53

Springer Series in Chemical Physics

Edited by Fritz P. Schäfer

Springer Series in Chemical Physics

Editors: Vitalii I. Goldanskii   Fritz P. Schäfer   J. Peter Toennies

Managing Editor: H. K. V. Lotsch

Volume 40 **High-Resolution Spectroscopy of Transient Molecules**
By E. Hirota

Volume 41 **High Resolution Spectral Atlas of Nitrogen Dioxide 559–597 nm**
By K. Uehara and H. Sasada

Volume 42 **Antennas and Reaction Centers of Photosynthetic Bacteria**
Structure, Interactions, and Dynamics
Editor: M.E. Michel-Beyerle

Volume 43 **The Atom-Atom Potential Method.** Applications to Organic Molecular Solids
By A.J. Pertsin and A.I. Kitaigorodsky

Volume 44 **Secondary Ion Mass Spectrometry SIMS V**
Editors: A. Benninghoven, R.J. Colton, D.S. Simons, and H.W. Werner

Volume 45 **Thermotropic Liquid Crystals, Fundamentals**
By G. Vertogen and W.H. de Jeu

Volume 46 **Ultrafast Phenomena V**
Editors: G.R. Fleming and A.E. Siegman

Volume 47 **Complex Chemical Reaction Systems**
Mathematical Modelling and Simulation
Editors: J. Warnatz and W. Jäger

Volume 48 **Ultrafast Phenomena VI**
Editors: T. Yajima, K. Yoshihara, C.B. Harris, and S. Shionoya

Volume 49 **Vibronic Interactions in Molecules and Crystals**
By I.B. Bersuker and V.Z. Polinger

Volume 50 **Molecular and Laser Spectroscopy**
By Zu-Geng Wang and Hui-Rong Xia

Volume 51 **Space-Time Organization in Macromolecular Fluids**
Editors: F. Tanaka, M. Doi, and T. Ohta

Volume 52 **Multiple-Photon Laser Chemistry**
By R. V. Ambartzumian, C. D. Cantrell, and A. Puretzky

Volume 53 **Ultrafast Phenomena VII**
Editors: C. B. Harris, E. P. Ippen, G. A. Mourou, and A. H. Zewail

Volumes 1–39 are listed on the back inside cover

Ultrafast Phenomena VII

Proceedings of the 7th International Conference, Monterey, CA, May 14–17, 1990

Editors: C.B. Harris, E.P. Ippen, G.A. Mourou, and A.H. Zewail

With 435 Figures

Springer-Verlag
Berlin Heidelberg New York London
Paris Tokyo Hong Kong Barcelona

Professor Charles B. Harris, Ph.D.
Department of Chemistry, University of California,
Berkeley, CA 94720, USA

Professor Erich P. Ippen, Ph.D.
Department of Electrical Engineering and Computer Science,
Massachusetts Institute of Technology, Cambridge, MA 02139, USA

Professor Gerard A. Mourou, Ph.D.
Electrical Engineering and Computer Science Department,
University of Michigan, Ann Arbor, MI 48109, USA

Professor Ahmed H. Zewail, Ph.D.
Department of Chemistry, California Institute of Technology,
Pasadena, CA 91125, USA

Series Editors

Professor Dr. Fritz Peter Schäfer
Max-Planck-Institut
für Biophysikalische Chemie
D-3400 Göttingen-Nikolausberg, FRG

Professor Vitalii I. Goldanskii
Institute of Chemical Physics
Academy of Sciences, Kosygin Street 4
Moscow, 117334, USSR

Professor Dr. J. Peter Toennies
Max-Planck-Institut für Strömungsforschung
Böttingerstrasse 6–8, D-3400 Göttingen, FRG

Managing Editor: Dr. Helmut K. V. Lotsch
Springer-Verlag, Tiergartenstrasse 17,
D-6900 Heidelberg, Fed.Rep.of Germany

ISBN 3-540-53049-5 Springer-Verlag Berlin Heidelberg New York
ISBN 0-387-53049-5 Springer-Verlag New York Berlin Heidelberg

Printing: Weihert-Druck GmbH, D-6100 Darmstadt
Binding: J. Schäffer GmbH & Co. KG, D-6718 Grünstadt
2154/3140-543210 – Printed on acid-free paper

Preface

This volume contains contributions presented at the 1990 Topical Meeting on Ultrafast Phenomena. The meeting, held in Monterey, California, from May 14 to 17, was the seventh in a biennial series whose purpose is to provide a forum for discussion of the latest advances in ultrafast optics and their application to fundamental scientific and engineering studies. The 362 registered participants, including 115 students, represented a wide range of scientific and engineering disciplines and came from 18 different countries. Their enthusiasm, the originality and quality of the papers that were presented, and the beautiful coastal setting combined to produce a very successful and enjoyable conference.

The reports herein reflect the vitality of the conference and the field in general. Novel methods for generating picosecond and femtosecond pulses continue to be developed. Even as pulse generation and measurement methods test the limits of visible optics, new techniques extend the technology to different laser media, new wavelength regimes, and ultrahigh power levels. The range of applications, too, continues to expand into new areas of physics, chemistry, biology, material science, and engineering. Although the space available for each contribution does not permit a very detailed discussion of any particular topic, this volume provides an impressively complete, up-to-date record of current research in the field of ultrafast phenomena.

Many people contributed to the success of the conference. Members of the international program committee not only provided general guidance, they reviewed all of the papers and organized the program. The staff of the Optical Society of America took care of the rest of the conference arrangements with their usual efficiency. Special thanks are due to the Optical Society of America for the sponsorship that makes that help possible and the Air Force Office of Scientific Research, which provided the major part of the funding. Additional grants from the following companies are also greatly appreciated: Burleigh Instruments, Clark Instruments, Coherent, Hamamatsu, Klinger Scientific, Lambda Physik, Medox Electro-optics, Metalaser Technologies, Newport Corporation, Quantel, Quantronix, and Spectra Physics.

Berkeley, CA — *Charles B. Harris*
Cambridge, MA — *Erich P. Ippen*
Ann Arbor, MI — *Gerard A. Mourou*
Pasadena, CA — *Ahmed H. Zewail*
August 1990

Contents

Part II High Intensity Sources and Interactions

Part IV **Nonlinear Effects in Optical Fibers**

Part VII Semiconductors; Solid-State Physics

Part VIII **Coherent Phonons**

Part X Interfaces and Surfaces

Part XI Electron or Proton Transfer and Solvation

Part XIII Biology

Part I

Ultrashort Pulse Generation

Hybrid and Passive Mode Locking in Coupled-Cavity Lasers

W. Sibbett

J.F. Allen Physics Research Laboratories, Department of Physics and Astronomy, University of St. Andrews, North Haugh, St. Andrews, Fife KY16 9SS, UK

1. Introduction

The idea of exploiting a coupled-cavity configuration for improved active mode-locking was first reported for linear external cavities (1-3). With the low-gain He-Ne lasers involved there was an obvious practical advantage in minimising insertion loss by siting the relevant modulator in the external cavity. The alternative concept of exploiting an optical nonlinearity in an external coupled cavity was first described by Mollenauer and co-workers in respect of the soliton laser (4,5). Subsequent experimental (6-10) and theoretical (11) studies then showed that the generation of bright optical solitons was not necessary to achieve enhanced mode-locking using nonlinear coupled cavities.

These schemes involving coupled-cavity configurations which have been variously described as coupled cavity (6-8), additive-pulse (9) and interferential (10) mode-locking are now being applied to an increasing number of laser types. The nonlinear medium in the external cavity has most frequently been a monomode optical fibre (6) but a semiconductor diode amplifier (6,8) or a nonlinear crystal (10,12) has also been demonstrated as suitable in particular laser systems. When substantial nonlinear phase shifts can be induced during the coupled-cavity mode-locking (CCM) process then it is possible to obtain pulse durations in the sub-100 fsec region (13). This is in contrast to the case of *empty* coupled cavities where some phase-locking enhancement can be obtained (14) but the degree of pulse refinement and shortening is less spectacular.

In this paper the emphases have been directed towards a selected subset of coupled-cavity mode-locking aspects. These relate to the assessment of (i) nonlinear Fabry-Perot and Michelson interferometric laser arrangements, (ii) exploitable fibre and semiconductor chip optical nonlinearities, (iii) hybrid and passive mode-locking and (iv) frequency-chirp compensation in CCM lasers.

2. Hybrid Coupled-Cavity Mode Locking

The dominating passive role of the control-fibre nonlinearity in both the soliton laser (4) and the generalised coupled-cavity colour centre laser (15) has been experimentally confirmed. It has also been shown that for an arrangement such as that illustrated in figure 1, subpicosecond pulses can reliably evolve from initial noise bursts (15). It is only required that the pump laser generates colour centre laser pulse fluctuations that have intensities sufficient to induce significant self-phase modulation in the optical fibre. The resultant feedback of the self-phase-modulated control-cavity signal then essentially dictates the nature of the pulse shaping and shortening processes which relate to the parameters of the main cavity (ie gain medium, bandwidth-limiting elements etc). Because of the intensity-dependent passive nature of the control cavity nonlinearity it is valid to describe this technique as *hybrid* coupled-cavity mode locking (15). The implementation of this hybrid mode-locking scheme in a KCl:Tl colour centre laser is reviewed here for the cases of nonlinear Fabry-Perot and Michelson cavity configurations.

2.1 Nonlinear Fabry-Perot Cavity Configuration

For an arrangement of the type shown in figure 1, there is a considerable choice for the fibre that is incorporated within the control cavity. In addition to the possible avoidance of

Springer Series in Chemical Physics, Vol. 53 **Ultrafast Phenomena VII**
Editors: C.B. Harris · E.P. Ippen · G.A. Mourou · A.H. Zewail

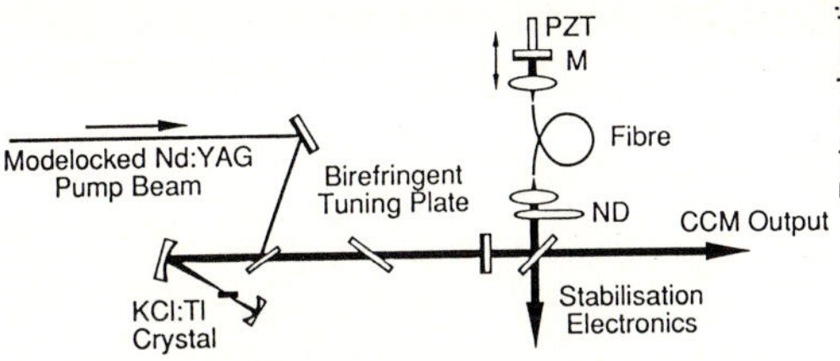

Fig 1: CCM laser with nonlinear Fabry-Perot cavity

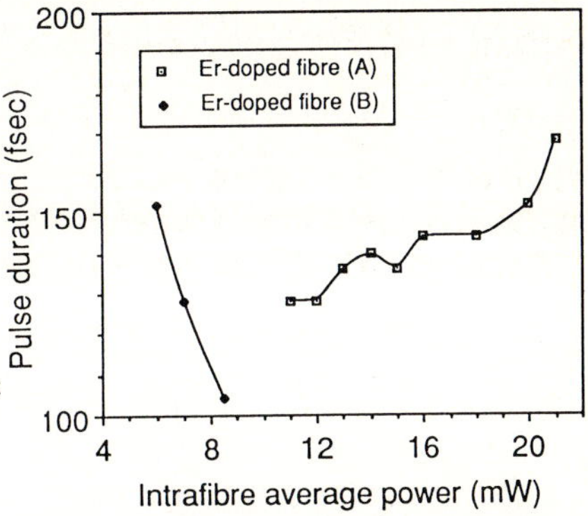

Fig 2: CCM pulse duration vs power in Er-doped control fibres

deliterious effects arising from modulational instability in anomalously-dispersive fibres (16), the normally-dispersive fibres also offer attractively large operational latitudes in terms of intrafibre optical powers. A good example of this is shown in figure 2 where the selected erbium-doped fibre (type A, doping concentration of 6 x 10^{18} cm^{-3}) enables subpicosecond pulses to be generated for average intrafibre powers from 11 mW to the maximum available power of 21 mW. Despite a rather more limited operational latitude, an anomalously-dispersive fibre can be chosen (see figure 2, Er-fibre (B), doping concentration of 1.6 x 10^{17} cm^{-3}) which enables the coupled-cavity mode locking to be initiated and sustained at substantially lower intrafibre power levels.

The coupled nonlinear Fabry-Perot cavity scheme has already been employed in our laboratory for a number of fibre types that have included dispersion-shifted, erbium-doped, and erbium-free normally-dispersive fibres and anomalously-dispersive standard (or soliton), erbium-doped (low concentrations), neodymium-doped and dispersion-flattened fibres. Moreover, this hybrid technique has also been shown to lead to enhanced mode-locking when coupled to a *passively* mode-locked Li:F_2^+ colour centre laser (17) as distinct from the more usual actively or synchronously mode-locked lasers.

2.2 Nonlinear Michelson Cavity Configuration

Although investigated experimentally for a pulsed CO_2 laser (10) and studied theoretically (10,18), the nonlinear Michelson cavity configuration shown schematically in figure 3 has been implemented less frequently than the Fabry-Perot counterpart discussed in the previous subsection. It is quite distinct in that it offers the potential for greater physical compactness and it is free from the optical ringing effect that is associated with a coupled Fabry-Perot interferometer.

In an assessment of the mode-locking performance of the synchronously-pumped KCl:Tl colour centre laser having a nonlinear Michelson cavity as illustrated it has been found that this scheme performs extremely well and, as expected, possesses greater tolerance to perturbations than the physically more extended Fabry-Perot alternative. With an 18 cm long erbium-doped, normal-dispersion fibre (type A) in the nonlinear arm, ~ 90 fsec duration, near bandwidth-limited ($\Delta\nu\Delta\tau \sim 0.35$) $sech^2$ pulses have been produced using this approach (see figure 4). Subpicosecond pulses have also been produced when an InGaAsP semiconductor amplifier chip was used as the nonlinear medium.

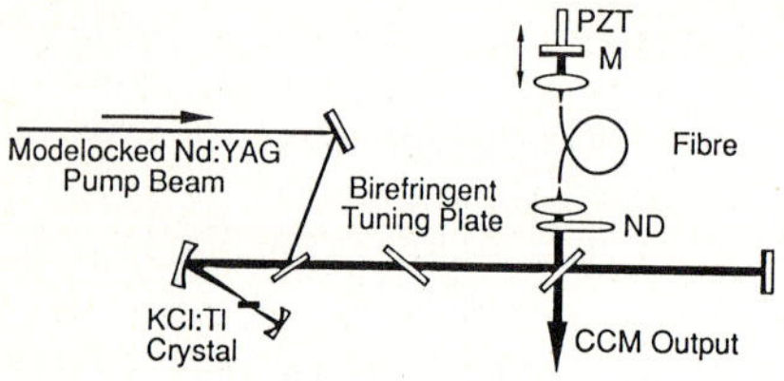

Fig 3: CCM laser with nonlinear Michelson cavity

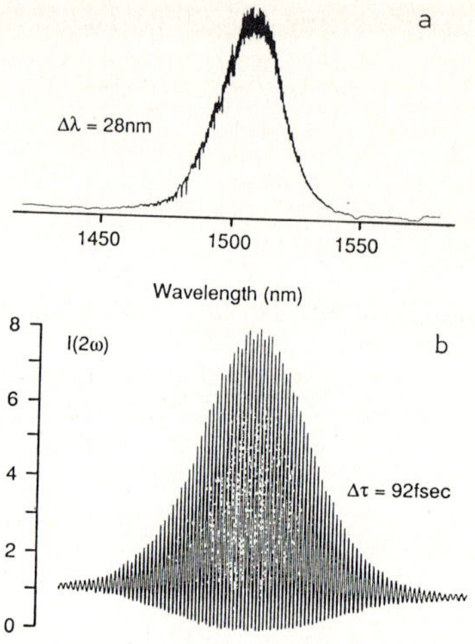

Fig 4: Spectral and temporal data for laser of Figure 3

3. Passive Coupled-Cavity Mode Locking

As already pointed out in Section 2 of this paper, coupled-cavity mode-locking in the KCl:Tl laser having a fibre-based nonlinearity in the control cavity can be readily and reliably established from periodic noise bursts. However, observations to date imply that this CCM process cannot be initiated or sustained in anything other than the short term, when the active modulation is removed. In contrast, a self-starting and entirely passive CCM operation has been observed for the $Ti:Al_2O_3$ laser (19). [This has the obvious advantage that the provision and control of a suitable modulator having appropriate RF drive electronics is no longer required.] The reason for the ease with which a self-starting and sustained passive mode-locking operation can be achieved in a coupled-cavity $Ti:Al_2O_3$ laser configuration (see figure 5) is most probably due to the observed strong mode-beating that occurs in the main cavity. Also, in a theoretical analysis of the self-starting condition for such a laser system the relatively low emission cross-section of the $Ti:Al_2O_3$ gain medium has been shown to be a favourable parameter (σ (KCl:Tl) = 1.3×10^{-17} cm^2, σ ($Ti:Al_2O_3$) = 2.7×10^{-19} cm^2) (20). Experimental aspects relating to intracavity frequency chirp compensation in a passive CCM, $Ti:Al_2O_3$ laser together with the conditions under which self-initiating/sustaining passive CCM operation can be demonstrated for a KCl:Tl colour centre laser are presented in the following subsections.

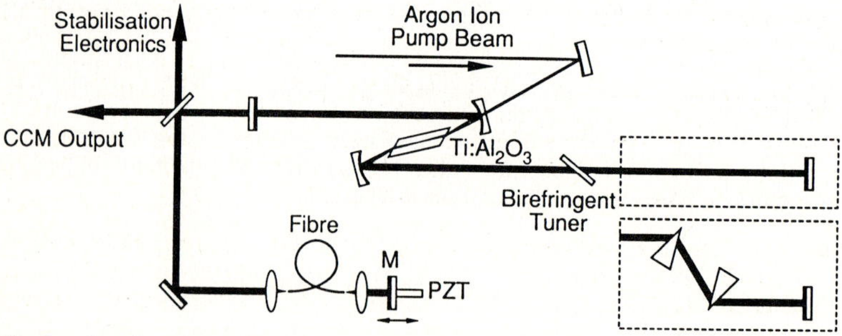

Fig 5: Passive CCM, $Ti:Al_2O_3$ laser (Inset: with intracavity GVD compensation)

3.1 Titanium-Sapphire Laser

For the particular data presented here, an adapted Spectra Physics Model 3900 $Ti:Al_2O_3$ laser having an extended resonator (f_{cavity} ~ 70 - 190 MHz) has been used. The typical all-line argon-ion pump power was ~ 6 W and the $Ti:Al_2O_3$ laser was mostly operated in the 800 - 900 nm tuning range. The length (~ 30 cm) of optical fibre which was incorporated within the control cavity (see figure 5) was monomode in this spectral region (Andrew Corporation Inc - type 48280-1-P elliptical core fibre) and has previously been exploited with coupled-cavity $LiF:F_2^+$ colour-centre laser configurations (6,17). The typical optical power coupled into this control fibre was ~ 50-100 mW and when passively mode-locked the average output power was ~ 100 mW.

A distinctive feature of the coupled-cavity $Ti:Al_2O_3$ laser is that significant self-phase-modulation (SPM) and group velocity dispersion (GVD) effects arise in the gain medium as well as in the nonlinear medium within the control cavity. It is therefore not surprising that the pulses produced by the CCM, $Ti:Al_2O_3$ exhibit a considerable degree of frequency chirp. This is clearly illustrated by the interferometric SHG autocorrelation trace included as figure 6(a) for the case where a 1.6 mm thick birefringent filter was employed within the main cavity and the laser tuned to 850 nm. By propagating the pulses via an extracavity diffraction grating pair (1200 l/mm, 30 mm separation) it can be seen from figure 6(b) that much of the frequency chirp can be removed and the pulse durations are reduced from 2.7 psec ($\Delta\nu\Delta\tau$ = 3.7) to 290 fsec ($\Delta\nu\Delta\tau$ ~ 0.35) (durations deduced from intensity autocorrelation data).

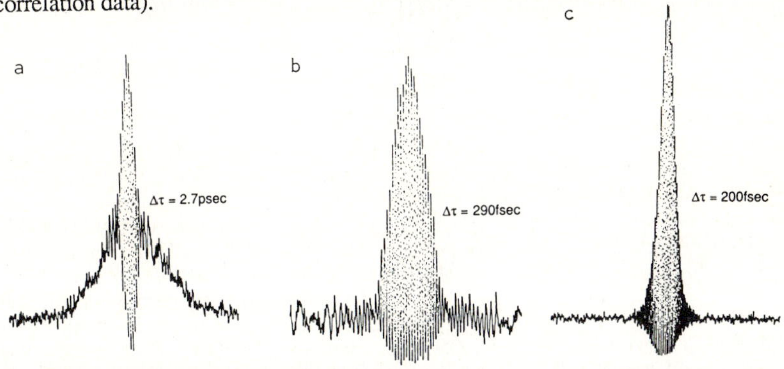

Fig 6: Autocorrelation data for CCM, $Ti:Al_2O_3$ laser having (a) no, (b) extracavity and (c) intracavity GVD compensation

An alternative approach to frequency-chirp compensation involves the cavity refinement as indicated in the inset of figure 5. The inclusion of the pair of high-dispersion, flint-glass, Brewster-angled prisms provides a convenient and efficient means of controlling the net group velocity dispersion. For a prism separation of 40 cm it can be seen from figure 6(c) that the pulses that are directly produced by this GVD-compensated, CCM laser are essentially free from excess frequency chirp. The bandwidth ($\Delta\nu$ = 1.74 THz)-duration (200 fsec) products for typical pulses are ~ 0.35 which indicate that they are close to the Fourier transform limit if $sech^2$ intensity profiles are assumed. Under these operating conditions the average and peak powers from our passive CCM $Ti:Al_2O_3$ laser were 50 mW and 3 kW respectively.

3.2 KCl:Tl°(1) Colour Centre Laser

In earlier publications (6,8) experimental data for a CCM, KCl:Tl colour centre laser were presented for both fibre-based and semiconductor-amplifier-based optical nonlinearities in the control cavity. The latter nonlinearities have the appeal that they can be accessed at relatively low propagating optical powers and the GVD contributions can be small. Moreover, by supplying dc injection current to the semiconductor chip the dynamic nonlinearities are

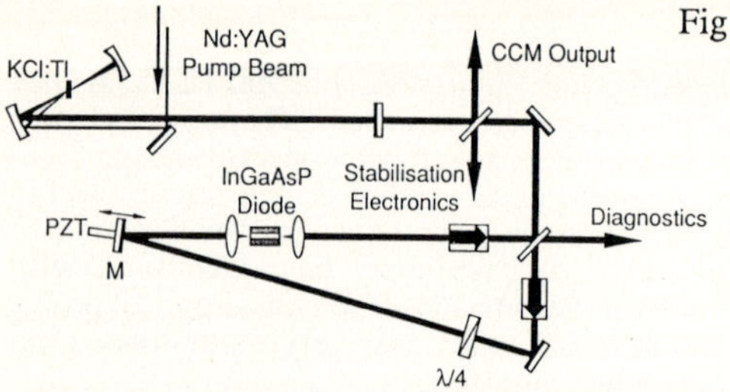

Fig 7: Passive CCM, KCl:Tl laser with nonlinear InGaAsP chip

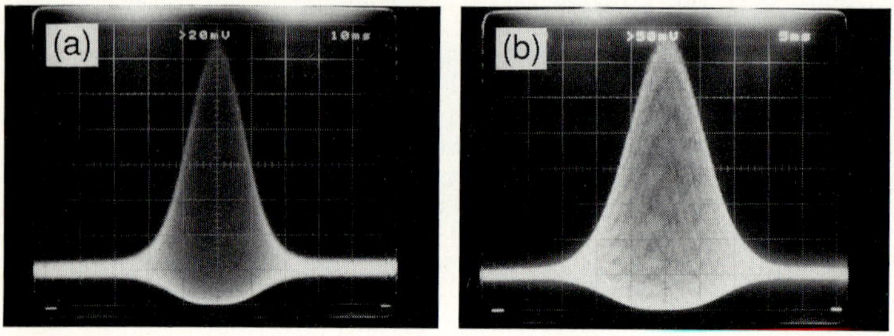

Fig 8: Autocorrelations for saturable (a) absorption, (b) gain regimes; $\Delta\tau_p$ (a), (b) = 480, 280 fsec.

induced by the intracavity radiation such that their influence has a flux-dependent passive nature.

The mode-locking performance of the KCl:Tl laser has been investigated for the coupled-cavity configuration, shown schematically in figure 7. An anti-reflection coated, 350 μm long, double-channel planar buried heterostructure InGaAsP semiconductor was employed as the nonlinear element in the control cavity. At injection currents below the transparency current of 6 mA in the 1.5 μm spectral region the chip exhibits net absorption while at currents in excess of 6 mA it provides net amplification. The induced carrier dynamics due to a propagating optical field can give rise to nonlinear phase changes which are associated with the saturable absorption/gain regimes. Subsequent feedback to the main cavity of such phase-modulated signals can therefore lead to enhanced mode-coupling and thus to the generation of ultrashort pulses through the CCM influence.

It is especially noteworthy that whereas self-initiating coupled-cavity mode-locking was not observed for the KCl:Tl colour centre laser when fibre-related nonlinearities were involved, this has not been the case when the optical nonlinearity was provided by the semiconductor device. Indeed, we have observed, for the first time, self-initiating,self-sustaining and stable coupled-cavity mode locking of the KCl:Tl laser under the conditions of both saturable absorption and saturable gain in the control-cavity InGaAsP medium (21). Representative interferometric autocorrelation traces for this laser (pumped by an unmode-locked Nd:YAG laser) are reproduced in figure 8(a), (b) for both of these novel operating regimes. It can be readily seen from these data that the pulses generated by this passive-CCM KCl:Tl colour centre laser are essentially free from any excess frequency chirp. Significantly it has also been estimated that the optical powers that are coupled into the semiconductor waveguide and fed back into the main cavity are as low as 700 μW and 5 μW respectively. Thus this represents an attractively practical and convenient mode-locking technique where adequate nonlinear phase can be induced at low powers such that unchirped femtosecond pulses can be reliably obtained.

4. Conclusions

These hybrid and passive mode-locking schemes which involve a variety of nonlinear coupled-cavity configurations offer practical and versatile means by which femtosecond laser pulses can be reliably produced. From the examples described above it can be seen that particular nonlinear media can be incorporated into the control cavity such that enhanced mode-locking can be achieved for a relatively wide range of laser parameters. Moreover, these observations are further complemented by the other related work that has been reported in the *Short Pulse Lasers* session of this conference. It can therefore be expected that ongoing active research in this topic will lead to yet more variants in the exploitable nonlinearities and in the laser types for which the coupled-cavity mode-locking technique will be shown to be especially appropriate.

5. Acknowledgements

The contributions of experimental data from my colleagues, Robert Grant, David Spence and Xiaonong Zhu are gratefully acknowledged. Overall financial support for the work described here has been provided by the UK Science and Engineering Research Council.

References

[1]. L C Foster, M D Ewy and C B Crumly, Appl Phys Lett, **6**, 6 (1965).
[2]. M Di Domenico Jr and V Czarniewski, Appl Phys Lett, **6**, 150 (1965).
[3]. L B Allen, R R Rice and R F Mathews, Appl Phys Lett, **15**, 416 (1969).
[4]. L F Mollenauer and R H Stolen, Opt Lett, **9**, 13 (1984).
[5]. F M Mitschke and L F Mollenauer, IEEE J Quantum Electron **QE-22**, 2242 (1986).
[6]. P N Kean, X Zhu, D W Crust, R S Grant, D Burns and W Sibbett, Paper PD-7, Tech Dig CLEO '88; also Opt Lett, **14**, 39 (1989).
[7]. K J Blow and B P Nelson, Opt Lett, **13**, 1026 (1988).
[8]. A Miller and W Sibbett, J Mod Opt, **35**, 1871 (1988).
[9]. J Mark, L Y Liu, K L Hall, H A Haus and E P Ippen, Opt Lett, **14**, 48 (1989).
[10]. F Oullette and M Piché, Can J Phys, **66**, 903 (1988).
[11]. K J Blow and D Wood, J Opt Soc Amer B, **5**, 629 (1988).
[12]. J R M Barr and D W Hughes, Appl Phys B, **49**, 323 (1989).
[13]. X Zhu, P N Kean and W Sibbett, Opt Lett, **14**, 1192 (1989).
[14]. S M J Kelley, Opt Commun, **70**, 495 (1989).
[15]. A Finch, X Zhu, P N Kean and W Sibbett, IEEE J Quantum Electron, to be published in June 1990.
[16]. X Zhu, P N Kean and W Sibbett, IEEE J Quantum Electron, **25**, 2445 (1989).
[17]. C I Johnston, D E Spence, R S Grant and W Sibbett, Opt Commun, **73**, 370 (1989).
[18]. M Morin and M Piché, Opt Lett **14**, 1119 (1989).
[19]. J Goodberlet, J Wang and J G Fujimoto, Opt Lett, **14**, 1123 (1989).
[20]. E P Ippen, L Y Liu and H A Haus, Opt Lett, **15**, 183 (1990).
[21]. R S Grant, P N Kean and W Sibbett, to be published.

Generation of Ultrashort Nd:YLF Pulses by cw Passive Mode Locking with a Nonlinear External Cavity

J.K. Chee, M.N. Kong, E.C. Cheung, and J.M. Liu

Department of Electrical Engineering, University of California at Los Angeles, Los Angeles, CA 90024, USA

Abstract. We present experimental results on the cw passive mode-locking operation of a flashlamp-pumped Nd:YLF laser with a nonlinear external cavity. We have obtained nearly transform-limited pulses which are as short as 3.2 ps at usable output power levels of up to 10 W. Cavity length detuning characteristics are also discussed.

1. Introduction

Since the first demonstration of pulse shortening in the soliton laser[1] using the feedback from an external cavity, it has been demonstrated that the additive-pulse mode-locking is an effective method to further shorten the actively mode-locked laser pulses[2–6]. Recently, it has been demonstrated in a $Ti:Al_2O_3$ that the nonlinear external coupled-cavity acted as a passive mode-locker to generate a stable train of pulses without synchronous pumping or active mode-locking[7].

To date, all the demonstrations of coupled-cavity mode-locking using a nonlinear external cavity are in lasers with gain media which have extremely broad bandwidths, such as color center crystals or $Ti:Al_2O_3$. Moreover, the average output power of the pulse train is at the relatively low power level of a few hundred milliwatts or much less. In this paper, we report cw passive mode locking operation of the Nd:YLF laser at high power with a nonlinear external coupled-cavity using an optical fiber as the nonlinear medium. We have obtained nearly transform-limited pulses as short as 3.2 ps at usable average output power up to 10 W.

2. Experimental Results

Figure 1 shows our experimental setup. The main cavity formed by M_1 and M_2 was a Quantronix 416 Nd:YLF laser at 1.053 μm wavelength which delivered up to 14 W cw at the highest pumping level. The external cavity was formed by M_2 and M_3 and consisted of an optical fiber as the nonlinear medium. The end mirror, M_3, of the external cavity had a 5-cm radius of curvature and was coated for maximum reflectivity at 1.053 μm. The

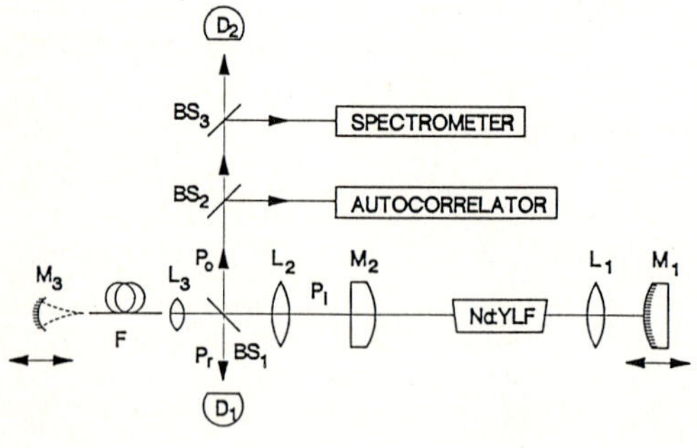

Fig. 1. Experimental setup.

Springer Series in Chemical Physics, Vol. 53 **Ultrafast Phenomena VII**
Editors: C.B. Harris · E.P. Ippen · G.A. Mourou · A.H. Zewail

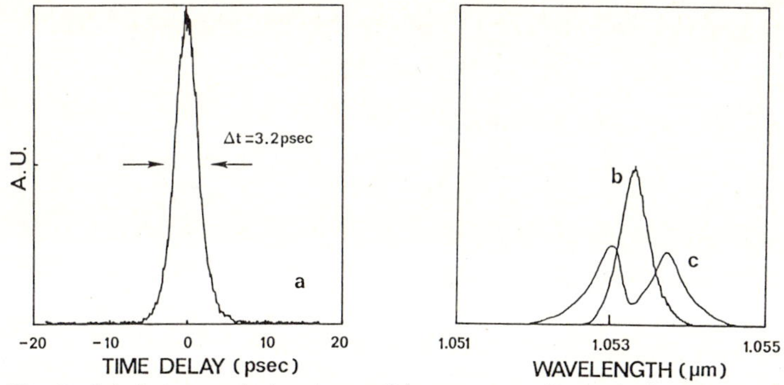

Fig. 2. (a) Autocorrelation trace, (b) spectrum of a typical output pulse at $P_o = 3.5$ W, and (c) spectrum of self-phase modulated pulse after traveling through the fiber in the external cavity.

lens L_2 inside the external cavity was used to correct the divergence of the beam exiting the main cavity for efficient coupling into the fiber with an NRC F-L10B objective lens, L_3. The usable output power from the system was obtained through the beam splitter BS_1. The typical coupling efficiency into the fiber was 60%, while the retroreflective coupling efficiency of the mirror M_3 was nearly 100% except the reflection loss at the fiber tip.

The performance characteristics of this laser depend on the values of the various parameters, including the fiber length, the beam splitter reflectivity, the power coupled into the fiber, the laser output power level, and the relative cavity detuning. Typical lengths of fiber used were 30, 60, and 90 cm. The typical beam splitter reflectivity ranged from 50% to 90%. The output power, P_o, ranged from 1 to 10 W and the power coupled into the fiber was varied from 300 mW to 2 W. Pulses of widths ranging from 3 to 10 ps at a repetition rate of 76 MHz were obtained when the laser was mode locked. Stable mode-locking operation was observed over a large range of the relative cavity detuning.

Figures 2(a) and (b) show the background-free second-harmonic generation autocorrelation trace and corresponding spectrum of a typical output pulse at $P_o = 3.5$ W when a fiber length of 30 cm and a beam splitter of 70% reflectivity were used. This autocorrelation trace was fitted almost perfectly with a sech^2 pulse of FWHM pulse width $\Delta t = 3.2$ ps. The FWHM spectral width $\Delta\lambda = 4.2$Å corresponds to a time-bandwidth product of 0.36 and the pulse was nearly transform limited. Figure 2(c) shows the spectrum of the pulse after traveling through the fiber. This broadened spectrum corresponds to a maximum nonlinear phase shift of 1.5π due to self-phase-modulation, which is consistent with the calculated value using the measured power in the fiber. At a higher pumping level, output power of 6 W could be obtained at an expense of lengthening the pulse to 4 ps. Moreover, it was necessary to reduce the power coupling efficiency into the fiber in order to maintain a stable train of pulses at a higher pumping level.

With proper choices of the external cavity parameters, the pulse width could be broadened while maintaining stable passive mode locking operation by detuning the cavity length. Although the shortest pulse width did not depend much on the power level, the broadest mode-locked pulse width and the range of cavity detuning attainable through this depended largely on the output power.

Figures 3(a) and (b) show the output pulse width as a function of relative cavity detuning when a fiber length of 60 cm was used. Shown in Fig. 3(a) are the cavity detuning characteristics corresponding to the output power levels of 4 W and 6.5 W

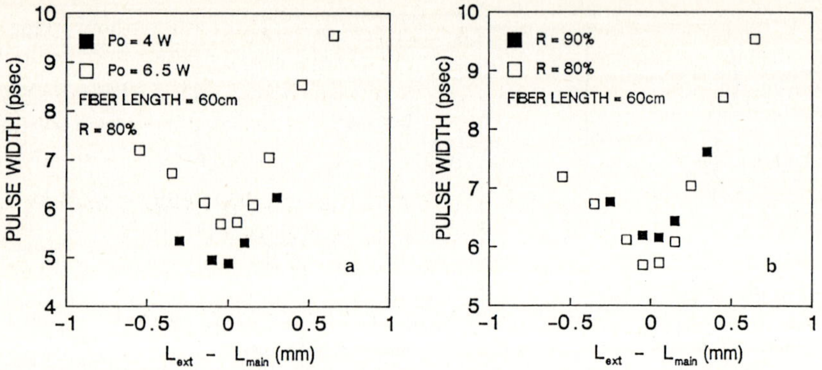

Fig. 3. Output pulse width as a function of relative cavity length detuning for (a) two output power levels and (b) two beam splitter reflectivities.

when a beam splitter of 80% reflectivity was used. The output pulse width becomes broadened as two cavities are mismatched in length. It is also shown that the detuning range increases at a higher output power level. Typically, stable operation of passive mode locking was observed with a cavity detuning up to ±1 mm. From the measured spectra, it was found that the pulses stayed nearly transform-limited over the entire detuning range before mode locking ceased. In Fig. 3(b), the cavity detuning characteristics are compared for different beamsplitter reflectivities.

3. Conclusion

We have demonstrated cw passive mode locking of the Nd:YLF laser with a nonlinear external cavity. A stable train of nearly transform-limited pulses as short as 3.2 ps at a repetition rate of 76 MHz is obtained. Stable passive mode locking operation could be maintained with average output powers ranging from 1 to 10 W. By detuning the relative cavity length over a range of ±1 mm, the pulse width could be varied between 3 and 10 ps before mode locking stopped.

This research was supported by the Academic Senate at the University of California, Los Angeles under Grant No. 4258.

References

1. L.F. Mollenauer and R.H. Stolen, Opt. Lett. **9**, 13(1984).
2. K.J. Blow and B.P.Nelson, Opt. Lett. **13**, 1026(1988).
3. P.N. Kean, X.Zhu, D.W. Crust, R.S. Grant, N.Langford, and W. Sibbett, Opt. Lett. **14**, 39(1989).
4. C.P. Yakymyshyn, J.F. Pinto, and C.R. Pollock, Opt. Lett. **14**, 621(1989).
5. X.Zhu, P.N. Kean, and W. Sibbett, IEEE J. Quantum Electron. **25**, 2445(1989).
6. P.M.W. French, J.A.R. Williams, and J.R. Taylor, Opt. Lett. **14**, 686(1989).
7. J. Goodberlet, J. Wang, J.G. Fujimoto, and P.A. Schulz, Opt. Lett. **14**, 1125(1989).

Ultrashort Pulse Generation with Additive Pulse Modelocking in Solid State Lasers: $Ti:Al_2O_3$, Diode Pumped Nd:YAG and Nd:YLF

J. Goodberlet[1], *J. Jacobson*[1], *J. Wang*[1], *J.G. Fujimoto*[1], *T.Y. Fan*[2], *and P.A. Schulz*[2]

[1]Department of Electrical Engineering and Computer Science and Research Laboratory of Electronics, Massachusetts Institute of Technology, Cambridge, MA 02139, USA
[2]Lincoln Laboratories, Lexington, MA 02173, USA

Abstract. Additive pulse modelocking provides a powerful technique for passive modelocking a wide variety of solid state lasers. The additive pulse modelocking technique uses a laser coupled with a nonlinear external cavity to produce modelocking by a fast saturable absorber like mechanism without the need for active modulation. Ultrashort pulse generation is demonstrated in $Ti:Al_2O_3$, diode pumped Nd:YAG and Nd:YLF.

The development of ultrashort pulse solid state lasers is important for a wide range of ultrafast studies. The $Ti:Al_2O_3$ laser has an extremely broad tuning range from 670 nm to greater than 1000 nm and thus is a natural choice for femtosecond generation and spectroscopy. The diode pumped Nd:YAG and Nd:YLF have the potential for developing a compact ultrashort pulse source suitable for applications in electro-optic sampling and high speed signal processing.

Recent theoretical and experimental studies have demonstrated that a nonlinear external cavity can be used to enhance pulse shortening and modelocked performance in actively modelocked lasers [1-3]. The pulse shaping mechanism has been termed additive pulse modelocking (APM) because the pulse shaping depends on coherent addition of pulses from the main laser and external cavity [3]. The APM technique has recently been extended to achieve self starting additive pulse modelocking in $Ti:Al_2O_3$ [4]. Recent demonstrations of self starting APM in the diode pumped Nd:YAG as well as the lamp pumped Nd:YAG, Nd:YLF, and laser pumped Nd:Glass establish the generalizability of these techniques to other solid state laser systems [5-8]. Self starting APM achieves ultrashort pulse generation without the need for active amplitude or phase modulation and thus provides a significant reduction in cost and complexity over actively modelocked laser systems.

In this paper we describe investigations of self starting additive pulse modelocking in $Ti:Al_2O_3$ and diode pumped Nd:YAG and Nd:YLF. Figure 1 shows a schematic diagram of the $Ti:Al_2O_3$ and the diode pumped Nd:YAG and Nd:YLF lasers. The lasers consist of a main laser and an external cavity composed of a beamsplitter, optical fiber, and retro-reflecting mirror. The relative cavity lengths are set so that pulses returned from the external cavity will interferometrically recombine with pulses in the main laser.

Viewed from the main laser, the output mirror of the main laser and the external cavity may be considered as a Fabry-Perot which has a phase dependent reflectivity. The nonlinear refractive index of the optical fiber produces an intensity dependent phase shift so that the external cavity has a nonlinear intensity dependent reflectivity. Depending on the relative cavity length settings, the external cavity functions as fast saturable absorber or fast saturable gain. The laser can also be viewed as a system which incorporates all optical switching to perform pulse shaping.

Springer Series in Chemical Physics, Vol. 53 **Ultrafast Phenomena VII**
Editors: C.B. Harris · E.P. Ippen · G.A. Mourou · A.H. Zewail

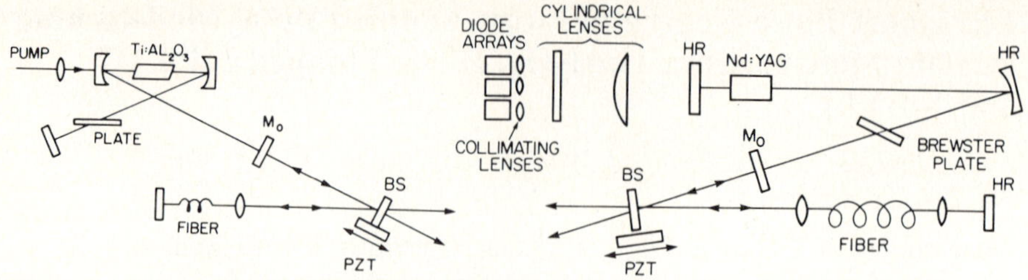

Figure 1. Schematic of self starting APM $Ti:Al_3O_3$ and diode pumped Nd:YAG and Nd:YLF lasers. The nonlinear external cavity produces a fast saturable absorber modelocking. Pulse generation is achieved without active amplitude or phase modulation.

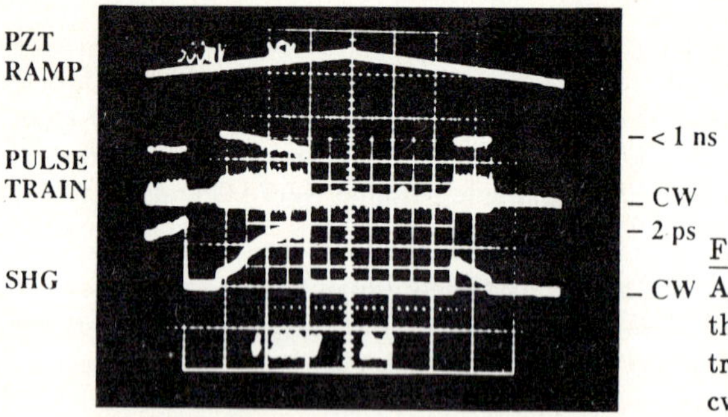

Figure 2. Dependence of the APM $Ti:Al_2O_3$ laser operation on the relative cavity lengths. District phase settings produce either cw or modelocked behavior.

The relative cavity lengths or phase setting is a key parameter for obtaining self starting additive pulse modelocking since it determines the magnitude and sign of the nonlinear reflectivity. Figure 2 shows the typical behavior of the $Ti:Al_2O_3$ laser as a function of external cavity phase. The external cavity length was varied by ramping a PZT while monitoring the ouput pulse train and its second harmonic. Note that the laser behavior is interferometrically periodic in cavity length. The transition from cw to short pulse operation exhibits hysteresis and bistability behavior. Different phase regimes produce optimum self starting versus steady state short pulse operation.

The main laser cavity and external cavity parameters will determine the strength of the nonlinear reflectivity. In $Ti:Al_2O_3$, the output mirror reflectivity is 85% and the external cavity beamsplitter is 60% reflectivity. The length of the optical fiber in the external cavity is 12 cm. For these laser parameters, the threshold for self starting APM is approximately 100 mW of power coupled into the optical fiber. The laser is typically operated with $\sim$ 300 mW in the fiber.

Since our objective is to minimize pulse durations, we have chosen to use shorter fiber lengths to reduce dispersion at the expense of higher modelocking threshold intensities. In $Ti:Al_2O_3$ using a single plate birefringent tuning plate, chirped pulses of 1.4 ps duration with output powers of 400 mW are generated. Pulses as short as 200 fs can be produced with a diffraction grating pair to compensate for dispersion. By restricting the bandwidth using a two plate birefringent filter and using a 7 cm fiber length to reduce dispersion, pulses as short as 700 fs may be generated directly from the laser.

In the diode pumped Nd:YAG, larger nonlinearities were required to obtain self starting APM. The output mirror reflectivity was 85% and the external cavity beamsplitter was 80% or 90%. In Nd:YAG, using a fiber length of 83 cm, the threshold for self starting APM was 80 mW in the fiber. The pulse duration was 1.7 ps with a spectral bandwidth of 0.67 nm and an output power of 25 mW. We believe that these are the shortest pulses generated to date in a Nd:YAG laser.

Preliminary results were also obtained in diode pumped Nd:YLF. Using a 155 cm length of fiber, 2.0 ps chirped pulses were generated with a 0.8 nm bandwidth and an output power of 10 mW. The Nd:YLF had a lower output power, thus it was necessary to using longer fiber lengths to obtain modelocking at the expense of increased dispersion and longer pulses.

These results demonstrate the applicability of additive pulse modelocking to a range of solid state laser systems. The inherent scalability of this technique as well as the possibility of explicitly optimizing the modelocking nonlinearity can provide a powerful approach for ultrashort pulse generation.

We wish to thank E. P. Ippen and H. A. Haus for helpful discussions. The work at MIT campus was supported in part by the Air Force Office of Scientific Research grant F49620-88C-0089, the Joint Services Electronics Program contract DAAL03-86-K-002, the National Science Foundation PYI program 85-52701, and the AT&T Foundation. The work at Lincoln Laboratory was sponsored by the Department of the Navy.

References

[1] K.J. Blow and D. Wood, J. Opt. Soc. Am. B **5**, 629 (1988).

[2] P.N. Kean et al., Opt. Lett. **14**, 39 (1989).

[3] J. Mark et al., Opt. Lett. **14**, 48 (1989).

[4] J. Goodberlet et al., Opt. Lett. **14**, 1125 (1989).

[5] J. Goodberlet et al., Opt. Lett. **15**, 504, (1990).

[6] L.Y. Liu et al., Opt. Lett. **15**, 553, (1990).

[7] J. K. Chee et al., in Ultrafast Phenomena VII, C. B. Harris, E. P. Ippen, G. A. Mourou, A. Zewail, Eds., Springer Verlag (1990).

[8] Ch. Spielmann et al., in Ultrafast Phenomena VII, C. B. Harris, E. P. Ippen, G. A. Mourou, A. Zewail, Eds., Springer Verlag (1990).

Chirp in Actively Modelocked Diode Lasers

A.S. Hou [1], *R.S. Tucker* [2], *and E.P. Ippen* [1]

[1]Department of Electrical Engineering and Computer Science, Research Laboratory of Electronics, Massachusetts Institute of Technology, Cambridge, MA 02139, USA

[2]Department of Electrical and Electronic Engineering, University of Melbourne, Parkville, Vic. 3052, Australia

Pulses from actively modelocked diode lasers exhibit a positive wavelength chirp [1] and recently have been compressed externally using normal linear dispersion [2,3]. In this paper, we present time-resolved chirp measurements of a modelocked external-cavity InGaAsP laser. These measurements reveal a wide range of chirp values for different laser operating conditions. A self-consistent modelocking theory which incorporates both carrier-induced index (phase) modulation and group velocity dispersion has been used to explain the observed chirping.

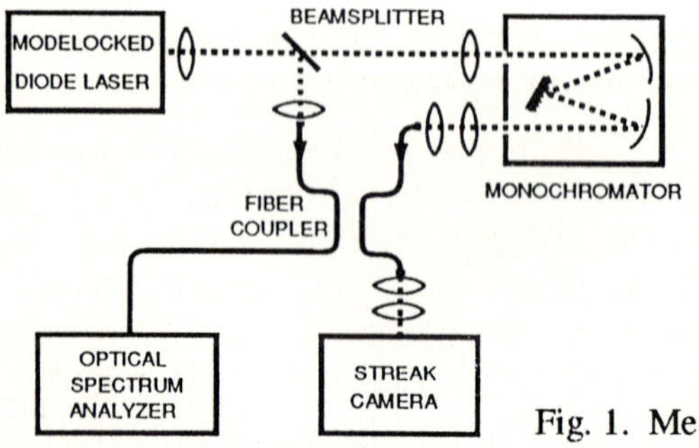

Fig. 1. Measurement setup

The experimental setup is illustrated in Figure 1. Our measurements were performed with a 15 cm long external-cavity laser, consisting of an anti-reflection-coated 1.3 μm wavelength InGaAsP diode and a 1200 lines/mm gold diffraction grating for bandwidth selection and feedback. Under modelocked operation, the laser is biased below or near threshold and driven with a sinusoidal RF current. Typical pulses were 10-15 ps at FWHM, and the average optical power was approximately 100 μW. For a time-resolved spectral measurement, the modelocked pulses are filtered through a monochromator and then recorded using a syncroscan streak camera with 3 ps resolution. Part of the unfiltered pulse train is tapped to establish a precise time reference for each streak camera trace. An optical spectrum analyzer is used to pinpoint the filter wavelength and bandwidth to within 0.2 Å. Figure 2 shows a sample time-resolved spectrum for a 15.5 ps pulse. The chirp can be deduced directly from the instantanous peak wavelength of such a spectrum [4].

By adjusting the tilt of the intracavity diffraction grating, we could tune our modelocked laser from 1.26 μm to 1.32 μm. In general, the spectral width of the pulses increased with longer wavelength operation, but the pulse widths did not vary much. Time-resolved measurements indicate a clear trend of increasing chirp with

Springer Series in Chemical Physics, Vol. 53 **Ultrafast Phenomena VII**
Editors: C.B. Harris · E.P. Ippen · G.A. Mourou · A.H. Zewail

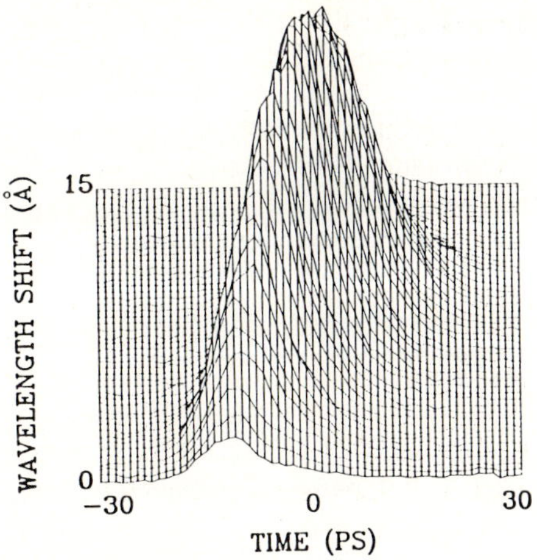

Fig. 2. Time-resolved spectrum of a chirped modelocked diode pulse.

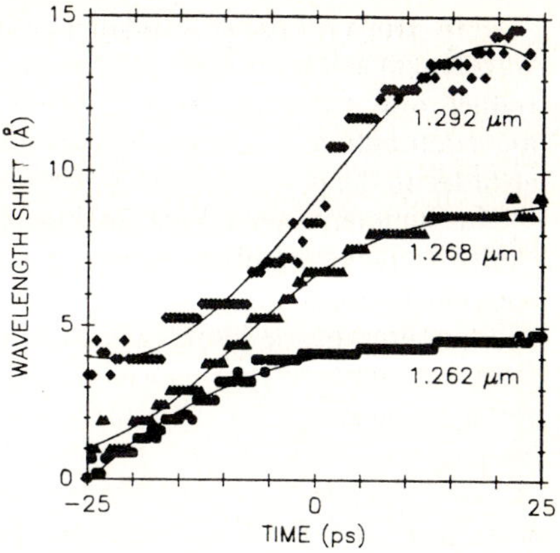

Fig. 3. Measured chirp at 3 operating wavelengths of the laser. The pulses are centered at t=0.

increasing wavelength. A comparison of chirping for three operating wavelengths is displayed in Figure 3. From these plots, the linear component of chirp is estimated to be 0.2, 0.4 and 1.0 Å/ps at 1.262, 1.268, and 1.315 μm, respectively.

The effect of bandwidth selectivity on chirping was also investigated. There are two stable resonator configurations for the external-cavity laser; one with the cavity beam focussed on the grating and a second with the beam collimated onto the grating. The two yield dramatically different results. In the focussed configuration, the laser chirped as much as 1.3 Å/ps. In the collimated configuration, the increased efficiency of the grating greatly diminished the chirping, and nearly-transform-limited ($\Delta\nu\Delta\tau \approx 0.56$) pulses were generated. Similarly, we could reduce the chirping by using an intracavity etalon. We conclude that bandwidth selectivity is a key determinant of chirp in a modelocked diode laser. With ample bandwidth, the laser pulse can undergo a large degree of chirping. Tightening the bandwidth can reduce the amount of chirp without broadening the pulses.

The chirp in modelocked diode lasers can be analytically modeled using standard active modelocking theory [5]. In addition to operators for round-trip gain and loss, one needs to include an operator $e^{i\phi(t)}$ that modifies the phase of the circulating pulse.

$$\phi(t) = \alpha g(t)L + D\frac{d^2}{dt^2}, \qquad (1)$$

where α is the linewidth enhancement factor, g(t) is the electric field gain waveform, L is the length of the diode, and D is dispersion. The solution to the modified steady-state modelocking equation is a chirped Gaussian pulse [4]. The α parameter couples diode gain modulation to phase modulation. A linear chirp is established when the gain possesses a quadratic curvature. Dispersion also contributes to the chirping by redistributing the spectral contents of the pulse in time. Both phase modulation and dispersion are necessary to give rise to spectral broadening in excess of $\sqrt{2}$ times the transform-limit, as was observed.

The proposed theory offers a possible explanation for the variation in chirp with wavelength and bandwidth selectivity. At shorter wavelengths, the laser is operating farther away from the bandgap, where α is calculated to be much smaller [6]. Thus, the contribution to chirp by gain modulation is reduced. With less bandwidth afforded to the pulse, the effect of dispersion is substantially moderated.

In summary, time-resolved measurements were conducted with an external-cavity modelocked diode laser. We observed a positive wavelength chirp whose magnitude ranged from 0.05 to 1.3 A/ps depending on the operating conditions and configuration of the laser. The results support the conclusion that chirp originates from two physical processes: (1) carrier-induced index changes that accompany variations in gain and (2) group velocity dispersion.

This work was supported in part at M.I.T. under NSF Grant No. EET-8815834. A. S. Hou and R. S. Tucker were with AT&T Bell Laboratories, Crawford Hill Laboratory, Holmdel, NJ.

References

1. P. P. Vasil'ev, V. N. Morozov, G. T. Pak, Yu. M. Popov, and A. B. Sergeev, Sov. J. Quantum Electron., 15, 859 (1985).
2. J. M. Wiesenfeld, M. Kuznetsov, and A. S. Hou, IEEE Photon. Tech. Lett., 2, 319, (1990).
3. A. S. Hou, R. S. Tucker, and G. Eisenstein, IEEE Photon. Tech. Lett., 2, 322, (1990).
4. A. S. Hou, M.S. Thesis, MIT, (1990).
5. H. A. Haus, IEEE J. Quantum Electron., 11, 736, (1975).
6. M. Osinksi and J. Buus, IEEE J. Quantum Electron., 23, 9, (1987).

Femtosecond Color Center Lasers near 2.8 μm with HgCdTe Multiple Quantum Well Saturable Absorber

C.L. Cesar[1], *M.N. Islam*[1], *C.E. Soccolich*[1], *R.D. Feldman*[1], *R.F. Austin*[1], *and K.R. German*[2]

[1]AT&T Bell Laboratories, Room 4D-437, Holmdel, NJ 07733, USA
[2]Burleigh Instruments, Burleigh Park, Fishers, NY 14453, USA

We have used a HgCdTe multiple quantum well saturable absorber to passively modelock KCl:Li and RbCl:Li color center lasers and generated pedestal free pulses as short as 120 fs near 2.8 μm.

We have used HgCdTe multiple quantum well (MQW) saturable absorbers to passively modelock two color center lasers (CCL) near 2.8 μm. The KCl:Li CCL generates pulses as short as 120 fs and the RbCl:Li CCL generates pulses as short as 190 fs, which are the shortest pulses in the mid-infrared wavelength range. These CCL are the only direct laser source of sub-picosecond pulses in this wavelength range. Our results extend the semiconductor passive modelocking technique previously applied to the NaCl CCL system (1) to different wavelength range, saturable absorber material and laser systems. Both pulse durations are close to the 150 fs estimated LO phonon collision thermalization time for HgCdTe MQW (2) and are shorter than the 275 fs pulses obtained with the NaCl CCL (1). The NaCl CCL used an InGaAs/InAlAs MQW saturable absorber that has a thermalization time of 200 fs. These results are consistent with the proposal that the pulses are generated by the fast saturable absorption due to the thermalization, through LO phonon collision, of a cold carrier distribution (1).

Figure 1 shows the experimental set up. The saturable absorber is an anti-reflection coated HgCdTe MQW grown by molecular beam epitaxy. The MQW consists of 50 periods of 30 Å thick $Hg_{0.82}Cd_{0.18}Te$ wells between 100 Å thick

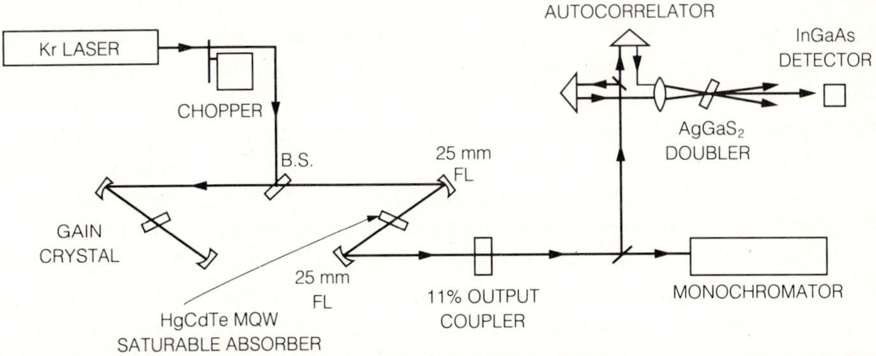

Fig. 1 . Experimental set up for the CCL lasers and diagnostics.

Springer Series in Chemical Physics, Vol. 53 **Ultrafast Phenomena VII**
Editors: C.B. Harris · E.P. Ippen · G.A. Mourou · A.H. Zewail

$Hg_{0.15}Cd_{0.85}Te$ barriers. Figure 2 shows the MQW room temperature absorption spectrum at low-level excitation and the wavelength where the lasers operate. The CCL's are pumped with a chopped (50% duty cycle) all lines Krypton laser. For the KCl:Li the pump power varies between 3.7 to 5.6 W, with the shortest pulse obtained at 3.7 W. The gain crystal is inside of a cryostat and the external part of the cavity is purged with nitrogen. Without the nitrogen purge the pulses are broadened by the water vapour absorption lines to about 5 ps. For the RbCl:Li the pump power is 4.2 W and the cavity can be open to the air, since the wavelength falls out of the water absorption band. The shortest pulses are obtained without any tuning element in the cavity. The pulse width is measured using a noncollinear autocorrelator with an anti-reflection coated, 1 mm thick $AgGaS_2$ doubling crystal and the second harmonic is detected with an InGaAs detector.

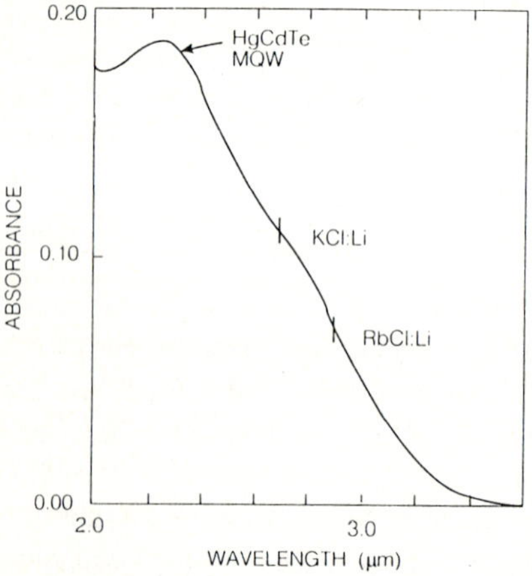

Fig. 2. The HgCdTe MQW room temperature spectrum.

Figure 3 shows the autocorrelation and spectrum of pulses from both lasers. The KCl:Li pulse is 120 fs and the RbCl:Li is 190 fs, if we assume a secant hyperbolic pulse shape. In the inserts the autocorrelations in the -20 to +20 ps range show no pedestal. The 120 fs pulse at 2.7 μm corresponds to 13.3 optical cycles. Some of the features of the spectra come from a cw contribution at the beginning of the chopper cycle, and others come from the water absorption lines that are not fully removed by the purge. When the lasers work with birefringent tuning plates, the cw sharp features are resolved from the pulse spectrum and appear at the long wavelength side. Without the birefringent plates the cw features are no longer sharp because of the superposition of several modes and overlap with the pulse spectra.

From the average output power and repetition rate we calculate the output peak power of 470 W for the KCl:Li and 415 W for the RbCl:Li. When the KCl:Li is pumped with 5.2 W the peak power is 720 W and the pulse width is 150 fs. For both lasers the energy densities in the absorber are 2 to 5 times larger than the small signal saturation energy density for the MQW. The energy densities in both gain crystals are about 10% of the gain saturation energy densities. Therefore, unlike the NaCl system (1), these lasers work with little gain saturation, which is possible only

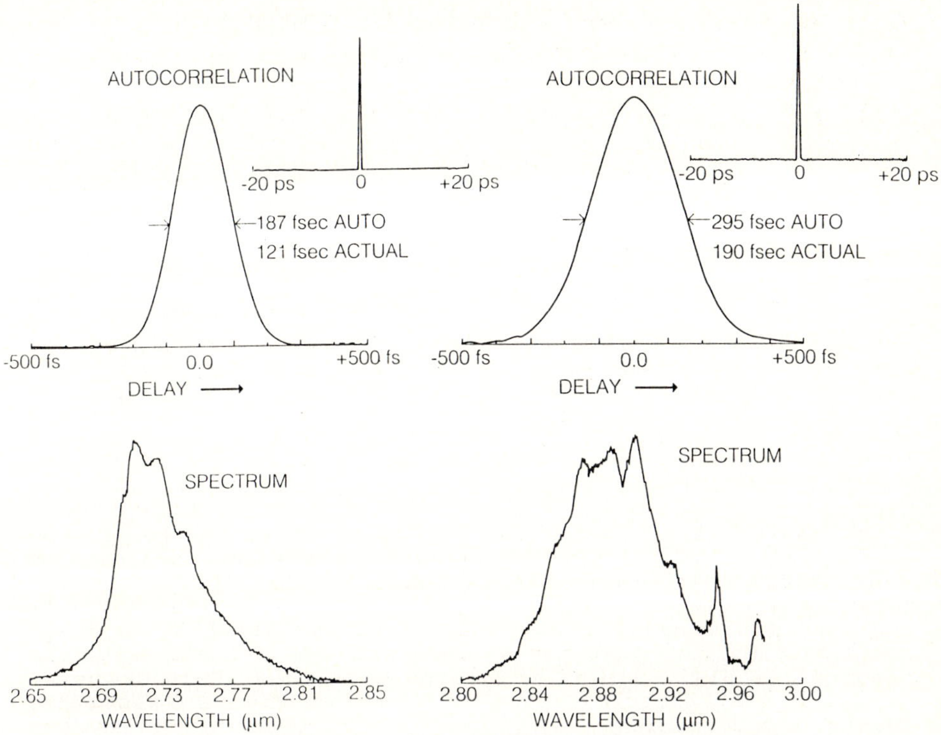

Fig. 3. Top : laser pulse autocorrelation. Insert : extended range autocorrelation showing no pedestal. Bottom : laser pulse spectrum. Left : KCl:Li. Right : RbCl:Li.

if we have a fast saturable absorption component. We believe that this fast component comes from the thermalization of a cold carrier distribution through LO phonon collisions (1).

In conclusion we obtained the shortest pulses in the mid-infrared region by passive modelocking the KCl:Li and the RbCl:Li CCL to generate pulses as short as 120 fs. Together with the previous NaCl CCl (1) work our results stress the role of the LO phonon collision thermalization time and point out the possibility of modelocking without gain saturation. C.L.Cesar is on leave from UNICAMP and has a fellowship from CNPq, Brazil.

1. M.N.Islam, E.R.Sunderman, C.E.Soccolich, I.Bar-Joseph, N.Sauer, T.Y.Chang and B.I.Miller, IEEE J. Quantum Electron. 25, 2454 (1989).
2. C.L.Cesar, M.N.Islam, R.D.Feldman, R.F.Austin, D.S.Chemla, L.C.West and A.E.DiGiovanni, Appl. Phys. Lett. 56, 283 (1990).

820 nm Colliding-Pulse Modelocked Dye Laser

W.H. Knox and F.A. Beisser

AT & T Bell Laboratories, Holmdel, NJ 07733, USA

The 620 nm CPM laser has become the most common laser for femtosecond pulse generation, with perhaps as many as several hundred units around the world in operation. The CPM laser, of course, provides direct conversion of CW light to femtosecond modelocked pulses. While it is desirable to extend this technique to other wavelengths, each new wavelength requires a painful and lengthy materials research program, with no guarantee of success. A number of conditions have been investigated [1], but the 620 nm CPM laser is still the most common laser, and one wonders whether the R6G-DODCI pair is the most "magic pair" in existence. We have developed a new CPM laser which operates in single pulse trains of high stability and find performance comparable to 620 nm CPM lasers. A six mirror cavity with gain jet of Rhodamine-800 dye and saturable absorber of HITCI dye is completed with four SF-10 brewster prisms in a standard cavity configuration. Figure 1 shows the relevant spectroscopy. We choose R-800 because it has the longest wavelength emission of the high-quantum efficiency family of rhodamine dyes; this may represent the longest wavelength attainable for CW-pumped CPM dye lasers. We find that direct pumping with an active-stabilized CW krypton laser yields better stability than when pumping with a Ti:Sapphire laser [2], because of the improved mode quality, noise and pointing stability.

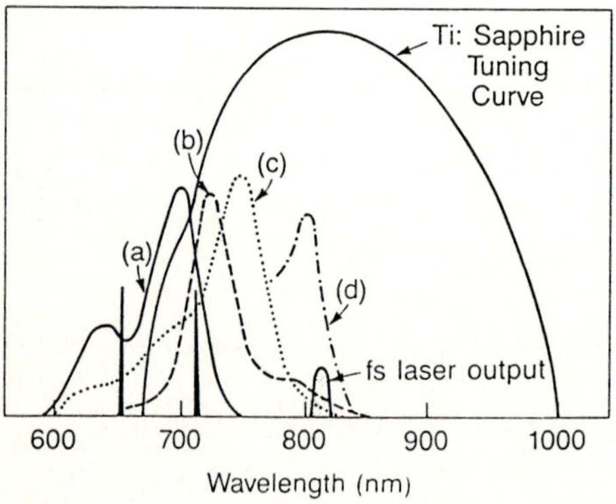

Figure 1. (a) R-800 absorption spectrum, (b) R-800 emission spectrum, (c) HITCI ground-state absorption, (d) HITCI photoisomer absorption, and Ti:Sapphire gain curve. The krypton pump line at 647 nm and the 710 nm pump wavelength used previously [2] are also shown.

Springer Series in Chemical Physics, Vol. 53 **Ultrafast Phenomena VII**
Editors: C.B. Harris · E.P. Ippen · G.A. Mourou · A.H. Zewail

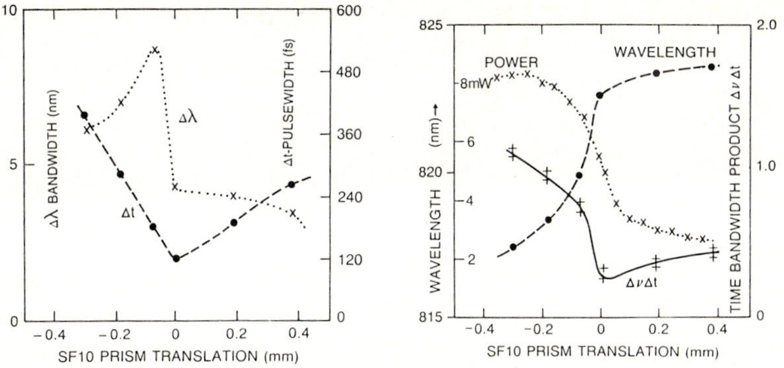

Figure 2. Scaling of laser parameters with relative intracavity glass length.

In a standard 10-12 ns cavity, double pulse trains are observed which cannot be eliminated. This is presumably due to the fact that although we measure for R-800 a 3 ns lifetime in ethylene glycol, which is comparable to the R6G lifetime, the larger emission cross-section of the IR rhodamine leads to a higher stimulated emission rate. We find that the CW pump threshold with the absorber jet out of the cavity is only 200 mw, and the operating point is at only 1.7 W average power. In a 7 ns short cavity this instability vanishes, but we find a different instability which results from the long lifetime of the HITCI (1.2 ns). We eliminated this strong instability by reducing the gain mirror radii to 5 cm radius (equal to the absorber radii), thus increasing the gain saturation level, but decreasing the S-parameter for modelocking. The strong gain saturation and lower S-parameter results in a complex behavior with respect to intracavity dispersion (Figure 2a). Minimum pulsewidths are about 100 fs, with transform-limited pulsewidths obtained for in the net positive dispersion limit. A strong variation in the mean wavelength and output power is obtained (Figure 2b). Interestingly, at low absorber concentrations, stable modelocking at positive GVD is not obtained.

In order to obtain a stable spectrum we use a mirror for the back high reflector which stabilizes the spectrum on the long wavelength end. We use a Newport corporation DM.4 mirror, which reflects about 98-99% at the lasing wavelength, but drops to 50% at about 845 nm. This mirror also provides a net negative group velocity dispersion, making it possible to run the laser even without prisms in the cavity. In this mode, extremely stable trains of pulses are obtained, with 15 mW per beam in four separate beams. The pulsewidth is about 140 fs directly from the cavity, but with several cm of extra glass added outside the cavity, 85 fs pulses are obtained. This mode of operation is similar to that of early CPM lasers. There are at least three reasons why the prisms create some problems for our infrared CPM laser. First, they lengthen the cavity by about 1 ns, moving closer to the unstable region. Second, prisms provide scattering losses which are undesirable, and third, higher order dispersion eventually becomes a limiting factor.

We find that the dye stability is excellent; the R-800 lasts at least 5 months when operated in a refrigerated pump station, and the HITCI is also stable for several months, after some

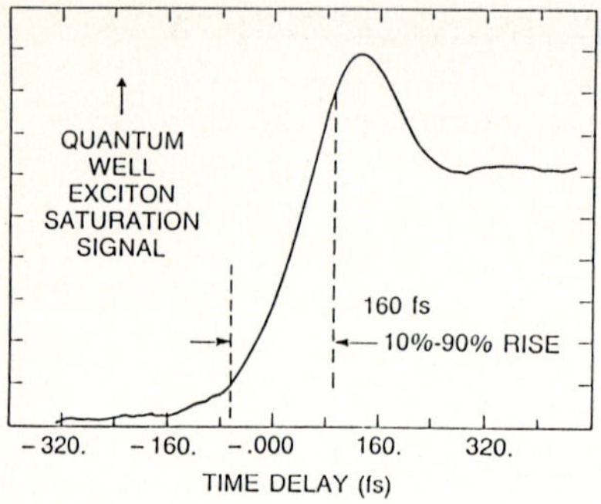

Figure 3. Exciton saturation dynamics in room-temperature GaAs quantum wells.

initial small degradation. We use nitrogen purges in both pump stations. We have tested this new laser in electroabsorption sampling experiments [3]. Figure 3 shows one result. A simple exciton saturation experiment shows a 160 fs risetime, indicating pulses of 100 fs duration.

The wavelength of 820 nm is close to gain peak of LDS-821 emission. We have amplified these pulses to several microjoules energy at 8 kHz repetition rate in a 5-pass copper-vapor laser-pumped dye jet amplifier, obtaining 33,000x net energy gain. Preliminary pulse compression results in short single-mode fibers indicate chirped bandwidth equivalent to a 7 fs pulse, although imperfect phase compensation in preliminary experiments produced compressed pulses of 10-12 fs duration.

Clearly, amplification in Ti:Sapphire is a desirable application for this source (Figure 1), where the clean satellite-free CPM laser operation and lack of cavity length drift will be important considerations.

[1] P.M.W. French and J.R. Taylor, Laser Focus Feb. 1989 gives a recent review, and P. Georges, F. Salin and A. Brun, Opt. Lett. 14, 785 (1989).

[2] W.H. Knox, Postdeadline paper PD31, CLEO 1989, Baltimore, MD.

[3] W.H. Knox, G.E. Doran, B. Tell, K.W. Goossen, G. Hasnain, J. Cunningham and D.S. Chemla, IEEE JQE 25, 2586 (1989).

Clean Subpicosecond Pulses by Coupling Resonantly to a Linear High Loss Cavity

P. Beaud, J.Q. Bi, J. Schütz, W. Hodel, and H.P. Weber

Institute of Applied Physics, University of Berne, Sidlerstr. 5, CH-3012 Berne, Switzerland

Abstract. *Extremely low feedback from an external linear cavity coupled to a synchronously pumped dye laser drastically stabilizes the laser operation on a microsecond time scale and significantly improves the mode locking. Nearly Fourier transform limited pulses result.*

In recent years coupling of a mode locked laser to a nonlinear external cavity has been established as an excellent technique for the generation of ultrashort light pulses [1-4]. In general, the mode locking improves and shorter pulses are generated. The simpler case of a linear external cavity is not expected to lead to the generation of shorter pulses. However, we experimentally discovered that extremely low feedback from the external cavity leads to a significant improvement of the mode locking and transform limited pulses are generated.

The mode locked laser investigated is a conventional linear dye laser, synchronously pumped by compressed Nd:YAG laser pulses. It delivers subpicosecond pulses tunable in wavelength between 1.25 and 1.35 µm [5]. The length of the external cavity (formed by the output coupler of the dye laser and an end mirror) is matched to that one of the dye laser. A beam splitter inside this external cavity serves to couple light out of the laser system and together with an additional variable attenuator allows to adjust the feedback level (over a large range down to less than 10^{-10}).

When the laser is coupled to the external cavity the pulse width measured by autocorrelation slightly increases but the pronounced wings of the pulse present in absence of any feedback are clearly suppressed. The autocorrelation closely fits that one of a $sech^2$ pulse (see Fig.1). In addition, the laser stability dramatically improves on a microsecond time scale. The examples shown in the Figures are for a feedback level of 10^{-10}. For levels lower than 10^{-10} the influence of the feedback rapidly reduces. On the other hand increasing the feedback level up to 10^{-6} the pulse duration gradually increases to approximately 700 fsec and the rapid part of the laser noise further decreases. For feedback levels higher than 10^{-6} the pulse duration approaches that of the pump pulse (4 psec) and satellites appear.

The pronounced wings of the autocorrelation trace in absence of any feedback from the external cavity clearly indicates the presence of a random substructure within the pulse envelope [6] and it is the random nature of this substructure that causes the rather large noise band observed in Fig. 2 [7]. The action of the feedback is therefore to suppress the pulse substructure and consequently improves mode locking. This interpretation is strongly supported by the fact that the

Springer Series in Chemical Physics, Vol. 53 **Ultrafast Phenomena VII**
Editors: C.B. Harris · E.P. Ippen · G.A. Mourou · A.H. Zewail

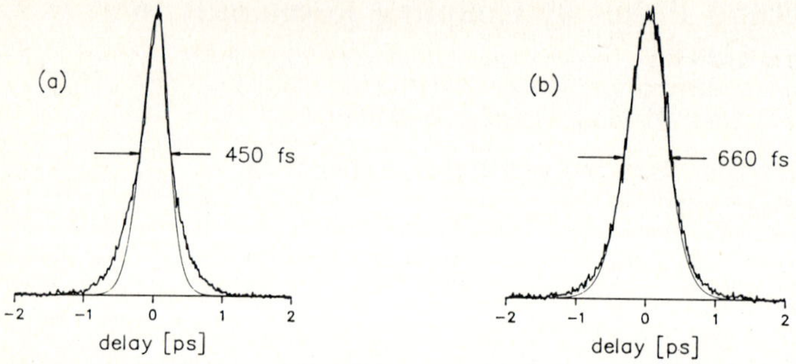

Fig. 1: Autocorrelation of the laser pulses (a) without and (b) with external cavity for a feedback of 10^{-10}. Assuming a $sech^2$ pulse shape the corresponding pulse durations are 290 and 430 fsec, respectively. For comparison the calculated traces of a $sech^2$ pulse are also shown as weak solid lines.

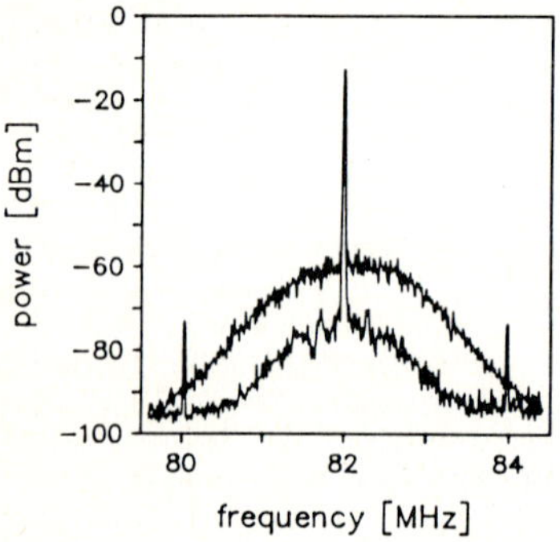

Fig. 2: Power spectrum of the dye laser centered at the repetition frequency of the laser system without (upper) and with external cavity (lower trace) for a feedback level of 10^{-10}. The resolution was 30 kHz.

time bandwidth product which is without feedback typically 0.75 now is reduced into the range from 0.25 to 0.35. This is close to 0.31, the value for a perfectly Fourier transform limited $sech^2$ pulse.

This result is quite surprising considering that a feedback level of 10^{-10} roughly corresponds to three photons per roundtrip which are coupled back into the main laser cavity. How can some single photons significantly influence the mode locking of the dye laser? Using a numerical model Catherall and New have shown [8] that spontaneous emission plays a crucial role in the dynamics of mode locking by synchronous pumping. At the setting of the cavity length, where optimum pulses are generated, the effective velocity of the pulse envelope is slightly different from that one of the signal velocity. Spontaneous noise initially located just ahead of the pulse moves therefore into the pulse envelope and causes severe perturbations due to the random nature of the spontaneous light. These perturbations which are then progressively amplified until they have reached the pulse maximum, are responsible for the relatively poor mode locking usually observed in synchronously pumped dye lasers [8]. We believe that the feedback pulse acts as a coherent seeding signal at the (very early) leading edge of the laser pulse and suppresses the influence of the spontaneous emission by gain competition. Note that the power of the seeding pulse at the 'feedback threshold' of 10^{-10} and the power of spontaneous light emitted into the laser mode are both in the same order of magnitude (10^{-6} W).

It should be mentioned that the long term stability of the laser system is not improved by the external cavity. On a millisecond time scale we observe a hopping of the laser frequency and discontinuities of the laser intensity of up to 10 %. These features are probably due to random fluctuations of both cavity lengths which causes the weak feedback pulse to interfere either constructively or destructively with the main laser pulse. In order to keep the optimum phase matching between the pulses, the laser (based on the group velocity dispersion of both cavities) self-adjusts its effective optical path length through a shift in wavelength. This is possible because - in contrast to the case of a nonlinear external cavity - the spectral bandwidth of the laser pulse is much narrower than the bandwidth supported by the tuning element. To improve the laser stability in this time scale an active stabilization of the cavity lengths similar to that reported in Ref. [1] would be required.

In conclusion we have shown that extremely low feedback from a linear external cavity strongly improves the mode locking performance of a synchronously pumped dye laser. In general practically Fourier transform limited pulses result. Besides its usefulness this phenomena allows a deeper insight into the dynamics of mode locking by synchronous pumping.

[1] F.M. Mitschke, and L.F. Mollenauer: IEEE J. Quantum Electron. **QE-22**, 2242 (1986).

[2] K.J. Blow, and D. Wood: J. Opt. Soc. Am. **B5**, 629 (1988).

[3] P.N. Kean, X. Zhu, D.W. Crust, R.S. Grant, N. Langford, and W. Sibbet, Opt.Lett. **14**, 39 (1989).

[4] J. Mark, L.Y. Liu, K.H. Hall, H.A. Haus, and E.P. Ippen: Opt. Lett. **14**, 48 (1989).

[5] P. Beaud, B. Zysset, A.P. Schwarzenbach, and H.P. Weber: Opt. Lett. **11**, 24 (1986); P. Beaud, B. Zysset, and H.P. Weber: SPIE **710**, 446 (1987).

[6] D.B. MC Donald, J.L. Rossel, and G.R. Fleming, IEEE J. Quantum Electron. QE-17, 1134 (1981).

[7] J. Kluge, D. Wiechert, and D. Von der Linde: Opt. Comm. **51**, 271, (1984).

[8] J.M. Catherall, and G.H.C. New, IEEE J. Quantum Electron. QE-22, 1593 (1986).

Multireflection Effects in Gain-Switched and Q-Switched Semiconductor Lasers

B.J. Thedrez and C.H. Lee

Joint Program of Advanced Electronic Materials, Electrical Engineering Department, University of Maryland, College Park, MD 20742, USA

Abstract. A theoretical study of a one-section and a two-section laser driven by extremely fast risetime signal is presented. Pulses of 1 ps are predicted with reduced multireflection effects.

1. Introduction

In order to get short pulses from semiconductor lasers, an extremely short and intense pumping signal has to be applied to the device. Optical switches have proved to be efficient sources for providing electrical pulses with risetimes as short as the incident optical beam and thus on the order of 1 ps or less [1]. The exponential decay time depends on the recovery time of the switch and ranges from the nanosecond to the picosecond time scale. We have theoretically investigated the optical response of a semiconductor laser to short electrical pulses of this type (<25 ps) and demonstrated that stable and repetitive pulses of 1 ps full width at half maximum (FWHM) can be generated.

2. Description of the device.

For applications where Q-switching is employed, a multisection laser is required. One section at least provides gain, while another section is used to absorb the light and increase the loss. Large changes in loss due to the quantum confined Stark effect (QCSE) have already been demonstrated in a multiquantum-well two-section laser by applying a reverse voltage on the modulator section [2]. Since the QCSE has an extremely fast response time, it is possible to achieve adequate losses in times as short as 1 ps, with an optical switch integrated into the laser. For our simulation, we have investigated a 1 ps loss change in a 30 μm long modulator section. The gain section in this case is 270 μm long. On the other hand, the device chosen for simulating gain-switching is a one section 300 μm long laser. Parameters used for our calculations are given in Table 1. To generate stable and repetitive pulses, the laser must be biased strictly above threshold [4]. For reason of power consumption, we will only consider the case of threshold.

3. Multireflection effects

An extension of the common rate equations have been used where forward and backward travelling beams are separately calculated [3]. The approximation that distributed loss is valid on time scales larger than the roundtrip time T_r

Springer Series in Chemical Physics, Vol. 53 **Ultrafast Phenomena VII**
Editors: C.B. Harris · E.P. Ippen · G.A. Mourou · A.H. Zewail

Table 1: Parameters of the semiconductor laser.

Parameters	value
confinement factor	0.2
spontaneous emission factor	10^{-5}
differential gain	5 10^{-16} cm^{-2}
carrier lifetime	2 ns
index of refraction	3.6
internal loss	12 cm^{-1}
wavelength	830 nm

is not needed, and extremely fast phenomena can be studied both in time and space. This is important since a spatial dissymmetry exists in the multisection laser already described. If the change in the modulator section occurs in a time smaller than half the roundtrip time T_r, then we *locally* change the intensity inside the cavity. This gives rise to a propagating pulse whose FWHM is smaller than $T_r/2$. This pulse experiences many roundtrip excursions and gives rise to a train of separate pulse with final energy which is limited by the saturation of the gain (fig.1-a).

In the case of gain-switching, if the laser is uniformly pumped at or above threshold, the field is uniform and no travelling pulse is generated inside the cavity. The pulse is created at the end of the process by the saturation of the gain. However, even if the saturation is fast, it takes at least half a roundtrip-time to extract the light from the cavity, and multireflections tend to broaden the envelope of the emitted pulse (fig.1-b).

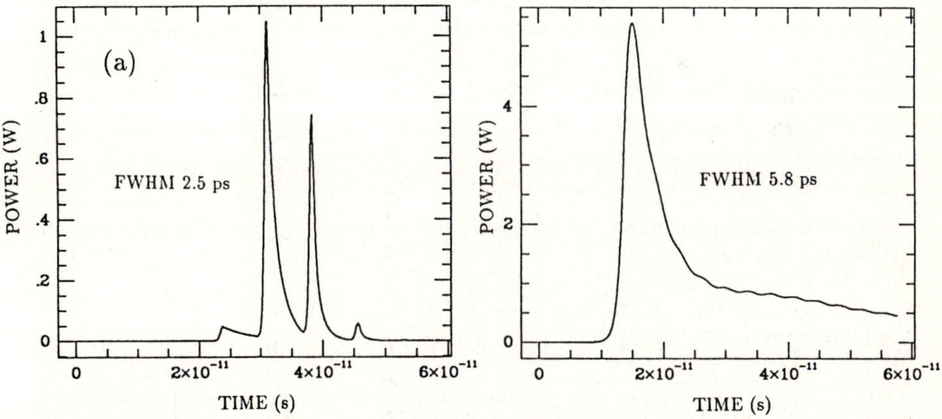

Fig.1(a): Gain-switching in a 300 μm long laser biased at threshold. Pump pulse is 25 ps wide with 1 ps risetime. Decay is exponential. Increase of carriers is 5 x 10^{18} cm $^{-3}$.

Fig.1(b): Q-switching in a 300 μm long laser with a 30 μm long modulator section. Loss decreases linearly from 4500 cm^{-1} to 100 cm^{-1} in 1 ps and recovers exponentially with a 25 ps FWHM . The laser is biased at threshold.

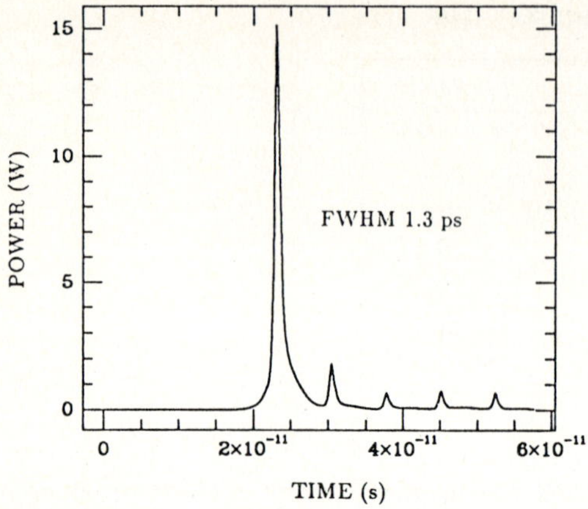

Fig.2: Output pulse from a 300 μm long laser with a 30 μm long modulator section. The driving signals are the same as in fig.1(a) for the gain section and fig.1(b) for the loss section.

Both an increase in power compared to Q-switching and a reduction of the FWHM compared to gain-switching with reduced reflection effects can be observed if both techniques are applied together. The gain-switch process provides the build-up of the pulse. If the loss is switched off after a proper delay, an effective Q-switch follows, and gives rise to a narrow pulse. Due to the large increase of carriers created by pumping, a fast saturation of the gain occurs. As a result, the second reflection of the pulse only experiences loss and is immediately quenched (fig.2). Therefore,the resulting FWHM of the laser pulse depends mainly on the risetime of the electrical pulse, not on its duration.

4. Conclusion

Ultrashort pulses have been predicted when fast risetime electrical signals are used to drive a two-section laser. A new technique based on gain-switching and Q-switching processes should result in pulses of 1 ps FWHM with reduced multireflection effects.

5. References

[1] Chi H.Lee ed.,Picosecond Optoelectronics Devices (Academic, Orlando, 1984)

[2] Y.Arakawa,A.larsson,J.Palaski,and A.Yariv, Appl.Phys.Lett. 48, 561 (1986).

[3] S.W.Corzine,J.E.Bowers, G.Przybylek,U.Koren,B.I.Miller, and C.E. Soccolich, Appl. Phys. lett. 52, 348 (1988).

[4] A.Mecozzi, P.Spano, and A.D'Ottavi, Appl.Phys.Lett. 55, 769 (1989).

Dramatic Reduction of the Amplitude Noise of Colliding-Pulse-Modelocking Dye Lasers

M.C. Nuss, U. Keller, G.T. Harvey, M.S. Heutmaker, and P.R. Smith

AT&T Bell Laboratories, Crawfords Corner Road,
Room 4C-330, Holmdel, NJ 07733, USA

The colliding pulse modelocking (CPM) dye laser has established itself as the work horse of femtosecond optics and spectroscopy owing to the ease of generating stable optical pulses of less than 100 fsec duration. The probably most important improvement of the CPM laser has been the addition of a four-prism sequence to adjust the intracavity group velocity dispersion [1]. Although the CPM laser has been object of extensive studies to understand the complex pulse forming mechanisms, the characterization of its amplitude noise has not received much attention. We show in this manuscript that the strong amplitude noise of the CPM laser at frequencies of several MHz can be significantly reduced by simply inserting a single frequency etalon in the cavity of the (argon-ion) pump laser.

It is well known that the noise spectrum of a CPM laser does not roll off at a 1/f rate as expected, but shows distinct maxima of considerable width and amplitude at spacings of several MHz [2]. Figure 1 (ΔΔΔ) shows the rf power spectrum of the CPM laser in a 0-5 MHz range obtained with a shot noise limited detector and a rf-spectrum analyzer. The shot noise of the detector in this plot is at -96 dBc, whereas the instrument noise (dotted curve) of the detector-amplifier-spectrum analyzer com-

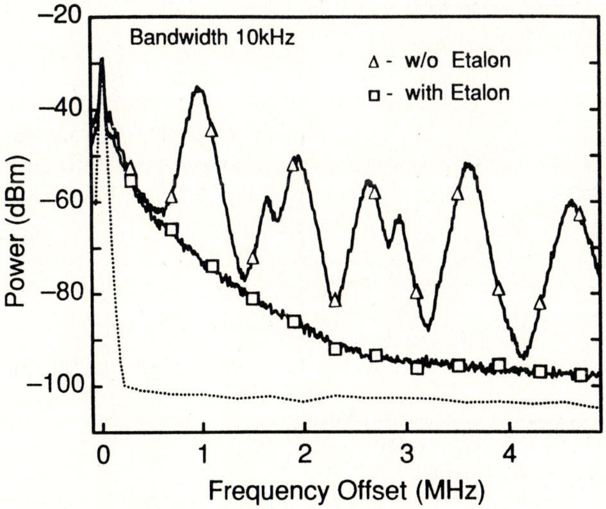

Fig. 1 CPM amplitude noise spectrum with and without etalon in the argon cavity.

Springer Series in Chemical Physics, Vol. 53 **Ultrafast Phenomena VII**
Editors: C.B. Harris · E.P. Ippen · G.A. Mourou · A.H. Zewail

bination is at -104 dBc. The detector was ac-coupled for this measurement, so that the amplitude noise at low frequencies (<100 kHz) is not plotted in perspective. Although the peak of the sidebands is well below the carrier, their amplitude rises typically 50 dB from the 1/f-like background noise. The origin of these spurious sidebands has been completely obscure so far. We find that the position of these sidebands moves as the cavity length of the CPM laser is varied. From the analysis of the peaks and their shift with cavity length we argue that the noise peaks originate from mixing of the amplitude noise of the argon laser due to longitudinal mode beating with the CPM laser [3,4]. With a cavity frequency of the argon laser $f_{AR} = 76.37\,MHz$ and $f_{CPM} \sim 100\,MHz$, many mixing products $f_{mix} = |Kf_{AR} - Jf_{CPM}|$ are possible that yield mixing frequencies at few MHz. Here, K and J are the K-th and J-th harmonic of the cavity frequency of argon and CPM laser, respectively.

Therefore, operating the argon pump laser in a single longitudinal mode should remove harmonics of the cavity roundtrip frequency from the output of the argon laser and hence the spurious amplitude noise of the CPM laser. The curve with the square symbols (□□□) in Fig. 1 shows the rf spectrum with the etalon in place with the same output power and photocurrent as before. Indeed, the spurious modulations have disappeared completely and a clean noise spectrum that rolls off as roughly 1/f is obtained. The amplitude noise reaches the shot noise of the detector at roughly 3.5 MHz.

With the etalon in place, we can perform a thorough analysis of the amplitude noise of the CPM laser over a wide frequency range. A detailed description of the principle of the noise characterization can be found in reference [5]. Fig. 2 displays the amplitude noise of the CPM laser at a photocurrent of 1 mA and compares it to

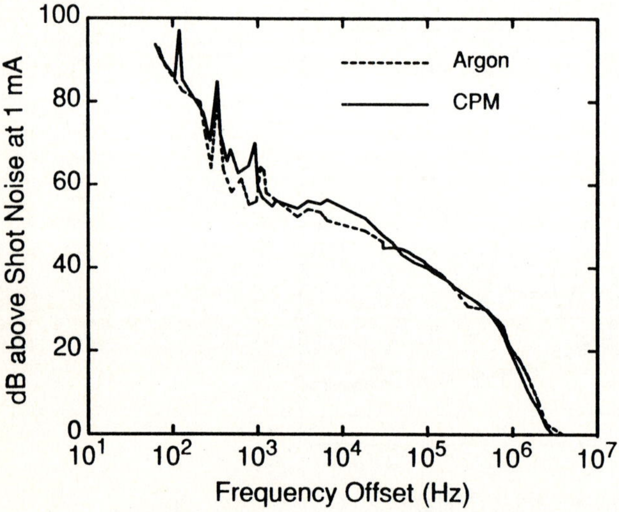

Fig. 2 Amplitude noise of the CPM laser as well as the argon-ion pump laser with etalon in the pump laser cavity.

the amplitude noise of the argon-ion pump laser. Both noise spectra are essentially featureless except for harmonics of the power line frequency at frequencies below 1 kHz. It is apparent from Fig. 2 that over the entire frequency range from 50 Hz to 3.5 MHz the CPM noise is limited by the noise of the argon laser, which presumably results from plasma instabilities. The AM noise reaches the shot noise of the detector at roughly 3.5 MHz. This compares to ~100 kHz for a Nd:YAG laser and suggests that the amplitude noise of the CPM laser could be further reduced if a solid-state cw pump laser in the visible spectral range was available.

The reported noise reduction due to intermodulation of the harmonics of the cavity roundtrip frequency of pump laser and dye laser is most prominent at frequencies of several MHz although we have observed mixing sidebands at frequencies of only few kHz for certain cavity lengths. The reported noise reduction can be exploited by MHz modulation and detection of the laser beam. For example, we observe a 60 times improvement in signal-to-noise ratio in electro-optic sampling experiments when using a single frequency etalon in the argon laser cavity and a modulation frequency of 3.5 MHz. A similar increase in signal-to-noise ratio can be expected in any pump-probe experiments when a shot-noise limited detector and chopping above 3 MHz (e. g. with an acousto-optic modulator) is used.

1. J. A. Valdmanis, and R. Fork, Opt. Lett. **10**, 131 (1985)
2. D. von der Linde, Appl. Phys. **B39**, 201 (1986)
3. G. T. Harvey, M. S. Heutmaker, P. R. Smith, J. A. Valdmanis, and M. C. Nuss, Proceedings of the Picosecond Electronics and Optoelectronics Conference, Salt Lake City, March 1989
4. M. C. Nuss, U. Keller, G. T. Harvey, M. S. Heutmaker, and P. R. Smith, Optics Lett., August 1990 issue
5. U. Keller, K. D. Li, M. Rodwell, and D. M. Bloom, IEEE Journ. Quant. Electr. **25**, 280 (1989)

Kilohertz Amplification of Femtosecond Pulses near 1.55 μm, to Microjoule Energies

G. Sucha, M. Wegener, S. Weiss, and D.S. Chemla

AT&T Bell Laboratories, Room 4B-401, Holmdel, NJ 07733, USA

There is much interest in femtosecond spectroscopy of semiconductor heterostructures in materials such as InGaAs or InAlAs which have a bandgap wavelength near 1.5μm . For a long time, the time resolution of these studies was limited to several picoseconds, which is characteristic of synchronously pumped F-center lasers used to do spectroscopy in these materials. Recently, stable sources of 100 fsec pulses around 1.5μm have been developed, using F-center lasers. The soliton laser [1] and subsequently, the additive-pulse-modelocked laser (APM) [2,3] and coupled-cavity-modelocked laser [4] are capable of producing 100-200 fsec pulses routinely, with up to 1.5 nJ of energy. However, for some applications, higher pulse energies are needed. In particular, femtosecond continuum generation (desirable for spectroscopy) requires pulse energies of a few microjoules. This requires amplification of the existing sources by a factor of 1000 or more. We report here the amplification of femtosecond pulses to energies of 1-2 μJ using a multi-pass amplifier with NaCl F-centers as the gain medium. F-centers in KCl:Li have been used previously to amplify nanosecond pulses from another F-center laser operating at $\lambda \approx 3\mu$m. [5]

The NaCl F-center system (actually an impurity stabilized F_2^+ center with oxygen as the impurity [6]), is an attractive medium for femtosecond pulses because of its large gain bandwidth ($\gtrsim$200nm). The gain spectrum of the NaCl F_2^+ is shown in figure 1 along with the spectrum of a transform-limited 70 fsec pulse for comparison. Also, it has a gain cross section ($\lesssim 10^{-16}$cm^2) which is comparable to dyes. Thus, it has a saturation fluence on the order of a few mJ/cm^2. Energy storage densities are typically on the order of a few mJ/cm^3. This center also has a gain storage lifetime of 150 nsec which is over 10 times longer than for dyes. This allows us to use a pump source with pulses as long as 100 nsec, which reduces the cost and complexity of the system.

The multi-pass amplifier, shown in Fig. 2, is fashioned after the multi-pass "bowtie" amplifier of Knox, et al. [7]. We use a 7 mm thick NaCl crystal which is held at 77K in a cryostat. This crystal is pumped by a Q-switched Nd:YAG laser which produces pulses at λ=1.06μm , of 2 mJ energy with 80 nsec duration at 1 kHz repetition rate. A pump spot size of ≈ 2 mm is used. Both the pump pulse and the injected femtosecond pulse were horizontally polarized. Blue light from a filtered mercury-arc lamp was used to illuminate the crystal in order to reverse the orientational bleaching effect which is common to many F-center lasers.

Springer Series in Chemical Physics, Vol. 53 **Ultrafast Phenomena VII**
Editors: C.B. Harris · E.P. Ippen · G.A. Mourou · A.H. Zewail

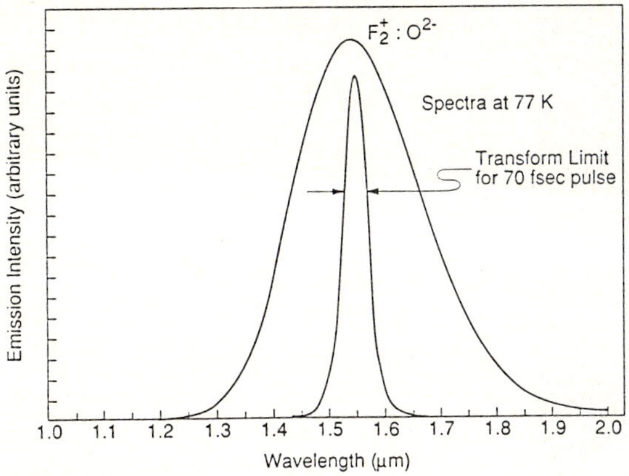

Figure 1.
Gain spectrum of F_2^+ centers in NaCl (upper curve) and theoretical spectrum of a transform limited 70 fsec pulse (lower curve).

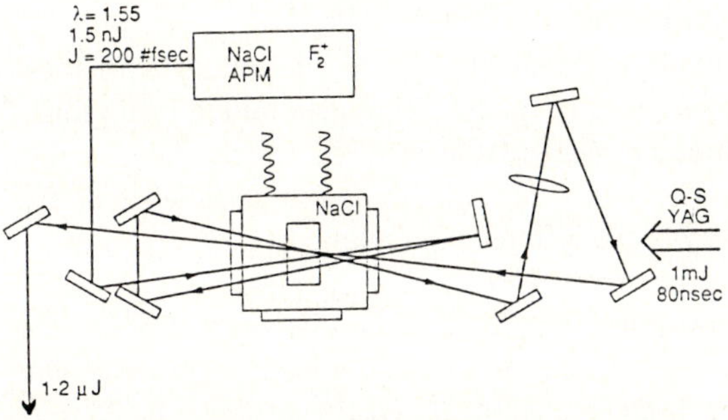

Figure 2. Schematic of 4-pass "bowtie" amplifier.

Due to the large crystal thickness, all the beams are sent through the crystal in a very nearly collinear fashion for multi-pass operation. The pulses to be amplified are produced by an NaCl APM laser [3]. Interestingly, this laser is self-starting [8]. It produces 200 fsec pulses at $\lambda \approx 1.55 \mu m$ with energies of 1.5 nJ.

The crystal in this amplifier has a single-pass gain of nearly 7, which is to be expected for nearly total inversion, while the entire 4-pass amplifier has a 50% linear loss when unpumped. Thus, the amplifier has a net gain of over 1000. The measured ASE energy is at least a factor of 100 below the amplified pulse energy. The amplified pulse train is shown in figure 3. The peak pulse of this train has 1 to 2 μJ energy while the pulse train contains well over 10 μJ . This energy is not limited by gain saturation. When the beam is reinjected for a fifth pass through the amplifier, the single pass gain of the fifth pass is nearly 6 --almost the small signal gain. Thus, we should be able to extract over 50 μJ without serious gain saturation. However, we are presently limited to 4 passes through the crystal due to lensing from the curvature of the crystal surfaces which occurred during hand polishing.

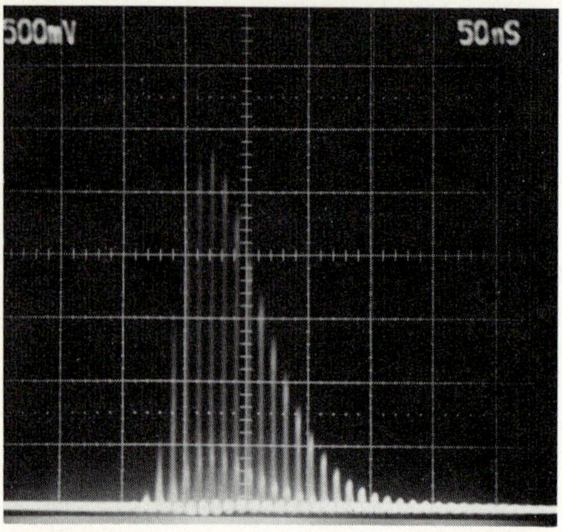

Figure 3. Amplified pulse train after 4-pass amplifier. Pulse train contains over 10μJ .

Increasing the repetition rate of the Q-S YAG laser to 5 kHz causes a decrease in the pump pulse energy which results in a decreased gain by about a factor of 5. This should be easily compensated by another pass through the crystal.

Preliminary attempts were made to generate infrared continuum with these pulses by focusing into a 3 mm thick piece of quartz. The unamplified background from the oscillator made direct measurement of the infrared continuum difficult. However, visible continuum was clearly produced [9]. It is not clear whether this visible continuum is the tail of a very broad infrared continuum, or if it is a narrower infrared continuum which has been upconverted by the fundamental.

In conclusion, we have demonstrated that femtosecond pulses at 1.55μm can easily be amplified to microjoule energies using F_2^+ centers in NaCl. This was done at a 1 kHz repetition rate. This system should be easily scalable to tens of microjoules and/or several kHz rep rate.

REFERENCES

1. L.F. Mollenauer et al., Opt. Lett. (9), 13(1984)
2. J.Mark et al., Opt. Lett. (14), 48(1989)
3. C.Yakymishyn et al., Opt. Lett. (12), 621(1989)
4. P.N.Kean et al., Opt. Lett. (14), 39(1989)
5. K.R.German, Proc. International Conference on Lasers '79 p. 374.
6. E.Georgiou et al., Phys. Rev. B (35) 7636(1987)
7. W.H.Knox et al., Opt. Lett. (9) 552(1984)
8. G.Sucha et al., to be published.
9. G.Sucha et al., to be published.

A Broadly Tunable cw Femtosecond Source for Ultrafast Spectroscopy

E.S. Wachman[1], *D.C. Edelstein*[2], *W.S. Pelouch*[1], *and C.L. Tang*[1]

[1]Materials Science Center, Cornell University, Ithaca, NY 14853, USA
[2]IBM T.J. Watson Research Center, Box 218, Yorktown Heights, NY 10598, USA

Abstract. Preliminary measurements of hot carrier relaxation at energies near the L valley minimum in InGaAs are reported using a broadly tunable cw femtosecond optical parametric oscillator.

1. The Femtosecond Optical Parametric Oscillator

As ultrafast laser technology becomes available to researchers in increasingly diverse fields, there is a growing interest in versatile subpicosecond sources that can be tuned broadly throughout the near- to mid-infrared. Much of this region remains inaccessible to all but a few researchers. In addition, most femtosecond tunable systems do not provide the cw tunability desirable for many experiments.

We have developed a broadly tunable, high repetition rate femtosecond source by directly pumping an optical parametric oscillator (OPO) with a continuous wave femtosecond laser. The fs OPO, shown in figure 1, is based on a 1.4 mm piece of $KTiOPO_4$ located at an intracavity focus of a standard Rh6G-DODCI CPM dye laser.

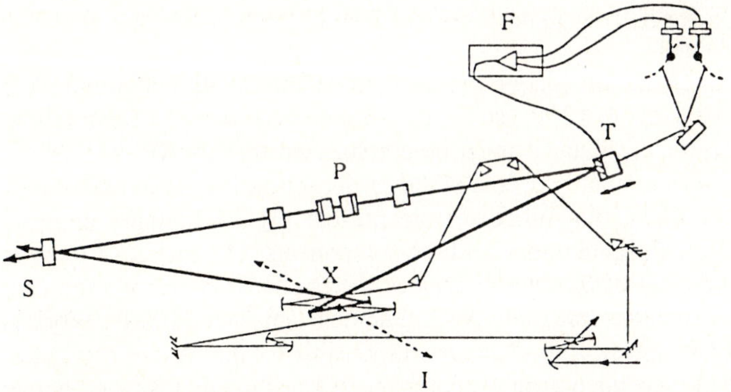

Figure 1: Configuration of the femtosecond OPO. Signal optics and beam (S) are indicated by thick lines. The idler beams (I) are shown by dotted lines, and exit the KTP crystal (X) as shown. Four prisms (P) are included for GVD compensation in the OPO cavity. The back mirror is mounted on a piezoelectric transducer (T) for length adjustment, and is controlled by the feedback circuit (F). The OPO is pumped by a CPM dye laser (thin lines).

Springer Series in Chemical Physics, Vol. 53 **Ultrafast Phenomena VII**
Editors: C.B. Harris · E.P. Ippen · G.A. Mourou · A.H. Zewail

The CPM non-collinearly pumps the singly-resonant oscillator, resulting in milliwatt outputs at a 100 MHz repetition rate that are potentially tunable from 0.72 to 4.5 μm [1]. A four-prism sequence inserted in the oscillator cavity provides adjustable group velocity dispersion (GVD) compensation, permitting chirp-free pulse generation with pulsewidths of 105 fs obtainable between 800 and 900 nm. Active cavity stabilization enables stable operation for periods of hours [2].

2. Hot Carrier Relaxation Measurements in InGaAs

The femtosecond OPO is well-suited for optical measurements of carrier relaxation times in semiconductors, since it allows the selective excitation of carrier populations throughout a continuous range of energies, and its cw output is particularly beneficial for high signal-to-noise measurements.

Although the ternary III-V semiconductor InGaAs has shown great promise for a number of device applications, relatively few measurements of carrier relaxation times have been reported in the material, and the bulk of these have been made near the band-edge (0.74 eV). For many electronic device applications, however, energies closer to the L valley minimum (~1.29 eV) are of the greatest interest. In particular, it should be possible to isolate the effect of Γ-L intervalley scattering on carrier relaxation at these energies by exciting carriers in the central Γ valley both above and below the energy of the L valley minimum. The fs OPO is the only high repetition rate femtosecond source sufficiently tunable in this wavelength range to permit such an investigation.

Transmission correlation measurements of this type are now in progress to study samples of high quality MBE-grown InGaAs etched to a thickness of ~4500 Å (roughly two absorption lengths at 1.29 eV). Using a novel galvonometer-based optical delay generator to scan over delays greater than 7 ps at a rate of 10 Hz, and averaging the results of many successive scans, signal to noise ratios well in excess of 60 dB may be obtained.

Typical data at an excitation energy 185 meV above the L valley minimum and an excitation density of 4×10^{18} cm^{-3} are shown in figure 2. Despite the extraordinary quality of the data, it must be emphasized that careful analysis is required in order to extract reliably details of the carrier relaxation, since both multi-component and non-exponential processes are present [3]. Preliminary analysis indicates a relaxation with several distinct time components. The fastest of these is apparent in the broadening of the central peak of the transmission correlation data relative to the autocorrelation trace and has a time constant under 100 fs. Work is underway to identify the physical mechanisms responsible for the various processes, and to investigate their dependence on excitation energy as the OPO is tuned below the threshold for intervalley scattering.

3. Conclusion

We have described a cw mode-locked optical parametric oscillator providing subpicosecond tunability throughout the near- to mid-infrared with transform-limited

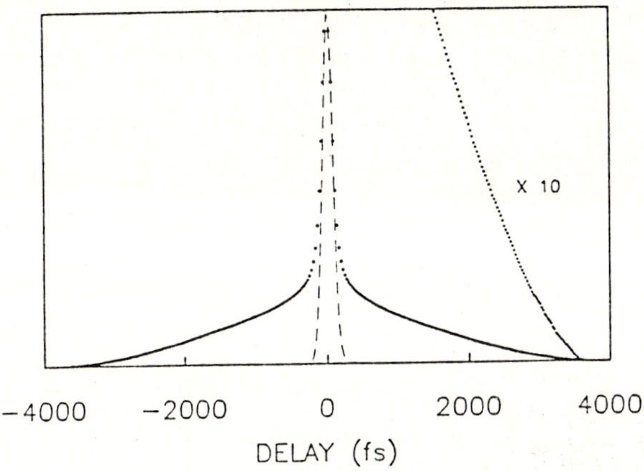

Figure 2: Typical transmission correlation data (symbols) in bulk InGaAs at 185 meV above the L valley at a carrier density of 4×10^{18} cm^{-3}. The pulse autocorrelation (dashed line) is shown for reference. The data is also shown on an expanded (x10) scale.

pulses and milliwatt outputs. Preliminary transmission-correlation measurements of hot carrier relaxation in bulk InGaAs in the vicinity of the L valley using the OPO are presented that reveal the presence of a relaxation with multiple time components.

4. References

[1] D.C. Edelstein, E.S. Wachman, C.L. Tang: Appl. Phys. Lett. **54**, 1728 (1989)
[2] E.S. Wachman, D.C. Edelstein, C.L. Tang: Optics Letters **15**, 136 (1990)
[3] F.W. Wise, M.J. Rosker, G.L. Millhauser, C.L. Tang: IEEE J. Quantum Electron. **23**, 1116 (1987)

Passive Feedback for Stable Short Pulse Generation

A. Agnesi[1], *J.-C. Diels*[2], *P. Di Trapani*[1,2], *V. Kubecek*[3], *J. Marek*[3], *G. Reali*[1], *and C.Y. Yeh*[2]

[1]Department of Electronics, University of Pavia, Italy
[2]Physics Department, University of New Mexico, Albuquerque, NM87131, USA
[3]Faculty of Nuclear Science and Physical Engineering, Technical University of Prague, Czechoslovakia

Abstract Self defocusing associated with intracavity two-photon absorption in mode-locked solid state laser is shown to extend and stabilize the pulse train, and compress the pulses.

1. Motivation

Hybrid pumping of fs lasers requires a train of at least 100 pulses with high amplitude stability. Therefore, continuously mode-locked (YAG or argon) lasers have been the most commonly used sources. Q-switched lasers would be preferable, because they provide shorter pulses of higher peak power, but the train consists only of a few pulses of widely varying amplitude. It is shown here that inclusion of a properly apertured two-photon absorbing semiconductor, in a hybridly Q-switched and mode-locked solid state laser, results in the generation of roughly 100 pulses of equal amplitude, bandwidth limited and high energy (as compared to the cw mode-locked laser) pulses.

2. The Role of an Intracavity Slab of GaAs

In a mode-locked laser, optimal pulse shortening in the saturable absorber can be achieved if the peak intensity is maintained close to the saturation intensity of the dye. An electronic method [1] has been demonstrated, which uses an electro-optic modulator and a negative feedback loop to limit the individual pulse energy. A two-photon absorber would provide a similar function, with the additional advantage of acting -- jointly with the saturable absorber -- directly on the pulse intensity. By limiting the peak intensity, the "inducible absorption" [2] reduces the depletion rate of the inversion, thereby extending the duration of emission. However, two photon absorption is known to cause pulse broadening [2], and to require peak powers close to the damage threshold.

Our measurements show that a two photon absorbing semiconductor contributes instead to pulse **shortening** inside the laser cavity, at power densities below those required for the observation of direct two photon attenuation. The photoinduced carriers create a negative lens that defocuses the beam.(Fig 1) This self-defocusing is an **integrated** effect, since the recombination time of the two-photon generated carriers is 400 ps. It is mainly the tail of the pulse that is deflected, (Fig. 1b) resulting in a pulse shortening that can be as large as 50% in

Springer Series in Chemical Physics, Vol. 53 **Ultrafast Phenomena VII**
Editors: C.B. Harris · E.P. Ippen · G.A. Mourou · A.H. Zewail

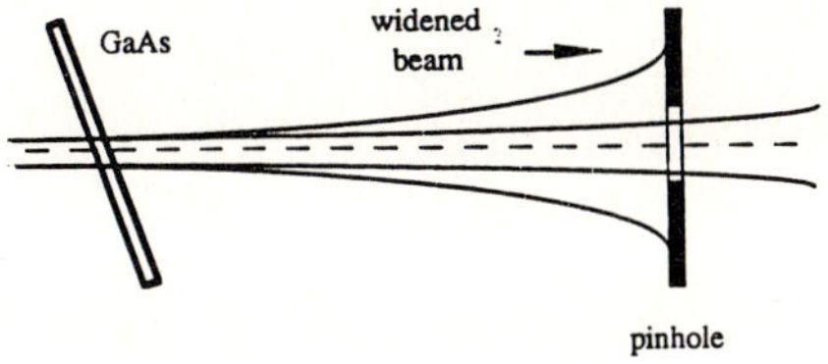

Figure 1a: The limiter

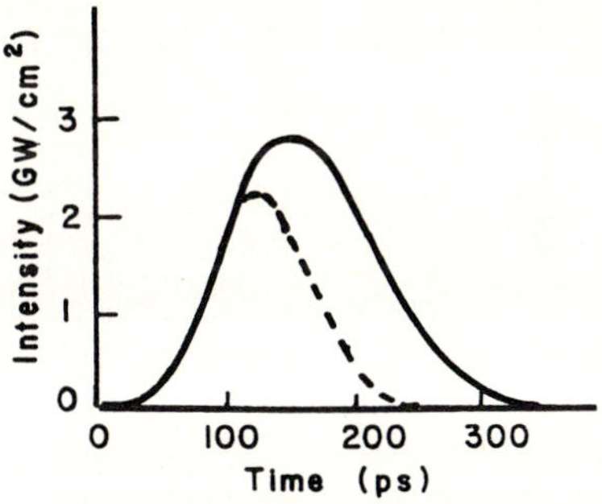

Figure 1b: Pulse compression

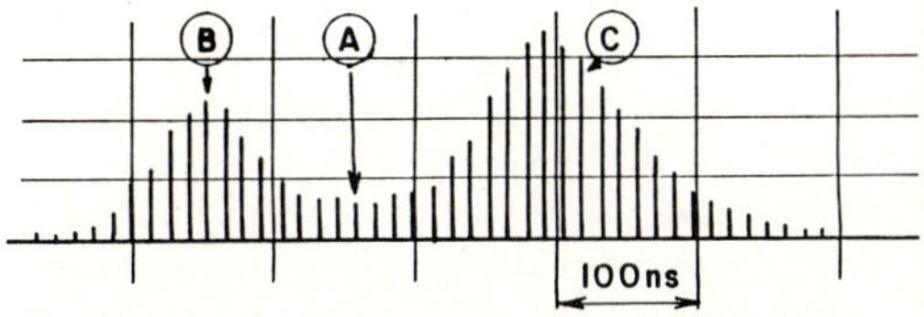

Figure 2: Pulse train transmitted through the pinhole.

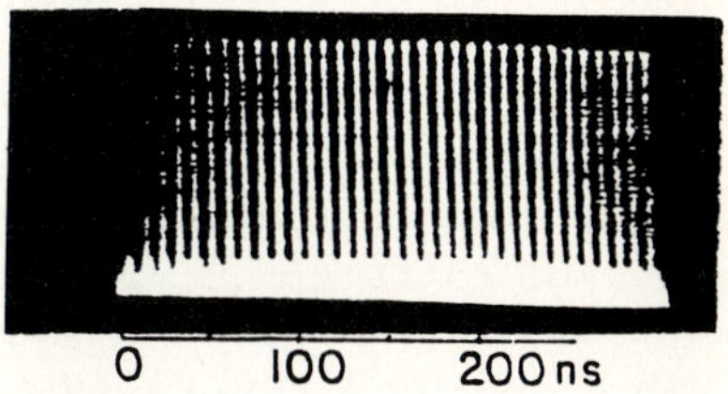

Figure 3: Pulse train with intracavity negative feedback.

a single path. Indeed, measurements of the transmission of a Q-switched-mode-locked train through a 400 μm thick GaAs sample in the geometry of Fig. 1a clearly show enhanced losses near the peak of the train.(A in Fig. 2) Autocorrelation measurements of the successive pulses from the train indicate a reduction in pulse duration from 120 ps to 60 ps for the central pulse A. The duration of pulses B and C (Fig. 2) changes from 125 to 105 ps, and 105 to 115 ps, respectively. The broadening of C, together with an increase in transmission (Fig. 2) indicate self focusing due to the cumulative heating by successive pulses and the positive temperature dependence of the index of refraction of GaAs.(dn/dT)

3. The laser with Passive Negative Feedback

The arrangement of Fig. 2 combines the action of energy limiter and pulse tail clipper. The latter effect complements the front edge steepening of an intracavity saturable absorber. After insertion of the pinhole-GaAs combination of Fig. 1a in the cavity of a Nd laser, the pulse train is extended to include over 100 pulses of nearly equal amplitude, following a initial transient of 20 to 30 pulses. Autocorrelation measurements recorded on a transient digitizer enabled us to select a portion of the train, showing the pulse duration to be 10, 5 and 1 ps in the flat portion of the train, for Nd:YAG, Nd:YAP and Nd:glass, respectively. The individual pulse energies are 2μJ and 10 μJ, with a shot to shot reproducibility better than 2.75 and 1.5 % for the Nd:YAP and Nd:YAG lasers, respectively.

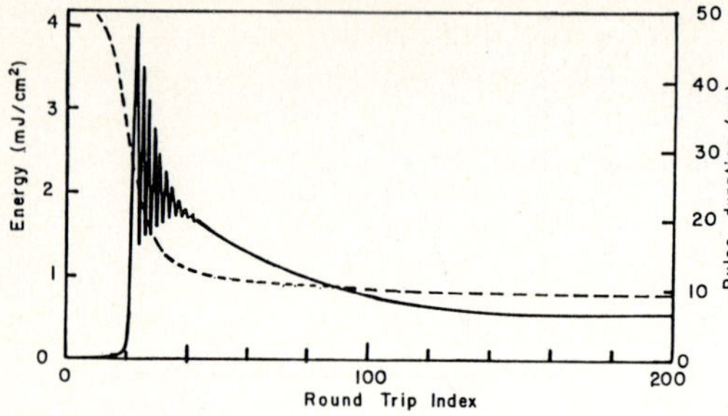

Figure 4
Pulse energy (solid line) and duration (dashed line) versus round-trip index.

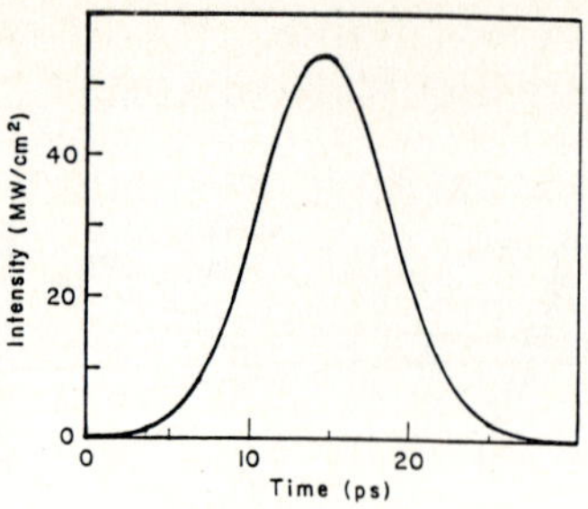

Figure 5
Steady state pulse (simulation)

For Q-switched solid state lasers, modeling of the pulse evolution with differential equations [3] is inappropriate, because the pulse modifications at each round trip cannot be assumed to be small. An infinitesimal approach is particularly questionable in the present situation, because the desired result (i.e. a stabilized pulse train) may be implicit in the assumption of small changes per round-trip. We computed successively for each element of the cavity the pulse evolution, no assumption being made as to the magnitude of change per round-trip or per element. The standard quantum mechanical transient response to a two level system is used to model the absorber. The transmission through the gain medium is calculated in the limit of negligible spontaneous emission during the pulse, taking into account non-uniform depletion of the gain at each passage. The simple Drude model is used for the negative contribution to the refractive index due to carriers created by two-photon absorption. The main results of the simulation of the pulse evolution in the cavity of a Nd:YAG laser are illustrated in Figs. 4 and 5.

This work was supported in part by the National Science Foundation under grant Nb 88 02530.

REFERENCES

1. K. Burneika, R. Grigoris, A. Piskarskas, G. Sinkyavichyus, and V. Sirutkaitis, Kvantovaya Elektron. **15**, 1658 (1988).
2. J. Schwartz, W. Weiler, and R. K. Chang, IEEE J. of Quantum Electron. **QE-6**, 442-450 (1970).
3. W. Rudolph and B. WIlhelmi, Appl. Phys. **B35**, 37-44 (1984).

A New Method of Ultrashort Pulse Generation: Modified Fabry-Perot Electrooptic Modulator

T. Kobayashi, A. Morimoto, Bong Young Lee, and T. Sueta

Engineering Science, Osaka University, Toyonaka, Osaka 560, Japan

Abstract. A new ultrashort optical pulse generator, the "modified Fabry-Perot electrooptical modulator" is proposed. From theoretical analysis it is derived that this method makes it possible to convert cw light to subpicosecond high peak optical pulses with near 100 % energy conversion efficiency regardless of laser linewidth.

1. Introduction

To generate ultrashort optical pulses, modelocking of lasers has normally been used. Even using this excellent method, however, we cannot generate subpicosecond pulses from narrow-linewidth lasers because the pulse width is limited by the gain-linewidth of the laser. On the other hand, a Fabry-Perot electrooptic modulator (FP modulator) [1] is applicable to almost all kinds of lasers and can produce subpicosecond pulses without any linewidth limitation. In fact, 0.8 ps pulses have been obtained from an Ar laser (linewidth: 4 GHz) using this method [2]. Unfortunately, the FP modulator has the disadvantage of low power efficiency. Since the peak value of the power transmittance is at most unity, the energy efficiency (average output power/average input power) becomes smaller as the pulsewidth becomes shorter as shown in Fig. 1a.

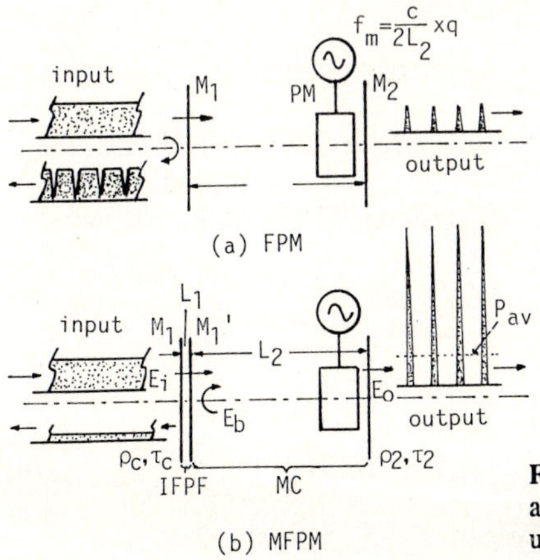

Fig. 1. Fabry-Perot electrooptic modulator and modified Fabry-Perot electrooptic modulator as optical pulse generators

In this paper, we propose a new ultrashort optical pulse generator, the "modified Fabry-Perot electrooptic modulator", which makes it possible to convert cw light to subpicosecond optical pulses, regardless of laser linewidth, with near 100 % energy efficiency. This new method has the advantages of good controllability and applicability, and may take the place of active modelocking in the future.

2. Basic Considerations

The basic arrangement of the proposed modified FP modulator (MFPM) is shown in Fig. 1b, where the input mirror of the FP modulator (Fig. 1a) is replaced by a Fabry-Perot filter with high finesse and a wide free spectral range (FSR). Accordingly, the MFPM is a composite optical cavity consisting of two cavities: an input Fabry-Perot filter (IFPF) and a modulation cavity (MC) including an electrooptic phase modulator as shown in the figure. A resonant pass frequency of the IFPF is just tuned to the input optical frequency ω_0. The internal phase modulation in the MC is synchronized to the optical round trip in the MC, that is, the modulation frequency $f_{\mathrm{m}} = qc/2L_2$ (q: integer). Under these conditions, a large part of the input light power is injected into the MC through the IFPF and converted to optical sideband components whose frequencies are $\omega_0 + 2n\pi f_{\mathrm{m}}$ ($n = \pm 1, \pm 2, \dots$) by the phase modulation. Since the sidebands produced cannot pass back through the IFPF, they are stored in the MC and the phase modulation is multiplied. Finally, the number of sidebands increases and the spectral width of them is greatly broadened. Accordingly, the fraction of the optical energy taken by the carrier component becomes low and optical power returning back through the IFPF becomes small. Consequently, most of the input light energy can be taken out as sidebands forming a pulse train through the output mirror M_2.

3. Analysis of Pulse Shape, Pulse Width, and Efficiency

The complex reflectance and transmittance of the IFPF are given by

$$\varrho_{\mathrm{c}n} = \varrho_1 - \varrho_1 T_1 \exp(-\mathrm{j}2\theta_{1n}) / \left[1 - R_1 \exp(-\mathrm{j}2\theta_{1n})\right] , \tag{1}$$

$$\tau_{\mathrm{c}n} = T_1 \exp\left[-\mathrm{j}(\omega_n L_1/c - 2\arg\tau_1)\right] / \left[1 - R_1 \exp(-\mathrm{j}2\theta_{1n})\right] , \tag{2}$$

where $R_k = |\varrho_k|^2$, $T_k = |\tau_k|^2$, ($k = 1, 2$), and $\omega_n = \omega_0 + 2\pi n f_{\mathrm{m}}$. Under the tuning condition $\theta_{1n} = \omega_n L_1/c - \arg\varrho_1 = 2\pi n f_{\mathrm{m}} L_1/c$, $|\varrho_{\mathrm{c}0}| \sim 0$, and $|\varrho_{\mathrm{c}n}(n \neq 0)| \sim 1$. The backward traveling electric field in the MC on the IFPF side can be written as

$$E_{\mathrm{b}}(t) \exp(\mathrm{j}\omega_0 t) = \exp(\mathrm{j}\omega_0 t) \sum b_n \exp(\mathrm{j}2\pi n f_{\mathrm{m}} t) \tag{3}$$

and it satisfies the self-consistent equation

$$\begin{aligned} E_{\mathrm{b}}(t) =& \varrho_2 \left[\varrho_{\mathrm{c}} : E_{\mathrm{b}}(t - 2L_2/c) + \tau_{\mathrm{c}0} E_{\mathrm{i}}\right] \\ & \times \exp\left\{-2\mathrm{j}\left[\omega_0 L_2/c + \Delta\theta \sin\left(2\pi f_{\mathrm{m}}(t - L_2/c)\right)\right]\right\} , \end{aligned} \tag{4}$$

where E_{i} is the input field and $[\varrho_{\mathrm{c}} : E_{\mathrm{b}}(t)]$ means $\sum \varrho_{\mathrm{c}n} b_n \exp(\mathrm{j}2\pi n f_{\mathrm{m}} t)$.

If we make the approximation that all $\varrho_{cn}(n \neq 0)$'s are constant, i.e., take them as ϱ_{cs} (~ 1), $[\varrho_c : E_b(t)]$ becomes $\varrho_{cs}E_b(t) - b_0(\varrho_{cs} - \varrho_{c0})$. Substituting this into (4) and after an intricate calculation we obtain the output field as

$$E_0(t) = \left(\frac{\tau_{c0}E_i}{1 + A(1 - \varrho_{c0}/\varrho_{cs})}\right)\frac{\tau_2 \exp[-j(\Delta\theta \sin 2\pi f_m t + \omega_0 L_2/c)]}{1 - R\exp[-2j(\theta_0 + \Delta\theta \sin 2\pi f_m t)]}, \quad (5)$$

where $R = |\varrho_{cs}\varrho_2|$, $\theta_0 = \omega_0 L_2/c - (1/2)\arg(\varrho_{cs}\varrho_2)$, $A = \{R\exp[-2j\theta(t)]/[1 - R\exp(-2j\theta(t))]\}_{\text{time average}}$, and $\theta(t) = \theta_0 + \Delta\theta \sin(2\pi f_m t)$, respectively. This output has a similar waveform to that of the Fabry-Perot modulator. The term in large parentheses corresponds to the forward traveling carrier component in the MC and its amplitude is much larger than that of an ordinary FP modulator. Consequently, the energy efficiency is very much improved. From (5) the pulse width is approximately given by [1]

$$\Delta\tau \sim \frac{(1-R)}{\pi\sqrt{R}} 2\Delta\theta f_m, \quad \text{for } \theta_0 = p\pi \ (p: \text{integer}). \quad (6)$$

Precisely speaking, the amplitudes and phases of ϱ_{cn} are not constant and depend on each sideband. Accordingly, the actual output field may slightly differ from (5). We also calculated the stationary solution of (4) rigorously. The pulse width, peak and average output powers calculated by (4) are shown in Fig. 2 as functions of modulation indices $\Delta\theta$. The approximation equations (5) and (6) differ from the rigorously calculated results by only a few percent. Accordingly, (5) is useful for a quick estimate. Anyway, it is seen from the figure that ultrashort pulses whose peak power is much higher than the input power are obtainable with high conversion efficiency.

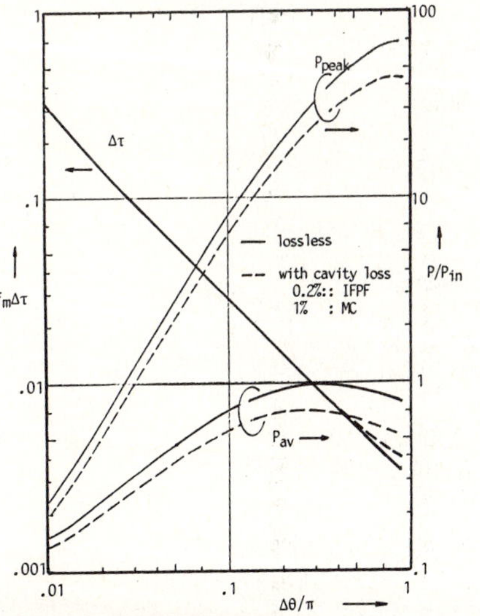

Fig. 2. Calculated pulse width, peak and average output power. $R_1 = R_1' = 0.98$, $R_2 = 0.9$

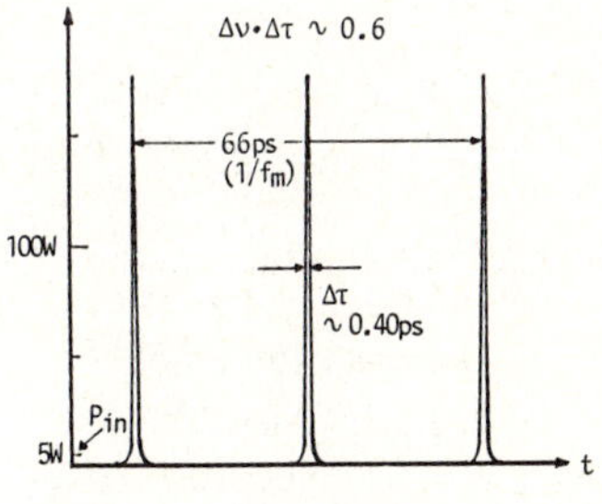

Fig. 3. Typical example of output pulse shape (for 5 W cw input)

Realistic numerical example. Light source: cw 5 W laser with single longitudinal mode oscillation. $R_1 = R_1' = 0.98$, $T_1 = 0.018$, $R_2 = 0.9$, $T_2 = 0.09$, $L_1 = 0.14$ mm, $L_2 = 2$ cm, losses: 0.2 % (= $1 - R_1 - T_1$) in the IFPF, and 1 % (= $1 - T_2 - R_2$) in the MC. The FSR and finesse of IFPF: 1.5 THz and 156. Modulation index $\Delta\theta = 0.5\,\pi$, $f_m = 15\,\text{GHz} = 2 \times c/2L_2$.

The calculated pulse shape is shown in Fig. 3. Here, $\Delta\tau = 0.40$ ps, repetition rate: 30 GHz (= $2f_m$). Peak and average output power: 175 W, 3.2 W (64 % efficiency).

4. Conclusion

A new ultrashort optical pulse generating method, the modified Fabry-Perot electrooptic modulator, has been proposed and its performance has been discussed theoretically. This method has the following excellent features and the possibility of taking the place of the modelocking method:

1. Subpicosecond pulses can be generated from almost all types of lasers, whatever their linewidth.
2. High energy efficiency and high peak power are obtainable.
3. Pulse repetition can be varied over a wide range without any modification of the laser cavity.

References

1. T. Kobayashi, T. Sueta, Y. Cho, Y. Matsuo: Appl. Phys. Lett. **21**, 341 (1972)
2. T. Kobayashi, A. Morimoto, T. Fujita, T. Uemura, K. Amano, T. Sueta: In *Ultrafast Phenomena V*, ed. by G.R. Fleming, A.E. Siegman, Springer Ser. Chem. Phys., Vol. 46 (Springer, Berlin, Heidelberg 1986)

Subpicosecond Argon Pumped Fibre Raman Laser

R. Leonhardt and J.D. Harvey

Department of Physics, University of Auckland, Private Bag, Auckland, New Zealand

Abstract. An argon pumped mode-locked Raman laser has been constructed which produces subpicosecond pulses at 526nm wavelength.

Recently it has been demonstrated that a modelocked argon laser operated in the superfluorescent regime, can generate a train of pulses with peak power in the kilowatt regime[1,2,3]. These pulses, having an order of magnitude higher peak power than can be obtained with a conventionally modelocked argon laser, can be exploited to generate a variety of nonlinear effects. Whilst in theory the high peak power could be exploited to generate subpicosecond pulses by conventional fibre grating compression techniques[4,5], it is not possible to generate pedestal free compressed pulses (which require lengthy fibres) due to the strong stimulated Raman scattering and the high attenuation of single mode fibres at this wavelength. In this paper we show that the stimulated Raman emission can be harnessed to generate subpicosecond pulses in a dispersion compensated fibre Raman laser.

The laser is shown schematically in Fig.1. The pump pulses are derived from a modelocked cavity dumped argon laser with a pulse repetition rate of 9.5MHz, having a pulse width of ~40ps and an average power of 180mW. The grating combination with the prism compensate for the dispersion in the fibre, and the variable delay line is used to synchronise the intracavity round trip time with the repetition rate of the pump laser. The choice of fibre used in the laser is critical to the performance of the system. Whilst conventional Germania doped fibres exhibit stronger Raman scattering, early attempts to use these fibres were hampered by damage induced by the high peak powers. The

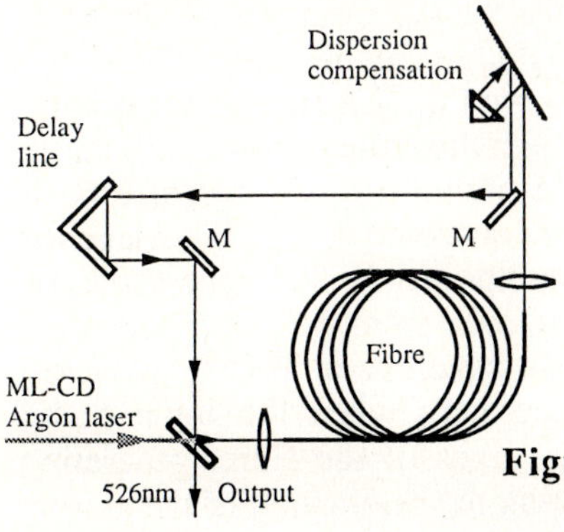

Figure 1 Schematic diagram of the fibre Raman ring laser

Springer Series in Chemical Physics, Vol. 53 **Ultrafast Phenomena VII**
Editors: C.B. Harris · E.P. Ippen · G.A. Mourou · A.H. Zewail

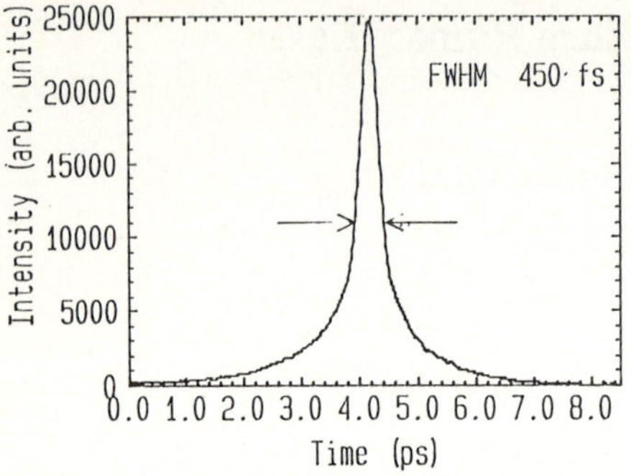

Figure 2
Autocorrelation function of 526 nm pulse train, as measured by background free autocorrelation in a scanning autocorrelator using a 1mm thick BBO crystal

increased attenuation of Germania doped fibres exposed to intense blue/green radiation is generated by colour centre formation via two photon absorption in the fibre[6]. Although non polarisation preserving fused silica core fibres avoid the damage problems, stable output was obtained only with a polarisation preserving fused silica core fibre. The fibre used in these experiments had a length of 20m and an attenuation of ~140db/km. The maximum single pass transmission through the fibre was only 25% due to the high Rayleigh scattering losses at 514.5nm and an input coupling efficiency of ~50%. The laser operates at a wavelength of 526nm and an autocorrelation function of the pulse train is shown in Fig.2. The observed FWHM of this autocorrelation function corresponds to a pulse duration of 320fs for a Gaussian pulse and 290fs for a sech squared shape. The corresponding spectrum (FWHM 1.3nm) of the pulse train is shown in Fig.3a which indicates that the pulses are nearly bandwidth limited. The observed lasing spectrum is relatively symmetrical, in contrast the spectrum of the (1st Stokes) stimulated Raman scattering for a single pass of the fibre (Fig.3b) has a FWHM of 2.6nm and is asymmetric. The output power of the laser was 4mW corresponding to a peak power of 0.56kW. Some limited tunability (± 2nm) was obtainable by adjustment of the mirror M (or equivalently the output coupler) since the single pass dispersion compensation produces an elliptical spot with spatially separated frequency components. Both the pump and the laser beam had p-polarisation corresponding to a high transmission (97%) of the output coupler.

Whilst other fibre Raman lasers have been reported in the infra red[7,8,9], this laser operates in a region of the visible spectrum which is not readily accessed by modelocked dye lasers. By adjusting the orientation of the grating and the delay time, we have also observed laser action on the second Stokes Raman component (at ~538nm), although this output was significantly less stable than that of the first Stokes. In addition it has recently been shown[10,11] that modulation instability can be observed in high birefringence fibres in the normal dispersion regime characterised by the appearance of upshifted and downshifted sidebands a few nm away from the pump frequency and an oscillation in the intensity of the pulse envelope at THz frequencies. This effect is very strong in the fibre used here when the input pump polarization is oriented at 45° to the major axes of the fibre, generating sidebands at ±2nm from 514.5nm. By adjusting the orientation of the grating

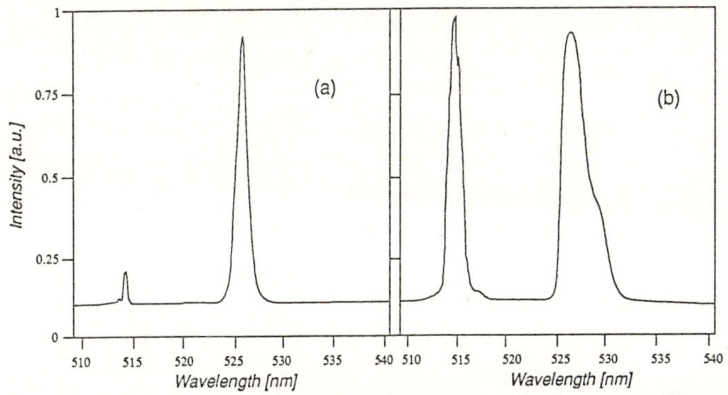

Figure 3(a) Spectrum of the pulse train from the fibre Raman laser with a wavelength of $\lambda = 526$nm and a FWHM of 1.3nm. The small peak is resolved light from the Argon laser reflected by the output coupler

(b) Spectrum for a single pass through the fibre. On the left side you can see the spectrum of the self-phase modulated Argon pulses, on the right the asymmetric Raman spectrum of the fibre with a FWHM of 2.6nm

to optimise back coupling at the pump wavelength, we have also seen laser action in the modulation instability sidebands. This laser is currently the subject of active investigation.

We would like to thank L. Poyntz-Wright for providing the experimental high birefrigence single mode optical fibre used in this work and the Alexander von Humboldt-Stiftung for financial support.

References

1. J.D. Harvey, M.J. Proctor, C. Steed, Appl.Phys. Lett. 52, 688 (1988)
2. J.D. Harvey, R. Leonhardt, S. Carter & P.D. Drummond, Phys.Rev., A40, 4789 (1989)
3. J.D. Harvey, J. Dudley, R. Leonhardt, S. Carter and P.D. Drummond, Quantum Optics V, Springer Proceedings in Physics 41, 214 (1989)
4. D. Grischkowsky and A.C. Balant, Appl.Phys.Lett. 41, 1 (1982)
5. R.H. Stolen, C.V. Shank and W.J Tomlinson, *Ultrafast Phenomena IV* (Springer-Verlag, Berlin Heidelberg 1984) p.46
6. L.J. Poyntz-Wright, M.E. Fermann and P.St.J. Russell, Opt.Lett. 13, 1023 (1988)
7. J.D. Kafka, D.F. Head and T. Baer in *Ultrafast Phenomena V* (Springer-Verlag, Berlin Heidelberg 1986) p.51
8. E.A. Golovchenko, E.M. Dianov, P.W. Mamyshev, A.M. Prokhorov and D.G. Fursa in *Ultrafast Phenomena VI* (Springer-Verlag, Berlin Heidelberg 1988) p.40
9. R. Schulz, M. Kuckartz and H. Harde, Opt.Commun. 70, 239 (1989)
10. J.E. Rothenberg, Paper MC1 Topical Meeting on Integrated Photonics Research, Hilton Head, March 1990; and Phys.Rev.A, in press.
11. P.D. Drummond, T.A.B. Kennedy, J.M. Dudley, R. Leonhardt and J.D. Harvey, Opt.Commun. (in press) (1990)

Generation and Amplification of Femtosecond Pulses at 800 nm

P. Georges [1], *F. Salin* [2], *F. Estable* [3], *G. Roger* [2], *and A. Brun* [2]

[1] Ecole Polytechnique Féminine, 3 bis rue Lakanal, F-92230 Sceaux, France

[2] Institut d'Optique Théorique et Appliquée, CNRS UA 14, Université Paris-Sud, Bât. 503, BP 147, F-91403 Orsay Cedex, France

[3] B.M. Industries, 7 rue du Bois Chaland, CE 209 Lisses, F-91029 Evry Cedex, France

Abstract. We report the generation of femtosecond pulses in the near infrared by passive mode-locking of a cw Rh700/HITCI dispersion-compensated ring dye laser. Pulses as short as 50 fs at 800 nm have been produced for the first time to our knowledge. Titanium sapphire multipass amplifiers have been studied to increase the energy of the pulses to the millijoule level.

Passive mode-locking with an intracavity group velocity dispersion (GVD) compensation is the most interesting technique for producing femtosecond pulses [1] and pulses as short as 27 fs have been obtained. However, passive mode-locking was, until recently, restricted to the single active/passive dye combination of rhodamine 6G/DODCI, yielding pulses in the spectral region around 620 nm, whereas the study of GaAs semiconductors needs femtosecond pulses in the near infrared (around the band gap energy). Furthermore there is now a great interest in the production of very high peak power (> 1 TW) pulses for X-ray generation and multiphoton ionization. With the recent development of new solid state materials such as $Ti{:}Al_2O_3$ having a large amplification bandwidth, the generation of femtosecond pulses in the near infrared is very attractive.

Following preliminary research on the generation of femtosecond ultrashort pulses at different wavelengths [2], we have studied the generation of pulses shorter than 100 fs around 800 nm. Taking into account that, in the passive mode-locking technique, the amplifier is cw pumped, it is necessary to use a very efficient dye, while it is not so important in active mode-locked dye lasers. The active medium is rhodamine 700, which provides an efficiency similar to rhodamine 6G when pumped with a cw krypton ion laser. Figure 1 shows the cavity configuration based on a design of *Valdmanis* et al. [1]. It is a classical six-mirror ring cavity with 150 mm radius of curvature mirrors around the amplifier medium and 50 mm radius of curvature mirrors around the saturable absorber jet. A sequence of four prisms controls the intracavity GVD. Due to the decrease of the negative GVD introduced when the wavelength increases, the distance between the prisms is increased to 380 mm

Two saturable absorbers, DDI and HITCI, were used to produce ultrashort pulses. Although we have obtained 36 fs pulses at 775 nm [3] with DDI, we have found that HITCI is better adapted to the rhodamine 700 fluorescence spectrum. With the dye combination of rhodamine 700 and HITCI, very stable and reliable

Springer Series in Chemical Physics, Vol. 53 **Ultrafast Phenomena VII**
Editors: C.B. Harris · E.P. Ippen · G.A. Mourou · A.H. Zewail

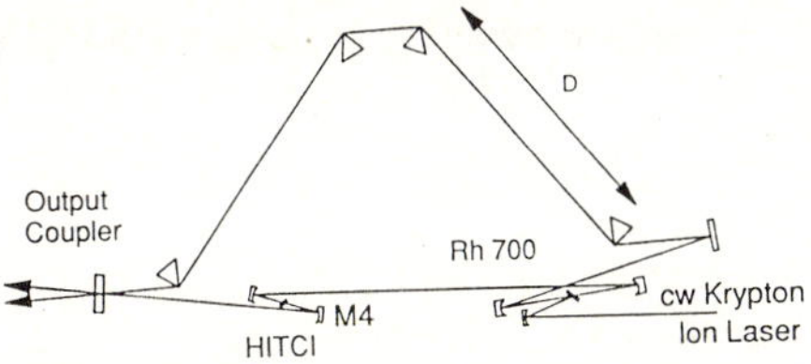

Fig. 1. Laser configuration

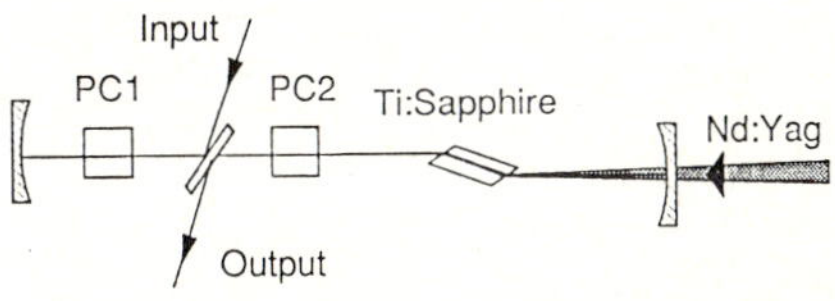

Fig. 2. Regenerative amplifier configuration

pulses as short as 50 fs near 800 nm have been produced with a spectrum width of 14 nm [4]. In order to avoid a multiple-pulse regime, we reduced the cavity round trip time to 7.2 ns. This behavior can be explained by the relatively fast gain recovery time of rhodamine 700.

If one considers the generation of very high peak power pulses, it is important to use an oscillator which produces pedestal-free pulses. Actual high peak power systems [5] are based on an actively mode-locked Nd:YLF oscillator followed by a fiber pulse compressor which introduces some wings in the temporal pulse shape. We think that this kind of infrared femtosecond laser could be the seeding oscillator of a high peak power Ti:Al_2O_3 amplifier using the chirped pulse amplification technique [5].

In order to model the amplification process of Ti:Al_2O_3, we first measured the saturation fluence of this solid-state amplifier medium [6]. Using the Frantz and Notwik model, we found a saturation fluence of 1.2 J/cm^2 at 770 nm and an "effective quantum efficiency" of nearly 0.5. We then modeled Ti:Al_2O_3 amplification with this value and determined that efficient extraction of the energy can be obtained in a relatively small number of passes in the amplifier due to the very large gain. For example, the model predicted that 76 % of the stored energy could be transferred to the signal in only 9 passes if we considered a small gain of 12 and losses of 10 % per pass.

To increase the energy of the ultrashort pulses produced, we have developed a regenerative Ti:Al_2O_3 amplifier [7] pumped by a Q-switched Nd:YAG laser (6 ns, 50 Hz). The regenerative amplifier (Fig. 2) consists of two highly reflecting mirrors (1 and 0.5 m radius of curvature separated by 1.2 m), the laser medium, a polarizer and two Pockels cells. In order to avoid self-focusing in the amplifier, we stretched the femtosecond pulses to a few tens of picoseconds by coupling the pulses into 30 m of single mode optical fiber. After expansion, the pulses are injected in the regenerative amplifier through the polarizer. The first Pockels cell is then polarized at its quarter wave voltage to close the cavity. When the pulse's energy reaches the saturation level, a quarter wave voltage is applied to the second Pockels cell to eject the pulse. Taking into account the losses through the stretcher stage, the input pulse's energy is around 50 pJ while the amplified pulse's energy is around 0.5 mJ (gain: 10^7). The pump pulse's energy is on the order of 6 mJ and the absorption of the crystal is 80 % at 532 nm. We have found experimentally that the saturation level is reached after 20 passes in the crystal, which is in good agreement with the theoretical predictions taking into

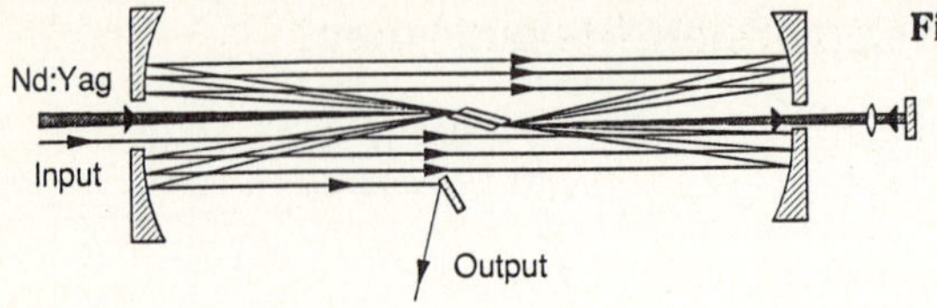

Fig. 3. Experimental setup of the passive multipass amplifier

account a small-signal gain of around 4. No spectral narrowing is observed in the amplification process and the spectrum of the amplified pulses is as large as the input pulse spectrum. This would allow us to compress the pulses back to nearly their original duration by using a standard double pass grating compressor.

Taking into account the relatively small number of passes in the amplifier, a passive multipass $Ti:Al_2O_3$ amplifier using curved mirrors has also been studied. The design is based on a scheme presented recently for a dye amplifier [8]. It consists of a two mirrors of slightly different radii of curvature (980 and 1050 mm) placed in a nearly confocal arrangement. The propagation of the beam illustrates the principle of an unstable cavity. The number of passes (six in our experiment) can be adjusted by translating a plane mirror as shown in Fig. 3. The pump beam is focused by a 1 m focal length lens leading to a 1 mm diameter. To avoid a laser effect between the two facing mirrors we have cut each of them into two parts. Furthermore, the alignment of the amplifier is easier with four mirrors than with only two. This multipass amplifier has been first studied to amplify a cw single-mode ring dye laser operating at 780 nm.

A single pass gain of 10 has been measured with a pump energy of 30 mJ. After six passes we obtained a gain of 2×10^6 when we injected the transmitted pump beam back into the crystal. The amplified pulse duration is 20 ns and the pulse energy is about 0.7 mJ [9]. The shortening of the amplified pulse duration is due to the gain saturation in the crystal because we reached the saturation fluence of titanium sapphire (1 J/cm^2). We think this kind of very simple amplifier may be used to amplify ultrashort pulses to the millijoule level and we are now working on this topic.

Acknowledgements. This work was partially supported by the Agence Nationale de Valorisation de la Recherche and the Ministère de la Recherche.

References

1. J.A. Valdmanis, R.L. Fork, J.P. Gordon: Opt. Lett. **10**, 131 (1985)
2. P.M.W. French, J.A.R. Williams, J.R. Taylor: Rev. Phys. Appl. **22**, 1651 (1987)
3. P. Georges, F. Salin, G. Roger, A. Brun: Opt. Lett. **14**, 940 (1989)
4. P. Georges, F. Salin, G. Le Saux, G. Roger, A. Brun: Opt. Lett. **15**, 446 (1990)
5. P. Maine, D. Strickland, M. Pessot, J. Squier, P. Bado, G. Mourou, D. Morter: In *Ultrafast Phenomena VI*, ed. by T. Yajima, K. Yoshihara, C.B. Harris, S. Shionoya, Springer Ser. Chem. Phys., Vol. 48 (Springer, Berlin, Heidelberg 1988) p. 2
6. F. Estable, F. Salin, M. Allain, P. Georges, A. Brun: Opt. Commun. **72**, 235 (1989)
7. P. Bado, M. Bouvier, J.S. Coe: Opt. Lett. **12**, 319 (1987)
8. E.V. Khoroshilov, I.V. Kryukov, P.G. Kryukov, A.V. Sharkov: In *Ultrafast Phenomena VI*, ed. by T. Yajima, K. Yoshihara, C.B. Harris, S. Shionoya, Springer Ser. Chem. Phys., Vol. 48 (Springer, Berlin, Heidelberg 1988) p. 22
9. P. Georges, F. Estable, F. Salin, J.P. Poizat, P. Grangier, A. Brun: High efficiency multipass Ti:sapphire amplifiers for continuous wave single mode laser. Submitted to Opt. Lett.

Femtosecond Laser Cavity Dispersion Measurement Using Soliton Properties

F. Salin[1], *P. Grangier*[1], *P. Georges*[2], *G. Roger*[1], *and A. Brun*[1]

[1]Institut d'Optique Théorique et Appliquée, CNRS UA 14, Université Paris-Sud, Bât. 503, BP 147, F-91403 Orsay Cedex, France
[2]Ecole Polytechnique Féminine, 3 bis rue Lakanal, F-92230 Sceaux, France

Abstract. Near-zero second-order and nonnegligible third-order cavity dispersion have been measured using the soliton properties of pulses generated in a CPM laser.

We present in this paper two experimental measurements of the intracavity dispersion of a colliding pulse mode-locked (CPM) dye laser. This parameter is of crucial importance for this kind of laser and, until now, has never been experimentally measured. First, we will show that using soliton properties of the pulses one can obtain a good approximation of the second-order dispersion. Next we will show that the third-order dispersion is not negligible in our CPM laser and that, indeed, it is the factor preventing shorter pulses from being obtained.

All these experiments were conducted with a classical CPM laser using rhodamine 6G and DODCI dyes. All the mirrors were single stack coated for 100 % reflectivity at normal incidence at 620 nm. With freshly prepared dyes, pulses as short as 31 fs (assuming a sech^2 profile) are obtained.

1. Second-Order Dispersion

Measuring the GVD and the SPM experimentally is a complicated problem which can be addressed by using soliton-like properties of the pulses. We have recently observed in our femtosecond laser what we think is an $N = 2$ soliton-like pulse with characteristics very similar to the usual $N = 1$ soliton ones. For some adjustment of the laser, a $\sim$ 100 kHz sinusoidal modulation appears on the pulse train envelope when the pump power is increased from 2.5 W to 2.8 W. From autocorrelations and spectra we found that, except for this modulation, the pulse shape and wavelength do not change with pump power. We think that these pulses correspond to $N = 2$ solitons having one pole much smaller than the other. Such a soliton is very similar to an $N = 1$ soliton. The only difference is a small periodic variation of its shape. In the case of a cavity the soliton period N_0, in terms of the number of cavity round trips, and the soliton intensity I are given by

$$N_0 = \tau^2/1.98\phi'' \quad \text{and} \tag{1}$$

$$I = \lambda/4n_2N_0l_{\mathrm{NL}}\ , \tag{2}$$

where τ is the pulse FWHM, ϕ'' is the GVD per round trip (in fs^2), n_2 and

Springer Series in Chemical Physics, Vol. 53 **Ultrafast Phenomena VII**
Editors: C.B. Harris · E.P. Ippen · G.A. Mourou · A.H. Zewail

l_{NL} are respectively the nonlinear index of refraction and the thickness of the nonlinear medium in the cavity, and λ is the laser wavelength. The energy E of the pulse is proportional to the product $I\tau$. Thus for a pulse with a nearly constant energy, the combination of (1) and (2) shows that the pulse duration and the soliton period have to be linearly proportional to the cavity GVD ϕ'':

$$N_0 \propto \phi''/E^2 \text{ and } \tau \propto \phi''/E\,. \tag{3}$$

We have checked the soliton-like character of these pulses by plotting the pulse duration and the modulation period in units of cavity round trips versus the variation of the GVD introduced by the translation of one of the intracavity prisms. The origin of the Ox axis is arbitrary. The experimental points are reasonably fitted by straight lines as expected from (3). However, if a perfectly constant energy is assumed, these two straight lines should intercept the Ox axis at the same point. In spite of the discrepancy between these two values we can deduce that the shortest pulses (40 fs) are obtained, in a CPM laser, for a small negative GVD ($-30\,\text{fs}^2 < \phi'' < -5\,\text{fs}^2$).

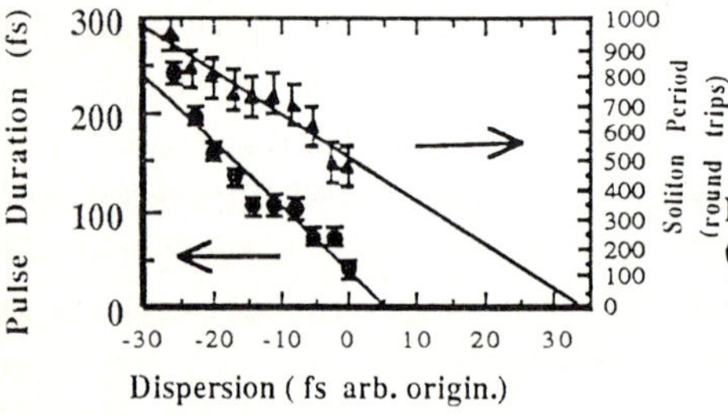

Fig. 1. Pulse duration (•) and modulation frequency (▲) versus intracavity group velocity dispersion. Least squares fits are also presented

2. Third-Order Dispersion

Starting from the alignment which gives the shortest pulses, we increased the DODCI jet thickness by translating it vertically. The estimated thickness is then greater than 50 μm. When one of the prisms is translated, introducing a little bit more glass into the cavity than needed in order to get the shortest pulses, a complicated spectral and temporal behavior is observed. A second peak appears in the pulse spectrum at a wavelength which depends on the intracavity GVD. The autocorrelation trace evolves from a clean near-sech^2 shape to a complicated multiple-peaked shape. Figure 2 gives an evolution of the pulse spectrum as a function of intracavity GVD variation. The upper curve corresponds to the shortest "sech^2" pulses. A second spectral peak is clearly visible on the right and its position changes until it merges with the principal part of the spectrum. Such a splitting of a pulse spectrum has already been observed in an optical fiber [1] and in a CPM laser [2]. It has been attributed to the third-order dispersion of the fiber or the laser. When a pulse is launched near the zero-dispersion (for example in the negative region), a part of its spectrum lies in the positive dispersion region. If the pulse is a soliton its shape stays approximately constant while the energy lying

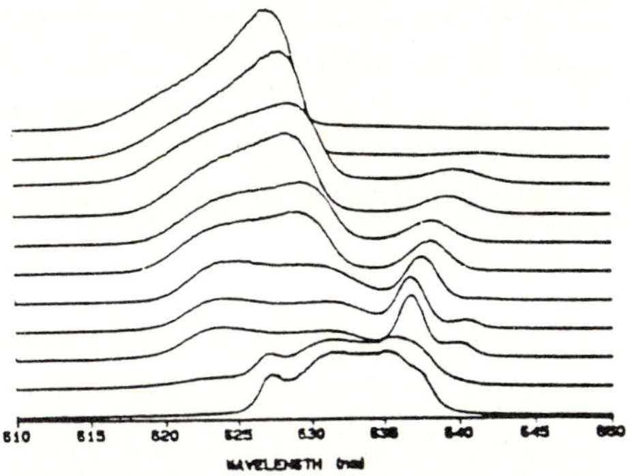

Fig. 2. Evolution of the pulse spectrum with GVD. GVD becomes less negative from top to bottom

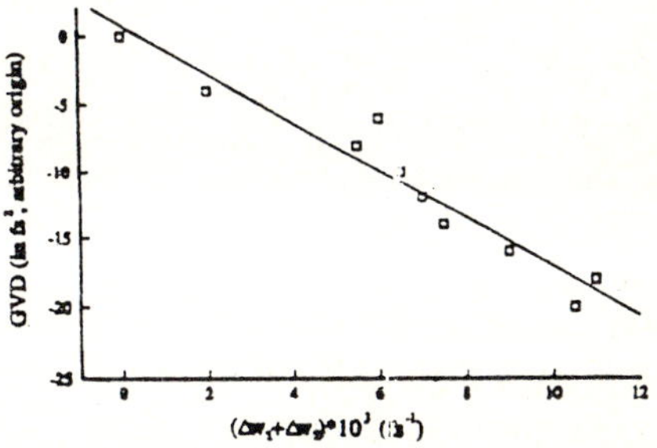

Fig. 3. Sum of the relative variation of the two peak frequencies versus GVD

in the positive dispersion does not travel at the same speed and is progressively radiated. But, with third order there is a wavelength in the positive region where the group velocity is the same as the soliton group velocity. A part of the energy is then trapped in the soliton and leads to spectral resonance. The exact position of this spectral resonance can be obtained by looking for two wavelengths ω_1 and ω_2 having the same group velocity. Using a second-order development of the group velocity around a wavelength ω_0, we find

$$(\omega_1 - \omega_2)\left[\phi_0'' + \tfrac{1}{2}\phi_0'''(\omega_1 - \omega_2 - 2\omega_0)\right] = 0\ , \tag{4}$$

where ϕ'' and ϕ''' are the second- and third-order dispersion at ω_0. We have recorded the frequencies ω_1 and ω_2 of the maxima of the two spectral peaks shown in Fig. 2. In Fig. 3 we have plotted the variation of the quantity $\omega_1 + \omega_2 - 2\omega_0$ versus the variation of ϕ_0''. The frequency ω_0 was arbitrarily chosen to correspond to 634 nm. The experimental points are quite aligned as predicted from (4). At least squares fit gives a slope of $-1800\,\mathrm{fs}^3$, leading to a third-order intracavity dispersion of $-3600\,\mathrm{fs}^3$. This large negative value shows that the prisms overcompensate the third-order dispersion of the rest of the cavity.

In conclusion, we have studied the dispersion of a CPM laser cavity. We found that the shortest pulses are obtained for a very small and negative second-order dispersion. Furthermore, we show that the third-order dispersion plays a predominent role. We think that the minimum pulse duration achieved in a CPM laser is not actually limited by the laser bandwidth but by high-order dispersion effects.

References

1. A.S. Gouveia-Neto, M.E. Faldon, J.R. Taylor: Opt. Commun. **69**, 173 (1988)
2. F.W. Wise, I.A. Walmsey, C.L. Tang: Opt. Lett. **13**, 129 (1988)

Physical Properties of Coupled Cavity Lasers Used for Ultrashort Pulse Generation

M. Morin, Y. Champagne, F. Martin, R. Vallée, N. McCarthy, and M. Piché

Equipe Laser et Optique Guidée (COPL), Département de Physique, Université Laval, Québec, Canada G1K 7P4

Coupling a mode-locked laser to a nonlinear external cavity has led [1-8] to the emission of ultrashort pulses whose duration can be reduced below 100 fs. The physical process leading to pulse narrowing has been identified [2,9-11]: chirping of the pulse circulating in the coupled cavity produces an ultrafast modulation after interferometric recombination with the pulse circulating in the main cavity. The first analyses [2,10-11] of this type of mode-locking have relied on restrictive assumptions (such as a smooth pulse shape) which cannot fully describe the variety of experimental observations. In this paper, we study the response of nonlinear interferometric reflectors and present the results of numerical modelling.

The nonlinear reflectors used in mode-locked coupled cavity lasers involve either multiple-beam or two-beam interference effects. Typical examples are the nonlinear Fabry-Perot (NFPI) and Michelson (NMI) interferometers, as shown in Fig. 1. The laser with a NMI has the advantage that the interference is made between a reference arm and a nonlinear arm; hence, the interferometric condition is insensitive to the thermal fluctuations of the gain medium. The response of the NMI exhibits a sinusoidal variation as a function of total phase-shift, as shown in Fig. 2a. The total phase-shift includes a linear part ϕ_1 (which is fixed by length adjustment) and a nonlinear part ϕ_{nl} which is proportional to incident laser intensity. The phase response of the NMI is delayed by $\pi/2$ with respect to its amplitude response. As a consequence, the points where the reflectivity is maximum ($0, 2\pi, \ldots$) have the largest slope of the phase response, leading to maximum phase modulation; minimum phase modulation occurs where the derivative of the coefficient of reflection is maximum.

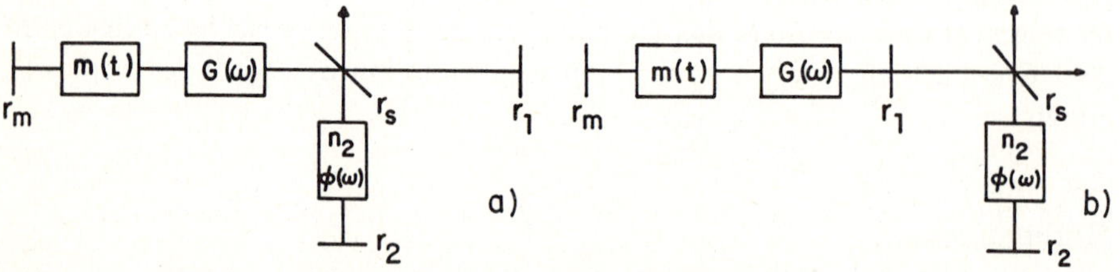

Fig. 1. Laser with a nonlinear Michelson interferometer (a) or a nonlinear Fabry-Perot interferometer (b). G(ω) and m(t) represent gain broadening and external modulation; n_2 and $\phi(\omega)$ stand for nonlinearity and dispersion.

Springer Series in Chemical Physics, Vol. 53 **Ultrafast Phenomena VII**
Editors: C.B. Harris · E.P. Ippen · G.A. Mourou · A.H. Zewail

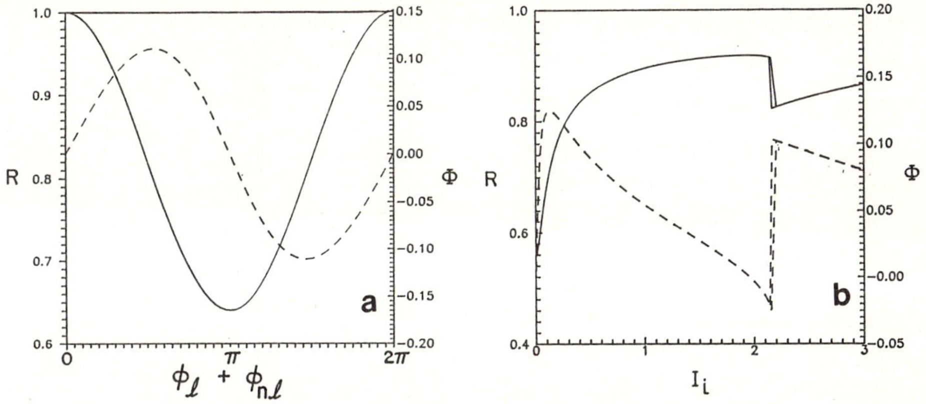

Fig. 2. Reflectivity (——) and phase-shift (---) of: a) a nonlossy NMI with $r_s = 0.32$; b) a NFPI with $r_1 = 0.89$, $r_2 = 0.71$, $r_s = 0.95$, $\phi_1 = -\pi$, assuming a coupling efficiency of 70% in the nonlinear medium.

The response of the NFPI is not sinusoidal, except for extremely lossy situations. In general, the steady-state reflectivity exhibits a rapid rise and a smooth plateau as a function of incident intensity, as shown in Fig. 2b where the incident intensity is normalized such that $I_i = 1$ produces a nonlinear phase-shift of π in one round-trip. One immediate consequence is that nonlinear phase-shifts larger than π are possible with such a device, as observed by Yakymyshyn et al. [3]. For certain conditions, the response exhibits hysteresis, bistability, period doubling and chaos.

Extensive numerical simulations with the two configurations shown in Fig. 1 have allowed to identify the possible regimes of emission of mode-locked coupled cavity lasers. In the following lines, we discuss the results relevant to a NaCl: OH^- laser equipped with a nonlossy NMI having a non-dispersive optical fibre of length L.

Smooth pulses are obtained when the pumping, and the nonlinearity are weak, as shown in Fig. 3a. The shape of the pulses is well approximated by an hyperbolic secant; the pulses have negligible chirp since the interferometer is tuned for minimum phase modulation ($\phi_1 = -\pi/2$). At higher nonlinearities (or pumping rates), the total phase-shift reaches a value where the reflectivity is maximum; as a result, the pulse intensity is clamped at a maximum value and the pulse shape becomes square, as shown in Fig. 3b. Increasing the nonlinearity or the fiber length leads to proportionally longer pulses with the same peak intensity. An unstable regime of emission called "compression to zero" occurs when the nonlinearity allows to reach the peak of the reflectivity curve but the pumping is insufficient to sustain the pulse. Under such conditions, the peak intensity remains fixed but the pulse duration decreases at each round-trip up to a point where the pulse carries no energy. This opens the possibility of generating pulses significantly shorter than the inverse bandwidth of the gain medium, at least on a transient basis.

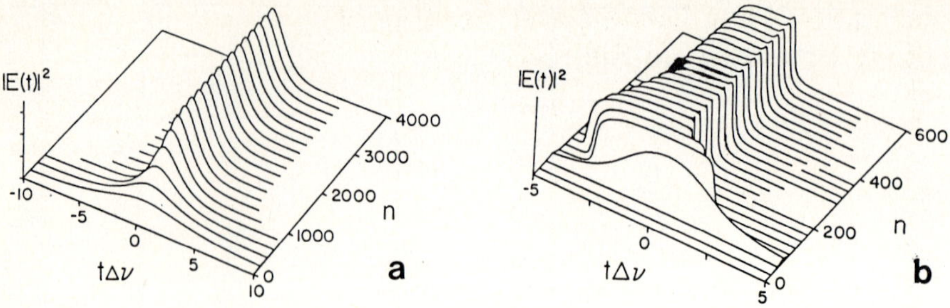

Fig. 3. Evolution of pulse shape with $\phi_1 = -\pi/2$ and pumping 50% above threshold. a) L = 0.1 m; b) L = 1.0 m.

References

[1] P.N. Kean, X. Zhu, D.W. Crust, R.S. Grant, N. Langford and W. Sibbett, Opt. Lett. 14, 39 (1989).

[2] J. Mark, L.Y. Liu, K.L. Hall, H.A. Haus and E.P. Ippen, Opt. Lett. 14, 48 (1989).

[3] C.P. Yakymyshyn, J.F. Pinto and C.R. Pollock, Opt. Lett. 14, 621 (1989).

[4] X. Zhu, P.N. Kean and W. Sibbett, IEEE J. Quantum Electron. QE-25, 2445 (1989).

[5] P.M.W. French, J.A.R. Williams and J.R. Taylor, Opt. Lett. 14, 686 (1989).

[6] J. Goodberlet, J. Wang, J.G. Fujimoto and P.A. Schulz, Opt. Lett. 14, 1125 (1989).

[7] J. Goodberlet J. Jacobson, J.G. Fujimoto, P.A. Schulz and T.Y. Fan, Opt. Lett. 15, 504 (1990).

[8] L.Y. Liu, J.M. Huxley, E.P. Ippen and H.A. Haus, Opt. Lett. 15, 553 (1990).

[9] F. Ouellette and M. Piché, Opt. Comm. 60, 99 (1986).

[10] M. Morin and M. Piché, Opt. Lett. 14, 1119 (1989).

[11] E.P. Ippen, H.A. Haus and L.Y. Liu, J. Opt. Soc. Am. B-6, 1736 (1989).

Continuous Wave Diode Laser Seeding of Tunable Femtosecond Soliton Pulses

E.J. Greer, D.M. Patrick, J.I. Vukusic, P.G.J. Wigley, and J.R. Taylor

Femtosecond Optics Group, Physics Department, Imperial College, Prince Consort Road, London SW7 2BZ, UK

Abstract. Using a seed derived from a wavelength-tunable cw diode laser, tunable fundamental solitons with durations as short as 200 fs have been generated via pulsed stimulated Raman amplification from a mode locked Nd:YAG laser.

1 Introduction

Over the past few years there has been a considerable number of theoretical and experimental investigations of the soliton-Raman generation mechanism [1-7]. The most simple experimental realisation is that provided by the single pass scheme [3,5-7], which has been shown to give rise to a wide spectral continuum made up of several solitons. The actual number and the spectral extent of the solitons are not particularly controllable, although there is a pump power dependence.

For pump pulses in the region of anomalous dispersion, the pulse formation mechanism can proceed via modulational instability. However, modulational instability is not a prerequisite for the process although it does lower the required threshold powers substantially. The solitary waves generated via amplification of the modulational instability at the peak of the pump pulse experience intrapulse Raman scattering, down-shifting in frequency. Due to the associated change in the group velocity, the down-shifted solitary waves exhibit a relative motion to the rear of the pump pulse, where soliton collisions take place with slower moving modulations. This generates higher power pulses and gives additional spectral broadening.

Since the process is initiated from noise, the soliton-Raman continuum generated is the sum of several solitons with random temporal separation and undefined central wavelengths. Spectral filters can be used to select a single soliton of defined wavelength, however this gives rise to considerable overall system loss. Through seeding the process at a specific central wavelength, fundamental solitons at precisely controllable wavelengths can be generated. Here we describe the experimental characterisation of this system.

2 Experimental

The continuous seed signal was obtained from a 500μm-long InGaAsP laser diode in an external cavity arrangement, with one of the plane facets anti-reflection coated, and wavelength tuned using a diffraction grating, over the range 1.33 to 1.38 μm. Typically, the average output power was ~1mW. A cw-pumped, actively mode locked Nd:YAG laser operating at 1.32μm generating 100ps pulses at a 100MHz repetition rate, with an average power of 2W provided the pump pulses. Both pump and seed signal were directed into a 1.98km length of low loss, single mode fibre, which had a dispersion zero at 1.34μm. The calculated fundamental soliton power for a 250fsec pulse at 1.36μm was ~62W. Temporal and spectral measurements of the radiation

Springer Series in Chemical Physics, Vol. 53 **Ultrafast Phenomena VII**
Editors: C.B. Harris · E.P. Ippen · G.A. Mourou · A.H. Zewail

coupled out of the fibre were taken using a scanning autocorrelator and a 0.5m monochromator with 20fs and 0.2nm resolutions respectively.

3 Results and Discussion

The average power of the pump was maintained at a level below which spontaneous evolution of the soliton-Raman continuum did not take place. Figure 1(a) shows a typical spectrum recorded for an average input power of 160mw (approximately 16W peak). A low level signal in the 1st Stokes Raman band was apparent, but no short pulse temporal feature could be detected attributed to this band.

On introduction of 70μW of continuous radiation at 1.36μm from the cw diode laser a substantial amplification took place around this central wavelength, as can be seen in figure 1(b), and a similar behaviour occurred for other input wavelengths. For higher than 180mW, spontaneous soliton evolution occurred, and although amplification took place for seeded signals, the spectral development was not as simple since soliton collisions gave rise to spectral broadening in this band.

A representative autocorrelation trace is shown in figure 2 of the pulses obtained at the launched signal wavelength, in this case 1.36μm, when the pump power was maintained below the spontaneous soliton generation level. One pulse per pump pulse was generated. As can be seen a 250fs pulses was recorded on top of a pedestal which was approximately 3% of the autocorrelation intensity. Approximately 28% of the total energy was in the pulse which corresponded to an average power of 1.8mW, 72.7W peak in the 250fs pulse. Consequently the power corresponded to that of a single fundamental soliton, within experimental error. Similar operation was obtained by tuning throughout the range available from the cw diode.

The principal contribution to the pedestal component was from the stimulated Raman signal arising from the 100ps pump pulses not required for soliton formation, which appeared as the dispersive wave. Through using intensity dependent polarisation rotation effects, it should be possible to totally remove the pedestal. On increasing the pump power, the pedestal component increased in intensity. This characteristic, however, was due to spontaneous evolution of other solitons and a contribution to the pedestal from cross-correlation of these randomly timed pulses,

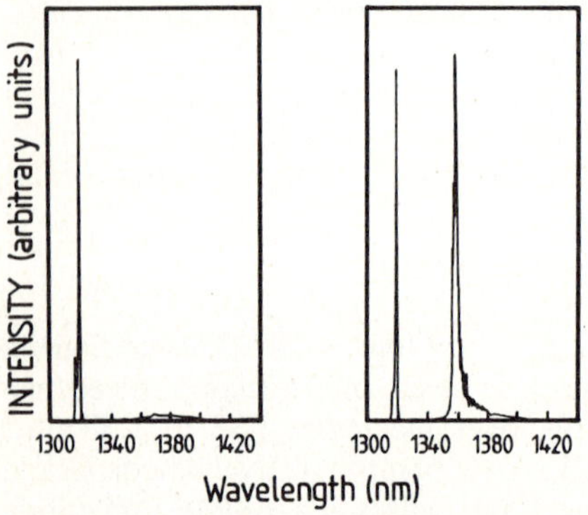

Fig 1 Output spectra from fibre for 160mW average power at 1.32 μm a) without and b) with 70μW cw diode laser signal at 1.36 μm.

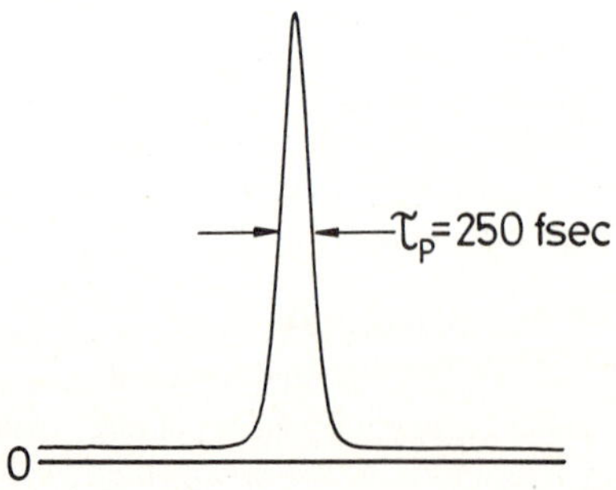

Fig 2 Background free autocorrelation trace of 250fs fundamental soliton generated at 1.36μm using the seeding mechanism.

such as is observed in the single pass generation scheme. With increased power, the duration of these pulses also increased, further indicating a breakdown of the purely seeded pulse formation mechanism to the spontaneous formation of many solitons in the soliton-Raman continuum.

4 Conclusion

We have shown that by seeding a low level cw diode laser signal, single pulse trains of fundamental solitons can be generated at predetermined wavelengths, by maintaining the modulated pump power below the level for self noise-seeded soliton generation. Elimination of the temporal jitter associated with the spontaneous evolution process should also be possible through use of pulsed semiconductor laser inputs.

Acknowledgement

The financial support for this work by British Telecom & the UK SERC is gratefully acknowledged.

References

[1] V.A. Vysloukh and V. N. Serkin, JETP Letts 28,199(1984)

[2] K.J. Blow, N.J. Doran & D. Wood, J, Opt. Soc. Am. B. 5,381(1987)

[3] M. N. Islam, G. Sucha, I. Bar-Joseph, M. Wegener, J.P. Gordon & D.S. Chemla, J. Opt. Soc. Am. B. 6, 1149(1989)

[4] K. J. Blow & D. Wood, IEEE J. Quant. Elect. QE25, 2665(1989)

[5] E. M. Dianov, A.Y. Karasik, P. Mamyshev, A. M. Prokhorov, V. N. Serkin, M. F. Stelmalch & A.A. Fomichev, JETP Lett. 41, 242(1985)

[6] B. Zysett, P. Beaud, W. Hodel & H. P. Weber, Ultrafast Phenomena V. 54(1986)

[7] A.S. Gouveia-Neto, A.S.L. Gomes & J.R. Taylor, Opt Lett . 12, 1035 (1987)

Gain Compression and Recovery Dynamics in Multiple Quantum Well Optical Amplifiers

J.M. Wiesenfeld, G. Eisenstein, M. Wegener, D.S. Chemla, G. Raybon, U. Koren, and S. Weiss*

AT&T Bell Laboratories, Holmdel, NJ 07733, USA
*Present address: Department of Electrical Engineering, Technion, Haifa 32000, Israel

Abstract: The gain of a multiple quantum well optical amplifier recovers with fast (1ps, 7ps) and slow (200 - 400 ps) components subsequent to gain compression by an intense pump pulse. The 7 ps time constant is caused by replenishment of carriers in the wells by carriers from the barriers and can be made to dominate the slower recovery process.

Semiconductor optical amplifiers with multiple quantum well (MQW) active regions have broad gain bandwidth and high saturation output power, at the expense of small-signal gain.[1,2] The high saturation output power is caused, presumably, by a rapid gain recovery time. Here we directly measure the gain recovery dynamics in MQW optical amplifiers.

The active region of the device studied contained four 8 nm wide InGaAs quantum wells separated by 10 nm wide 0.95 eV bandgap-InGaAsP barriers. On both sides of the MQWs are 100 nm wide InGaAsP cladding layers for establishing the waveguide, and further buffer and cladding layers of InP.[3] The amplifier studied most extensively was 600 μm long and had small-signal gain of 19 dB at 1.50 μm wavelength. The gain recovery dynamics were measured by a pump-probe technique using 7 ps duration pulses from a modelocked color center laser or 0.4 ps pulses from an additive pulse modelocked (APM) color center laser. The pump pulse was polarized perpendicular to the junction plane of the amplifier (TM polarization) and orthogonal to the probe pulse (TE polarization). The pump beam was at least 100 times more intense than the probe.

The intense pump pulse removes carriers by stimulated emission, thereby reducing the gain for the probe pulse. Gain recovery measured using 7 ps duration pulses is shown in Figs. 1 and 2 for several conditions. (Note that gain compression is the difference in dB between the small signal gain and the measured gain at a particular time delay, and is positive for reduced gain.) The initial recovery is unresolved by the 7 ps probe pulse. The slower recovery depends only on current bias applied to the amplifier and varies from 200 ps at 275 mA bias to 420 ps at 100 mA bias. The slow decay component is consistent in its magnitude and bias dependence with recovery dynamics dominated by Auger recombination.[4] For the case of 275 mA bias in Fig. 2, the gain overshoots (gain compression is negative) after the initial rapid recovery

Springer Series in Chemical Physics, Vol. 53 **Ultrafast Phenomena VII**
Editors: C.B. Harris · E.P. Ippen · G.A. Mourou · A.H. Zewail

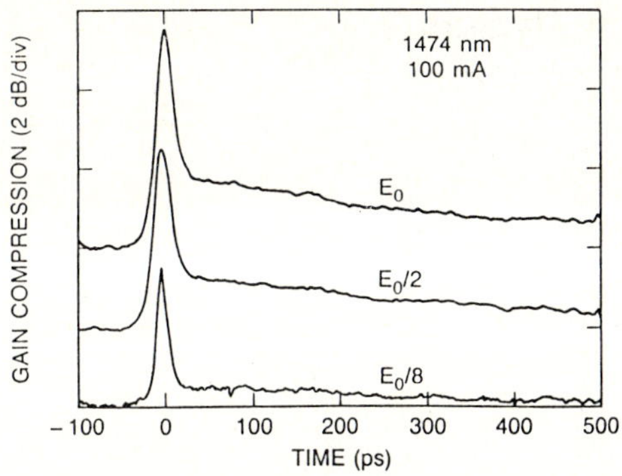

Figure 1: Gain compression at 1474 nm and 100 mA bias. Relative pump pulse energy is 1, 0.25, and 0.125 from top to bottom.

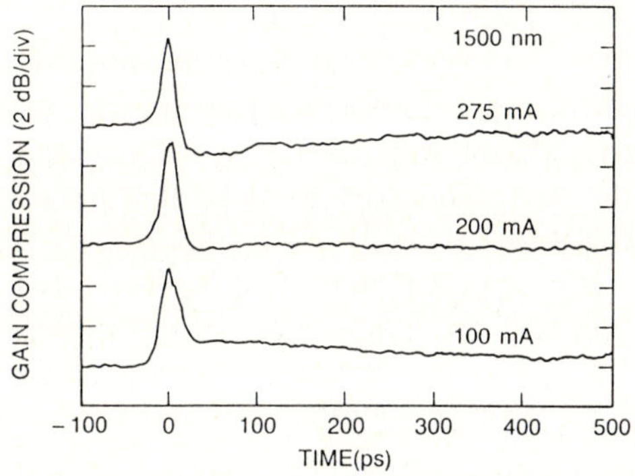

Figure 2: Gain compression for fixed pump energy and varying bias.

and then recovers with the slower time constant. To explain the gain overshoot it is necessary to consider carrier-density dependent refractive index effects.[5] As the carrier density decreases, in some cases the refractive index in the active region can increase, leading to a transient increase in the value of the mode confinement factor and thus, in the gain.

To resolve the faster gain recovery time it was necessary to use 0.4 ps pulses from the APM laser. The gain recovery (plotted as the logarithm of the gain compression) measured with such pulses is shown in Fig. 3. Two decay times are evident. The faster decay time of ~1 ps is caused by cooling of the carrier distribution after removal of carriers by the pump pulse and is consistent with previous measurements of this cooling time in active devices.[6] The 7 ps decay component is gain recovery caused by replenishment of carriers in the wells by carriers from a charge storage reservoir,[4] which for the MQW structure is the InGaAsP barrier and cladding regions. For an average distance l to the cladding region of 50 nm, the diffusion time in InGaAsP materials is about 6 ps. Further

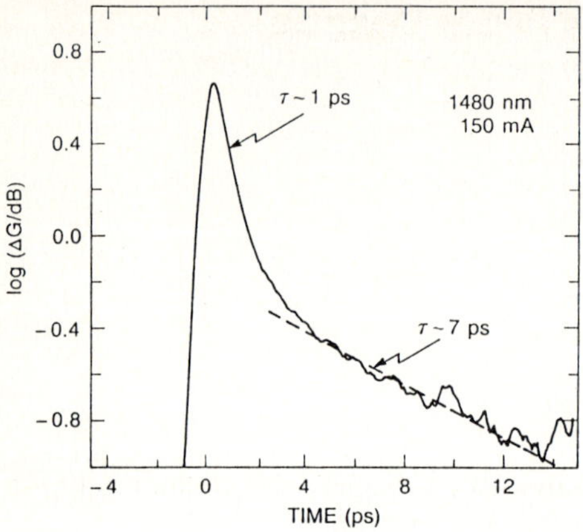

Figure 3: Gain recovery measured with 0.4 ps pulses.

support for the diffusion mechanism is found in Fig. 1. There are a finite number of carriers in the reservoirs (InGaAsP regions). As the energy of the pump pulse is reduced, fewer carriers are removed from the InGaAs wells by stimulated emission and, hence, more complete recovery can be effected by the finite number of carriers in the reservoirs. Thus, the amplitude of the slow component decreases relative to that of the fast component as the pump energy is decreased.

The results show that gain recovery can be very fast because the barriers and cladding regions can function as charge storage reservoirs. By proper design of the dimension and energy of the barriers in the MQW structure, very rapid and complete gain recovery may be achieved.

References:

[1] G. Eisenstein, U. Koren, G. Raybon, T. L. Koch, J. M. Wiesenfeld, M. Wegener, R. S. Tucker, and B. I. Miller, Appl. Phys. Lett.,**56**, 1201 (1990).

[2] M. Bagley, G. Sherlock, D. M. Cooper, L. D. Westbrook, D. J. Elton, H. J. Wickes, P. C. Spurdens, and W. J. Devlin, Electron. Lett., **26**, 512 (1990).

[3] U. Koren, B. I. Miller, Y. K. Su, T. L. Koch, and J. E. Bowers, Appl. Phys. Lett., **51**, 1744 (1987).

[4] G. Eisenstein, R. S. Tucker, J. M. Wiesenfeld, P. B. Hansen, G. Raybon, B. C. Johnson, T. J. Bridges, F. G. Storz, and C. A. Burrus, Appl. Phys. Lett., **54**, 454 (1989).

[5] C. Ell, H. Haug, and S. W. Koch, Opt. Lett. **14**, 356 (1989).

[6] K. L. Hall, J. Mark, E. P. Ippen, and G. Eisenstein, Appl. Phys. Lett., **56**, 1740 (1990).

Generation of Blue-Green 10 fs Pulses Using an Excimer-Pumped Dye Amplifier

R.W. Schoenlein, J.-Y. Bigot, M.T. Portella, and C.V. Shank

Lawrence Berkeley Laboratory, University of California,
1 Cyclotron Road, Berkeley, CA 94720, USA

Abstract. A CPM laser at 620 nm is used to generate 22 fs pulses at 500 nm using an excimer-pumped dye amplifier and continuum generation. The blue-green pulses are compressed to 10 fs using a fiber and grating/prism combination.

1. Introduction

We report the generation of blue-green optical pulses as short as 10 fs duration. The system is based on a CPM dye laser operating at 620 nm[1] and incorporates a two-stage excimer-pumped dye amplifier with continuum generation. The system produces 22 fs pulses at 500 nm which are compressed to 10 fs using a fiber with gratings and prisms for phase correction. This represents an important extension of pulse compression techniques which have been previously demonstrated at longer wavelengths in the visible at 620 nm[2], and in the near infrared at 800 nm[3].

2. Excimer-Pumped Dye Amplifier

The 50 fs pulses from the CPM laser are initially amplified in a three-pass dye amplifier pumped at 400 Hz by a XeF excimer laser (351 nm). The gain medium is Sulforhodamine 640, dissolved in a 30% solution of ammonyx LO, flowing through a quartz cell with a 2 mm interaction length[4] The cell is AR coated to prevent lasing off the glass/air interface which clamps the gain. The excimer pump beam is delivered to the gain cell via multimode silica fiber with a 1 mm core diameter. Though the ~30% coupling efficiency results in reduced pump pulse energies (~7.5 mJ), the fiber delivery system homogenizes any beam filamentation, and mode-filters the excimer beam such that only the useful uv power is delivered to the gain cell. This helps to minimize thermal blooming as well as uv photodegradation of the dye.

The three pass amplifier geometry is shown in figure 1. Pulses from the CPM are focused to a spot size of ~150 μm on the first pass. Subsequent passes are re-imaged to successively larger spot sizes on the gain cell in order to avoid gain saturation effects. excimer laser pulses from the fiber are imaged on the cell to a spot size of ~1 mm, and are nearly colinear with the CPM pulses. This provides a gain of ~27 per pass with a total gain of 2 x 10^4 and output pulse energies of ~4 μJ. A pair of flint glass prisms is employed after the amplifier to compensate group velocity dispersion introduced by the optical elements. The output pulse duration is ~50 fs at 620 nm.

Springer Series in Chemical Physics, Vol. 53 **Ultrafast Phenomena VII**
Editors: C.B. Harris · E.P. Ippen · G.A. Mourou · A.H. Zewail

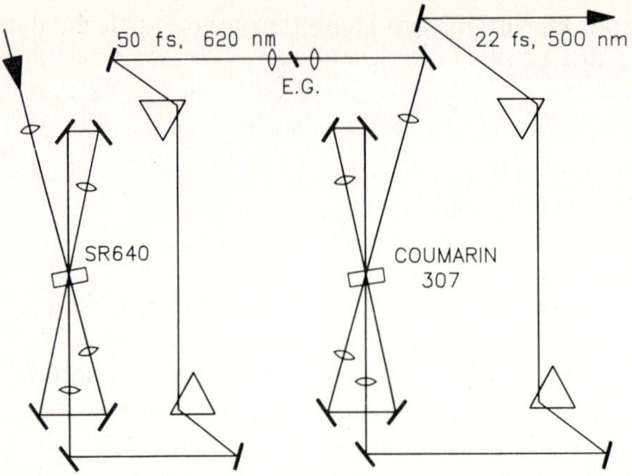

Figure 1. Schematic of two stage dye amplifier pumped by a XeF excimer laser.

Following the first stage of amplification, a white-light continuum is generated by focusing the amplified pulses onto a 1.2 mm jet of ethylene glycol. A portion of this continuum near 500 nm is selected using an induced-transmission filter with a bandwidth of 40 nm. This allows us to exploit the full gain-bandwidth of the dye in the second amplifier.

The second amplification stage employs the same three-pass geometry used in the first stage. The gain medium is Coumarin 307 dissolved in an ethanol/ethylene glycol mixture. The gain per pass is ~10, yielding a total gain of ~10^3. The reduced gain is attributed to the efficiency of the Coumarin dye, as well as decreased pump fluence due to attenuation in the optical fiber. A pair of BK7 prisms are used to compensate the pulse chirp from the continuum generation as well as group velocity dispersion from the optical elements. Using this approach, we are able to generate pulses as short as 22 fs at 500 nm, with pulse energies of ~0.5 μJ. Figure 2 shows a background-free autocorrelation of a 22 fs blue-green pulse, assuming a $sech^2$ pulse shape. The pulse spectrum (shown inset) is 27 nm FWHM, indicating that the pulses are nearly transform-limited. Recently, we have attained pulse

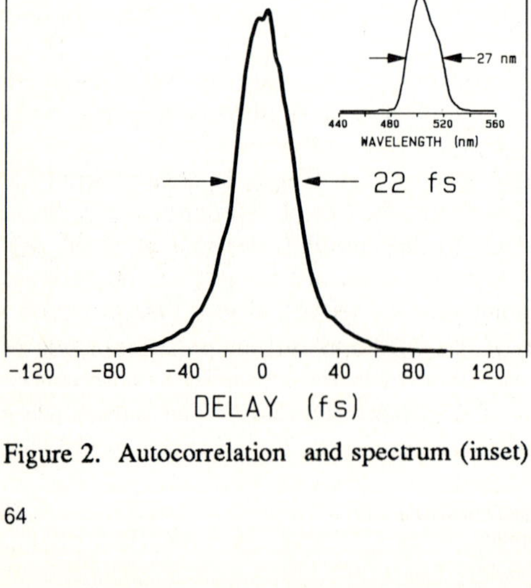

Figure 2. Autocorrelation and spectrum (inset) of 22 fs pulse at 500 nm.

durations as short as 27 fs in the blue at 450 nm using this system with the dye Coumarin 2. This demonstrates that the chirp associated with continuum generation in a jet[5] can be compensated over a sufficient spectral range to shorten the pulses by a factor of two or more. The final pulse duration is limited primarily by the gain bandwidth of the dye.

3. Pulse Compression

The femtosecond pulses at 500 nm are compressed to 10 fs duration using a single-mode pm fiber (~9 mm in length) followed by a pair of gratings and a sequence of four prisms in order to compensate both quadratic and cubic dispersion[2]. Figure 3 shows an autocorrelation corresponding to a 10 fs pulse (assuming a $sech^2$ pulse shape). This was measured using a background-free geometry with a 200 μm thick crystal of BBO. The corresponding spectrum (shown inset) is 59 nm FWHM. The measured pulses are nearly transform-limited, suggesting that these compression techniques may be effective at even shorter wavelengths. Generation of femtosecond pulses with higher photon energies may have important applications for the investigation of ultrafast phenomena in large band-gap semiconductors, metals, and organic molecules.

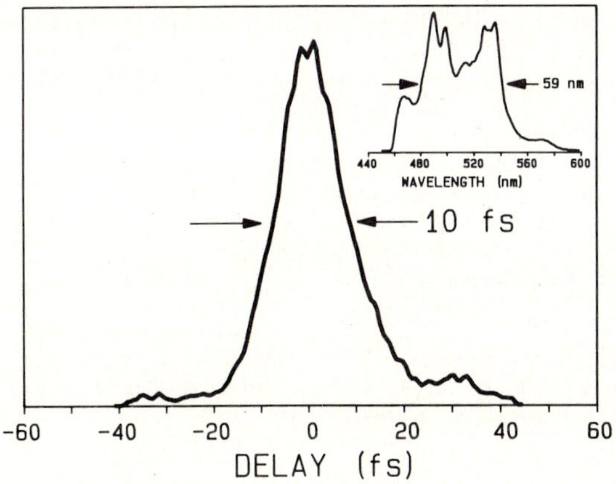

Figure 3. Autocorrelation and spectrum (inset) of 10 fs pulse at 500 nm.

References

[1] J. Valdmanis, R. Fork, and J. Gordon, Opt. Lett 10, 131 (1985)

[2] R.L. Fork, C.H. Brito Cruz, P.C. Becker, and C.V. Shank, Opt. Lett. 12, 483 (1987)

[3] P.C. Becker, H.L. Fragnito, R.L. Fork, F.A. Beisser, and C.V. Shank, Appl. Phys. Lett. 54, 41 1 (1989)

[4] D. Nickel, D. Kuhlke, and D. von der Linde, Opt. Lett 14, 36 (1988)

[5] R.L. Fork, C.V. Shank, C. Hirliman, R. Yen, and W.J. Tomlinson, Opt. Lett. 8, 1 (1983)

Work supported by U.S. Department of Energy under Contract No. DE-AC03-76SF00098.

Pulse Compression of a Mode-Locked Ti:Sapphire Laser

J.D. Kafka, M.L. Watts, D.J. Roach, M.S. Keirstead, H.W. Schaaf, and T. Baer

Advanced Product Technology Group, Spectra-Physics, Inc.,
1250 West Middlefield Road, Mountain View, CA 94042, USA

A mode-locked Ti:sapphire laser has the potential to produce ultrashort pulses while preserving the high power and tunability inherent in this gain media. Several mode-locking techniques have produced pulses of less than 10 psec duration from a Ti:sapphire laser [1-7], however in most of the configurations the output powers have been limited. We have previously reported on 6 psec pulses with 750 mW of average power generated by active mode-locking with an acousto-optic modulator [8]. In this paper, we use active mode-locking to generate picosecond pulses with 2.0 W of average power and pulse compression to generate tunable pulses shorter than 50 fsec with 750 mW of average power.

Our Ti:sapphire oscillator is pumped by the all-lines output of an argon ion laser as shown in Fig. 1. The basic cavity design is similar to the Spectra-Physics model 3900 cw Ti:sapphire laser. It uses a 2 cm Ti:sapphire rod in an astigmatically compensated cavity with the length extended to provide a repetition rate of 82 MHz. A high Q acousto-optic modulator is located near one end of the cavity and the tuning element is a one-plate birefringent filter. A Gires-Tournois interferometer [9] is used to compensate dispersion in the cavity resulting in shorter pulses. With 9 W of pump power, typical performance is 2.0 W of output power at a pulsewidth of 1.3 psec. The autocorrelation is shown in Fig. 2 and is fit well by a sech squared pulse shape. The spectrum of this pulse is shown in Fig. 3 and has a full width half maximum of 0.49 nm which yields a time bandwidth product of 0.31.

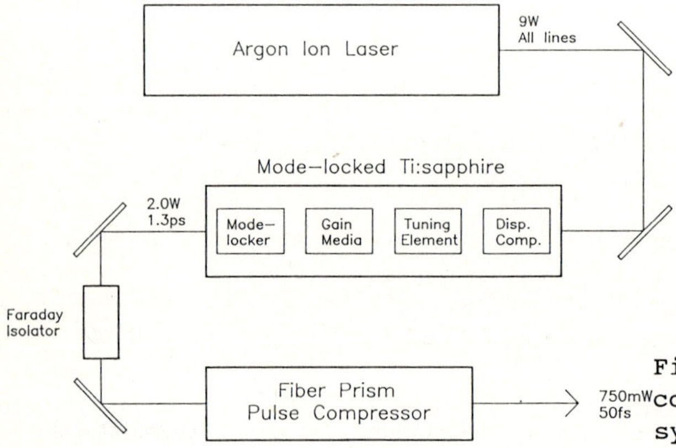

Fig. 1. The experimental configuration showing key system components.

Springer Series in Chemical Physics, Vol. 53 **Ultrafast Phenomena VII**
Editors: C.B. Harris · E.P. Ippen · G.A. Mourou · A.H. Zewail

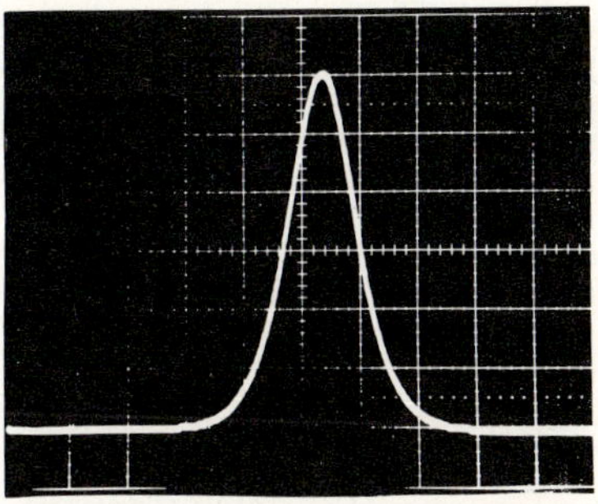

Fig. 2. Autocorrelation trace of a 1.3 psec pulse from the mode-locked Ti:sapphire laser.

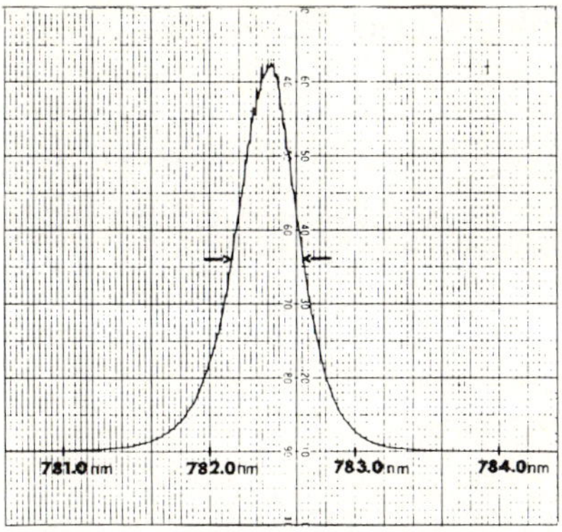

Fig. 3. Spectrum of the mode-locked pulse with a bandwidth of 0.49 nm.

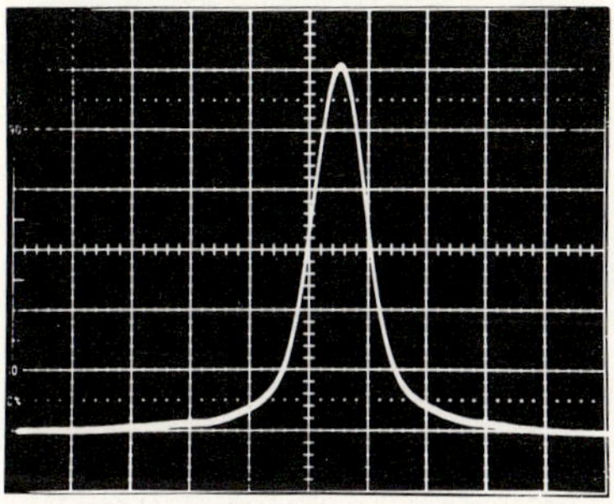

Fig. 4. Autocorrelation trace of a 50 fsec pulse from the pulse compressor (assuming a sech squared pulse shape).

We have constructed a fiber-prism pulse compressor which generates 50 fsec pulses while preserving the high power of the oscillator. The fiber is a single mode polarization preserving fiber with a length of 75 cm. A Faraday isolator is used to isolate the laser from back reflections from the fiber (Fig. 1). We observe 60% throughput for the fiber and isolator combination. We have constructed a novel dispersive element out of 3 closely spaced SF-10 Brewster prisms configured so that their dispersive power is additive. The dispersive delay line consists of this single dispersive element and a series of folds which is equivalent to the usual four prism sequence. The equivalent spacing between the first and second dispersive elements is approximately 60 cm and the throughput of the dispersive delay line is 60%.

The output from the pulse compressor is 750 mW of average power with 50 fsec pulsewidths at 780 nm. The input polarization to the fiber is adjusted for optimal pulse quality and the autocorrelation is shown in Fig. 4. The compressed pulses are within a factor of two of the transform limit, and the laser is tunable from 745 nm to 810 nm with the tuning range being limited by mirror coatings. Over this entire tuning range the pulse width was 50 fsec and the output power ranged from 600 to 800 mW. The stability of the compressed pulses is similar to that of the oscillator and the amplitude fluctuations are less than 1% rms measured from 10 Hz to 2 MHz.

The compressed, actively mode-locked Ti:sapphire laser provides a new method of generating ultrashort pulses near 800 nm. Moreover, we believe that the pulse widths and powers that we have demonstrated make this laser system the highest average power source of sub-100 fsec pulses at any wavelength.

We gratefully acknowledge the contributions of J. B. Clark.

References

1. P. M. W. French, J. A. R. Williams and J. R. Taylor, Opt. Lett. 14, 686 (1989).
2. J. Goodberlet, J. Wang, J. G. Fujimoto and P. A. Schulz, Opt. Lett. 14, 1125 (1989).
3. N. Sarukura, Y. Ishida, H. Nakano and Y. Yamamoto, Appl. Phys. Lett. 56, 814 (1989), Y. Ishida, N. Sarukura and H. Nakano, in Digest of Meeting on Ultrafast Phenomena (1990), postdeadline paper PD11, and N. Sarukura, Y. Ishida and H. Nakano, in Digest of Conference on Lasers and Electro-Optics (1990), paper CWF12.
4. U. Keller, W. H. Knox and H. Roskos, in Digest of Meeting on Ultrafast Phenomena (1990), postdeadline paper PD9.
5. J. Squier, F. Salin, C. Rouyer, S. Coe and G. Mourou, in Digest of Conference on Lasers and Electro-Optics (1990), postdeadline paper CPDP9.
6. D. E. Spence, P. N. Keen and W. Sibbett, in Digest of Conference on Lasers and Electro-Optics (1990), postdeadline paper CPDP10.
7. P. F. Curley and A. I. Ferguson, in Digest of Conference on Lasers and Electro-Optics

Coupled-Cavity Resonant Passive Modelocked (RPM) Ti:Sapphire Laser

U. Keller, W.H. Knox, and H. Roskos

AT&T Bell Laboratories, Room 4B-523, Crawfords Corner Road, Holmdel, NJ 07733, USA

Coupled cavity resonant passive modelocking (RPM) is a new simple technique to passively modelock laser media with a small gain saturation cross section such as Ti:Sapphire, Nd:YAG etc. lasers. For these lasers the gain saturation is weak making intracavity saturable absorber modelocking problematic. We simply added to a cw pumped Ti:Sapphire laser an external cavity with a nonlinear quantum well reflector (Fig. 1). We call this laser "resonant passive modelocked" because of the near-resonant nonlinearity used in the external cavity. Without any dispersion compensation and any active feedback control system, we generated pulses as short as 2 ps and less than 10ps over a 50 nm wavelength tuning range.

The general concept of coupled-cavity modelocking has been introduced by the soliton laser [1] where the nonlinear external cavity medium supports soliton pulse formation when the net group velocity dispersion is negative. This technique has been extended to an external nonlinear coupled cavity with positive dispersion, now referred to as APM [2] (additive pulse modelocking) or CCM [3] (coupled cavity modelocking), where the pulse formation is based on a nonlinear phase effect with constructive interference at the peak of the pulse and destructive interference in the wings. Recently, an APM Ti:Sapphire laser [4] generated 1.4 ps pulses which was dispersively compensated to 200 fs.

We introduce a new embodiment of the general concept of coupled-cavity modelocking, in which a cw pumped gain medium is placed in a high Q main cavity, and the external coupled cavity contains a near-resonant nonlinearity provided by a semiconductor device. We built a coupled cavity (Fig. 1) where the main laser is Spectra Physics #3900 Ti:Sapphire laser pumped by a cw argon ion laser which is coupled to an external cavity with a quantum well reflector [5] as the end mirror. We used a 5x microscope lens to focus the beam down to a 30 μm spot diameter. Either a pellicle or a folding mirror inside the external cavity was used as the output coupler. The nonlinear reflector was a GaAs/GaAlAs quantum well structure, which was AR coated, and incorporated a AlAs/GaAlAs dielectric mirror centered around 850 nm [6]. The undoped MQW region consists of 75 periods of 95 Å wide GaAs quantum wells and 45 Å wide $GaAl_{0.3}As_{0.7}$ barriers for which the exciton saturation intensity at zero bias was measured to be 5 to 10 kW/cm^2 [7].

Figure 2 shows a non-collinear, real-time (not averaged) autocorrelation trace obtained with a Femtochrome autocorrelator giving 2.1 ps FWHM pulse duration, assuming $sech^2$ pulse shape. The pulses are in single transverse TEM_{00} mode (also verified with an rf spectrum analyzer), the repetition rate was ≈ 250 MHz, and the average output power for the stronger beam was 90

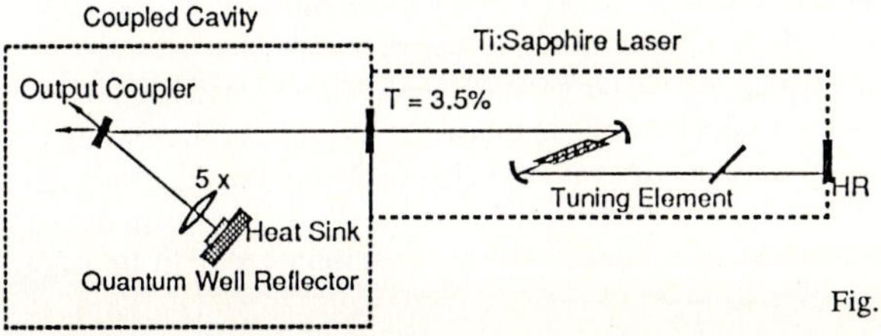

Fig.1: Cavity design

Springer Series in Chemical Physics, Vol. 53 **Ultrafast Phenomena VII**
Editors: C.B. Harris · E.P. Ippen · G.A. Mourou · A.H. Zewail

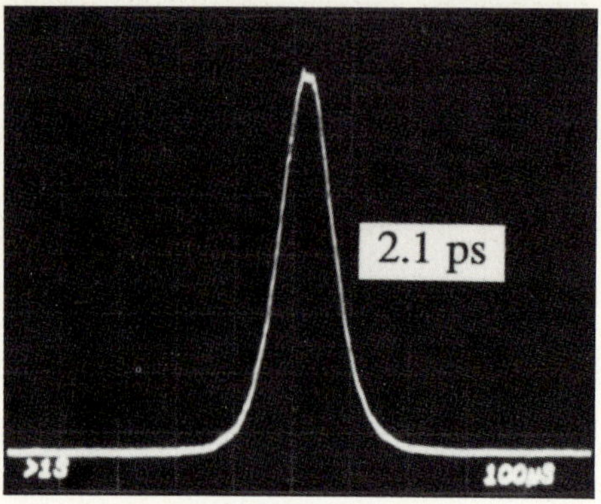

Fig. 2: Autocorrelation

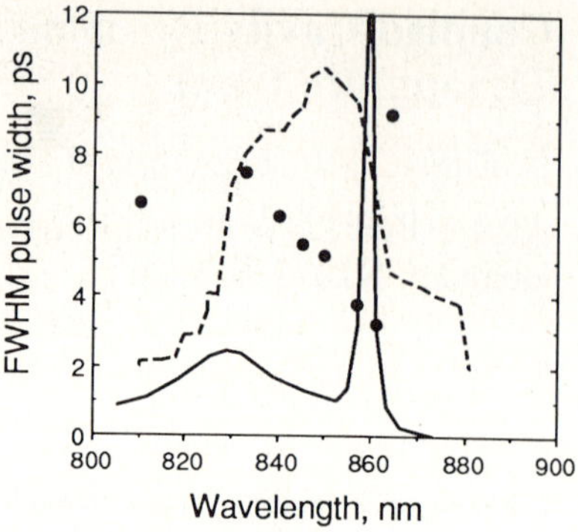

Fig. 3: Tunability over 50nm

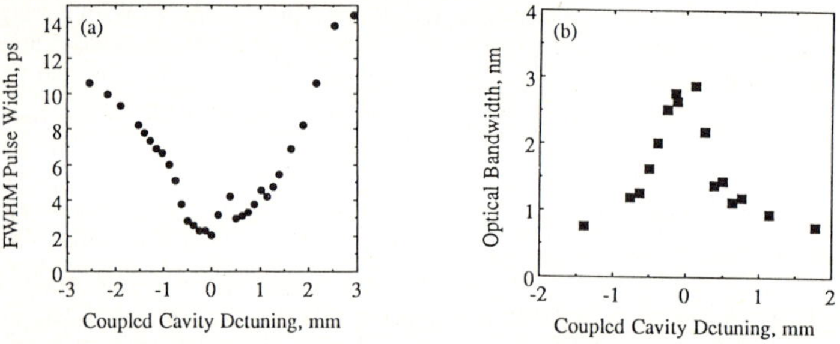

Fig. 4: (a) Pulse width and (b) optical bandwidth versus external cavity length

mW using a 10% output coupler. The 50 nm tunability is shown in Fig. 3, which also displays the normalized output power of the cw Ti:Sapphire laser cavity alone (dotted line), and the in-situ measured luminescence from the quantum well reflector which also includes the scattered lasing wavelength at around 860 nm (solid line). The 50 nm tunability was primarily limited by the tuning range of the Ti:Sapphire mirror set. We note that the shortest pulse durations are obtained at wavelengths slightly below (a few tens of meV) the luminescence peak. The intensity on the reflector is estimated to be ≈ 36 kW/cm^2 for the 2 ps pulses which is well above the exciton saturation intensity, as in the case of intracavity cw passive modelocked color center laser[8].

We were able to detune the external cavity length over ± 2 mm and still achieve pulse duration below 10 ps (Fig. 4a). The measured bandwidth at 846 nm at around zero detuning was 3.4 nm and reduced to 0.8 nm at a detuning of 1.4 mm (Fig. 4b). A 2.9 nm time-bandwidth limited pulse (assuming sech^2 pulses) would be 260 fs long. The pulse width was not very sensitive to the argon ion pump power. By varying the pump power to the Ti:Sapphire from 6 W to 10 W the pulse duration first decreased slightly to a minimum of 5 ps at ≈ 7 W power and increased to 7.5 ps at ≈ 10 W pump power at a cavity detuning of + 0.5 mm. The pulse repetition rate is determined by the external cavity length provided it is matched within a few millimeters to an integer fraction or multiple of the main cavity length. By decreasing the external cavity length to half the main laser's length, we were able to double the repetition rate. In this case, the fundamental mode beating of the main laser was suppressed by more than 55 dB with respect to the doubled repetition rate signal. Higher harmonics should be easily obtained.

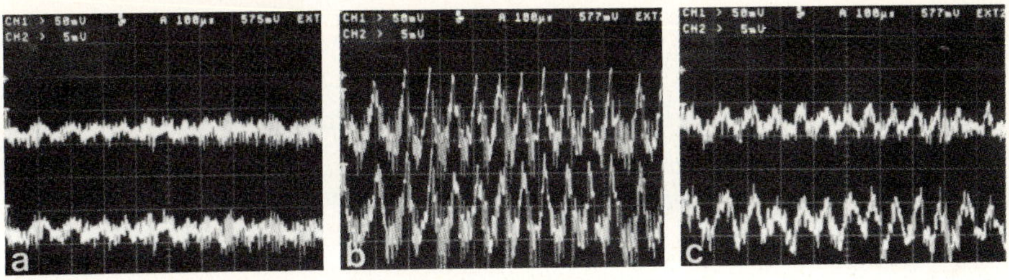

Fig. 5: Non-APM mode of operation

The near resonant nonlinearity is distinctively different from the non-resonant kerr-type nonlinearity used in the previous techniques. We have demonstrated that the pulse formation is non-APM and is based on a large nonlinear amplitude signal which, when re-injected into the main cavity, strongly modelocks the laser. The high Q of the main cavity is important for the short pulse formation. We proved the non-APM operation of the laser by repetitively varying the external cavity length with a piezo transducer over about 2 μm, while monitoring the intensity of both the fundamental and the second harmonic of the output power [2]. Figure 5 shows the periodic modulation for the output power (top trace) and the second harmonic of the output power (lower trace) at different cavity length detuning: +1.3 mm (a), matched (b), and -1.3 mm (c). In both cases the dc level is at about 3.5 divisions. At exactly matched cavity length, we observed up to 50% intensity modulation in the longer wavelength tuning range for both the fundamental and the second harmonic intensity due to interference between coupled cavities (Fig. 5b). This intensity modulation strongly reduces at detuned cavity lengths where, however, picosecond pulses were still observed (Fig. 5a and c). That these intensity modulations are similar for both the fundamental and the second harmonic signal and even disappear at larger detuning proves that the dominant pulse formation mechanism is not phase-coherent re-injecting (i.e. non-APM), although the details of the spectrum do exhibit some dependence on the phase.

In conclusion, we introduce a completely new and extremely simple technique of short pulse generation which can be applied to any kind of low gain laser media.

[1] L. F. Mollenauer, and R. H. Stolen, *Optics Lett.* **9**, 13, 1984

[2] C. P. Yakymyshyn, J. F. Pinto, and C. R. Pollock, *Optics Lett.* **14**, 621, 1989, and J. Mark, L.Y. Liu, K.L. Hall, H.A. Haus, and E.P. Ippen, *Optics Lett.* **14**, 48, 1989

[3] P. N. Kean, X. Zhu, D. W. Crust, R. S. Grant, N. Langford, and W. Sibbett, *Optics Lett.* **14**, 39, 1989

[4] J. Goodberlet, J. Wang, and J. G. Fujimoto, and P. A. Schulz, *Optics Lett.* **14**, 1125, 1989

[5] G. D. Boyd, D. A. B. Miller, D. S. Chemla, S. L. McCall, A. C. Gossard, and J. H. English, *Appl. Phys. Lett.* **50**, 1119, 1987

[6] The sample was provided by K. W. Goossen, J. E. Cunningham, and W. Y. Jan

[7] A. M. Fox, D. A. B. Miller, G. Livescu, J. E. Cunningham, J. E. Henry, and W. Y. Jan, submitted to *Appl. Phys. Lett.* 4/4/90

[8] M. N. Islam, E. R. Stunderman, C. E. Soccolich, I. Bar-Joseph, N. Sauer, T. Y. Chang, and B. I. Miller, *IEEE J. Quantum Electr.* **25**, 2454, 1989

Self-Starting Additive Pulse Mode-Locking of a Nd:Glass Laser

Ch. Spielmann, F. Krausz, E. Wintner, and A.J. Schmidt

Abteilung für Quantenelektronik und Lasertechnik,
Technische Universität Wien, Gusshausstr. 27, A-1040 Wien, Austria

Self starting continuous-wave mode-locked operation of a Nd:glass laser has been achieved by using a nonlinear external resonator. The coupled cavity incorporating an optical fiber initiates and sustains ultrashort pulse generation resulting in a stable train of 380 fs pulses.

Additive pulse mode-locking (APM) has become a powerful technique for femtosecond pulse generation in solid state lasers.[1] Recently, APM was found to be self-starting in a $Ti : Al_2O_3$ laser.[2] Motivated by this discovery we have developed a cw passively mode locked Nd:glass laser using a nonlinear external cavity. The laser generates stable subpicosecond pulses of 380 fs duration.

It has been shown that continuous wave operation of a Nd:glass laser is possible with low threshold and high slope efficiency.[3] Thermal problems can be avoided by careful design and proper host selection. Owing to its broad gain profile Nd:glass is able to amplify pulses shorter than 100 fs. Additionally, it offers the advantage to be pumped efficiently around 0.8 μm by diode lasers.[4] These properties make Nd:glass an attractive material for ultrashort pulse generation and amplification.

A schematic diagram of the coupled nonlinear cavity Nd:glass laser is shown in Fig. 1. The main cavity consists of mirrors M1-M3 and the output coupler OC.

It was designed to compensate for astigmatism caused by the active material. A Brewster-angle-oriented 2 mm thick plate of Nd:glass (LG-760, Schott) with 4% Nd^{3+} concentration by weight was inserted at the cavity mode waist in the Z-shaped resonator.

Using a 5% output coupler the pumping threshold was 50 mW when pumped simultaneously at the 752 nm and 799 nm lines of a krypton laser. At 800 mW of absorbed pump power the laser delivered nearly 300 mW output power at 1054 nm. For this experimental setup the critical pump powers indicating the onset of undesirable thermal effects lay only above 1W.[3]

The auxiliary resonator was formed by the output coupler OC of the main cavity, a 80% reflecting beam splitter BS, and a high reflecting mirror M4. The nonlinearity in

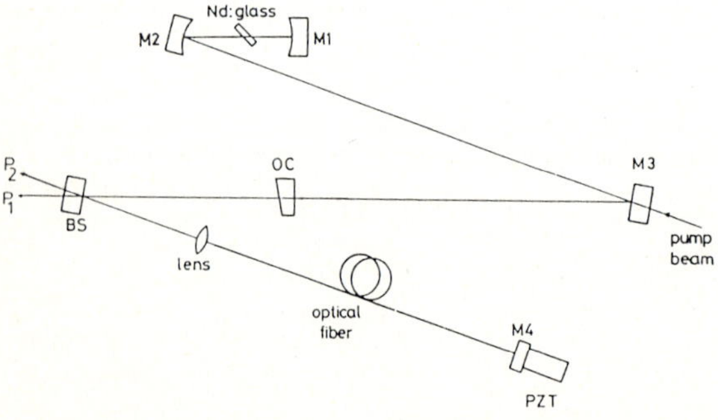

Fig.1: Schematics of the passively mode-locked Nd:glass laser. M1-M2, high reflecting mirrors with radii of curvature of 5 cm and 10 cm, respectively; M3, dichroic flat mirror; OC, output coupler with 80% reflectivity; BS, beam splitter with 80% reflectivity; M4, high reflecting flat mirror.

Springer Series in Chemical Physics, Vol. 53 **Ultrafast Phenomena VII**
Editors: C.B. Harris · E.P. Ippen · G.A. Mourou · A.H. Zewail

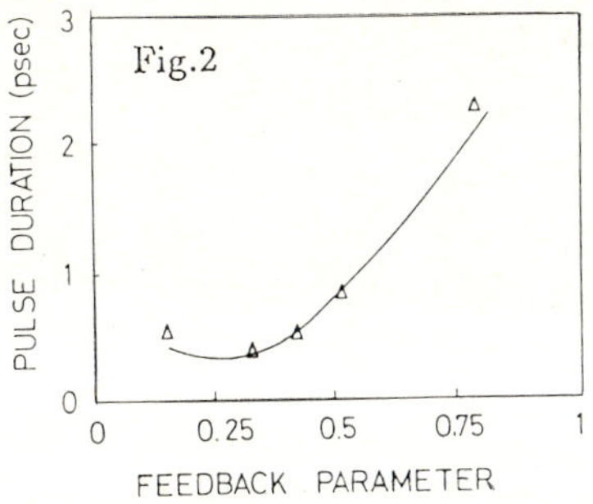

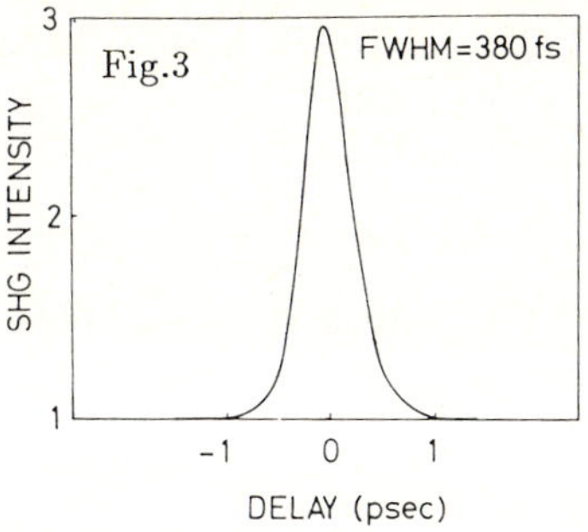

Fig.2: Pulse durations from the passively mode-locked Nd:glass laser as a function of the feedback parameter P_r/P_i

Fig.3: Collinear autocorrelation trace of the laser output.

the coupled cavity was provided by a single mode non-polarisation-preserving fiber of 90 *cm* length. Typical fiber coupling efficiency was 75%. This choice of high beam splitter reflectivity and long fiber provides a strong feedback and high nonlinearity facilitating self-starting of the mode-locking process.

With a 20% output coupler and the external cavity unblocked the pumping threshold was about 100 mW (absorbed power). It was found in order to get APM self-started it was necessary to eliminate any source of intracavity etalon effects in the *main* cavity. By contrast, etalon effects in the *auxiliary* cavity turned out not to impair the build-up of mode-locked pulses from the initial random noise.

In the pumping region of 100-400 mW only free-running lasing was observed. At about 400 mW transient mode locking occured. Microsecond trains of 40-50 ps pulses appeared and disappeared on a 100 μs time scale. Stable self-starting APM was achieved at pump powers above 500 mW. With an absorbed pump power of 600 mW typical average power levels at important locations in the coupled laser system were as follows: primary laser output (see Fig. 1) $P_1 = 60\ mW$, incident power into the fiber $P_i = 240\ mW$, power reflected back from the fiber by the mirror M4 $P_r = 180\ mW$, secondary laser output $P_2 = 36\ mW$.

With the external cavity tuned to resonance mode-locking turned on and off due to drift of the relative phase between the two cavities. The mode-locked train generally remained stable typically for hundreds of milliseconds. Now P_r could be continuously varied by removing mirror M4 from the fiber end. Mode-locked pulsewidths are shown in Fig.2 as a function of the ratio of P_3 to P_i. The shortest pulses were generated when the feedback was reduced by $\approx 30\%$.

The collinear autocorrelation trace shown in Fig.3 yields a pulse width of 380 fs assuming a sech^2 pulse shape, at a feedback parameter P_r/P_i of 0.28.

The output spectrum was measured to be within 1.2 times the transform-limited bandwidth. Second-harmonic fluctuations of less than 10% were observed on a long time scale when the average output power was stabilized with a servo loop. We expect further improvement of mode-locking performance by optimization of the external cavity parameters. Diode laser pumping of such a coupled-cavity Nd:glass laser will then allow the construction of an all-solid-state source of femtosecond optical pulses.

We gratefully thank Prof. H. A. Haus for stimulating conversations. We acknowledge discussions with Prof. D. C. Hanna, Dr. M. E. Fermann and T. Brabec. This research was supported by the Fonds zur Förderung der wissenschaftlichen Forschung in Österreich grant P 7282 and the Hochschuljubiläumsstiftung der Gemeinde Wien.

References

[1] E. P. Ippen, H. A. Haus, and L. Y. Liu, J. Opt. Soc. Am. B **5**, (1989)
[2] J. Goodberlet, J. Wang, J. G. Fujimoto, and P. A. Schulz, Opt. Lett. **14**, 1125 (1989)
[3] F. Krausz, E. Wintner, A. J. Schmidt, and A. Dienes, IEEE J. Quantum Electron. **QE-26**, 158 (1990)
[4] F. Krausz, T. Brabec, E. Wintner, and A. J. Schmidt, Appl. Phys. Lett. **55**, 2386 (1989)

Soliton-like Pulse Shaping in a cw Passively Mode-Locked $Ti:Al_2O_3$ Laser

Y. Ishida, N. Sarukura, and H. Nakano

NTT Basic Research Laboratories, 3-9-11 Midori-cho, Musashino-shi, Tokyo 180, Japan

We have generated stable pulses less than 170 fs around 750 nm from a cw passively mode-locked $Ti:Al_2O_3$ laser with a prism-pair dispersion compensator inside the cavity. The pulse shaping mechanism mainly originates from the combined actions of self-phase modulation (SPM) and group velocity dispersion or soliton-like pulse formation.

Recently, titanium-doped sapphires ($Ti:Al_2O_3$) have been receiving a lot of attention as a new type of solid state laser medium alternative to organic dyes because broadly tunable femtosecond pulses can be realized in the near-infrared region. At present, two different types of mode-locking have been reported for generating ultrashort pulses less than several picoseconds from a $Ti:Al_2O_3$ laser. One method, additive pulse mode-locking coupled with a nonlinear external fiber, produces pulses less than 0.8 ps and 200 fs in duration, after external dispersion compensation using a grating pair[1,2]. The other, cw passive mode-locking, uses a saturable absorber to produce pulses of about 4 ps for a DDI dye absorber, and 2.7 ps for a semiconductor-doped glass absorber[3,4].

In this paper, we describe the generation of tunable femtosecond pulses in a cw passively mode-locked $Ti:Al_2O_3$ laser based on soliton-like pulse shaping. The laser cavity consists of two sets of folded focusing mirrors for the laser rod and for absorber dye jet, which is basically the same as previously reported in Ref. 3. The saturable absorber dye used here is HITCI (3.6×10^{-5} M in ethyleneglycol). The output coupling is 5% around 800 nm. The $Ti:Al_2O_3$ laser is pumped by all of the lines (< 20 W) from a cw Ar ion laser.

Without a prism pair, threshold pumping power was 3.8 W. When pumped by 14 W, average output power was 1.1 W and the laser tuned from 750-810 nm. Typical pulse width was ~10 ps at about 780 nm for a single-plate birefringent filter and spectral width was ~0.35 nm. This spectral width is about 5.5 times transform limit assuming sech^2 shape and indicates that the pulses are chirping. It was found that the occurrence of positive linear chirp over the entire envelops strongly depends on the peak intensity of the pulse[5]. As pumping power increases, the SHG autocorrelation trace of the pulse shows a triangle shape, indicating that the rectangular shape of the actual pulse, and the spectral width broadens. This behavior is very similar to the SPM process when coupled with the linear dispersion of the optical fiber. The chirped pulses were compressed into a 1.6-ps pulse, indicating an almost transform-limited pulse (assuming sech^2 shape), by using a grating pair. On the other hand,

Springer Series in Chemical Physics, Vol. 53 **Ultrafast Phenomena VII**
Editors: C.B. Harris · E.P. Ippen · G.A. Mourou · A.H. Zewail

when the pumping power was less than 5 W, that is, at a low output power level (~100 mW), no spectral broadening ($\leq$0.1 nm) or pulse shortening was observed.

From these results, the mechanism of positive linear chirping can be explained according to SPM coupled with the linear dispersion inside the cavity. The SPM process of the sapphire rod plays a key role in spectral broadening and chirping because its peak intensity is high enough to induce nonlinear processes ($\geq$0.1 GW/cm^2) and its length is relatively long (20 mm). In order to control group velocity dispersion and to further generate short pulses, a Brewster prism pair (SF 57 glass) having a high dispersion was incorporated into the cavity. The values for $dn/d\lambda$ and $d^2n/d\lambda^2$ for SF 57 at 775 nm are -0.0725 μm^{-1} and 0.283 μm^{-2}, respectively. The dispersion, $d^2\phi/d\omega^2$ (negative or positive) can be continuously adjusted by translating one prism nomal to its base. Threshold pumping power with a prism pair was almost the same as with no prism pair, and a stable mode-locking was maintained until pumping rose over 10 W.

We observed dramatic reductions in pulse width and broadening in spectral width by placing a prism pair inside the cavity. When separation between the two prisms was adjusted to about 20 cm, stable pulses with no wings were obtained. The pulse trains are also stable with a small ripple of ~3 %. Under these conditions, the single pass dispersion $d^2\phi/d\omega^2$ for prism pair (SF 57) was calculated to be $-1{,}800$ fs^2 at 775 nm. Therefore, this negative dispersion can be compensated the positive chirp due to the SPM coupled with the group velocity dispersion of the sapphire rod. From the SHG autocorrelation trace shown in Fig. 1 a, pulse width was 165 fs for an assumed sech^2 shape. The spectrum measured had a symmetric shape, as shown in Fig. 1 b. For purposes of comparison, the calculated autocorrelation trace (open circles) and the cal-

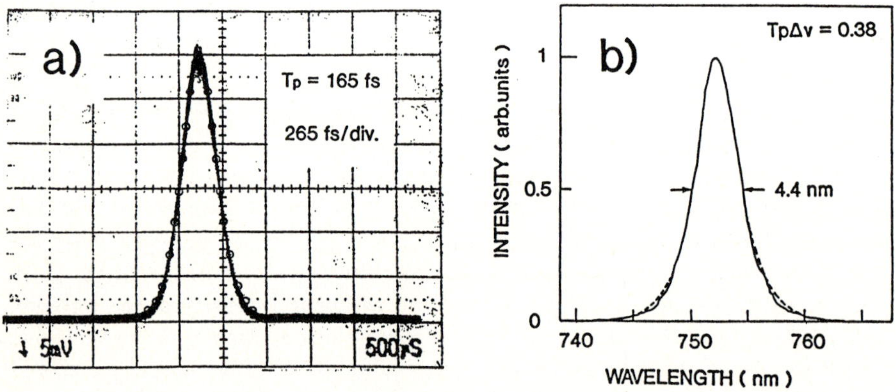

Fig. 1 a) Rapid-scanned SHG autocorrelation trace of the pulses. The correlation trace is best fitted by a 165-fs (FWHM) sech^2 shape (open circles). b) Spectrum. The trace is best fitted by a 4.4-nm (FWHM) sech^2 shape (dashed curves).

culated spectrum (dashed curves) for $sech^2$ shape are indicated in the Fig. 1 a and b, respectively. If spectral width is 4.4 nm (FWHM), the time-band width product is 0.38 at 752 nm, which indicates approximately transform-limited pulse. In this case, the compression ratio (correlation width between no prism and prism) is close to 40. With an 8 W pumping power, the average output power was ~450 mW at a repetition rate of 101 MHz. The wavelength was also tunable in the range of 750-805 nm. Tunability at the short wavelength side is limited by the reflectance of the cavity mirrors. Finally, the pulses with the peak power of 27 kW obtained from the cavity is compressed into 50-fs pulse with no wings by using a fiber prism pair (SF 59) compressor.

These experimental results indicate that the remarkable pulse shortening and high stability of the output pulses are mainly due to soliton-like pulse formation or dispersive pulse compression, which is the same as in the case for passively mode-locked ring dye or fiber soliton lasers.

In conclusion, we have demonstrated the generation of stable femtosecond pulses in a cw passively mode-locked, soliton-like $Ti:Al_2O_3$ laser. This laser can serve as a practical light source in studying various types of femtosecond spectroscopy or in high power laser amplifier systems in the near-infrared region (or near-ultraviolet region by frequency doubling) because of its excellent long term stability and simplicity.

References

1. P. M. W. French, J. A. R. Williams, and J. R. Taylor, *Opt. Lett.* **14**, 686(1989).

2. J. Goodberlet, J. Wang, J. G. Fujimoto, and P. A. Schulz, *Opt. Lett.* **14**, 1125(1989).

3. N. Sarukura, Y. Ishida, H. Nakano, and Y. Yamamoto, *Appl. Phys. Lett.* **56**, 814(1990).

4. N. Sarukura, Y. Ishida, T. Yanagawa, and H. Nakano, to be published in *Appl. Phys. Lett.*

5. N. Sarukura, Y. Ishida, and H. Nakano, to be published in *Opt. Comm.*

Sub-picosecond 38 W Optical Pulses from a Hybrid Mode Locked Semiconductor Laser System

P.J. Delfyett[1], *C.-H. Lee*[1], *L. Florez*[1], *N. Stoffel*[1], *N. Andreadakis*[1], *T. Gmitter*[1], *G.A. Alphonse*[2], *and J.C. Connolly*[2]

[1]Bell Communications Research, Red Bank, NJ 07701, USA
[2]David Sarnoff Research Center, Princeton, NJ 08543, USA

Abstract Ultrashort optical pulses of 0.56 picoseconds in duration with over 38 watts of peak power have been generated from an all semiconductor laser diode system. These optical pulses are the shortest to date and also represent the highest peak power produced from a mode locked semiconductor laser system.

1) Introduction

Compact and efficient sources of high repetition rate, high peak power optical pulses are needed to replace the standard ultrashort optical pulse generation schemes due to their relatively large size and poor wall plug efficiency [1]. These compact laser systems will find applications in areas spanning laser radar, applied nonlinear optics, such as optical computing and photonic switching, and optical time domain reflectometry. In this paper, we describe the first external cavity hybrid mode locked semiconductor laser system, which can produce optical pulses of 0.56 psec in duration with 38 watts of peak power. These pulses are the shortest optical pulses generated from a mode locked semiconductor laser system. The peak powers obtained from this system also represent the highest ever obtained from an all semiconductor laser system.

2) Experiment

The experimental setup is shown in Fig. 1. The mode locked laser was formed by placing a semiconductor traveling wave amplifier (TWA) inside of an external cavity. The TWA was connected to a bias tee, which supplied d.c. and R.F. power to the diode. The rear reflector is a multiple quantum well (MQW) structure which has been epitaxially removed and placed on a silver mirror reflector. The purpose of the MQW is that it serves as a saturable absorber which initiates passive mode locking. The output of the laser was passed through an isolater and TWA1. This TWA serves as a single pass amplifier which raises the peak power of the mode locked pulses to a level so that 'real time' autocorrelation techniques can be used to monitor the pulse shape. Once optimum mode locking was obtained, the output pulse from TWA1 was directed into a double pass dual grating pulse compressor which utilizes a 1:1 telescope to allow access to both positive and negative dispersion regimes. Due to the very low diffraction efficiency

Springer Series in Chemical Physics, Vol. 53 **Ultrafast Phenomena VII**
Editors: C.B. Harris · E.P. Ippen · G.A. Mourou · A.H. Zewail

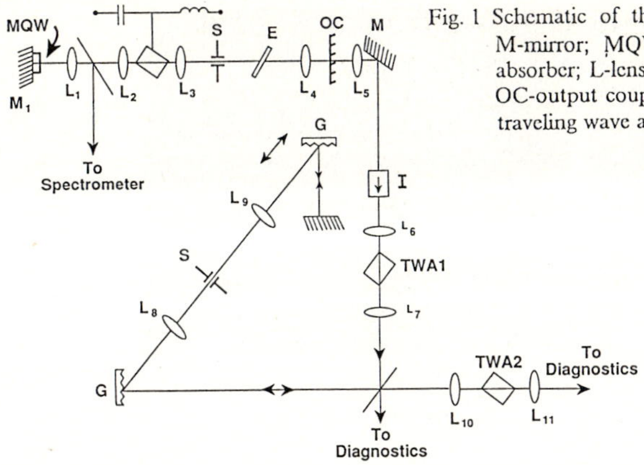

Fig. 1 Schematic of the experimental setup. Legend: M-mirror; MQW-Multiple quantum well saturable absorber; L-lens; S-adjustable slit; E-intracavity etalon; OC-output coupler; I-isolator; TWA-semiconductor traveling wave amplifier; G-grating.

of the gratings, the output of the compressor was subsequently reamplified by TWA2 before directing the pulses into the autocorrelator.

3) Results

Passive mode locking was obtained by tuning the laser to operate in the excitonic absorption region of the MWQ with an intracavity etalon. The pulses which were observed on the autocorrelator were typically 5 psec (FWHM) in duration after deconvolution, with noticeable satellite pulses [2]. Applying 0.5W of RF power with a frequency that matches the longitudinal mode spacing of the external cavity (~ 302 MHz), the satellite pulses were suppressed, emitting single pulses of 5 psec in duration, with a spectral width of 3.5 nm (FWHM). This gives a time bandwidth product of 8.7. The output power from TWA1 contained 32 mW of amplified signal power, which translates into a pulse energy of over 100 pJ and peak powers over 20 watts.

The large time-bandwidth product implies that a frequency sweep exists on the mode locked pulses. In order to take advantage of this frequency sweep, pulse compression techniques were employed to reduce the pulse duration. The compressor utilizes two 1200 l/mm gratings with 2 singlet lenses with 15 cm focal lengths. An adjustable slit was inserted in the Fourier plane of the 1:1 telescope to allow spectral windowing techniques to be used. The position of the second grating was located 15 cm away from the zero dispersion position towards the negative dispersion regime. By adjusting the grating position, the R.F. modulation frequency, and the amount of spectral windowing, optical pulses with a measured intensity FWHM of 0.56 psec were observed on the autocorrelator. In Fig. 2(a,b) an autocorrelation trace along with a low resolution spectrum of the optical pulse is shown. The spectral windowing reduced the laser linewidth to 3.5 nm., which gives a time-bandwidth product of 0.85, which is 2.7 times the transform limit for hyperbolic secant squared pulses. The amplified signal output power from TWA2 was 6.5 mW, which translates to 38 watts of

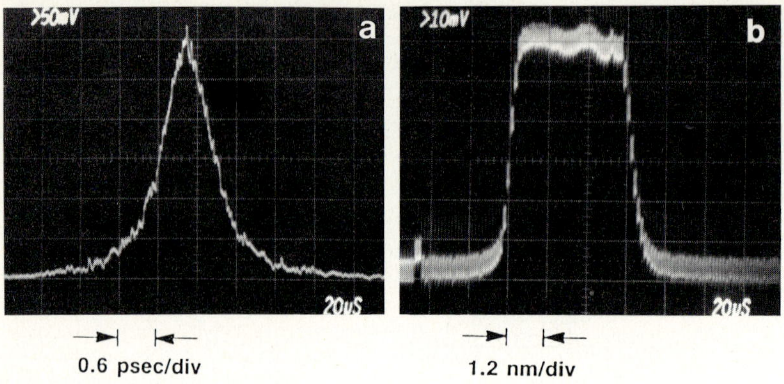

Fig. 2(a) :Real time autocorrelation trace of the compressed and re-amplified pulse. The deconvolved pulse width is 0.56 picoseconds.

Fig. 2(b) :Low resolution spectraum of the pulse after spectral windowing and compression. The spectral bandwidth is ~ 3.5 nm., which gives a time-bandwidth product of 0.85. The sharp edges of the spectrum are due to the adjustable slit located in the Fourier plane of the pulse compressor.

peak power. Higher peak powers were not obtainable due to the limited amount of injected signal power into TWA2, typically 100 μW, as compared to the optimum value of 1 mW. With an optimum value of injected optical signal power, the pulses should be able to be amplified up to the saturation energy level of the amplifiers, which is higher than 100 pJ. This implies that optical pulses with peak powers approaching 200 watts should be obtainable from this type of system.

4) Summary

In conclusion, we have demonstrated high peak power optical pulse generation from an all semiconductor laser diode system. By using spectral windowing and pulse compression techniques, optical pulses have been produced with a pulse duration of 0.56 psec, which are the shortest single optical pulses generated from a semiconductor diode laser system. In addition, the peak powers generated from this system represent the highest ever obtained from an all semiconductor laser system. Utilizing an optimized pulse compressor in order to obtain an optimum value of optical injected signal power into TWA2, optical pulses with peak powers approaching 200 watts should be obtainable. The peak powers obtained from this type of system would then make photonic switching applications possible with a compact laser source.

5) References

[1] P.J. Delfyett, C.-H. Lee, G.A. Alphonse, and J.C. Connolly, submitted to Appl. Phys. Lett.

[2] P.W. Smith, Y. Silberberg, and D.A.B. Miller, J. Opt. Soc. Am. B, 2, 1228, (1985)

Femtosecond Generation of Mid Infrared Pulses at 8.5 kHz via Parametric Mixing in $AgGaS_2$

P.C. Becker, D. Gershoni, and A. Prosser

AT&T Bell Laboratories, 600 Mountain Ave., Murray Hill, NJ07974, USA

There is currently a need for a stable and widely wavelength tunable femtosecond laser source for time resolved spectroscopic investigations. One approach is to amplify CPM pulses and generate a broad spectral continuum. Due to the high intensities available with the amplified CPM and continuum, there is potential for parametric mixing in a variety of nonlinear crystals so as to obtain femtosecond pulses in the 1-20 μm wavelength range. We report here preliminary results of the successful mixing of a femtosecond amplified CPM pulse and a femtosecond continuum so as to generate pulses in the mid infrared, at a repetition rate of 8.5 kHz. "Seeded" parametric amplification has been used previously by Jedju and Rothberg to generate pulses in the 2-5 μm wavelength range using 350 fs pulses at a 10 Hz repetition rate, and similarly by Moore and Schmidt at 100 Hz.[1,2] A recent advance has used a traveling wave cell.[3]

A schematic of the experiment is shown in Figure 1. The output of a CPM laser producing pulses of roughly 80 fs duration at 620 nm is amplified in a copper vapor laser pumped dye amplifier to energies of a few μjoules per pulse, at an 8.5 kHz repetition rate. These pulses are then divided in two by a 50/50 beamsplitter. One beam is focused with a 2.5 cm focal length lens into a flowing water cell of 3 mm path length, generating a broad spectral continuum. This continuum, filtered by a 590 nm long pass filter, is recombined with the remaining portion of the amplified CPM beam and another 50/50 beamsplitter directs the two beams in a collinear geometry towards the $AgGaS_2$ crystal. The beams are loosely focused onto the crystal with a 13 cm focal length lens. The detection system consists of a cooled HgCdTe detector followed by a lockin amplifier. A germanium flat was placed in front of the detector to filter out the visible and near infrared light.

$AgGaS_2$ is well known for its nonlinear properties.[4] It has been used for the generation of pulses covering the 1-10 μm wavelength range from the mixing of picosecond pulses.[5] Two crystals were used in the present experiment. One is 0.5 mm long and cut at 60° from the optic axis, which is the proper angle for the phase

Springer Series in Chemical Physics, Vol. 53 **Ultrafast Phenomena VII**
Editors: C.B. Harris · E.P. Ippen · G.A. Mourou · A.H. Zewail

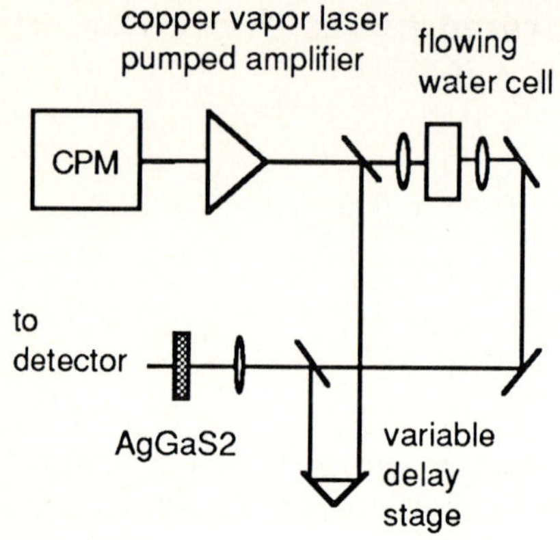

Figure 1: Experimental setup.

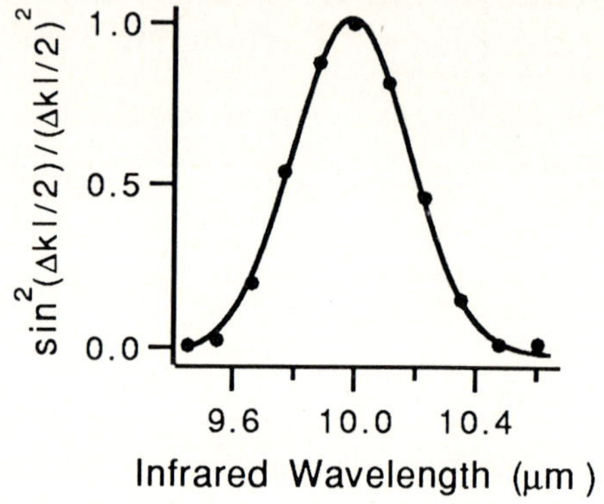

Figure 2: Calculated infrared intensity phase matching factor.

matched generation of 10 µm radiation from the difference frequency mixing of the fundamental light at 620 nm and 660 nm light from the continuum. The other is 3 mm long and cut at 70°. The phase matching is Type I with the fundamental at 620 nm polarized along the extraordinary axis and the continuum and mid infrared beams propagating as ordinary rays.

The bandwidth of the phase matchable infrared light, in a crystal of length l=0.5 mm, was calculated from the half-width of the quantity $\sin^2(\Delta kl/2)/(\Delta kl/2)^2$, plotted in Figure 2, where $\Delta k = n_1\omega_1/c - n_2\omega_2/c - n_3\omega_3/c$. ω_1 is the frequency corresponding to 620 nm, ω_2 that of the continuum, and ω_3 that of the mid infrared. The n_i are the corresponding indices of refraction, calculated from the Sellmeier equation given by Fan et al.[6] From Figure 2, the infrared bandwidth available in a 0.5 mm crystal length is seen to be about 50 cm^{-1}, which can support a pulsewidth of approximately 300 fs at 10 µm.

An infrared signal, very sensitive to both crystal orientation and temporal overlap between the pulses, was detected with both crystals. The amount of light collected by the detection system is estimated to correspond to an energy, per infrared pulse, on the order of a picojoule. The intensity of the infrared signal is plotted in Figure 3a as a function of the time delay between the fundamental 620 nm pulse and the continuum pulse, for the 0.5 mm crystal. The cross correlation has a FWHM of 135 fs. In a 0.5 mm crystal, it can be calculated that the group velocity dispersion causes the 620 and 660 nm pulses to acquire a relative temporal separation of 20 fs. Thus Figure 1 is essentially the cross correlation of the two visible pulses, measured by the generated infrared light. In a 3 mm crystal, the

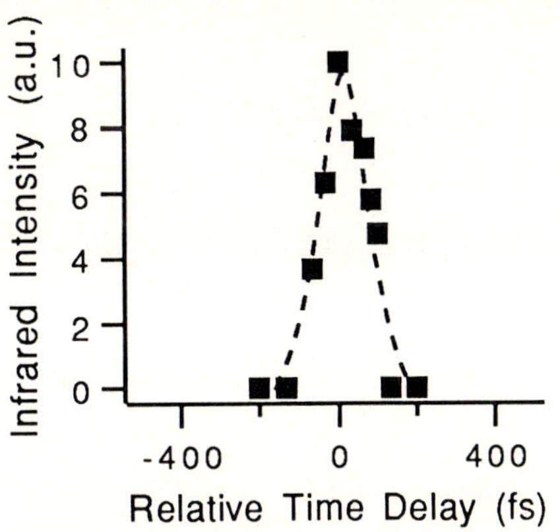

Figure 3a: Intensity of the infrared pulse, as a function of time delay between the 620 nm and continuum pulses, for a 0.5 mm crystal.

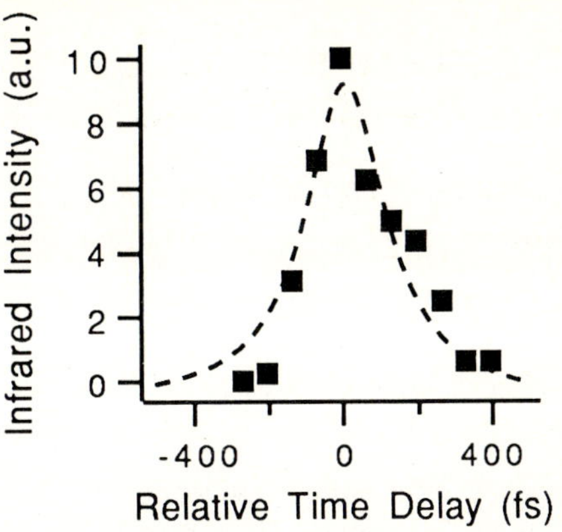

Figure 3b: Intensity of the infrared pulse, as a function of time delay between the 620 nm and continuum pulses, for a 3 mm crystal.

relative delay is calculated to increase by 120 fs. Thus we expect the cross correlation of the two pulses, as measured by the infrared generation, to lengthen by this amount. This is confirmed in Figure 3b, where the cross correlation half width has increased to 225 fs. A similar calculation of the walkoff between the visible and infrared pulses leads to an estimate of 400 fs for the infrared pulse duration.

In conclusion, we have shown that an 8.5 kHz repetition rate amplified CPM system operating at 620 nm, followed by a continuum generation cell, can be used to generate pulses in the mid infrared via parametric mixing in $AgGaS_2$. We expect that it will be possible to increase the energy per pulse in the infrared by several orders of magnitude by reamplifying the necessary portion of the continuum. These pulses should have applications in investigating the dynamics of many interesting physical systems such as intersubband transitions in quantum wells and molecular vibrations.

1. T.M. Jedju and L. Rothberg, *Appl. Opt.*, **27**, 615 (1988).
2. D.S. Moore and S.C. Schmidt, *Opt. Lett.*, **12**, 480 (1987).
3. T. Elsaesser and M. Nuss, Ultrafast Phenomena VII, Postdeadline paper PD14.
4. D.S. Chemla, P.J. Kupecek, D.S. Robertson, and R.C. Smith, *Opt. Commun.*, **3**, 29 (1971).
5. T. Elsaesser, A. Seilmeier, W. Kaiser, P. Koidl, and G. Brandt, *Appl. Phys. Lett.*, 44, 383 (1984).
6. Y.X. Fan, R.C. Eckartdt, R.L. Byer, R.K. Route, and R.S. Feigelson, *Appl. Phys. Lett.*, **45**, 313 (1984).

Femtosecond Pulses in the Medium Infrared by Down-Conversion of a Traveling Wave Dye Laser

T. Elsaesser and M.C. Nuss*

AT&T Bell Laboratories, Crawfords Corner Road, Holmdel, NJ 07733, USA
*Permanent address: Physik-Department E 11, Technische Universität München, D-8000 München 2, Fed. Rep. of Germany

Parametric frequency conversion in nonlinear crystals represents a widely applied technique to generate ultrashort pulses in the wavelength range between 1 and 10 μm [1]. Subpicosecond infrared pulses have been obtained by difference frequency mixing of amplified pulses from femtosecond dye lasers with part of a white light continuum [2]. In this paper, we present a novel parametric device to produce femtosecond pulses around 5 μm with a repetition rate of 8 kHz and a higher efficiency and stability than previously reported.

The experimental setup is depicted schematically in Fig. 1. The output of a colliding-pulse-modelocked dye laser (λ=620 nm) is amplified in a 6 pass dye amplifier pumped by a copper vapor laser. Part of the amplified pulses pump a traveling wave arrangement (TWDL) consisting of a cylindrical lens, a prism and a dye cell (length 0.4 cm). The optical path length in the prism varies for different parts of the pumping beam introducing a continuous temporal delay across the pumping wavefront. This delay is adjusted to the traveling time of stimulated emission along the pumped volume [3]. The femtosecond amplified spontaneous emission of the dye solution (oxazine 725 in methanol/water, concentration 1g/l) and a second part of the pulse at 620 nm are focused to a 3mm $LiIO_3$ crystal to generate the difference frequency.

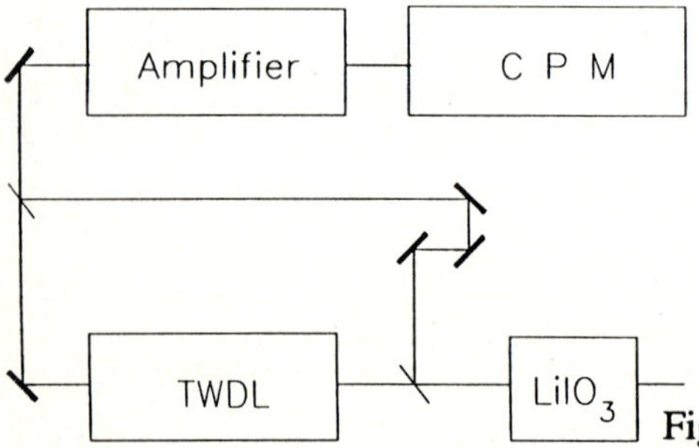

Fig. 1 Schematic of the experimental setup

In the traveling wave laser, approximately 15 percent of the pumping energy of 3 μJ are converted to a pulse at 690 nm with a spectral bandwidth of 30 nm. The fluctuations of the output power are less than ±5 percent. Autocorrelation and cross-correlation measurements indicate a duration of 250 fs for the pulses at 690

Springer Series in Chemical Physics, Vol. 53 **Ultrafast Phenomena VII**
Editors: C.B. Harris · E.P. Ippen · G.A. Mourou · A.H. Zewail

nm and at 620 nm with a negligible temporal jitter between them. Presently, the pulses at 620 nm are not recompressed after amplification.

Difference frequency mixing in the $LiIO_3$ crystal gives intense pulses in the wavelength range around 5 μm. For an energy of 3 μJ at 620 nm, approximately 8 percent of the photons around 690 nm are down-converted corresponding to an energy of 5 nJ per infrared pulse. This conversion efficiency is substantially higher than in other frequency mixing schemes for femtosecond pulses [2]. The infrared intensity fluctuates by less than ±10 percent and depends linearly on the intensity of the pulses at 620 and 690 nm.

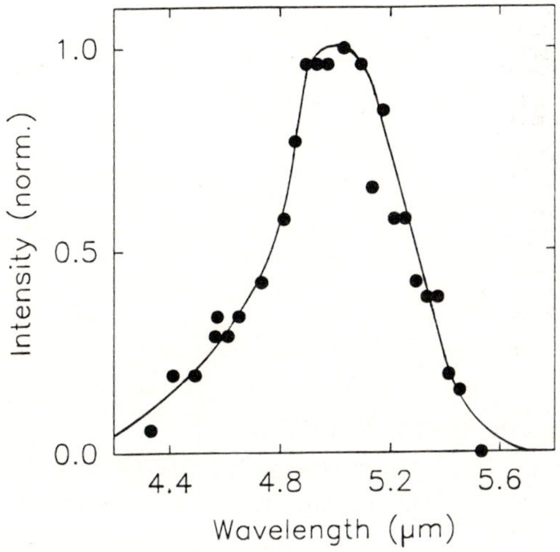

Fig. 2 Spectrum of the infrared pulses at 5 μm

The infrared energy distribution plotted in Fig. 2 has a spectral width (FWHM) of approximately 170 cm^{-1}. The long wavelength tail of the spectrum is determined by the absorption of the $LiIO_3$ crystal in this region. In the range from 4.3 to 4.8 μm, the decrease of the traveling wave output between 700 and 720 nm leads to a decrease of infrared energy.

In Fig. 3, we present the cross correlation trace of the infrared and the pulse at 620 nm used in the down-conversion process. The latter is broadened to approximately 430 fs by dispersion and self-phase modulation in the nonlinear crystal. From the data of Fig. 3, a duration of 350 ±50 fs is estimated for the infrared pulses.

We expect that with recompression of the amplified CPM pulses even shorter pulses at 5 μm can be generated. By using other dyes in the traveling wave cell [3], our technique should also be useful to generate pulses at mid-infrared wavelengths other than 5 μm.

We would like to thank Fred Beisser for the expert preparation of the traveling wave dye cell.

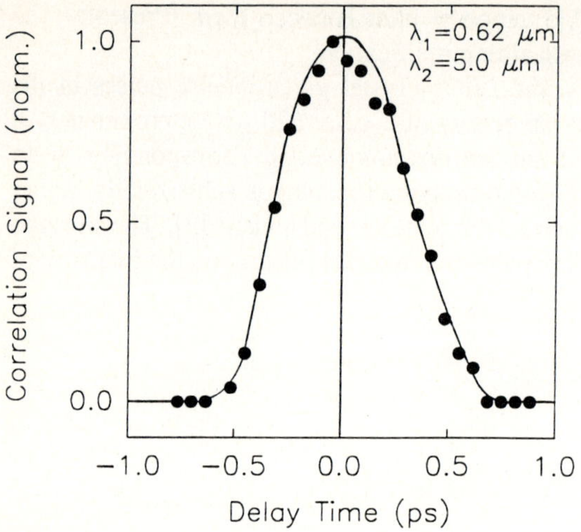

Fig. 3 Cross-correlation trace of pulses at 5 μm and at 0.62 μm.

[1] T. Elsaesser, H. Lobentanzer, and A. Seilmeier, Opt. Commun. 53, 355 (1985)

[2] D.S. Moore, and S.C. Schmidt, Opt. Lett. 12, 480 (1987);
T.M. Jedju, L. Rothberg, and A. Labrie, Opt. Lett. 13, 961 (1988)
P.C. Becker, D. Gershoni, and A. Prosser, Ultrafast Phenomena VII, Postdeadline paper PD 13

[3] J. Hebling, and J. Kuhl, Opt. Lett. 14, 278 (1989)

Pulse Evolution in a Coupled-Cavity Mode-Locked KCl:Tl Color-Center Laser

X. Zhu, W. Sleat, D. Walker, and W. Sibbett

J.F. Allen Physics Research Laboratories, Department of Physics and Astronomy, University of St. Andrews, North Haugh, St. Andrews, Fife KY169SS, UK

In recent years when considerable research interest has been concentrated upon the generation of femtosecond laser pulses the exploitability of coupled nonlinear cavities has been especially highlighted [1-7]. Although some numerical simulations on the pulse evolution in this type of mode-locked laser have been reported [2] relevant experimental information regarding the actual pulse build-up time has not been available to date. In this paper we present the principal operational data relating to pulse evolution in our coupled-cavity mode-locked (CCM) KCl:Tl color center laser.

The measurements were made initially by switching "off" and "on" a quarter-wave voltage applied to a Pockels cell inserted into the control cavity and simultaneously monitoring the change of the second harmonic (SH) intensity produced by the laser pulses. It is known that when the control cavity is at a state of "off" the ouput pulses from the main laser have durations of several tens of picoseconds and the corresponding SH signal is very low [3,4,7]. However, when the control cavity is switched to the "on" state the coupled-cavity mode-locking process starts and leads to a dramatic decrease in the output pulse duration and a ~100 times increase of the SH signal level is observable [3-5, 7]. One example which shows the rapid increase in SH signal after the control cavity is turned from "off" to "on" (at a switching speed of less than 10 ns) is reproduced in Fig. 1. It can be seen that the risetime is ~4.5 μs, which means that it takes about 360 cavity roundtrips for the steady state CCM laser operation to be established.

During the experiments it was generally observed that after the switching action of the Pockels cell there was a variation in the time delay before the SH signal started to increase and also the risetimes were seen to vary substantially during the measurements. Such an observation implies that the degree of phase match at the initial stage of pulse evolution plays a crucial role in determining the actual pulse build-up time. If the phases are matched when the control cavity is switched "on" the CCM process will start immediately and the SH intensity will have a very rapid rise. In contrast, if the initial phases are not properly matched when the control cavity is switched to "on" the CCM process will not start until some adjustment of the

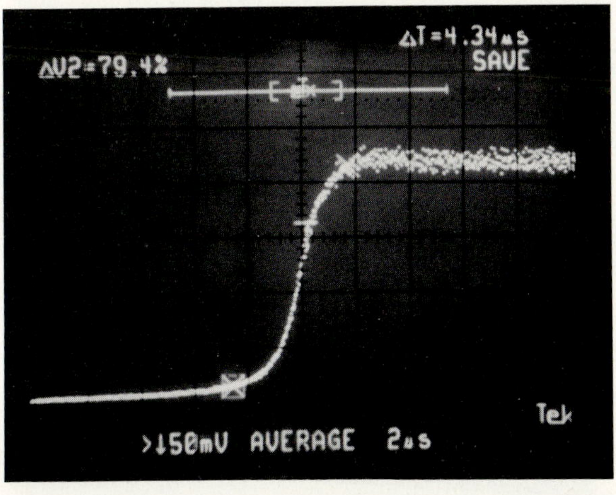

Fig. 1 Build up of the SHG intensity as the CCM laser pulses evolve.

Springer Series in Chemical Physics, Vol. 53 **Ultrafast Phenomena VII**
Editors: C.B. Harris · E.P. Ippen · G.A. Mourou · A.H. Zewail

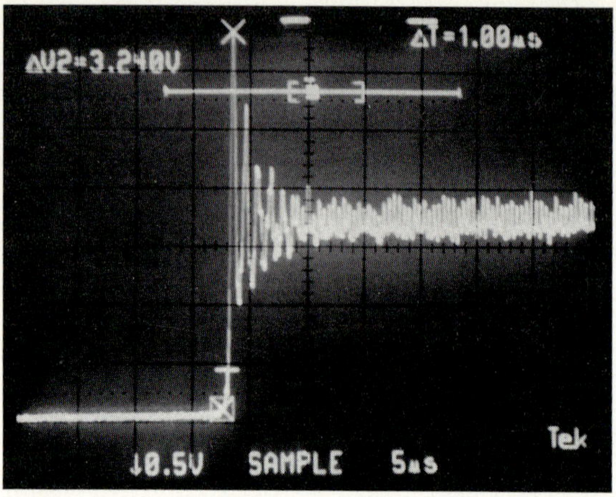

Fig. 2 SHG signal characteristic associated with the rapid evolution of the CCM laser pulses.

cavity lengths occurs such that the phases become properly matched. Such a phenomenon (the dependence on the initial "phase jitter" of the pulse build-up process) represents a very important feature of the pulse evolution kinetics involved in the coupled-cavity mode-locked laser. This can be regarded as a particular characteristic of the passive nature of this type of mode locking.

Because of the random time delay which was often observed to vary from 10 μs to 100 μs it was concluded that it was not essential to incorporate a fast switch to observe the pulse build-up process in the laser system. Therefore, instead of using the fast electro-optical switch to turn "off" and "on" the optical intensity of the feedback pulses we have also employed the scanning PZT-driven retroreflector of the control cavity to achieve an effective "phase-switching" operation. In this case the phase of the returned pulse was switched "off" and "on" at a lower speed with respect to the main cavity pulse, and similar intensity risetimes have been recorded to those for the Pockels-cell approach. Furthermore, the effect of the initial pulse characteristics on the pulse build-up time has also been examined. With different bandwidth control the initial pulse could be either a relatively clean pulse or just a noise burst. We found that a discrete initial pulse did not lead to a faster evolution to the CCM steady state. In contrast, because of the larger bandwidth tolerance associated with the noise-burst pulses the evolution of femtosecond CCM laser pulses was much more rapid.

Complementary measurements were taken by removing the Pockels cell from the control cavity and placing it in the pump beam. A representative single-shot record is shown in Fig. 2 where it can be seen that the pulse build-up time is 1 μs, which is equivalent to approximately 80 cavity periods. However, compared with the trace given in Fig.1 this reduction of the pulse build-up time was not attributed to the method of switching but to the increase of the optical power level fed back to the main cavity. Since the Pockels-cell switch is quite lossy the maximum intensity ratio of the return pulse to the main cavity pulse in the previous circumstance is only 1:2000. Without the Pockels cell in the control cavity this ratio can be increased by several times. The value of intensity ratio relating to the data in Fig. 2 is 3:2000. There is evidence that as the coupling power into the fiber increases the CCM laser pulses build up more rapidly. The shortest risetime observed at the highest available intrafiber power of 26 mW was only 200 ns.

A primary reason for applying the method of pump-beam switching was to investigate whether there are any significant differences in the pulse evolution process of the CCM laser in two particular instances. One is where the main cavity laser starts operation first and then the control cavity is added to it and the other is where the two cavities initiate together from the outset. Interestingly, our experimental results (e.g. see Fig. 3) indicate that for the latter situation the laser system seems to be initially oblivious to the existence of the control cavity and the CCM process does not start until a main cavity laser pulse has been established. Therefore there are no observable distinctions for the pulse evolution kinetics between pump-beam switching and the control-cavity switching.

In conclusion, the timescale of the pulse evolution process in a coupled-cavity mode-locked laser has been measured for the first time. Two different switching methods: control-cavity

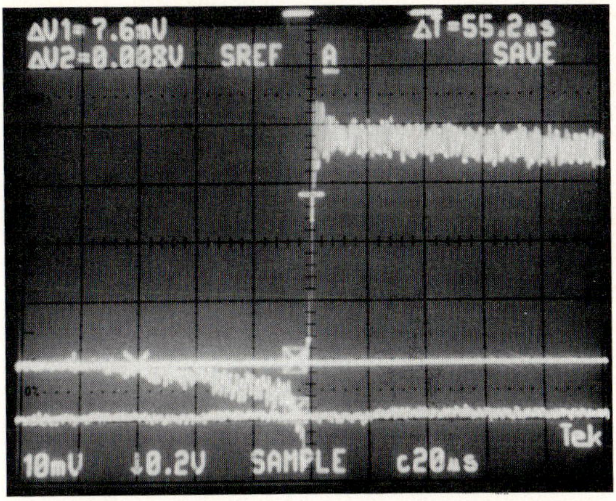

Fig. 3 SHG signal risetime characteristics obtained during pump-beam switching. The bottom line, which represents the signal level obtained when the control cavity was blocked, and the trace immediately above it are inverted and at a magnified scale (10 mV/div.).

switching and pump-beam switching have been employed. In both cases a random time delay between the time when the shorter pulses start to build up and that of each switching action is observed. The recorded pulse build-up times, we believe, are much faster than those predicted in current theory. Thus these results will provide further insight into the mechanisms involved in the coupled-cavity mode-locking process.

References

[1] L. F. Mollenauer and R. H. Stolen, Opt. Lett. **9**, 13, (1984).
[2] K. J. Blow and D. Wood, J. Opt. Soc. Am. B **5**, 629, (1988).
[3] P. N. Kean, X. Zhu, D. W. Crust, R. S. Grant, N. Langford and W. Sibbett, Opt. Lett. **14**, 39, (1989).
[4] W. Sibbett, Paper FQ-1, Tech. Digest of CLEO'89 (April 24-28, 1989, Baltimore USA).
[5] J. Mark, L. Y. Liu, K. L. Hall, H. A. Haus and E. P. Ippen, Opt. Lett. **14**, 48, (1989).
[6] M. Morin and M. Piché, Opt. Lett. **14**, 1119, (1989).
[7] X. Zhu, P. N. Kean and W. Sibbett, IEEE J. Quantum Electron. **25**, 2445, (1989).

Part II

High Intensity Sources and Interactions

Ultrahigh Peak Power Generation: Present and Future

P. Bado, J.S. Coe, J. Squier, F. Salin, C.-Y. Chien, J.-P. Tapié, and G.A. Mourou

University of Michigan, EECS, Ultrafast Science Laboratory,
2200 Bonisteel, Ann Arbor, MI48109, USA

Abstract. Compact solid-state laser sources based on "Chirped Pulse Amplification" are capable of generating intensities in excess of 10^{18} W/cm^2 and make possible the investigation of a new class of relativistic experiments in a physical regime previously inaccessible.

1. Introduction

New short-pulse technology based on Chirped Pulse Amplification (CPA) [1] has made possible the production of extremely bright laser sources. Such laser sources are capable of producing field strengths in excess of an atomic unit (e/a_0^2) and intensities superior to 10^{18} W/cm^2. These sources make possible the investigation of a new class of relativistic experiments in a physical regime previously inaccessible.

First demonstrated in the optical regime in 1985, the CPA approach is conceptually simple (see Fig. 1) and is now the standard technique to amplify ultrashort pulses in solid-state media.

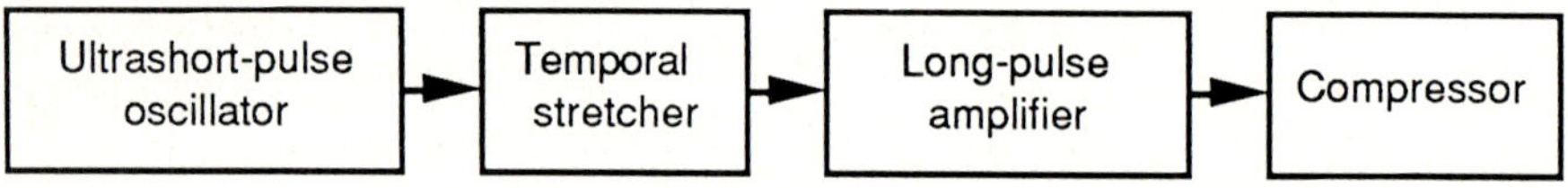

Fig.1: Standard Chirped Pulse Amplification Architecture

In order to avoid undesirable non-linear effects (such as self-focusing) in the amplifying medium, a short optical pulse is initially frequency-chirped and temporally stretched, allowing it to be amplified to saturation while maintaining relatively low peak power. After amplification, an optical compressor is used to restore the original pulsewidth, producing a pulse with short duration and large energy.

The CPA concept is very flexible and has been installed on numerous high-peak power solid-state sources using Nd:glass [1], alexandrite [2], and/or Ti:sapphire as gain media. Several Nd:glass CPA sources are already operating in the 1-10 Terawatt range (1-10 Joule, 0.7-1 ps) [3],[4], and present research is directed towards the

Springer Series in Chemical Physics, Vol. 53 **Ultrafast Phenomena VII**
Editors: C.B. Harris · E.P. Ippen · G.A. Mourou · A.H. Zewail

development of systems capable of generating petawatt pulses (1 kJ, 1 ps) [5] with intensities approaching 10^{21} W/cm^2. In parallel with the development of high-energy/high peak power sources, a large effort is direct towards the amplification of even shorter pulses. Sources based on Ti:sapphire and/or alexandrite are presently capable of amplifying 100-fs pulses to the 10-mJ level [2], and should reach the 100 mJ to Joule level in the near-future [6].

In order to take full advantage of these CPA sources, two key design issues must still be successfully addressed: (i) One must increase the peak-to-background intensity contrast ratio, and (ii) one must reduce the gain narrowing in the amplifiers.

2. Peak-to-Background Intensity Contrast Ratio

Pulse stretching and pulse compression are the two pulse manipulation operations that have made the CPA concept so successful. To first order, the increase in extractable peak power from a given source is equal to the ratio of the duration of the stretched pulse over the duration of the original unstretched pulse. Presently this ratio is of the order of 10^3 and work is in progress to push it to 10^5. The pulse manipulations that are at the heart of CPA deserve special attention since they are generally associated with limitations in the peak-to-background intensity contrast ratio.

The fundamental importance of this parameter is easily understood. First generation CPA sources operate at up to 10^{16}-10^{18} W/cm^2 [7]. The second generation sources are designed to operate at up to 10^{21} W/cm^2. For many experiments, the background intensity must be kept below 10^{11} W/cm^2. Thus, to be fully effective, an intensity contrast ratio as high as 10^{10} is required from these ultrahigh peak power sources.

Existing Nd:glass CPA sources fall short of this requirement. The (sub)picosecond seed pulse is not generated directly by an oscillator, but through a combination involving a modelocked oscillator, followed by a self-phase modulation stage (i.e. a relatively long single-mode fiber). The use of a fiber for spectral broadening and (partial) chirping, and of a grating pair for compression imposes certain constraints and limitations on the achievable peak-to-background intensity contrast ratio [1]. The phase law arising from self-phase modulation and group velocity dispersion in fiber can be compensated by a grating pair to first-order only. Residual higher-order terms reduce the intensity contrast ratio. This phase law can be modified by generating the spectrum in a short fiber, and by temporally stretching the pulse with a positive group velocity grating stage [8]. However, even after optimization of the fiber length, the intensity contrast ratio is still limited to 10^4:1.

To improve the intensity contrast ratio of our Nd:glass Terawatt source, we are redesigning its front-end. First, we have eliminated the need for a fiber-stage by generating directly (through active mode-locking) sub-picosecond pulses of the proper wavelength. Second, we are testing a non-linear device that improves the intensity contrast ratio of any (ultrashort) pulse.

In order to generate directly a sub-picosecond seed pulse for our CPA Nd:glass chain, a large bandwidth medium lasing in the 1.05-1.06 micron region is required.

We have selected Ti:sapphire since it satisfies these requirements. The sub-picosecond capability of lasers built with this material had been established earlier in the 800 nm region using the coupled-cavity technique [9]. However, in order to avoid the stability concerns associated with the coupled-cavity approach, we have opted for a more traditional actively mode-locked oscillator design. Our Ti:sapphire oscillator (derived from a Spectra-Physics unit) is pumped by a cw-Argon laser. Pulses shorter than 2 ps are routinely generated (centered at 1054 nm). These pulses carry a residual frequency-chirp, that can be partially eliminated with an extra-cavity grating pair to yield stable sub-picosecond pulses.

Our second approach is based on devices (birefringent fibers [10] and non-linear interferometers [11]) offering a transmission that is a non-linear function of the instantaneous intensity. These devices can be used to improve the intensity contrast ratio by 2 to 3 orders of magnitude. Our two approaches are compatible and will be combined to generate sub-picosecond pulses, with an intensity contrast ratio superior to 10^7:1.

3. Gain Narrowing

The second problem associated with our high-peak power sources is not specific to the CPA architecture, but is of concern with all laser sources. The final pulse duration is limited by the gain narrowing associated with the amplification process. The regenerative amplifier provides a net energy gain of 10^6 to 10^7. The total gain, taking into account the cavity loss, can be as high as 10^{17}. This very large gain is accompanied by a substantial gain narrowing. In order to avoid this effect, one can: (1) increase the gain per pass (or reduce the loss), (2) increase the gain medium bandwidth.

We have built a pulse-pumped Ti:sapphire regenerative amplifier, operating in the 1.0 -1.1 micron range. Ti:sapphire offers a very broadband gain centered around 800 nm. As we are operating far from the central wavelength, the gain profile is (almost) flat and should be able to support very short pulses. Additionally, one notes that the thermal birefringence that is the dominating loss factor in our Nd:glass amplifiers is of no-concern in a Ti:sapphire source, as this material is naturally birefringent. Thus, our new Ti:sapphire regenerative amplifier has a slightly higher net gain, and a much flatter gain profile than our Nd:glass regenerative amplifier. We have demonstrated the regenerative amplification of 7.2 nm pulses without noticing any gain narrowing. In fact, the gain bandwidth limit is no longer imposed by the gain medium, but is dominated by the cavity optical components (Pockels cell, thin-film intracavity polarizer, etc.).

4. Future Development

Several high-energy Nd:glass chains are now being upgraded to operate in the CPA mode. These sources will be able to amplify short pulses to the 100 to 1000 Joule level. The Nd:glass gain medium will limit the pulse duration to ~1 ps if a conventional front-end is used. A lower limit (~500 fs) and a much higher intensity contrast ratio will be achievable with a Ti:sapphire front-end (oscillator and regenerative amplifier).

Beside this high-energy approach, numerous groups are working on smaller, lower energy sources based only on broadband gain media, such as Ti:sapphire and alexandrite. Ti:sapphire offers a high-gain and a very broad bandwidth. However, its energy storage capability is limited. Ti:sapphire based sources should be able to amplify very short pulses (20-30 fs) to the 10-100 mJ level. Alexandrite is a much better storage medium, but suffers from a somewhat lower bandwidth. A system based on alexandrite should be able to amplify 100 fs pulses to the Joule level. A compact source combining these two materials, with Ti:sapphire being used for the oscillator and the pre-amplifier, and alexandrite being used in the final amplifier(s) should be able to amplify 50 fs pulses to the Joule level.

In summary, multi-terawatt Nd:glass CPA sources are already operating in numerous laboratories. In order to move to the petawatt regime, one needs to generate higher contrast ratio pulses and to eliminate the gain narrowing effects. Our new Ti:sapphire based front-end successfully addresses these two issues. In the near-future, extremely-compact CPA sources based on Ti:sapphire and alexandrite should reach the terawatt level (100 mJ @100 fs), and could reach the 20 terawatt mark (1 Joule @ 50 fs) within a few years.

5. References

[1] P. Maine, D. Strickland, P. Bado, M. Pessot, and G. Mourou, IEEE J. Quantum Electron., **QE-24**, 398 (1988).

[2] M. Pessot, J. A. Squier, G. A. Mourou, and D. J. Harter, Optics Letters, **14**, 797 (1989).

[3] M. D. Perry and F. G. Patterson, J. Weston, Optics Lett., **15**, 383 (1990).

[4] M. Ferray et al., Optics Commun., **75**, 278 (1990).

[5] CLEO/IQEC, Anaheim 1990, Paper JMA2.

[6] D. J Harter, M. Pessot, J. A. Squier, J. Nees, P. Bado, and G. A. Mourou, SPIE Los Angeles 1990.

[7] D. Normand, M. Ferray, L. Lompre, O. Gobert, A. L'Huillier, and G. Mainfray, submitted to Optics Letters.

[8] O. E. Martinez, IEEE J. Quantum Electron., **QE-23**, 59 (1987).

[9] J. Goodberlet, J. Wang, J. G. Fujimoto, and P. A. Schulz, Opt. Letters, **14**, 1125 (1989).

[10] N. J. Halas and D. Grischkowsky, Appl. Phys. Lett., **48**, 823 (1986).

[11] F. Ouelette and M. Piche, Optics Commun. , **60**, 99 (1986).

Very High Order Harmonic Generation

L.A. Lompré, A.L'Huillier, M. Ferray, and G. Mainfray

Service de Physique des Atomes et des Surfaces, CEN Saclay,
F-91191 Gif-sur-Yvette Cedex, France

Abstract. The main experimental and theoretical results on VUV harmonic generation in a gaseous medium exposed to a strong 1064 nm laser field will be reviewed. Experimental results show for example up to the 33rd harmonic in Ar at $3x10^{13}$ Wcm^{-2}. The main parameters which can influence the conversion efficiency will be discussed.

1. Introduction

With the exception of two experiments using a 1064 nm [1] and a 1315 nm laser radiation [2], most of the previous experiments on harmonic generation in a gaseous medium use short incident wavelengths in order to generate the shortest wavelengths [3-6]. But the conversion efficiency strongly depends on the medium.

The purpose of the present paper is to discuss experimental measurements of VUV light emission (355 - 30 nm), in particular harmonic generation in a rare gas medium irradiated by an intense 1064 nm laser field in the 10^{13} Wcm^{-2} intensity range. Recent experiments [7,8] have shown the production of high-order harmonics in a rare gas medium, with photon energies much above the ionization energy.

These results give rise to two essential questions: (1) What is the influence of multiphoton ionization on the conversion efficiency ? (2) What is the role of the propagation effects, and all the parameters related to high pressure ? Both questions are of the greatest interest in the perspective of the development of very high intense lasers, which would enable the study of harmonic generation in very intense laser fields.

2. Experimental Arrangement

The experimental set-up has been described in detail elsewhere and only a brief description will be given here [7,8]. The laser used is a 40 ps mode-locked Nd-YAG laser operated at 10 Hz. It is focused into the vacuum chamber by a lens with a focal length which can be varied from 75 to 300 mm. The gaseous target is provided by a pulsed gas jet. At 0.5 mm from the 1 mm diameter nozzle, the gas density in the jet has been measured to be 15 Torr, for a backing pressure of 150 Torr. The low ratio between the backing pressure and the pressure measured at the output of the gas jet is very important in order to minimize the contribution of dimers. The complete description of the gas jet and the method to measure the pressure in the gas jet and the jet profile has been described elsewhere [9]. The knowledge of the jet parameters is essential in order to take into account these characteristics in the calculations. Furthermore, the use of a gas jet limits the reabsorption of the VUV light and enables the detection of very short wavelengths. The VUV radiation (from 350 to 10 nm) emitted along the laser axis is

Springer Series in Chemical Physics, Vol. 53 **Ultrafast Phenomena VII**
Editors: C.B. Harris · E.P. Ippen · G.A. Mourou · A.H. Zewail

analyzed by a monochromator. It is composed of a toroidal grating which forms an aberration free image of the interaction region on the output slit for the spectral range of the grating. The advantage of this device is that the collection solid angle is imposed by the grating size and not by the entrance slit. The VUV radiation is detected by a photomultiplier or a windowless electron multiplier depending on the spectral range investigated.

3. Experimental Results

Figure 1a) shows typical results obtained in Xe and Ar at a pressure of 15 Torr for a laser intensity of $3x10^{13}$ Wcm^{-2}. The most salient features are the following: the harmonic intensity distribution first decreases rather steeply from the third to the fifth or seventh harmonic, remains almost flat from the seventh to a high-order harmonic, which depends on the atom and on the laser intensity, and finally decreases rapidly. Several major points come out: (1) only odd harmonics are observed. That is due to the fact that the harmonics are created in an isotropic gaseous medium with inversion symmetry which excludes even harmonics. (2) the maximum number of harmonics created at the maximum laser intensity is strongly dependent on the rare gas investigated; up to the 21st harmonic in Xe and up to the 33rd harmonic in Ar. (3) Furthermore, over the whole spectrum investigated (10-355 nm), a small number of lines are observed, corresponding to transitions from excited states to the fundamental of the neutral atom and also of the singly charged ion.

Figure 1b) shows the number of photons produced at each harmonic in xenon for several laser intensities from $5x10^{12}$ to $3x10^{13}$ Wcm^{-2}. The vertical scale of Fig.1a) and 1b) has been calibrated in absolute value by estimating the collection efficiency, in

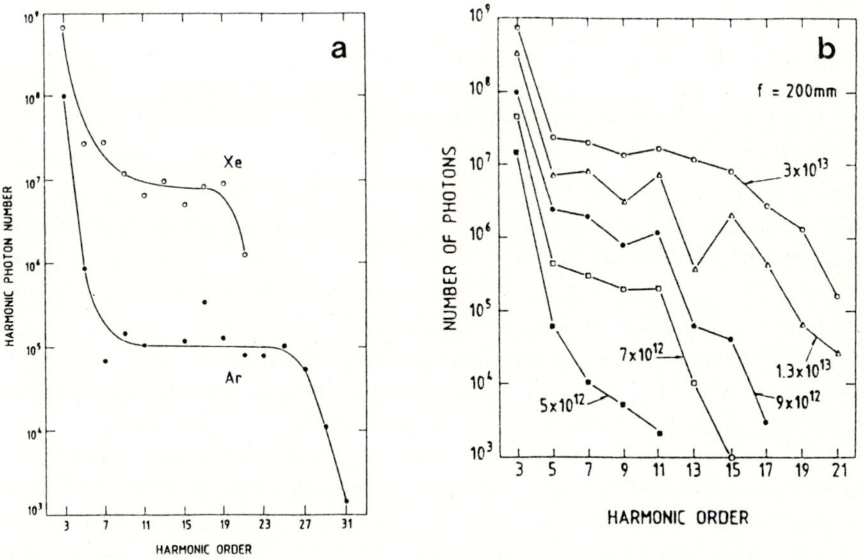

Fig.1 a) Harmonic intensity distributions generated in Xe and Ar at $3{\cdot}10^{13}$ Wcm^{-2}, b) Number of photons in each harmonic at different laser intensities, in Xe.

particular by taking into account the spectral responses of the grating and the photon detectors for the different harmonics. The experimental uncertainties are estimated to a factor of 5. Most of the modulations in Fig.1 are therefore not significant. As the intensity increases, it can be seen than a plateau appears at about 7×10^{12} Wcm^{-2}. Beyond 1.3×10^{13} Wcm^{-2}, the distribution becomes smoother, and the number of harmonics which can be detected remains unchanged, 21st in Xe.

4. Discussion

These experimental results have stimulated a number of theoretical works [10-14] either within the framework of perturbation theory and going beyong it, using Floquet method in H or performing time-dependent calculations [15]. All calculations are based on a single atom response and calculate the square of the Fourier transform of the atomic dipole moment induced by the laser field, $|d(q\omega)|^2$. A very good review of these theoretical works and others can be found in a special issue of J.O.S.A. B 7, April 1990. They lead to the conclusion that the main features are reproduced by time dependent calculations and by non perturbative theories when the laser intensity is high enough. But is it meaningful to compare in this way experimental and theoretical results ? What is the relationship between N_q and $|d(q\omega)|^2$.

It is well known that an electromagnetic field propagating in a dielectric medium characterized by an electronic polarization must satisfy the Maxwell propagation equation. If the incident laser field is assumed to be monochromatic, and considering only the lowest-order in the interaction with the radiation field, then, for a gaussian beam focused into a nonlinear medium of finite length, and separating single-atom and many-atom contributions, the number of photons N_q detected in an experiment may be written as :

$$N_q = K\,(|\chi^{(q)}|^2 I^q) * (b^3 \mathcal{N}^2 |F_q|^2/q^{1/2})$$

where K is a constant. The main parameters which govern harmonic generation are the *microscopic* one, the qth order susceptibilty $\chi^{(q)}$ as well as the *macroscopic* ones, the confocal parameter b, the atomic density $\mathcal{N}$ and the phase matching factor, $|F_q|^2$.

The influence of the different macroscopic parameters on the harmonic production have been studied. It clearly shows that the number of photons produced in each harmonic follows a b^3 and a $\mathcal{N}^2$ power laws. The effect of the phase matching factor $|F_q|^2$ is not clear yet and is still under consideration [16-17].

One important parameter is the laser intensity. Figure 2 shows the intensity dependences of the 9th, 13th and 19th harmonic frequency in Xe in a double logarithmic plot. The behaviour of the harmonic power as a function of the laser intensity is very similar to the ion behaviour [18]. A saturation of the number of ions produced is observed at the same laser intensity (I_S). This observation is very important because it shows that the harmonic production is strongly influenced by the presence of ions in the interaction volume. This observation has two possible explanations, either the harmonic production efficiency is much smaller in an ionic medium than in an atomic vapor, or the presence of free electrons in the interaction volume seriously affects phase

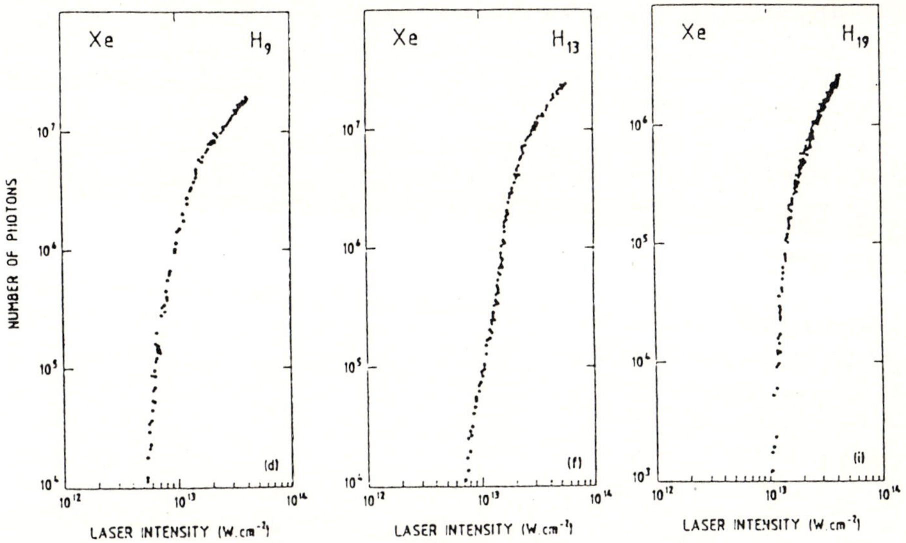

Fig.2 Intensity dependence of the 9th, 13th and 19th harmonics in Xe.

mismatching conditions. But the experiment does not allow to conclude which effect is the major one.

These intensity dependences show another interesting feature. The harmonics produced below the ionization limit (3, 5, 7, 9) closely follow a I^q power law for laser intensity lower than I_S. The 11th and 13th harmonics have a much more complicated behavior. Finally, all the higher order harmonics (15-21) vary approximately as I^{12} for laser intensities below I_S. These results show that data must be taken for laser intensities below or close to the saturation intensity for comparison with theoretical calculations, but in any case not for laser intensities larger than I_S. Most of the high order harmonics are created in an intensity such that the medium is not or weakly ionized. The maximum number of harmonics, 21 in Xe, 29 in Kr, 33 in Ar is directly related to the saturation intensity, more harmonics can be generated when the atom is lighter because the saturation intensity is higher. In the case of Ne and He with saturation intensities of the order of 3 to 5×10^{14} Wcm^{-2} respectively, we may expect the creation of the 59th and the 65th harmonics of the incident laser radiation at 1064 nm. This would correspond to generated radiations at 18.3 and 16.8 nm.

The interpretation of these results are very promising for the near future. With the development of new high power lasers working in the terawatt regime laser intensities of the order of 10^{18} Wcm^{-2} can be reached and a new field of investigation is opened [19]. At such a high laser intensity, a great number of electrons can be removed from the atom, leading to the production of very highly charged ions. We may expect that similar processes may occur in an ionized gas, leading to the production of higher order harmonics, and shorter wavelengths.

The authors would like to thank Dr. C. Manus for many stimulating discussions, X.F. Li and A. Sanchez for their help in the experiments.

5. References

1. M.G. Groseva, D.I. Metchkov, V.M. Mitev, L.I. Pavlov and K.V. Stamenov, Opt. Commun. **23**, 77 (1977).
2. J. Wildenauer J. Appl. Phys. **62**, 41 (1988).
3. J. Bokor, P.H. Bucksbaum and R.R. Freeman, Opt. Lett. **8**, 217 (1983).
4. J. Reintjes, C.Y. She and R. Reckardt, IEEE J. Quatum Electron. QE-**14**, 581 (1978).
5. J. Reintjes, L.L. Tankersley and R. Christensen, Opt. Commun. **39**, 334 (1981).
6. A. McPherson, G. Gibson, H. Jara, U. Johann, T.S. Luk, I. McIntyre, K. Boyer and C.K. Rhodes, J. Opt. Soc. Am. B **4**, 595 (1987).
7. M. Ferray, A. L'Huillier, X.F. Li, L.A. Lompré, G. Mainfray and C. Manus, J. Phys. B **21**, L31 (1988).
8. X.F. Li, A. L'Huillier, M. Ferray, L.A. Lompré and G. Mainfray, Phys. Rev. A **4**, 5751 (1989).
9. L.A. Lompré, M. Ferray, A. L'Huillier, X.F. Li and G. Mainfray, J. Appl. Phys. **63**, 1791 (1988).
10. Y. Gontier and M. Trahin, IEEE, J. Quantum Electron. QE-**18** ,1137 (1982).
11. B. Gao and A.F. Starace, Phys. Rev. A **39**, 4550 (1989).
12. L. Pan, K.T. Taylor and C.W. Clark, Phys. Rev. A **39**,4894 (1989).
13. J.H. Eberly, Q. Su and J. Javanainen, Phys. Rev. Lett **62**, 881 (1989).
14. M. Dörr, R.M. Potvliege and R. Shakeshaft, J.O.S.A. B **7**, 433 (1990).
15. K. Kulander and B.W. Shore, J.O.S.A. B **7**, 502 (1990).
16. A. L'Huillier, X.F. Li and L.A. Lompré, J.O.S.A. B **7**, 527 (1990).
17. L.A. Lompré, A. L'Huillier, M. Ferray, P. Monot and G. Mainfray, J.O.S.A. B, in press (Mai 1990).
18. A. L'Huillier, L.A. Lompé, G. Mainfray and C. Manus, J. Phys. B **16**, 1363 (1983).
19. M. Ferray, L.A. Lompré, O. Gobert, A. L'Huillier, G. Mainfray, C. Manus, A. Sanchez and A. Gomes, Opt. Comm. **75**, 278 (1990).

Time-Resolved Spectral Shifts of Femtosecond Pulses in Atmospheric Density Plasmas

W.M. Wood, W.C. Banyai, J.L. Trisnadi, and M.C. Downer

Physics Department, University of Texas at Austin, Austin, TX78712, USA

Abstract. Femtosecond pump-probe measurements of spectral shifts provide novel diagnostics of collisional dynamics in atmospheric density laser-produced plasmas under intense field conditions.

Atmospheric density plasmas produced by intense femtosecond pulses have recently received a great deal of interest because of their possible role as a high intensity source of soft x-rays[1] and as media for particle acceleration[2]. One method for achieving such an x-ray source involves the creation of an inverted population of ionized core states on an ultrashort time scale. Multiphoton ionization of gases with high intensity ultrafast light pulses can result in a cold plasma of highly stripped ions at atmospheric density. The electrons recombining into the deeper core states emit soft x-rays. With 100 fs, mJ energy, 620 nm pulses, we are able to both create such a plasma on an ultrashort time scale, and study the rapid collisional dynamics of the plasma.

We have shown recently that the ionization of atmosphere pressure gases during the first few tens of cycles of a 100 fs laser pulse results in a large blue shift of the pulse spectrum[3]. The microscopic collisionless ionization of a single atom has now been modelled quantitatively with the Keldysh theory, and incorporated into a macroscopic Drude model to obtain an *ab initio* theory of the blue shift which agrees well with most experimental observations[4]. However, we have also observed two new effects: anomalously large blue shifts in Ar and N_2 at pressures above several atm. and time-resolved red shifts of probe pulses. Neither of these effects is adequately described by our collisionless Keldysh ionization model. We attribute these effects to the collisional dynamics of the plasma.

The basis for the spectral shifts lies in the rapid modulation of the index of refraction of the plasma:

$$n^2 = 1 - \omega_p^2/(\omega_o^2 + \gamma^2) \qquad (1)$$

Here, ω_p is the plasma frequency and γ is the electronic collision frequency. A rapid decrease in n causes a blue shift, while a rapid increase in n causes a red shift:

$$\Delta\omega = -\omega_o\,(z/c)\,(dn/dt) \qquad (2)$$

z represents the confocal beam parameter of our focal geometry. In Fig. 1 we

Springer Series in Chemical Physics, Vol. 53 **Ultrafast Phenomena VII**
Editors: C.B. Harris · E.P. Ippen · G.A. Mourou · A.H. Zewail

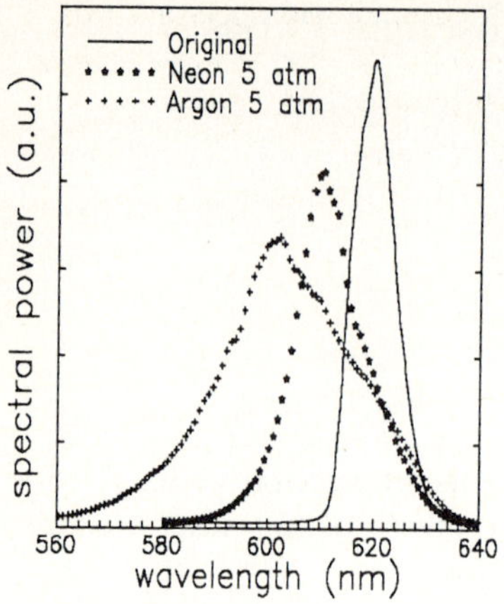

Fig. 1. Blue shifted spectra at 10^{16} W/cm^2.

show the input spectrum and the blue shifted spectra in 1 atm. Ne and 5 atm. Ar for 10^{16} W/cm^2. The former (and numerous similar spectra) are explained quantitatively by our collisionless Keldysh model, which describes a rapid increase in ω_p resulting from the ionization of non-interacting single atoms in a strong light field[4]. Note, however, the anomalously greater shifting and broadening for Ar (also observed for N2). We now introduce the additional mechanism of impact ionization. Under the stated experimental conditions, the quiver amplitude ($r = eE/m\omega^2 \cong 120$ Angstroms) begins to exceed the average interatomic spacing ($l \cong 40$ Angstroms at 5 atm.). Furthermore, the wiggle energy ($E_{quiver} \cong 350$eV) corresponds to nearly the peak electron impact ionization cross-section, σ[5]. Consequently, wiggling electrons efficiently impact ionize neighboring atoms in cases (Ar, N_2) where σ is large enough[5]. The average time between ionizing collisions is:

$$T_i = 1 / (N_i \sigma v) \tag{3}$$

Using v = the electron quiver velocity and σ from Ref. [5], we find $T_i \cong 15$ fs $<< t_{pulse}$ for both Ar and N_2 at 5 atm., whereas $T_i \cong 150$ fs. $> t_{pulse}$ for Ne and He, which do not exhibit the anomalous effect. This additional rate of increase in ω_p from collisional ionization quantitatively explains the extra blue shifting we observe at higher pressures of Ar and N_2 when incorporated into a comprehensive model with Keldysh ionization.

Time-resolved experiments show a red shift of the probe pulse spectrum in the trailing edge of the ionizing pulse, which can be explained as a rapid modulation of the collision frequency γ in eq. (1). For a classical Debye screened plasma, the effective collision frequency scales as the electron density, and inversely as the cube of the mean electron velocity[6]. Therefore, γ is small in the leading edge of the ionizing pulse because N_e is small and remains small because the electrons are ionized with a large quiver velocity. As the ionizing pulse departs, the collision frequency increases rapidly to an equilibrium value determined by the thermal velocity. From eq. (1), a rapid increase in refractive index and red shifted spectrum will occur if γ approaches ω. This places upper and lower limits on the

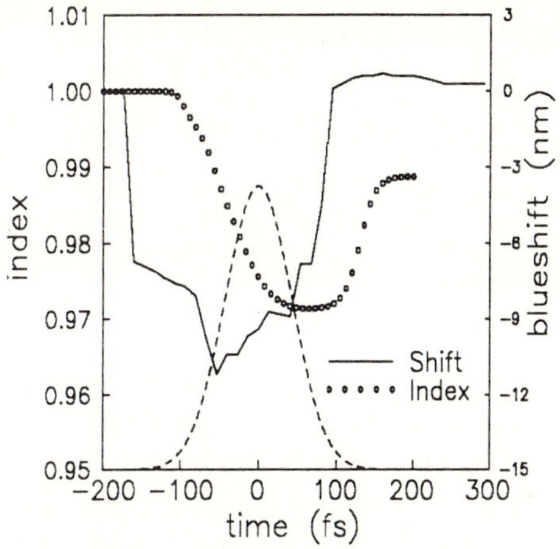

Fig. 2. Measured peak spectral shifts, calculated refractive index and original spectrum.

final thermal velocity of the plasma required to produce a red shift. If the final temperature > 10 eV, then no red shift occurs because γ/ω remains small. On the other hand, if the final temperature is less than 1/2 eV, then a larger red shift than observed would occur because γ would dominate the denominator in eq. (1). The calculated effects of plasma creation and thermalization are shown in Fig. 2. The refractive index change versus time was calculated by combining collisionless Keldysh ionization with collision frequency modulation. The slope of the index change becomes positive at a time delay of 120 fs which is where the observed red shift begins to turn-on, as shown in Fig. 2. The refractive index relaxes to a value less than unity because we have assumed a final plasma temperature of 5 eV, which corresponds to a red shift between 2 and 5 nm.

In summary, in addition to collisionless ionization, ionizing collisions between electrons and ions during the peak part of the pulse and Coulomb collisions during the trailing edge of the pulse contribute to the observed spectral shifts.

Research was supported by NSF Grant DMR-8858388 and Welch Foundation Grant F-1038.

References:

1. N. H. Burnett and P. B. Corkum, J. Opt. Soc. Am. B **6,** 1195. (1989).
2. *Advanced Accelerator Concepts*, ed. C. Joshi, AIP Conference Proceedings No. 193 (American Institute of Physics, New York 1989).
3. W. M. Wood, G. Focht, and M. C. Downer, Opt. Lett. **13**, 984 (1988); M. C. Downer, G. Focht, D. H. Reitze, W. M. Wood, and T. R. Zhang, *Ultrafast Phenomena VI*, eds. T. Yajima, K. Yoshihara, C. B. Harris, and S. Shionoya (Springer-Verlag, 1988).
4. M. C. Downer, W. C. Banyai, W. M. Wood, D. C. Anacker and J. L. Erskine, Proc. SPIE **1209**, 166-173 (1990).
5. H. Tawara and T. Kato, Atomic Data and Nuclear Tables **36**, 167. (1987).
6. D. V. Sivukhin, *Reviews of Plasma Physics vol.4*, ed. M.A. Leontovich, Consultants Bureau, N.Y. 1966.

Characterization of a High-Intensity, Subpicosecond XeCl Laser System

A.J. Taylor, T.R. Gosnell, J.P. Roberts, D.C. MacPherson, and C.R. Tallman

Los Alamos National Laboratory, Los Alamos, NM 87545, USA

Abstract. We describe here a terawatt-class laser system based on the amplification of subpicosecond pulses in XeCl discharge amplifiers. For this system we have made a direct measurement of the output beam quality and carried out extensive measurements of the pulsewidth throughout the entire optical train.

Recent advances in ultrafast lasers and large-aperture optical amplifiers have spurred the development of terawatt-class laser systems capable of delivering focal-spot intensities in excess of 10^{19} W/cm^2. At these extremely high intensities, the optical field strength is more than twenty times larger than the Bohr electric field, e/a_0^2, permitting for the first time investigations of the optical properties of matter in a previously unexplored intensity regime.

Our laser system consists of a front-end which generates 308-nm seed pulses by frequency-doubling the output of a synchronously-pumped subpicosecond dye oscillator-amplifier system[1], followed by an amplifier chain comprised of a small-aperture (1-cm^2) XeCl preamplifier, a beam-expanding vacuum spatial filter, and a large-aperture (100-cm^2) final amplifier[2]. This system generates 250-mJ, 275-fsec pulses at 308 nm and operates at a sustained repetition rate of 1 Hz. The ultrashort output pulse lies on a 20-nsec, 50-mJ ASE pedestal, but only half of the ASE is included in the solid angle subtended by the output beam. Using f/3.7 optics we measure FWHM focal-spot dimensions of 3.4 x 4.1 μm^2 for the full-energy final output beam and therefore demonstrate a true focal-volume intensity of 6.5 x 10^{18} W/cm^2.

Autocorrelations of the pulse and its corresponding spectrum as it traverses the amplifier chain are presented in Figure 1. The autocorrelations, obtained using two-photon ionization of DABCO vapor, all exhibit the 3:1 contrast ratio indicative of well-mode-locked pulses.

The 190-fsec, 308-nm seed pulse with its relatively smooth 0.96-nm wide frequency spectrum (1a) broadens temporally to 270 fsec (1b) after

Springer Series in Chemical Physics, Vol. 53 **Ultrafast Phenomena VII**
Editors: C.B. Harris · E.P. Ippen · G.A. Mourou · A.H. Zewail

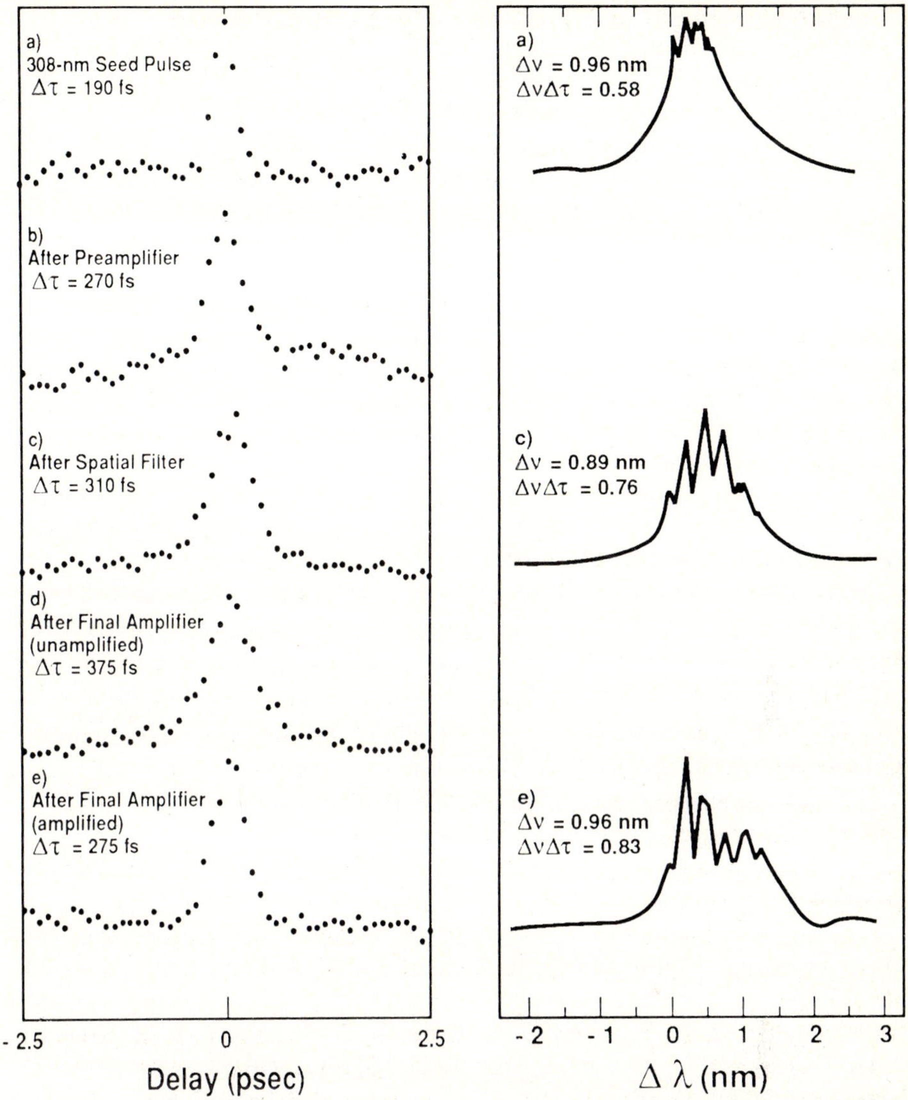

Figure 1. Temporal and spectral evolution of the pulse through the laser.

amplification in the relatively low gain ($g_0l=9$) preamplifier. This type of broadening has been observed previously in XeCl[3] and is caused by modulation of the pulse's spectrum by the XeCl gain profile (1c). After traversing the spatial filter, this pulse is broadened to 310 fsec (1c). An experimental investigation of this broadening reveals that it is caused not by the aperture in the spatial filter but rather dispersion in the spatial filter optics: fused silica lenses (0.5 cm) and windows (1.5 cm). The effects of the final amplifier are twofold: With the amplifier turned off and evacuated, the pulse broadens from 310 to 375 fsec (1d), as a result of group velocity dispersion in the amplifier's four 0.95-cm quartz windows. On the other hand, gain saturation during the amplification process in the final amplifier broadens the pulse spectrum and compresses the pulse to a final width of 275 fs (1e).

The beam quality of the fully amplified output pulse is determined by measuring its transmission through a calibrated pinhole. The focusing optic is a high-quality on-axis parabolic mirror with a 14.8-cm-diameter clear aperture and a focal length of 28.55 cm. At a wavelength of 633 nm, this optic produces a diffraction-limited focal spot. With the output beam dimensions of 6.2 cm by 9 cm, the effective f number for the focusing optic is f/3.7. Using this optic, the measured transmission of the output beam through a circular pinhole 10 μm in diameter and carefully centered in the focal plane is 82%. Since the intensity profile of the unfocused beam is fairly flat-topped, we assume that the intensity profile in the focal plane is given by the Fraunhofer diffraction integral for a uniformly illuminated rectangular aperture with an aspect ratio of 1.23. With this assumption, the measured transmission of the 10-μm pinhole implies FWHM dimensions for the central diffraction spot of $3.4 \times 4.1\ \mu m^2$. These dimensions are 3.1 times larger than would be obtained for a diffraction-limited beam.

With these focal-spot dimensions, our estimate for the mean intensity obtained with a 250-mJ, 275-fsec pulse is $6.5 \pm 1.4 \times 10^{18}$ W/cm^2. With f/1 focusing optics, this system will produce an intensity of 8.9×10^{19} W/cm^2, which corresponds to an electric field in the focal volume 39 times the binding field of the hydrogen atom.

References

[1]A. J. Taylor, J. P. Roberts, T. R. Gosnell and C. S. Lester, Opt. Lett.**14,** 444 (1989).
[2]A. J. Taylor, C. R. Tallman, J. P. Roberts, C. S. Lester, T. R. Gosnell, P. H. Y. Lee, and G. A. Kyrala, Opt. Lett. **15**, 39 (1990).
[3]Bernhard Dick, Sandor Szatmari, Bela Racz and Fritz Schafer, Opt. Commun. **62**, 277 (1987).

Soft-X-Ray Generation from 100 fs Irradiation of Thin Film Targets

H.W.K. Tom[1], *O.R. Wood, II*[1], *G.D. Aumiller*[1], *and M.D. Rosen*[2]

[1]AT&T Bell Laboratories, Room 4B-429 Crawfords Corner Road, Holmdel, NJ 07733, USA

[2]Lawrence Livermore National Laboratory, Livermore, CA 94550, USA

In this paper we show that the soft-x-ray yield from laser-exciting solid carbon with 100 fsec, 620 nm pulses, is increased 10-fold by using 100 Å film targets in which electronic diffusion is limited. We also show that the optical properties of the highly excited C films are well-described by the Drude model even when electron densities and temperatures exceed $6X10^{23}$ and 300 eV respectively. We also show that the electronic diffusion is well-described by the Conductivity and Spitzer models without the often used flux-limiter of 1/30.

The current work was stimulated by previous studies of soft-x-ray generation by irradiating solid Ta targets with 100 fsec, 620 nm pulses [1]. There, we measured surprisingly low x-ray yields and deduced that rapid electronic diffusion carried the optically deposited energy from the front 100 Å skin depth of the target into the bulk even before the majority of the x-rays were emitted. Several other groups [2,3,4,5] have also recently begun to explore the physics of laser-heating solid density targets and the subsequent plasma evolution because the soft-x-ray pulses so generated can be used to pump x-ray lasers and to do time-resolved x-ray measurements directly.

1. ENHANCED X-RAY YIELD

In Fig. 1 we show the x-ray yield in four energy bands from 100 and 300 Å free-standing carbon films as a function of incident fluence. For x-ray emission energies > 750 eV, the difference in yield is > 10 X. Each data point in Fig. 1 was obtained in a single shot from the excitation laser. The free-standing C films, provided by Lebow, Goleta, CA, were supported on a wire grid in a vacuum chamber. The

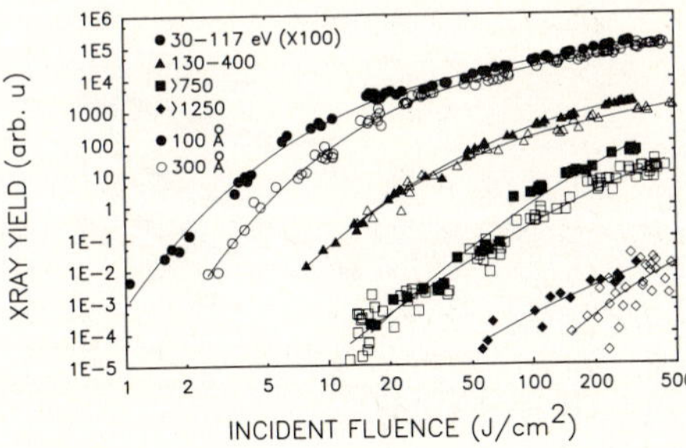

Fig. 1. X-ray yield vs incident fluence

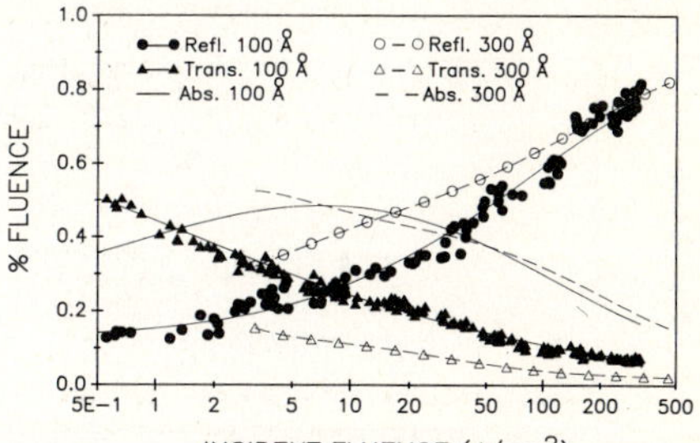

Fig. 2. % reflected, transmitted, and absorbed fluence vs incident fluence

Springer Series in Chemical Physics, Vol. 53 **Ultrafast Phenomena VII**
Editors: C.B. Harris · E.P. Ippen · G.A. Mourou · A.H. Zewail

excitation pulses (620 nm CPM laser pulses, amplified to 1 mJ at 10 Hz, dispersion-compensated to 100 fsec with prisms, < 10^{-3} ASE) were focussed with a lens into the vacuum chamber to be incident on the film targets at 3 degrees with a 14 μm diameter. On every shot (10 Hz) of the laser the target was translated to an intact part of the film between different grid wires. The soft-x-rays were detected at 60 degrees incidence to the film with an MCP and gated electronics. We used 4 spectral filters in transmission: Acton filters A and C (300 nm Al, 27 nm C, 3 nm Si, 30-117 eV), 200 nm Ag (130-400 eV), 5 μm Be (>750 eV), and 25 μm Be (>1250 eV).

2. OPTICAL PROPERTIES OF HOT PLASMA

In Fig. 2, we show the fraction of the incident energy reflected from and transmitted through the 100 and 300 Å (smooth curve only) films as a function of the incident fluence. As the target gets hotter the reflectivity increases toward unity. Despite the differences in R and T the absorbed energy (1-R-T) seems to be about the same for the 2 films.

Because the excitation is uniformly distributed over the 100 Å film by direct optical deposition or by diffusion, we can easily compare the optical properties of the film to that predicted from the Drude model. Assuming the optical energy is deposited slowly enough that the plasma is always in local thermal equilibrium (LTE), the instantaneous reflectivity, ρ(t), only depends on the absorbed fluence at time t. The function ρ(*absorbed fluence*) that is self-consistent with the 100 Å reflected energy data in Fig. 2 is shown in Fig. 3 as the filled circles.

The solid line is the function ρ(*absorbed fluence*) calculated using the Drude model for the dielectric function with the electron density and temperatures determined by Saha equilibrium at the absorbed energy density. The collisional frequency is calculated from Spitzer. The Drude model gives a very good approximation of the optical properties of this plasma even though the solid density carbon is nearly 6 times ionized and the electron temperatures are as high as 300 eV.

3. DIFFUSION IN HOT PLASMA

In Fig. 4, we replot the x-ray data of Fig. 1 as a function of absorbed excitation fluence. The xray yields are extremely sensitive to the peak electron temperatures.

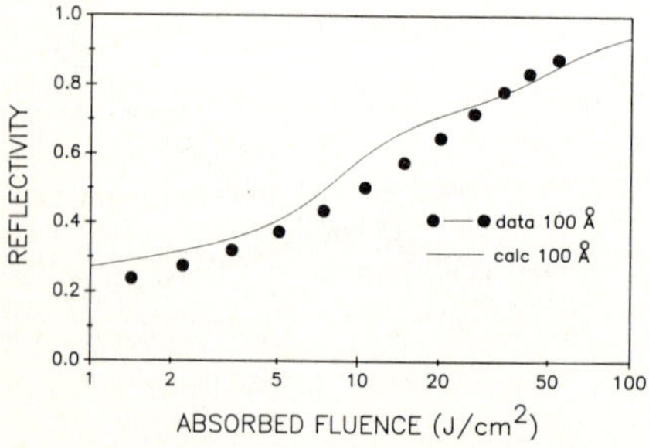

Fig. 3. Reflectivity vs absorbed fluence

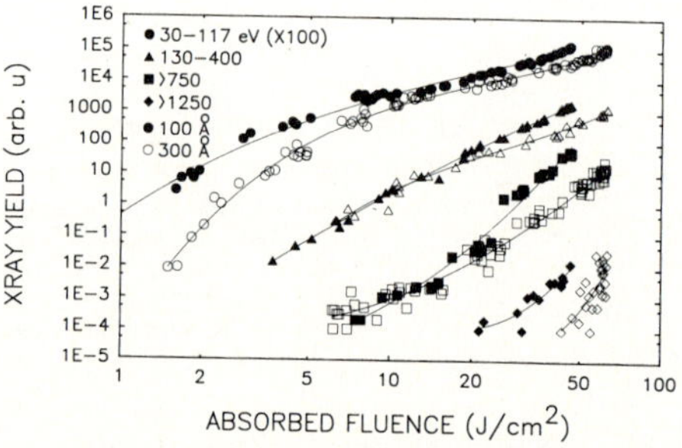

Fig. 4. X-ray yield vs absorbed fluence

The data at the highest excitation levels shows that the 300 Å film requires 1.4 times more absorbed fluence to reach the same temperature as the 100 Å film.

For T_e near 100 eV the Drude model gives an optical penetration depth of 40 Å. Solving the ordinary one-dimensional heat flow equation during the excitation pulse, the peak temperatures would differ by 1.4 in the two films if the diffusion constant $D=10cm^2/sec$. This value agrees with the conductivity model and Spitzer models without any flux limiter! and implies that previous analysis of experiments that have required flux-limiters are incorrect.

We also expect the x-ray pulses from the films to be extremely short because expansion cools the plasmas rapidly. Detailed hydrodynamic simulations were performed using the LASNEX code in which the diffusion rates were those calculated following the conductivity model of Lee and More [6] using no flux-limiters. The calculations predict that for 100 and 300 Å films heated with 10 J/cm^2 absorbed energy, emission at 50 eV has pulse durations of 200 and 360 fsec, respectively, while emission at 500 eV has pulse durations of 120 and 120 fsec, respectively. The similar times for the 500 eV emission are due to the expansion cooling in the 100 Å film being the same rate as the initial diffusion cooling in the 300 Å film.

In summary, we have shown that electronic diffusion on the 100 fsec time-scale is an important loss mechanism during the laser-heating of solid density targets. The optical and diffusion properties of the excited plasma are well-described by the standard models. The thin film geometry can produce ultrashort x-ray pulses.

REFERENCES

[1] H.W.K. Tom and O.R. Wood, II, *Appl. Phys. Lett. 54* , 517 (1989).

[2] R.W. Falcone and M.M. Murnane, in *Short Wavelength Coherent Radiation*, ed. D.T. Attwood and J. Bokor (American Institute of Physics, New York, 1986), p. 81.

[3] O.R. Wood, II, *et al*, *Appl. Phys. Lett. 53*, 654 (1988).

[4] M.M. Murnane, H.C. Kapteyn, and R.W. Falcone, *Phys. Rev. Lett. 62*, 155 (1989).

[5] H.M. Milchberg, R.R. Freeman, S.C. Davey, and R.M. More, *Phys. Rev. Lett. 61*, 2364 (1988).

[6] Y.T. Lee and R.M. More, *Phys. Fluids 27*, 1273 (1984).

Ultrafast Microscopy of Exploding Solids

D.D. Dlott

School of Chemical Sciences, University of Illinois at Urbana Champaign, Box 37 Noyes Lab., 505 S. Mathews Ave., Urbana, IL61801, USA

Ultrafast microscopy augments the high temporal resolution provided by ultrashort light pulses with the high spatial resolution provided by optical microscopy. Spatial resolution is required whenever the process of interest is spatially inhomogeneous. One example, interfacial phenomena, includes explosions occurring at a gas-solid interface. In this paper, I wish to point out that it is now possible to add ultrafast imaging capabilities to almost any ultrafast pump-probe experiment at a modest cost, often less than $1,000. I will then illustrate the capabilities of our ultrafast microscope by reviewing briefly some experiments on exploding polymers: the laser surface ablation of PMMA (poly methyl methacrylate).

Some of the spatially distinct phenomena one expects to observe during surface ablation are shown in Fig. 1. A tightly focused (25 μm diam.), 100 ps, 532 nm pulse of up to 675 μJ energy obtained from a Nd:YAG regenerative amplifier [1] was the ablation source [2-4]. Via self-focusing [3] and nonlinear absorption [4], this pulse causes the surface of the solid to vaporize, and explode into the air-solid interface at supersonic velocity, producing a blast wave which initially propagates at up to Mach 200 (ca. 60 km/s). The recoil from the blast launches a monopolar (compressive) shock wave which propagates back into the solid target, while the transient heating at the surface produces bipolar acoustic waves [4]. These waves propagate at roughly the acoustic velocity of PMMA, ca. 2.7 km/s [2]. The propagation of the heat conduction zone caused by thermal diffusion and motion of the melt front is far slower. After ca. 0.5 s, the surface has resolidified, during which time the melt front has moved only a few μm [3].

Figure 2 shows a "sideways" imaging geometry where the large diameter probe pulse propagates at right angles to the ablation pulse. Because our polymer samples are transparent, an "in-line" geometry where ablation and probe pulses counterpropagate through the sample was also used [2-4]. In either case, diffraction-limited achromatic lenses are used to expand the probe pulse before it is incident on the camera photocathode. The inexpensive camera used (Panasonic WV 1550) requires 0.3 μJ for a good image. Of

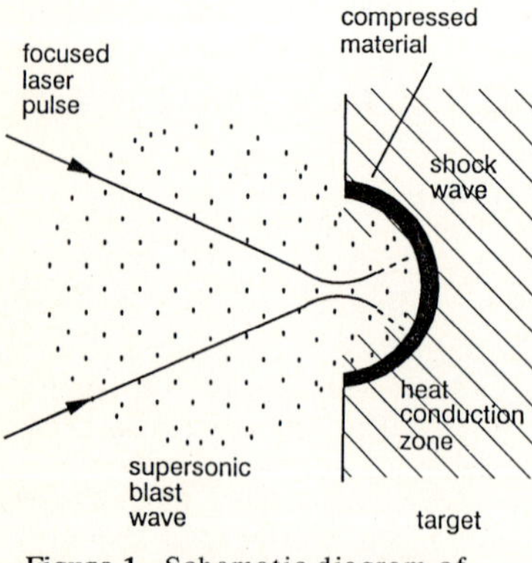

Figure 1. Schematic diagram of some of the spatially-dependent processes which have been observed using the ultrafast microscope on ablating polymer samples [4].

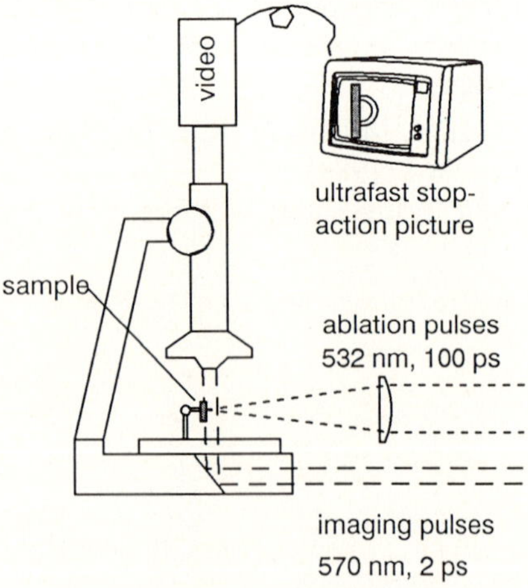

Figure 2. Schematic of the ultrafast microscope apparatus using the sideways geometry.

Springer Series in Chemical Physics, Vol. 53 **Ultrafast Phenomena VII**
Editors: C.B. Harris · E.P. Ippen · G.A. Mourou · A.H. Zewail

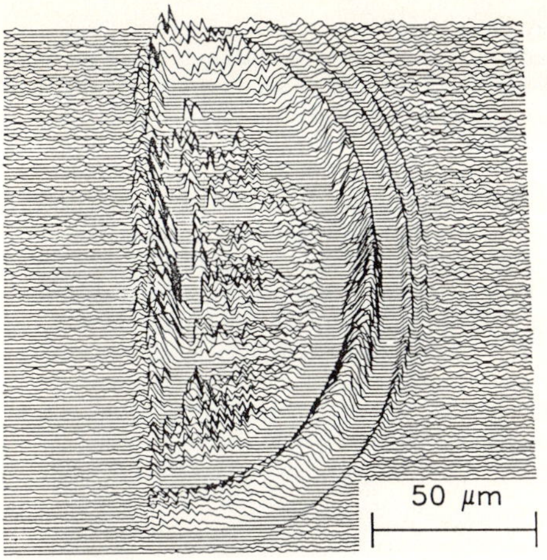

Figure 3. An ultrafast image of ablation at the polymer surface taken 10 ns after the ablation pulse. The sample surface-air interface is at the left. The blast wave hemisphere is observed via its attenuation of the probe pulse.

Table I. Blast wave parameters

$E_{pulse} = 285\ \mu J$	$E_{blast} = 60\ \mu J$	$m_{blast} = 5 \times 10^{-10} g$
V(at 600 ps) = 22 km/s	P(at 600 ps) = 10,000 atm	$\Delta H_{blast} = 120$ kJ/g

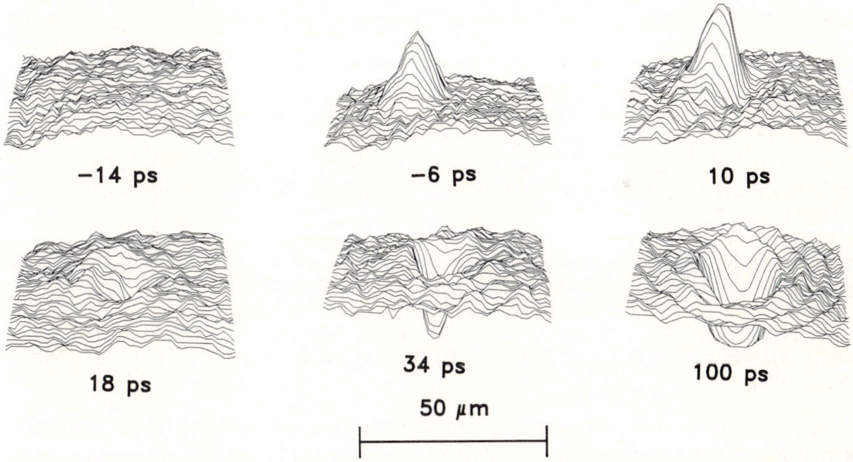

Figure 4. Ultrafast images of the ablation initiation process obtained during the 100 ps ablation pulse with a 2 ps probe. At -6 ps a filament produced by self-focusing appears. About 20 ps later, the darkening produced by the appearance of the blast wave is observed. The blast wave then propagates as in Fig. 3.

course more sensitive cameras with image intensifiers could be used if necessary. The image is retained on the cathode long enough to permit it to be captured by a video frame grabber and personal computer.

A blast wave at 10 ns delay is seen in Fig. 3. The blast front is then roughly 80 μm from the surface, giving an average velocity of 8 km/s (Mach 27). However the propagation is nonlinear; as time increases the blast slows down. The radius of the blast wave is proportional to $t^{0.4}$, so that the initial velocity is much faster than the average [2]. Properties of a typical blast wave [4] are summarized in Table I. The table shows that the laser ablation blast is quite powerful. The energy density of the blast wave, 120 kJ/g, is thirty times greater than the heat of explosion of TNT, 4.3 kJ/g [5].

The early stages of ablation initiation were observed to involve time-dependent self-focusing [3]. Figure 4 shows some images during the first few ps. The image at -14 ps (relative to the peak of the ablation pulse at t=0) shows the undistorted center of the probe pulse. The spatial profile is flat over this 50 μm x 50 μm region. At -6 ps, a small bright spot (indicated by the height above the plane) is observed. The probe pulse is focused by the transient lens formed by the intense ablation pulse, which undergoes filamentation. The filament irradiates the sample for roughly 20 ps, at which time an expanding dark region (indicated by depth below the plane) appears. The darkening is caused by attenuation of the probe by the blast wave.

Once the filament appears, laser heating of the polymer is greatly accelerated. We have conducted a detailed series of intensity-dependent experiments on PMMA and discovered an interesting effect [3]. The intrinsic time delay between the appearance of the filament and the appearance of the blast wave is 20 ± 4 ps, irrespective of the intensity of the ablation pulse or the energy of the blast wave. That is not what one expects for thermally induced decomposition of a polymer. There the rate of decomposition is expected to increase exponentially with laser energy. This observation leads us to conclude that the appearance of the blast wave is controlled by physical, rather than chemical kinetics. The blast wave is not produced until the optical excitations relax, through nonradiative processes, into phonons and ultimately into dissociative states with large translational quantum numbers [3].

Acknowledgement.

This research would not have been possible without the efforts of Drs. Hackjin Kim, Taehyoung Zyung and Jay C. Postlewaite. Financial support from the National Science Foundation, through grant NSF DMR 87-21243, and the US Army Research Organization, through grant DAAL03-90-G-0300 is gratefully acknowledged.

References

1. "Picosecond YAG Laser Amplifier and Electrooptic VFET Pulse Switch for High Power, High Repetition Rate Operation", J. C. Postlewaite, J. B. Miers, C. Reiner, and D. D. Dlott, IEEE J. Quantum. Electron. **QE-24**, pp. 411-417 (1988).

2. "Ultrafast Imaging Study of Optical Damage Dynamics and Laser-produced Wave Propagation in Polymethyl Methacrylate", H. Kim, J. C. Postlewaite, T. Zyung, and D. D. Dlott, J. Appl. Physics **64**, pp. 2955-2958 (1988).

3. "Time Dependent Self-Focusing and a 20 Picosecond Delay in Laser Ablation of Polymers". H. Kim, T. Zyung, J. C. Postlewaite, and D. D. Dlott. Appl. Phys. Lett., **54**, 2274-2276 (1989).

4. "Ultrafast Imaging of 0.532 μm Laser Ablation of Polymers: Time Evolution of Surface Damage and Blast Wave Generation", T. Zyung, H. Kim, J. C. Postlewaite, and D. D. Dlott. J. Appl. Phys. **65**, pp. 4548-4563 (1989).

5. "The Relationship Between Kinetic Data of the Low-Temperature Thermolysis and the Heats of Explosion of Organic Polynitro Compounds", S. Zeman, M. Dimun, and S. Truchlik, Thermochem. Acta., **78**, pp. 181-209 (1984).

Femtosecond Melting of Diamond

D.H. Reitze, H. Ahn, X. Wang, and M.C. Downer

Physics Department, University of Texas at Austin, Austin, TX78712, USA

Abstract. We have performed time-resolved reflectivity and transmission studies of crystalline diamond in order to understand the complex nature of the solid-liquid phase transition in carbon. Above a well defined melting critical fluence of 0.55±0.1 J/cm^2, diamond undergoes a transition into a metallic liquid state, in agreement with both theoretical predictions of the liquid carbon and previous experiments conducted on graphite. Below the critical fluence, we observe the creation of a photogenerated electron-hole plasma.

1. Introduction

The electrical properties of liquid carbon have remained a subject of interest for almost a century, ever since Ludwig reported increased resistance in carbon rods melted at high pressure[1]. Although solid carbon sublimes when heated at atmospheric pressure, several recent experiments have shown that a transient liquid phase is created when heated by a short laser or electrical pulse. Despite the numerous experimental studies [2-5], the question remains: Is liquid carbon metallic or insulating? We recently reported the first time-resolved femtosecond reflectivity experiments on graphite melting[6] in which an initial high reflectivity phase (caused by a phase transition into a liquid metallic state) decays into a final low reflectivity phase (see Fig. 1). The origin of the low reflectivity phase can be interpreted as a second phase transition into an insulating liquid phase based on time resolved ellipsometry measurements of the optical constants[6], although we have not ruled out the possibility that ablated material is responsible for the delayed reflectivity drop, based on post-mortem analysis of the irradiated regions. In this paper we report on time-resolved optical reflectivity and transmission studies of crystalline diamond excited with intense 620 nm, 90 fs pulses. The transparency of diamond at visible wavelengths permits probing from the back surface of the sample. By probing the reflectivity of the melted region at the liquid carbon/diamond interface rather than the liquid carbon/air interface, ablation is eliminated as a cause of optical response.

2. Experimental Procedure

In the front-surface (back surface) reflectivity measurements, 90 fs, 0.62 μm pump pulses from a colliding-pulse dye laser were amplified to 0.2 mJ at a 10 Hz repetition rate and then focused at normal incidence to a 60 μm. spot on the (100) surface of a 250 μm thick diamond sample. The probe beam was split from the pump beam using a glass wedge, passed through a variable optical delay line, and focused to 20 μm. diameter waist centered on the pump spot at a 15° (45°) angle of incidence. The pump and probe beams were orthogonally polarized in order to discriminate against scattered pump light. Between each shot the sample was translated to ensure that the pump excited an undamaged region. Each shot was normalized to discriminate against pulse-to-pulse energy fluctuations.

Springer Series in Chemical Physics, Vol. 53 **Ultrafast Phenomena VII**
Editors: C.B. Harris · E.P. Ippen · G.A. Mourou · A.H. Zewail

3. Results

In Fig. 1, we present results of front surface reflectivity measurements of diamond excited at 10x the critical melting fluence F_m. The long time response is shown in Fig. 1a (solid squares), in which the reflectivity initially rises ($\Delta t < 3$ ps) above and then decays ($\Delta t > 3$ ps) to well below the solid diamond reflectivity value. The fast reflectivity response is detailed in Fig. 1b (solid squares). The discrepancy in peak reflectivities is an artifact of the coarse time delay between data points in Fig. 1a. This pattern of a fast initial rise followed by slower decay to low reflectivity values does not change as the pump fluence is varied, although the decay to low reflectivity becomes longer as the fluence is reduced. For comparison, we plot front surface reflectivity data for highly oriented pyrolytic graphite (HOPG) in each panel (open squares). In HOPG, the initial reflectivity rise is caused by the formation of metallic liquid phase[6]. The similarity of the responses provides strong evidence that the dynamics of the solid-liquid phase transition is independent of the initial lattice structure and in each case a metallic liquid film is created.

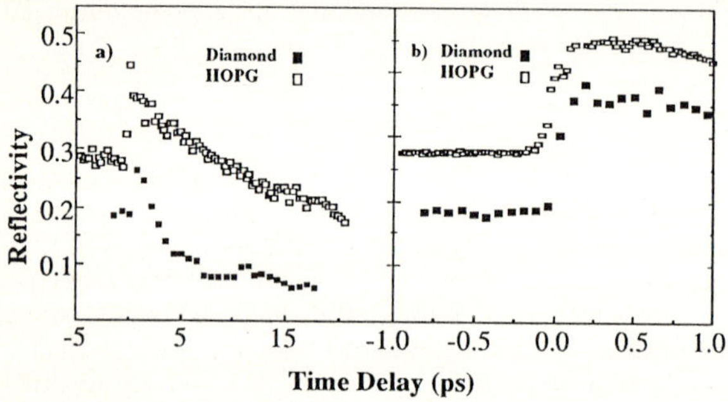

Fig. 1 - Front surface time-resolved reflectivity on diamond (100) and HOPG using 620 nm, 90 fs pulses. a) long time scale dynamics (Diamond: $F = 10F_m$, HOPG: $F = 20F_m$) b) risetime dynamics (Diamond: $F = 10F_m$, HOPG: $F = 6F_m$)

In Fig. 2, we present the results of probing the back surface reflectivity (Fig. 2a) and transmission (Fig. 2b) of diamond when excited at $1.25F_m$ (filled squares). In both cases, a large initial decrease occurs followed by a slight recovery toward the solid reflectivity and transmission levels after 20 ps. This behavior is seen at any fluence above F_m. The transmission drop can be intuitively explained in terms of the creation of a thin, highly absorptive metallic layer, consistent with the front surface reflectivity results. The drop in reflectivity, however, seems inconsistent with this explanation since a metallic layer should highly reflective, as observed in front surface reflectivity. Further,

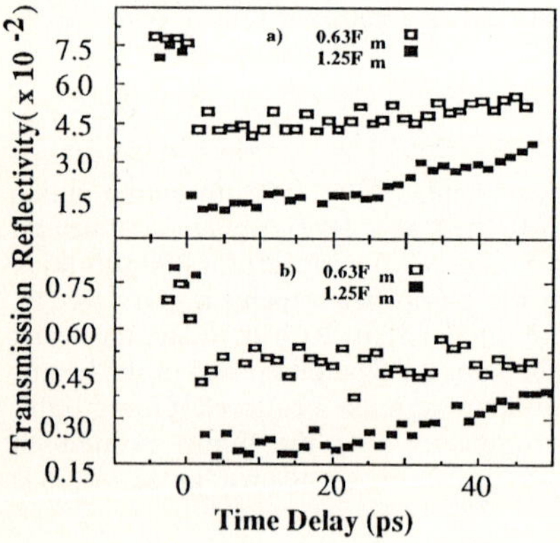

Fig. 2 - Back surface time resolved a) reflectivity, and b) transmission on diamond (100) taken below($0.63F_m$) and above ($1.25F_m$) the critical melting fluence using 620 nm, 90 fs pulses.

the reflectivity and transmission responses of diamond when excited *below* the critical melting fluence ($0.63F_m$, open squares) are qualitatively identical to those observed above F_m.

The signs and magnitudes of the initial reflectivity and transmission drops below F_m can be explained in terms of the generation of free carriers in diamond (E_{gap}=5.48 eV) via three-photon absorption. As the pump pulse is absorbed, a graded electron-hole plasma forms below the front surface. Neglecting the effect of the gradient, we can model the plasma as a slightly absorptive layer ($n=n_{eh}+ik_{eh}$) at the front surface. Using the Fresnel equations, we calculate theoretical reflectivity and transmission changes equal to the observed changes for n_{eh}~2.1, k_{eh}~0.0004 and a plasma thickness d_{eh} = 60 μm. In order to model the reflectivity and transmission changes above F_m, we must include the effects of both the thin metallic liquid layer and the electron-hole plasma. We use the thin film equations[7] to model our data. Briefly, we use as inputs to the thin film equations n_{eh},k_{eh}, and d_{eh} (determined from the analysis above) and the complex index of liquid carbon n_{lc} and k_{lc} (determined from time-resolved ellipsometry experiments[6]). We adjust only the value of the liquid film thickness d_{lc} to calculate theoretical reflectivity and transmission values. We obtain the values T = 0.25, R = 0.02 for d_{lc} ~ 20 nm, near the observed experimental values of T = 0.2, R=0.015. Although this model is crude in that it neglects spatial variations in the optical constants, it qualitatively accounts for the observed responses.

The prolonged decrease in transmission in Fig. 2b above F_m is indicative of a persistent metallic film. As a basis for comparison, we performed back surface experiments on a 1.0 μm silicon film on sapphire. Silicon is known to metallic in the liquid state. We find, for the same reason as in diamond, that the reflectivity and transmission both decrease sharply for positive time delays. The transmission in silicon does not show any observable recovery over a 60 ps period, in contrast to diamond. However, the partial recovery in diamond can be attributed to the e-h plasma rather than a transformation in the liquid film.

This research was supported by NSF Grant DMR-8858388, AFOSR Contract F9620-89-C-0044, and the Robert A. Welch Foundation Grant F-1038.

4. References

1. A. Ludwig, Z. Elektrochem. 8, 273 (1902).
2. T. Venkatesan, D.C. Jacobson, J.M. Gibson, B.S. Elman, G. Braunstein, M.S. Dresselhaus, and G. Dresselhaus, Phys. Rev. Lett. 53, 360 (1984).
3. J. Steinbeck, G. Braunstein, M.S. Dresselhaus, T. Venkatesan, D.C. Jacobson, J. Appl. Phys. 58, 4374 (1985).
4. J. Heremans, C.H. Olk, G.L. Eesley, J. Steinbeck, and G. Dresselhaus, Phys. Rev. Lett. 60, 452 (1988).
5. A.M. Malvezzi, N. Bloembergen, and C.Y. Huang, Phys. Rev. Lett. 57, 146 (1986).
6. D. H. Reitze, X. Wang, H. Ahn, M. C. Downer, Phys. Rev. B **40**, 11986 (1989).
7. O.S. Heavens, *Optical Properties of Thin Solid Films* (Dover, New York, 1985).

Femtosecond Photoemission and Reflectivity Measurements of High Temperature Electron Dynamics in Solid Density Plasmas

W.C. Banyai, D.C. Anacker, X.Y. Wang, D.H. Reitze, G.B. Focht, M.C. Downer, and J.L. Erskine

Physics Department, University of Texas at Austin, Austin, TX78712, USA

Abstract. The energy distribution of electrons photoemitted from a hot, solid-density plasma created by photoexciting a Ag surface with two collinear, *p*-polarized, 90 fs pulses was measured via photoelectron time-of-flight spectroscopy. At zero time delay the average photoelectron energy corresponds to several 10^5 K and drops to half this value when the time delay exceeds 180 fs. Time-resolved reflectivity studies, under similar excitation conditions, indicate electron heating is pulsewidth limited, but recovery caused by electron cooling and surface expansion occurs over several picoseconds.

1. Introduction

Metal surfaces have been excited with low and high-intensity ultrafast optical pulses in order to study the relaxation dynamics of the transient disequilibrium between the electron and lattice temperatures[1], the temperature-dependent collisionality of the material[2], resonant absorption[3] and the radiation of soft x-rays from high-temperature plasmas[4]. At high intensities ($>10^{12}$ W/cm^2), the modulation of the absorption at a metal surface can occur, according to the Drude model, by either increasing the collision frequency, as in the case of an increasing electron temperature, or tuning the plasma frequency to the optical frequency, as in the case of the critical surface in a steep plasma gradient in front of a rapidly-heated metal surface. Increased collisional absorption affects both *s* and *p*-polarizations while collisionless absorption in steep plasma gradients affects only the *p*-polarization. In either case, a large amount of energy can be deposited within a few skin depths when 100 fs pulses are used to excite a surface, leading to a high-temperature, solid-density plasma. Furthermore, high ionization states can be achieved through the combined effects of pressure and multiphoton ionization, leading to x-ray emission from free-free, free-bound and bound-bound transitions. The generation of ultrafast x-ray pulses is contingent on rapid cooling of a high-temperature plasma by electron-ion interaction, electronic thermal conduction[5] and plasma expansion.

We have time-resolved the high-intensity, ultrafast excitation and recovery of a metal surface in order to address the questions of collisional versus resonant absorption and the cooling dynamics of the excited hot-electron distribution. Time-resolved reflectivity measurements were used to study the development of plasma gradients from short to long scale-lengths while the evolution of the average electron temperature was studied with time-resolved photoemission spectroscopy.

2. Experiments

The intensity-dependent reflectivity of a polished Ag(111) surface was measured by focusing a 90 fs, 620 nm pulse to a $1/e^2$ spot size of 40 μm at a 45° angle of incidence. Here we emphasize the material as Ag(111) since the reflectivity measurements were single shot with the sample moved to a fresh part of the surface for each data point. The normalized reflectivity curves for both *s* and *p*-polarizations are shown in Fig. 1. The reflectivity remains close to unity until the intensity reaches 10^{12} W/cm^2. The reflectivity

Springer Series in Chemical Physics, Vol. 53 **Ultrafast Phenomena VII**
Editors: C.B. Harris · E.P. Ippen · G.A. Mourou · A.H. Zewail

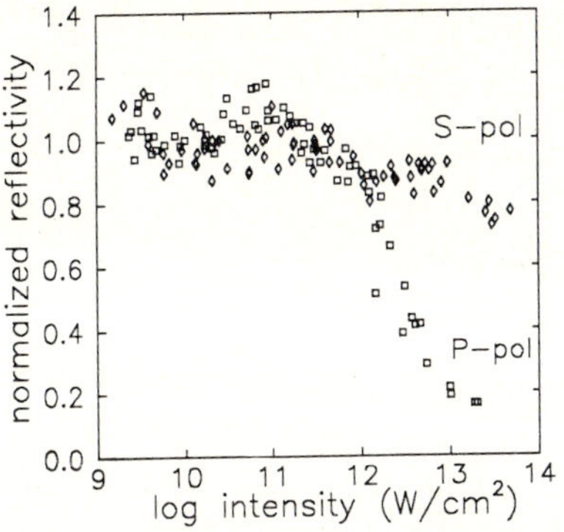

Fig. 1. Intensity-dependent reflectivity. at a 45° incident angle.

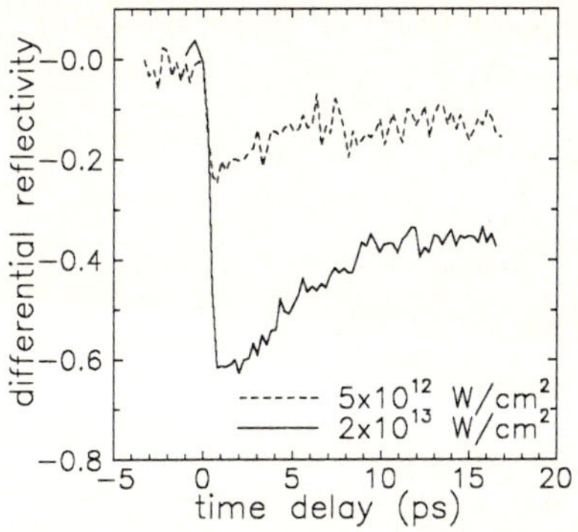

Fig. 2. Time-resolved differential reflectivity of a weak probe.

then decreases for both *s* and *p*-polarizations, but more dramatically for *p*-polarized light. This effect, observed at 45° incident angle, can be explained, using the Drude-Fresnel model for reflection from a plasma, as an increase in the collision frequency of the plasma, neglecting the expanding plasma at the surface.

The time-evolution of the intensity-dependent change in the reflectivity was examined for *p*-polarized light in order to distinguish between collisional and resonant absorption. A pump pulse was focused on a Ag(111) surface at a 45° incident angle to a 60 μm x 80 μm spot while a weak probe pulse, incident at -68°, sampled the pump-induced change in reflectivity. The time-resolved reflectivity reveals a rapid decrease in the initial 500 fs, followed a slower, but not complete, recovery on the order of tens of picoseconds, as shown in Fig. 2. Note that pre-pulse heating is negligible in view of the rapid, pump-induced change in the reflectivity. The turn-on time is approximately the same for the two intensities studied and, when the geometry of the experiment is taken into consideration, follows the leading edge of the 90 fs pump pulse. This suggests that the initial increase in absorption is due to a rapid increase in the collision frequency of the plasma, while resonant absorption, which has a slower turn-on time, contributes to the sustained decrease in the reflectivity. The smaller initial absorption observed at the lower pump intensity clearly reflects the lower electron temperature and therefore reduced collisional absorption. However, the slower recovery of reflectivity reveals a complicated combination of a cooling plasma tending to reduce collisional absorption and the development of a plasma scale-length conducive to absorption. Similar reflectivity transients are observed in Al and W, although the detailed recovery pattern is material dependent.

In order to gain further insight into the electron dynamics, a photoemission experiment was performed under similar excitation conditions. A Ag sample was placed in a vacuum chamber at 10^{-10} Torr and cleaned, as indicated by LEED and Auger measurements. Two equal-energy, collinear, *p*-polarized pulses were incident on the sample at 68°. Electron energy distributions were recorded with a time-of-flight photoelectron detector for successive time delays between the two pulses. Each distribution resulted from a 256-shot average. The five distributions shown in Fig. 3 were obtained by illuminating a single spot on the sample. The emitted energy distributions are approximately Maxwellian, which can be seen by comparing the curves in Fig. 3 to the calculated time-of-flight distributions[6], using a Maxwellian velocity distribution with temperatures ranging from $2x10^5$ - $4x10^5$ Kelvin, shown in Fig. 4. The measured temperatures are consistent with temperatures calculated for an electron gas that absorbs 60% of a 90 fs, 2 x 10^{12} W/cm^2 pulse in a 100 Å skin-depth as shown in Fig. 1, although the substantial effect of space-charge is not yet included in the calculated time-of-flight distributions. It is apparent, from a comparison of Figs. 3 and 4, that the temperature decreases by half when the time delay

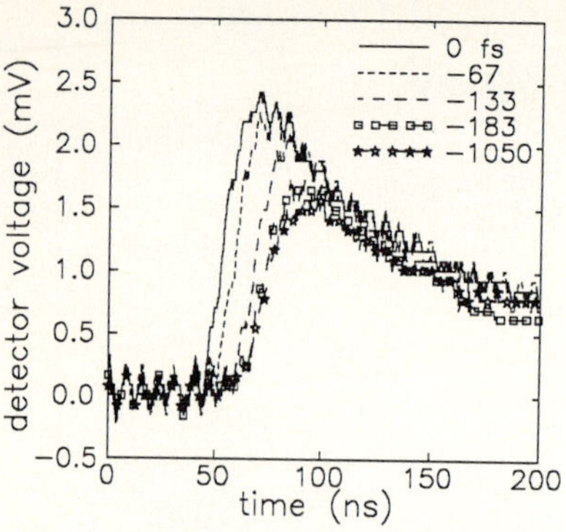

Fig. 3. Time-of-flight distributions for equal-energy pump-probe.

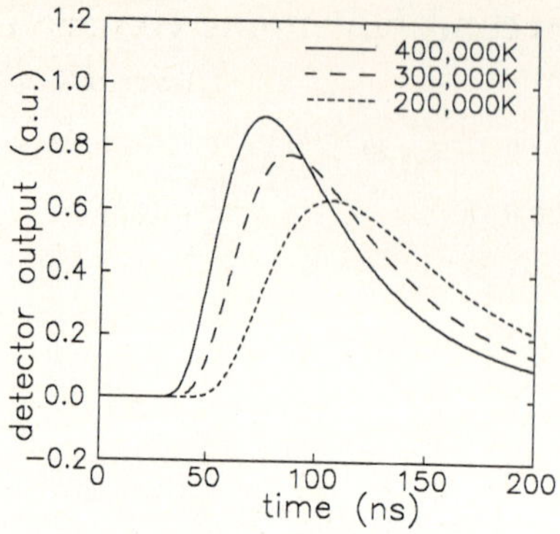

Fig. 4. Calculated time-of-flight distributions.

between the pulses is greater than 180 fs. It is important to note that when one of the beams was blocked at zero time delay, the electron energy distribution corresponding to a 180 fs time delay was obtained. This means that after the first 180 fs the second pulse does not cause any additional heating or photoemission.

This research was supported by ONR Contracts N00014-88-K-0063 and N00014-88-K-0754, NSF Grant DMR-8858388, AFOSR Contract F9620-89-C-0044 and Robert A. Welch Foundation Grant F-1038.

3. References

1. H. E. Elsayed-Ali, T. B. Norris, M. A. Pessot and G. A. Mourou, Phys. Rev. Lett. **58**, 1212 (1987); R. W. Schoenlein, W. Z. Lin, J. G. Fujimoto and G. L. Eesley, Phys. Rev. Lett. **58**,1680 (1987).
2. H. M. Milchberg, R. R. Freeman, S. C. Davey and R. M. More, Phys. Rev. Lett. **61**, 2364 (1988).
3. H. M. Milchberg and R. R. Freeman, J. Opt. Soc. Am. B **6**, 1351 (1989); J. C. Kieffer, P. Audebert, M. Chaker, J. P. Matte, H. Pepin, T. W. Johnson, P. Maine, D. Meyerhofer, J. Delettrez, D. Strickland, P. Bado and G. Mourou, Phys. Rev. Lett. **62**, 760 (1989); R. Fedosejevs, R. Ottmann, R. Sigel, G. Kuhnle, S. Szatmari and F. P. Schafer, Phys. Rev. Lett. **64**, 1250 (1990).
4. D. Kuhlke, U. Herpers and D. von der Linde, Appl. Phys. Lett. **50**, 1785 (1987); H. W. K. Tom and O. R. Wood, II, in *Laser Interaction and Related Plasma Phenomena*, H. Hora and G. H. Miley, eds. (Plenum Press, N.Y.,1989), p.183; M. M. Murname, H. C. Kapteyn and R. W. Falcone, Phys. Rev. Lett. **62**, 155 (1989).
5. S. D. Brorson, J. G. Fujimoto and E. P. Ippen, Phys. Rev. Lett. **59**, 1962 (1987).
6. N. G. Utterback, S. P. Tang and J. F. Friichtenicht, Phys. Fluids **19**, 900 (1976).

Second Harmonic Generation in Femtosecond Laser-Produced Plasmas

T. Engers, H. Schüler, W. Fendel, H. Schulz, and D. von der Linde

Institut für Laser- und Plasmaphysik, Universität-GHS-Essen, D-4300 Essen 1, Fed. Rep. of Germany

Abstract. Reflected second harmonic with a conversion of 10^{-4} is observed. The spectra show a multicomponent structure dependent on the angle of incidence. In pump probe experiments we detect maxima of the probe pulse second harmonic a few picoseconds after the pump pulse.

We have studied laser plasmas produced by very intense subpicosecond pulses focussed on both metallic and insulating solid targets. One of the unique aspects of subpicosecond laser-plasma interaction is that there is not enough time for substantial hydrodynamic expansion of the plasma. The laser pulse interacts with matter of approximately solid state density, and the scale length of the generated plasma is much less than the wavelength of light [1-3].

Here we present initial observations of second harmonic generation in subpicosecond laser-produced plasmas which reveal a rich variety of interesting phenomena, likely to carry useful information about the plasma dynamics.

In these experiments we used 350 fs laser pulses at 624 nm with energies up to 0.5 mJ. The pulses are focussed on target using an achromatic objective lens (f = 2.5 cm). With a spot diameter of 3.6 μm the maximal peak intensity is estimated to be 1.4×10^{16} W/cm^2. The amplified spontaneous emission integrated over the total emission cone was a few times 10^{-2} of the total pulse energy.

In our experiments we observed strong second harmonic radiation both from plasmas produced on 0.1 μm aluminium films on glass substrates, and on bulk glass samples. The second harmonic is generated as a well-collimated beam in the direction of the reflected fundamental. In the following the main properties of the reflected second harmonic (SH) are summarized.

1) SH generation increased strongly with the angle of incidence. For example, the SH at 70 degrees was 50 times stronger than that at 35 degrees. The angular dependence is qualitatively similar to surface SH generation on metals[4].

2) For 70 degrees (35 degrees) angle of incidence p-polarization was about 1000 times (10 times) more effective than s-polarization. The second harmonic was always p-polarized.

3) The SH increased with the fundamental energy faster than a square law. For example, on glass at 50 degrees the SH was measured to increase proportional to $E^{2.6}$, where E is the energy of the incident fundamental pulse.

4) Energy conversion to SH as high as 10^{-4} was measured. The SH beam can be readily observed with the bare eye on a fluorescent white paper.

We note that the observed second harmonic conversion efficiency is comparable to that expected from metal surfaces [4].

The frequency spectra of the reflected SH light show very interesting structure. The solid curve in Fig. 1 depicts an example of a SH spectrum from glass at 42 degrees angle of incidence. The dashed line shows a SH spectrum from a 0.1 mm KDP-crystal for comparison. The peak of the plasma SH is red-shifted by $\Delta\lambda = +1.4$ nm. However, there is also a weaker blue-shifted component ($\Delta\lambda = -2$ nm).

Springer Series in Chemical Physics, Vol. 53 **Ultrafast Phenomena VII**
Editors: C.B. Harris · E.P. Ippen · G.A. Mourou · A.H. Zewail

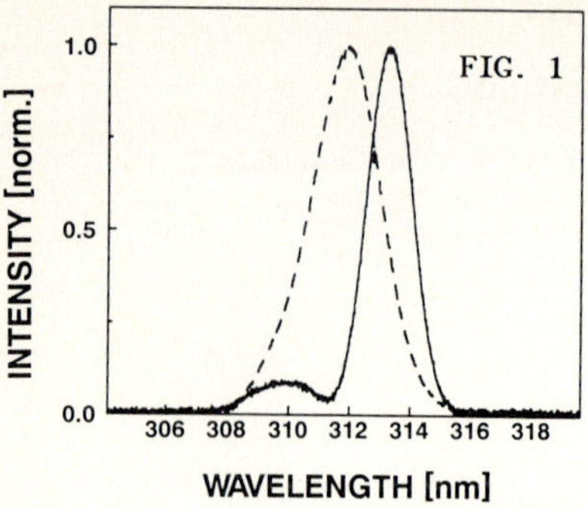

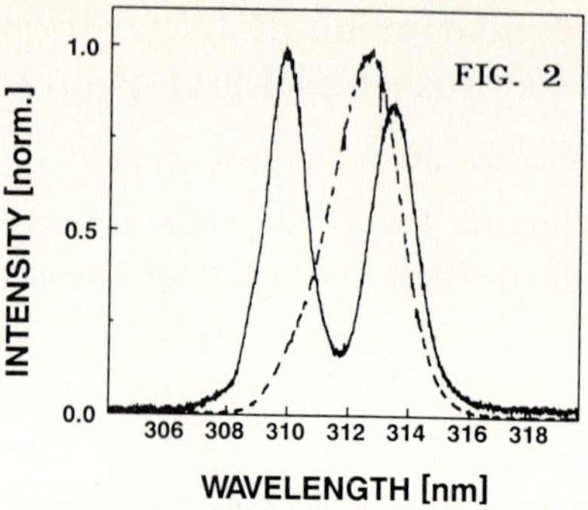

Spectra of the reflected second harmonic from laser plasmas produced on glass targets. The dashed curves are the reference second harmonic spectra from a KDP crystal. Angle of incidence: 42 degrees (Fig. 1); 35 degrees (Fig. 2).

Figure 2 shows an example of a SH spectrum measured at 35 degrees. In this case a red-shifted and a blue-shifted component of comparable magnitude was measured. We note that the spectra were also critically dependent on the focussing conditions. In our measurements of the angular dependence great care was required to ensure the same focussing when the angle of incidence is changed. Thus we believe that our data indicate a distinct variation of the SH spectra with the angle of incidence of the fundamental beam.

The observed polarization properties of the second harmonic suggest that the mechanism of second harmonic generation is dominated by the following process. The p-polarized incident light drives some modulation of the electron density at the fundamental frequency, and light is scattered from these coherent electron density fluctuations at a frequency close to the second harmonic. However, the observed second harmonic from s-polarized light for smaller angles of incidence indicates that there are also contributions from other mechanisms.

The plasma dynamics during and after subpicosecond laser excitation lead to interesting changes of the second harmonic with time. Pump-probe experiments were performed with s-polarized pump pulses and collinear p-polarized probe pulses at the same frequency. The angle of incidence was 50 degrees.

A collinear pump geometry was chosen in these initial experiments in order to minimize the problems of maintaining spatial overlap of the very tightly focussed beams when the sample surface is raster-scanned during data recording. However, a disadvantage of the collinear geometry is that a relatively high probe-to-pump energy ratio (1 : 4) had to be used to enable detection of the probe pulse second harmonic on the background of the second harmonic generated by the pump pulse.

An example of pump-probe second harmonic measurements on glass is shown in Fig. 3. The p-polarized SH of the probe pulses is plotted as a function of the delay time between pump and probe pulses. The upper and the lower mark on the right indicate the SH signals produced by the probe pulse only, and the pump pulse only, respectively.

The data reveal a rather complicated variation of the second harmonic with time. A first maximum is reached at 2 ps, followed by a decrease, and a second maximum at 5 ps. The probe pulse SH is down to the background after 10 ps. A close inspection of the minimum near zero delay time reveals some subpicosecond fine structure (see insert of Fig. 3). A maximum of the second harmonic was also observed for negative time delay at −2 ps.

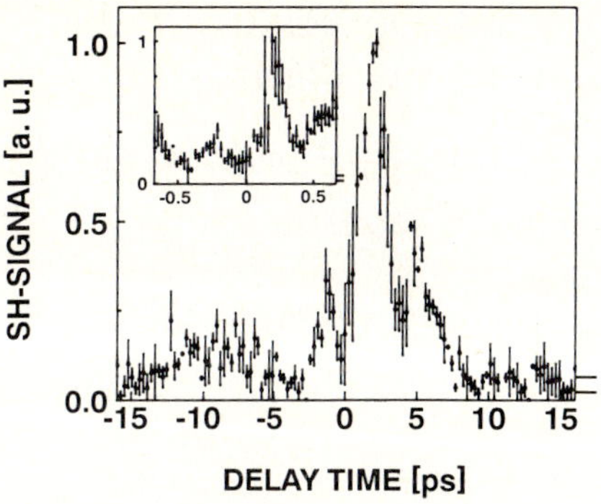

FIG. 3

Second harmonic of the probe pulse as a function of delay time for a plasma produced on a glass target. The insert shows the structure near zero delay time on an expanded time scale.

The peak of the second harmonic at negative delay times signifies a situation in which pump and probe pulses change their roles. For negative delay the p-polarized pulses interact with the sample first. We note that at 50 degrees p-polarized light is expected to be absorbed more efficiently than s-polarized light. Thus the p-polarized "probe" is likely to produce substantial excitation, although its energy is only 25 % of the s-polarized "pump" pulse. The data show that substantial second harmonic (p-polarized) is generated by the s-polarized pulses after a delay time of 2 ps.

Very similar results were also obtained from pump-probe experiments on aluminium targets, except that it appears that in this case the second maximum for positive delay is absent.

In summary, we have observed strong second harmonic generation from subpicosecond laser plasmas. The explanation of the observed multicomponent structure of the spectra and of the time dependence of the reflected second harmonic require detailed analysis of the plasma dynamics and of the various second harmonic generation mechanisms.

Support of this work by the Deutsche Forschungsgemeinschaft is gratefully acknowledged.

References

1. H. M. Milchberg and R. R. Freeman, Opt. Soc. Am. **B 6**, 1351 (1989).
2. J.-C. Kieffer, J.-P. Matte, S. Bélair, M. Chaker, P. Audebert, H. Pépin, P. Maine, D. Strickland, P. Bado, and G. Mourou, IEEE J. Quant. Electr. **QE-25**, 2640 (1989).
3. R. Fedosejevs, R. Ottmann, R. Sigel, G. Kühnle, S. Szatmari, and F. P. Schäfer, Appl. Phys. **B 50**, 79 (1990).
4. J. E. Sipe and G. I. Stegman, in "*Surface Polaritons*", ed. by V. M. Agranovich and D. L. Mills (North Holland, Amsterdam, 1982), p. 661.

Enhanced Absorption and ASE Pedestal Suppression in the Generation of Ultrashort-Pulse Solid-Density Plasmas

H.C. Kapteyn[1], *M.M. Murnane*[1], *A. Szoke*[2], *A. Hawryluk*[2], *and R.W. Falcone*[1]

[1]Department of Physics, University of California at Berkeley, Berkeley, CA 94720, USA

[2]Lawrence Livermore National Laboratory, Livermore, CA 94550, USA

Abstract. Two techniques are demonstrated which will facilitate the production of solid-density ultrashort-pulse plasmas at very high (>10^{19} W/cm^2) intensities. Grating targets are shown to dramatically enhance the energy absorption of an incident 150 fs laser pulse over that of a flat target, from <20% to >90%. Also, the technique of self-induced plasma shuttering is demonstrated to be an effective means of reducing the pedestal energy accompanying the ultrashort pulse. Thus, solid-density plasmas can now be produced even with a laser pulse of low contrast ratio.

1. Introduction

A high-intensity femtosecond laser pulse, when focused onto a solid target can produce a high-density, short-lived plasma, which emits a sub-picosecond burst of x-rays. This has been demonstrated experimentally in our research[1], using a 3 mJ, 150 fs laser pulse focused to 10^{16} W/cm^2. Laser systems of peak intensity 10^2 to 10^3 times this value will soon become available, making it possible to create hotter and more intense x-ray emitting plasmas. However, two potential obstacles to creating such plasma have existed. First, in order for the plasma to cool on a sub-picosecond time scale, the intense laser pulse must be absorbed at near solid densities. If a nanosecond duration ASE pedestal of sufficient intensity accompanies the short pulse, it can vaporize the target and the femtosecond pulse will be absorbed in the low-density vapor, resulting in dramatically reduced cooling rates and increased x-ray pulse width. Since the threshold prepulse energy is independent of the main pulse intensity, the contrast ratio requirement increases with increasing peak intensity. The second problem is that as the temperature of the plasma increases beyond a few hundred eV, electron-ion collision rates reduce, and the absorption of the incident radiation decreases, resulting in a situation of diminishing returns in terms of attainable temperature. Herein we present an experimental demonstration of two techniques which dramatically reduce these two problems.

2. Self-induced plasma shuttering for prepulse suppression

When a high-intensity laser pulse is incident on a flat target, most of the incident energy is reflected specularly[1], since the pulse is reflected from the surface before the plasma can expand significantly. Consider a laser beam focused onto a glass target, at an intensity such that any pedestal energy in the pulse is below the damage threshold of the glass, but the femtosecond pulse has an intensity above the damage threshold. The prepulse will be reflected at the surface reflectivity value of the

Springer Series in Chemical Physics, Vol. 53 **Ultrafast Phenomena VII**
Editors: C.B. Harris · E.P. Ippen · G.A. Mourou · A.H. Zewail

glass(which can be minimized by placing the target at Brewster angle), while the main pulse will reflect off a plasma, at near unity reflectivity. The pedestal of the reflected pulse is thus dramatically suppressed.

The setup used to experimentally demonstrate this scheme consists of a glass target in a vacuum chamber. A 3 mJ, 150 fs pulse is focused onto the target at an incident angle of 45° using a 15 cm lens (to 10^{15} W/cm^2) and is then recollimated using another lens. The recollimated beam is then refocused in another vacuum chamber to create a solid density plasma. X-ray emission from this plasma is monitored using an x-ray streak camera. Our previous work has demonstrated that the x-ray pulse duration is very sensitive to prepulse energy incident on the target-- a pedestal energy of <10 µJ in 5-10 ns is sufficient to create observable pulse broadening due to coupling of the laser energy into a low-density vapor. For these experiments, a ~500 µJ ASE pedestal was introduced into the beam by removing saturable absorbers from the amplifier chain. A portion of the glass target was coated with aluminum; when the target is rotated to the aluminum portion, the prepulse is reflected as well as the main pulse, and the resulting x-ray emission was of long duration, as is illustrated in figure 1(a). In contrast, when the laser was reflected off the glass target, the x-ray emission was short (figure 1(b)), and resolution-limited when compared with a pulse incident with no prepulse (figure 1(c)). Indeed, the duration of the x-ray pulse was resolution-limited even for an ASE level of 50% (~1.5 mJ). The reduced intensity of the short-pulse plasma is a result of the limited time resolution of this data (8 ps) compared with the x-ray pulsewidth of ~1 ps. The integrated reflectivity of the pulse focused onto the glass was 50±5%, and of the pulse focused onto aluminum was 67±6%. This compares with a low-intensity reflectivity of the glass of 1.1%; thus a prepulse suppression ratio of ~50 is expected for this configuration. The x-ray yield of the prepulse-suppressed plasma compared to the yield of a plasma created (with no ASE) without focusing onto the prepulse target was 30±2%. This indicates that the focusability of the reflected beam was not affected, since x-ray yield is expected to vary approximately as the intensity to the 1.6 power. This technique is universally applicable for removing pedestals in high-energy ultrashort pulses due to ASE in the laser amplifiers, or pedestals resulting from fiber-grating compression or spectral distortion in chirped-pulse amplification. Brewster angle or antireflection-coated targets can further increase the contrast ratio.

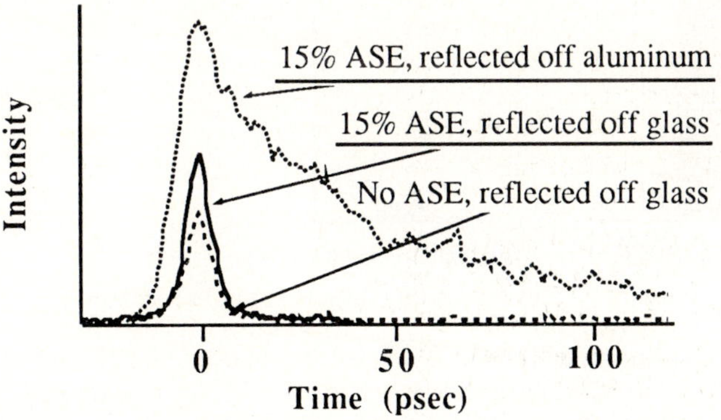

Figure 1. X-ray pulse length for differing target conditions.

3. Structured targets for enhanced laser-light absorption

In a separate experiment, diffraction grating structured targets were used to dramatically increase the laser energy absorbed by a target. The gratings used were lithographically produced with a period of 300 nm and a depth of 150 nm; there is no first-order reflection from these gratings. We observed that rough-surface targets generally have increased x-ray yields; the gratings were used as a well-characterized rough target. As illustrated in Fig. 2, the x-ray yield of the grating targets (for normal incidence) increased by ≈12 over that of a flat target, and was strongly polarization dependent. KeV x-rays, observed through a 5 μm Be filter, were seen using the grating targets but were never present using flat targets. Target reflectivity measurements (the reflected energy was entirely specular) indicate that the enhanced emission is due to enhanced absorption of the laser radiation. The reflectivity of a flat target is approximately 80%, while the reflectivity of the grating was 10% when the incident polarization was perpendicular to the grooves, and 50% for the other polarization. The enhanced absorption can entirely account for the enhanced emission; 4-5 times more laser energy is absorbed by the target, and there is a 1.6 power dependence of x-ray yield on intensity. The enhanced absorption mechanism may be explained as follows. The grating reflectivity can be analyzed in a polarization-dependent effective index approximation. For perpendicular polarization, the effective index of the grating layer is near unity (≈ 1.426) and the surface reflectivity is ~4%. Therefore, the laser energy penetrates into the structure, but the electric field is excluded from the grooves and must therefore be enhanced in the voids. An approximate doubling of the electric field due to a ~50% fill factor corresponds to an effective 4x increase in intensity and deposited energy, assuming constant target area.

In conclusion, we have demonstrated two techniques that will allow the efficient production of solid-density plasmas at intensities of $>10^{18}$ W/cm^2. This may result in plasma temperatures of several keV and x-ray conversion efficiency approaching unity. Such a source can be used for condensed matter, atomic and molecular physics.

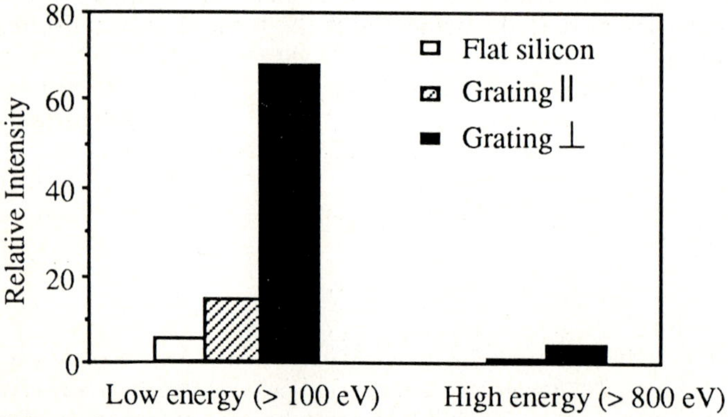

Figure 2. X-ray yield for flat targets and for grating targets for both polarizations.

Acknowledgements. This work is funded by the US AFOSR, and through a collaboration with Lawrence Livermore National Laboratory. M. M. acknowledges support from a UC President's Postdoctoral Fellowship.

Reference. 1. M.M. Murnane, H.C. Kapteyn, and R.W. Falcone, IEEE J. Quan. Electron. **25** (12) p. 2417 (1989).

Proposed Source of Sub-picosecond Far Infrared Radiation

H. Hamster and R.W. Falcone

Department of Physics, University of California at Berkeley,
Berkeley, CA94720, USA

Abstract
We propose that far infrared radiation (FIR) with gigawatt peakpower can be generated from a plasma produced at the focus of a high-intensity, ultra-short laser pulse as a result of the coherent acceleration of electrons through ponderomotive forces. A well defined vacuum-plasma interface will optimize the generated FIR.

Large ponderomotive forces created by a focused, short duration laser pulse, can cause the spatial redistribution of charge carriers in plasmas. This effect has been studied previously [1]. Of particular relevance to this proposal is the case of a quasi-neutral, collisionless plasma which is illuminated by a laser pulse with a duration short enough that the inertia of the massive, positive ions prevents them from moving significantly. The ponderomotive force of focused laser will then cause only electrons to be accelerated and expelled from the focal region. Together with the relativistic mass increase of the electrons, such effects can lead to self focusing [1] of the intense laser light. In this proposal we are concerned with the coherent expulsion of charges from a laser focus leading to intense, broadband far infrared radiation (FIR). The generated FIR field will have an amplitude proportional to the intensity of the laser pulse, and frequencies on order of the inverse of the duration of the laser pulse. Key to this proposal is the observation that the far-field radiation pattern of all accelerated electrons constructively interfere if the geometry of the focused laser pulse and plasma is chosen correctly.

To obtain an estimate for the FIR energy output, we first equate the ponderomotive force with the force due to the space charge field generated from electron-ion separation. For a geometry where the laser pulse is incident on a vacuum-solid interface, with exponential decay of intensity into the surface, the surface charge density, σ, is given by

$$\sigma = \frac{U_{pond}}{4\pi e \delta} \quad (1) \qquad \text{where} \qquad U_{pond} = \frac{2\pi e^2}{m c \omega^2} I \quad (2) \ .$$

U_{pond} is the ponderomotive potential, e is the electron charge, m is the electron mass, c is the speed of light, δ is the absorption depth, I is the intensity of the laser pulse, and ω is the laser frequency. Equation (1) requires that the absorption depth, δ, be much smaller than the radius of the laser focal spot, r_0. The effective charge, q_{eff}, accelerated within the focal region is

$$q_{eff} = \sigma \pi r_0^2 \quad (3) \ .$$

Assuming equilibrium between ponderomotive and space charge forces, we can write the resulting charge acceleration, a, as

Springer Series in Chemical Physics, Vol. 53 **Ultrafast Phenomena VII**
Editors: C.B. Harris · E.P. Ippen · G.A. Mourou · A.H. Zewail

$$a = \frac{2\delta}{[\tau/2]^2} \qquad (4)$$

where τ is the duration of the focused optical pulse. The radiated far infrared power, P, can be obtained from Larmor's formula. If we write the laser intensity, $I = E / (\pi r_o^2 \tau)$, where E is the energy of the incident laser pulse, we then obtain

$$P = \frac{2^5}{3} \frac{e^2}{c^5 m^2} \frac{E^2}{\omega^4 \tau^6} \quad \text{or} \quad P\,(W) = 10^2 \left[\frac{E\,(J)\,\lambda\,(\mu m)^2}{\tau\,(psec)^3} \right]^2 \qquad (5)$$

Assuming a laser with an energy of 1 J, a pulse length of 60 fsec, and a wavelength of 800 nm we obtain a predicted FIR power of 8.5×10^8 W and energy of 60 μJ . A laser capable of producing this power is currently being constructed in our laboratory [2]. Note that an ultra short pulse length is very crucial in achieving high output power.

To obtain the angular, temporal and spectral distributions of the FIR we developed a more sophisticated model. However, calculation of the ponderomotive forces normal to the surface of a solid target is complex and depends on the details of plasma expansion during the laser pulse [3]. We therefore assumed a model with a spatially and temporally Gaussian focused laser pulse incident on a layer of freely moving charges. The total radiative output was calculated by summing the (retarded) potentials from the accelerated charges within the target. The charge distribution along the axis of pulse propagation was chosen to be Gaussian to allow an analytical solution. The results derived from this model are given in Figure (1) - (3) : They show the spatial, temporal, spectral distribution of the FIR.

The use of a thin gas target is appealing from an experimental viewpoint, but does not appear to be feasible since the ultrahigh intensity laser focal volume would then be surrounded by a much larger volume of partially ionized gas due to the spatial wings of the focused laser pulse. This surrounding plasma could be penetrated by the optical pulse, but would be dense enough to absorb the generated FIR. We therefore

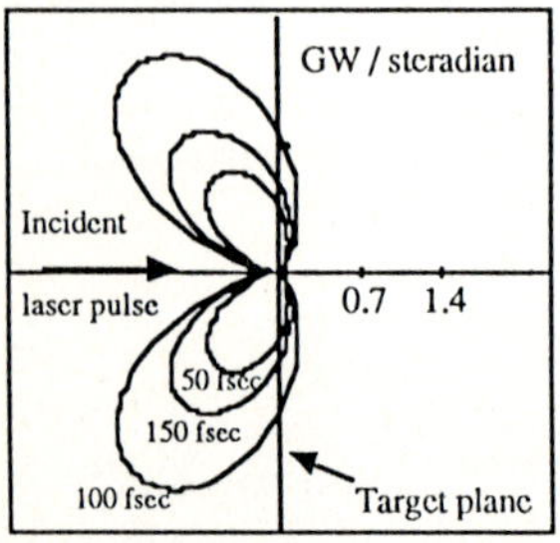

Figure (1) Angular distribution of FIR power. A 60-fsec, 1 J pulse is assumed. The intensity is shown 50, 100 and 150 fsec before or after the pulse center hits the target.

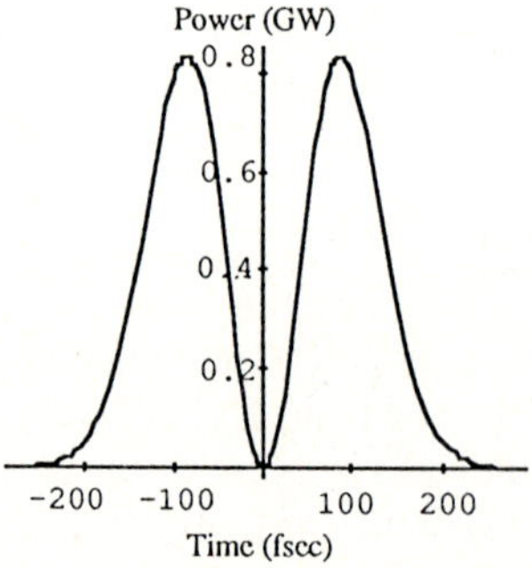

Figure (2) Temporal shape of the same pulse. The FIR field goes as t exp [- (t / τ)^2]

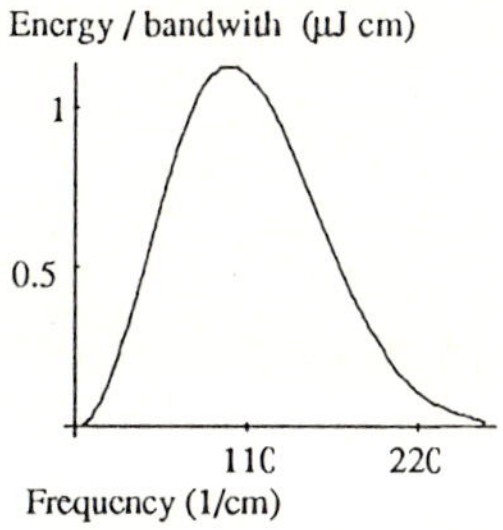

Figure (3) Power spectrum for a 60 fsec, 1 J pulse.

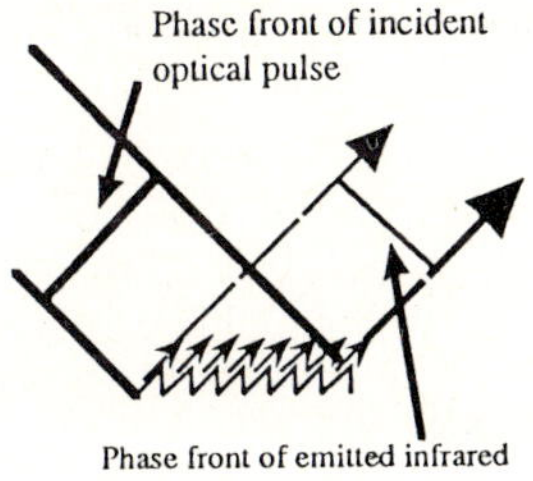

Figure (4) Coherent addition of radiation in a grooved target geometry.

assume that the most optimum geometry will involve an abrupt interface with a solid target. Note that this mechanism does not require a large energy deposition in the target. The large fields that occur in the very steep density profile close to the interface [3] are sufficient for the proposed mechanism.

The predicted pulse energy output of our source is comparable to the energy from infrared free electron lasers [4]. Our proposed source has a pulse duration several orders of magnitudes shorter and will have a synchronized optical pulse for femtosecond cross-correlation experiments.

It appears possible to control the direction of the emitted radiation by coherent addition of fields along a line focus. If one chooses a target which has a grooved structure, as shown in Fig. 4, the radiation will add coherently in a direction parallel to the groove tilt. The phase front of the exciting pulse can also be tilted. It should therefore be possible to direct most of the power into a small solid angle and achieve $dP / d\Omega$ in excess of a gigawatt per steradian. A further spectral and spatial enhancement by control of the source geometry may also be possible.

Acknowledgements
This work was supported by the U.S. Air Force Office of Scientific Research and through a collaboration with Lawrence Livermore National Laboratory under auspices of the U.S. Department of Energy under contract No. W-7405-Eng-48.

References

1. G. Sun, E. Ott, Y. Lee and P. Guzdar: Phys. Fluids 30, 526 (1987); W. Mori, C. Joshi, J. Dawson, D. Forslund, J. Kindel: Phys. Rev. Letters 60, 1298 (1988).
2. H.C. Kapteyn, A. Sulivan, H. Hamster, R.W. Falcone in: Proceedings of the SPIE, Vol. 1229 (1990).
3. R. Fedosejevs, et al: Appl. Physics B 50, 79 (1990) and Phys. Rev. Letters 64, 1250 (1990); J.C. Kieffer, et al: J. Quantum Elect. 25, 2640 (1989); H.M. Milchberg and R.R. Freeman: J. Opt. Soc. Amer. B6, 1351 (1989).
4. Report of the Workshop on Scientific Opportunities for Infrared Free Electron Lasers, Vol.1;Lawrence Berkeley Laboratory, Oct. 31 - Nov. 1, 1988; LBL-26783 1/2 ; CONF-8810126.

Part III

Spectroscopic Techniques and Nonlinear Optics

Picosecond Stimulated Raman Measurement of Enhanced Optical Nonlinearities in a Conjugated Polymer

G.J. Blanchard and J.P. Heritage

Bellcore, Red Bank, NJ 07701, USA

We measure optical Stark effects in a polydiacetylene using picosecond stimulated Raman scattering. Large signal enhancements occur when exciton sidebands are used as Raman intermediate states, in agreement with calculations of the overlap integrals for these transitions.

Conjugated polymers are attractive candidate materials for nonlinear optical signal processing applications. The crystalline polydiacetylene PTS, a model 1-D material, possess values of $\chi^{(3)}$ which compare with those of 2-D III-V multiple quantum well structures (MQWS) [1]. The on-resonance nonlinear optical response of both of these materials is described by phase-space filling and has been examined experimentally [2,3]. In contrast to resonant excitation, the non-resonant nonlinear response of polydiacetylenes and III-V MQWSs are not identical. The phonon-mediated nonlinear responses of each of these materials is determined by how strongly their phonons are coupled to their excitons.

We have used picosecond stimulated inverse Raman spectroscopy to study the fast off-resonance nonlinear response of PTS. In addition to inverse Raman scattering there are several optical Stark effects which contribute to the data [4]. We detect the optical Stark effect on the excitonic transition(s) and the phonon-mediated optical Stark effect on the ground-state phonon-to-exciton transitions. These different effects interfere with one another to produce excitation energy-dependent variations in phonon lineshape which can be modeled quantitatively using the theory of phonon-mediated optical nonlinearities [5]. We find that phonon coupling to a v=1 exciton is enhanced significantly over that for the v=0 exciton.

The spectrometer used in these studies has been described elsewhere [6]. For the light intensities used here $\chi^{(3)}$ provides an adequate first-level description of the data.

$$\chi^{(3)} \propto \frac{\langle 0|\mu^2{}_x|X\rangle\langle X|\mu^2{}_1|1\rangle\langle 1|\mu^2{}_1|X\rangle\langle X|\mu^2{}_x|0\rangle}{(\omega_x-\omega_{pr}-i\gamma_x)^2[\omega_1-(\omega_{pr}-\omega_p)-i\gamma_1]} \qquad (1)$$

The μ's are the transition moments, ω_p, ω_{pr} and ω_1 refer to the pump, probe and phonon frequencies, respectively, 0, 1 and X are the ground state, phonon and exciton wave functions and the γ's are the linewidths. Two limiting cases are instructive in understanding the excitation energy-dependent variations in phonon lineshape. Off resonance, $\omega_p+\omega_1 \ll \omega_x$ and $\omega_{pr}-\omega_p=\omega_1$. The second term in the denominator of Eq. 1 dominates and produces an absorptive resonance in $\mathrm{Im}\{\chi^{(3)}\}$. Near the double resonance, where ω_{pr} is tuned through ω_x and $\omega_p+\omega_1=\omega_x$, the real terms in the denominator of Eq. 1 approach zero yielding a transmissive response. Between these two limiting cases, both terms in the

Springer Series in Chemical Physics, Vol. 53 **Ultrafast Phenomena VII**
Editors: C.B. Harris · E.P. Ippen · G.A. Mourou · A.H. Zewail

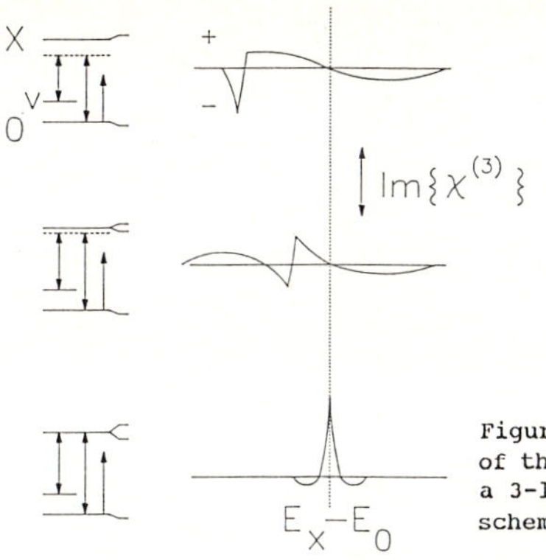

Figure 1: Schematic representation of the $Im(\chi^{(3)})$ responses of a 3-level system for experimental scheme used here.

Table 1: PTS transition cross-sections

phonon	ω (eV)	$\sigma_{0v\to x0}$ (eV)	$\sigma_{0v\to x1}$ (eV)
1+1	0.517	0.09	0.65
2+3	0.333	-	0.20
2+4	0.302	0.06	0.55
2+2	0.368	0.12	0.55

denominator contribute to produce dispersive lineshapes, as schematized in Fig. 1. While Eq. 1 is useful in understanding the data intuitively, we have chosen the more complete theory due to Schmitt-Rink for modelling purposes [5]. This theory calculates the <u>total</u> susceptibility experienced by the probe due to the presence of the pump and accounts for many-body effects. We include the complex dielectric response of PTS and multiple phonons and excitons (via linear superposition) in our calculations [6]. Initially, this model has two adjustable parameters; the sample thickness and the cross-section for the phonon-to-exciton transition. The sample thickness is determined using transitions for which the cross-sections are known [7]. The cross-sections for other transitions are then fitted using this same sample thickness. The cross-sections for the ω_{2+2}, ω_{2+3} and ω_{2+4} phonon-to-exciton transitions are substantially larger for the ω'_2 exciton than for the fundamental (v'=0) exciton. These data are presented in Fig. 2 and their best-fit cross-sections in Table 1.
The dash in the table indicates that the cross-section was too small to be measured. X_0 indicates the v'=0 exciton and X_1 the ω'_2 exciton. The phonon numbering scheme follows Ref. 7. These results can be understood using a harmonic oscillator model. Assuming Franck-Condon transitions, the ratio of the phonon-to-exciton cross-sections is given by the square of the ratio of the overlap integrals for these transitions. We express this ratio as a function of the displacement between the ground state and excitonic potential wells,

$$\frac{\sigma_{0v-X_1}}{\sigma_{0v-X_0}}=\left(\frac{[z^2-v]}{z}\right)^2$$

for v = 1,2. Z is a dimensionless quantity related to the well displacement, D the phonon energy, f, and the reduced mass, m;

$$z=\left(\frac{2\pi^2mf}{h}\right)^{1/2}D$$

For modest values of z the cross-section ratio is expected to exceed unity. It is thus possible to determine the well displacement, D, from the experimental values of the cross-section ratios. For the ω_{2+2} phonon , D = 0.0024±0.0002 nm; D = 0.0021±0.0001 nm for ω_{2+4} and 0.0022±0.0002 nm for ω_{1+1}. These values are very similar to those measured for different phonons in PTS using spontaneous resonance Raman spectroscopy [7].

Stimulated inverse Raman scattering is a very sensitive probe of several fundamental spectroscopic properties. We

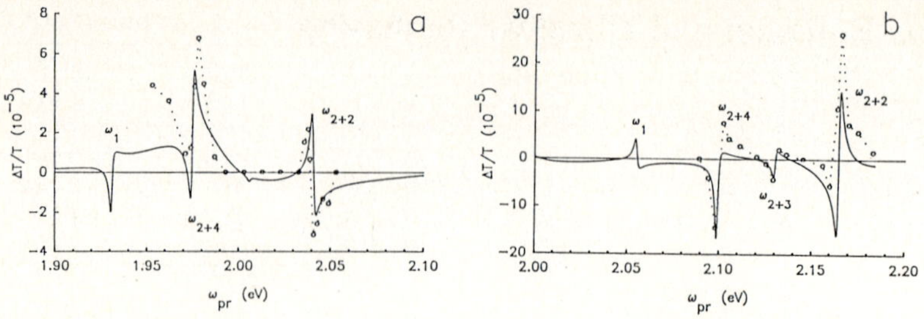

Figure 2: Comparison of experiment (points) to the theory of phonon-mediated optical nonlinearities (lines) [5]. (a) ω_{pump}=1.673 eV, (b) ω_{pump}=1.797 eV.

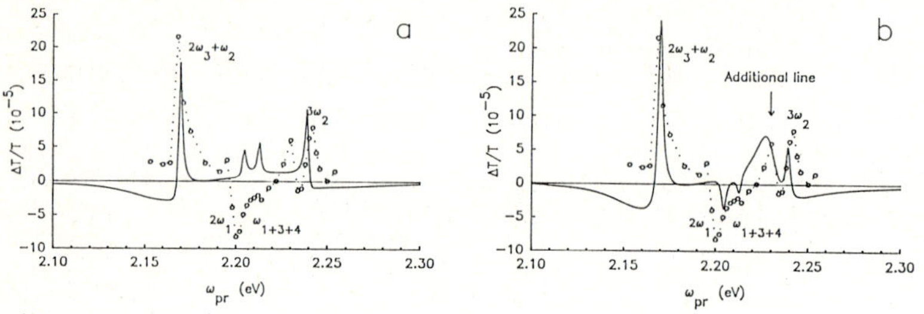

Figure 3: Comparison of experiment (points) to theory. (a) Calculation including 11 sidebands of the ^1B_u exciton. (b) Same as (a) but including a narrow line (γ=8 meV) at 2.23 eV, believed to be an exciton different from the ^1B_u exciton.

obtain quantitative agreement between theory and experiment except near 2.2 eV (see Fig. 3). There is disagreement between theory and experiment despite the use of 11 exciton sidebands to model the (^1B_u) exciton manifold (Fig. 3a). We recover quantitative agreement with the data only when an additional, narrow line centered at 2.23 eV is introduced (Fig. 3b). PTS is known to have an excitonic feature near 2.2 eV [8], thought to be of ^1A_g symmetry - the same as the ground state. We believe this is the feature we are sensing, but that it is not of pure ^1A_g symmetry. We should not be able to detect the presence of this state if it is of pure ^1A_g symmetry. Perhaps it is mixing with states from the ^1B_u exciton manifold. The fact that ground state phonons are coupled to this exciton indicates that it occupies a spatial region overlapping the ^1B_u exciton. Further investigation is in order to better understand the origin of this state.

References

1. A. Von Lehmen, D. S. Chemla, J. E. Zucker, J. P. Heritage, Opt. Lett., 11, 609 (1986).
2. S. Schmitt-Rink, D. S. Chemla, D. A. B. Miller, Advances in Physics, in press.
3. B. I. Greene, J. Orenstein, R. R. Millard, L. R. Williams, Phys. Rev. Lett., 58, 2750 (1987).
4. D. Frolich, A. Nothe, K. Reimann, Phys. Rev. Lett., 55, 1335 (1985).
5. B. I. Greene, J. F. Mueller, J. Orenstein, D. H. Rapkine, S. Schmitt-Rink, M. K. Thakur, Phys. Rev. Lett., 61, 325 (1988).
6. G. J. Blanchard, J. P. Heritage, A. Von Lehmen, M. K. Kelly, G. L. Baker, S. Etemad, Phys. Rev. Lett.,63, 887 (1989).
7. D. N. Batchelder, D. Bloor, J. Phys. C, 15, 3005 (1982).
8. D. Bloor, D. J. Ando, P. A. Norman, J. S. Obhi, P. V. Kolinsky, B. Movaghar, Physica Scripta, T19, 226 (1987).

Spontaneous Emission and Ultrafast Dephasing in Front of a Mirror

K.E. Drabe, G. Cnossen, and D.A. Wiersma

Department of Chemistry, Ultrafast Laser and Spectroscopy Laboratory, University of Groningen, Nijenborgh 16, 9747 AG Groningen, The Netherlands

Abstract. We show that the emission spectrum of Nile Blue in front of a mirror, exhibits Wiener–type fringes that decay with a 100 femtoseconds coherence lifetime.

We report a new technique to probe femtosecond dephasing processes in solids, requiring only low excitation intensities. The method is based on the measurement of the *spontaneous emission* spectrum of a thin sample (about 2μm) in front of a mirror, excited by 40 fs pulses from a colliding pulse modelocked (CPM) laser. Fig. 1A shows the simplified level scheme of a prototype molecule, and fig.1B gives the schematics of the experimental arrangement.

The presence of a mirror has two effects. Firstly, it leads to the creation of a transient Wiener fringe pattern [1]; secondly, it leads to an interference effect in spontaneous emission. It can be shown [2] that for an emission spectrum that is predominantly inhomogeneously broadened, the combination of the two effects leads to a spectrum of the form

$$I(\omega) \propto I^{e}_{inhom}(\omega)\ [1 + \tfrac{1}{2}\exp\{-(\Gamma_{31}+\Gamma_{21})t_{21}\}\cos\kappa(L_1+L_2)\,\mathrm{sinc}K(L_2-L_1)] \qquad (1)$$

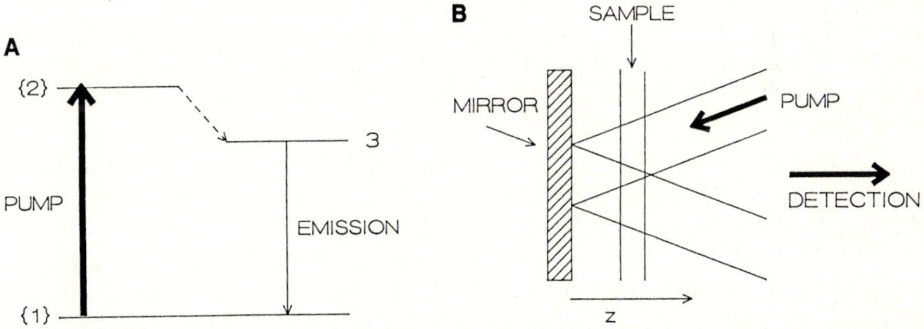

Fig.1 (A) Relevant level scheme for Nile Blue. {1} and {2} denote vibrational manifolds in the ground and excited state, while **3** stands for the vibrationless excited singlet state.
(B) Schematics of the experimental arrangement. Note that the emission is detected in a direction perpendicular to the mirror.

Springer Series in Chemical Physics, Vol. 53 **Ultrafast Phenomena VII**
Editors: C.B. Harris · E.P. Ippen · G.A. Mourou · A.H. Zewail

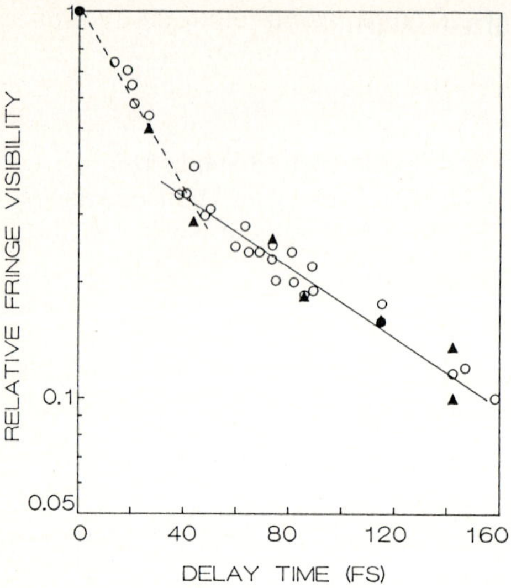

Fig.2 Relative fringe visibility in the emission of Nile Blue doped in a polymer film as a function of delay between the incident and reflected pulse from the mirror. The delay times were calculated from the modulation period in emission and use of eq. (1). The open circles are data points obtained using the CPM laser; triangles with a stochastic excitation source.

Here $I^e_{inhom}(\omega)$ is the inhomogeneously broadened emission spectrum, $\kappa = (n/c)[(\omega_p+\omega-\Omega^e_0)\cos\nu_p-\omega]$, $K = k_{zp}-k$, and sincx=sinx/x. $k_{zp} = (\omega_p \cos\nu_p)n/c$ is the projection of the wavevector of the pump beam on the normal to the mirror and k is the absolute value of the wave vector in emission. Γ_{21} is the homogeneous linewidth of the excited **2←1** vibronic transition and Γ_{31} the homogeneous linewidth of the **3←1** zero–phonon transition. L_1 represents the distance between the sample and the mirror and (L_2-L_1) the thickness of the sample. For a "homogeneously" broadened emission spectrum we obtain an expression of the same form as eq. (1) but with a damping term $\exp(-\Gamma_{21}t_{21})$ instead of $\exp\{-(\Gamma_{31}+\Gamma_{21})t_{21}\}$ and with $\kappa \equiv K = k_{zp}-k_z$.

The experiments were performed on thin (about 2μm thick) polymer films, prepared by spincoating a solution of poly(4– vinyl)pyridine doped with the fluorescent molecule Nile Blue.

Fig. 2 shows a plot (open circles) of the *relative* fringe intensity (I_{CPM}/I_{HeNe}) in emission as a function of distance between the emitting layer and the mirror. For pulse delay times greater than 40 fs , the decay of the fringe intensity is fit to an exponential with a dephasing time constant (T_2) of 100 ± 10 fs (solid line in fig. 2). Since the detected emission arises from an incoherently populated state, the measured decay is not complicated by quantum beats as in a fs photon echo decay experiment [3]. The same decay was obtained using a stochastic excitation source, obtained from a broad–band ns dye laser pulse.

Regarding the question of what the relationship is between the measured optical dephasing time of 100 $\pm$ 10 fs and a "relaxation time" of 7 fs, calculated from the 1450 cm^{-1} "homogeneous" width in the emission spectrum, we propose that the apparent "homogeneity" of the emission spectrum is caused by overlap of spectra from many different initial vibronic states.

References

1. O.Wiener, Ann. d.Physik, **40**,203(1890).
2. K.E.Drabe, G.Cnossen and D.A.Wiersma, submitted for publication.
3. P.C.Becker, H.L.Fragnito, J.Y.Bigot, H.Brito Cruz, R.L.Fork and C.V.Shank, Phys.Rev.Lett., **63**, 505(1989).

Polariton Effects in Transient Grating Experiments

J. Knoester

Department of Chemistry, University of Groningen, Nijenborgh 16,
9747 AG Groningen, The Netherlands

Abstract. A theory for the transient grating signal in low temperature crystals is presented that microscopically accounts for the role of polariton transport.

1. Introduction

The traditional approach for calculating nonlinear optical signals is based on performing response theory of the material system with respect to the external laser (electric) fields. The optical response is then characterized by the eigenstates of the material system with instantaneous Coulomb interactions. For a molecular crystal, these eigenstates are the Coulomb excitons. This approach, however, neglects the transverse internal radiation field (retardation) [1], which is a quantum degree of freedom and cannot be switched off experimentally. Therefore, in low temperature pure crystals, the proper eigenmodes are those of the coupled material–radiation system: polaritons [2–4]. The polaritons are coupled to the external fields through the boundary conditions on the crystal's surface and in a proper description of the optical experiment, we should consider the response of the polariton system to these external fields [1]. The optical signal then contains characteristics of polaritons rather than excitons. Recently, such explicit signatures of retardation have received much experimental and theoretical attention [3–7]. In the present study, we focus on the role of polaritons in transient grating (TG) experiments on molecular crystals.

2. Theory and Results

The conventional interpretation of the TG is that it probes the motion of excitons during the time delay (τ) between creation of the grating by two pump pulses (with wave vectors $\mathbf{k}_1$ and $\mathbf{k}_2$) and detection of the grating by a third pulse ($\mathbf{k}_3$). As during the time interval τ no external electric field exists, the conventional microscopic description of the grating's decay is based on solving the evolution of the

Springer Series in Chemical Physics, Vol. 53 **Ultrafast Phenomena VII**
Editors: C.B. Harris · E.P. Ippen · G.A. Mourou · A.H. Zewail

isolated exciton system furnished with some scattering model [8]. Following the observation of abnormally high diffusion constants in TG's on Anthracene crystals [5], Agranovich was the first to note that at low temperatures the grating decay is probably better understood by considering propagation of polaritons [6], as these may have much larger group velocities than excitons. However, a theory for the TG signal that incorporates polaritons from first principles has so far not been published.

To give such a theory, we apply a recently proposed hierarchy of equations of motion for polariton operators [1]. To explain nonlinear effects, it is essential that we use the full *non-bosonic* commutation relations for the polaritons when deriving this hierarchy. These commutators are not trivial, because molecular excitons are not bosons. In fact, the polariton commutators are proportional to the exciton population operator. Using a factorization approximation to close the hierarchy, it then follows that, as in the traditional theory, the transient grating signal at wave vector $\mathbf{k}_1-\mathbf{k}_2+\mathbf{k}_3$ is determined by the expectation value of the *exciton* population operator at the grating wave vector $\mathbf{K}\equiv\mathbf{k}_1-\mathbf{k}_2$ [9]. The important differences with the traditional approach are the following. (i) The initial condition for the grating is a coherence between two polariton states rather than two exciton states. (ii) During the delay time τ, the internal radiation field is not neglected and the evolution of the full polariton system is considered. An important step in our theory is then to solve coupled equations of motion for the coherences $\langle\xi^\dagger(\mathbf{q},t)\xi(\mathbf{K}+\mathbf{q},t)\rangle$ during the delay time. Here $\xi^\dagger(\mathbf{k})$ [$\xi(\mathbf{k})$] is the creation [annihilation] operator for the polariton with wave vector $\mathbf{k}$ and $\langle..\rangle$ indicates the expectation value. The coupling between coherences at different $\mathbf{q}$ values results from scattering of the polaritons on phonons; the crystal's translational symmetry guarantees that only coherences of the above type (with total wave vector $\mathbf{K}$) couple to each other. In absence of scattering, the different coherences are decoupled, the polariton motion is coherent, and the grating signal does not decay.

In a simple model, we assume equal scattering rates (cf. the Haken–Strobl model for dephasing [8]) between all polaritons with an exciton component greater than 1/2, whereas the other polaritons are assumed not to take part in the scattering process. The rationale for this is that it is the exciton component of the polaritons that couples them to the phonons and induces the scattering. In the limit of strong scattering, we find diffusive behavior leading to a transient grating signal decay

$$S(\tau) = S(0)\exp[-K^2 v_{eff}^2 \tau/\Gamma], \tag{1}$$

with Γ the total scattering rate out of any polariton coherence (dephasing rate) and $K\equiv|\mathbf{K}|$. The most important parameter in Eq. (1) is

v_{eff}, which can be expressed in terms of the polariton dispersion relation [9] and may be interpreted as an effective polariton group velocity. Rough estimates for aromatic crystals yield values for v_{eff} of 10^5cm/s and 10^6cm/s for the polariton belonging to the lowest singlet **a**–excitons of Naphthalene and Anthracene, respectively. These values are much higher than typical exciton group velocities, thus predicting an acceleration of the TG decay.

An alternative way to model the scattering process, is to assume that the initially created polaritons rapidly relax towards a bottleneck region in the lower polariton curve. This idea was put forward by Agranovich [6] and is probably of great relevance in the case of Anthracene. Our microscopic theory can be applied to this model and it is possible to find an exact solution for the TG signal at arbitrary scattering strength [9]. In the limit of strong scattering, the solution is independent of the initial condition right after the relaxation to the bottleneck and we recover Eq.(1), where v_{eff} is now the polariton group velocity in the bottleneck region. This result gives microscopic support to Agranovich's interpretation of the TG's on low temperature Anthracene [5,6].

We conclude that the polariton hierachy proposed in Ref.[1], is a useful tool in evaluating optical response in low temperature crystals. We have applied this approach to describe transient gratings and shown from first principles how the polariton dispersion curve (group velocity) enters the observed signal. In general, polaritons will accelerate the transient grating decay.

References

[1] J. Knoester and S. Mukamel, Phys. Rev. A**41**, 3812 (1990).
[2] J.J. Hopfield, Phys. Rev. **112**, 1555 (1958).
[3] S.H. Stevenson, M.A. Connolly, and G.J. Small, Chem. Phys. **128**, 157 (1988).
[4] J. Knoester and S. Mukamel, J. Chem. Phys. **91**, 989 (1989).
[5] T.S. Rose, R. Righini, and M.D. Fayer, Chem. Phys. Lett. **106**, 13 (1984).
[6] V.M. Agranovich, A.M. Ratner, and M. Salieva, Solid State Comm. **63**, 329 (1987).
[7] E. Hanamura, Phys. Rev. B**39**, 1152 (1989).
[8] See, e.g., D.K. Garrity and J.L. Skinner, J. Chem. Phys. **82**, 260 (1985); R.F. Loring and S. Mukamel, J. Chem. Phys. **83**, 4353 (1985); R.F. Loring and S. Mukamel, J. Chem. Phys. **84**, 1228 (1986).
[9] J. Knoester, J. Chem. Phys. (to be submitted).

Atomic Wave Packets and Optical Pulses

L.D. Noordam[1], *A. ten Wolde*[1], *H.G. Muller*[1], *A. Lagendijk*[1,2], *T.F. Gallagher*[3], *and H.B. van Linden van den Heuvell*[1]

[1]FOM Institute for Atomic and Molecular Physics,
Kruislaan 407, 1098 SJ Amsterdam, The Netherlands
[2]Natuurkundig Laboratorium of the University of Amsterdam, The Netherlands
[3]University of Virginia, Charlottesville, VA22901, USA

Abstract. In this contribution we will consider the similarities between atomic electron wave packets and short optical pulses. It will be shown that well known effects in optics, such as dispersion, modulation of the group velocity, and shortening of pulses, can also occur for matter waves.

1. Introduction

In a short pulse laser a light packet is constructed by coherent superposition of eigen modes of the laser cavity. An atomic electron wave packet is a coherent superposition of atomic eigenstates. As in the optical case the electronic wave packet will broaden if the group velocity is not constant for all frequencies (Sec. 2). If the index of refraction is constant the wave packet will be dispersion free (Sec. 3). Modulation of a wave packet can be realized by modulating the atomic frequencies (Sec. 4). Finally, in Sec. 5, we show that also for matter waves there are nonlinear effects opening the possibility to shorten the wave packet.

2. Rydberg wave packet [1-3]

We created a radial wave packet by coherent excitation of several Rydberg states of atomic Rb by a ps laser pulse. The wave packet starts to oscillate in the radial direction between the classical turning points. The period of this oscillation is determined by the spacing between the energy levels. The wave packet can be probed by ionization, since ionization of an electronic wave packet is most likely near the core. This offers the possibility of performing a time-resolved pump-probe experiment. The first pulse creates a radial wave packet, which can be ionized by a second laser pulse. In this way we observed two oscillations of the wave packet. After that the dispersion of the wave packet became too large, due to the fact that the spacing between adjacent Rydberg states varies with the principal quantum number n: $\Delta E \approx 1/n^3$.The different frequency components have different oscillation times and the wave packet disperses.

3. Parabolic wave packet [4-5]

In order to create a dispersion free wave packet, the spacing between adjacent atomic eigenstates must be constant. This is the case in an electric field. If a hydrogen atom is placed in a dc electric field, each degenerate n-manifold splits into n parabolic levels labeled with k. The spacing between subsequent k-states is constant: $\Delta E = 3Fn$. This means that a dispersion-free wave packet can be created by coherent excitation of several parabolic states. The created parabolic wave packet will have an oscillation time of $\tau = 2\pi/3Fn$. The superposition leads to localization of the angular momentum l. This angular momentum of the electron wave packet oscillates between $l = 0$ and $l = n$-1. The oscillation can also be probed by ionization, since the ionization probability is strongly l-dependent. In a pump-probe experiment in Rb we now observed ten oscillations of the parabolic wave packet (fig. 1), showing that it has much less dispersion than the radial wave packet. In the case of a parabolic wave packet all the frequencies have the same oscillation time and there is almost no spreading.

Springer Series in Chemical Physics, Vol. 53 **Ultrafast Phenomena VII**
Editors: C.B. Harris · E.P. Ippen · G.A. Mourou · A.H. Zewail

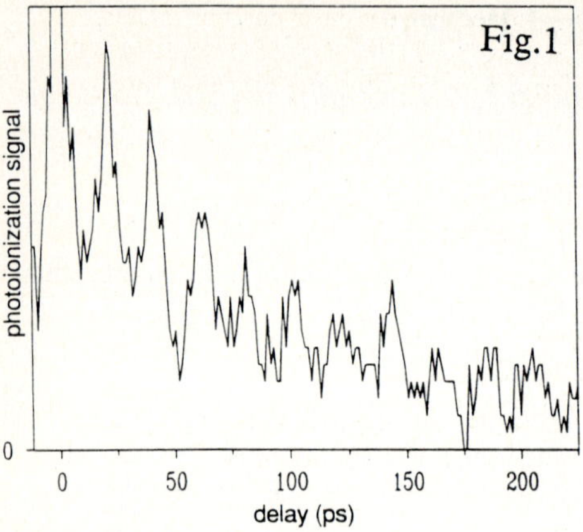

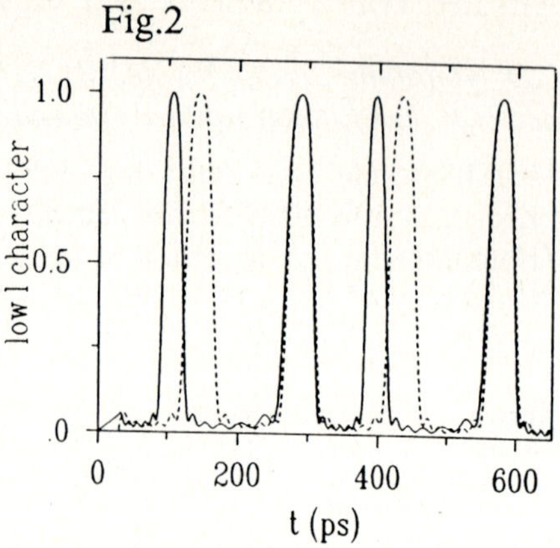

Fig.1: Oscillations of a parabolic wave packet of n=19 in an electric field of 645 V/cm. There is almost no dispersion of this wave packet.
Fig.2: Oscillations of a parabolic wave packet of n=9 in a static electric field of 200 V/cm, modulated with a field of 100 V/cm.The oscillation time and width of the wave packet are no longer constant.

4. Modulation of a parabolic wave packet [6]

Consider the propagation of a short optical pulse in a cavity. By modulating the refractive index of the dielectric, the oscillation time and the length of the optical pulse can be varied. The same type of behavior can occur with matter waves. Modulating the electric field in the case of a parabolic wave packet, gives rise to variations in oscillation time and width of the matter waves. (Fig.2)

5. Shortening of matter waves [7]

The two ingredients on which compression schemes are based are a nonlinear process to increase bandwidth and a dispersive delay to bring the thus generated frequencies into phase. We calculated that a Rydberg wave packet can be used for both increase of bandwidth and compression. If the wave packet is excited by an intense laser pulse the states shift and more states will be excited than one would assume on basis of the bandwidth alone. The bandwidth of the matter waves exceeds the laser bandwidth. This chirped wave packet compresses itself since different frequencies have different oscillation times. We study these effects by calculating the ionization probability of the wave packet: this is equivalent to the width of the wave packet after returning to the core. Figure 3 shows

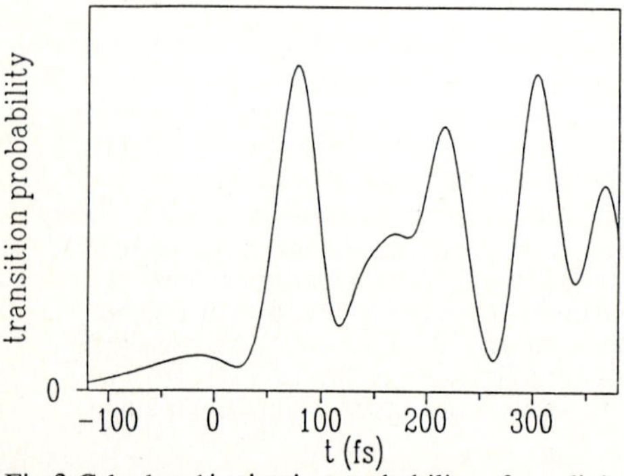

Fig 3:Calculated ionization probability of a radial wave packet as a function of time. The peak intensity of the excitation pulse is $1.5 \cdot 10^{13}$ W/cm^2. The duration of the first return to the core of the wave packet (44 fs) is six times shorter than the excitation pulse (275 fs).

that the first return of the wave packet is six times shorter than the pulse duration. Due to the AC Stark shift the response of the wave packet depends on the light intensity, inducing additional frequencies (for optical pulses this is realized by the well known self phase modulation in a fiber). Compression of the chirped matter waves is realized by the varying spacing of the Rydberg states (the Tracy grating pair for optical pulses).

[1] L.D. Noordam et al, J. Phys. B **21**, L533 (1988);
[2] A. ten Wolde et al, Phys. Rev. Lett. **61**, 2099 (1988);
[3] J.A. Yeazell et al, Phys. Rev. A. **40**, 5040 (1989);
[4] A. ten Wolde et al, Phys. Rev. A. **40**, 485 (1989);
[5] L.D. Noordam et al, Phys. Rev. A. **40** , 6999 (1989);
[6] L.D. Noordam and T.F. Gallagher, submitted for publication;
[7] L.D. Noordam et al, to be published in J. Phys. B.

High Time Resolution and Coherence Effects with Incoherent Light in the Raman Fringe Decay

M. Müller, K. Wynne, and J.D.W. van Voorst

Laboratory for Physical Chemistry, University of Amsterdam, Nieuwe Achtergracht 127, 1018 WS Amsterdam, The Netherlands

Abstract. Using a new technique - Raman Fringe Decay - ultra high time resolution is obtained with incoherent light and detuning oscillations are observed for narrow bandwidth detection.

Since 1984, many theoretical and experimental studies have shown that the time resolution in certain four-wave-mixing experiments (e.g. degenerate four-wave-mixing, coherent Stokes Raman scattering) is determined by the coherence time of the pulses rather than their widths [1]. In this paper it is shown that this concept is also valid for a new experimental technique: the Raman Fringe Decay (RFD) [2]. Furthermore it is shown that the interpretation of the results becomes more complex when the detection bandwidth is small compared to the inverse dephasing time due to coherence effects.

In a RFD experiment the probe pulse is an exact replica of the excitation pulse and both consist of a Laser pulse (frequency ω_L) and a Stokes pulse (frequency ω_S), the frequency difference of which equals the Raman transition frequency ($\Omega = \omega_L - \omega_S$). The signal is measured at the anti-Stokes frequency (ω_{AS}) as a function of the time delay between excitation and probe. It is found that the signal shows oscillations (Raman fringes) as a function of the delay with a frequency corresponding to the Raman transition (i.e. with a period of approximately 50 fs for the experiments described here). In experiments on the 524 cm^{-1} mode of liquid methyliodide and the 656 cm^{-1} mode of liquid carbondisulfide, it is found that in the case of broadband detection (detection bandwidth $\gamma >> 1/T_2$) and resonant excitation ($\omega_L - \omega_S = \Omega$) the amplitude of the fringes decays, as a function of the delay, with the vibrational dephasing time T_2 (see figure 1). When - instead of parallel polarization of excitation and probe - the polarization of the probe pulse pair is made orthogonal to that of the excitation pulse pair, the signal becomes asymmetric and the effect of the limited pulse width (Γ_p^{-1}) upon the decay, when $T_2 > \Gamma_p^{-1}$, disappears. Both in extensive theoretical calculations and in the experiments it is found that the time resolution for cross-polarization experiments is determined by the correlation time of the pulses only, both for dephasing times much longer

Springer Series in Chemical Physics, Vol. 53 **Ultrafast Phenomena VII**
Editors: C.B. Harris · E.P. Ippen · G.A. Mourou · A.H. Zewail

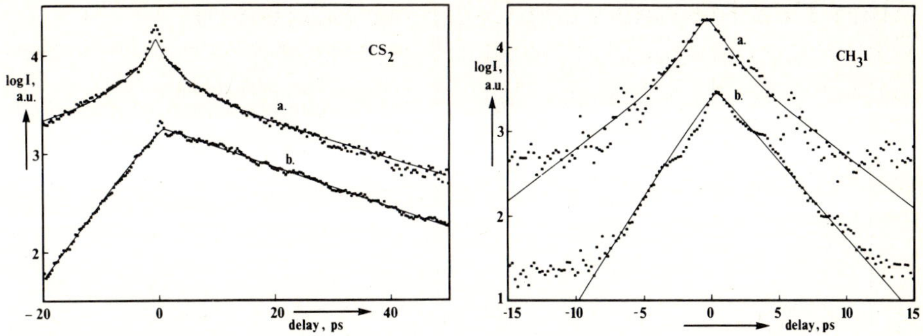

Figure 1. Experimental data for parallel and orthogonal excitation in RFD experiments with broadband detection on CS_2 (left) and CH_3I (right). The solid curves are least-squares fits of the theoretical calculations to the data. The parameters are: for CS_2 (left): (a) $\Gamma_p = 0.23(2)$, $T_2^{-1} = 0.0487(7)$ and (b) $\Gamma_p = 0.141(1)$, $T_2^{-1} = 0.0469(2)$; and for CH_3I (right): (a) $\Gamma_p = 0.09(5)$, $T_2^{-1} = 0.39(5)$; the data of (b) have not been fitted because of the significant departure from exponential decay for small τ, due to a malfunctioning of the analog-to-digital converter. The parameters used for the solid curve are $\Gamma_p = 0.19$, $T_2^{-1} = 0.42$. All values are given in ps^{-1}.

and much shorter than the pulse width. No coherence spike (as in coherent Stokes Raman scattering) is found, enabling a straightforward determination of the dephasing time.

In the case of white detection (detection bandwidth $\gamma >> 1/T_2$) and detuned excitation ($\omega_L - \omega_S = \Omega + \Delta$), detuning oscillations are observed in the signal, with a beat frequency equal to the detuning Δ. For monochromatic detection ($\gamma << 1/T_2$) these detuning oscillations become much more pronounced and become dependent on the detection detuning ($\delta = 2\omega_L - \omega_S - \omega_{det}$). In this case the beat frequency of the signal is given by the relation $2\delta + \Delta$ (see figure 2). Indeed, detuning oscillations have been observed even for resonant excitation ($\Delta = 0$; $\delta \neq 0$).

All the observed features discussed above are neatly described by third order perturbation theory, analytically solved including the effects of finite pulse widths.

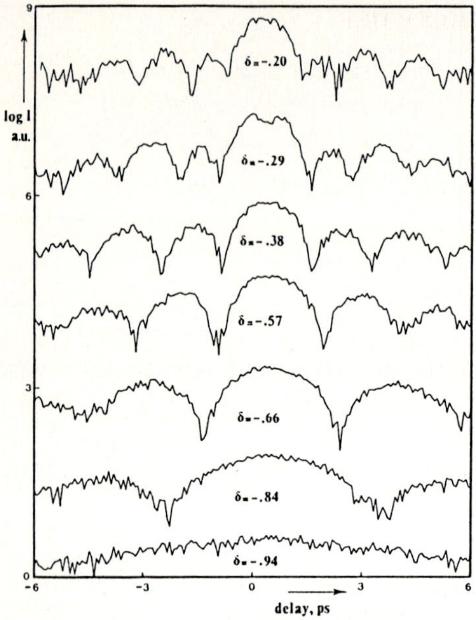

Figure 2. Raman fringe amplitude as function of the delay between excitation and probe for the 656 cm^{-1} mode in carbondisulfide with pseudo-monochromatic detection (γ = 0.04 ps^{-1}; T_2= 0.05 ps^{-1}) for different values of the detection detuning δ (in ps^{-1}). The Raman detuning Δ was set at 0.42 ps^{-1}.

Acknowledgement. The research reported here was partially supported by SON (Stichting Scheikundig Onderzoek Nederland) and NWO (Stichting Nederlands Wetenschappelijk Onderzoek).

References:

1. See among others:
   N. Morita, T. Yajima, Phys. Rev. **A30**, 2525 (1984)
   R. Beach, D. DeBeer, S.R. Hartmann, Phys. Rev. **A32**, 3467 (1985)
   T. Hattori, A. Terasaki, T. Kobayashi, Phys. Rev. **A35**, 715 (1987)

2. K. Wynne, M. Müller, J.D.W. van Voorst, Phys. Rev. Lett. **62**, 3031 (1989)
   K. Wynne, M. Müller, J.D.W. van Voorst, Accepted for publication in Phys. Rev. **A**;
   M. Müller, K. Wynne, J.D.W. van Voorst, Accepted for publication in J. Opt. Soc. Am. **B**.

Fluorescence Detection of Femtosecond Accumulated Photon Echo

K. Uchikawa[1], *H. Ohsawa*[1], *T. Suga*[1], *A. Furusawa*[1], *and S. Saikan*[2]

[1]Nikon Corporation, Nishi-Ohi, Shinagawaku, Tokyo 140, Japan
[2]Department of Physics, Faculty of Science, Osaka University, Toyonaka, Osaka 560, Japan

The photon echo measurement in opaque or scattering materials is generally difficult since the scattered pump light overloads the detector. In this sense, a different modulation technique which does not induce the intensity modulation of the pump beam is desirable. Recently, Saikan et al. demonstrated that phase modulation method is also applicable to the measurement of accumulated photon echo[1]. This technique has been much improved by using a piezo-electric actuator as the phase modulator, and fluorescence detection of the accumulated photon echo has been succeeded.

In the heterodyne detection of accumulated photon echo, the probe light monitors the rephased nonlinear polarization after the excitation of three optical pulses. In this case, the probe light also induces the population changes in the ground and excited states. Strictly speaking, these population changes correspond to the fourth-order diagonal elements of the density matrix. Therefore, if we observe the fluorescence associated with the population change in the excited state, the fluorescence detection of photon echoes is possible. Based on this principle, Zewail et al. previously succeeded the fluorescence detection of the transient photon echo in a gaseous sample[2]. However, their technique is not available to condensed matter due to the ultrafast phase relaxation time. The present phase modulation technique enables fluorescence detection of femtosecond accumulated photon echo even in opaque samples.

As is well known, the heterodyne detected accumulated photon echo can be interpreted in terms of write-in and read-out processes[3]. Here, the term write-in process means the second-order population change induced by the time-delayed two optical fields. On the other hand , the read-out process is associated with the heterodyne detection of the nonlinear polarization. The accumulation means that the former process should be time-averaged over the life time of the bottleneck state. Hence, so long as the modulation frequency of the pump light is higher than the decay rate of the bottleneck state, the write-in process is not affected by the phase modulation. However, the read-out process is fast enough to follow the phase modulation. Considering the above difference between the two processes, we find that the phase modulation of the pump beam induces the intensity modulation in the fluorescence through the fourth-order population change of the excited state. Usually, we have detected the second harmonic component in the lock-in detected fluorescence signal although the even-order higher harmonic components are also available for the measurement of the photon echo.

A kiton red dye laser, which was synchronously pumped by a mode-locked Nd:YAG laser, was used in this experiment. The field correlation time of this laser was about 40 fsec. The sample employed in this experiment was oxazine-4 doped in polyvinylbutyral. The absorption spectrum of this sample and the laser power spectrum are shown in Fig.1(a) and 1(b), respectively. Fluorescence spectrum suffering from the scattering of the laser light is also shown in Fig.1(c). In the fluorescence detection of accumulated photon echo, the scattered laser light was eliminated by using a sharp cut filter (HOYA R66) which was positioned in front of the photo-

Springer Series in Chemical Physics, Vol. 53 **Ultrafast Phenomena VII**
Editors: C.B. Harris · E.P. Ippen · G.A. Mourou · A.H. Zewail

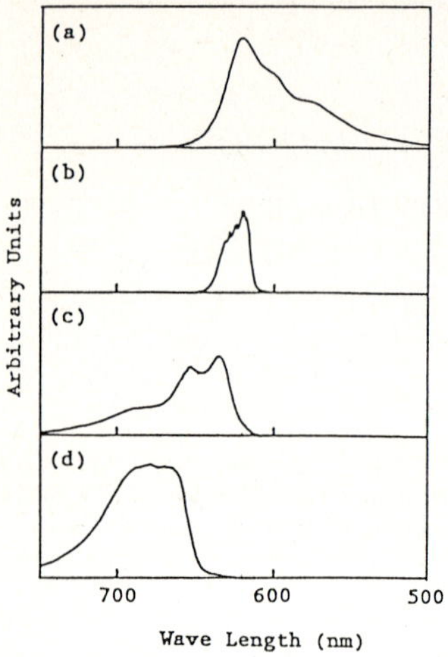

Fig.1 (a) Absorption spectrum of Oxazin-4 doped in PVB; (b) Laser power spectrum; (c) Fluorescence spectrum suffering from the scattered laser; (d) Fluorescence spectrum observed behind the sharp cut filter

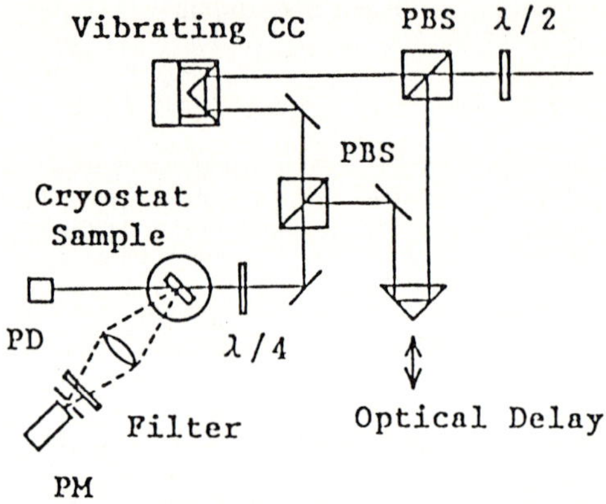

Fig.2 Experimental arrangement

multiplier. Filtered fluorescence spectrum is shown in Fig.1(d). Using this laser, we measured the fluorescence decay time and the absorption recovery time in this sample, which were found to be 5 nsec and 10 msec, respectively. This result indicates that the phase modulator should be driven at the frequency ranging from 1 KHz to 10 MHz for the fluorescence detection of the photon echo. Before the experiment, two types of phase modulators had been tested, one type was an electro-optical modulator consisting of $LiTaO_3$ crystal, and the other type was a vibrating corner cube prism glued on a piezo-electric actuator. Although the former modulator works at higher modulation frequency, which is useful to suppress the laser noise, the so-called residual amplitude modulation (RAM) remained substantially[4]. On the other hand, detectable RAM was absent in the latter modulator. In this experiment, the vibrating corner cube prism was adopted as the phase modulator. The modulation frequency was 20 KHz.

The experimental set-up is depicted in Fig.2. The two echo signals generated in the transmitted light and the fluorescence are simultaneously detected. In order to eliminate the influence of the polarization dependent fluorescence yield, a quarter wave plate has been inserted in the collinear pump and probe beams to convert the orthogonal linear polarization into counter rotating circular polarization. This was necessary to obtain the symmetrical decay curves of the photon echoes, as shown in Fig.3. Figures 3(a) and 3(b) show, respectively, the fluorescence detected and transmitted light detected photon echoes. Temporal behavior of the echo signals measured by two different methods almost agreed as was expected from the theoretical analysis. The temperature dependence of the dephasing time in this sample is summarized in Fig.4. The observed power law dependence is consistent with the previous results which were obtained by amplitude modulation method[5].

In summary, fluorescence detection of femtosecond accumulated photon echo has been demonstrated. This technique is especially useful for the study of phase relaxation in scattering materials such as polycrystal and biological materials.

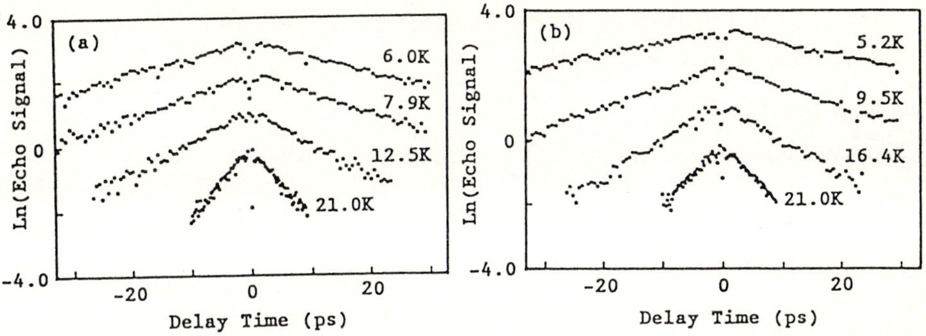

Fig.3 Echo signals plotted as a function of delay time at various temperatures. (a) Fluorescence detection (b) Transmitted light detection

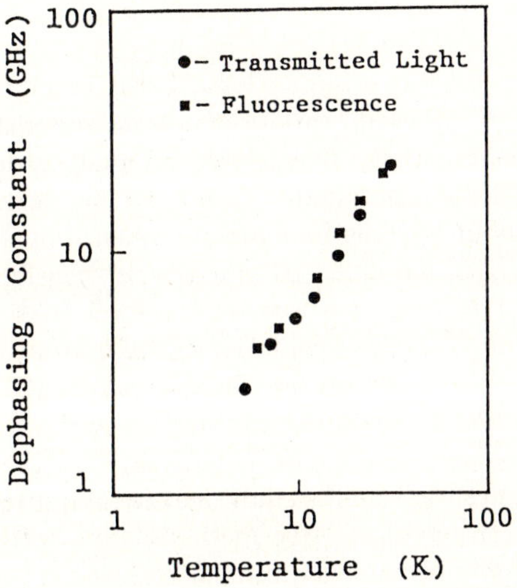

Fig.4 Temperature dependence of optical dephasing time of Oxazin-4 doped in PVB.

References

1, S.Saikan, H.Miyamoto and T.Nakabayashi, IQEC'88 Technical Digest Tokyo, 558(1988)
2, A.H.Zewail, T.E.Orlowski, K.E.Jones and D.E.Godar, Chem.Phys.Lett.48,256 (1977)
3, S.Saikan, H.Miyamoto, Y.Tosaki and A.Fujiwara, Phys.Rev.B36,5074(1987)
4, E.A.Whittaker, M.Gehrtz and G.C.Bjorklund, J.Opt.Soc.Am.B2,1320(1985)
5, R.M.Macfarlane and R.M.Shelby, J.Lumin.36,179(1987)

Analysis of Femtosecond Pulse Systems Using Ray-Pulse Matrices

A.G. Kostenbauder

Ginzton Laboratory, Stanford University, Stanford, CA 94305, USA

Abstract. This paper gives a canonical treatment for the analysis of optical systems in which all rays make small angles to the optical axis and where all frequencies present are close to some center frequency.

1. Background

The last decade has seen the development of a wide variety of optical systems for the generation and shaping of ultrashort pulses. The design and analysis of these systems has usually been done in an ad hoc manner. For example, Fork et al. [1] computed the dispersion of the ubiquitous four prism sequence by a difficult geometric argument. Martinez [2, 3] has since given a prescription for modeling time-invariant systems with $ABCDEF$ matrices and several auxiliary phases. In this paper, we show how to model a wide variety of systems, both time-invariant and time-varying, with the use of four by four "ray-pulse matrices." This treatment is correct to first order in the coordinates and momenta and to second order in the phases, just as are $ABCD$ matrices and the Huygens integral. Using classical mechanics, treatments equivalent to ray-pulse matrices have been developed for electron optics. These methods have been carried out to nth order using Lie algebraic methods.[4] Similar extensions to nth order for optical systems are also possible but will not be discussed here.

2. Ray-Pulse Matrices

The first step in the construction of ray-pulse matrices is the erection of reference planes. Fire a reference pulse into the system with a center frequency that corresponds to a wavelength of λ_0. The path of this pulse defines the optical axis. At each place of interest, erect a reference plane perpendicular to the optical axis and set the local clock to zero at the time of arrival of the reference pulse. We now measure the quantities x, θ, t, and f as the changes in position, slope, time of arrival, and center frequency of other beams. For simplicity in later results, we assemble these four into a vector $(x, t, \theta, -\lambda_0 f)^T$. Provided that the four displacements are small, we expect that the output vector will be a matrix (the ray-pulse matrix of the system) times the input vector.

Springer Series in Chemical Physics, Vol. 53 **Ultrafast Phenomena VII**
Editors: C.B. Harris · E.P. Ippen · G.A. Mourou · A.H. Zewail

$$\begin{bmatrix} x \\ t \\ \hline \theta \\ -\lambda_0 f \end{bmatrix}_{\text{out}} = \left[\begin{array}{c|c} \mathbf{A} & \mathbf{B} \\ \hline \mathbf{C} & \mathbf{D} \end{array}\right] \begin{bmatrix} x \\ t \\ \hline \theta \\ -\lambda_0 f \end{bmatrix}_{\text{in}} . \tag{1}$$

The sixteen elements of these matrices are not all independent due to the geometry of phase space. The allowed matrices have ten independent elements and make up the symplectic group Σ^2.

Space limitations prevent a tabulation of ray-pulse matrices here. With such a tabulation [5], it is possible to find the matrix for a complicated system by multiplying the matrices for the elements. This systematic approach allows one to rapidly compute the behaviour of a wide variety of optical systems used with femtosecond pulses.

3. Narrowband Huygens Integral

For monochromatic systems, the output wave may be found by integrating the input wave against $\exp(-j2\pi L(x_{\text{in}}, x_{\text{out}})/\lambda_0)$ where L is the path length or eikonal through the system. With reference planes perpendicular to the optical axis and small angles, L is a constant plus a quadratic form in x_{in} and x_{out}. Similarly, for systems that involve only small angles and small detunings, the output wave can be found by integrating $\exp(-j2\pi L(x_{\text{in}}, t_{\text{in}}, x_{\text{out}}, t_{\text{out}})/\lambda_0)$ against the input wave where L is again a constant (that we will ignore) plus a quadratic form. The coefficients in the quadratic form are easily expressed in terms of the block matrices

$$\mathcal{E}(x_{\text{out}}, t_{\text{out}}) \propto \iint \exp\left\{-\frac{j\pi}{\lambda_0}\left[\begin{pmatrix} x_{\text{in}} \\ t_{\text{in}} \end{pmatrix}^T \mathbf{B}^{-1}\mathbf{A}\begin{pmatrix} x_{\text{in}} \\ t_{\text{in}} \end{pmatrix} - 2\begin{pmatrix} x_{\text{out}} \\ t_{\text{out}} \end{pmatrix}^T \mathbf{B}^{-1}\begin{pmatrix} x_{\text{in}} \\ t_{\text{in}} \end{pmatrix}\right.\right.$$
$$\left.\left. + \begin{pmatrix} x_{\text{out}} \\ t_{\text{out}} \end{pmatrix}^T \mathbf{D}\mathbf{B}^{-1}\begin{pmatrix} x_{\text{out}} \\ t_{\text{out}} \end{pmatrix}\right]\right\} \mathcal{E}(x_{\text{in}}, t_{\text{in}})\, dx_{\text{in}}\, dt_{\text{in}}. \tag{2}$$

This is identical to the form for the Huygens integral in two spatial dimensions given by Nazarathy.

4. Gaussian Beam-Pulses

Conventional gaussian beams are easily propagated through $ABCD$ systems via the rule $\tilde{q}_{\text{out}} = (A\tilde{q}_{\text{in}} + B)/(C\tilde{q}_{\text{in}} + D)$. Similarly, we define a "gaussian beam-pulse" to be a signal whose electric field satisfies

$$\mathcal{E} \propto \exp\left\{-\frac{j\pi}{\lambda_0}\begin{pmatrix} x \\ t \end{pmatrix}^T \mathbf{Q}^{-1}\begin{pmatrix} x \\ t \end{pmatrix}\right\} \tag{3}$$

Such fields are gaussian in both space and time, and may involve tilted pulses where the right end of the pulse is out in front of the left end due to the off diagonal elements in $\tilde{\mathbf{Q}}^{-1}$. ($\tilde{\mathbf{Q}}$ is symmetric.) Using the results of the last section, we find that

$$\tilde{\mathbf{Q}}_{\mathrm{out}} = [\mathbf{A}\mathbf{Q}_{\mathrm{in}} + \mathbf{B}]\,[\mathbf{C}\mathbf{Q}_{\mathrm{in}} + \mathbf{D}]^{-1} \tag{4}$$

which is precisely analogous to the usual result. A theory of modelocking that simultaneously accounts for spatial and temporal shape changes can be developed from this relationship.

5. Summary

We have given treatments for narrowband paraxial systems for both the ray and wave pictures. The connection between the parameters that are used in the two approaches have been given. Gaussian beam-pulses have been introduced and shown to obey a simple and familiar transformation rule.

6. References

[1] R. L. Fork, O. E. Martinez, and J. P. Gordon, "Negative Dispersion Using Pairs of Prisms," Optics Letters, Vol. 9, pp. 150-152, 1984.

[2] Oscar Eduardo Martinez, "Matrix Formalism for Pulse Compressors," IEEE Journal of Quantum Electronics, Vol. 24, pp. 2530-2536, 1988.

[3] Oscar Eduardo Martinez, "Matrix Formalism for Dispersive Laser Cavities," IEEE Journal of Quantum Electronics, Vol. 25, pp. 296-300, 1989.

[4] A.J. Dragt, "Lectures in Nonlinear Orbit Dynamics," *Physics of High Energy Particle Accelerators*, eds. R. A. Carrigan, F. R. Huson, and M. Month, (American Institute of Physics, New York, 1982), pp. 147-313.

[5] A. G. Kostenbauder, "Ray-Pulse Matrices: A Rational Treatment for Dispersive Optical Systems," IEEE Journal of Quantum Electronics, to appear june 1990.

Picosecond Universal Optical Fluctuation in Strongly Disordered Solids

M. Tomita[1] *and M. Matsuoka*[2]

[1]Department of Physics, Faculty of Science, Shizuoka University, 836 Ohya, Shizuoka 422, Japan
[2]Institute for Solid State Physics, University of Tokyo, 7-22-1 Roppongi, Minato-ku, Tokyo 106, Japan

Propagation and scattering of a coherent wave in disordered media show inherently noise-like behavior due to random interference among multitudinously scattered waves [1]. The optical speckle and the universal conductance fluctuation are examples of this noise-like behavior. We consider this interference in time domain and report the first observation of picosecond universal optical fluctuation in the transmitted pulse profiles through a strongly disordered solid. The temporal fluctuation discussed here is essentially different from the one originating in the Brownian motion of scatterers, which is often discussed in QELS (quasi-elastic light scattering) or DWS (diffusing wave spectroscopy) [2].

The simplest way to understand the above new type of fluctuation is to consider "transient speckle patterns" by an incident coherent optical pulse. A certain set of photons from shorter trajectories in the sample forms a certain speckle pattern first on the observation plane, which stays for a moment of the order of the incident pulse duration. After this pattern goes out, another set of photons from longer trajectories forms a new pattern which has no correlation to the previous one. In this way, transmitted light of a short pulse has large fluctuation in time domain with a correlation time of the order of the incident pulse duration.

A schematic diagram of the experiment for the observation of the fluctuation is shown in Fig.1. We used the second harmonic of a cw mode-locked Nd^{3+}YAG laser which provides controlled optical pulses of 65, 110, and 150 psec duration. Corpuscles of $BaSO_4$ compacted to 680 μm thickness was used as a sample. The transmitted light from the sample was led to a synchro-scanning streak camera via two apertures in order to detect a single coherent cell of the transient speckle. Examples of the transmitted profile observed with 65 psec pulse are shown in Fig.2 where the polarization of the detected light is parallel (Fig.2(a)), and perpendicular (Fig.2(b)) to that of the incident light. We see that the transmitted pulse profiles through a strongly disordered medium have inherently large

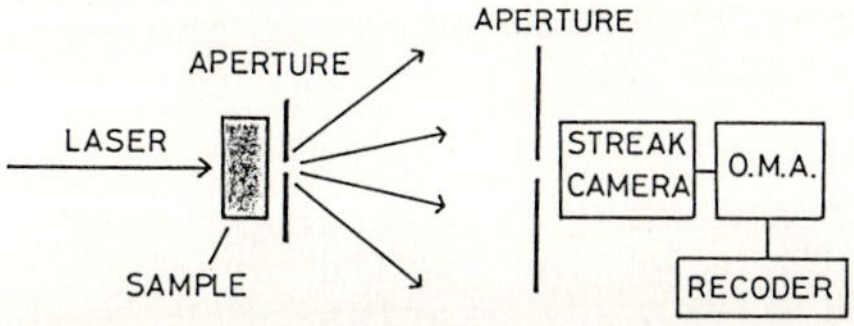

Fig.1 Schematic diagram of the experiment.

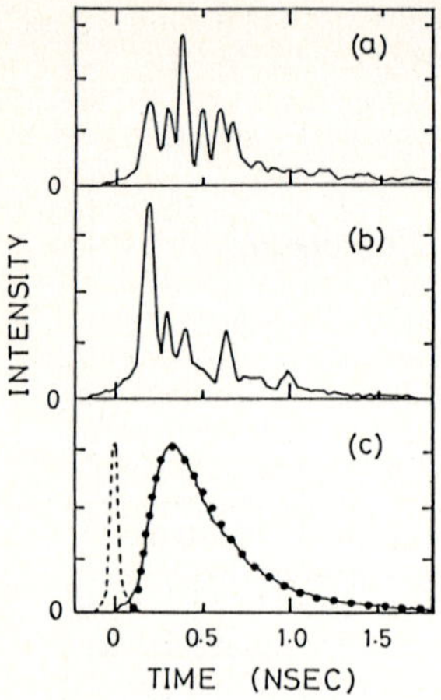

Fig.2 Time resolved transmitted pulse profile of 65 psec optical pulses. The broken line represents the instrument response function to the 65 psec input pulse.

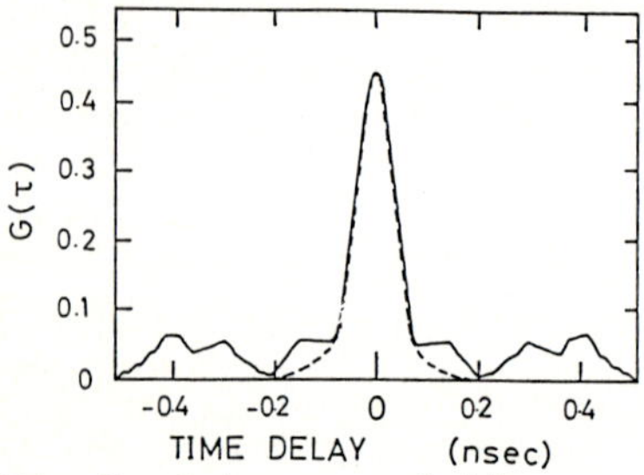

Fig.3 Auto-correlation functions.

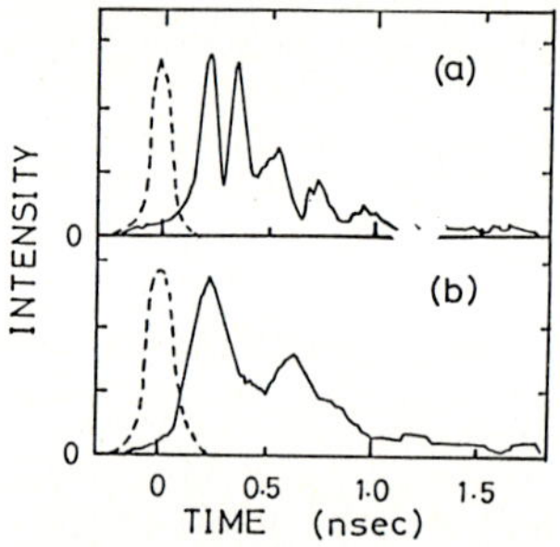

Fig.4 Time resolved transmitted pulse profile of (a) 110 psec and (b) 150 psec pulses.

temporal fluctuation of the order of the incident pulse duration. In Fig.3, an intensity correlation function calculated for ten sets of the measured profiles (solid line) is shown together with the intensity correlation function of the incident pulse (broken line). The width of the correlation peak, 92 psec is in good agreement with that of the auto-correlation of the incident pulse. In Fig.4, examples of the transmitted profiles observed with optical pulses of 110 and 150 psec duration are shown. We see clearly the broadening of fluctuation peaks with the incident pulse width.

The profile of this fluctuation is sample-specific and reproducible for every laser shot, thus we can recognize it as a "fingerprint in time-domain" for the specific arrangement of the disordered structure. The temporal fluctuation discussed here is "universal" in the sense that the time scale of the fluctuation is independent of the sample size and order of randomness.

Every small area of the output plane of the scattered light has a different temporal profile. Thus we can obtain an ensemble average of the fluctuation over many samples of fluctuation, if we observe a sufficiently large scattering area [3]. In Fig.2(c) a transmitted pulse profile taken with a corrective lens between the two apertures is shown, where we obtained a smooth profile as a result of an ensemble average of more than 10^4 samples of fluctuation. The filled circles in Fig.2(c) represent a calculated curve from the photon-diffusion approximation as a best fit to the ensemble averaged smooth temporal profile, where the diffusion constant is 0.14 mm^2/nsec, and the photon lifetime is 1.5 nsec. This value of the diffusion constant is small compared with the value of 0.67 mm^2/nsec estimated from the width of a coherent backscattering peak in the same sample. One of the possible explanations of this discrepancy may be a length-scale-dependent diffusion constant of the localized photons [4], because the shape of the backscattering peak is mainly determined by short trajectories, while the transmitted pulse profile is determined by longer trajectories.

A series of transient speckle patterns at various time points form "a speckle in time-space domain". When we observe the time integrated spatial scattering pattern of the transmitted picosecond pulse, we obtain a smooth profile as a function of the scattering angle because the transient speckle patterns are averaged out by the temporal fluctuation. We therefore recognize that time averaged measurement with coherent short pulses reveals a transition from the fluctuation region to one self-averaging intensity transport region by the shortening of the optical coherence length as is the case with temporally incoherent light [5].

References;

1. M.Kaveh,M.Rosenbluh,I.Edrei,and I.Freund; Phys.Rev.Lett., 57, 2049 (1986); S.Etemad, R.Thompson, and M.J.Andrejco, Phys. Rev. Lett., 57, 575, (1986).
2. B.J.Berne and R.Pecora, Dynamic Light Scattering, (Wiley, New York, 1976); D.J.Pine, D.A.Weitz, P.M.Chaikin, and E.Herbolzheimer, Phys. Rev. Lett. 60, 1134, (1988); X.L.Wu, D.J.Pine, P.M.Chaikin, J.S.Huang, and D.A.weiz, J. OPt. Soc. Am. B7, 15, (1990).
3. J.M.Drake and A.Z.Genack; Phys.Rev.Lett. 63, 259 (1989).
4. P.W.Anderson, Philos. Mag., B52, 505, (1985).
5. M.Tomita and H,Ikari, (submitted).

Wavepacket Theory of Femtosecond Dynamic Absorption Spectroscopy

W.T. Pollard and R.A. Mathies

Department of Chemistry, University of California at Berkeley, Berkeley, CA 94720, USA

By measuring the *spectrum* of the probe pulse, rather than just its energy, Shank and coworkers have demonstrated how to take advantage of the large spectral width of the fastest femtosecond pulses to obtain *complete time-resolved absorption spectra* of coherently evolving molecules *on a 6 fs timescale.* [1] To provide a quantitative analysis of these experiments, we have formulated the perturbative density matrix theory for the third order susceptibility of a multi-level system in terms of four-time correlation functions which can be interpreted as the time-dependent overlap of bra and ket vibrational wavepackets propagating independently on the ground and excited state electronic potential surfaces. [2] When vibrational motion during the excitation is minimal, or when the pump pulse is shorter than the optical T_2, the system will be prepared in an essentially pure vibrational state. The time-resolved differential absorption spectrum can then be treated as the *first-order* spectroscopy of the *non-stationary state* created by the pump pulse. For a non-stationary initial wavefunction $|\psi_1(t)\rangle$, the dynamic absorption spectrum obtained with a single ultrafast probe pulse $\mathbf{E}_{pr}(t)$ can be written as [2]

$$\sigma(\omega) \propto \omega \, \mathrm{Im}\{ \mathbf{P}_{\mathbf{pr}}^{(1)}(\omega)/\mathbf{E}_{\mathbf{pr}}(\omega) \} \tag{1}$$

$$P_{pr}^{(1)}(t) = \langle \psi_1(t)| \, \mu_{21}^{*} \, |\psi_2(t)\rangle \tag{2}$$

where

$$|\psi_2(t)\rangle = \frac{i}{\hbar} \int_{-\infty}^{t} dt' \, e^{-\gamma_{21}(t-t')} \, e^{-ih_2 (t-t')/\hbar} \, \mu_{21} E_{pr}(t') \, |\psi_1(t')\rangle \tag{3}$$

is the wavefunction created on the final state by the probe pulse, convolved with an exponential electronic dephasing function. Here, $\mathbf{E}_{\mathbf{pr}}(\omega)$ and $\mathbf{P}_{\mathbf{pr}}(\omega)$ are the Fourier transforms of the probe pulse electric field and of the system polarization induced by the probe; h_1 and h_2 are the vibrational Hamiltonians of the initial and final electronic states, which are coupled by the transition dipole μ_{21} and whose superposition is dephased at the rate γ_{21}. The correlation function in Eqn. 2 is simply the time-dependent matrix element of μ_{21} between the time-evolving initial wavefunction and the time-dependent wavepacket created on the final state by the probe pulse (Fig. 1). This result represents a natural generalization of the usual time-dependent picture of optical absorption [3] to the non-stationary state case, which cannot be treated by linear response theory. The time-dependent

Springer Series in Chemical Physics, Vol. 53 **Ultrafast Phenomena VII**
Editors: C.B. Harris · E.P. Ippen · G.A. Mourou · A.H. Zewail

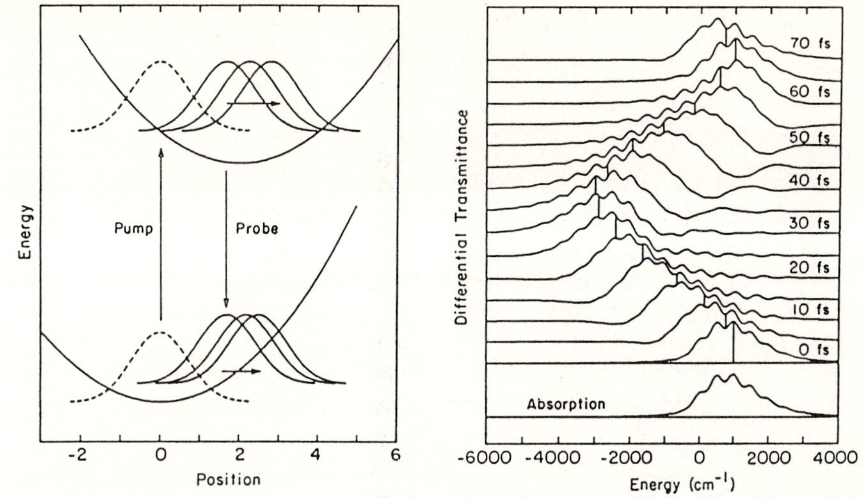

Figure 1. Ground and excited state wavepacket dynamics following the probe pulse for stimulated emission from the nonstationary excited state wavefunction $|\psi_1(t)\rangle$. The stimulated emission spectra calculated for a series of pump-probe delay times are shown for a system of 500 cm-1 harmonic surfaces and δ-function pulses.

formulation is particularly appropriate for femtosecond absorption studies, since these wavepackets are actually physically realized.

This approach provides both a quantitative and intuitive link between nuclear dynamics on the molecular potential surfaces and the time-evolving dynamic absorption spectra measured with femtosecond pulses. An initial application was to analyze the 60 fs pump/6 fs probe spectra of the excited state *trans-cis* isomerization of the retinal chromophore in bacteriorhodopsin. [4,5] Using a one-dimensional model for the isomerization coordinate, we have been able to accurately reproduce the evolution of both the transient excited state absorption and stimulated emission bands, demonstrating that the observed spectral behaviour can be explained in terms of coherent intramolecular dynamics. [2] This theory has also been used [6] to understand the 6 fs pump/6 fs probe spectroscopy of Nile Blue by Fragnito et al. [7] Our analysis shows that the oscillatory evolution of the absorption spectrum reflects coherent vibrations on both the ground *and* excited state surfaces, and that the two contributions may be distinguished by their unique spectral behaviour (Fig. 2). We have recently begun to investigate multi-dimensional models for Nile Blue that will allow us to accurately predict the effects of multiple-pulse excitation sequences on this system.

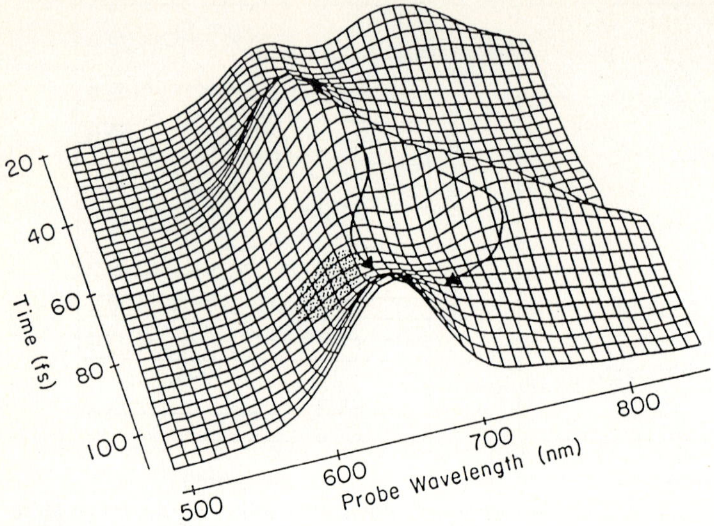

Figure 2. Calculated differential absorbance spectra of Nile Blue as a function of pup-probe delay time, covering a single 55 fs period of the impulsively excited 590 cm mode. The oscillatory shift of the stimulated emission band out to ~800 nm reflects excited state wavepacket motion, while the more limited oscillations of the ground state depletion band at ~640 nm reflect vibrational coherence generated by the pump pulse in the ground state.

References

1. C.V. Shank, *Science* **233**, 1276 (1986).
2. W.T. Pollard, S.-Y. Lee and R.A. Mathies, *J. Chem. Phys.* **92**, 4012 (1990).
3. E.J. Heller, *Acc. Chem. Res*, **14**, 368 (1981).
4. R.A. Mathies, C.H. Brito Cruz, W.T. Pollard and C.V. Shank, *Science* **240**, 777 (1988).
5. W.T. Pollard, C.H. Brito Cruz, C.V. Shank and R.A. Mathies, *J. Chem. Phys.* **90**, 199 (1989).
6. W.T. Pollard, H.L. Fragnito, J.-Y. Bigot, C.V. Shank and R.A. Mathies, *Chem. Phys. Lett.* (in press)
7. H.L. Fragnito, J.Y. Bigot, P.C. Becker and C.V. Shank, *Chem. Phys. Lett.* **160**, 101 (1989).

The Anti-Resonant Ring in Ultrafast Excite–Probe Spectroscopy

R. Trebino and C.C. Hayden

Combustion Research Facility, Sandia National Laboratories, Livermore, CA 94551, USA

Abstract. We show that the anti-resonant ring is a versatile detection arrangement in ultrafast spectroscopy, allowing very accurate *optical* removal of the probe-laser background light and slow effects, independent measurement of absorption and refractive-index transients through optical homodyning, and maximal signal-to-noise ratios.

1. Introduction

The anti-resonant ring (ARR), first introduced by Siegman[1], has found many applications, most recently in colliding-pulse mode-locked lasers[2]. Here, we introduce the use of the ARR as a detection arrangement for performing ultrafast excite-probe induced-absorption experiments, where it offers many advantages over current techniques. Specifically, it automatically subtracts off the probe-laser background and allows the removal of dc or slow (> ~100 psec) effects from the measured ultrafast transient both by an automatic *optical* subtraction method (more accurate than electronic subtraction). Also, by deliberately allowing some leakage from the ring, optical homodyne detection (OHD) is possible, independently yielding both the absorption and refractive-index transients. By choosing the appropriate leakage value in this mode, maximal signal-to-noise ratios can be achieved for a given amount of laser shot-to-shot jitter. Further, a continuum probe may be used because there is no phase-matching requirement.

2. The Anti-Resonant-Ring Technique

Figure 1 illustrates the ARR's use in an excite-probe experiment. The probe pulse is injected into the ARR, with the sample inside the ring, but *off-center*. In the absence of an excitation pulse, both clockwise-propagating (cwp) and counterclockwise-propagating (ccwp) probe pulses have identical fields on returning to the beam splitter, so that no light "leaks" out of the ring. If, however, an excitation pulse excites the sample a short time, τ, before the cwp pulse, then the cwp pulse probes both the long-lived or dc absorption (α_0) plus any induced transient increase or decrease in absorption $[\Delta\alpha(\tau)]$. Later, the ccwp pulse arrives at the sample and probes only the long-lived or dc absorption. Thus, the two pulse fields returning to the beam splitter are no longer identical. Consequently, the intensity leaking out of the ring, I_{out}, vs. relative delay, τ, yields the ultrafast sample transient, $\Delta\alpha(\tau)$:

$$I_{out} \propto |E_{cwp} - E_{ccwp}|^2 = |\tfrac{1}{2} E_{in} \exp\{-\tfrac{1}{2}[\alpha_0+\Delta\alpha(\tau)]\, L\} - \tfrac{1}{2} E_{in} \exp\{-\tfrac{1}{2}\alpha_0 L\}|^2 \quad (1)$$

$$= |E_{in}|^2 \exp\{-\alpha_0 L\}\, \{\tfrac{1}{4}\Delta\alpha(\tau)L\}^2 \quad (2)$$

where E_{cwp} and E_{ccwp} are the cwp and ccwp field strengths returning to the beam splitter after traversing the ring, E_{in} is the input field strength, and L is the sample length. In the last result, we have assumed that $|\Delta\alpha(\tau)L| \ll 1$. Note that the result is proportional to the intensity transmitted by the sample and to the square of the transient absorption, $\Delta\alpha(\tau)$.

On the other hand, deliberately allowing leakage by using a beam-splitter reflectivity of $\tfrac{1}{2}+\varepsilon$, provides a local oscillator for OHD, producing a result that is a constant plus a term that is linear in $\Delta\alpha(\tau)$:

$$I_{out} \propto |(\tfrac{1}{2}-\varepsilon)\, E_{in} \exp\{-\tfrac{1}{2}[\alpha_0+\Delta\alpha(\tau)]\, L\} - (\tfrac{1}{2}+\varepsilon)\, E_{in} \exp\{-\tfrac{1}{2}\alpha_0 L\}|^2 \quad (3)$$

Springer Series in Chemical Physics, Vol. 53 **Ultrafast Phenomena VII**
Editors: C.B. Harris · E.P. Ippen · G.A. Mourou · A.H. Zewail

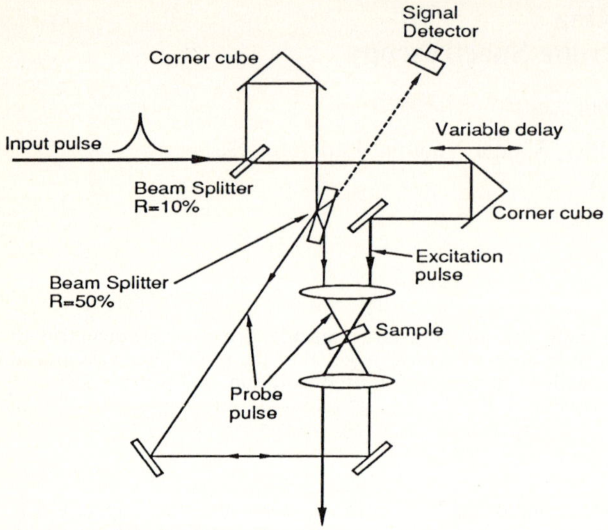

Fig. 1. Simplified experimental arrangement for ultrafast ARR spectroscopy. The clockwise-propagating probe pulse sees the sample just after the excitation pulse excites it. The counterclockwise-propagating probe pulse sees the sample long after the excitation pulse has gone. The detected signal, proportional to the squared modulus of the difference between these two fields, is the transient absorption only. A cw train of <100-fs pulses provided both the excite and probe pulses in our experiments. The actual ARR used in this study involved the ~50% beam splitter, two mirrors, two corner cubes, and a double-passed lens pair, avoiding both relative image inversions and relative-magnification mismatches at all optics.

$$\approx \; |E_{in}|^2 \exp\{-\alpha_0 L\} \; \{4\varepsilon^2 + \varepsilon\Delta\alpha(\tau)L\} \qquad (4)$$

where we have neglected the term that is second order in $\Delta\alpha(\tau)$. The first term (the "beam-splitter leakage") represents the leakage due to the deviation of the beam-splitter reflectivity from 1/2, while the second term is the leakage due to the transient-absorption signal. In this case, maximal signal-to-noise (S/N) ratios occur, as has been shown for OHD polarization spectroscopy[3]. When laser noise dominates, the improvement in S/N over the simple two-beam excite-probe technique is given by $(4\varepsilon)^{-1}$ or $(4\delta)^{-1/2}$, where $\delta \equiv 4\varepsilon^2$ is the ratio of the beam-splitter leakage and the input intensity transmitted by the sample cell. In addition, polarizers are not required, so material birefringence, optical phase drift, and phase uncertainty—problems in OHD polarization spectroscopy—are not significant problems here. Finally, the pulse polarizations are not constrained by the technique. Sensitivity is also very good: in preliminary experiments with femtosecond input pulses, we have obtained $\sim 10^{-4}$ leakage, using a 45°-incidence-angle, dielectric-coated beam splitter (limited by slight reflectivity variations across its surface). Lower leakage, and hence, higher sensitivity, appears possible using an uncoated, high-incidence-angle beam splitter, which can also exhibit zero dispersion ($dR/d\lambda$). In initial experiments, we used R $\approx$55%, yielding $\sim 10^{-2}$ leakage.

3. Experiment and Discussion

Figure 2 shows a comparison of the results typically obtained both in a two-beam excite-probe (2BEP) induced-absorption experiment and in an ARR experiment. The same apparatus, alignment, and integration time were used for both experiments. To convert an ARR to a 2BEP experiment without changing beam alignment—allowing valid S/N comparison—a photo-diode detector simply intercepted the cwp probe beam after the sample cell. The sample used was a jet of dilute (3%-absorbing) malachite green dissolved in ethylene glycol in order to simulate a weak-signal experiment. For a beam-splitter leakage of 1%, use of the ARR typically yielded a factor of approximately 5 improvement in the signal-to-noise ratio over the two-beam experiment, in agreement with the theoretical prediction of 5.

The ARR detection method should yield even more improvement in S/N when less beam-splitter leakage occurs; leakage values much less than 10^{-4} appear achievable. In

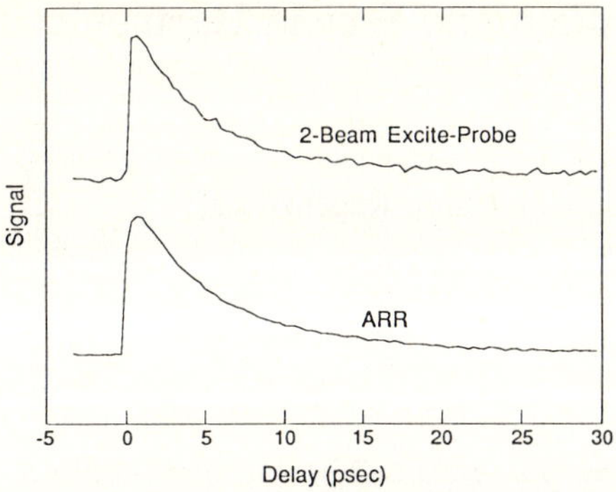

Fig. 2. Comparison of ultrafast transients using the standard two-beam excite-probe induced-absorption technique and the anti-resonant-ring method. Observe the decrease in noise in the transient obtained using the anti-resonant-ring method by a factor of ~5. Use of an ARR arrangement with less than the 1% leakage value used here should yield greater improvements in S/N.

addition, ARR-detection offers even more improvement for low-repetition-rate laser experiments where shot-to-shot jitter is generally greater. In summary, the ARR appears to be quite useful for obtaining sensitive ultrafast transient-absorption measurements.

We would like to thank Mark Kimmel for assistance in all experiments and A.G. Kostenbauder and A.E. Siegman for several excellent suggestions. This work was supported by the U.S. Department of Energy, Office of Basic Energy Sciences, Chemical Sciences Division.

4. References

1. A.E. Siegman, IEEE JQE **QE-9**, 247 (1973).
2. A.E. Siegman, Opt. Lett. **6**, 334 (1981).
3. G.L. Eesley, M.D. Levenson, and W.M. Tolles, IEEE JQE **QE-14**, 45 (1978).

Simultaneous Two-Photon Absorption and Second Harmonic Generation in CdSe

A. Penzkofer, M. Schäffner, and X. Bao

Naturwissenschaftliche Fakultät II–Physik, Universität Regensburg, D-8400 Regensburg, Fed. Rep. of Germany

Abstract. A hexagonal CdSe crystal is excited with picosecond Nd:glass laser pulses. Two-photon absorption and resonant non-phasematched second harmonic generation occur simultaneously. Using different crystal orientations all components of the second harmonic susceptibility tensor and some components of the two-photon absorption susceptibility tensor are determined.

1. Introduction

Two-photon absorption (TPA) and second harmonic generation (SHG) occur simultaneously when picosecond light pulses of a Nd:glass laser (wavelength $\lambda_L = 1.054$ µm, duration $\Delta t_L \approx 6$ ps) pass through a hexagonal CdSe crystal (wurtzite structure 6mm point symmetry).

Previous TPA studies are summarized in [1,2]. The simultaneous occurrence of TPA and SHG is studied scarcely [3,4]. Excited state absorption of the two-photon generated carriers was observed for nanosecond excitation pulses [4,5].

Intensity dependent transmission measurements of the pump laser and second harmonic conversion efficiency measurements are carried out to separate TPA and SHG. A detailed theoretical treatment is given elsewhere [6].

2. Experimental

Single picosecond light pulses separated from a passively mode-locked Nd:phosphate glass laser were applied in the experiments. The size of the CdSe crystal (from Cleveland Crystals Inc., selenium treated for high resistivity) was 12 × 12 × 10 mm with the c-axis parallel to a 12 mm edge. Energy transmission measurements of the pump laser (T_E) and SHG energy conversion efficiency measurements (η_E) were performed. Four different interaction schemes of the pump pulses with the crystal were studied (see Table 1) to determine all non-zero SHG susceptibility components d_{ij} [7] and some TPA coefficients $\alpha^{(2)}_{ijkl}$.

3. Results

The T_E and η_E results of the oo→e interaction (wavevector $\mathbf{k}_L \perp$ c-axis and electric field strength $\mathbf{E}_L \perp$c-axis) are presented in Fig.1. The pump pulse reduction by second harmonic generation is estimated to be $T_{E,SHG} \approx 1-\eta_E \ell \alpha_2$ where ℓ is the sample length and α_2 the linear absorption coefficient at the second harmonic frequency (ordinary ray o: $\alpha_{2o} \approx 1.2\times10^5$ cm^{-1}, extraordinary ray e: $\alpha_{2e} \approx 1.5\times10^5$ cm^{-1} [8,9]). η_E is found to be

Springer Series in Chemical Physics, Vol. 53 **Ultrafast Phenomena VII**
Editors: C.B. Harris · E.P. Ippen · G.A. Mourou · A.H. Zewail

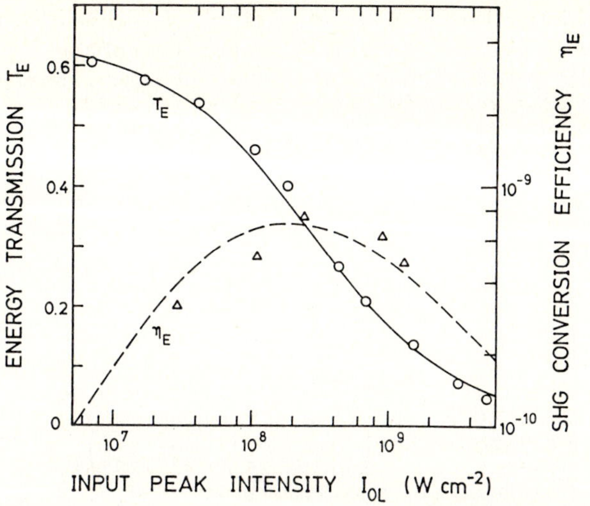

Fig.1: o——o: Energy transmission T_E. Δ---Δ: SHG energy conversion efficiency η_E. oo→e interaction. Sample length ℓ=1 cm. Fit parameters see Table 1.

Table 1: Interaction schemes and corresponding SHG and TPA coefficients. $T_{E,x}$ ($T_{E,z}$) is transmission of laser light polarized ⊥(∥) to c-axis.

Orientation ($\mathbf{k}_L$,**c**) (degrees)	($\mathbf{E}_L$,**c**) (degrees)	Inter-action	SHG d_{ij} (10^{-11}m V^{-1})	TPA $\alpha^{(2)}_{ijkl}$ (10^{-10}m W^{-1})
0	90			$\alpha^{(2)}_{xxxx}$:1.8±0.3
90	90	oo→e	d_{31}:2.5±0.3	$\alpha^{(2)}_{xxxx}$:1.8±0.3
90	0	ee→e	d_{33}:3.9±0.3	$\alpha^{(2)}_{zzzz}$:1.8±0.3
90	45	oe→o	d_{15}:2.3±0.3	
			$T_{E,x}$: $\alpha^{(2)}_{xxxx}$ and $\alpha^{(2)}_{xzzx}+\alpha^{(2)}_{xzxz}$	:0.3±0.3
			$T_{E,z}$: $\alpha^{(2)}_{zzzz}$ and $\alpha_{zxxz}+\alpha_{zxzx}$	:0.3±0.3

two orders of magnitude too small to influence the energy transmission T_E.

The experimental points in Fig.1 are fitted by theoretical curves [6]. The T_E curves include no excited state absorption and the good fit to the experimental data indicates that an absorption of the two-photon generated carriers is negligible at λ_L = 1054 nm.

In Table 1 the investigated crystal orientations relative to the laser pulses, the SHG interaction processes, and the fitted SHG susceptibility coefficients d_{ij} and TPA coefficients $\alpha^{(2)}_{ijkl}$ are listed.

4. Conclusions

In non-centrosymmetric crystals TPA and resonant non-phase-matched second harmonic generation occur simultaneously. Energy transmission and second harmonic conversion efficiency measurements are necessary to separate the contributions of the SHG and the TPA to the nonlinear transmission.

The measured energy transmissions of CdSe versus input pump pulse peak intensity may be used as calibration curves for the peak intensity detection of picosecond light pulse by energy transmission measurements [10].

References

1. W.L. Smith, in CRC Handbook of Laser Science and Technology, Vol.III, Optical Materials: Part 1, edited by M.J. Weber (CRC Press, Boca Raton, Florida, 1986) pp. 229.
2. E.W. VanStryland, H. Vanherzeele, M.J. Soileau, A.L. Smirl, S. Guha, and T.F. Boggess, Opt. Engng. 24, 613 (1985).
3. F. Bryukner, V.S. Dneprovskii, and V.U. Khattatov, Sov. J. Quantum Electron. 4, 749 (1974).
4. V.S. Dneprovskii and Sh.M. Ok, Sov. J. Quantum Electron. 6, 298 (1976).
5. D.C. Hanna and A.J. Turner, Opt. Quantum Electron. 8, 213 (1976).
6. A. Penzkofer, M. Schäffner, and X. Bao, Opt. Quantum Electron. (1990) to be published.
7. F. Zernike and J.E. Midwinter, Applied Nonlinear Optics (Wiley, New York, 1973).
8. X. Bao, M. Schäffner, and A. Penzkofer, to be published.
9. M. Cardona and G. Harbeke, Phys. Rev. A137, 1467 (1965).
10. W. Blau and A. Penzkofer, Opt. Commun. 36, 419 (1981).

Anomalous Dispersion of the Electronic Kerr Effect n_2

D.J. Hagan, Y.Y. Wu, E. Canto, M. Sheik-bahae, A.A. Said, J. Young, T.H. Wei, M. Junnarkar, and E.W. Van Stryland

Center for Research in Electro-Optics and Lasers,
University of Central Florida, Orlando, FL32816, USA

Measurements of the bound electronic nonlinear refractive index n_2, using picosecond DFWM and beam distortion, show a region of anomalous dispersion where n_2 eventually turns negative between the two-photon absorption edge and the fundamental edge.

We have performed picosecond Degenerate Four-Wave Mixing (DFWM) experiments and beam distortion experiments, including using the newly developed Z-scan technique,[1] to accurately measure the bound electronic nonlinear refractive index n_2 in a variety of materials at 1.06 and 0.53 μm. In semiconductors where two-photon absorption (2PA) is allowed, we find that n_2 decreases from its maximum value near the 2PA ($2\hbar\omega \simeq E_g$) edge to zero as the wavelength is increased, changing to negative nearer the fundamental absorption edge. In order to measure n_2, we experimentally separated the effects of 2PA excited carrier refraction (coefficient σ) from the electronic Kerr effect induced refraction. Using DFWM this is accomplished by observing that the carrier effects last for hundreds of picoseconds and vary as the fifth power of the incident irradiance I, while the $\chi^{(3)}$ effect follows the temporal variation of the pulse and varies as I^3. Such a signal is shown in Fig. 1.

The inset shows the order of the nonlinearity (all beam irradiances varied simultaneously) at zero delay and at long delays where only the free-carrier refraction remains. DFWM studies in CdTe at 1.06μm showed similar fast $\chi^{(3)}$ and slow fifth order signals.

To determine the nature of these unexpectedly large and fast $\chi^{(3)}$,s, beam distortion measurements were made at various irradiance levels and fitted numerically to a nonlinear beam

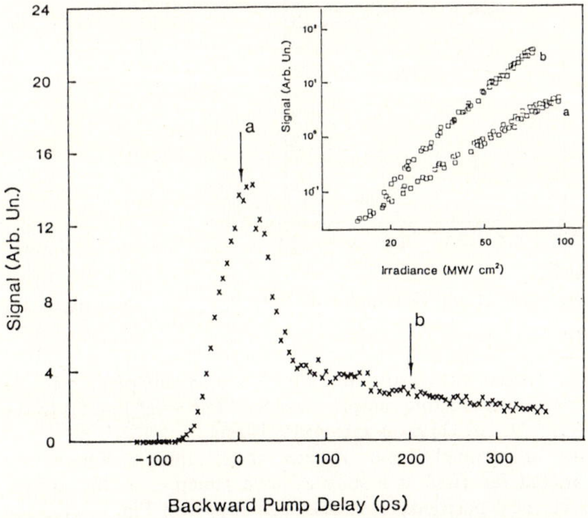

Fig. 1. The DFWM signal as a function of the temporal delay of the backward propagating pump beam (forward pump and probe temporally overlapped) for 30 ps, 0.53 μm pulses in ZnSe. The inset shows a log-log plot of the irradiance dependence at temporal delays of a) 0 ps and b) 200ps respectively (all three input beams varied simultaneously).

Springer Series in Chemical Physics, Vol. 53 **Ultrafast Phenomena VII**
Editors: C.B. Harris · E.P. Ippen · G.A. Mourou · A.H. Zewail

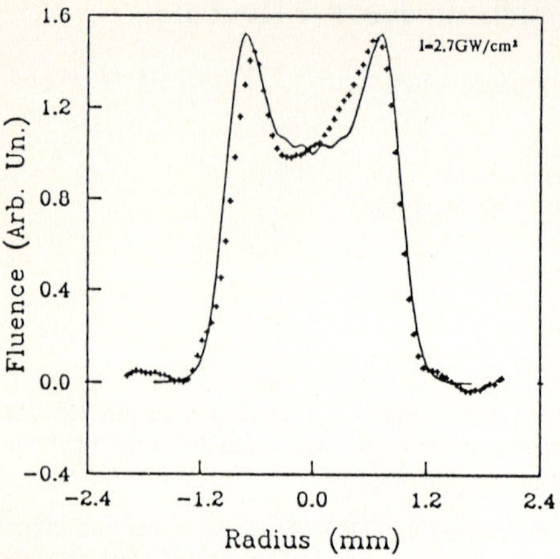

Fig. 2. The transmitted spatial energy distribution of a 30 ps 0.53 μm pulse after transmission and propagation through ZnSe as viewed on a vidicon. The solid line is the theoretical fit to the defocusing.

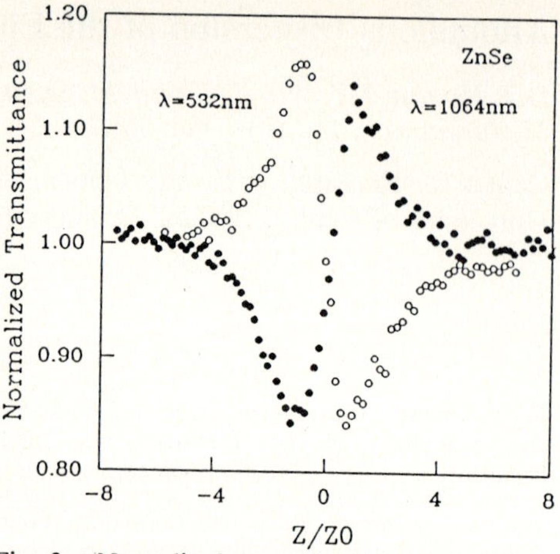

Fig. 3. Normalized transmittance versus sample position in units of $Z_0=\pi\omega_0^2/\lambda$ in a Z-scan experiment showing the sign change of n_2 from + at 1.06μm to - at 0.53μm.

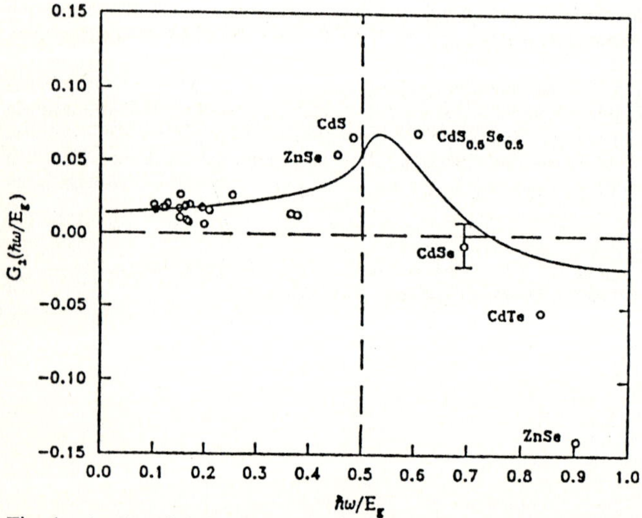

Fig. 4. A plot of $G_2(\hbar\omega/E_g)$, the dispersion of n_2. Data are scaled by $n_0E_g^4/K$, where n_0 is the linear index and K is a numerical constant.

propagation code. Figure 2 shows the spatial energy distribution of a 0.53 μm pulse $\simeq$10 cm after transmission through a 3 mm thick polycrystalline sample of ZnSe. The solid line fit to the data using β=5.5 cm/GW [2] and σ=0.8x10^{-21} cm^3 [1] gives $n_2 \simeq$-4x10^{-11} esu.

These numbers agree with those determined from Z-scan measurements where the transmittance of an aperture placed in the far field is monitored as a function of the sample position with respect to the linear optics focal position.[1] Z-scans, as shown in Fig. 3, give the sign of n_2 and show that for ZnSe the nonlinear refraction changes from positive at 1.06μm to negative at 0.53μm. We have performed similar experiments on CdTe and many other materials and they show similar behavior. The data for n_2 obtained by these methods along with the data from Ref. 3 for n_2 in wide gap dielectric materials fits a universal dispersion curve for n_2 as

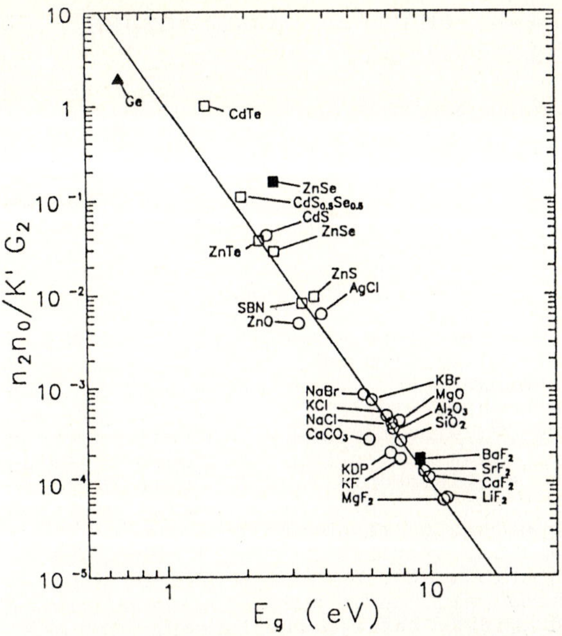

Fig. 5. Log-log plot of the scaled n_2 versus E_g showing the E_g^{-4} dependence of the straight line.

shown in Fig. 4. The solid curve in Fig. 4 was obtained from a nonlinear Kramers-Kronig transformation of the 2PA dispersion as given in Ref. 2. This theory, as well as the data, show a region of anomalous dispersion in n_2 between the 2PA edge and the fundamental edge.

1. M. Sheik-bahae, A.A. Said, T.H. Wei, D.J. Hagan, and E.W. Van Stryland, Jour. Quan. Elect., QE XX, April(1990).
2. M. Sheik-bahae, D.J. Hagan, and E.W. Van Stryland, submitted to Phys. Rev. Lett., (1990).
3. R. Addair, L.L. Chase, and S.A. Payne, Phys. Rev B*39*, 3337-3349 (1989).

Electronically Programmable Femtosecond Pulse Shaping Using a Multielement Phase Modulator

A.M. Weiner, D.E. Leaird, J.S. Patel, and J.R. Wullert

Bellcore, 331 Newman Springs Road, Red Bank, NJ 07701, USA

Considerable progress has been made in developing techniques for generating ultrashort pulses with arbitrarily controllable pulse shapes [1-3]. Nevertheless, it has not generally been possible to achieve both high resolution and precision [1] and electronic programmability [2] simultaneously. In this paper we describe a programmable, high resolution, fsec pulse shaping apparatus based on a multielement liquid crystal phase modulator [4]. Applications include fsec pulse position modulation and programmable fsec pulse compression.

Our pulse shaping apparatus, sketched in Fig. 1, consists of a pair of 1800 line/mm gratings at the focal planes of a unit-magnification, confocal lens pair, and a multielement modulator (MEM), which is placed midway between the lenses. The MEM, depicted in Fig. 2, is a linear array of 32 individual liquid crystal phase modulators on 100 μm centers. Further details are given in [4]. 75-fsec, 0.62-μm pulses from a CPM dye laser comprise the input for our experiments. The optical frequency components are spatially dispersed by the first grating and lens, operated on by the MEM, and then recombined by the second lens and grating. The output pulse shape is the Fourier transform of the spectrum as modified by the phase modulator. In our previous works pulse shaping was achieved by means of fixed, microlithographically defined spatial masks [1]. By using the MEM, we now achieve programmability and continuous (gray-level) phase control.

Figure 3 shows an example of real-time pulse shaping. For this measurement one half of the modulator elements were held at a fixed voltage V_1, while the other half were switched at a 20 Hz rate between voltages V_1 and V_2. V_2-V_1 was chosen to yield a relative phase shift of π. The resulting pulse intensity profiles were measured by using a real-time cross-correlator. The data demonstrate complete switching from a single pulse to a pulse doublet on a millisecond time scale.

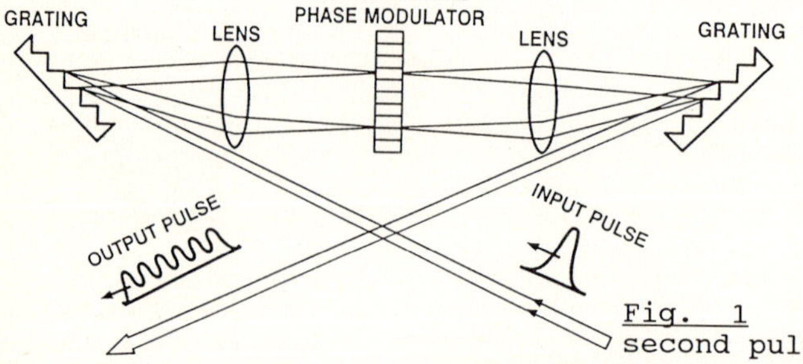

Fig. 1 Programmable femtosecond pulse shaping apparatus.

Springer Series in Chemical Physics, Vol. 53 **Ultrafast Phenomena VII**
Editors: C.B. Harris · E.P. Ippen · G.A. Mourou · A.H. Zewail

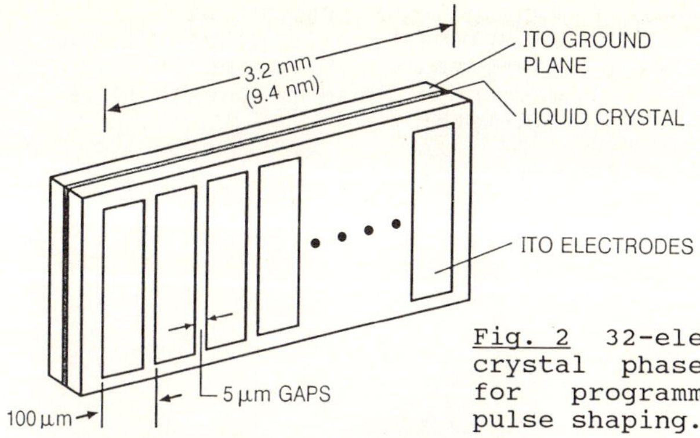

Fig. 2 32-element liquid crystal phase modulator used for programmable femtosecond pulse shaping.

Fig. 3

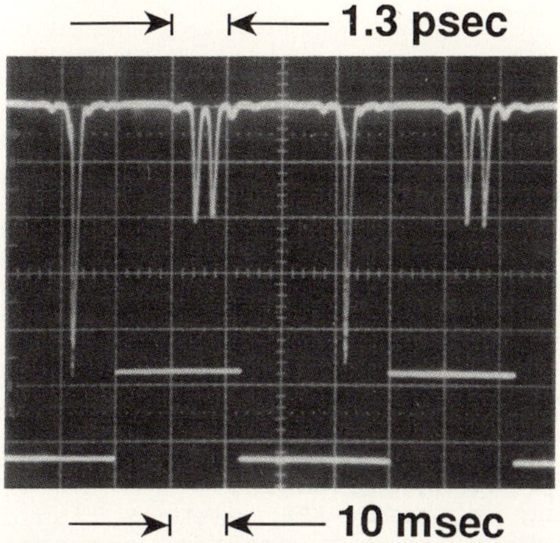

Fig. 4

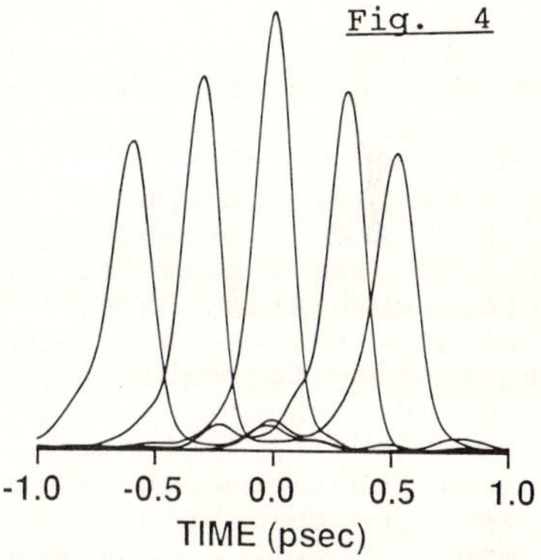

Fig. 3 Top trace: Cross-correlation trace of shaped pulses, showing switching from a single pulse to a pulse doublet at a 20 Hz rate. Bottom trace: Modulated drive signal switching between 1.9 and 5.0 volts at a 20 Hz rate.

Fig. 4 Intensity profiles of pulses shifted temporally by various amounts.

Gray level control of the optical phase makes possible further applications such as fsec pulse position modulation (PPM). We rely on the fact that if $f(t)$ and $F(\omega)$ are a Fourier transform pair, then the delayed signal $f(t-\tau)$ is the Fourier transform of $F(\omega)\exp(-i\omega t)$. Thus, a pulse can be retarded (or advanced) by imposing a linear phase sweep onto its spectrum. Figure 4 shows measurements of pulses emerging from the shaper for different linear phase sweeps. As seen, the pulse is shifted temporally over a range several times the pulse duration. Our fsec PPM technique also makes it possible to perform cross-correlation measurements without moving parts. We have demonstrated this fact by comparing cross-correlation measurements performed by using a standard stepper-motor-driven delay stage with measurements performed without moving parts by using fsec PPM. Excellent agreement is obtained [4].

As one further example, we have demonstrated programmable fsec pulse compression. In the spectral domain, linearly chirped pulses exhibit a phase response which varies

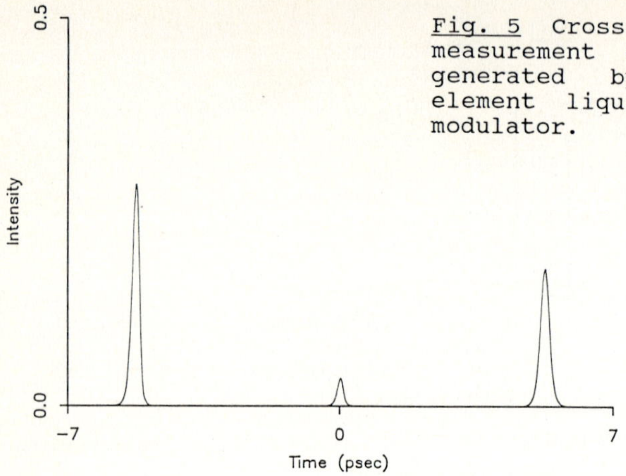

Fig. 5 Cross-correlation measurement of a pulse doublet generated by using a 128-element liquid crystal phase modulator.

quadratically with frequency. By using the MEM to compensate for the quadratic phase shift, linearly chirped pulses can be compressed down to the bandwidth limit. We have performed experiments which verify this point. By translating one of the gratings in the pulse shaper, we obtained a pulse whch was chirped and broadened by a factor of ten. Then, by using the MEM, we were able to quickly and conveniently dial in the proper quadratic phase to compress the pulse to its original duration [4]. Our technique holds out the possibility of compensating for cubic and higher order phase variations, which arise for example in pulse compression below 10 fsec or in propagation of femtosecond pulses through kilometer lengths of optical fiber near the zero dispersion wavelength.

We are currently evaluating a new, 128 element liquid crystal phase modulator, which should provide spectral resolution comparable to the best which we have obtained with fixed masks defined by microlithography. The device consists of a linear array of 128 individual phase modulators on 40 μm centers, with gaps of $\simeq$1.5 μm separating the electrodes of neighboring modulator elements. An example of pulse shaping with this device is shown in Fig. 5. The phases alternate from 0 to π from one pixel to the next, and this splits the original pulse into a pulse doublet separated by >10 psec. Already this simple example demonstrates the ability to program 128 separate phase features into the frequency spectrum. Ultimately we hope to use a 128-element modulator as a general laboratory tool which can replace prefabricated masks in many pulse shaping applications.

References

1. A.M. Weiner, J.P. Heritage, and E.M. Kirschner, J. Opt. Soc. Am. B5, 1553 (1988).
2. M. Haner and W.S. Warren, Appl. Phys. Lett. 52, 1458 (1988).
3. T. Kobayashi and A. Morimoto, in OSA Proceedings on Picosecond Electronics and Optoelectronics (OSA, Washington, D.C., 1989), 81-86.
4. A.M. Weiner, D.E. Leaird, J.S. Patel, and J.R. Wullert, Opt. Lett. 15, 326 (1990).

Pulsed Diffusing-Wave Spectroscopy: Pathlength Specific Observation of Speckle Fluctuation Spectra from Dense Colloids

A.G. Yodh[1], *P.D. Kaplan*[1], *and D.J. Pine*[2]

[1]Department of Physics, University of Pennsylvania, Philadelphia, PA 19104, USA

[2]Exxon Research & Engineering Co., Rt. 22 East, Annandale, NJ 08801, USA

Abstract. We introduce a probe of fluctuating dense random colloidal suspensions which uses phase fluctuations of optically gated photons to obtain pathlength specific speckle correlation functions.

1. Introduction

Nonlinear optical gating has enjoyed wide success over the years [1]. We describe a new class of gating experiment which we have developed and applied to study particle diffusion in colloids at high concentration. In contrast to previous work, the new scheme exploits the *phase fluctuations* of gated photons and enables the experimenter to access properties of dense colloidal suspensions that *multiply* scatter light.

Diffusing-wave spectroscopy (DWS) is a new method of light scattering [2,3] that exploits the diffusive nature of light transport in strongly scattering media to relate temporal intensity fluctuations of the scattered light to particle motion. Unlike traditional quasi-elastic light scattering methods, DWS probes particle motion in dense, *strongly scattering colloids* and over length scales much *shorter* than the wavelength of light.

In a typical DWS experiment a cw laser beam illuminates a colloidal suspension, and the temporal autocorrelation function of the multiply scattered output field is measured. Properties of the scatterers, such as their diffusion coefficient are extracted by analyzing this autocorrelation function. Microscopically one can envision each photon traveling ballistically between particles, and experiencing changes in propagation direction after each scattering event. Two length scales characterize photon transport in the media: (1) s, the total distance traveled by a photon, and (2) l^*, the transport mean free path of the photon. Physically, l^* ($l^* \ll s$) is the mean distance a photon travels before its propagation direction is randomized. Thus l^* is the random walk step size for the "diffusing photons." If we consider only the electric field due to photons that travel a length s through the media, then the time-averaged autocorrelation function, $g_1(\tau,s)$, of the scattered field takes on a simple form when the particles move independently [2,3],

$$g_1(\tau,s) = \exp[-k_0^2\langle\Delta \mathbf{r}^2(\tau)\rangle(s/l^*)/3] \ . \qquad (1)$$

Here $\langle\Delta \mathbf{r}^2(\tau)\rangle$ is the mean square displacement of a particle in time τ and k_0 is the wavenumber of the light in the media.

The first DWS experiments used cw lasers [3] to measure the *total* electric field autocorrelation, $G_1(\tau)$, function. his can be computed by incoherently summing the contributions of each path-dependent $g_1(\tau,s)$, weighted by P(s), the probability a photon will travel a distance s through the media. For Brownian motion, $\langle\Delta \mathbf{r}^2(\tau)\rangle = 6D\tau$, and $G_1(\tau)$ is the Laplace transform of P(s) with new scale factors.

In pulsed-DWS (PDWS) spectroscopy standard nonlinear optical gating methods [1] isolate the contributions of photons that travel paths of a single specified length through

Springer Series in Chemical Physics, Vol. 53 **Ultrafast Phenomena VII**
Editors: C.B. Harris · E.P. Ippen · G.A. Mourou · A.H. Zewail

the media. The upconverted photons are then further processed. Using photon correlation techniques [3] we perform an intensity-intensity autocorrelation measurement on the upconverted photons. Effects such as sample absorption and experimental geometry are rendered unimportant. In addition to studying particle diffusion with greater precision, we can test fundamental elements of the original DWS theories.

2. Pulsed Diffusing-Wave Spectroscopy (PDWS)

The basic ideas of PDWS are illustrated in Fig. 1. A 100 MHz pulse train from a mode-locked Nd:YAG laser is divided between a sample and a reference line of variable length. Sample pulses impinge on a dense colloidal suspension of 0.46-μm-diam. polystyrene spheres in water. A portion of the output "speckle" is collected, and recombined with the reference pulse in a KTP crystal where it is upconverted. The sample pulse is temporally broadened due to multiple scattering in the suspension, and the reference pulse gates [1] out a small portion of the broad pulse. Importantly, the fluctuations of the gated speckle field are impressed upon the upconverted field! Thus, the standard autocorrelation measurement can be performed on the upconverted photons.

The autocorrelation function of the SH field is proportional to $g_1(\tau,s)$. By changing the relative delay between reference and sample arms we can vary s. Note that the temporal behavior of the *upconverted* autocorrelation function, $g_1(\tau,s,2\omega)$, no longer depends on the shape of P(s), and for fixed s, a plot of $\ln[g_1(\tau,s,2\omega)]$ vs τ directly yields the time dependence of $\langle\Delta r^2(\tau)\rangle$. In DWS any process that affected P(s), modified the temporal decay of the measured autocorrelation function. Thus even processes that *do not* effect particle diffusion must be properly accounted for in analyzing DWS data. PDWS eliminates these types of problems.

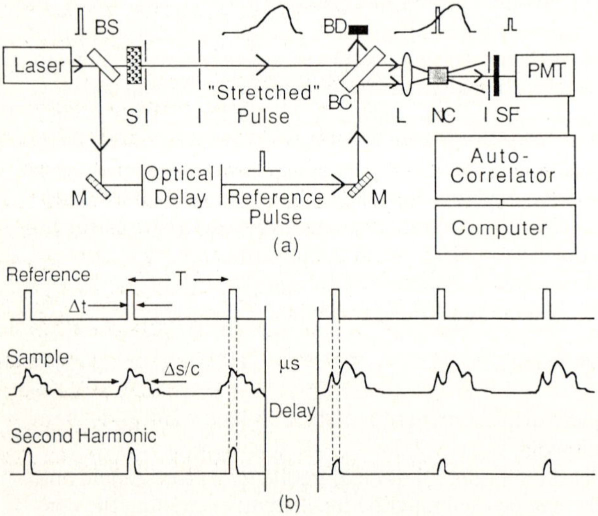

Fig. 1: (a) Schematic of experimental set-up: BC, beam combiner; BD, beam dump; BS, beam splitter; I, iris; L, lens; M, mirror; NC, nonlinear crystal (KTP); PMT, photomultiplier tube; S, sample; SF, second-harmonic spectral filter. (b) Sketch of reference, sample, and upconverted pulse intensities during two time intervals separated by several microseconds.

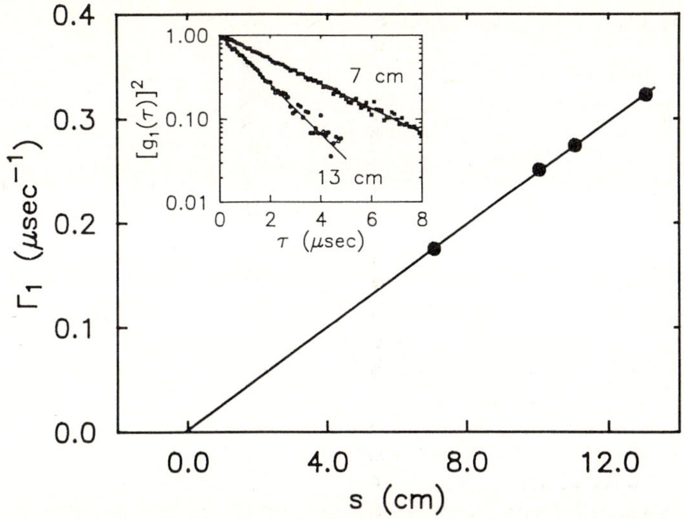

Fig. 2: Inset: Plot of the upconverted intensity autocorrelation function $[g_1(\tau,s,2\omega)]^2$ *vs* τ. Main: Plot of the temporal decay rate, Γ_1, of the upconverted field autocorrelation function vs reference arm delay *s*.

To illustrate the basic features of PDWS we have carried out measurements on the Brownian dynamics of dense colloidal suspensions. Typical photon count rates were between 80 and 600KHz, allowing normalized intensity autocorrelation functions to be obtained with 100nsec time resolution in ~15 minutes. The intensity and field correlation functions are related through the Siegert relation [3]. In the inset of Fig. 2 we plot typical SH intensity autocorrelation functions $[g_1(\tau,s,2\omega)]^2$ vs τ. The volume fraction of spheres was 0.3, the sample thickness was 2 mm, and the reference arm delays were s=7.0 cm and s=13.0 cm. In contrast to DWS measurements the curves decay exponentially. Using this sample we performed measurements at different optical delays s. In the main part of Fig. 2 we plot the temporal decay rate, Γ_1, of the upconverted field autocorrelation function vs. reference arm delay *s*. These measurements corroborate the primary result of DWS. That is, the path-dependent electric field autocorrelation function decays exponentially at a rate proportional to s.

This work was supported by the NSF through the MRL Program #DMR-8519059, through equipment loans from the MRL Laser Facility, and through a Presidential Young Investigator Award (AGY). We also received support from the PEW Foundation.

3. References

1. J.Shah, IEEE J. Quantum Electronics **24**, 276 (1988) and references therein; R. Vreeker, M.P. Van Albada, R. Sprik, and A. Lagendijk, Phys. Lett **A 132**, 51 (1988); K.M. Yoo, Y. Takiguchi, and R.R. Alfano, Applied Optics **28**, 2343 (1989).
2. M.J. Stephen, Phys. Rev. B **37**, 1 (1988).
3. G. Maret and P.E. Wolf, Z. Phys. B **65**, 409 (1987); D.J. Pine, D.A. Weitz, P.M. Chaikin, and E. Herbolzheimer, Phys. Rev. Lett. **60**, 1134 (1988).

Part IV

Nonlinear Effects in Optical Fibers

Ultrafast All-Optical Fiber Soliton Logic Gates

M.N. Islam, C.E. Soccolich, and D.A.B. Miller

AT&T Bell Laboratories, Room 4D-411, Holmdel, NJ 07733, USA

We demonstrate an all-optical fiber soliton dragging NOR-gate with a switching energy of 5.8pJ, a fanout of six, and a speed of up to 0.2THz. Cascadability and gain are proven by feeding the gate output back to the input. Soliton dragging can be represented as a Generalized Exclusive-OR module, two of which can be interconnected as broadcast and routing switches or NOR- and AND-gates.

We demonstrate a 5.8pJ switching energy all-optical NOR-gate with a fanout of six that is based on timing shifts from soliton dragging [1] in a fiber. This three-terminal, cascadable gate satifies all requirements for a clocked digital optical processor. Furthermore, we describe here an equivalent circuit for soliton dragging in terms of a generalized exclusive-or (GEO) module. By changing the interconnections between two GEO-modules, we can implement NOR and AND-gates or broadcast and routing switches.

The logic gate operates based on time shifts from soliton dragging in a clocked digital system. A "1" corresponds to a pulse that arrives within the clock window and a "0" either to no pulse or an improperly timed pulse. In soliton dragging two pulses interact through cross-phase modulation [1] and, consequently, chirp in frequency. The frequency shift is converted to a time shift by propagating in a dispersive delay line. For the NOR-gate the fiber length is trimmed so that in the absence of any signal the power supply or control pulse C arrives within the clock window and corresponds to a "1". When either or both signals are incident, they interact with the control pulse through soliton dragging and pull C out of the clock time window.

The insert in Fig. 1 shows a schematic of the NOR-gate that consists of two birefringent fibers connected through a polarizing beam splitter with the output filtered by a polarizer. The control pulse C provides gain and logic level restoration, propagates along one principal axis in both fibers and corresponds to A NOR B at the output. The two signal pulses A and B are polarized orthogonal to C and are blocked by the polarizer at the output. The signals are timed so that A and C coincide at the input to the first fiber and B and C coincide (in the absence of A) at the input to the second fiber.

Springer Series in Chemical Physics, Vol. 53 **Ultrafast Phenomena VII**
Editors: C.B. Harris · E.P. Ippen · G.A. Mourou · A.H. Zewail

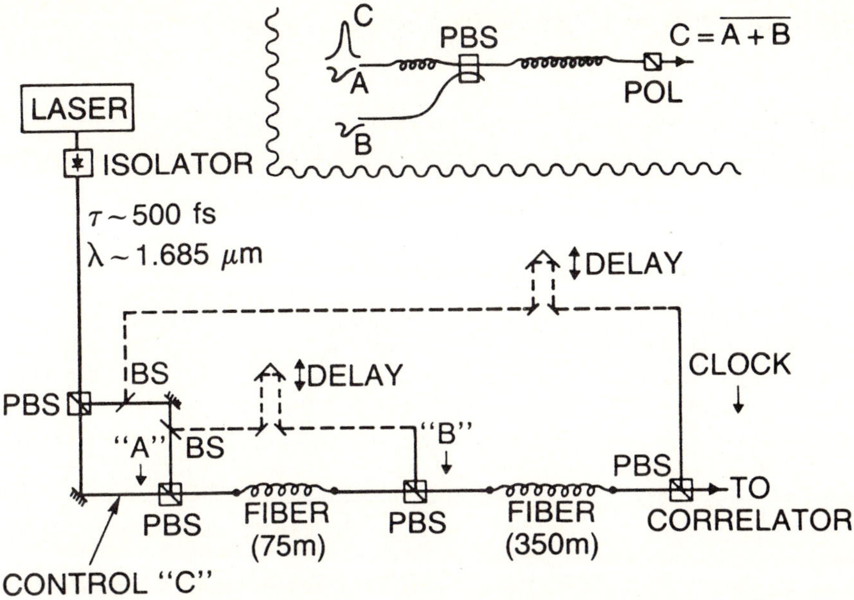

Fig. 1. Experimental configuration for testing an all-optical NOR-gate. The insert shows a simplified schematic of the NOR-gate.

Figure 1 shows the experimental apparatus for testing a single NOR-gate. We obtain $\tau\sim$ 500fsec pulses near 1.685μm from a passively mode locked NaCl color center laser in which a 2mm thick quartz birefringent plate limits the bandwidth and, thus, intentionally broadens the pulses [2]. The input stage separates the control C, signals A and B, and clock beams, and stepper motor delay stages are used to time properly signal B and the clock. The two fibers are 75m and 350m long, have a polarization dispersion of about 80psec/km, and exhibit a polarization extinction ratio better than 14:1. The control pulse output and the clock are directed to a correlator to measure the time shifts.

The correlation of the clock with the NOR-gate output is illustrated in Fig. 2. The dotted box corresponds to the clock window, and we see that C arrives within this window when no signal is present. When A=1 or B=1, C shifts between 2 to 3 psec out of the clock window; the shift from A is larger since C can time shift in both fibers. When A=B=1, C shifts by about 4psec. In this example the signal energies are 5.8pJ each and the fanout or gain (control out / signal in) is six. The control pulse energy in the first fiber is 54pJ and is reduced to 35pJ in the second fiber because of coupling losses.

To prove the cascadability and fan-out of the logic gate, we connected the NOR-gate as an inverter and fed the output back to the input

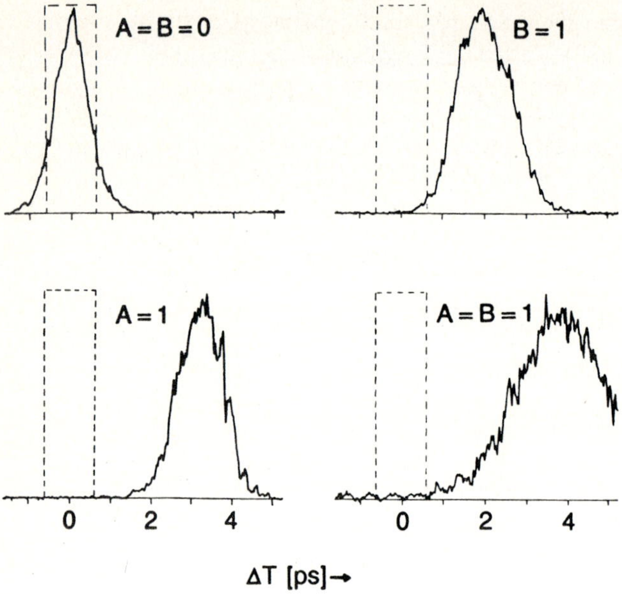

Fig. 2. Correlation of clock with NOR- gate output.

(A = 0, B = previous output from gate). We placed a 50:50 beam splitter at the output and sent half of the output through a delay line to the B input. The correlator was set to the center of the clock time window. As Fig. 3 shows, with the feedback blocked the output is a string of 1's. When the feedback is added, the output becomes an alternating train of 1's and 0's whose period is twice the fiber latency (1.75 μsec).

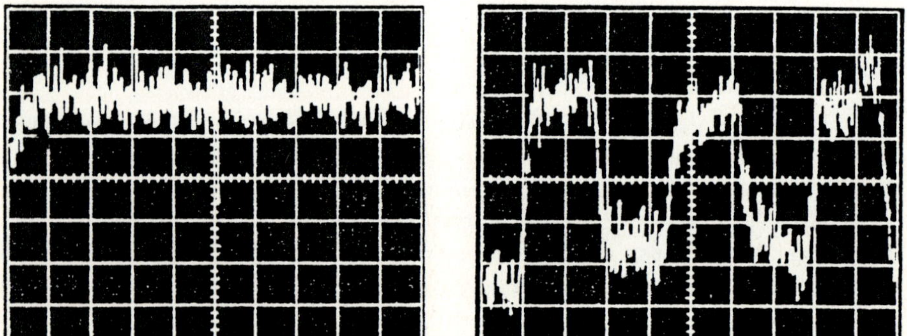

Fig. 3 Output from a NOR-gate configured as an inverter whose output is fed back to the input. Feedback blocked on the left and feedback added on the right.

We now switch gears from the physical configuration to an idealized logical construct that can be used for design purposes. Consider the basic building block consisting of a birefringent fiber surrounded by two polarizing beam splitters (Fig. 4a). We treat the general case of two equal amplitude inputs a and b (which may be signal or control) and outputs a' and b'. From a logic viewpoint, this building block can be replaced by a four-port generalized exclusive-or (GEO) module where $a' = a \cdot \bar{b}$ and $b' = \bar{a} \cdot b$.

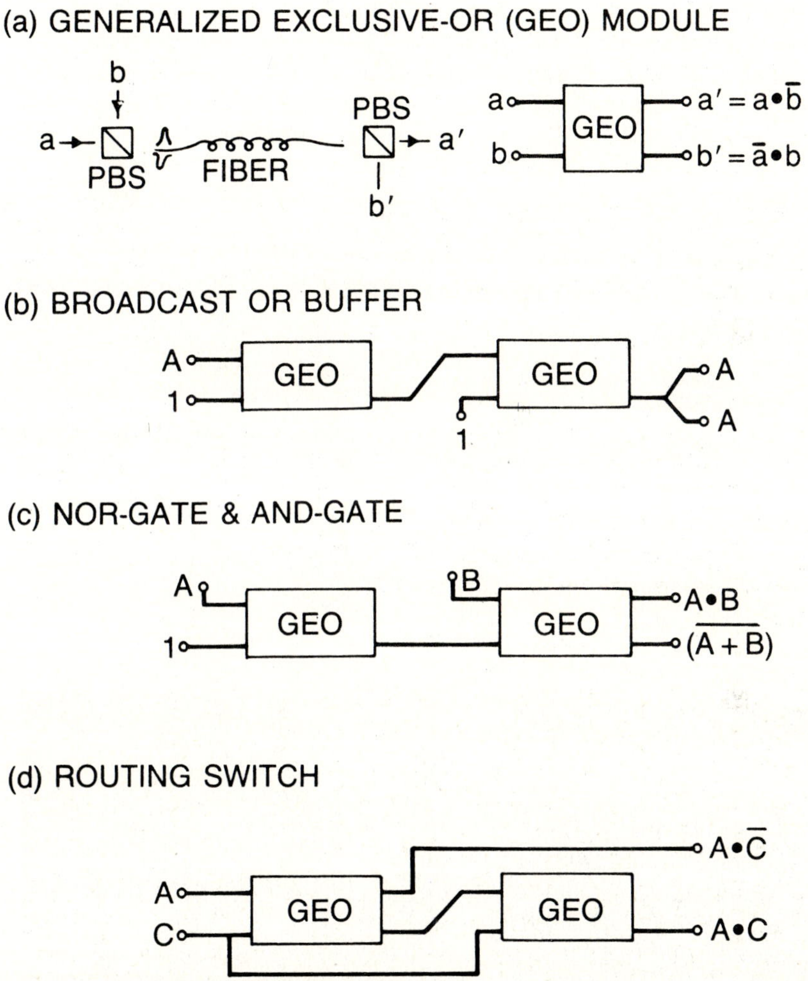

Fig. 4. (a) The basic building block consisting of a birefringent fiber surrounded by two polarizing beam splitters (PBS) can be represented by a four-port generalized exclusive-OR (GEO) module. Two GEO-modules can be configured as (b) broadcast or buffer switches, (c) NOR and AND-gates, and (d) routing switches.

Two such GEO-modules can be interconnected to provide high functionality. Figure 4b shows that by using each GEO-module as an inverter, we can implement a broadcast or buffer switch to provide multiple copies of the input. The NOR-gate described above is shown in Fig. 4c, where the control pulse is replaced by a power supply "1". Along the control axis we find A NOR B and along the orthogonal axis we find A AND B. Furthermore, for the same hardware used in the NOR-gate, we can also implement a routing switch (Fig. 4d). Depending on the value of the control C, signal A will appear at one of two separate ports.

In summary, we have presented an all-optical cascadable NOR-gate appropriate for a digital optical processor. This three-terminal gate has a switching energy of 5.8pJ, fanout of six and restores logic levels and timing at the output. Two fiber GEO-modules can be configured as broadcast or buffer switches, NOR and AND-gates, or routing switches.

[1] M.N. Islam, Opt. Lett. **15** , 417 (1990).

[2] M.N. Islam, E.R. Sunderman, C.E. Soccolich, I. Bar-Joseph, N. Sauer, T.Y. Chang and B.I. Miller, IEEE J. Quantum Electron. **25** , 2454 (1989).

Propagation and Amplification of Ultrashort Optical Soliton Pulses in Erbium-Doped Fibers for Very High Speed Communication

M. Nakazawa

NTT Transmission System Laboratories, Lightwave Communication Laboratory, Tokai, Ibaraki-Ken 319-11, Japan

Abstract Picosecond~femtosecond soliton pulse amplification and transmission in erbium-doped optical fiber amplifiers are presented, where adiabatic soliton narrowing is observed for femtosecond soliton amplification. Data transmission experiments at 5 Gbit/s-400 km and 10 Gbit/s -300 km are described, that use a new method, so called dynamic soliton communication, for sending solitons over long distances.

1. Introduction

Recent progress on erbium-doped fiber amplifiers (EDFA) has been very rapid since they have a great potential for opening a new field in high-speed optical communication [1]~[3]. EDFAs operate in the 1.5 μm region where the fiber loss in an ordinary silica-based single-mode optical fiber is minimized. Their typical advantages are a high gain of more than 30 dB, low noise, wide bandwidth, polarization insensitive gain, and high saturation output power. Because of these excellent characteristics, EDFAs are very useful not only for ordinary high speed communication, but also for optical soliton communication .

2. Erbium-doped fiber amplifiers

Figure 1 shows a block diagram of the EDFA[3]. The pump beam at 1.48 μm is coupled into an erbium fiber through a WDM fiber coupler and a non-polarized optical isolator. The WDM fiber coupler passes the 1.55 μm signal and the pump beam into the erbium fiber. The pump source is an InGaAsP laser diode (LD) and the spectral width is over 20 nm, but all spectral components can effectively contribute to the generation of the population inversion. The erbium fiber installed in the EDFA module has a doping (erbium ion) concentration of 100 ppm and a length of approximately 100 m. The 1.53 μm~1.55 μm signal is fed to the input port and the amplified signal exits from the output port. A narrowband optical filter which has a passband of 1~3 nm is installed to remove amplified spontaneous emission, resulting in a high signal to noise ratio. The pump power of 50 mW is enough to obtain a gain of more than 30 dB.

Currently, a gain as high as 46.5 dB at 1.535 μm has been successfully obtained for a pump power of 133 mW using 1.48 μm LDs [4]. The gain coefficient is estimated to be 0.76 dB/m, and the saturation output power was about + 11 dBm, which means that the EDFA can operate as a power amplifier. It should be noted that a bandwidth of approximately 30 nm from 1.53 μm to 1.56 μm enables the amplification of pico~femtosecond pulses. We have shown that, with the advantage of the high saturation output, optical soliton pulses can be amplified with EDFAs [5].

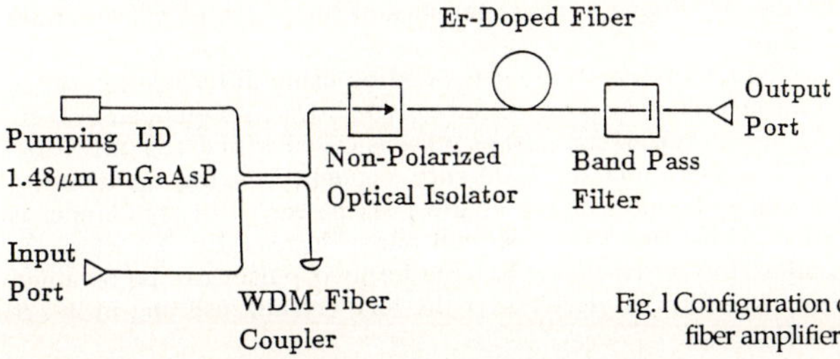

Fig. 1 Configuration of erbium-doped optical fiber amplifier (EDFA).

Springer Series in Chemical Physics, Vol. 53 **Ultrafast Phenomena VII**
Editors: C.B. Harris · E.P. Ippen · G.A. Mourou · A.H. Zewail

3. SUBPICOSECOND SOLITON AMPLIFICATION

When applying EDFAs to high-speed optical soliton communication systems, it is important to reveal their pico ~ subpicosecond response characteristics [6]. The pump power dependence of the gain of the EDFA for subpicosecond pulses is shown in Fig. 2. Open and closed circles show the gain characteristics for input peak powers of 40 W (open circles) and 12 W (closed circles), respectively. The signal wavelength was 1.552 μm and the pulse width was 0.9 ps. When the coupled pump power was 90 mW, gains for input peak powers of 40 W and 12 W were 3.8 dB and 5.5 dB, respectively. The gain for an input power of 40 W is lower than that for one of 12 W due to gain saturation.

The triangles in Fig. 2 show the gain for a signal wavelength of 1.535 μm. The pulse width was set at 0.7 ps and the input peak power was 43 W. In the present case, a gain of 2.8 dB was obtained for a pump power of 90 mW. Since EDFAs can amplify pulses with peak powers of a few tens of watts with a relatively low pump power, it is superior to stimulated Raman scattering when it is used as an optical amplifier in soliton transmission systems.

It is found that the net gain at 1.535 μm is less than that at 1.552 μm. However, the fluorescence curve of erbium-doped fiber shows that the small signal gain at 1.535 μm is larger than that at 1.552 μm. This is because a gain bandwidth of several nm is required to amplify subpicosecond soliton pulses. For example, assuming a hyperbolic secant squared waveform, a soliton with a pulse width of 0.7 ps at 1.535 μm has a spectral bandwidth of 3.6 nm. The 3 dB gain bandwidth of erbium-doped fiber is 5.2 nm around 1.535 μm and 14.3 nm around 1.552 μm. Since the gain bandwidth at 1.535 μm is almost equal to or less than the spectral width of subpicosecond pulses, the effective gain for the subpicosecond soliton is small although the narrow band gain at 1.535 μm is large. However, the gain bandwidth centered on 1.552 μm is broad enough to amplify subpicosecond solitons. Thus the gain for subpicosecond pulses in the 1.552 μm band is larger than that in the 1.535 μm band.

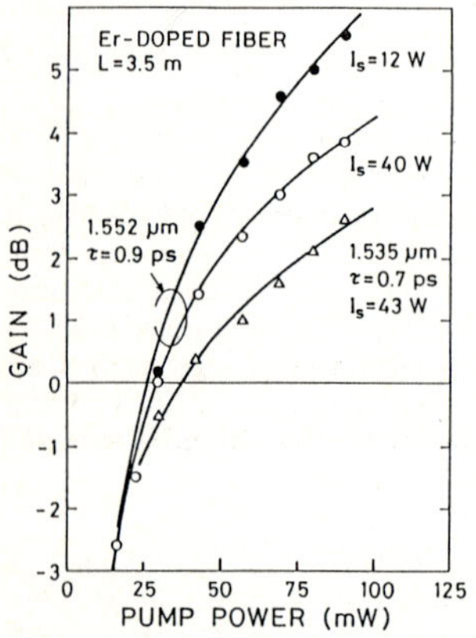

Fig. 2 Pump power dependence of EDFA gains for subpicosecond pulses. 1.535 μm (triangles) and 1.552 μm (open and closed circles).

4. FEMTOSECOND SOLITON AMPLIFICATION WITH ADIABATIC NARROWING

Recently, Zhu et al. reported the application of an erbium-doped fiber to the coupled-cavity mode locking of a KCl:Ti femtosecond laser [7], and Ainslie et al. reported a few dB gain for a 200 fs soliton pulse [8]. If the full bandwidth of approximately 30 nm can be utilized, a 100 fs pulse could be amplified. Here, we describe the amplification characteristics of femtosecond pulses under various pumping conditions for EDFAs.

Femtosecond infrared pulses in the 1.5 μm region are generated by using difference frequency mixing in a KTP crystal [9] Here, a synchronously pumped femtosecond dye laser pumped by a cw mode-locked 1.064 μm YAG laser is used. A 1.064 μm , 100 ps pulse is mixed with a 0.63 μm visible femtosecond pulse in a 3 mm KTP crystal, and the difference frequency mixing generates a femtosecond pulse in the 1.5 μm region. To increase the 1.5 μm peak power, a cavity damper is installed inside the dye laser cavity. Thus, the pulse repetition rate is decreased to 3.8 MHz. By rotating a birefringent filter installed in the dye laser, tunable infrared pulses can be obtained between 1.4 μm and 1.6 μm. An optical filter is placed after the KTP crystal, resulting in 250 fs pulses with peak powers of 2 ~ 4 kW.

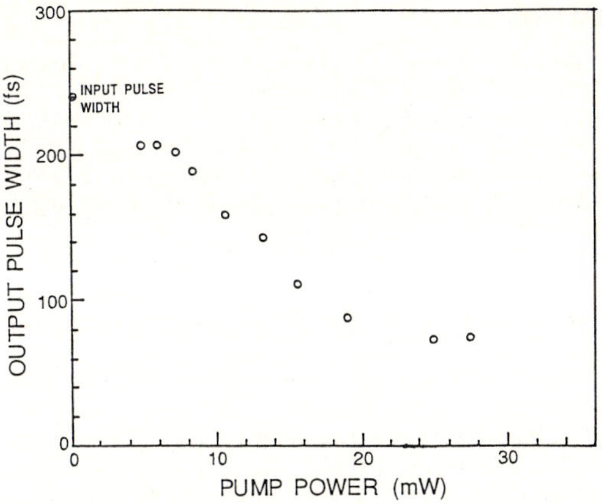

Fig.3 Changes in the soliton pulse width vs the pump power.

When erbium fibers about 100 m long and with a doping concentration of 100 ppm or 20 m long with a concentration of 470 ppm were used for the EDFA, the soliton self-frequency shift (SSFS) [10] became quite large and the continuum spectrum broadened to 1.75 μm from 1.55 μm. In this case, the spectral component of the soliton has already moved outside the EDFA gain profile, so high gain was not obtained. The key to achieving efficient femtosecond pulse amplification is to use a shorter erbium fiber with a relatively high doping concentration of between 1000 ppm and 2000 ppm.

Changes in the pulse width of the output soliton as a function of the pump power thus obtained are shown in Fig. 3. As seen, the pulse width is linearly shortened with an increase in the pump power and the spectral width is broadened corresponding to this pulse shortening. It is important to note here that the output pulse width can be controlled by using adiabatic soliton narrowing in a gain medium. For pump powers smaller than 5 mW, the EDFA acts as a loss medium, so that the output power is small and the pulse width can not be measured with the autocorrelator. A gain of 9 dB has been obtained for 260 fs input pulses with a peak power of 30 W and an average power of 30 μW (-14.2 dBm).

For a - 7 dBm input, the gain was about 5 dB. According to the perturbation theory for soliton pulses [11], the soliton pulse is shortened to $1/(1 + 2\Delta)$ when a perturbation Δ exists. Here Δ for the gain medium is given by $2\Delta = G/(10\log e)$. G is the power gain in dB. Therefore, when G is equal to 5 dB, 2Δ is 1.15 and hence the pulse width is shortened to 0.46 (=1/2.15) that of the input pulse (240 fs x0.46=110 fs). 60~70 fs pulses result from SSFS rather than from the adiabatic narrowing. It should be noted that soliton trapping occurs within a gain bandwidth of the EDFA when adiabatic gain narrowing occurs. However, when the pump power is increased to more than 20 mW, soliton narrowing is stopped at a pulse width of 60~80 fs, and pedestal components appear on the wing of the soliton pulse. At this stage, SSFS dominates and the center wavelength of the soliton spectrum is far from the gain bandwidth of the EDFA, resulting in no further adiabatic soliton narrowing.

5. GHZ SOLITON GENERATION USING A GAIN-SWITCHED DFB LASER DIODE WITH SPECTRAL WINDOWING

To date, solitons have been mainly generated by high power lasers such as color center lasers and infrared dye lasers. From a practical point of view, soliton generation from a laser diode is most desirable, however, it is well known that the generation of transform-limited pulses from a gain-switched laser diode is very difficult since a frequency down-chirp inevitably occurs due to carrier density modulation. Even in a distributed feedback laser diode (DFB LD) transient spectral changes exist, which cause the non-linear chirp induced dispersion penalty. Here we generate transform-limited solitons from a gain-switched DFB LD with spectral windowing and EDFAs [12]. The laser source is a bulk type DFB LD, which is driven by sinusoidal current modulation in the GHz region. With this gain switching technique, we can easily generate a picosecond pulse train with a GHz repetition rate. To remove the nonlinear chirp, we use a Fabry-Perot resonator (FPR) as a narrow-band spectral filter which can select a limited portion of the whole spectral component from the DFB LD. With this technique, although the transmitted average power through the FPR is reduced to 1/10 of the input power, the pulse becomes transform-limited. Then, EDFAs are used as booster amplifiers to increase the peak power up to the soliton power level.

6. 5 GBIT/S~10 GBIT/S DYNAMIC SOLITON COMMUNICATION

Here soliton transmission with lumped gain media is described [13]. Rather than the Raman process, we use a new method for transmitting solitons named "Dynamic soliton Communication", in which we use a combination of an N=1.2-2.0 soliton, lumped gain generated by an erbium-doped fiber and low-loss fiber [14]. The key to success is to use a relatively large amplitude soliton input instead of an exact N=1 soliton. Since we assume that the fiber loss is small, of the order of 0.2-0.3 dB/km, soliton narrowing occurs near the beginning of the fiber for $1<N<2$ solitons. Then the soliton starts to broaden since the GVD becomes dominant because the fiber loss weakens the intensity dependent nonlinearity. In these soliton dynamics, there exists a characteristic distance L_c at which the pulse width recovers to that of the input soliton although the intensity is attenuated by the fiber loss. The decrease of the soliton amplitude at the fiber output can be compensated for by a lumped gain generated by erbium-doped fiber. The nonlinear phase change along the fiber, which is caused by fiber loss, should be taken into account when the total transmission distance including many lumped amplifiers exceeds 1000 km.

The experimental setup for soliton transmission thus constructed is shown in Fig. 4, in which the peak power of the output soliton pulse from EDFA 2 was set in the N=2 soliton regime to extend the repeater spacing, resulting in L_c of 50 km. The laser source for soliton generation is a DFB LD. Here, we use an FPR with a bandwidth of 0.18 nm, a cavity spacing of about 200 µm, and a finesse of about 30. As a result, the output pulse width from the LD shortens to 20 ps from 27 ps. This transform-limited Gaussian pulse is coupled into a $LiNbO_3$ light intensity modulator to switch the soliton train. The modulator is a Mach-Zehnder type and the V_π= 6.6 V. The output from the modulator is amplified with two EDFAs 1 and 2 to increase the peak power in the soliton power regime.

The soliton transmission fiber (STF) comprises eight 50-km dispersion shifted fibers with zero dispersions at 1.490-1.500 µm and -2.3 - -2.4 ps/km/nm dispersion at 1.545 µm. The fiber loss is 0.24 dB/km and the mode field diameter is 5.6 µm-6.0 µm. The N=1 soliton peak power for the 20 ps pulse is as small as 6-10 mW .

After propagation over 400 km, the output soliton is detected with a high speed InGaAs PIN photo-diode with a bandwidth of 15 GHz. Thus, the modulated soliton pulse train is converted into an electrical pulse train and is amplified with a wideband electrical amplifier. Then, error rates are measured at 5 Gbit/s, where the clock pulse is lead from the pulse pattern generator. The output soliton is also measured with a synchroscan streak camera.

Digitally coded soliton trains at 400 km are shown in Fig. 5 (a)~(c). Traces (a)~(c) were coded as <11101>, <11001>, and <10001>, respectively. As can be seen clean solitons could be transmitted and no pulse broadening was observed over 400 km. The error rate thus obtained was less than 1×10^{-9}, which means that the present transmission is error free. It is reasonable to expect error free characteristics because the amplitude of the soliton is more than 20 dB larger than that of ordinary IM/DD signals.

A 10 Gbit/s soliton transition experiment was also undertaken over 300 km. Transmitted data codes of <011100110>, <110010011>, and <100110100> are shown in Fig. 6 (a)~(c), respectively. It is clearly seen that no distortion appears on the transmitted soliton waveforms. Although there was a small overlapping on the wing between the adjacent solitons, nonlinear pulling or pushing was not observed since the transmission distance was as short as 300 km.

So far there has been no report on soliton transmission in a single-pass long fiber with optical repeaters. The present results show that such a transmission is possible with the Dynamic soliton communication technique using erbium amplifiers and repeaters. By increasing the bit-rate, a few tens Gbit/s soliton transmission will be possible over 3000 km.

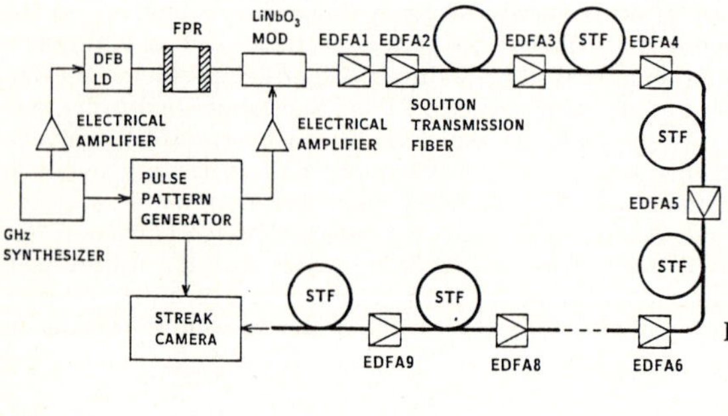

Fig. 4 Experimental setup for Dynamic Soliton Communication.

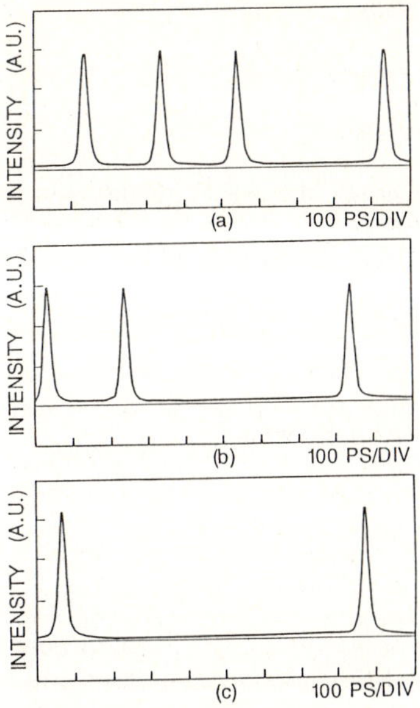

Fig. 5 5 Gbit/s digitally coded soliton transmission over 400 km.

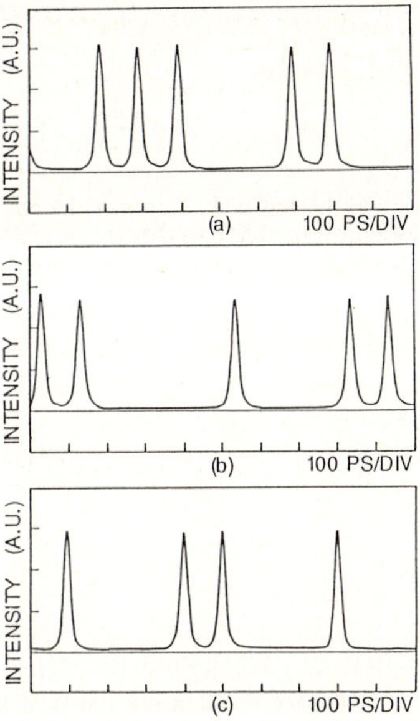

Fig. 6 10 Gbit/s soliton transmission over 300 km.

References

[1] E. .Desurvire, J. R. Simpson, and P. C. Becker, Opt. Lett., 12, 888 (1987).
[2] R. J. Mears, L. Reekie, J. M. Jauncy, and D. N. Payne, Electron. Lett., 23, 1026 (1987).
[3] M. Nakazawa, Y. Kimura, and K. Suzuki, Appl. Phys. Lett., 54, 295 (1989).
[4] Y. Kimura, K. Suzuki, and M. Nakazawa, Electron. Lett., 25, 659 (1989).
[5] M. Nakazawa, K. Suzuki, and Y. Kimura, Electron. Lett., 25, 199 (1989).
[6] K. Suzuki, Y. Kimura, and M. Nakazawa, Opt. Lett., 14, 865 (1989).
[7] X. Zhu, P. N. Kean, and W. Sibbet, Opt. Lett., 14, 865 (1989).
[8] B. J. Ainslie, K. J. Blow, A. S. Gouveia-Neto, P. G. J. Wigley, A. S. B. Sombra, and J. R. Taylor, Electron. Lett., 26, 186 (1990).
[9] K. Kurokawa and M. Nakazawa, Appl.Phys. Lett., 55, 7 (1989).
[10] J. P. Gordon, Opt. Lett., 11, 662 (1986), see also, F. M. Mitschke and L. F. Mollenauer, Opt. Lett., 11, 659 (1986).
[11] J. Satsuma and N. Yajima, Suppl. Prog. Theoret. Phys., 55, 284 (1974).
[12] M. Nakazawa, K. Suzuki, and Y. Kimura, Opt. Lett., May Issue (1990);See also, ibid, June Issue (1990).
[13] H. Kubota and M. Nakazawa, IEEE, J. Quantum Electron., April Issue (1990).
[14] K. Suzuki, M. Nakazawa, E. Yamada, and Y. Kimura, Electron. Lett., 26, 551(1990); See also, M. Nakazawa, K. Suzuki, and Y. Kimura, IEEE, Phot. Tech. Lett., 2, 216 (1990).

Observation of Two-Color Soliton Collisions in Optical Fiber

S.R. Friberg [1], *W. Jiang* [1], *Y. Sakai* [1], *and R.J. Hawkins* [2]

[1] Basic Research Laboratories, Nippon Telegraph and Telephone Corporation, 3-9-11 Midor-cho, Musashino-shi, Tokyo 180, Japan
[2] Lawrence Livermore National Laboratory, Livermore, CA 94550, USA

Abstract: Collisions of two optical solitons, with center frequencies at 1461.3 nm and 1465.6 nm, were observed using autocorrelation and spectral measurements.

1. Introduction

The interaction of two optical pulses, a process that takes place only in nonlinear media, is invariably described using CW models modified to account for pulse intensity variations. Such descriptions, even with generous use of computer simulations, are often unenlightening, relying as they do on liberal use of approximations. Soliton interactions are on firmer ground. Individual solitons propagate and collide with no change in shape, acquire no frequency chirp, and are transform limited. They are described by a nonlinear Schrödinger equation with exact analytic solutions [1]. This gives a combination of ideal pulse interaction properties and elegant mathematical characterization that has interesting applications, including quantum non-demolition measurement of photon number [2],[3] and ultrafast all-optical logic devices [4]. It also offers a new model for the interaction of two optical pulses. Here, we describe the experimental observation of a two-color soliton collision.

2. Experimental description

Observation of soliton collisions requires two solitons with jitter-free time and frequency separation. Two solitons with identical frequencies are easily made by a beamsplitter [5], but if the frequencies are different it becomes more complicated. Our solution is to use a modified pulse-compression grating-pair that produces two different frequency pulses without timing and frequency jitter.

The experimental apparatus is diagrammed in Fig. 1. Nine picosecond color-center laser pulses at 1463 nm are frequency-chirped and broadened by self-phase modulation in a polarization-preserving, single-mode, dispersion-shifted PANDA fiber ($\lambda_o = 1.53\ \mu m$) and focused into the diffraction grating pair (600 gr/mm). This gives a spectrum about 8 nm wide, 25 times that of the color center pulses. Slits directly in front of two mirrors select two separate frequency ranges to make two pulses. These are reflected back through the slits and the grating-pair and into 400 meters of single-mode polarization-preserving fiber (λ_o = 1.32 μm, $|D|$ = 11.8

Springer Series in Chemical Physics, Vol. 53 **Ultrafast Phenomena VII**
Editors: C.B. Harris · E.P. Ippen · G.A. Mourou · A.H. Zewail

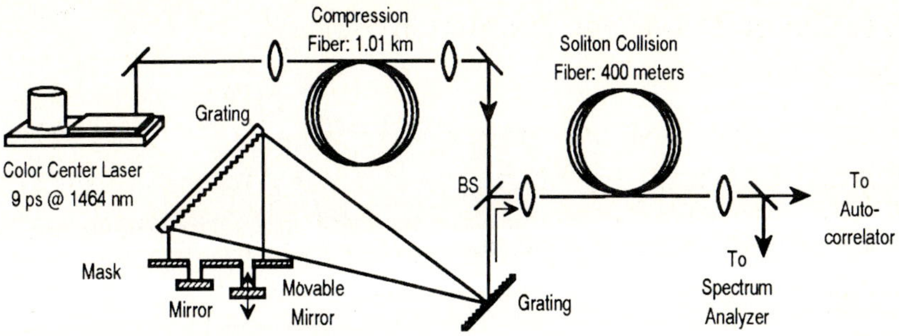

Figure 1: Soliton collision apparatus. A grating-pair pulse-compressor with two masks gives two separate frequency pulses. They collide in the fiber at a position determined by the movable mirror.

psec/km-nm). Time translation of one of the pulses is accomplished by mounting one mirror on a translator. Pulses leaving the fiber are observed using an autocorrelator and a spectrum analyzer.

Figure 2 shows autocorrelation and spectral traces for the two solitons used for the soliton collisions. These were obtained by adjusting the intensity of the pulse going into the fiber while monitoring the pulse time-bandwidth product ($\Delta\tau\Delta\nu$) at the exit. The solitons have values of $\Delta\tau\Delta\nu$ very close to the ideal value of 0.315, and are separated by 4.3 nm, or 4.3 times their bandwidth. Energies are 13 pJ each and the pulse widths deconvolve to 2.26 and 2.35 ps.

The solitons collide in the fiber if the shorter-wavelength, faster-moving soliton catches up with the slower-moving, longer-wavelength soliton. We observe that part of the collision which occurs at the end of the fiber using the autocorrelator and spectrum analyzer. Figure 3 shows a series of experimental autocorrelation traces with the center point of the collision at various places in the fiber. The traces, shown as solid lines, are labelled by the mirror position in the pulse compressor, the collision center is at 250 μm. Numerical solutions of the nonlinear Schrödinger equation [3] are next to the experimental data, offset slightly for clarity. For these, we have assumed only that the two solitons are equal in intensity, and that they are separated by 4.3 times their bandwidth. There is almost perfect agreement between the experimental and numerical results, probably because of the good solitons.

3. Discussion and Conclusions

Our results show explicitly some of the features of soliton collisions. Clearly, we have a nonlinear coherent interaction of two coherent pulses. There is no phase or timing jitter between the pulses, and the pulses are not chirped. This is implied by detailed structure of the autocorrelation traces, which otherwise would be washed out. The autocorrelation traces correspond roughly to what you would see for pulses colliding in a vacuum. However, both pulses have acquired timing shifts and phase shifts because of the nonlinear interaction [1]-[4].

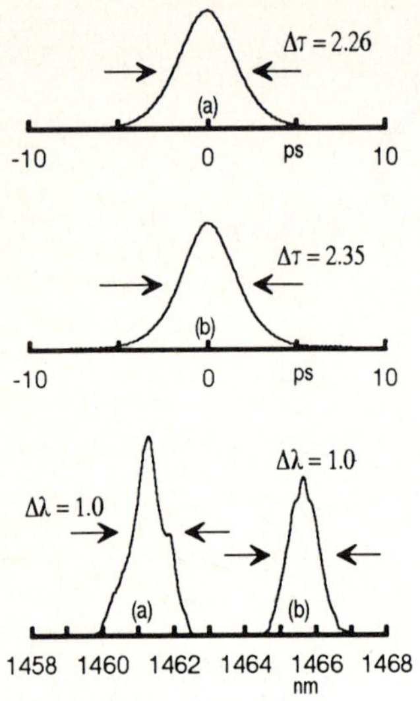

Figure 2: Autocorrelations & spectra of colliding solitons. Soliton (a): λ_c=1461.3 and $\Delta\tau\Delta\nu$=0.316. Soliton (b): λ_c=1465.6 and $\Delta\tau\Delta\nu$=0.328.

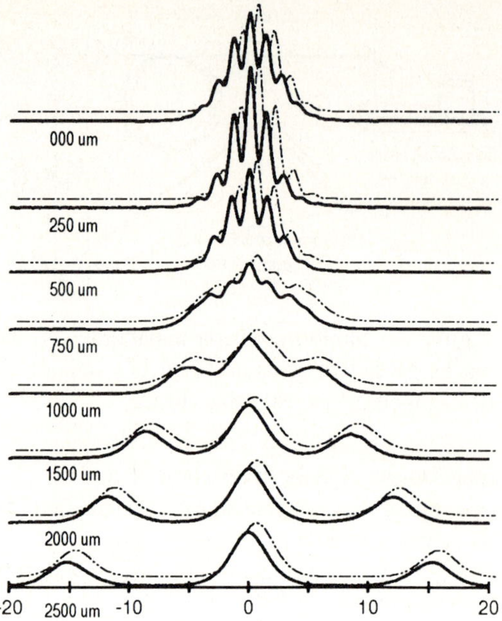

Figure 3: Soliton collision autocorrelation traces; experimental (thick lines) and numerical results (broken lines). The axis is in picoseconds, the labels give the position of the movable mirror.

References

1. V. E. Zahkarov and A. B. Shabat, Sov. Phys. JETP 34, 62 (1972).
2. H. A. Haus, K. Watanabe, and Y. Yamamoto, J. Opt. Soc. Am. B, 1138 (1989).
3. Y. Sakai, R. J. Hawkins and S. R. Friberg, Opt. Lett. **15**, (1990).
4. S. R. Friberg and W. Jiang, in **Photonic Switching II**, K. Tada and H. S. Hinton, eds., (Springer, Berlin, Heidelberg, 1990).
5. F. M. Mitschke and L. F. Mollenauer, Opt. Lett. 12, 355 (1987).

Modulational Instability for Normal Dispersion

J.E. Rothenberg

IBM Watson Research Center, P.O. Box 218,
Yorktown Heights, NY 10598, USA

Abstract: Modulational instability has been observed for the first time with visible waves. Intensity modulation at 3.5 THz is observed on two orthogonally polarized waves copropagating in a birefringent fiber.

Optical modulational instability (MI), [1] the spontaneous generation of intensity modulation on a continuous wave, has received much attention recently as a novel nonlinear optical process and as a source of high repetition rate subpicosecond pulses. In a detailed analysis [2] it was pointed out that MI would only be seen for anomalous group velocity dispersion (GVD). This effect was first observed in the nonlinear propagation of infrared light in optical fibers. [3] It has been known for some time that nonlinearly coupled waves can exhibit MI even for normal dispersion. [4] The situation of two orthogonally polarized waves nonlinearly coupled via cross-phase modulation in a birefringent fiber was treated [5] for anomalous GVD. Copropagating waves of one polarization but different frequencies was subsequently suggested to exhibit MI for normal dispersion, [6] however, the coherent interaction between the two waves tends to inhibit MI in this case. [7] The system of [5] does not suffer from this difficulty, and an extension of these analyses shows that MI occurs for normal dispersion in birefringent fibers. In the time domain both waves develop intensity modulation, and in the spectral domain the slow wave develops a single sideband frequency down-shifted, whereas for the fast wave it is up-shifted. This behavior is in contrast to previously observed optical MI [3] where the pump develops dual sidebands, and thus the present system represents a new class of MI.

The experimental apparatus used for this measurement is shown in Fig 1. A Michelson type interferometer supplies two orthogonally polarized 600 nm, 9 psec pulses which are focused into a 18m birefringent fiber. The pulse in the fast mode of the fiber is delayed at the fiber input by 25 psec with respect to the slow mode to allow for maximum interaction during propagation. The relative group delay of the two fiber modes is 1.6 ps/m. The output wave is selected by a polarizer and was either autocorrelated or measured with a spectrometer.

Figure 2 shows the spectra near and above threshold for the slow and fast waves. The pump is clearly separated from the generated single modulational sideband, which is shifted by the modulation frequency. Note also the exponential threshold behavior of the sidebands, characteristic of MI. It has been

Springer Series in Chemical Physics, Vol. 53 **Ultrafast Phenomena VII**
Editors: C.B. Harris · E.P. Ippen · G.A. Mourou · A.H. Zewail

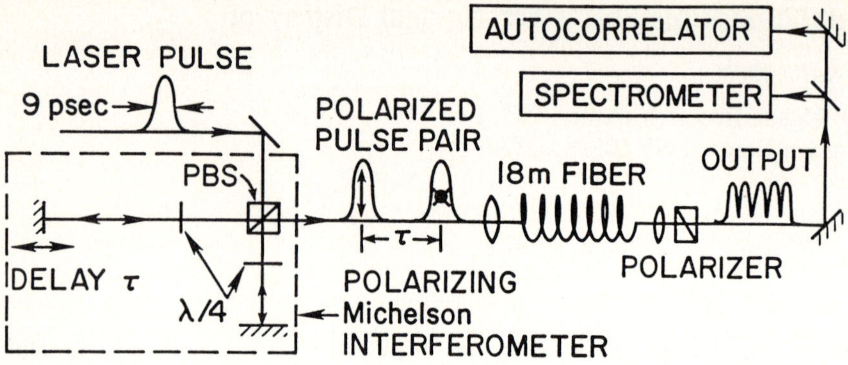

Figure 1: Apparatus used to observe MI. Polarizing beam splitter (PBS) in interferometer creates two orthogonally polarized pulses aligned to fiber axes.

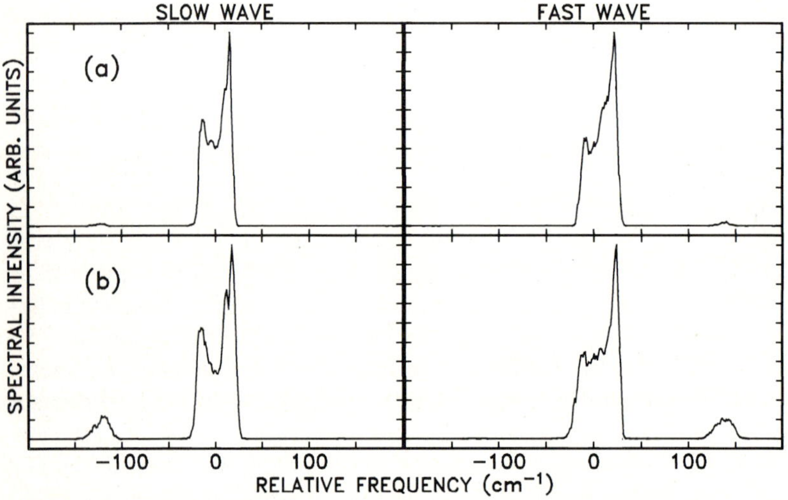

Figure 2: Measurements of the spectra of the fiber output of the slow (first) and fast (second column) waves at peak input powers of (a) 90 and (b) 110 W.

pointed out that MI can be understood in terms of four wave mixing. [2] In this new type of MI the phasematching is assisted by the fiber birefringence. [8] Figure 3 shows the autocorrelation of the slow wave at increasing power levels. The modulation frequency's variation with power is in good agreement with theory. The structure and reduction in the width of the envelope of modulation is partially understood in terms of an increasing MI bandwidth with power. Raman scattering is also shown to have a significant effect on MI. Note that the autocorrelation of a 100% modulated cw sinusoid would only have a peak to minimum ratio of 3:1, and the observed ratio approaches this value, indicating that the peak intensity modulation approaches 100%.

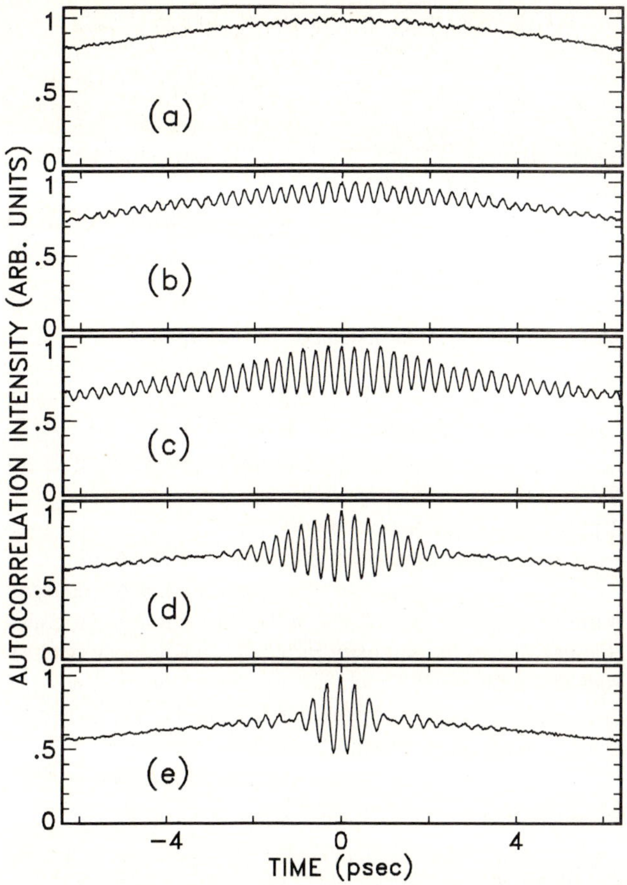

Figure 3: Background free autocorrelations of the slow output wave for peak input powers of (a) 90, (b) 110, (c) 140, (d) 170, and (e) 220 W. The intensity baseline is the true zero.

References

1. L.A. Ostrovskii, Sov. Phys. JETP **24**, 797 (1967).
2. A. Hasegawa and W.F. Brinkman, IEEE J. Quant. Elect. **QE-16**, 694 (1980).
3. K. Tai, A. Hasegawa, and A. Tomita, Phys. Rev. Lett. **56**, 135 (1986).
4. A.L. Berkhoer and V.E. Zakharov, Sov. Phys. JETP **31**, 486 (1970).
5. C.R. Menyuk, IEEE J. Quant. Elect. **QE-23**, 174 (1987).
6. G.P. Agrawal, Phys. Rev. Lett. **59**, 880 (1987).
7. J.E. Rothenberg, Phys. Rev. Lett. **64**, 813 (1990).
8. R.K. Jain and K. Stenersen, Appl. Phys. B **35** 49 (1984).

Ultrafast All-Optical Switching in a Polarization-Maintaining Fiber

H. Vanherzeele [1] *and B.K. Nayar* [2]

[1] E.I. Du Pont de Nemours & Company, CR & D Department, Experimental Station, P.O. Box 80356, Wilmington, DE 19880, USA

[2] British Telecom Research Laboratories, Martlesham Heath, Ipswich IP5 7RE, UK

Ultrafast all-optical switching has been demonstrated in various optical fiber interferometric devices using the optical Kerr effect [1,2]. The ultrafast nonlinearity of silica results in incomplete optical switching with non-square pulses. Complete pulse switching can be achieved using soliton-like pulses and we recently demonstrated its implementation in a nonlinear fiber loop mirror [3]. In many applications it is not possible to be in the soliton regime. For these applications, we propose complete switching of a short signal pulse by a high power broad control pulse using a two-wavelength operation of a Kerr modulator/switch consisting of a polarization-maintaining fiber and an analyzer [4]. In the absence of the control pulses the analyzer blocks the signal pulses. In the presence of the control pulses the output polarization state of the signal pulses changes due to the intensity induced birefringence change. The transmission, T, through the analyzer is dependent on the phase change, $\Delta\phi$, induced by the control power, P, and is given by $T=\sin^2(\Delta\phi/2)$. Dual-wavelength operation is generally considered unattractive due to the walk-off associated with the group velocity dispersion (GVD). However, the walk-off can generate the uniform phase change required for complete optical switching and is given by [4,5]

$$\Delta\phi = (8\pi n_2/3\,\lambda A_{eff}) \int_0^L P(t-\Delta\beta' z)dz$$

where, λ is the signal wavelength, L is the fiber length, A_{eff} is the effective mode area, n_2 is the Kerr coefficient, and $\Delta\beta'$ is the relative group delay between the control and signal pulses. Here we demonstrate experimentally that GVD can be exploited to achieve complete optical switching for a range of temporal delays such that the fast pulse (the longer of the two wavelengths) can walk-through the peak of the slow pulse. A minimum switching power results when the initial temporal offset is one-half of the relative group delay. The results on the completeness of optical switching are presented as a function of the initial temporal delay and the ratio of the signal and control pulse widths. In addition, broadband operation of the device is demonstrated.

In this work a broadly tunable laser system was used [6]. The control pulses are derived from a dual-head Nd:YLF laser, producing a 100 MHz train of 52 ps (FWHM) pulses at 1.053 µm. A synchronously mode-locked dye laser was used to generate signal pulses continuously tunable in the range 0.78-1.0 µm. The duration of the dye laser pulses can be selected in the range 2-6 ps or 30-50 ps (FWHM). The relative temporal delay between the control and signal pulses is adjustable by an optical delay line. The basic experimental arrangement is an optical Kerr

Springer Series in Chemical Physics, Vol. 53 **Ultrafast Phenomena VII**
Editors: C.B. Harris · E.P. Ippen · G.A. Mourou · A.H. Zewail

modulator/switch and consists of a YORK polarization-maintaining fiber and an analyzer at the output. The signal pulses (average power 1 mW) and the control pulses (average power 0-1 W) are co-propagating and have their polarizations at 45° and along one of the fiber axes respectively. The short dye laser pulses, have a polarization dispersion walk-off length of 2.2 m. In order to overcome this, we used two approximately equal lengths (1.6 m) spliced such that the fiber axes were orthogonal at the splice and this resulted in 100:1 polarization extinction ratio. At the output of the fiber, the control beam is rejected by a dichroic beam splitter, while the signal beam is passed through a Soleil-Babinet compensator to obtain a linearly polarized beam. This beam was analyzed with a Glan-Taylor polarizer.

In a first set of experiments we investigated the signal switching ratio for short (3 ps) and broad (30 ps) signal pulses as a function of the initial delay between the control and signal pulses. The control pulse energy was selected to give about 50% switching in order to operate near the maximum sensitivity of the switching characteristic and was slightly different in each experimental run. The results at two signal wavelengths, 0.85 μm and 0.95 μm, are shown in Fig.1. Clearly, a maximum switching ratio results when the control pulses are optimally delayed with respect to the signal pulses. This occurs at 0.85 μm (0.95 μm) when the control pulses are lagging by 18 ps (8 ps). The difference in the optimum delay is governed solely by the GVD and as expected is independent of the signal pulse width. The relative group delay between the signal and control pulses was independently measured using a long length of a similar fiber : $\Delta\beta'=13.8$ ps/m (at 0.85 μm) and $\Delta\beta'=5.3$ ps/m (at 0.95 μm). Thus the one-half relative group delays in a 3.2 m long fiber are 22 ps and 9 ps at 0.85 μm and 0.95 μm respectively and are in reasonable agreement with the measured optimum temporal offset values. Hence these results clearly demonstrate that a minimum switching power results for an initial temporal offset equal to one-half of the relative group delay. In a second set of experiments we investigated the completeness of optical switching for both short (3 ps) and broad (30 ps) signal pulses at 0.85 μm and 0.95 μm as a function of the control pulse energy at the appropriate optimum delay settings. The results are plotted in Fig.2. It can be seen that for the short signal pulses complete switching occurs. The solid line is a theoretical fit at 0.85 μm. In addition, it is significant to note that even for the broad signal pulses (0.6 x control pulse width) nearly complete switching results. The control pulse energy for complete switching is 4.8 ± 0.5 nJ and was nearly the same over the 0.83 μm (fiber's higher mode cut-off) to 1.0 μm wavelength range within the measurement accuracy. The above results motivated the investigation of the switching ratio as a function of

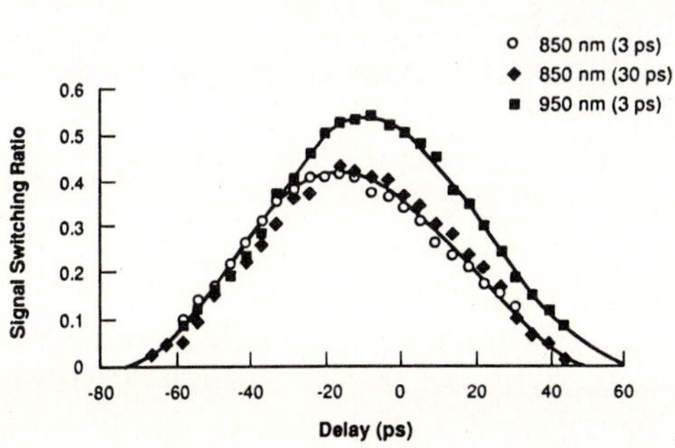

Fig.1 : Signal pulse switching ratio as a function of the initial delay between the signal and control pulses. For negative delays, the control pulses are lagging the signal pulses.

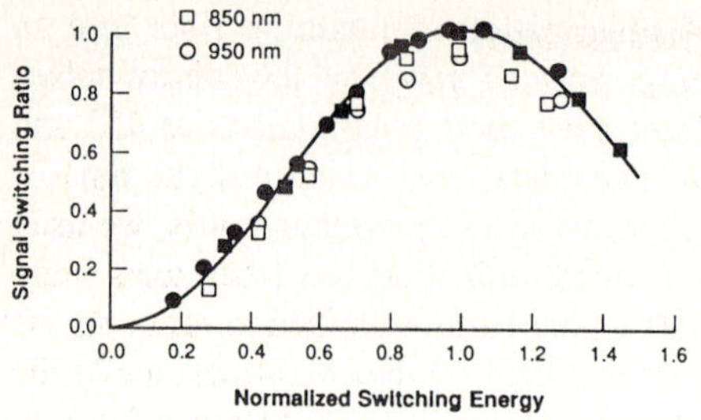

Fig.2 : Signal pulse switching ratio for optimum initial delay settings at 0.85 μm and 0.95 μm as a function of the normalized control pulse energy at 1.053 μm. Open (filled) dots correspond to 30 ps (3 ps) signal pulses. The normalization factor for the control pulse energy is 4.8 nJ.

the signal to control pulse width ratio, r. For this purpose, the control pulse energy was set at 4.8 nJ to achieve complete switching of the shortest signal pulses (at 0.85 μm), and the switching ratio was measured as a function of signal pulse width. The results show that the switching ratio rapidly decreases for r >0.6. This is to be expected as for the fiber length used, insufficient walk-through occurs between the pulses. However, a longer fiber length can be used to ensure an adequate walk-through and hence yield complete switching.

In conclusion, we have demonstrated for the first time to our knowledge complete optical switching of the shortest, 2 ps non-soliton pulses. It has been experimentally shown that the GVD can be exploited to achieve complete switching of signal pulses comparable in duration to the control pulses. A more general conclusion follows that the relative ratio of the pulse widths is not critical as long as the fast pulse can walk-through the slow one. As an example, complete switching of a signal pulse can be accomplished using a very short control pulse. Furthermore it has been shown that this switching scheme is broadband. The operation of this device in the near IR will result in a lower switching energy as the smaller GVD will enable the use of longer fibers. This switching concept is quite general and can be used for cross-correlation experiments, multiplexing/demultiplexing, optical routing, sampling and logic applications.

REFERENCES

1. J.M. Dziedzic, R.H. Stolen, and A. Ashkin, Appl. Opt. 20, 1403 (1981).
2. N.J. Doran, D.S. Forrester, and B.K. Nayar, Electron. Lett. 25, 267 (1989).
3. K.J. Blow, N.J. Doran, and B.K. Nayar, Opt. Lett. 14, 754 (1989).
4. T. Morioka and M. Saruwatari, Opt. Eng. 29, 200 (1990).
5. K.J. Blow, N.J. Doran, B.K. Nayar, and B.P. Nelson, Opt. Lett. 15, 248 (1990).
6. H. Vanherzeele, Appl. Opt. 29 (20 May 1990).

Observation of Induced Phase Modulation of Femtosecond Pulses in Glass and Organic Fibers

M. Yamashita[1], *K. Torizuka*[1], *and T. Uemiya*[2]

[1]Laser Section, Electrotechnical Laboratory, Umezono 1-1-4, Tsukuba, Ibaraki 305, Japan
[2]New Chemical Laboratory, Sumitomo Electric Industries, Shimaya 1-1-3, Konohana, Osaka 554, Japan

Since spectral broadening of a weak, picosecond pulse (I_2 pulse) was observed due to phase-modulation induced (IPM) by a second, independent and intense pulse (I_1 pulse) at a different wavelength in condensed media [1], picosecond or femtosecond pulse compressions using IPM in fibers were proposed and analyzed [2-7]. However, IPM phenomena in the femtosecond time-region in fibers have not been yet made clear experimentally. In this paper, we report the first observation of IPM for femtosecond pulses propagating in a fused-silica glass fiber (glass fiber) and a DAN (4-(N, N-dimethylamino)-3-acetamidonitrobenzene) single crystal-cored fiber (DAN fiber) [8]. Furthermore, it is shown that the latter's ultrafast nonlinear refractive index $n_{2.DAN}$, whose value has been so far unknown to our knowledge, is much larger than the former's $n_{2.G}$.

A 1-kW, 100-fs pulse around 625 nm from a CPM laser was amplified to 50 MW by means of two-stage dye-amplifiers pumped by a 1-kHz, 10-mJ XeCl laser. The amplified pulse of the 250-fs duration was split into intense I_1 and weak I_2 pulses (9:1) as a modulation-inducing pulse and a pulse to be modulated, respectively. After the linear polarization of the two pulses was set to be perpendicular to each other, they were copropagated to fibers by an optical delay line including polarizing beam splitters. The fiber-output spectra of I_2 and I_1 pulses for observation of IPM and self-phase modulation (SPM), respectively, were detected by means of OMA gated by 10-ns electric pulses in synchronization with the 1-kHz amplified pulses.

In Figs.1-A) and -B), the observed spectra of the I_2 pulses which were and were not copropagated with a 500-kW I_1 pulse in an isotropic single-mode glass fiber of the 5-cm length, respectively, are shown. We find that the former spectral width is wider than those of the latter and of the input pulse (Fig.1-C)). In addition, it was confirmed that the former spectral width was almost equal to that of the I_1 output pulse broadened by dispersive SPM. Furthermore, with the decrease of the I_1 pulse power to 70 kW, it was observed that the spectral width of the I_2 output pulse decreased from 23.9 to 15.5 nm. These results suggest that the former spectral broadening is due to dispersive IPM in the fiber.

A similar experiment was carried out for a DAN fiber with dense flint glass (SF4) cladding. The DAN single crystal in the fiber was grown in the glass capillary by the Bridgeman-Stockbarger method. One dielectric axis (y-axis) of the crystal is perpendicular to the fiber propagation-axis and another one (x-axis) is inclined at 36° against its fiber axis, so that the DAN fiber shows the birefringence property. At first, in order to

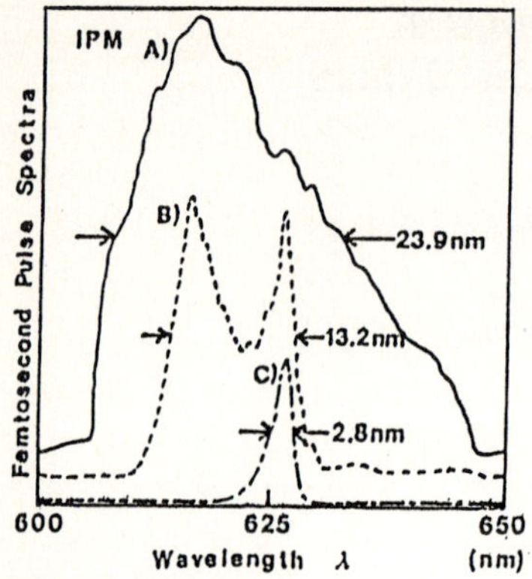

Fig.1. A) and B) are the spectra of the I_2 output pulses modulated and non-modulated by a 500-kW I_1 pulse in a 5-cm glass fiber, respectively. C) is the input pulse spectrum.

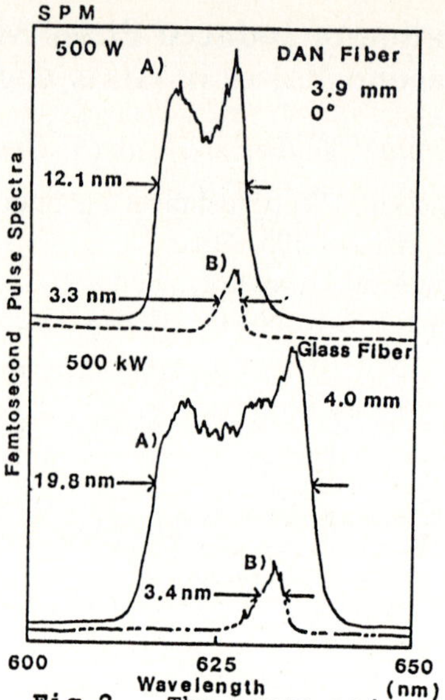

Fig.2. The upper and lower A) are the SPM spectra of the I_1 outpulses of the 500-W input power in a DAN fiber and of the 500-kW input power in a glass fiber, respectively. B) is the input pulse spectrum.

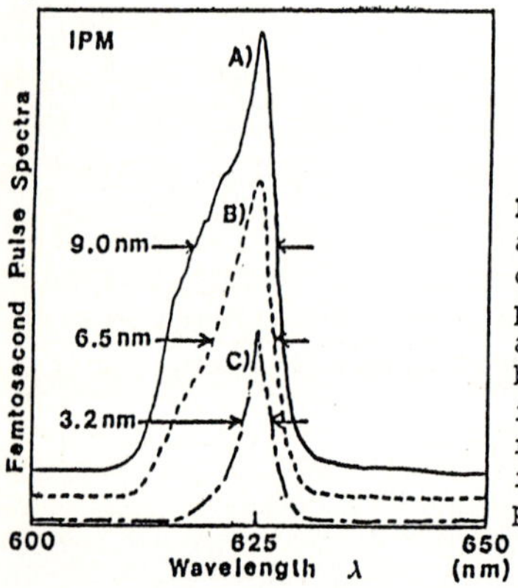

Fig.3. A) and B) are the spectra of the I_2 output pulses modulated and non-modulated by a 10-kW I_1 pulse in a DAN fiber, respectively. C) is the input pulse spectrum.

evaluate the value of the $n_{2.DAN}$ we measured spectral broadening due to dispersive SPM in the fiber of the 3.9-mm length and 4.6-μm core-diameter and in a glass fiber of similar length and core-diameter. The angle of the linear polarization of the input pulse was adjusted to the dielectric y-axis of the DAN fiber so that its pulse propagated effectively. The upper and lower A) in Fig.2 show the spectrum of the output pulse observed for the DAN fiber at the low input power $I_{p.DAN}$ of 500 W and that for the glass fiber at the much high input power $I_{p.G}$ of 500 kW, respectively. From the comparison between the two spectra and input powers, we find that the DAN fiber broadens the spectrum of the propagation pulse by dispersive SPM much more effectively than the glass fiber. We can determine the $n_{2.DAN}$ value using an equation [9] of the ratio of the spectral width $\Delta\nu_{DAN}$ broadened in the DAN fiber to that $\Delta\nu_G$ in the glass fiber,

$$\Delta\nu_{DAN}/\Delta\nu_G = \{ [(n_{2.DAN} \times I_{p.DAN})/(n_{0.DAN} \times k_{DAN})] / [(n_{2.G} \times I_{p.G})/(n_{0.G} \times k_G)] \}^{1/2}$$

and dispersion equations of the DAN crystal [10]. Here, the $n_{0.DAN}$ and $n_{0.G}$, and the k_{DAN} and k_G are the linear refractive indexes and the group velocity dispersions of the DAN crystal and fused-silica fibers, respectively. Consequently, $n_{2.DAN} = 1.7 \times 10^4 \times n_{2.G} = 2.2 \times 10^{-9}$ esu and $\chi^{(3)}_{DAN}(-\omega:\omega,-\omega,\omega) = 1.0 \times 10^{-10}$ esu in the femtosecond time-region. We find that this very large value is reasonable, when considering the already known n_2 value due to the electronic nonlinear polarization of liquid nitrobenzene [11, 12] of the same family as DAN, the larger molecular-number per volume in the crystal, and the large molecular packing and parallel effects in the crystal on the nonlinear polarization.

Finally we observed the output spectra (Figs.3-A) and -B)) of the I_2 pulses which were and were not copropagated with a 10-kW I_1 pulse in the DAN fiber of the 3-mm length and 4 μm core-diameter. Here, the two linear polarizations of the I_1 and I_2 input pulses, perpendicular to each other, were at an angle of 45° against one dielectric y-axis of the DAN fiber. From Fig.3 we find that the spectrum of the I_2output pulse at the I_1-pulse-on is broadened more widely than those of the I_2output pulse at the I_1-pulse-off and of the input pulse, like the result for the glass fiber. In addition, it was confirmed that the spectral width of the former I_2output pulse was almost equal to that of the I_1 output pulse. Therefore, we conclude that the former spectral broadening is mainly due to dispersive IPM. We think that, at this experimental condition, the spectral broadening is hardly affected by the practical difference of the wavenumbers, group velocities and group velocity dispersions between the effective dielectric axes. A detailed examination of this DAN fiber is in progress.

References

1. R.R.Alfano, Q.X.Li, T.Jimbo, J.T.Manassah and P.P.Ho, Opt. Lett. 11, 626 (1986).
2. D.Schadt and B.Jaskorzynska, Electron. Lett. 23, 1090 (1987).
3. B.Jaskorzynska and D.Schadt, IEEE J. Quantum Electron. QE-24, 2117 (1988).
4. M.Yamashita, K.Torizuka, T.Shiota and T.Sato, in Ultrafast Phenomena VI, Ed. by T.Yajima et al. (Springer-Verlag, Berlin, 1988) p.50.
5. J.T.Manassah, Opt. Lett. 13, 755 (1988).
6. G.P.Agrawal, P.L.Baldeck and R.R.Alfano, Phys. Rev. A40, 5063 (1989).
7. M.Yamashita and K.Torizuka, Japanese J. Appl. Phys. 29, 294 (1990).
8. T.Uemiya, N.Uemishi, Y.Shimizu, S.Okamoto, K.Chikuma, T.Tohma and S.Umegaki, SPIE vol.1148, 207 (1989).
9. J.Herrmann and B.Wilhelmi, in Lasers for Ultrashort Light Pulses, (Norh-Holland, Amsterdam, 1987) p.253.
10. P.Kerkoc, M.Zgonik, K.Sutter, Ch.Bosshard and P.Gunter, Appl. Phys. Lett. 54, 2062 (1989).
11. P.P.Ho and R.R.Alfano, Phys. Rev. A20, 2170 (1979).
12. W.T.Lotshaw, D.McMorrow, C.Kalpouzos and G.A.Kenney-Wallace, Chem. Phys. Lett. 136, 323 (1987).

Part V

Optoelectronics

Optically Induced Femtosecond Ele
from Semiconductor Surfaces

X.-C. Zhang, B.B. Hu, J.T. Darrow, S.H. Xin,

Electrical Engineering Department, Microelec
and Center for Telecommunications Research,
New York, NY 10027, USA

Abstract. Terahertz bandwidth electromagnetic ra
semiconductor surfaces [1]. This diffraction-limite
free space and can be optically steered by varying th
addition, from the analysis of measured wavefor
concentration, carrier mobility, crystal symmetry , o
field near the semiconductor surfaces can be estimate

Most semiconductor surfaces have surface states. The
and bend the energy bands near the surfaces or int
region and the built-in surface electric field, as shown in Fig. 1. The depletion field is directed normal to the surface, and the strength of the field depends on the Schottky barrier potential V, the doping concentration N and the thermal energy kT/e, as described by the expression:

$$E(x) = \frac{eN}{\varepsilon}(W - x) \quad ; \quad W = \sqrt{\frac{2\varepsilon}{eN}\left(V - \frac{kT}{e}\right)}$$

where e is the electron charge, ε is the dielectric constant, W is the depletion width and x is the distance measured from the surface. When a femtosecond laser beam illuminates on a bare semiconductor surface with the photon energy greater than the bandgap of the semiconductor, photons are absorbed, creating free carriers. The natural bias (the built-in electric field) drives the surface photocarriers to form a transient photocurrent. The current pulse has a risetime on the order of the laser pulse duration and a fall time comparable with the sweeping time of the free carrier across the depletion width. The transient photocurrent radiates inward (transmitted) and outward (pseudoreflected) femtosecond electromagnetic pulses. If the optically illuminated area on the semiconductor is larger than or comparable with the wavelength of the electromagnetic wave, the radiation is directional and diffraction-limited. A duration of 50 fs laser pulse from CPM laser was used to optically induce and detect electromagnetic radiation. The radiated field is detected by a 100 μm photoconducting dipole antenna fabricated on radiation-damaged silicon-on-sapphire . The temporal measurement is achieved by varying the time-delay between the excited laser pulse illuminated on the semiconductor surface and the trigger laser pulse focused on the photoconducting detector.

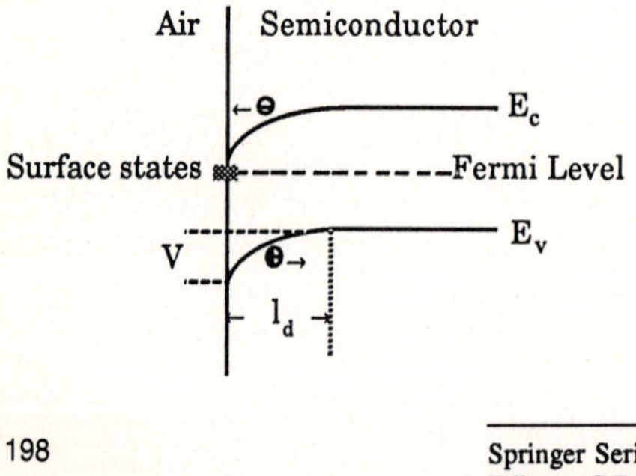

Fig. 1: Band diagram of semi-insulating InP. Surface states near the conduction-band edge cause Fermi "pinning" at the interface. Photocarriers are swept across the depletion width l_d to form a photocurrent pulse.

Springer Series in Chemical Physics, Vol. 53 **Ultrafast Phenomena VII**
Editors: C.B. Harris · E.P. Ippen · G.A. Mourou · A.H. Zewail

The radiated field of the electromagnetic wave can be derived from the Hertz vector potential with a plane wave approximation and the boundary condition of a current normal to the semiconductor surface. For a weak laser beam, the amplitudes of outward optically induced radiated field E_{out}, inward field E_{in} and surface current J_s can be expressed as [1]

$$E_{out} = -n_s E_{in} = \eta_s J_S \frac{\sin\theta_1}{\cos\theta_1 + n_s \cos\theta_2}; \qquad J_S = e\int_0^{\infty} n(x)\mu(E_d,x)E_d(x)\,dx$$

where n_s and η_s are the index of refraction and the characteristic impedance of the sample, θ_1 and θ_2 are the outward and inward angles of the radiated fields relative to normal, e is the electron charge, n(x) is the photocarrier density, $E_d(x)$ is the static built-in field, and $\mu(E_d, x)$ is the transient carrier mobility which is a function of $E_d(x)$. The incident optical angle θ_{op}, propagating direction of outward radiated field θ_1 and inward radiated field θ_2 satisfy a generalized Fresnel law [3].

A large variety of samples as radiation sources have been selected from III-V, II-VI and group IV semiconductors. For bandgaps larger than the CPM photon energy ($\hbar\omega$ = 2 eV), no radiation process is expected, and this has been verified by ZnSe (E_g = 2.4 eV) and GaP (E_g = 2.2eV). The normalized amplitude of radiated fields from several samples are listed in Table 1. Si and GaSe show moderate signals, because the absorption lengths in Si and GaSe at an optical wavelength of 620 nm are approximately 10 times longer than that in InP and GaAs, causing few carriers in the depletion layer to contribute to the radiation. We also have found that different semiconductors show different waveforms of radiated fields. For example, under the same experimental conditions, the waveforms of InP are close to bipolar and that of GaAs are close to unipolar.

Table 1

Sample	InP	GaAs	CdTe	CdSe	InSb	Ge	GaSb	Si	GaSe
Signal	100	71	33	11	8	7	2	0.5	<0.1

Figure 2(a), (b) plot the temperature dependent time-integrated radiation powers and the radiated fields (peak value) from an undoped GaAs (n= $10^{16}/cm^3$) and a undoped InSb (n= $8 \times 10^{15}/cm^3$ at 77 K), respectively.

The radiated field is proportional to the surface static field, resulting in opposite polarities of the radiated fields from n-type and p-type samples. Figure 3 shows the opposite radiated fields from a n-type and a p-type GaSb (n,p$\approx 10^{18}$ /cm^3) samples. The differences of the waveforms are due to the different values of the surface built-in potentials. A similar result has been achieved from GaAs samples (n,p$\approx 10^{16}$ /cm^3) also.

Metal/semiconductor Schottky barriers have been used to determine and estimate the sign and the strength of the surface field. Thin gold films were evaporated on n-type and p-type GaAs (111) wafers (n, p $\approx 10^{16}$ /cm^3). The thickness of the gold films were varied

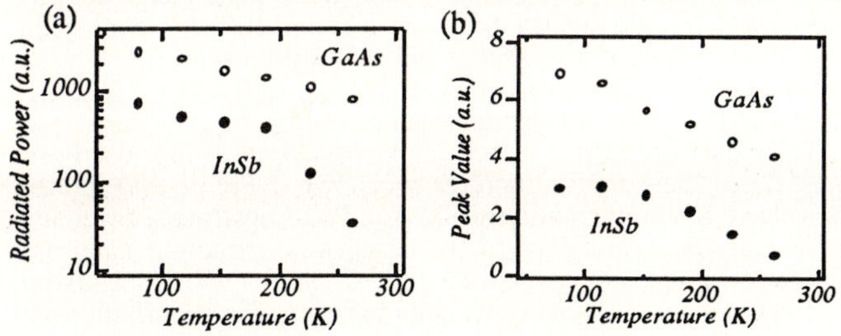

Fig.2: (a) Time-integrated powers and (b) peak values vs temperature from GaAs and InSb.

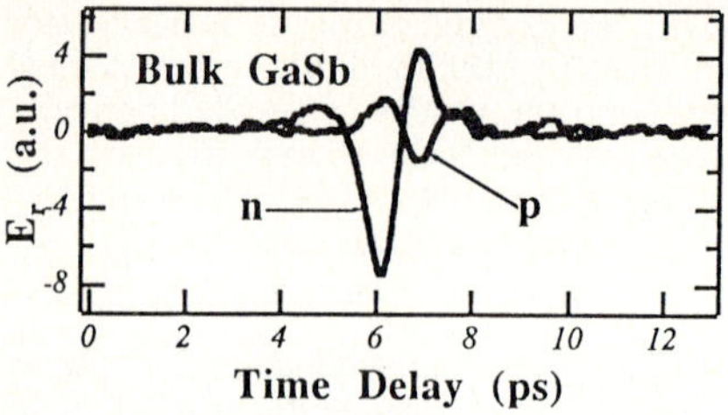

Fig.3: Radiated fields from n-type and p-type GaSb bulk samples.

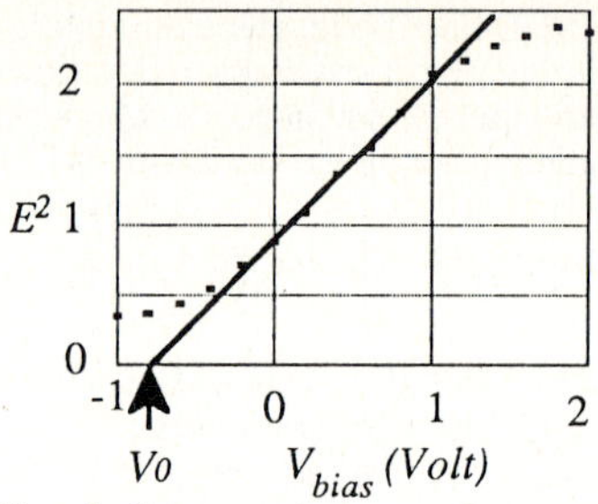

Fig. 4: Square of the radiated field (peak value) as a function of bias voltage from n-type GaAs. Schottky barrier height is given by *V=kT/e-Vo.*

from 40 to 80 angstroms. An indium contact on the back of the sample was annealed to provide an ohmic contact. The Schottky samples show good rectifying behavior in current-voltage measurements. For a moderate bias voltage the external potential is mainly applied across the Schottky barrier and changes the depletion width *W*. For high forward bias, the high current causes the bulk resistance to reduce the potential affecting the Schottky barrier and the built-in field can not be completely eliminated. For high reverse bias, the field across the depletion region causes breakdown of the Schottky diode, again limiting the effect of the surface field. The externally controlled surface field can be monitored by the radiated electromagnetic signal.

For GaAs with the doping concentration greater than 10^{16} /cm^3, the radiated field E_{out} is proportional to the depletion width *W* [2]. The radiated field of n-type GaAs at zero bias is about 5 times stronger than that of p-type. Figure 4 shows the amplitude square of the radiated signal from n-type GaAs with a 40 angstrom Au film vs the applied bias voltage. The radiated field increases at a moderate reverse bias on the n-type sample and saturates at higher reverse bias. Basically, the reverse bias enhances the internal (built-in) field and increases the depletion width, while the forward bias cancels the internal field. The direction of reverse bias (as well as forward bias) of n-type and p-type Schottky barriers is opposite. Since $E^2 \propto W^2 \propto (V + V_{bias} - kT/e)$, the static built-in potential *V* can be estimated from the intersection point where $V_{bias} = kT/e - V$ in Figure 4. We have estimated that the built-in field *V* has a value of 0.81±0.05 eV for an n-type Au/GaAs Schottky sample. Similar to the analysis of the n-type Au/GaAs Schottky barrier, we have estimated *V* =0.32±0.1 eV for a p-type Au/GaAs Schottky barrier. The increased experimental error in p-type sample is from the larger leakage current in our p-type Schottky diode.

The strain generated electric field in a <111> oriented misfit superlattice provide a suitable bias for generating optically induced femtosecond electromagnetic radiation. The radiated fields from strained layer superlattices (InGaAs/GaAlAs, GaSb/AlSb and InGaSb/InAs) have been measured and analyzed. The direction of the piezoelectric field in the layer depends on the type of distortion of the layer (contraction or expansion) and the grown face of the substrate. For the layer which has a small lattice constant and suffers a expansion, the piezoelectric field from a A-face (B-face) superlattice points inward (outward).

The results of one pair of GaSb/AlSb strained layer superlattices and one pair of GaSb thin film samples grown on <100> and <111>B oriented GaAs wafers by molecular beam epitaxy (MBE) are discussed here. The sample structures are shown in Figure 5(a) and (b). The GaSb/AlSb superlattices have 20 multiquantum wells with 400 Angstroms of well and barrier width. The GaSb films have a thickness of 0.5 μm. The conduction band and valence band offsets in the superlattices are approximately 0.5 and 0.4 eV, respectively, giving a strong spatial confinement for both electrons and the holes. The superlattice and

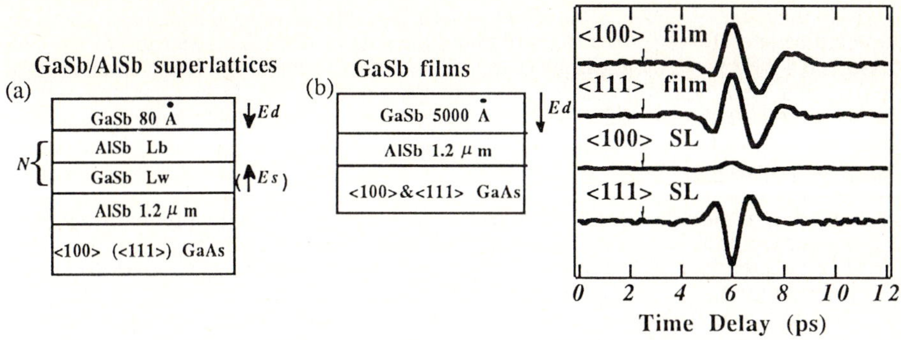

Fig. 5: (a) Structures of GaSb/AlSb superlattices, (b) GaSb films and (c) their waveforms of the radiated fields.

film samples are not intentionally doped; Hall measurements show that all samples are p-type and have a typical impurity level of $8.2 \times 10^{15}/cm^3$.

The Fermi level in these p-type samples "pins" the energy bands downward near the surface and forms the depletion field having a direction inward to the substrates. The samples should show the same p-type behavior in the electromagnetic waveforms if the only contribution to the radiation is from the surface depletion layer. A larger radiation amplitude from the GaSb films than that from the superlattice GaSb cap layer is expected due to the greater number of photocarriers in this region. When the strain-induced electric field contributes to the electromagnetic radiation, the waveforms from the <111> oriented GaSb/AlSb superlattices should be very different from the others. For a free-standing <111> oriented strained layer superlattice, piezoelectric fields exist both in the wells and the barriers, and their directions are opposite. With a thick (1.2 μm) AlSb as a buffer layer for the superlattice, the lattice distortion in the AlSb barriers is much less than in the GaSb wells and the strain in the AlSb barriers is negligible compared with that in the GaSb wells. Furthermore, in <111> oriented GaSb/AlSb superlattices with the optical excitation energy of 2 eV, the radiation from the quantum barriers is much weaker than that from the quantum wells because of the moderate field strength and a smaller absorption coefficient (α(AlSb) is 288 times less than α(GaSb)) in the barriers. Since the <111> oriented superlattices were grown on B-face substrates, the direction of the strain-induced field in the GaSb quantum wells, which have a smaller lattice constant and suffer compression, points outward, opposite to the surface depletion field for p-type samples. The strain-induced field overrides the surface fields, resulting in an outward macroscopic electric field.

The samples were mounted in a cryostat at a temperature of about 80 K, with their front surfaces aligned to the same plane to equalize the timing for the temporal waveform measurement. The incident angle of the laser beam was 55 degrees and the optical power density on the sample was 3 mW/cm^2. The radiated electromagnetic pulses at the specular angle were collected by two gold-coated off-axis parabolic mirrors and were focused on a photoconducting antenna by an attached sapphire lens. Fig. 5 (c) shows the radiation waveforms from the GaSb films and the superlattices at 80 K. The waveforms (p-type samples) from the surface depletion fields of the two GaSb films and the <100> superlattice are similar, while the waveform from the <111> superlattice is opposite from the others, as we expect from the opposite direction of the strain-induced field. The radiation from the piezoelectric field in the <111> superlattice clearly overrides the contribution from the surface depletion field. The amplitude (absolute value) from the <111> superlattice is 8.6 times larger than that from the <100> superlattice.

A pulse duration of 450 fs (full width at half maximum) is measured for the <111> superlattice and 800 fs for the films (without deconvolution). The spectrum of the film peaks at 0.4 THz while the spectrum of the <111> superlattice is completely limited by the detector response. We also observed that the spectra from the <111> superlattices with narrower well width have richer frequency components in the high frequency region than the spectra from wider wells, but lower amplitudes due to the poor antenna efficiency at the higher frequencies.

In summary, femtosecond electromagnetic radiation can be optically induced from semiconductor surfaces. In addition, the radiated field E has the following behaviors:

(1) The radiated field E is TM mode, the measured ratio of TM/TE is over 20:1. The polarities of the inward and the outward fields are opposite.
(2) The amplitude and sign of E depend on the incidént angle of the laser beam. The E_{max} was achieved near the Brewster's angle.
(3) The amplitude of E decreases as $N^{-0.4}$ where N is the impurity doping concentration ($10^{15}/cm^3 < N < 10^{20}/cm^3$) in GaAs.
(4) The polarity of E from n-type samples shows an opposite sign compared with that of p-type samples. It reflects the opposite direction of the surface fields of n-type and p-type samples.
(5) The amplitude of E varied as the crystal rotated around its normal axis. Three-fold rotation symmetry on (111)-orientated CdTe, GaAs and InP has been directly observed from the radiated fields.

This research was supported by the Office of Naval Research under grant number #N00014-86-K-0694.

References:

[1] X.-C. Zhang, B.B Hu, J.T. Darrow and D.H. Auston, Appl. Phys. Lett, **56,** 1011, (1990)
[2] X.-C. Zhang, J.T. Darrow, B.B Hu, D.H. Auston, M.T. Schmit, P.Tham and E.S. Yang Appl. Phys. Lett, **56,** 2228, (1990)
[3] J.T. Darrow, B.B Hu, X.-C. Zhang and D.H. Auston, Opt. Lett, **15**, 323, (1990)

Femtosecond Excitonic Sampling in Submicron Thickness Coplanar Striplines

W.H. Knox, G.E. Doran, B. Tell, K.W. Goossen, G. Hasnain, J. Cunningham, and D.S. Chemla

AT&T Bell Laboratories, Holmdel, NJ 07733, USA

Excitonic electroabsorption in thin GaAs-AlGaAs free-standing films is a novel approach to femtosecond optoelectronics in which limits of femtosecond electrical pulse generation, propagation and detection can be investigated [1]. The use of ultrathin substrates as a propagation medium for femtosecond electrical signals appears to be a interesting new feature. As the substrate becomes thinner, the TE cutoff frequency increases, until the field lines ultimately propagate in air above and below the lines. This drastically reduces the modal dispersion and phonon losses [1,2]. The ultimate limit for this technique is achieved when the substrate thickness is only one optical absorption length. By reducing the barrier thickness to only 3.0 nm to obtain optimum absorption per unit thickness, we have fabricated a sample which is only 0.75 microns thick, and transparent to visible light. The theory [2] predicts a 30 THz TE cutoff frequency and propagation speed close to that of light in vacuum. As a result of the thin barriers, we obtain a superlattice-like absorption structure with a very broad differential electroabsorption (Figure 1). This is well-suited to modulation of broadband femtosecond laser pulses.

For femtosecond measurements, we use a newly developed infrared CPM laser producing 100 fs pulses at 820 nm wavelength [3]. We first obtain an excitonic saturation signal with resonant excitation with the pump and probe spots overlapped. A

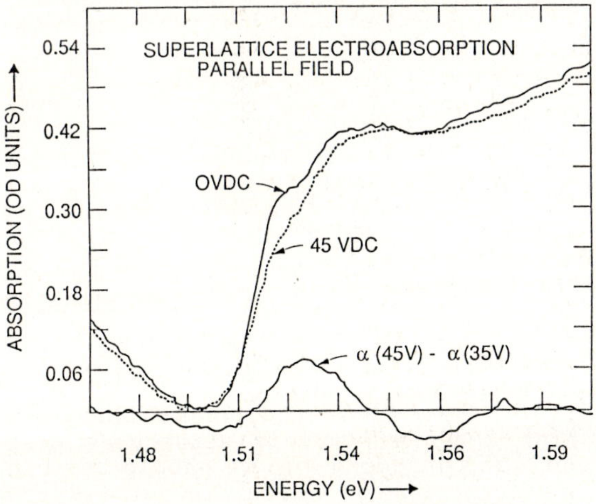

Figure 1. Absorption of ultra-thin free-standing sample at 300 K, and change in absorption at high bias (scaled 6x).

Springer Series in Chemical Physics, Vol. 53 **Ultrafast Phenomena VII**
Editors: C.B. Harris · E.P. Ippen · G.A. Mourou · A.H. Zewail

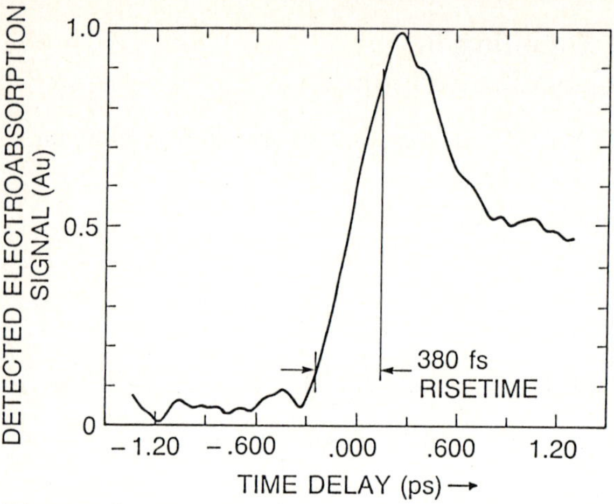

Figure 2. Propagating signal at 45 V bias, after 250 microns of propagation distance.

risetime of 160 fs is obtained [3]. We then separate the pump and probe beams and measure propagating electrical signals. Figure 2 shows the electroabsorption signal which is obtained at 45 V bias after 250 microns of propagation in the stripline. We obtain a risetime of 380 fs, with a rapid initial decay which is due to field-induced carrier heating from k=0 states ("velocity overshoot"). This result is confirmed by Monte-Carlo simulations [4].

The dispersion at high frequencies (above 1 THz) contains not only modal contributions, but also a strong TO phonon resonance at 8 THz. We have carried out a full-wave analysis with material dispersion [2]. For submicron thickness substrates, the effective index is relatively flat up to the phonon frequency at 8 THz. Our model indicates that such structures should be capable of supporting electrical pulses of less than 50 fs risetime over macroscopic distances. In the thin substrate limit, the propagation velocity should approach the speed of light. Figure 3 shows that we obtain propagation at 81% the speed of light, which is in good agreement with our model [2].

In summary, coplanar quantum well striplines of macroscopic dimensions (10 microns) can be fabricated in an ultrathin substrate limit by selective etching. We have pushed this technique to the limit where the sample is only one optical absorption depth thick, and we find that these structures have interesting high speed properties. These results, together with the recent realization that the excitonic time response is of the order of the inverse field-induced linewidth [5] (which is very short for our superlattice sample) suggests that significantly higher time resolution should be obtained with this approach. Recent advances in direct production of ultrashort optical pulses at 820 nm [2] should facilitate such investigations. In addition, we have the possibility to incorporate interesting high speed electronic structures into our measurement environment, and in doing so, we should be able to elucidate fundamental response speeds which are not limited by dielectric discontinuities or bonding pad capacitances or inductances.

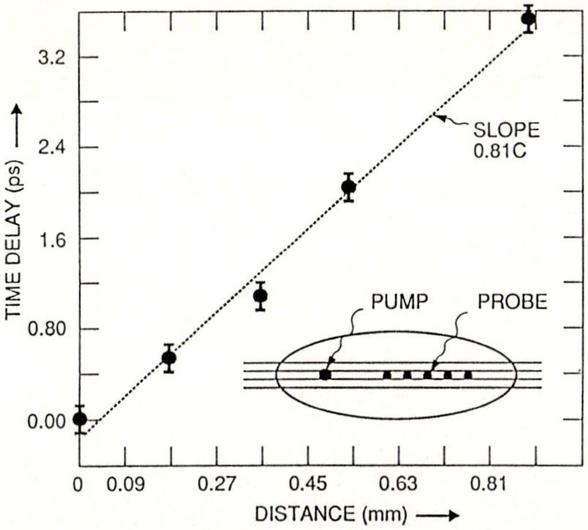

Figure 3. Measurement of propagation speed of 0.75 micron thickness 10 micron spacing coplanar stripline. We obtain propagation at 81% the speed of light. The calculated TE cutoff frequency of this structure is 30 THz. No significant distortion of the pulse is observed up to 1 mm propagation distance.

[1] W.H. Knox, J.E. Henry, K.W. Goossen, K.D. Li, B. Tell, D.A.B. Miller, D.S. Chemla, A.C. Gossard, J. English and S. Schmitt-Rink, IEEE JQE 25, 2586 (1989).

[2] G. Hasnain, K.W. Goossen and W.H. Knox, Appl. Phys. Lett. 56, 515 (1990).

[3] W.H. Knox and F.A. Beisser; this conference, to be published.

[4] S.M. Goodnick, private communication; to be published.

[5] S. Schmitt-Rink, D.S. Chemla, W.H. Knox and D.A.B. Miller, Opt. Lett. 15, 60 (1990).

Picosecond Pulse Generation Circuits Using a GaAs Nonlinear Transmission Line

C.J. Madden[1], *R.A. Marsland*[1], *D.M. Bloom*[1], *and M.J.W. Rodwell*[2]

[1]Edward L. Ginzton Laboratory, Stanford University, Stanford, CA 94305, USA
[2]Department of Electrical and Computer Engineering, University of California at Santa Barbara, Santa Barbara, CA 93106, USA

Circuits for generation picosecond step and pulse signals have many applications in wide bandwidth time-domain and microwave instrumentation. The silicon step-recovery diode is widely used in sampling oscilloscopes and network analyzers for gating sampling diodes but has limited utility in optoelectronic and ultrafast applications due to its relatively slow speed (30-40 ps). Conversely, photodiodes and photoconductors produce much narrower electrical pulses [1,2] but the real pulse generator is a femtosecond or picosecond mode-locked laser. Using only RF drive, we have measured a 6 V, 1.8 ps falltime step output as well as a 2.4 V, 4.4 ps full width at half maximum (FWHM) pulse output from monolithic GaAs nonlinear transmission line (NLTL) circuits.

1. Picosecond Shock-wave Generation

In the past, we have demonstrated shock-wave generation measured at as low as 2.5 ps falltime [3,4] using an NLTL. The 1.8 ps step function, shown in Fig. 1, demonstrates a refinement in our ability to measure these waveforms using the direct electrooptic (EO) sampling [5] technique. Our EO sampling system uses a pulse-compressed, CW mode-locked, Nd:YAG laser to directly probe the electrooptically induced birefringence in the GaAs substrate

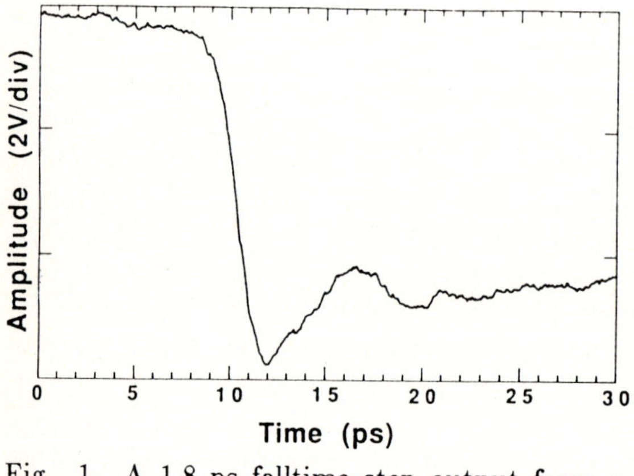

Fig. 1. A 1.8 ps falltime step output from a Nonlinear Transmission Line (NLTL).

Springer Series in Chemical Physics, Vol. 53 **Ultrafast Phenomena VII**
Editors: C.B. Harris · E.P. Ippen · G.A. Mourou · A.H. Zewail

on which the circuits are fabricated. Employing a two stage fiber-grating pulse compressor [6] we have reduced the optical pulsewidth at 1.06 μm to 400 fs FWHM, thus improving the overall EO sampling system time resolution to a 1.0 ps risetime. This yields a deconvolved falltime of 1.5 ps (10%-90%) for our NLTL circuit. Further circuit speed improvements are possible, but are not worthwhile until after another advance in EO time resolution.

2. Picosecond Pulser Circuit

In the pulser circuit, Fig. 2, the NLTL is a high-impedance coplanar waveguide transmission line periodically loaded with reverse-biased Schottky varactor diodes. The voltage variable capacitance of the diodes causes the wave propagation velocity on the 50Ω impedance NLTL to increase with increasing reverse bias so that the falltime of a negative-going transition will decrease as it propagates. If the input is a 7.5 GHz sinewave (40 ps 10%-90% falltime) the output is a sawtooth wave with a 2 ps falltime. To produce a pulse output, the NLTL step signal is differentiated using a pair of shorted transmission line stubs. In Fig. 2, the NLTL output is applied across the slot lines, Z_S, (through the terminating resistor R_O and coupling capacitor C_O) where the round trip delay on the stubs determines the output pulsewidth coupled to the load, R_L=50Ω, through C_L. Equivalently, since the slotline delay is set to the NLTL falltime, the stubs are inductors and the differentiator is simply an RL high-pass filter. The circuit's pulse output waveform is shown in Fig. 3, it was electrooptically measured at the output bondpad and has a 2.4 V peak amplitude and a pulsewidth of 4.4 ps FWHM. This result is comparable to that for photoconductors but is done all electrically and is well suited to driving active mode-locked laser diodes [7] or other applications where picosecond electrical drive pulses are desired.

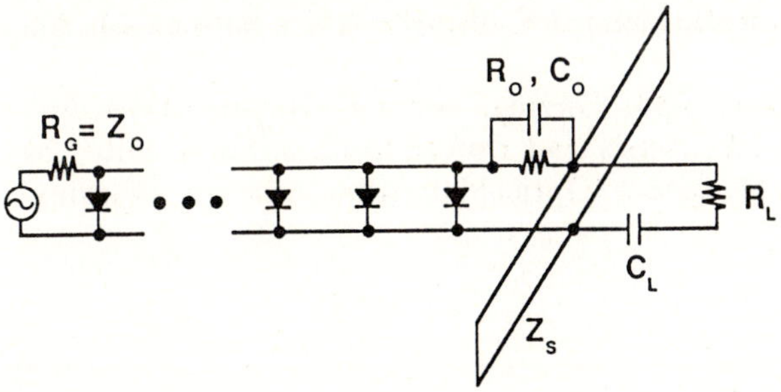

Fig. 2. Scematic diagram for the NLTL pulser circuit.

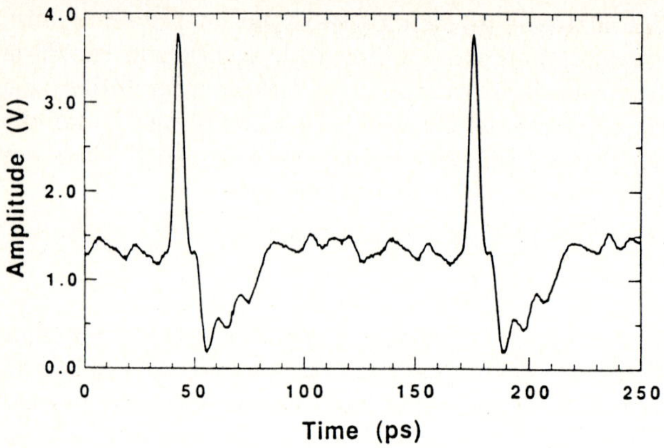

Fig. 3. A 4.4 ps FWHM pulse output from the NLTL pulser circuit.

References

[1] F.W. Smith, H.Q. Le, V. Diadiuk, M.A. Hollis, A.R. Calawa, S. Gupta, M. Frankel, D.R. Dykaar, G.A. Mourou, and T.Y. Hsiang, Appl. Phys. Lett., **54**, No. 10, p. 890 (1989).

[2] B.J. Van Zeghbroeck, W. Patrick, J.-M. Halbout, and P. Vettiger, IEEE Electron Device Letters, **EDL-9**, No. 10, p. 527 (1988).

[3] C.J. Madden, M.J.W. Rodwell, R.A. Marsland, D.M. Bloom, and Y.-C. Pao, IEEE Elect. Dev. Lett., **EDL-9**, No. 6, p. 303 (1988).

[4] C.J. Madden, R.A. Marsland, M.J.W. Rodwell, D.M. Bloom, and Y.-C. Pao, Appl. Phys. Lett., **54**, No. 11, p. 1019 (1989).

[5] K.J. Weingarten, M.J.W. Rodwell, and D.M. Bloom, IEEE J. Quantum Electron., **QE-24**, No. 2, p. 198 (1988).

[6] B. Zysset, W. Hodel, P. Beaud, and H.P. Weber, Optics Letters, **11**, No. 3, p. 156 (1986).

[7] P.A. Morton, R.J. Helkey, S.W. Corzine, and J.E. Bowers, *OSA Proceedings on Picosecond Electronics and Optoelectronics*, Vol.4, edited by T.C.L.G. Sollner and D.M. Bloom (Optical Society of America, Washington, D.C., 1989), p. 87.

Microwave Properties of Anisotropic Materials Characterized with Picosecond Transient Radiation

G. Arjavalingam, Y. Pastol, G.V. Kopcsay*, and J.-M. Halbout*

IBM Research Division, T.J. Watson Research Center, P.O. Box 218, Yorktown Heights, NY 10598, USA
*Present address: IBM – France, Corbeil-Essonnes, France

Abstract. The anisotropic microwave dielectric properties of uniaxial crystals and stretch-oriented conducting polymers are characterized using the coherent microwave transient spectroscopy technique.

The recently developed coherent microwave transient spectroscopy (COMITS) technique is used to characterize the complex dielectric properties of anisotropic materials, over a wide frequency range. COMITS is based on picosecond duration electromagnetic transients radiated and received by optoelectronically pulsed antennas [1,2]. The experimental setup is shown schematically in figure 1. The exponentially tapered coplanar stripline antennas are fabricated on implanted silicon-on-sapphire substrates [3]. Optical pulses from a mode-locked, pulse-compressed, and frequency-doubled, Nd:YLF laser are used in a standard pump/probe configuration to excite the dc-biassed transmitter and photoconductively sample the received waveforms. Hemispherical fused silica lenses collimate the radiation diverging from the transmitter and focus it onto the receiver. The measured signal has a 7 ps wide central peak, and frequency components from 0 GHz to 150 GHz. Since it is proportional to the field (and not power) incident on the receiver, phase information is preserved. The dielectric properties of a material are determined by making measurements with two different sample thicknesses. The corresponding Fourier spectra are divided in order to eliminate the effect of the sample surfaces and extract the net contribution due to the bulk [1,2]. Since COMITS is in effect a phase-sensitive microwave spectrophotometer, both real and imaginary parts of dielectric constants are obtained simultaneously, over a broad bandwidth.

The two-dimensional flow of current in the antenna structures gives rise to strongly plane-polarized microwave radiation. This was characterized with 100-μm-pitch wire-grating polarizers built with 50 μm thick tungsten wires. The polarization sensitivity of the experimental apparatus was found to be 60 times greater for the electric field parallel to the plane of the antennas $E_{\parallel}$, than for fields perpendicular to it. This property of the COMITS setup was used to characterize the

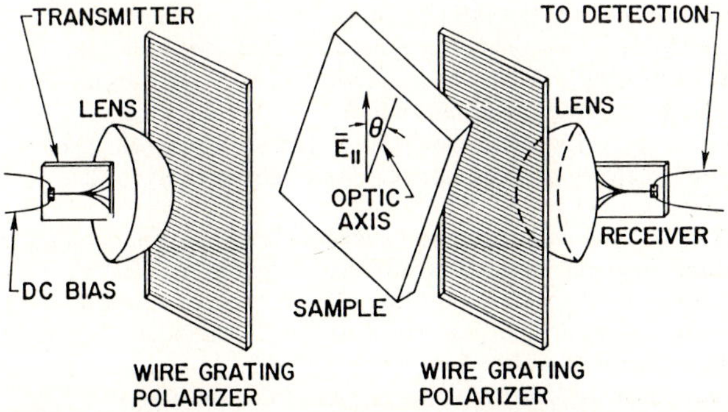

Figure 1. Schematic of the COMITS setup for polarized microwave experiments. θ is the orientation angle between the optic axis and the electric field.

Springer Series in Chemical Physics, Vol. 53 **Ultrafast Phenomena VII**
Editors: C.B. Harris · E.P. Ippen · G.A. Mourou · A.H. Zewail

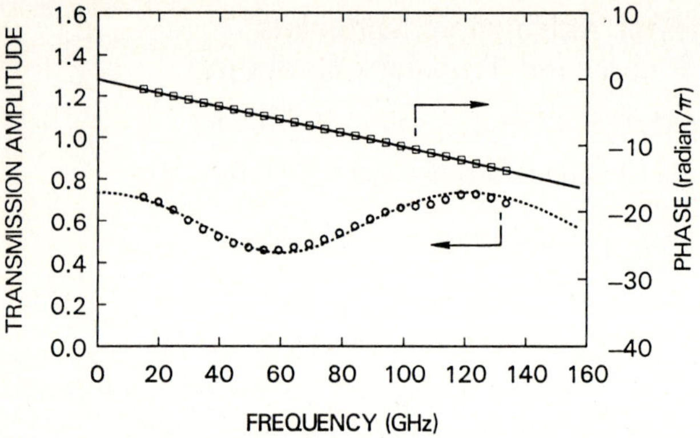

Figure 2. Amplitude and phase of the the transmission function of a 7.5 mm thick sapphire crystal for $\theta = 120°$. The circles and the squares correspond to the experimental results for the amplitude and the phase, respectively. The continuous and dotted lines, prediction of theory.

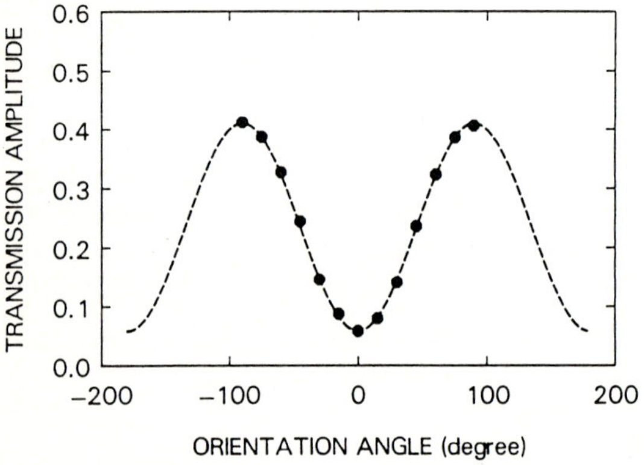

Figure 3. Amplitude transmission of a polyacetylene sample, at 60 GHz, plotted as a function of orientation angle. The dots are experimental results; the dashed line is calculated from the polarizer equation.

anisotropic dielectric constants of uniaxial crystals and the orientation dependent conductivity in stretch-oriented and doped polymers. For instance, the measured ordinary microwave index of sapphire was found to be $n_o = 3.06 \pm 0.03$, and the extra-ordinary index $n_e = 3.39 \pm 0.03$, while the loss in both directions was negligible. The orientation dependent electric-field transmission function of a 7.5 mm thick piece of sapphire was also characterized, and a typical result for an angle of 120 ° between electric-field and optic axis is shown in Fig. 2. The agreement between theory and experiment was found to be very good as shown in the figure. The results depicted in Fig. 2 correspond to the microwave analog of the single-plate birefringent or Lyot filter used in many ultrafast laser systems [4].

The orthogonal conductivities of stretch-oriented and doped polyacetylene and polyaniline were also characterized. We observed about 10 times greater conductivities parallel to the direction of elongation compared with the perpendicular direction. This is believed to be due to the increased mobility of electrons along the oriented polymer chains [5]. The angle dependent transmission for a polyacetylene sample is shown in Fig. 3 together with the theoretically expected curve given by the polarizer equation. The agreement is again excellent [6].

In summary, the microwave dielectric properties of anisotropic materials were characterized using the coherent microwave transient spectroscopy technique. The ordinary and extra-ordinary complex dielectric constants of uniaxial crystals, and the orientation dependent conductivities in stretch-oriented polymers were measured. We have also used this feature of COMITS to characterize lithographically fabricated gratings which are an inexpensive substitute for wire-gratings.

REFERENCES

1. Y. Pastol, G. Arjavalingam, J-M. Halbout, G.V. Kopcsay, Appl. Phys. Lett. 54, 307 (1989).
2. G. Arjavalingam, Y. Pastol, J.-M. Halbout, and G.V. Kopcsay, IEEE Trans. Microwave Theory Tech., 38, 615 (1990).
3. A.P. DeFonzo and C.R. Lutz, Appl. Phys. Lett. 51, 212 (1987); and Y. Pastol, G. Arjavalingam, J.-M. Halbout, and G.V. Kopcsay, Electron. Lett. 24, 1318 (1988).
4. Y. Pastol, G. Arjavalingam, J-M. Halbout, G.V. Kopcsay, Appl. Phys. Lett. 55, 2277 (1989).
5. N. Basescu et al. , Nature, 327, 403 (1987).
6. G. Arjavalingam, N. Theophilou, Y. Pastol, G.V. Kopcsay, and M. Angelopoulos, Jour. Chem Phys., July 1, 1990.

Microwave Phase Locking Using an Electro-optic Harmonic Mixer

Ming G. Li[1], *E.A. Chauchard*[1], *Chi H. Lee*[1], *and Hing-Loi A. Hung*[2]

[1]Electrical Engineering Department, University of Maryland, College Park, MD 20742, USA
[2]COMSAT Laboratories, Clarksburg, MD 20871, USA

Abstract. A phase locked loop (PLL) which synchronizes the microwave source with laser pulses has been developed. An electro-optic broadband harmonic mixer has been used in the PLL loop.

The electro-optic (E-O) sampling technique has potential applications in microwave measurement, including MMIC characterizations [1,2]. However, in the particular method to measure the wave form at microwave frequency by the E-O sampling used in [2], there is a requirement for stable synchronization between the microwave source and the laser pulses. Even in the case of amplitude measurement alone (excluding phase), the signal-to-noise ratio is proportional to the frequency stability of the microwave source. In order to achieve high sensitivity measurement, the synchronization is essential. One approach to meeting this requirement is to use a sophisticated frequency synthesizer as the microwave source and to synchronize the laser acoustic-optic (A-O) modulator driver (another frequency synthesizer) with it [2]. This type of system is very complex and expensive, especially when applied to the millimeter wave frequency range requiring multiple signal sources.

In this work, we report a new phase locking technique which can potentially lock any MMIC oscillator for frequencies up to 100 GHz at the wafer level before the wafer is diced and the device is RF packaged. This technique can be accomplished without requiring mechanical contacts to supply microwave/millimeter wave signal to the wafer.

A schematic of the phase locked loop (PLL) is shown in Fig. 1. The harmonic mixer consists of a GaAs microstrip line, a polarizer (optional), and a photodetector. The E-O effect can be detected by a conventional scheme, which measures the rotation of laser beam polarization with a polarizer [1], or by a new scheme, which utilizes the multiple beam interference effect from the substrate without using a polarizer [3]. The electro-optic effect results in the mixing of the microwave signal on the stripline with one of the harmonic frequencies of the laser repetition rate, thus producing an intermediate frequency, f_{IF}. The upper frequency limit of this mixer is determined by the thickness dimension of the GaAs substrate, and the laser pulse width. The output signal from the photodetector is amplified by a 10 MHz amplifier with a gain of 80 dB, and then mixed with a 10 MHz reference signal derived from the driver of the A-O modulator. The resultant error signal is applied to the voltage controlled oscillator (VCO),

Springer Series in Chemical Physics, Vol. 53 **Ultrafast Phenomena VII**
Editors: C.B. Harris · E.P. Ippen · G.A. Mourou · A.H. Zewail

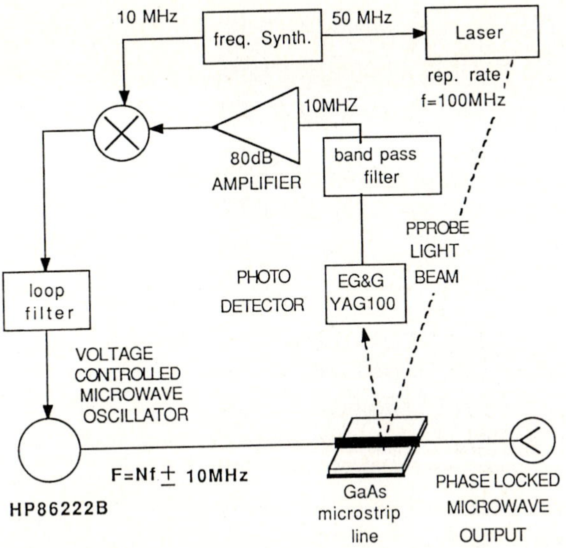

Fig. 1. Schematic of microwave phase lock loop. The optical-microwave intermixer consists of a GaAs microstrip line and a polarizer (optional)

currently represented by the signal generator HP86222B which can be controlled by an external dc voltage, through an active loop filter. The presence of the loop filter is essential in the PLL. The locked frequency of the VCO has to be $f = (100N \pm 10)$ MHz, where N is an integer and the laser pulse repetition rate is 100 MHz.

The phase locked condition can be determined by monitoring the output waveform of the intermediate frequency amplifier and the output of the 10 MHz reference oscillator using an oscilloscope. The time base of the scope is triggered by one of the signals. If phase lock is achieved, very clean waveforms from both signals can be observed as shown in Fig. 2. If the output signal is not phase locked, only one trace stays clear. During the experiment, the output spectrum of the microwave source was also monitored. Without phase lock, the output of the microwave source has a wide spectrum and is very unstable. After the signal has been phase locked, the measured spectrum becomes narrow and stable. Figure 3 shows a phase locked 90 MHz signal. A phase locked signal up to 1.3 GHz has been obtained.

In conclusion, a new microwave phase lock scheme which synchronizes the microwave output with picosecond laser pulses has been demonstrated. With this synchronized microwave source, it is possible to measure the microwave waveform. Since this technique allows the conversion of RF phase signal to dc error signal without requiring a RF probe physically contacting (or connecting) the microwave circuit, this makes on-wafer phase locking and characterization of a MMIC circuit possible, prior to the wafer being diced and individually MMIC packaged.

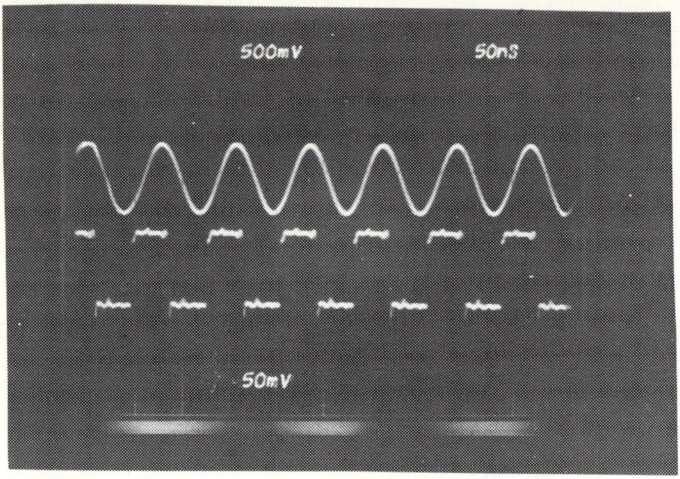

Fig. 2. *Upper trace:* 10 MHz reference oscillation. *Lower trace:* 10 MHz intermediate frequency output

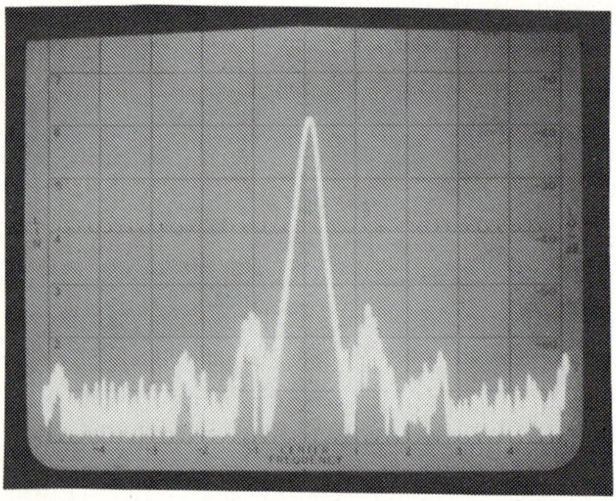

Fig. 3. Spectrum of phase locked microwave signal. Resolution band width: 10 Hz; frequency span: 50 Hz/div; center frequency: 90 MHz; amplitude scale: 10 dB/div

References

1. K.J. Weingarten, M.J.W. Rodwell, D.M. Bloom: Picosecond optical sampling of GaAs integrated circuits. IEEE J. QE-**24**, 198–220 (1988) and previous papers
2. M.J. Rodwell, M. Riaziat, K.J. Weingarten, B.A. Auld, D.M. Bloom: Internal microwave propagation and distortion chracteristics of traveling-wave amplifiers studied by electrooptic sampling. IEEE Trans. MTT-**34**, No. 12 (1986)
3. M.G. Li, E.A. Chauchard, C.H. Lee, H.A. Hung: "Microwave Modulation of Optical Signal by Electro-optic Effect in GaAs Microstrips" in IEEE MTT-S Intl. Microwave Symp. Digest, May 1990

Broadband Microwave Reflection Experiments with Picosecond Transient Radiation

W.M. Robertson, G. Arjavalingam, G.V. Kopcsay, and J.-M. Halbout

IBM Research Division, T.J. Watson Research Center, P.O. Box 218, Yorktown Heights, NY 10598, USA

Coherent broadband reflection spectroscopy using picosecond transient radiation from optoelectronically pulsed antennas is demonstrated. The complex reflection function (amplitude and phase) of Fabry-Perot interferometers, the dispersion from gratings, and the reflection from material samples are presented.

Recently there has been much interest in broadband spectroscopy techniques based on optically generated ultrashort electromagnetic pulses [1-3]. Auston et. al. used Cherenkov-like radiation produced by optical rectification in non-linear crystals to perform contact reflection measurements [1]. We have demonstrated the Coherent Microwave Transient Spectroscopy (COMITS) technique to measure the complex dielectric properties of materials in the 15-130 GHz range [2]. More recently this technique has been extended to higher frequencies by Grischkowsky et al. [3]. Here we present results of broadband microwave reflection spectroscopy with optoelectronically generated, freely propagating, picosecond transient radiation. The technique is illustrated with the measured complex reflection function of Fabry-Perot interferometers, the dispersion from amplitude gratings, and reflection from materials. The experimental results are compared with theory and in all cases very good agreement is found.

The experimental configuration for reflection studies is shown schematically in figure 1. The transmitter and receiver are identical coplanar antennas fabricated on ion-implanted silicon-on-sapphire substrates [4]. The transmitter is excited, and the receiver photoconductively sampled, by 1.5ps wide, 527nm wavelength optical pulses from a mode-locked, pulsed compressed, and frequency doubled Nd:YlF laser. The planar antennas leads to microwave radiation that is strongly polarized with the E-field in the plane of reflection (p-polarized). The broadband microwave radiation is collimated by a hemispherical fused-silica lens, and the reflected radiation is focused onto the receiver by an identical lens. The sample can be rotated about an axis providing an additional variable: the orientation angle θ (see fig. 1).

The Fabry-Perot was constructed from two 0.254mm thick fused-silica plates mounted parallel to each other; the plate separation could be accurately varied. The Fourier transforms of the time-dependent signals recorded with several different plate separations were divided by the Fourier transform of the signal recorded with a gold mirror. The experimentally obtained complex reflection function for a plate separation of 4mm is shown in

Springer Series in Chemical Physics, Vol. 53 **Ultrafast Phenomena VII**
Editors: C.B. Harris · E.P. Ippen · G.A. Mourou · A.H. Zewail

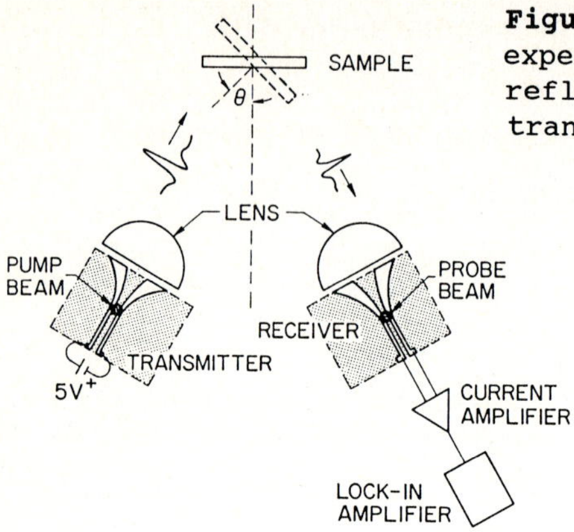

Figure 1. Schematic diagram of the experimental configuration for reflection studies of picosecond transient radiation.

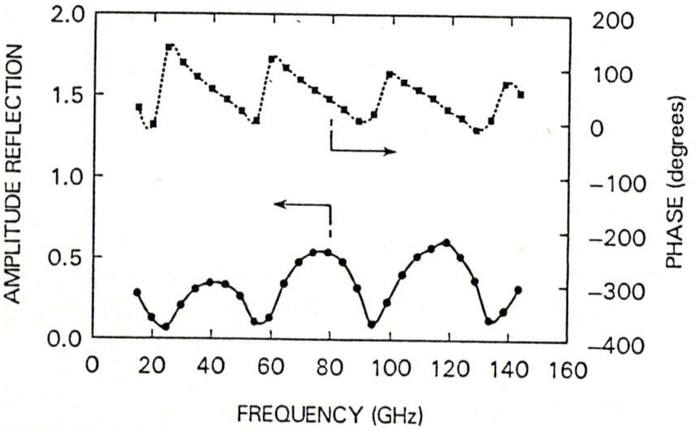

Figure 2. Complex reflection, amplitude (lower trace, circles) and phase (upper trace, squares) of a Fabry-Perot with plate spacing of 4 mm. The lines are guides to the eye.

figure 2. The measured results were found to be in very good agreement with theoretically predicted results.

Diffraction of microwave radiation was investigated using amplitude gratings to disperse the broadband transient electromagnetic radiation from the transmitter into its frequency components. Gratings were made by cutting strips out of a sheet of aluminum foil and mounting the periodic structure on a hard backing. Two gratings were used: one with a periodicity of 10mm (5mm Al/5mm spacer) and the other with a periodicity of 15mm (7.5mm Al/7.5mm spacer). The specularly reflected signal from the 15mm grating ($\theta=0°$) and the signal for $\theta=20°$ are shown in figure 3. The specular reflection was essentially the same as the reflection from a gold mirror but with an amplitude reduced by about 50%. The Fourier transform of the diffracted signal consisted of a narrow band of frequencies. The peak frequencies of the diffracted spectra were found to be in very good agreement with the diffraction equation.

We also measured reflected signals from material samples including fused silica, the orthogonal orientations of crystalline sapphire, and silicon substrates with a range of conductivities. These results were also found to be in agreement with theoretical predictions.

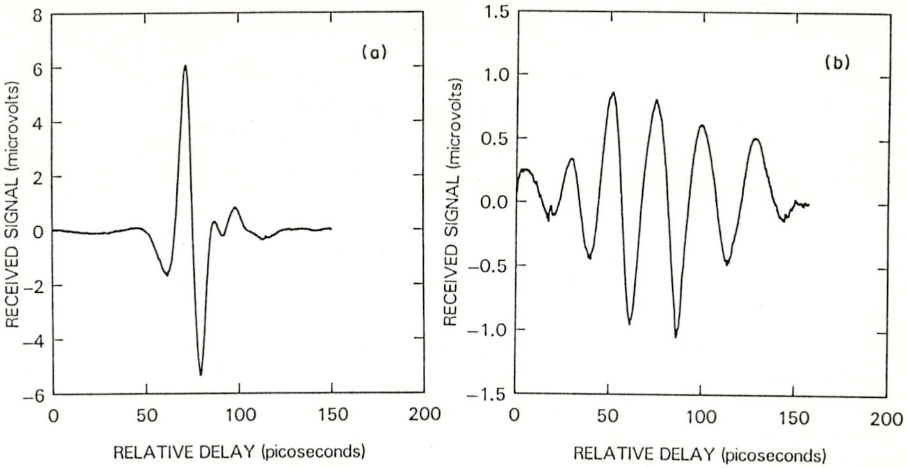

Figure 3. (a) Measured specularly reflected (θ=0°) transient from a grating with period of 15 mm. (b) Diffracted signal recorded with θ=20°.

In summary, we have demonstrated coherent, broadband microwave reflection spectroscopy with optoelectronically generated, freely propagating, ultrafast electromagnetic transients. In combination with the transmission experiments described previously, these results illustrate the possibility of optical experiments at GHz frequencies. Finally, the experimental apparatus and techniques described here can be applied to picosecond, time-resolved ellipsometry in the microwave spectrum.

References

1. K. P. Cheung and D. H. Auston, Infrared Phys. **26**, 23 (1986).
2. Y. Pastol, G. Arjavalingam, J.-M. Halbout, and G. V. Kopcsay, Appl. Phys. Lett. **54**, 309 (1989); ibid **55**, 2277 (1989).
3. M. van Exter, C. Fattinger, and D. Grischkowsky, Opt. Lett. **14**, 1128 (1989).
4. A. P. DeFonzo and C. Lutz, Appl. Phys. Lett. **51**, 212 (1987).

Ultrafast Coplanar Air-Transmission Lines

D.R. Dykaar, A.F.J. Levi, and M. Anzlowar

AT&T Bell Laboratories, 600 Mountain Ave., Murray Hill, NJ 07974, USA

We demonstrate subpicosecond electrical pulse propagation using coplanar air-transmission lines. Risetimes of 0.8 ps after 2.8 mm of propagation are achieved with a velocity of propagation which is 86% of the speed of light in vacuum. Our results suggest that intrachip communication in digital circuits with risetimes as short as one picosecond is readily achievable using electrical signals.

1. Introduction

A limit to ultrafast electronics with single picosecond switching time is the ability to rapidly propagate undistorted electrical signals from one transistor to another. We demonstrate experimentally that ultrafast Coplanar Air-Transmission (CAT) lines may be used to propagate subpicosecond electrical signals across typical chip dimensions of a few mm, at very high speeds and with little distortion. Our results suggest that electrical propagation (as opposed to optical interconnects) is a practical method of intrachip (~mm) communication in digital circuits requiring risetimes as short as 1 ps.

Subpicosecond electrical transients can be propagated for short distances ($\lesssim$100μm) without significant loss in fidelity (risetime) [1,2]. However, there has been little attempt to propagate subpicosecond electrical signals large distances (e.g. a few mm) because of dispersion mechanisms involving dielectric discontinuities and materials related losses [2-3]. We show that a practical solution to this problem is the formation of CAT lines in which the bulk of the substrate dielectric is removed around the conductors. This technique is also ideally suited to the study of ultrafast devices which can be integrated into the present structure.

2.Experiment

Gold coplanar transmission lines 1 μm thick by 5 μm wide with a 5 μm gap are fabricated on a 200 nm layer of SiO_2 on semi-insulating GaAs substrates. The layer of SiO_2 acts both as an etch mask and as a physical cantilevered support for the lines. In addition, the electrically insulating SiO_2 layer reduces leakage and dark currents in the system. The inset in Fig. 1(b) shows the device geometry. Note that the electrodes overhang the SiO_2 layer by

Springer Series in Chemical Physics, Vol. 53 **Ultrafast Phenomena VII**
Editors: C.B. Harris · E.P. Ippen · G.A. Mourou · A.H. Zewail

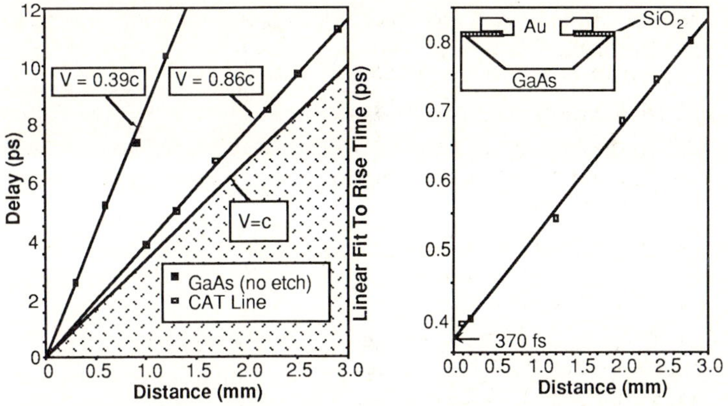

Figure 1(a). Plot of delay (as measured by midpoint of linear fit to risetime) versus propagation distance. Velocity, V, is found from a linear fit to the data. Shaded area is bordered by V = c. (b) Plot of risetime (as measured by midpoint of linear fit to risetime) versus propagation distance. Measured risetime after 2.8 mm of propagation is 0.8 ps. Intercept of 370 fs is a measure of system response.

about 2 μm, so the region of highest field strength experiences a uniform air dielectric.

Electrical transients are generated at the voltage biased photoconductive switch using the sub-100 fs full width at half maximum optical pulses from a colliding pulse mode locked laser. An optical fiber is used to deliver the excitation pulse thereby guaranteeing that the laser timing delays remain fixed. Electrical signals are detected using electro-optic sampling [2] and sub-millivolt detection electronics [4].

3. Results

Propagation times are measured for distances from 200 μm to 2.9 mm. The results are shown in Fig. 1(a) for both the CAT line and a conventional (unetched) transmission line. The solid lines are least-squares linear fits to the data. A naive calculation of the velocity, V, which uses an average value of the dielectric constant $\varepsilon_{av} = (\varepsilon+\varepsilon_{air})/2$, gives $V = c/\sqrt{\varepsilon_{av}} = 0.38c$, where $\varepsilon = 12.85$ ($\varepsilon_{air} = 1$) is the static dielectric constant of GaAs [5] (air) and c is the speed of light in vacuum. This is in good agreement (<3%) with the measured value of 0.39c for the unetched sample. In contrast, a CAT line with an etch depth of 12 μm and an etched width of 20 μm gives a velocity of propagation of 0.86c. This is more than a factor of two faster than the unetched transmission line. Calculating the effective dielectric of the etched substrate yields $\varepsilon = 1.7$ and an average dielectric $\varepsilon_{av} = 1.35$. This value is so close to a uniform air dielectric that electrical signals should propagate with little dispersion entirely in a TEM mode.

The effect of dispersion was investigated by measuring the risetime as a function of distance propagated along the CAT line. In order to measure the risetime, the optical fiber was removed to prevent dispersion of the fs excitation pulse. A summary of risetime measurements versus distance from 100 μm to 2.8 mm is given in Fig. 1(b). The zero distance intercept of 370 fs gives a measure of the overall system response which includes the effects of the sampling crystal, velocity mismatch between the electric and optical waves inside the $LiTaO_3$, etc. In addition, it is likely that further improvements could be achieved by operating a CAT line at low temperature, or by using superconducting electrodes [6].

4. Conclusion

We have developed a practical technique for fabricating ultrafast transmission lines with a demonstrated velocity of propagation of 0.86c and a measured risetime of 0.8 ps after 2.8 mm of propagation. Ease of fabrication and the practicality of our approach suggests that CAT lines are the most convenient way to implement intrachip communication over distances of a few mm in digital electronic circuits with rise times ~1 ps.

References

1. J.A. Valdmanis, and G.A. Mourou, IEEE J. Quantum Electronics, QE-22, 69 (1986). and
D.R. Dykaar, J. Chwalek, J.F. Whitaker, R. Sobolewski, S. Gupta, T.Y. Hsiang, and G.A. Mourou, IEEE Trans. Magn., MAG-25, 814, (1989).
2. D.H. Auston, in Picosecond Optoelectronic Devices, edited by C.H. Lee (Academic, London, 1984), pp. 73-116. and
W.H. Knox, J.E. Henry, K.W. Goossen, K.D. Lee, B. Tell, D.A.B. Miller, D.S. Chemla, A.C. Gossard, J. English, and S. Schmitt-Rink, IEEE J. Quant. Elec. **25**(12), 2586 (1989).
3. G. Hasnain, K.W. Goossen, and W.H. Knox, Appl. Phys. Lett. **56**, 515 (1990).
4. J.M. Chwalek and D.R. Dykaar, Rev. Sci. Ins., 61(4), 1273 (1990).
5. J.S. Blakemore J. Appl. Phys. **53**(10) R123 (1982).
6. D.R. Dykaar, J. Chwalek, J.F. Whitaker, R. Sobolewski, T.Y. Hsiang, G.A. Mourou, D.K. Lathrop, S.E. Russek and R.A. Buhrman, Appl. Phys. Lett., **52**, (17), 1444 (1988). and
D.R. Dykaar, R. Sobolewski, and T.Y. Hsiang, IEEE Trans. Magn., MAG-25, 1392, (1989). and
M.C. Nuss, P.M. Mankiewich, R.E. Howard, B.L. Straughn, T.E. Harvey, C.D. Brandle, G.W. Berkstresser, K.W. Goossen, and P.R. Smith, Appl. Phys. Lett., **54**, (22), 2265 (1989). and
C-C. Chi, W.J. Gallagher, I.N. Duling III, D. Grischkowski, N.J. Halas, M.B. Ketchen, A.W. Kleinsasser, IEEE Trans. Magn., MAG-23, 1666, (1987).

Part VI

Quantum Wells; Reduced Dimensionality

Ultrafast Studies of Resonant and Non-resonant Electron and Hole Tunneling in Quantum Well Structures

J. Shah[1], *D.Y. Oberli*[2], *K. Leo*[1,*], *T.C. Damen*[1], *D.A.B. Miller*[1], *J.P. Gordon*[1], *J.E. Cunningham*[1], *J.M. Kuo*[3], *and C.W. Tu*[4]

[1]AT&T Bell Laboratories, Holmdel, NJ 07733, USA
[2]Schottky Institute, Munich, Fed. Rep. of Germany
[3]AT&T Bell Laboratories, 600 Mountain Ave., Murray Hill, NJ 07974, USA
[4]University of California at San Diego, La Jolla, CA 92093, USA
*On leave from Max-Planck-Institut, Stuttgart, Fed. Rep. of Germany

ABSTRACT We review our comprehensive investigations of electron and hole tunneling times between quantum wells separated by a single barrier. These studies show strong electronic and phonon resonances, strong influence of collisions and relaxation on tunneling times and provide new insights into non-resonant and resonant tunneling in semiconductor microstructures.

1. INTRODUCTION

Tunneling is of fundamental interest and strongly influences high speed devices based on perpendicular transport of carriers in semiconductor microstructures. We have directly determined resonant and non-resonant tunneling of electrons and holes in asymmetric double quantum well structures (a-DQWS) using ultrafast luminescence spectroscopy [1-6]. For *non-resonant tunneling*, our results show that (i) tunneling times depend exponentially on the barrier thickness, (ii) impurity-assisted tunneling rate is small and (iii) optical-phonon-assisted tunneling dominates when the energy separation allows this process. For *resonant tunneling*, we have demonstrated strong reduction in tunneling times at resonance; however, tunneling times are considerably longer than expected because of destruction of coherence by relaxation and collisions. We have developed a unified theoretical framework for understanding tunneling in the presence of relaxation and collisions.

In a-DQWS, the two quantum wells have different thicknesses and hence distinct spectral signatures. Tunneling rates are directly determined by measuring the *decay rates* of luminescence of the well from which the carriers tunnel out. The use of isolated a-DQWS allows investigation of tunneling without the complicating effects of transport of carriers. Effects of various electronic and phonon resonances on the tunneling rates are investigated by applying reverse bias to different structures as illustrated in Fig. 1 and Fig. 2. Internal electric field was determined using quantum confined Stark effect. All measurements were made on GaAs/AlGaAs structures at low temperatures.

2. NON-RESONANT TUNNELING

When there is no resonance between the initial and the final energy levels, tunneling requires a momentum conserving transition. In this case, the tunneling process is the same as the inter-subband transition between the subbands of the entire DQWS. This process can be thought of as tunneling because initially the carrier is primarily in one well but ends up primarily in the other well. Two interesting cases have been investigated.

Springer Series in Chemical Physics, Vol. 53 **Ultrafast Phenomena VII**
Editors: C.B. Harris · E.P. Ippen · G.A. Mourou · A.H. Zewail

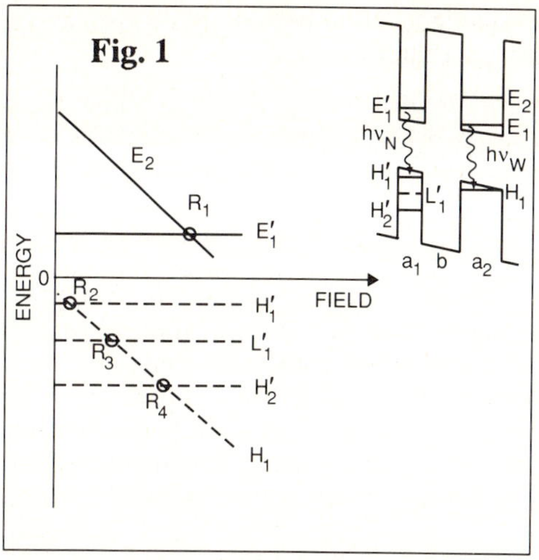

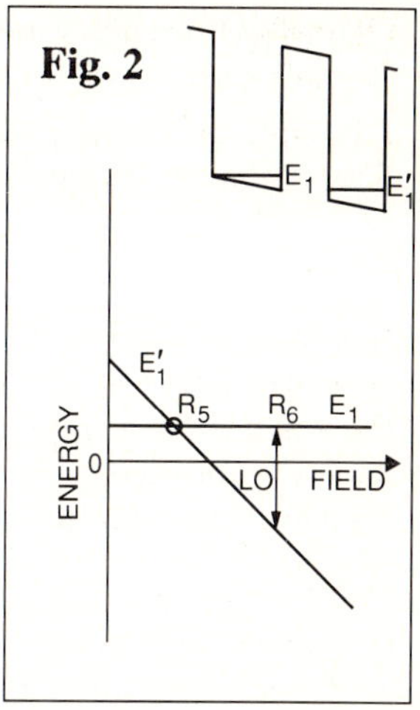

Fig. 1 and Fig. 2 : Schematic diagrams of a-DQWS and various resonances that can be explored using reverse bias.

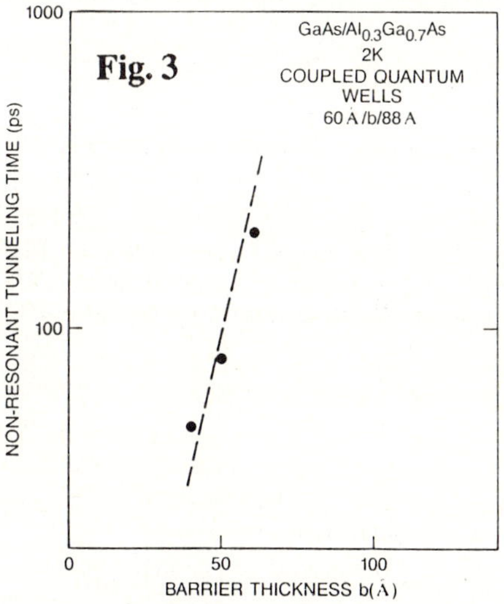

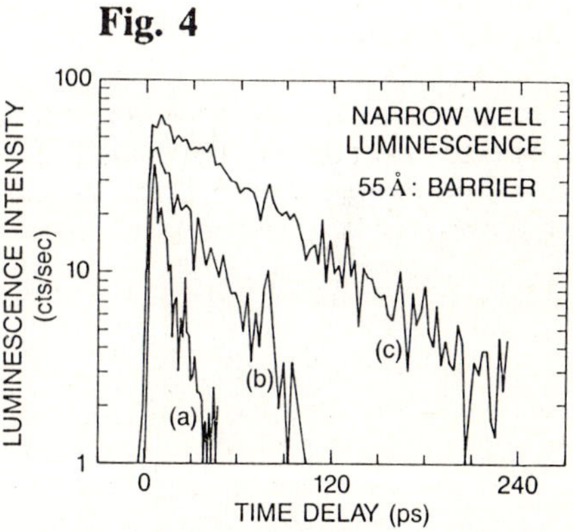

Fig. 3 : Measured electron tunneling time vs barrier thickness when $\Delta E > \hbar\omega_{LO}$.

Fig. 4 : Measured time evolutions of the wide well for three different electric fields which bring E_1' below E_1.

2.A Dependence on Barrier Thickness Fig. 3 shows the dependence of the electron tunneling time on the thickness b of the barrier separating the two quantum wells [3,4]. The separation between the initial and the final states was larger than the optical phonon energy ($\hbar\omega_{LO}$). The tunneling time increases exponentially with b, a result that is easily understood because the wavefunction overlap depends exponentially on the barrier thickness.

2.B Optical Phonon Resonance We have investigated how tunneling times depend on the energy separation ($\Delta E = E_1 - E_1'$) by applying electric field to a structure shown in Fig. 2 [1,2]. At the built-in bias, E_1' is slightly lower than the initial state E_1. The decay curves of the wide well luminescence for three different fields are shown in Fig. 4 and the decay time as a function of electric field is shown in Fig. 5. For ($\Delta E < \hbar\omega_{LO}$), the decay time is close to the normal recombination time, showing that impurity and acoustic-phonon assisted tunneling is not important in good quality samples. At the optical phonon emission threshold (R_6 in Fig. 2), the tunneling time *decreases sharply*, clearly showing the importance of optical phonons in tunneling. A comparison of the measured times with those calculated by Lary and Goodnick [2] using Monte Carlo techniques is also shown in Fig. 5. The results lead to interesting questions concerning the nature of phonon modes that participate in the tunneling process.

3. RESONANT TUNNELING

When two interacting levels are brought into resonance by applying an electric field, they exhibit an anti-crossing behavior. Excitation with an appropriate ultrashort pulse whose spectrum is broader than the energy separation between the two levels creates a linear superposition of the two eigenfunctions such that the carrier wavepacket is created primarily in one well. This wavepacket then

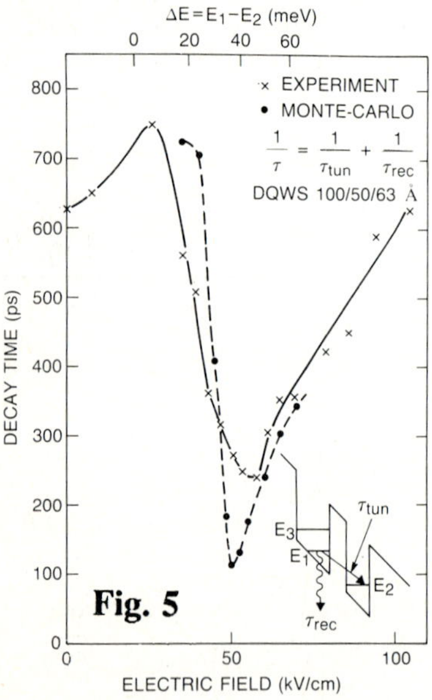

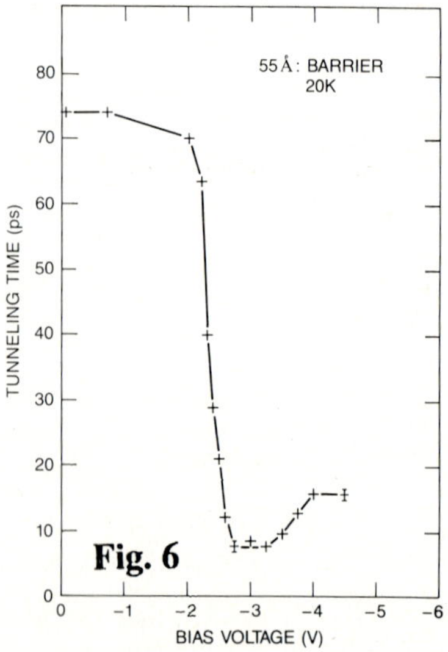

Fig. 5 : Measured decay time of the wide well as a function of reverse electric field as ΔE is increased through $\hbar\omega_{LO}$.

Fig. 6 : Measured decay times of the narrow well luminescence as a function of reverse field as E_2 is tuned through E_1'. The results clearly show the effect of electronic resonance.

oscillates between the two wells with a period determined by the energy separation. Although there has been considerable argument in the literature about the definition of tunneling times, there is no ambiguity about its definition in this case of coherent tunneling. It should be noted that excitation with longer pulses or monochromatic light source creates an eigenstate that is delocalized and experiments measuring ratios of luminescence from two wells as a function of electric field using such techniques primarily measure wavefunction overlaps rather than tunneling.

3.A Electron Resonance We find [3-5] that the tunneling times go through a minimum as we apply a reverse bias to the structure shown in Fig. 1. The results for the resonance R_1 of Fig. 1 are shown in Fig. 6. This was the first direct demonstration of resonance and determination of the resonant tunneling time. Effects of electron resonance have also been reported by other groups [7].

3.B Hole Resonance We have also made the first measurement of the resonant tunneling of holes [6]. A minimum in the tunneling times is observed for the case of holes (R_4 in Fig. 1). No resonances are observed when heavy hole and light hole levels are brought in resonance (R_3 in Fig. 1), even when the temperature is raised to 80 K and the density is increased. The most interesting aspect of the tunneling at resonance is that the tunneling times are longer than expected for coherent oscillations [8]. For electrons the tunneling time is 7 ps compared to about 1 ps expected for coherent tunneling, and for holes it is 1300 ps, about a factor of 40 longer than expected. We have shown [6] that this large increase in the tunneling time results from the destruction of coherence by collisions and relaxation. We have further shown that collisions, relaxation and tunneling cannot be considered as independent processes and must be considered in a unified manner [6].

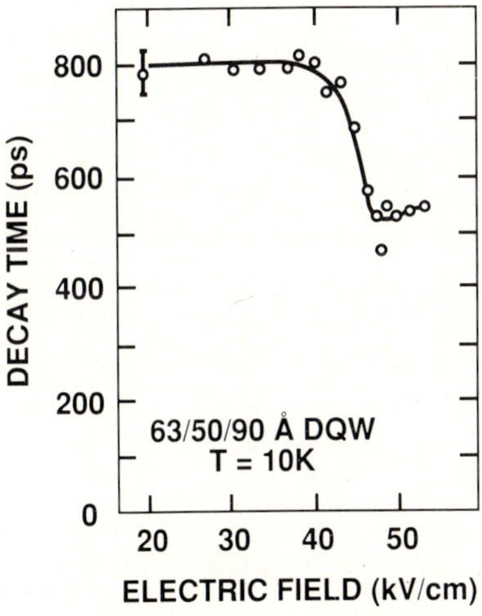

Fig. 7

Fig. 8

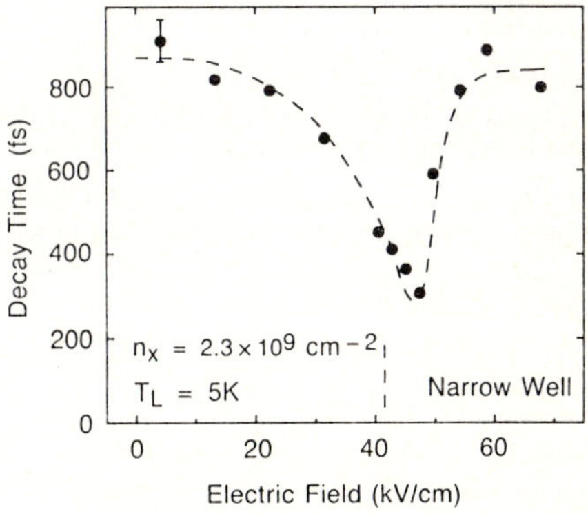

Fig. 7 : Measured decay time of the wide well luminescence as H_1 is tuned through L_1' and H_2'.

Fig. 8 : The measured decay time of the DFWM signal at the narrow well exciton as E_1' is tuned through E_1 by electric field.

3.C Investigation of Resonant Tunneling by DFWM We have introduced ttime-resolved degenerate four wave mixing as a powerful technique for investigating tunneling [9]. In Fig. 8 we show the measured decay time of the DFWM signal as a function of electric field when the laser is tuned to the lowest transition in the narrow well. We find that the dephasing time goes through a minimum when condition R_5 of Fig. 2 is fulfilled. This is a result of scattering of the wavepacket when it is in the wide well by carriers in the wide well. The tunneling time must be shorter than the minimum dephasing time measured in Fig. 8.

4. OBSERVATION OF COHERENT OSCILLATIONS

We have made the first observation of coherent oscillations of the carrier wavepacket, as discussed by Leo et al.at this Conference.

[1] D. Y. Oberli, Jagdeep Shah, T. C. Damen, R. F. Kopf, J. M. Kuo and J. E. Henry, OSA Proceedings on "Picosecond Electronics and Optoelectronics", Vol **4**, ed. by T. C. L. G. Sollner and D. M. Bloom, p. 111 (1989)

[2] D. Y. Oberli, Jagdeep Shah, T. C. Damen, J. M. Kuo and J. E. Henry, J. Lary and S. M. Goodnick, Appl. Phys. Letters **56**, 1239 (1990).

[3] D. Y. Oberli, Jagdeep Shah, B. Deveaud and T. C. Damen, Ref. 1; p. 94 (1989).

[4] Jagdeep Shah, in "Spectroscopy of Semiconductor Microstructures", edited by G. Fasol, A. Fasolino and P. Lugli, Proc. of NATO ARW, Venice, May 1989; published by Plenum Publishing; pp. 535-559 (1989).

[5] D. Y. Oberli, Jagdeep Shah, T. C. Damen, T. Y. Chang, C, W, Tu, D. A. B. Miller, J. E. Henry, R. F. Kopf, N. Sauer and A. E. DiGiovanni, Phys. Rev. B 40, 3028 (1989).

[6] Jagdeep Shah, K. Leo, J. P. Gordon, T. C. Damen, D. A. B. Miller, J. E. Cunningham and C. W. Tu, Bull. Am. Phys. Soc. **35**, 591 (1990); also to be published.

[7] T. Matsusue, M. Tsuchiya and H. Sakaki, Ref. 1, p. 254 (1989); T. Norris, N. Vodjdani, B. Vinter, C. Weisbuch and G. Mourou, Ref. 1, p. 106 (1989); M. G. W. Alexander, W. Rühle, R. Sauer and W. Tsang, Appl. Phys. Letters **55**, 885 (1989); B. Deveaud, F. Clerot, A. Chomette, A. Regreny, R. Ferreira and G. Bastard, Europhys. Lett. **11**, 367 (1990).

[8] S. Luryi, Sol. St. Commun. 65, 787 (1988).

[9] K. Leo, Jagdeep Shah, E. O. Göbel, T. C. Damen, K. Köhler and P. Ganser, Appl. Phys. Letters **56**, May 21 (1990).

Femtosecond Spectroscopy of Quasi-Zero-Dimensional Magneto-Excitons

J.B. Stark, W.H. Knox, and D.S. Chemla

AT&T Bell Laboratories, Crawfords Corner Road, Holmdel, NJ 07733, USA

Quantum confinement of electronic states in quantum wells is achieved by layered thin film growth techniques. These materials demonstrate that linear and nonlinear optical properties depend critically on dimensionality[1]. A magnetic field applied perpendicularly to the quantum well confines the electronic states within the plane, producing the bound and discrete magneto-exciton eigenstates. By increasing the field strength, the magnetic confinement length can be made smaller than the excitonic Bohr radius. In this way, the dimensionality of the quantum well states can be tuned continuously from quasi-two to quasi-zero dimensions. We report the first observation of femtosecond time resolved optical nonlinearities in quasi-zero dimensional magneto-excitons. Our results demonstrate that confinement into quasi-zero dimensions strongly modifies both linear and nonlinear optical response, as well as the kinetic processes governing thermalization and energy relaxation.

The sample used consists of 60 periods of 70 Å GaAs/AlGaAs quantum wells, mounted on a glass carrier, with the growth substrate removed by etching. The linear absorption spectrum is shown in Figure 1. At zero field, the *1s* heavy hole (*1shh*) and

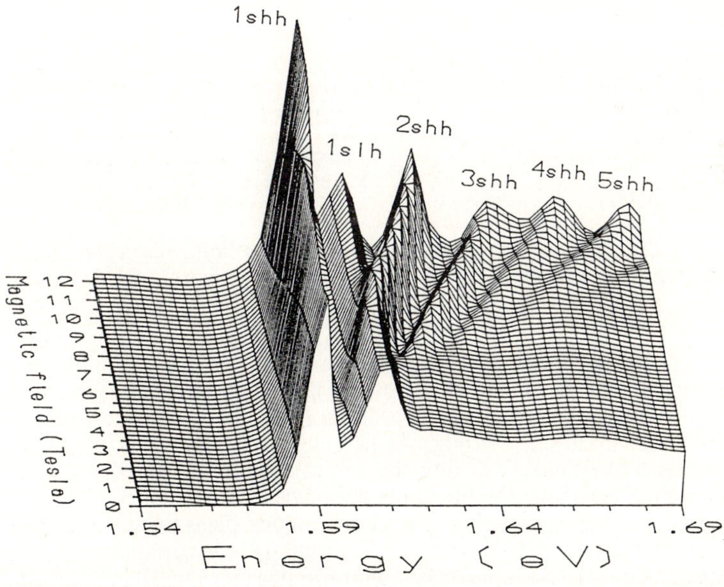

Figure 1. Magneto-exciton linear absorption spectrum, 70 Å quantum well, 4.2 K, *versus* magnetic field. The excited light hole states are weak, and not resolved.

Springer Series in Chemical Physics, Vol. 53 **Ultrafast Phenomena VII**
Editors: C.B. Harris · E.P. Ippen · G.A. Mourou · A.H. Zewail

1s light hole (*1slh*) peaks lie below the continuum absorption from the extended scattering states. As the magnetic field increases, the magneto-exciton states evolve smoothly from the bound states of the exciton. The optically active states of the magneto-exciton may therefore be classified as *ns*, where *n* is the principal quantum number. The *1shh* and *1slh* magneto-exciton states disperse slowly with field, whereas the *2shh* and higher excited states disperse much more rapidly.

To measure the dynamical nonlinear optical response of the sample, amplified pulses, at 805 nm with 1μJ energies and an 8 kHz repetition rate, were focused into an ethylene glycol jet to produce broadband continuum pulses of ~100*fs* duration[2]. The continuum was split into pump and probe, the pump beam passing through a variable delay stage, and a narrowband interference filter. Both beams were focused onto the sample, in a superconducting 12 Tesla magnet. The probe beam was collected by an optical fiber and delivered to an optical multichannel analyzer for parallel detection.

At zero magnetic field, a population was optically pumped into the *1shh* exciton state. The absorption spectrum (Fig. 2a) shows the optical density of the quantum wells, at 4.2 K, before and after the arrival of the pump. Creation of a population of *1shh* excitons causes that resonance to lose oscillator strength, due to phase space filling. The *1shh* resonance also shifts to higher energy. This can be understood by considering the excitons to interact via hard core repulsion[3], or in terms of a bandgap renormalization and decrease of the exciton binding energy[4].

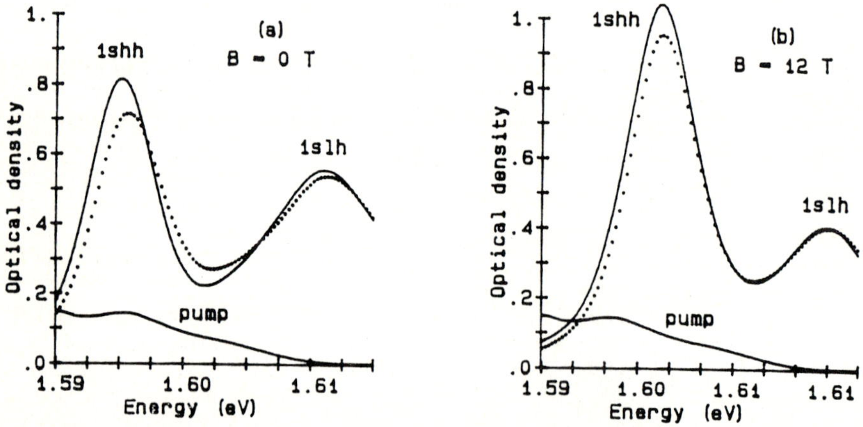

Figure 2. (a) Optical density, resonant pumping of *1s* at zero field, for probe 600 fs before (solid) and 600 fs after (dotted) pump. (b) Same, magnetic field 12 T.

The same experiment was performed at 12 Tesla (T), creating an occupation in the *1shh* magneto-exciton state. The absorption spectrum (Fig. 2b) again shows loss of oscillator strength due to phase space filling. At 12 T, however, there is no blue shift of the magneto-exciton resonance. Tuning the dimensionality of the states from quasi-two to quasi-zero dimensions, therefore, quenches the exciton-exciton Coulomb interactions. Because the magneto-exciton population renormalizes the gandgap, the *2shh* state (not shown) shifts to the red. Results of the analagous experiment, creating a population in the *2shh* magneto-exciton state, demonstrates this general conclusion: the occupation of a magneto-exciton state at high field causes a loss of oscillator strength of that state, with no shift, as well as a red shift of unpopulated states.

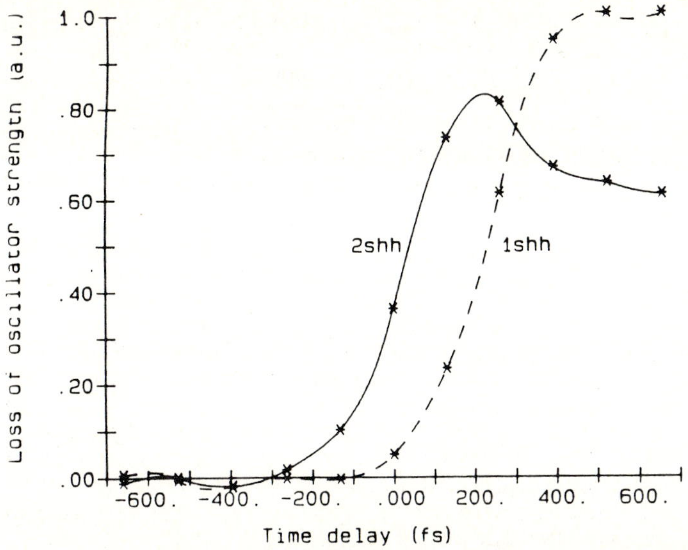

Figure 3. Loss of oscillator strength *versus* time delay at the *2shh* (solid) and *1shh* (dashed), for pumping resonant with *2shh*. Asterisks are data, line is interpolated.

The occupation of *2shh* magneto-exciton states also allows an observation of kinetic processes. The relaxation of carriers from the *2shh* to the *1shh* is observed (Fig. 3) as the delayed rise in the loss of oscillator strength at the *1shh* after the creation of a population of *2shh* magneto-excitons. This relaxation is much slower than that seen in zero field quantum wells, when excited at the same relative spectral position[5]. Because all states at high magnetic field are discrete and well separated in energy, the fastest carrier-carrier relaxation paths are eliminated, and relaxation proceeds at a much slower rate. A second relaxation process, presumably cooling to the lattice temperature, occurs on a time scale of 100 ps.

We have measured the femtosecond nonlinear optical dynamics of a quantum well system as its dimensionality is continuously tuned from quasi-two to quasi-zero dimensions by a perpendicular magnetic field. The exciton-exciton Coulomb interactions seen at zero field are quenched at high fields. In general, the bandgap is renormalized even at 12 Tesla, so that the unpopulated states red shift, while the populated states simply lose oscillator strength. Furthermore, the kinetic processes are seen to be slowed with respect to those in zero field, due to the elimination of low energy two-body scattering channels. Continuing studies of the optical Stark effect of magneto-excitons at 12 T show that optical pumping 40 meV below the *1shh* creates a virtual population in the lower lying magneto-exciton states, producing signals at the *1shh*, *1slh*, *2shh*, and *3shh*.

[1] S. Schmitt-Rink, D.S. Chemla, D.A.B. Miller, Adv. Phys. **38**, 89 (1989).
[2] W.H. Knox, J. Opt. Soc. Am. B **4**, 1771 (1987).
[3] N. Peyghambarian, et al, Phys. Rev. Lett. **53**, 2433 (1984).
[4] S. Schmitt-Rink, D.S. Chemla, H. Haug, Phys. Rev. B. **37**, 941 (1988).
[5] W.H. Knox, et al, Phys. Rev. Lett. **61**, 1290 (1988).

On the Lineshape of Time-Resolved Four-Wave Mixing

M. Wegener[1], *G. Sucha*[1], *E.-O. Göbel*[1], *D.S. Chemla*[1], *S. Schmitt-Rink*[2], *and W. Schäfer*[3]

[1]AT&T Bell Laboratories, Holmdel, NJ 07733, USA
[2]AT&T Bell Laboratories, 600 Mountain Ave., Murray Hill, NJ 07974, USA
[3]Forschungszentrum Jülich, P.O. Box 1913, D-5170 Jülich, Fed. Rep. of Germany

ABSTRACT: We study the effect of coherent interaction of polarizations in time-resolved Four-Wave-Mixing. The experimental results on excitons show novel effects. The physical origin of these effects is dicussed using simple models to emphasize their rather general nature.

Introduction: Experiments on time-resolved Four-Wave-Mixing (FWM) are usually interpreted in terms of the isolated two-level model in order to obtain information on the dephasing time T_2 [1,2]. Recent experiments involving excitonic resonances in semiconductor Quantum Wells show striking qualitative deviations from this picture. In particular an exponential tail for negative time-delays at low excitation is found to evolve into two distinct temporal maxima at high excitation [3,4]. We demonstrate that this can be understood by accounting for interactions between polarizations. This interaction in case of excitons is the coherent exciton-exciton interaction.

Experiment: In Fig. 1 we depict experimental results on time-resolved FWM on excitons at 5K in InGaAs QWs [4]. As the intensity is increased we find a gradual increase of the negative time-delay signal and the evolution of a lineshape that exhibits two distinct temporal maxima. This evolution can not be understood in terms of the isolated Two-Level model.

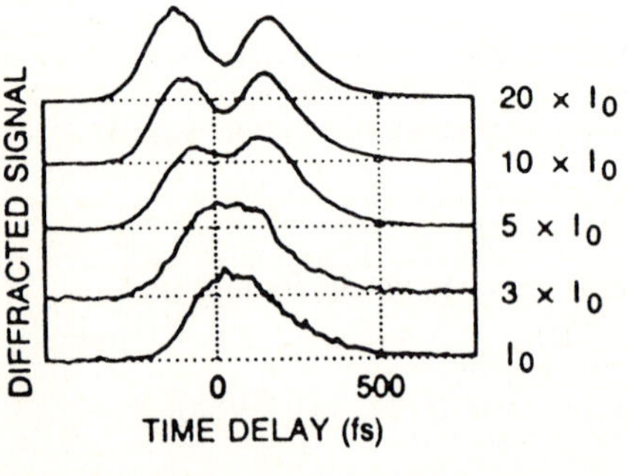

Fig.1 : Time-resolved FWM signal for increasing excitation. $I_0 \approx 3\text{MWcm}^{-2}$, 140fs pulses at 1500nm, 10meV detuning below the excitonic resonance.

Springer Series in Chemical Physics, Vol. 53 **Ultrafast Phenomena VII**
Editors: C.B. Harris · E.P. Ippen · G.A. Mourou · A.H. Zewail

Theory:

The experimental findings can be described in detail by numerically solving the kinetic equations for the semiconductor band edge [4]. It is, however, illustrative to first consider the much more simple model of local field interaction. It gives an intuitive understanding of the physics and shows the general nature of the effects.

Local field interaction: If a local field interaction $E_{loc} = E + L\,P$ (E is the external field, P the polarization) is incorporated into the usual Bloch equations of the isolated Two-Level model [1] we find the time-integrated signal I_{k_3} in direction $k_3 = 2k_2 - k_1$ as a function of time delay $T = t_2 - t_1$ for δ-shaped optical pulses to third order

$$I_{k_3}^{hom}(T) \sim \Theta(T)\, e^{-2T/T_2} + \frac{4\,V^2}{(2T_2^{-1} + T_1^{-1})\,(T_2^{-1} + T_1^{-1})} \times \left(\Theta(T)\, e^{-2T/T_2} + \Theta(-T)\, e^{4T/T_2}\right) \tag{1}$$

Θ is the Heaviside stepfunction, V is proportional to the Lorenz factor L and reflects the strength of interaction. For finite V the signal exhibits a finite rise time with a time constant $T_2/4$ and the decay as usual with time constant $T_2/2$. For V=0 the well known result for the isolated Two-Level model [1] is retained. This behavior can be interpreted quite intuitively. With or without local field interaction a grating is formed once both, pulse-1 and pulse-2 (for either time-ordering) have arrived in the sample. Without local field interaction only the applied fields can diffract from this grating. If pulse-2 comes first (negative time-delay), the grating is only set up after pulse-1 arrives. At this time, however, no photon from pulse-2 is available since pulse-2 has already passed the sample. Thus in the direction $2\times k_2 - k_1$ no signal is generated for negative time-delay. With local field interaction, however, it is possible to diffract the left-over of the first order polarization of beam-2 off the grating formed as soon as pulse-1 arrives. Since two first order polarizations from beam-2 are involved in this process and have both decayed with a time constant T_2 until pulse-1 arrives, the signal decays with twice the time constant for negative time delay. For an inhomogeneously broadened line we retain a photon-echo like response and the signal at negative time-delay disappears.

$$I_{k_3}^{inhom}(T) \sim \Theta(T)\, e^{-\frac{4T}{T_2}} + 4\,V^2\,T_1^2\,\Theta(T)\, e^{-\frac{4T}{T_2}}\left(1 - e^{-\frac{T}{T_1}}\right)^2 \tag{2}$$

Our numerical calculations for excitons in semiconductors show, that V is in fact so large, that the strength of the terms in eqs.(1,2) with prefactor V^2 are comparable to the ones without prefactor V^2.

Four-Wave-Mixing at the semiconductor band edge: The different states in the valence and conduction band of a real semiconductor can be decribed by a coupled

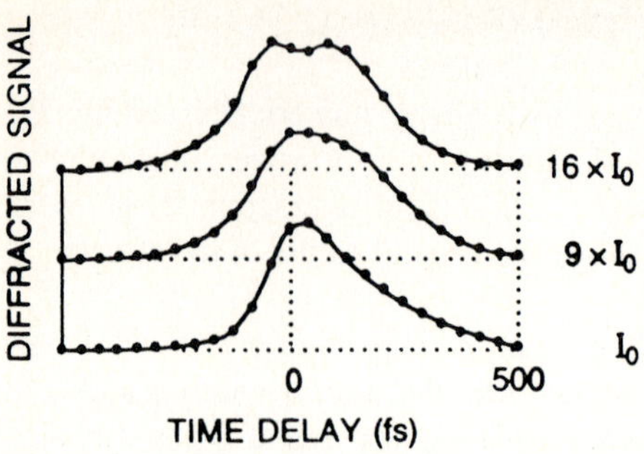

Fig. 2: Theory: Evolution of the diffracted signal as a function of time-delay for increasing intensity. For details see [4].

set of Two-Level systems. These states are interacting via the Coulomb-potential. This interaction can again be interpreted as a local field effect, which leads to coherent interaction of excitons. In this framework one has to solve the kinetic equations numerically. We have done this for parameters that are close to the experimental conditions. Results are depicted in Fig. 2. We obtain good qualitative agreement with the experiment. Again the signal for negative time-delay gains relative strength and two temporal maxima evolve for increasing excitation. This can be understood as a subtle interference of higher order polarizations with the third order polarization dicussed in the previous chapter. The lineshape thus reflects coherent and incoherent interaction of excitons. The coherent part changes the lineshape, the incoherent part merely reduces T_2.

Conclusions: The lineshape of time-resolved Four-Wave-Mixing on dense media can be qualitatively influenced by coherent interaction between polarizations. One can thus separate incoherent and coherent interaction of polarizations by time-resolved FWM.

References

[1] T. Yajima and Yoichi Taira, J. Phys. Soc. Jap. 47, 1620 (1979)
[2] L. Schultheis, M.D. Sturge and J. Hegarty, Appl. Phys. Lett. 47, 995 (1985)
[3] K. Leo et al., see proceedings of this conference
[4] M. Wegener, D.S. Chemla, S. Schmitt-Rink and W. Schäfer,
Phys. Rev. A, submitted

Stimulated Photon Echoes and Free Polarization Decay in GaAs Multiple Quantum Well Structures: Evidence for Localized and Delocalized Excitons

M.D. Webb, S.T. Cundiff, and D.G. Steel

The University of Michigan, Ann Arbor, MI48109, USA

Abstract. Interface disorder in multiple quantum wells leads to inhomogeneous broadening of excitons, and allows the study of the effects of disorder on exciton localization. Transient four-wave mixing studies illustrate that spectral diffusion via phonon assisted migration and thermal activation are the dominant mechanisms of localized exciton relaxation. Localized and delocalized excitons contribute to the nonlinearity, leading to the simultaneous observation of *both* stimulated photon echoes and free polarization decay from a semiconductor heterostructure.

1. Introduction

Recent experimental and theoretical work suggests that lowest energy excitons in GaAs/AlGaAs multiple quantum wells are localized by disorder, while higher energy excitons are quasi-delocalized, described by Bloch-type wavefunctions [1]. At low temperature, localized excitons are expected to be characterized by spectral diffusion and by long optical coherence times, while delocalized excitons have short coherence times and exhibit spatial mobility [1]. The short optical coherence time associated with delocalized excitons results in a linear absorption profile that is primarily homogeneously broadened, while the disorder associated with localized excitons leads to an inhomogeneously broadened contribution to the absorption. In this work, we investigate the energy and phase relaxation of HH1 excitons in GaAs/AlGaAs MQW's at temperatures below 15K. Specifically, we report the first observation of phonon assisted migration in GaAs multiple quantum wells. We also report the first simultaneous observation of free polarization decay and stimulated photon echo signals from semiconductor heterostructure materials. These measurements provide evidence that both localized and delocalized excitons contribute to the nonlinear optical response, and that the transition from localized to delocalized states may not be sharply defined in energy.

2. Scattering and relaxation studies of excitons.

To study the mechanisms responsible for the scattering and relaxation of excitons, the experimental configuration incorporates the method of backward, three-pulse transient four-wave mixing (FWM) studies on GaAs/$Al_{0.3}Ga_{0.7}As$ quantum wells with 65 periods of 96Å wells and 98Å barriers. The output signal is either time-integrated, or time resolved by mixing it with a reference pulse in a $LiIO_3$ crystal and detecting the background-free second harmonic. Under these experimental conditions, an inhomogeneously broadened exciton population produces a stimulated photon echo which is delayed in time from the third pulse by an amount approximately equal to the delay between the first and second pulses. A

Editors: C.B. Harris · E.P. Ippen · G.A. Mourou · A.H. Zewail

homogeneously broadened exciton population produces a free polarization decay signal, nearly coincident in time with the third pulse.

Compared to the 6 ps dephasing time (corresponding to an absorption linewidth of 0.2 meV) measured in high-quality single quantum wells [2], the HH1 exciton absorption width of 2.5 meV in our sample implies that the exciton is inhomogeneously broadened. This is most likely due to interface disorder present as a result of non-ideal growth conditions. Time-integrated transient FWM studies of exciton relaxation reveal that the scattering time is 68 ps (corresponding to a homogeneous linewidth of 0.02 meV); see Figure 1. This relatively slow phase relaxation time implies that the excitons are localized. The observed exciton lifetime in this sample is 34 ps, as determined by measuring the signal as a function of third pulse delay, and is much too fast for exciton recombination. This result, in combination with the dephasing time measurement, strongly suggest that spectral diffusion of localized excitons is the primary scattering mechanism. Recent theoretical work [3] predicts that phonon assisted migration of localized excitons may be the dominant mechanism of spectral diffusion below 10K, with a unique temperature dependence for the scattering rate given by $\Gamma = \Gamma_0 \exp(\beta T^{\alpha})$, where $1.6<\alpha<1.7$. At higher temperatures, thermal activation is expected to dominate the relaxation. Figure 2 shows the measurement of the scattering rate, with the data fitted to phonon assisted migration at low temperature and thermal activation at higher temperature.

A measurement of the dephasing rate as a function of excitation frequency shows the dephasing rate increases an order of magnitude near the absorption line center, confirming earlier measurements and suggesting evidence

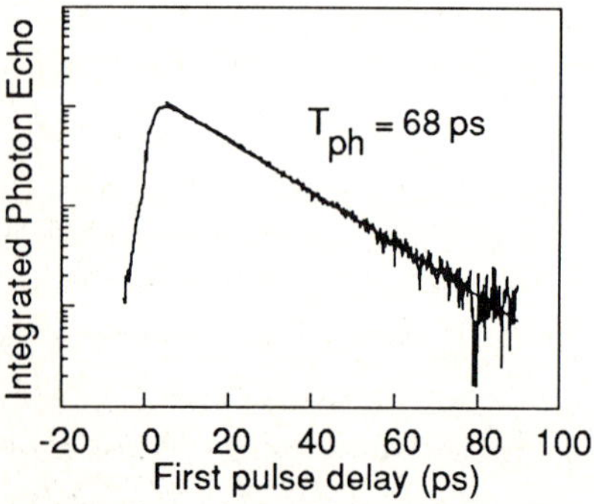

Figure 1. The exciton phase relaxation scattering curve, obtained for T = 2.3K, with an excitation density of 2.0×10^7 excitons/cm^2-layer, yielding a homogeneous linewidth of 0.02 meV.

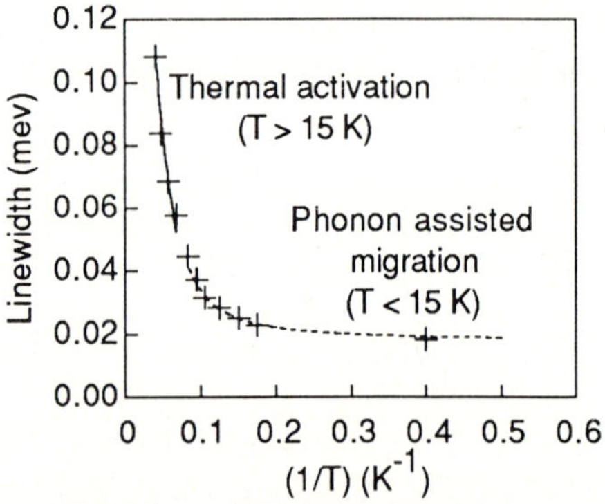

Figure 2. Homogeneous linewidth as a function of temperature; the dashed line is a fit to the theory of phonon assisted migration, with Γ_0 = 0.02 meV, β = 0.016, and α = 1.6. The solid line is a fit for thermal activation, with an activation energy of 3.4 meV.

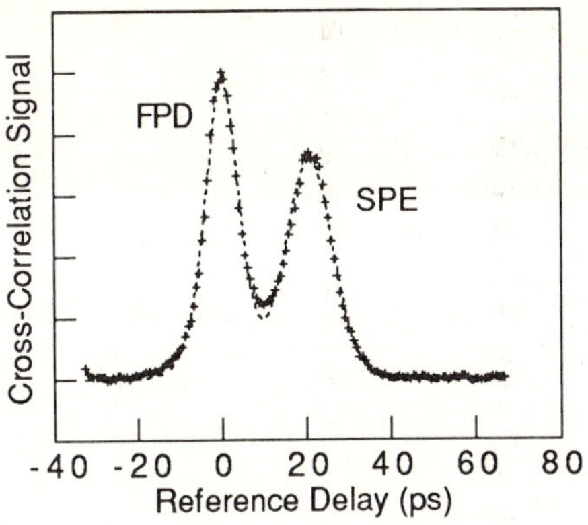

Figure 3. Stimulated photon echoes and free polarization decay, illustrating that both localized and delocalized excitons contribute to the response.

for the exciton mobility edge [1]. However, our time-resolved measurements of the emission signal show the interpretation may be more complex. Figure 3 shows the presence of both a prompt free polarization decay (FPD) signal, and a delayed stimulated photon echo (SPE) signal, representing contributions from a homogeneously broadened transition and an inhomogeneously broadened transition, respectively. Both features are resonant, have strong spectral overlap with each other, and are characterized by different dephasing and excitation relaxation rates. Indeed, the dephasing rate of the FPD signal is nearly pulsewidth limited, corresponding to a decay time shorter than 5 ps. The data suggest that the FPD signal arises from quasi-delocalized excitons, while the SPE signal is due to localized excitons and that both excitons can clearly exist at the same excitation energy.

The authors wish to acknowledge support for this research from ARO and to thank P. K. Battacharya for providing MBE samples.

References:

1. J. Hegarty, L. Goldner, M. D. Sturge, Phys. Rev. B, **30**, 7346, (1984).
2. L. Schultheis, A. Honold, J. Kuhl, K. Kohler, C. W. Tu, Phys. Rev. B, **34**, 9027, (1986).
3. T. Takagahara, Phys. Rev. B, **32**, 7013, (1985).

Γ-X Intervalley Electron Transfer Rates in Type-II $Al_xGa_{1-x}As/AlAs$ Superlattices

J. Nunnenkamp[1], *J. Kuhl*[1], *K. Ploog*[1], *J. Feldmann*[2], *E.O. Göbel*[2], *P. Dawson*[3], *and C.T. Foxon*[3]

[1]Max-Planck-Institut für Festkörperforschung, Heisenbergstr. 1, D-7000 Stuttgart 80, Fed. Rep. of Germany
[2]Fachbereich Physik & Zentrum für Materialwissenschaften, Philipps Universität, Renthof 5, D-3550 Marburg, Fed. Rep. of Germany
[3]Philips Research Laboratories, Redhill, Surrey RH1 5HA, UK

Reduction of the GaAs well width in $(GaAs)_m/(AlAs)_n$ (m, n = number of monolayers) superlattices (SL) permits tuning of the lowest Γ-conduction band state in the GaAs above the X-minimum in the AlAs layers for $m \leq 12$. In such a staggered type-II SL the highest valence band state is contained in the GaAs layers and the lowest conduction band state in the AlAs slabs. After photoexcitation of hot electrons and holes in the confined Γ conduction- and valence band states of the GaAs, the holes stay in the GaAs during relaxation and the electrons scatter to the energetically lower X-related states of the AlAs.

In spite of the spatial charge transfer from the GaAs to the AlAs associated with this intervalley electron scattering, the $\Gamma-X$ scattering rate in the type-II SL can be extremely fast and comparable to the respective events in a GaAs bulk crystal [1,2]. Ternary/binary $(Al_xGa_{1-x}As)_m/(AlAs)_n$ MQW show type-II behavior for much larger values of m [3,4].

In this paper we present a systematic study of the Γ-X intervalley scattering in $(Al_xGa_{1-x}As)_m(AlAs)_n$ type-II short period superlattices (SPS) grown by MBE with various values for x, m and n by means of pico- and femtosecond excite and probe spectroscopy. We demonstrate that the scattering time τ can systematically be varied from 100fs-22ps by changing the thickness and the composition of the well material.

Energy- and time-resolved differential transmission spectra (DTS) at T=10 K for a sample with m=11, n=24, are depicted in Fig. 1. The pronounced bleaching at the heavy-hole, light-hole and split off excitonic transition originates from screening and state filling effects. The plot of the peak differential transmission versus delay time for the hh-, hh- and s-o exciton transition shown in Fig. 1 (right part) demonstrates a partial recovery of the bleaching on a subpicosecond time scale in the type II structures opposite to the usual behavior in type I-MQW and SL [5]. No further appreciable change of the DTS signal is observed up to 20 ps which is the upper limit for τ in GaAs/AlAs SPS [6]. The initial rapid recovery of the exciton bleaching in type II structures can be attributed to intervalley scattering of electrons out of the Γ-conduction band in the GaAs into the X-conduction band of the AlAs and the corresponding reduction of state filling in the Γ-conduction band. From the spectra of Fig. 1 we obtain τ = 0.65ps for the (m = 11, n = 24) SPS. The remaining signal,

Springer Series in Chemical Physics, Vol. 53 **Ultrafast Phenomena VII**
Editors: C.B. Harris · E.P. Ippen · G.A. Mourou · A.H. Zewail

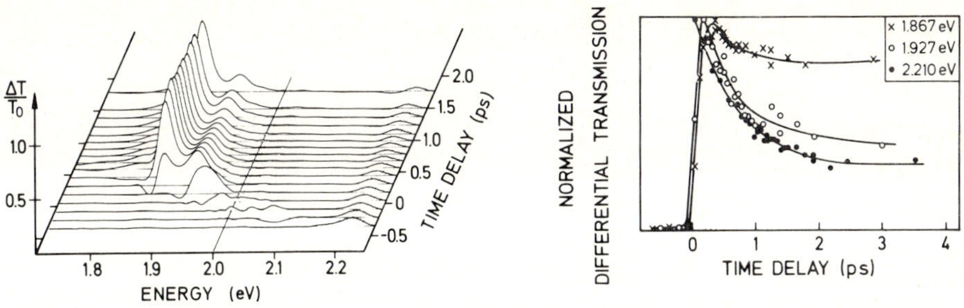

Fig. 1 Energy and time-resolved differential transmission spectrum (left) and time-resolved normalized differential transmission at the $hh, \ell h$ and split-off type-I transition of the [11, 24] SPS (right).

which decays on a timescale of 100 ps, originates from state filling of the still populated hole states of the Γ-valence band in GaAs and from screening of the excitonic enhancement by holes in the GaAs and electrons in the AlAs.

The measured Γ-X transfer times exhibit a rapid decrease with decreasing well width and cover the complete range between 100fs and 22ps. The dependence on AlAs barrier thickness is relatively weak. This pronounced variation of τ with the structure parameters directly follows the corresponding change in the spatial overlap of the initial and final state electronic wavefunctions, which have maximum amplitudes in the GaAs and AlAs, respectively. Figure 2 presents a plot of τ for 12 samples versus S_{xz} which is the sum over the squared overlap integrals of Γ and X_z Kronig-Penney envelope wavefunctions, including all possible final X_z states. The $\Gamma - X_z$ scattering does not require momentum conservation and is mediated via mixing of the Γ and X_z wavefunctions by the potential step at the (Al, Ga)As/AlAs interface. In contrast, $\Gamma - X_{x,y}$ scattering necessitates the participation of phonons or interface roughness in the scattering for conservation of the wavevector components parallel to the layers. Figure 2 demonstrates that the experimental data are fairly well described if only the overlap between the Γ and X_z envelope wavefunctions is considered.

Intensity and temperature dependent measurements of τ demonstrate that carrier/carrier interactions play no role for the Γ-X real space charge transfer and that phonon assisted scattering becomes important only for the thicker (Al,Ga)As/AlAs MQWs. As the transfer times exhibit no change with temperature between 295 and 10 K in the case of thin well layers ($m \leq 12$) (see Fig. 2) we have to conclude that Γ-X mixing and/or $\Gamma - X_{x,y}$ scattering induced by the interface roughness provide the dominant contribution to the interlayer scattering in these samples. On the other hand, $(Al_xGa_{1-x}As)_m/(AlAs)_n$ MQW with (m=68.3, n=36.3, x=0.37) and (m=39, n=33.5, x=0.36) reveal a decrease of τ from 22 to 2.5 ps and from 4.3 ps to 1.9 ps, respectively, if the temperature is raised from 10 K to 300 K. The detailed temperature dependence depicted in Fig. 3 for the (m= 68.3, n=36.3, x= 0.37)-sample can be explained by optical phonon assisted scattering if Stokes and anti-Stokes

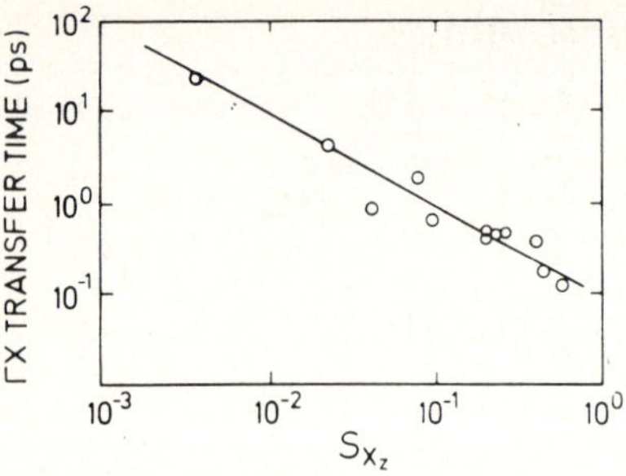

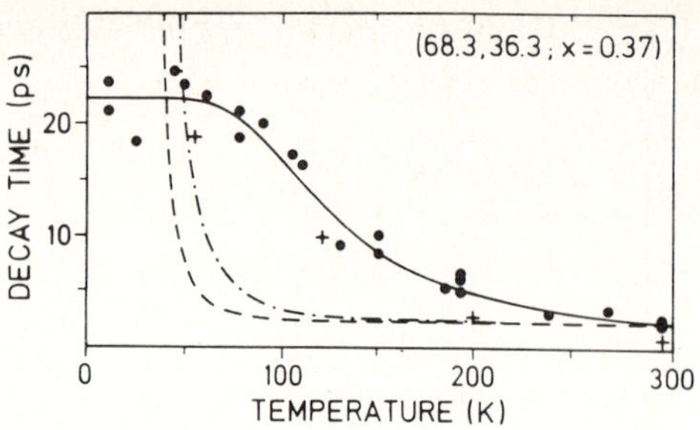

Fig. 2 (left) Plot of the experimental Γ-X transfer times versus the calculated squared overlap integrals for Γ and X_z envelope wavefunctions.

Fig. 3(right) Variation of the Γ-X transfer times for the (m=68.3, n=36.3, x=0.37) MQW with temperature.

scattering processes are taken into account. The solid line represents a fit to the experimental data corresponding to $\tau^{-1} = \tau_o^{-1}(J_{abs} \cdot N_{ph} + J_{em}(N_{ph}+1))$ with N_{ph} = Bose-Einstein phonon occupation probability. The first and second term account for processes associated with absorption and emission of phonons, respectively. J_{abs}, J_{em} and τ_o^{-1} have been used as fitting parameters. The dashed and the dashed/dotted lines describe the temperature dependence theoretically predicted for ionized impurity scattering assuming a carbon acceptor with a concentration of 10^{10}cm^{-2} and 10^{11}cm^{-2}, respectively. The result identifies electron-optical phonon interaction as the dominant intervalley scattering mechanism in MQW with well thicknesses $m > 30$.

References

1. J. Feldmann, R. Sattmann, E. Göbel, J. Kuhl, J. Hebling, K. Ploog, R. Muralidharan, P. Dawson, C.T. Foxon, *Phys.Rev.Lett.* **62**, 1892 (1989).
2. P. Saeta, J.F. Federici, R.J. Fischer, B.I. Greene, L. Pfeiffer, R.C. Spitzer, B.A. Wilson, *Appl. Phys. Lett.* **54**, 1681 (1989).
3. J. Feldmann, R. Sattmann, E.O. Göbel, J. Nunnenkamp, J. Kuhl, J. Hebling, K. Ploog, R. Muralidharan, P. Dawson, C.T. Foxon, *Sol.Stat.Electr.* **32**, 1713 (1989).
4. Y. Masumoto, T. Mishina, F. Sasaki, M. Adachi, *Phys.Rev.* **B40**, 8581 (1989).
5. W. Knox, C. Hirlimann, D.A.B. Miller, J. Shah, D.S. Chemla, and C.V. Shank, *Phys. Rev. Lett.* **56**, 1191 (1986).
6. G. Peter, E.O. Göbel, W.W. Rühle, J. Nagele and K. Ploog, *Superl. and Microstructures* **5**, 197 (1989).

Two-Dimensional Carrier–Carrier Screening Studied with Femtosecond Photon Echoes

J.-Y. Bigot[1], *M.T. Portella*[1], *R.W. Schoenlein*[1], *C.V. Shank*[1], *and J.E. Cunningham*[2]

[1]Lawrence Berkeley Laboratory, University of California, 1 Cyclotron Road, MS:70-110A, Berkeley, CA 94720, USA
[2]AT&T Bell Laboratories, Holmdel, NJ 07733, USA

Abstract. The two dimensional behavior of carrier-carrier screening is studied in a GaAs-GaAlAs quantum well structure with femtosecond photon echoes. The variation of the electron hole dephasing time with carrier density indicates that the screening is due to nearest neighbor interactions in a two dimensionnal electron-hole gas.

1. Introduction

The phase coherence time between an electron and the corresponding hole created during a band-to-band optical transition in a semiconductor diminishes with increasing carrier density. This electron-hole dephasing process, which is due to carrier-carrier interactions, is reduced however by the mutual screening between the electrons and holes. It is particularly interesting to study this carrier-carrier screening in quantum well (QW) structures where, due to the confinement of the carriers in the well, the system behaves like a two dimensional (2D) electron-hole gas.

We study here the influence of carrier-carrier interactions on the dephasing time of the electron-hole polarization induced by an ultrashort optical pulse on a GaAs-GaAlAs QW. Measurements performed at different carrier densities indicate a screening behavior that is characteristic of the two-dimensional structure. The 2D screening that we deduce from our measurements is well understood if we assume that the screened Coulomb interaction between carriers is determined by the average inter-carrier spacing.

Femtosecond optical spectroscopy now allows the dynamics of energy redistribution and dephasing of excited carriers in bulk semiconductors or 2D QW structures to be studied with 10 fs resolution. In bulk GaAs it is known [1] that, under high densities of excitation, carrier-carrier scattering plays an important role for the thermalization of hot carriers and it has been shown [2] that it is also the predominant mechanism for the momentum relaxation. In QW structures the carrier thermalization has been extensively studied in undoped [3,4] and doped [5] structures, where collisions between carriers was found to be the main process for the relaxation of energy through inelastic carrier-carrier scattering. Recent work [4,5] has also shown that phase space filling due to the Pauli exclusion principle prevails over the long range Coulomb screening for the bleaching of the exciton. Until now, there have been no direct temporal measurements of the carrier-carrier dephasing time in a 2D QW. The Coulomb screening is indeed expected to be important in the time scale when the carriers have not completely thermalized. The observation of this effect is investigated here with femtosecond photon echoes.

Springer Series in Chemical Physics, Vol. 53 **Ultrafast Phenomena VII**
Editors: C.B. Harris · E.P. Ippen · G.A. Mourou · A.H. Zewail

2. Femtosecond photon echoes in semiconductors

Experimentally, the technique of photon echoes produced by femtosecond pulses has already proved to be very powerful for measuring polarization dephasing times in molecules in solution as well as in semiconductors [6]. In this technique, the photon echoes are generated by a sequence of two-femtosecond laser beams with respective wave vectors k_1 and k_2. The energy of the echo produced in the direction $2k_2-k_1$ (or $2k_1-k_2$) is detected as function of the relative delay T_{12} between the two pulses. The resulting decay time T_{echo} of the echo signal is a direct measurement of the polarization dephasing time T_2 since $T_{echo}=T_2/4$ for an inhomogeneously broadened system (in the limit of infinite inhomogeneity).

Using this technique, dephasing times in the range 15 to 50 fs have been reported in bulk GaAs [2]. Moreover, the variation of the dephasing time with the three-dimensionnal carrier density N was shown to follow a power law : $T_{echo}\sim N^{-0.3}$, characteristic of a three dimensional environment.

The variation of T_{echo} in a 2D environment has been studied in a 9.6 nm-thick GaAs quantum well with 8 fs near infrared pulses. The large pulse spectrum (730-900 nm total bandwidth) covered only the bidimensional part of the structure. This is shown in Figure 1 where we have represented the pulse spectrum and the optical density of the QW.

The ultrashort infrared pulses have been generated as follows [7]. The 50 fs pulses coming from an Argon-pumped CPM cavity are amplified to a few microjoules by a copper vapor laser with 8 KHz repetition rate and focused in a jet of ethylene glycol to generate a white-light continuum. The infrared part of this spectral continuum is selected with a slit and amplified in a gain medium (LDS-821 diluted in ethylene glycol) with a second copper vapor laser. The resulting 50 fs infrared pulses, which can be tuned within the LDS-821 gain response curve, are

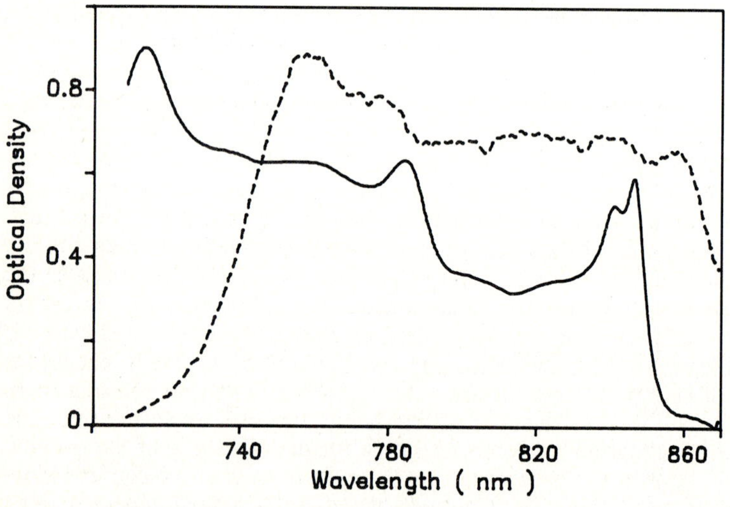

Figure 1: Optical density of the 9.6 nm-thick quantum well (full line) and pulse spectrum (dashed line).

further shortened to 10 fs using a fiber and a sequence of gratings and prisms [8]. The matching of the pulse spectrum to the 2D absorption region of the structure was accomplished by tuning the pulse wavelength to 805 nm before compression. In these conditions, the ratio between the density of photons exciting the 3D continuum of the QW structure and those exciting the well is estimated to be less than 2% (see fig.1). The photon echoes are produced by the compressed pulses which are split in two parts in a Michelson interferometer and focused on the QW structure to a spot of about 20 μm. The $2k_2$-k_1 echo signal is detected with a photomultipler, a box-car and a lock-in amplifier for different relative delays between the two incident pulses.

3. Two Dimensional screening in a Quantum Well

In figure 2 we have represented photon echo signals from the QW structure as function of the relative delay between the two infrared pulses. T_{echo} is obtained from the exponential decay of the corresponding echo signal using a least-square polynomial fit.

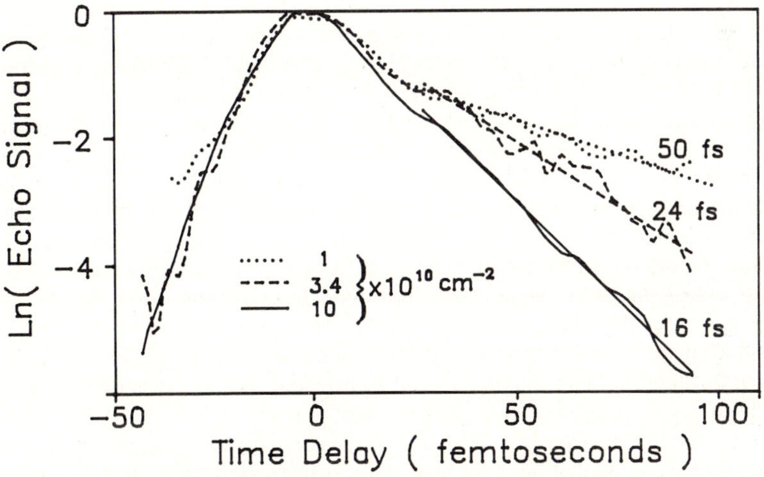

Figure 2: Photon echo signal in the quantum well structure as function of relative time delay between the two 8 fs infrared pulses.

Figure 3 shows the echo decay time T_{echo} measured for different carrier densities N. The experimental error bar for each point indicates the good signal-to-noise ratio. The solid curve is a fit to the data which gives a density dependence of the dephasing time on carrier density with an exponent -0.55 ±0.04.

The decrease of the relaxation time T_2 ($T_2=4T_{echo}$) with increasing density of carriers suggests that carrier-carrier scattering is the mechanism responsible for the electron-hole dephasing process observed in these data. Since the experimental technique that we use is sensitive to the phase relaxation, both elastic and inelastic

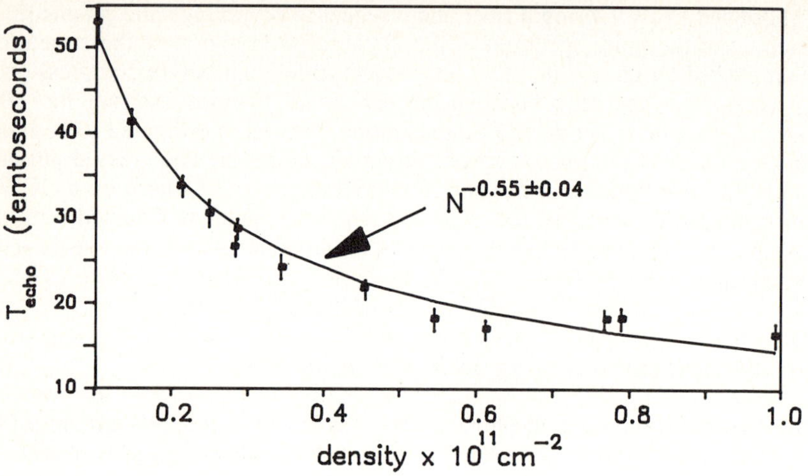

Figure 3: Variation of the echo decay time with the carrier density in a quantum well structure.

scatterings are involved. The fast decay times deduced from our data are consistent with the fact that the phase coherence between an electron and a hole is lost long before the carrier thermalization is reached, which is known to be of the order of 300 fs for similar 2D structures [4].

The factor -0.55 for the density law displayed in Figure 3 reflects the two-dimensional screening behavior of the carriers. A similar experiment performed on a bulk sample of GaAs gave a factor of -0.3 for the density law. Indeed, the Coulomb interaction between two charges is less screened in a 2D particle gas than in a 3D gas, since, due to the confinement of carriers in a layer, the environment is less effective in shielding the attractive or repulsive forces. The behavior observed here is consistent with a screened Coulomb interaction having a very short effective range. This is known to occur in systems with high density of carriers [9], where the range of the Coulomb interaction is much reduced due to the mutual screening of the charged particles.

All these observations suggest a description of carrier-carrier scattering with a small screening length. Indeed, our experimental results as well as those reported in [2] can be modelled with very simple geometrical arguments based on short range interactions. For a system with dimensionality n, considering nearest neighbor interactions, the scattering rate of a charge in a "sphere" containing N other charges scales as $N^{1/n}$. Therefore, the density law for the scattering time scales to -1/n, i.e., -0.33 for a 3D system and -0.5 for a 2D system. This simple nearest neighbor interaction picture gives a fairly good description of our data. Naturally, a complete theoretical description of our experimental results requires sophisticated many-body calculations taking into account the highly non-thermal carrier population created by the femtosecond pulses.

REFERENCES

[1] J. Shah, *IEEE J. Quantum Elect.* *22*, 1728 (1985).

[2] P.C. Becker, H.L. Fragnito, C.H. Brito-Cruz, R.L. Fork, J.E. Cunningham, J.E. Henry, C.V. Shank, *Phys. Rev. Lett. 61*, 1647 (1988).

[3] C.V. Shank, R.L. Fork, R. Yen, J. Shah, B.I. Greene, A.C. Gossard, C. Weisbuch, *Sol. State Commun. 47*, 981 (1983).

[4] W.H. Knox, C. Hirlimann, D.A.B. Miller, J. Shah, D.S. Chemla, C.V. Shank, *Phys. Rev. Lett. 56*, 1191 (1986).

[5] W.H. Knox, D.S. Chemla, G. Livescu, J.E. Cunningham, J.E. Henry, *Phys. Rev. Lett. 61*, 1290 (1988).

[6] C.V. Shank, P.C. Becker, H.L. Fragnito, R.L. Fork, *Ultrafast Phenomena VI*, p. 344, Springer-Verlag, 1988.

[7] P.C. Becker, H.L. Fragnito, R.L. Fork, F.A. Beisser, C.V. Shank, *Appl. Phys. Lett. 54,* 411 (1989).

[8] R.L. Fork, C.H. Brito-Cruz, P.C. Becker, C.V. Shank, *Opt. Lett. 12*, 483 (1987).

[9] see for instance: D. Pines, C. Nozieres, "The Theory of Quantum Liquids: Normal Fermi Liquids," Addisson-Wesley, 1966.

This work was supported by the U.S. Department of Energy under Contract No. DE-AC03-76SF00098

Femtosecond Spectroscopy of Highly Excited Cu_2O

A. Mysyrowicz[1], *D. Hulin*[1], *and E. Hanamura*[2]

[1]Laboratoire d'Optique Appliquée, ENSTA, Ecole Poytechnique, F-91120 Palaiseau, France

[2]Department of Applied Physics, University of Tokyo, Hongo, Bunkyo-ku, Tokyo, Japan

Abstract. Femtosecond time-resolved differential absorption spectroscopy of Cu_2O is performed following optical injection of up to 5.10^{19} cm^{-3} free carriers or n = 1 excitons. A collision-assisted absorption process is observed. The time of formation of excitons is found to be < 200 fs.

Cu_2O is a canonical exciton material. The lowest exciton series (the yellow series) exhibits up to 8 distinct absorption lines obeying a hydrogenic relation, starting with term n = 2. The n = 1 term is seen as a very weak quadrupole transition. It is also at the origin of a broad phonon-assisted absorption continuum below line n = 2. Using the familiar pump and probe detection technique, we have investigated the differential absorption spectra of highly excited Cu_2O with femtosecond time-resolution. Excitation of the crystal was performed by tuning the 200 fs pump pulse central wavelength to the n = 1 phonon continuum (direct optical injection of n = 1 excitons). The highest initial electron-hole pair density was estimated $n \sim 5.10^{19}$ cm^{-3}. Note that it still corresponds to a dilute excitation regime, well below the Mott critical density since the exciton Bohr radius $a_0 = 0.7$ nm ($na_0^3 = 0.035$ for $n = 10^{20}$ cm^{-3}).

Figure 1 shows two representative absorption spectra, recorded 2 ps after injection of n = 1 excitons together with the transmission spectrum of the unperturbed crystal. As the initial electron-hole pair density is increased, the different terms of the series disappear successively, starting with the highest terms. There is no appreciable shift of the lines accompaying this bleaching. However, an increase of absorption on the high energy of each discrete line is clearly seen.

We attribute these changes in absorption to a collision-assisted transition. In an elementary step of this process, an exciton of term n is created at large K vector with the simultaneous ejection of an exciton from the bottom of the n = 1 kinetic energy band to a higher momentum state. Because of energy and momentum conservation laws, this "indirect" (collision-assisted) transition gives rise to an assymetric broadening to the high energy side of each line as observed in the experiment. A complete model calculation of the expected lineshape and its particle density dependance, incorporating exciton-phonon broadening and the Fano interference between a discrete excitonic line and the underlying phonon sideband will be published elsewhere.

We have also compared the differential absorption spectra obtained under initial injection of either n = 1 excitons or free electron-hole pairs (pump laser central wavelength tuned to the ionization continuum above the band gap) using comparable electron-hole pair densities. Figure 2 shows differential spectra recorded sequentially under both pumping conditions at 200 fs time intervals. A striking feature is the similar evolution even at early times. We infer from this similitude that the free carriers have paired into excitons and have relaxed towards the n = 1 exciton band in a time comparable to the pump excitation pulse duration.

Springer Series in Chemical Physics, Vol. 53 **Ultrafast Phenomena VII**
Editors: C.B. Harris · E.P. Ippen · G.A. Mourou · A.H. Zewail

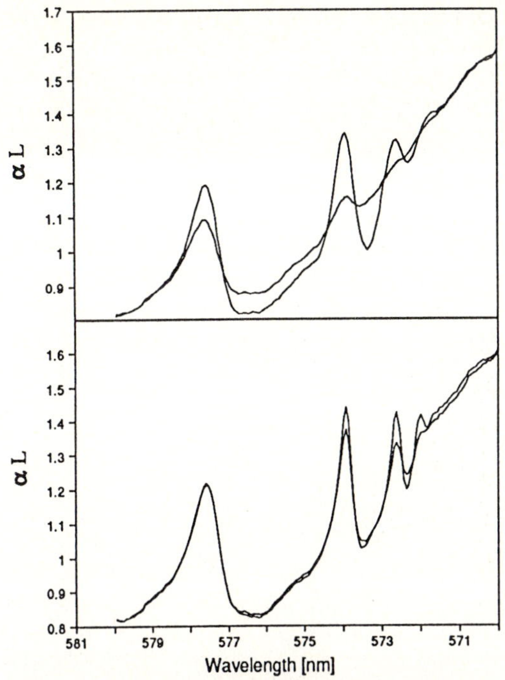

Figure 1. Transmission spectra of Cu_2O at 15 K measured 2 ps after injection of n = 1 excitons, for two injected densities differing by one order of magnitude. For comparison, the unperturbed spectrum is also shown in each case. Lines n = 2, 3, 4 and 5 of the yellow excitonic series are clearly apparent.

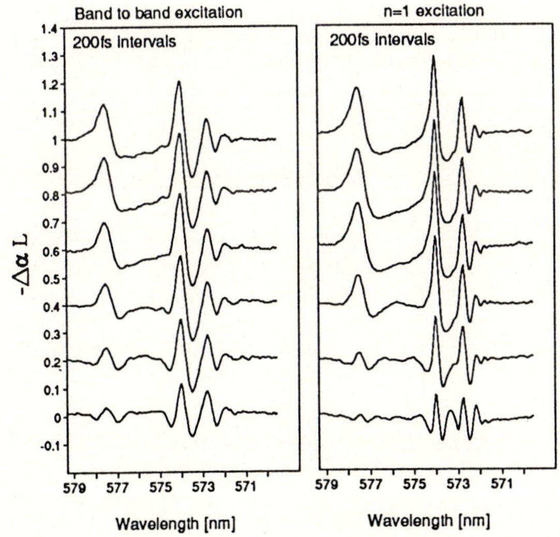

Figure 2. Differential transmission spectra of Cu_2O at 15 K recorded at 200 fs time intervals under injection of free carriers (left) or n = 1 excitons (right). For clarity, successive spectra have been displaced vertically by $-\Delta\alpha L = 0.2$.

In summary, a new absorption process (collision assisted exciton absorption) has been discovered in electronically excited Cu_2O. We expect this mechanism to be also effective in other materials at medium excitations levels. In addition, we deduce the exciton formation time from free carriers to be less than one picosecond under our experimental conditions .

* LOA is Laboratoire associé au C.N.R.S. (URA 1406).

Dynamics of Excitons in Conjugated Polymers

M. Yoshizawa, M. Taiji, and T. Kobayashi

Department of Physics, Faculty of Science, University of Tokyo, Hongo 7-3-1, Bunkyo-ku, Tokyo 113, Japan

Abstract. Time evolution of excitations in polydiacetylenes (PDAs) and polythiophenes (PTs) was investigated by time-resolved absorption and luminescence spectroscopies. The relaxation of photoexcited free excitons to self-trapped (ST) excitons was observed. The formation and decay kinetics of the ST excitons in conjugated polymers was discussed.

1. Introduction

In recent years, there has been great interest in optical and electrical properties of conjugated polymers because of their large and ultrafast optical nonlinearities and their role as model compounds for quasi-one-dimensional semiconductors. PDAs and PTs are among well studied conjugated polymers. PDAs show phase transitions with dramatic color changes and their properties can be modified by changing the side-groups. In PDAs photo-generated singlet and triplet excitons have been observed [1,2] and polarons and bipolarons are formed in PTs after photoexcitation or electrochemical doping [3]. The results of photoluminescence measurements have indicated that the decay time constants of PTs and red-phase PDAs are several picoseconds [4], while photoluminescence in blue-phase PDAs has not been observed.

2. Experimental

The time-resolved absorption spectroscopy was performed using the amplified pulse of a CPM dye laser [1]. A dye laser synchronously pumped by the second harmonic of a mode-locked Nd:YAG laser was used as the pump source of the luminescence spectroscopy. The time dependence of the luminescence was measured using a synchronously-scanning streak camera.

The cast films of a blue-phase PDA [PDA-3BCMU (3-butoxycarbonylmethyl-urethane] and a red-phase PDA [PDA-4BCMU (4-butoxycarbonylmethylurethane] were prepared on glass substrates. The thin films of poly(3-methyl-thiophene) (P3MT) and poly(3-dodecylthiophene) were prepared by electro-chemical polymerization on glass substrates.

3. Results and Discussion

The transient photoinduced absorbance changes in P3MT are shown in Fig. 1. The bleaching due to ground-state depletion appears around 2.0 eV and broad-band absorption below the band-gap is due to ST excitons. Similar absorbance changes were observed also in PDA-3BCMU and PDA-4BCMU. A small minimum at 1.8 eV can be explained in terms of Raman gain. The Raman shift is 1450 cm^{-1} and is assigned to the C=C stretching vibration. The absorbance changes due to several nonlinear optical processes, i.e. hole burning, Raman gain, resonant Kerr effect, and optical Stark effect were observed in PDA-3BCMU and P3MT at short delay times [1,5,6].

Figure 2 shows the time dependence of the absorbance change (ΔA) at 1.77 eV in PDA-3BCMU. The formation time of the ST excitons is estimated

Springer Series in Chemical Physics, Vol. 53 **Ultrafast Phenomena VII**
Editors: C.B. Harris · E.P. Ippen · G.A. Mourou · A.H. Zewail

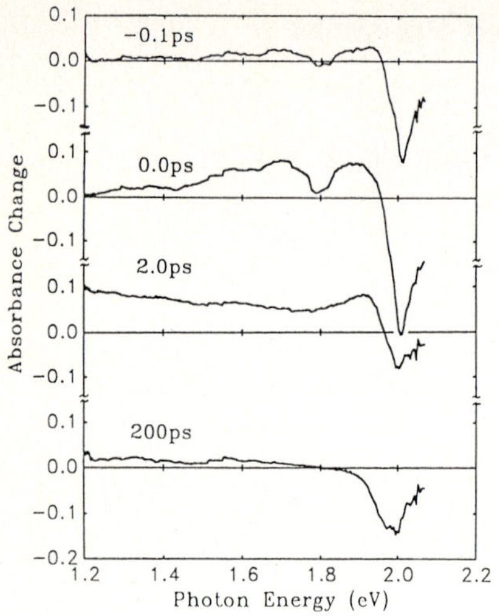

Fig. 1. Transient absorption spectra in P3MT at 10 K. The pump photon energy is 1.97 eV.

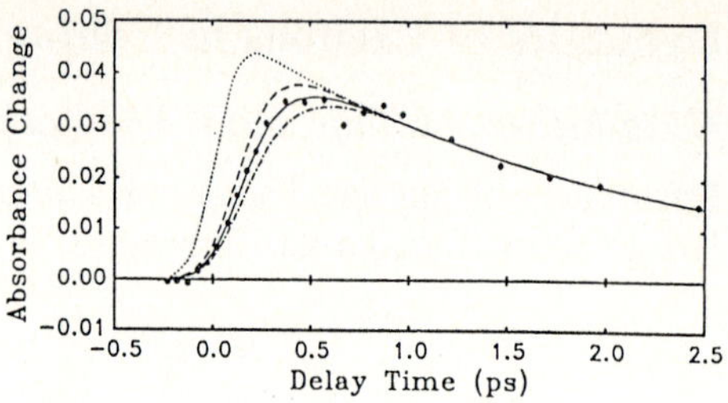

Fig. 2. The time dependence of the absorbance change of PDA-3BCMU at 1.77 eV. Curves are the calculated absorbance changes with the formation times of 0 fs (dotted curve), 100 fs (dashed curve), 150 fs (solid curve), and 200 fs (dash-dotted curve). The decay time is set at 2.0 ps in the calculation.

from the rise time of ΔA as to be 150±50 fs. After this fast relaxation the decay kinetics in PDA-3BCMU can be represented nearly by a single exponential function up to 5 ps with the time constant of 2.0 ps. The initial decay times in PDA-4BCMU and PTs are defined by the initial slope of the semilogarithmic plot of ΔA vs. delay time. They are 890±160 fs, 620±60 fs, and 450±50 fs at 10 K in PDA-4BCMU, P3MT, and P3DT, respectively. After the initial relaxation the induced absorption in PDA-4BCMU and PTs decays more slowly and the time constants defined by the slopes at 5 ps are longer than 10 ps.

The luminescence decay in PDA-4BCMU is plotted in Fig. 3 (log[I(t)] vs. $t^{1/2}$). The decay kinetics has two components. The fast component which could not be resolved by the 30 ps resolution time was observed at 1.9-2.0 eV. It is assigned to the hot excitons. The slow component is represented by $\exp[-(t/\tau)^{1/2}]$ with $\tau = 125$ ps. The spectrum of the slow component was broader than the fast component and it was observed down to 1.6 eV. The decay kinetics of the slow component is explained by the initial process of the trapping of the ST exciton by defects in fractal dimension [6].

Both the bleaching and the induced absorption with a peak around 1.4 eV have a decay time much longer than several hundred picoseconds in both PDAs and PTs (see Fig. 1). In PDAs this long-life component was assigned to be due to triplet excitons from the intensity and wavelength dependence of the pump pulse [1]. The dependence on the polarization of the pump pulse in oriented P3DT film shows that the component is due to polarons and/or bipolarons [7].

Since the formation process of ST excitons has no barrier in the one-dimensional system, the formation process is expected to take place within the coupled phonon cycle. The excitons in conjugated polymers are strongly coupled with the C-C stretching modes. The frequencies of the stretching modes are 1500-2100 cm^{-1} and correspond to the oscillation periods of 15-25 fs, while the trapping time constants of excitons are much longer than the phonon periods. It can be explained as follows. The binding energy remains as the kinetic energy of the lattice oscillation and the formation time of ST excitons is determined by the energy redistribution rate from the strongly coupled phonon to the other phonon modes.

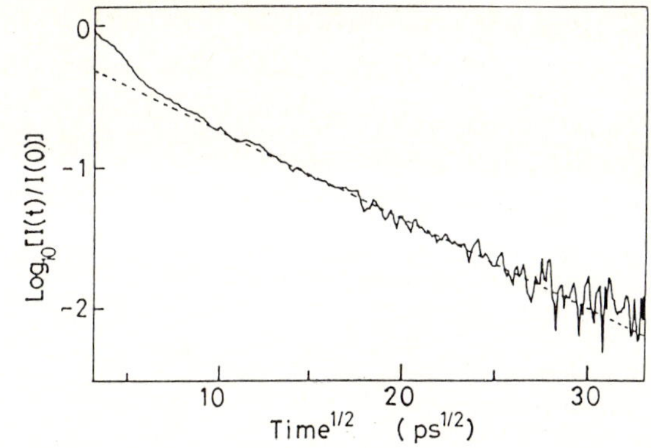

Fig. 3. The luminescence decay, I(t), in PDA-4BCMU at 10 K observed at 1.97 eV. The dashed curve is $\exp[-(t/\tau)^{1/2}]$ with τ = 125 ps.

The decay kinetics of the ST excitons can be explained as follows. The initial decay kinetics with the time constants of about 150-900 fs can be explained in terms of the hot ST excitons. The hot ST excitons can relax to the ground state by passing the crossing point between the ST-exciton and ground-state potentials. After the hot ST excitons relax to the lower states of the ST excitons, they relax by tunneling through and/or passing over the barrier. The decay time constant then becomes longer than the initial decay time. The remaining ST excitons are observed by the luminescence spectroscopy as the slow component and the decay kinetics is explained by the trapping model.

The difference between fluorescent and nonfluorescent polymers can be explained by the difference in the time constant. The potential barrier between the ST exciton and the ground state in PDA-3BCMU is expected to be lower than in PDA-4BCMU because of the lower energy of free and ST exciton. The decay kinetics of the relaxed ST excitons is faster and the number of the remaining ST excitons is smaller than in PDA-4BCMU. The decay kinetics of the ST excitons can be fitted nearly to a single exponential up to 5 ps and the luminescence is weaker and could not be observed.

References

1. M. Yoshizawa, M. Taiji, and T. Kobayashi, IEEE J. Quantum Electron, **QE-25**, 2532 (1989).
2. T. Kobayashi and H. Ikeda, Chem. Phys. Lett., **133**, 54 (1987).
3. K. Kaneto, F. Uesugi, and K. Yoshino, J. Phys. Soc. Jpn. **57**, 747 (1988).
4. K.S. Wong, W. Hayes, T. Hattori, R.A. Taylor, J.F. Ryan, K. Kaneto, K. Yoshino, and D. Bloor, J. Phys. **C18**, L843 (1985).
5. U. Stamm, M. Taiji, M. Yoshizawa, K. Yoshino, and T. Kobayashi, Mol. Cryst. Liq. Cryst. **182A**, 147 (1990).
6. T. Kobayashi, M. Yoshizawa, M. Taiji, U. Stamm, and M. Hasegawa, J. Opt. Soc. Am. B, in press.
7. M. Yoshizawa, K. Yoshino, and T. Kobayashi, to be submitted.

Femtosecond Transient-Grating Experiments in Quantum Wells

K. Leo[1,*], *E.O. Göbel*[1,**], *T.C. Damen*[1], *J. Shah*[1], *S. Schmitt-Rink*[2], *W. Schäfer*[2,***], *J.F. Müller*[3], *K. Köhler*[4], *and P. Ganser*[4]

[1]AT&T Bell Laboratories, Holmdel, NJ 07733, USA
[2]AT&T Bell Laboratories, Murray Hill, NJ 07974, USA
[3]Rutgers University, Piscataway, NJ 08855, USA
[4]Fraunhofer IAF, D-7800 Freiburg, Fed. Rep. of Germany
* On leave from Max-Planck-Institut FKF, D-7000 Stuttgart 80, Fed. Rep. of Germany
** Present address: Philipps-Universität, D-3500 Marburg, Fed. Rep. of Germany
*** Present address: KFA Jülich, D-5170 Jülich, Fed. Rep. of Germany

Abstract. Using degenerate four-wave mixing (DFWM) with subpicosecond time-resolution, we demonstrate the following phenomena: (i) a novel coherent interaction of excitonic polarizations and (ii) quantum beats of excitons in quantum wells due to the interference of light-hole (lh) and heavy-hole (hh) excitons or hh excitons of different confinement.

1. Introduction

Time-resolved DFWM has been proven to be a powerful technique to study the dephasing time constant T_2 of free carriers in bulk GaAs [1] and excitons in GaAs and GaAs/AlGaAs heterostructures [2]. In the self-diffracted geometry, a theory [3] describing the excitons as two-level systems predicts that the diffracted signal in the direction $k_3 = 2k_2 - k_1$ is zero for $T<0$, rises abruptly at $T=0$ and exponentially decays for $T>0$ with a time constant $T_2/2$ (homogeneous transition) or $T_2/4$ (inhomogeneous transition). In this paper, we show in the first part that the coherent interaction of the excitons leads to a signal at *negative delay times*, which rises with *half* the time constant of the decay, and in the second part that the diffracted signal of coherent exciton states shows interference effects (*quantum beats*).

2. Effects of coherent exciton-exciton interactions on the lineshape

Figure 1 shows the diffracted signal versus the delay time for a (nominally) 170 Å $GaAs/Al_{0.35}Ga_{0.65}As$ sample for lattice temperatures of 5, 40 and 70K excited with low intensity 500fs pulses. The signal for $T>0$ reflects the decay of the polarization due to incoherent scattering events; it is faster at higher temperatures due to exciton-phonon scattering [2]. For negative times, however, the signal has a rising wing, with a time-constant which is approximately half of the decay time, as recently predicted theoretically [4]. This signal is not predicted for noninteracting excitons [3]. A simple picture of this effect is that the coherent polarization left over by the earlier arriving pulse #2 interacts with the polarization grating created by the arrival of the pulse #1. This interaction leads to polarization with wavevector k_3, which is detected in the experiment. At higher densities, the signal for negative time increases, and, for excitation tuned to lower energies, a pronounced double-peak structure develops [5], similar to recent observations in InGaAs quantum wells by Wegener et al. [6].

Springer Series in Chemical Physics, Vol. 53 **Ultrafast Phenomena VII**
Editors: C.B. Harris · E.P. Ippen · G.A. Mourou · A.H. Zewail

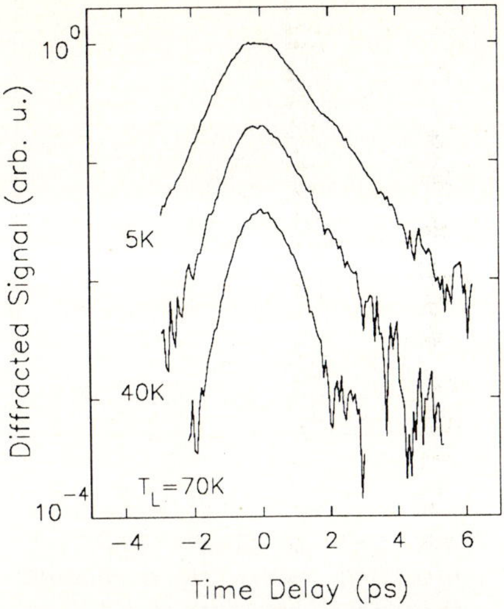

Figure 1: DFWM signals in a 170 Å QW versus time delay for different lattice temperatures and low excitation intensity.

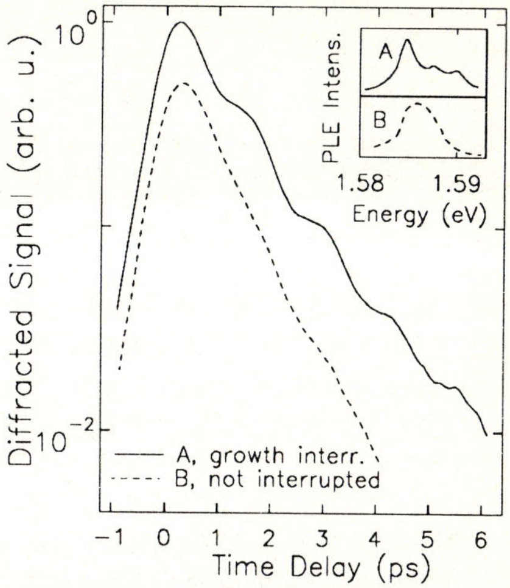

Figure 2: Diffracted signal of a QW with split hh exciton line (A, solid) and single (B, dashed) exciton line.

3. Quantum beats of excitons in quantum wells

In another experiment [7], we have studied two 70 Å GaAs/$Al_{0.35}Ga_{0.65}As$ QW samples grown with (sample A) and without (B) growth interruption. The photoluminescence excitation spectrum (inset of Fig.2) of sample A shows a split hh exciton line due to spatially varying confinement of the excitons. Sample B shows a broad line. Figure 2 shows the diffracted signal for sample A (solid line) and B (dashed line). Sample B shows the expected exponential decay. The signal for sample A, however, is strikingly different: The decay is modulated by an oscillatory structure with a period of about 1.3ps. The oscillation is caused by the interference of the polarization of the the split hh exciton lines. The period corresponds roughly to the splitting of the hh exciton lines of about 2.7meV. This seemingly surprising observation of *quantum beats from extended states in a solid* is possible due to the existence of sharp levels by exciton formation [7]. Our results are also the first observation of quantum beats resulting from *spatially separated states.* Recently, quantum beats have been observed in luminescence in AgBr [8] and in DFWM in BiI_3 [9].

The splitting of the exciton lines of the experiment described above is caused by extrinsic effects. The splitting of the lh and hh excitons is an intrinsic property of a quantum well. In this case, the quantum beats are due to the interference of the splitted levels in a three-level system. Figure 3 shows the first observation of such quantum beats between lh and hh excitons in a quantum well [10]. The dashed line shows the diffracted signal for excitation of the hh only, the solid line for simultaneous excitation of lh and hh. The beat period of 960fs agrees well with the lh-hh splitting of 4.2meV in the 170 Å well.

Recently, we have observed coherent oscillations in a double quantum well structure by DFWM [11], where the beats (Fig.4) correspond to an oscillatory motion of a

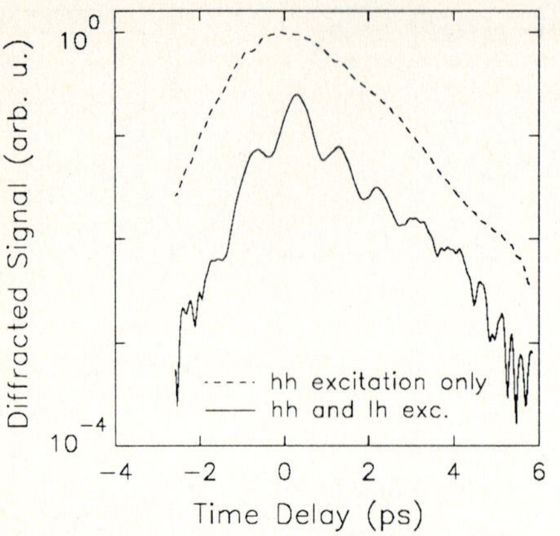

Figure 3: Diffracted intensity for hh excitation only (dashed line) and lh and hh excitation (solid line) in a 170 Å quantum well.

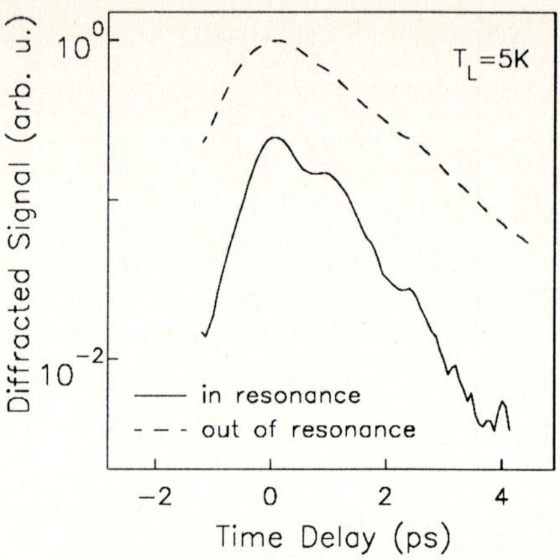

Figure 4: Quantum beats of anticrossed levels in a double quantum well structure (solid line), compared to the signal out of electronic resonance (dashed line).

wavepacket created out of the anticrossed eigenstates. These oscillations were also observed in pump-and-probe experiments [11].

4. References:

[1] P.C. Becker, H.L. Fragnito, C.H. Brito Cruz, R.L Fork, J.E. Cunningham, J.E. Henry, and C.V. Shank, Phys. Rev. Lett. **61**, 1647 (1988)

[2] L. Schultheis, M.D. Sturge, and J. Hegarty, Appl. Phys. Lett. **47**, 995 (1985); L. Schultheis, J. Kuhl, A. Honold, and C.W. Tu, Phys. Rev. Lett. **57**, 1635 (1986); **57**, 1797 (1986)

[3] T. Yajima and Y. Taira, J. Phys. Soc. Jpn. **47**, 1620 (1979)

[4] C. Stafford, S. Schmitt-Rink, and W. Schäfer, Phys. Rev. B, in press.

[5] K. Leo, M. Wegener, J. Shah, D.S. Chemla, E.O. Göbel, T.C. Damen, S. Schmitt-Rink, and W. Schäfer, submitted for publication

[6] M. Wegener et al., paper ME3, this conference

[7] E.O. Göbel, K. Leo, T.C.Damen, J. Shah, S. Schmitt-Rink, W. Schäfer, J.F. Müller, and K. Köhler, Phys. Rev. Lett. **64**, 1801 (1990)

[8] V. Langer, H. Stolz, and W. von der Osten, Phys. Rev. Lett. **64**, 854 (1990)

[9] T. Tokizaki et al., paper ThD6, this conference

[10] K. Leo, T.C. Damen, J. Shah, E.O. Göbel, and K. Köhler, Appl. Phys. Lett., July 9, 1990 issue.

[11] K. Leo, J. Shah, E.O. Göbel, T.C. Damen, S. Schmitt-Rink, W. Schäfer, and K. Köhler, submitted for publication

Femtosecond Quantum Beats in Excitons of a Layered Semiconductor BiI_3

T. Tokizaki[1], *A. Nakamura*[1], *Y. Ishida*[2,*], *T. Yajima*[2,**], *I. Akai*[3], *and T. Karasawa*[3]

[1]Department of Applied Physics, Nagoya University, Chikusa-ku, Nagoya 464-01, Japan
[2]Institute for Solid State Physics, Tokyo University, Minato-ku, Tokyo 106, Japan
[3]Department of Physics, Osaka City University, Sumiyoshi-ku, Osaka 558, Japan
* Present address: NTT Basic Research Lab., Musashino-shi, Tokyo 180, Japan
** Present address: College of Humanities and Science, Nihon University, Setagaya-ku, Tokyo 156, Japan

In recent years, ultrafast relaxation processes of excitons in solids have been widely studied with the development of femtosecond lasers and measuring technique. With a coherent transient spectroscopy we can measure not only the optical dephasing of excited states but also quantum beats among some states in multilevel systems. Interference beats were observed with a degenerate four-wave mixing (DFWM) for atomic Rb vapor[1]. However, such a coherent transient phenomenon has not been observed, to author's knowledge, for semiconductors.

Two-dimensional excitons trapped in stacking faults of a layered semiconductor BiI_3 are known to show several sharp absorption lines near the indirect bandgap at low temperatures. We have studied the optical dephasing of these excitons in the picosecond time region[2]. Since these excitons have a common initial state in valence band, this excitonic system is considered to be a multilevel system. This paper reports the first observation of femtosecond quantum beats among exciton levels in BiI_3 by DFWM using a femtosecond dye laser.

Figure 1 shows a transmission spectrum of the BiI_3 crystal at 4.2K. Absorption lines R, S and T are due to stacking fault excitons originating from the distortion of bulk excitonic states $\phi_{6,10}$, ϕ_3 and ϕ_4[3]. Absorption lines indicated by P and SB arise from "direct" excitons caused by the zone folding effect and excitons trapped at surfaces, respectively.

The conventional DFWM experiment was made using a hybridly mode-locked cavity-dumped R6G dye laser and an optical pulse compressor. The pulse duration was about 200 fs and the spectrum width was 5.3 nm, which sufficiently covered R, S, T absorption lines as shown by the dotted line in Fig.1. When two light beams with the wave vector $\vec{k}_1$, $\vec{k}_2$ were mutually delayed and focused on a sample, the signal beams were emitted to $\vec{k}_3=2\vec{k}_1-\vec{k}_2$ and $\vec{k}_4=2\vec{k}_2-\vec{k}_1$ directions. In the case of nontransform limited pulses, the time resolution is

Editors: C.B. Harris · E.P. Ippen · G.A. Mourou · A.H. Zewail

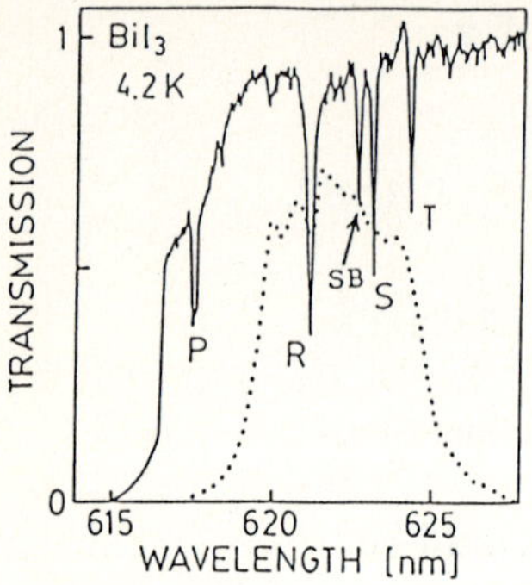

Fig.1 Transmission spectrum of BiI_3 and the laser spectrum (dotted line).

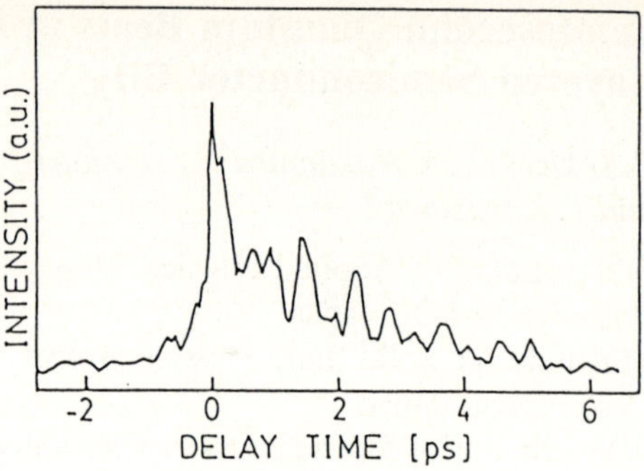

Fig.2 Correlation curves for the $\vec{k}_3$ output beam of DFWM.

governed not by the pulse duration but by the correlation time $\tau_c \simeq 100$ fs, that is almost the reciprocal of the spectral width[4].

Figure 2 shows the correlation trace for $\vec{k}_3$ signal. The signal intensity decays exhibiting a complicated oscillation, which suggests quantum beats superimposed with several beat-frequency components. Calculating the Fourier transform of the correlation curve, we have obtained the frequency components, 51, 32 and 83 cm^{-1}. These frequencies apparently correspond to the frequency differences between R and S lines, S and T lines, R and T lines, respectively. We have not observed any oscillation corresponding to the beats between the surface-related excitons and stacking fault excitons. This confirms that the quantum beats can occur in excitonic states with the same origin.

The oscillation in correlation traces of the conventional DFWM is considered to arise from the interference beat due to the following mechanisms: (A) interference of the polarizations of 2-level systems with two separated frequencies, Ω_a and Ω_b, (B) coherent superposition of electronic states in 3-level system. We have considered a new method to distinguish the two types of beats. A monochromator is added to the conventional DFWM method for resolving a frequency ω in the diffracted signal light. Therefore, we can measure the signal intensity as a function of the frequency as well as the time delay τ. Figure 3 shows the results of the numerical simulation to the systems (A) and (B) in the homogeneous broadening case. The vertical axis of correlation curves is normalized by the sum of the values of the exponential decay component and the beat amplitude at τ =0. In the case of (A), the normalized amplitude and the phase of the beats are dependent on the observed frequency ω. In the case of (B), however, the beat behavior is independent of ω except for the decreased intensity at off-resonance frequencies. Thus, the spectrum-resolved DFWM method allows us to distinguish the different beat phenomena. In the excitonic system of BiI_3, we can clearly observe the beat behavior even in the spectrum-resolved signal. This suggests that the observed beats are caused by the coherent super-

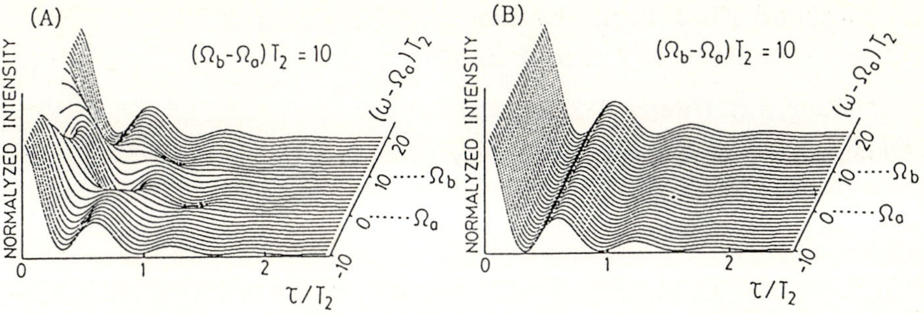

Fig.3 Simulated correlation curves of a new type of DFWM to the systems (A) and (B). Ω_a and Ω_b are resonant frequencies. T_2 is the dephasing time.

position of R, S and T excitonic states forming the 4-level system. Recently, Göbel et al.[5] have measured the quantum beats of the exciton system in GaAs quantum wells. They interpreted the beats in terms of the mechanism (A).

In summary, we have demonstrated quantum beats in two-dimensional excitons of BiI_3 by the DFWM method. We have proposed a new method to elucidate the beat phenomena observed in the pair of 2-level system and in the multilevel system.

References

[1] J. E. Golub and T. W. Mossberg: Opt. Lett. 11, 431 (1986)

[2] A. Nakamura, Y. Ishida, T. Yajima, T. Karasawa, I. Akai and Y. Kaifu: in Ultrafast Phenomena VI, ed. by T. Yajima, K. Yoshihara, C. B. Harris and S. Shionoya (Springer Verlag, 1988) p.266

[3] T. Komatsu , Y. Kaifu, S. Takeyama and N. Miura: Phys. Rev. Lett. 58, 2259 (1987)

[4] T. Yajima and N. Morita: in Method of Laser Spectroscopy, ed. by Yehiam Prior, A. Ben-Reuven and M. Rosenbluh (Plenum Publishing Corporation 1986) p.75

[5] E. O. Göbel, K. Leo, T. C. Damen, J. Shah, S. Schmitt-Rink, W. Schäfer, J. F. Müller, K. Köhler: Phys. Rev. Lett. 64, 1801 (1990)

Picosecond Photoluminescence of AlGaAs/GaAs Quasi-Quantum Wires and Boxes

H.Z. Wang, X.G. Hwang, W.Z. Lin, and Z.X. Yu

Ultrafast Laser Spectroscopy Laboratory, Zhongshan University, Guangzhou, China

Abstract. Picosecond photoluminescence measurements on AlGaAs/GaAs quasi-quantum wires and boxes reveal a significantly reduced carrier lifetime and an enhanced fluorescence peak intensity due to the quantum confinement as well as surface recombination.

The optical and electrical properties of quantum wires and quantum boxes have been the subject of growing interest not only for fundamental physics but also for modern device applications. Scattering suppression and high-mobility effects for these low dimensional superlattices have been described [1]. Striking structure in the photoluminescence of AlGaAs/GaAs quantum boxes has been observed and attributed to the spatial quantization of the low dimensional system [2]. In this paper we report time resolved photoluminescence (PL) studies on AlGaAs/GaAs quasi-quantum wires (QWWs) and quantum boxes (QWBs). Experimental results show an up-shifted frequency and enhanced peak intensity fluorescence and a significantly reduced carrier lifetime as the quantum dimension decreases from quantum well (QW) to QWW and QWB. The behavior of the PL of QWWs and QWBs may be attributed to the carrier localization and surface recombination in these low dimensional structures.

The quantum well samples used in this experiment were grown by MBE on Cr-doped GaAs substrates and consisted of a 0.1 μm GaAs buffer layer and a 0.1 μm $Al_{0.3}Ga_{0.7}As$ buffer layer, followed by four single quantum wells with GaAs well thickness of 30 Å, 50 Å, 70 Å, and 200 Å sandwiched between 100 Å $Al_{0.3}Ga_{0.7}As$ barriers, followed by a 0.1 μm $Al_{0.3}Ga_{0.7}As$ cap layer. The QWWs and QWBs were made from QWs using ion milling techniques. The lateral dimension for both the wires and boxes was about 1 μm. Thus each wire or box consists of four single quantum wells.

Picosecond photoluminescence experiments were performed at a sample temperature of about 80 K using a mode-locked and dumped dye laser which was pumped synchronously by a frequency doubled CW mode-locked Nd:YAG laser. The dye laser generated a pulse train with 6 ps pulse duration and 4 MHz repetition rate at a wavelength of 600 nm. The transient luminescence was detected by a synchroscan streak camera connected to a spectrometer with time resolution of 10 ps.

The PL spectra of the QWWs and QWBs are found to shift to high energy by as much as 10 meV. This energy up-shift is similar to that observed

Springer Series in Chemical Physics, Vol. 53 **Ultrafast Phenomena VII**
Editors: C.B. Harris · E.P. Ippen · G.A. Mourou · A.H. Zewail

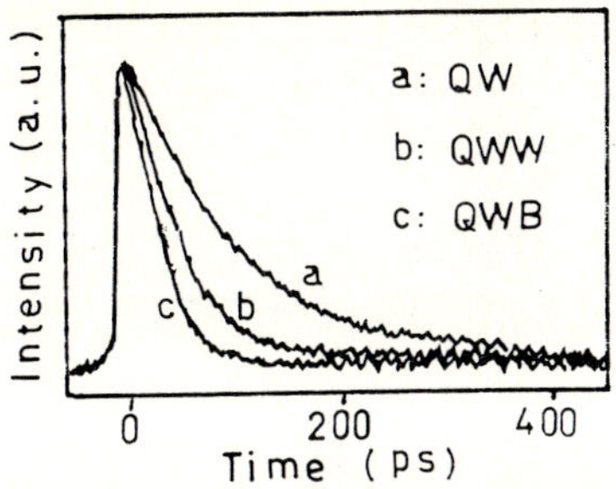

Fig. 1. Picosecond luminescence traces of a QW, QWW and QWB with $L_z = 30\,\text{Å}$

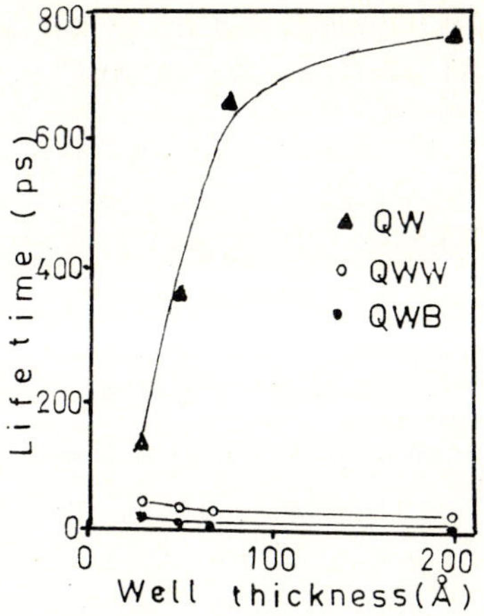

Fig. 2. Carrier lifetime versus quantum well thickness for a QW, QWW and QWB

by *Temkin* et al. [3] in InGaAs/InP QWWs and QWBs. It is attributed to low dimensional quantum confinement. The time behavior of the PL of QWWs and QWBs corresponding to the transition within a 30 Å well is displayed in Fig. 1.

The carrier lifetime decreases with decreasing quantum dimension from 154 ps for the QW to 28 ps for the QWW and 17 ps for the QWB. It is found that the PL peak intensity, which has been normalized in Fig. 1, increases as the quantum dimension decreases. Similar behavior of the PL is also observed for the cases with thicker wells. The carrier lifetime versus quantum well thickness L_z for the QWW, QWB and QW are shown in Fig. 2. For the QW the carrier lifetime reduces from 678 ps for $L_z = 200\,\text{Å}$ to 154 ps for $L_z = 30\,\text{Å}$. This reduction of the lifetime with decreasing well thickness has been observed previously and assigned to the quantum confinement [4]. As is shown in Fig. 2, for wells thinner than 70 Å the lifetime decreases dramatically. This indicates enhanced carrier localization in thin wells.

Several mechanisms may be responsible for the decreased lifetime of the carriers in QWWs and QWBs. As mentioned above, localization of carriers in a low dimensional quantum confined structure enhances the electron-hole radiative recombination, which results in a decrease of lifetime and an increase of peak intensity. Meanwhile, surface recombination may play an important role in the reduction of the lifetime in QWWs and QWBs. It is known that the GaAs system suffers from a relatively large surface recombination velocity. Any increase in the ratio of the surface to the volume will significantly enhance the recombination. Particularly, defects in the sample arising probably from imperfect ion milling processes can enhance the nonradiative recombination expected at layer sidewalls. This may explain the behavior that the peak intensity of QWWs and QWBs decreases with increasing well thickness and that the lifetime becomes shorter than the laser pulse duration in thicker wells.

The short lifetime and the enhanced peak intensity of the PL observed in QWWs and QWBs predict the possibility of developing ultrafast optoelectronic devices using these low quantum dimension structures.

References

1. H. Sakaki: Jpn. J. Appl. Phys. **28**, L314 (1989)
2. M.A. Reed, R.T. Bate, K. Bradshaw, W.M. Duncan, W.R. Frensley, J.W. Lee, H.D. Shih: J. Vac. Sci. Technol. B **4**, 358 (1986)
3. H. Temkin, G.J. Dolan, M.B. Panish, S.N.G. Chu: Appl. Phys. Lett. **50**, 413 (1987)
4. E.O. Gobel, H.Jung, J. Kuhl, K. Floog: Phys. Rev. Lett. **51**, 1588 (1983)

Absorption Saturation Dynamics in Capped CdSe Microcrystallites Exhibiting Quantum Confinement

W.L. Wilson, M.G. Bawendi, L. Rothberg, P.J. Carrol, T. Jedju, L. Brus, and M.L. Steigerwald

AT&T Bell Laboratories, 600 Mountain Ave., Murray Hill, NJ07974, USA

ABSTRACT. We report the use of subpicosecond transient bleaching techniques to characterize the spectral dynamics and isolate the single particle spectrum of 32 angstrom CdSe nanoclusters. We observe strong bleaching of the localized, particle in a box transitions as well as the onset of induced absorption to the blue of the homo-lumo transition. The holewidth, bleaching, and induced absorption dynamics suggest that the initially photogenerated electron-hole pair rapidly relaxes to a long lived, (>1ns), *trapped* carrier configuration faster than 500fs. The quality of our samples allows us to resolve the vibronic structure of the microcrystallite due to coupling of the *exciton* to the 200cm^{-1} LO phonon and observe it's effect upon the hole spectrum.

Restriction of carrier wave functions in semiconductor structures of reduced dimensionality results in increased oscillator strength in the excitonic absorption which can be exploited in the design of novel nonlinear optical devices[1]. We report transient bleaching studies of CdSe "quantum dots" whose size is small relative to the bulk material exciton radius [2]. Subpicosecond hole burning and steady-state luminescence spectroscopies are used to characterize the spectral dynamics and isolate the single particle spectrum of these 32 angstrom diameter CdSe microcrystallites.

These clusters, which are grown by a new synthetic method, have a well developed wurtzite crystal structure, [3], high quantum yield of band edge luminescence, low saturation intensities, (about ~100 microJ/cm^2) and long bleaching recovery times. Our measurements were performed at 10-20K in dilute glasses.

In this material the *exciton* band, i.e., the homo-lumo transition is shifted ~1 ev to higher energy by quantum confinement effects. The spectral positions of the localized states can be calculated using a simple particle in a sphere model, [4]. The optical absorption and transient hole burning spectra are shown in figures 1 and 2. There is strong bleaching of the, $1S_e$-

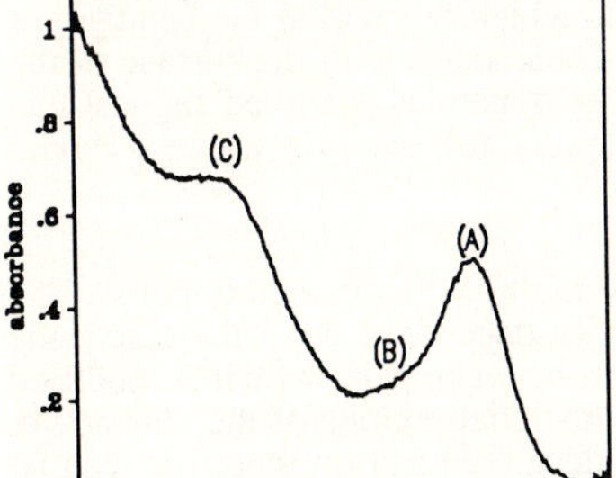

Figure 1: Absorption spectrum of CdSe nanoclusters, (T=20K). The spectral assignments are as stated in text.

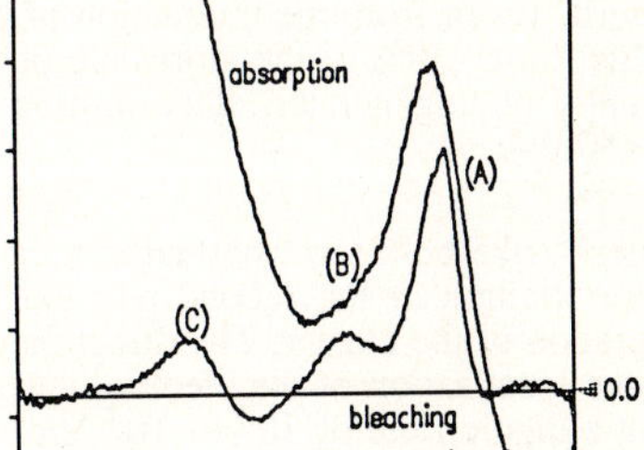

Figure 2: Transient hole spectrum,(T=20K).

Springer Series in Chemical Physics, Vol. 53 **Ultrafast Phenomena VII**
Editors: C.B. Harris · E.P. Ippen · G.A. Mourou · A.H. Zewail

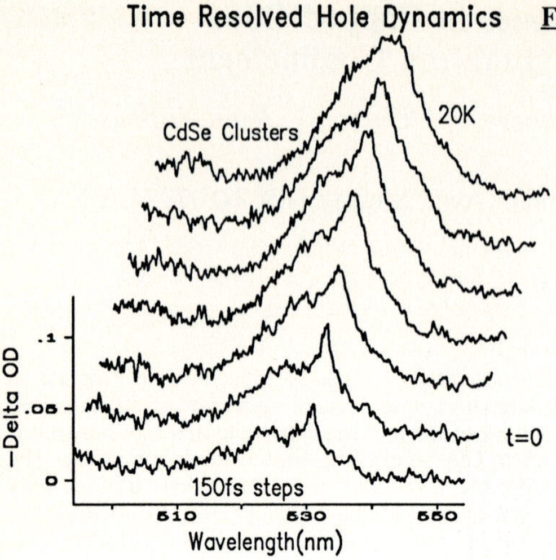

Figure 3: Time evolution of bleach.

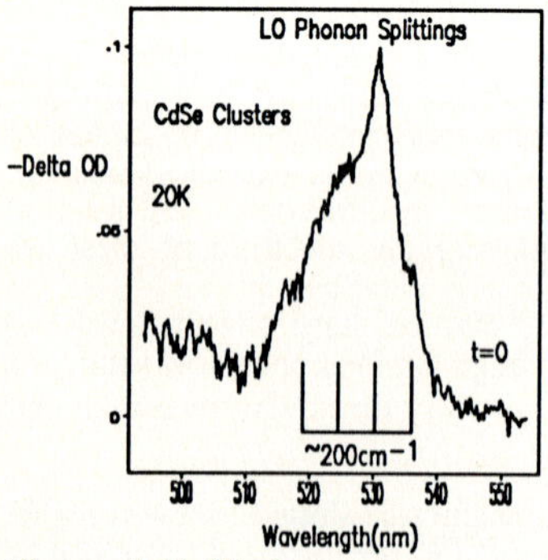

Figure 4: T=0 hole spectrum.

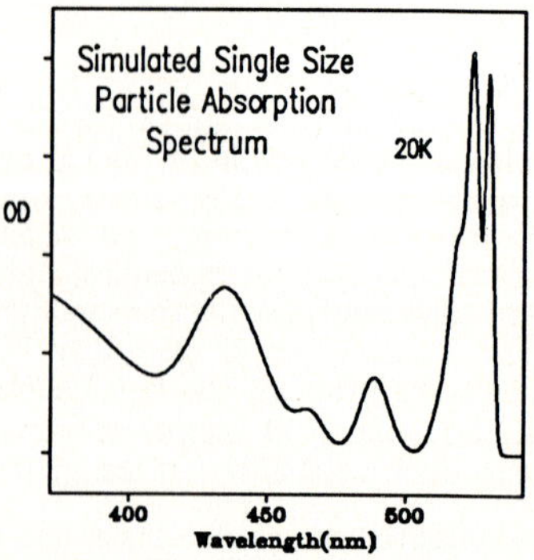

Figure 5: Simulated single cluster spectrum extracted.

$1S_h$, $1S_e$-$1P_h$ transitions as well as the spin-orbit split state, (A), (B), and (C) respectively. More importantly we resolve an instrumentally limited,(500fs), induced absorption to the blue of the homo-lumo transition which decays on a 10-100ns time scale. The long lifetime suggests that it is due to a "trapped" species. These states which fall within the bandgap of the material result from the interaction of the cluster electronic states with the surface of the crystallite, (note 30% of the atoms are on the surface). The dynamics observed suggest that the initially photogenerated electron-hole pair created rapidly relaxes to a trapped carrier state, (<500fs).

Clusters (like bulk semiconductors) can accommodate multiple electron-hole pairs. The oscillator strength of the second pair excitation is critically dependent upon the electronic configuration of the cluster. The direct coulomb interaction between pairs which is modified due to the localization of the electron and hole to different spatial regions of the cluster can result in a dipole field of 10^3 or 10^4 V/cm. The photoinduced absorption therefore, can be readily understood as a reduction in oscillator strength of the second electron-hole transition due to the presence of these effects. This description is consistent with recent theoretical work of Hu et al., [5], as well as early bleaching studies in CdS clusters, [6]. The evolution of the excitonic bleach is shown in figure 3. The sharp feature prominent at t=0 is a coherent coupling feature due to the 4-wave mixing interaction of the pump and probe fields. In addition, the high quality of our samples allows us for the first time to resolve the vibronic structure of the $1S_e$-$1S_h$ transition due to coupling to the $200cm^{-1}$ LO phonon and its effect on the emission and holeburning spectra. This is also clearly illustrated in figure 4 . Detailed analysis of the time-resolved and steady-state spectral data allows us to extract a $150cm^{-1}$ hole width for the exciton transition and construct the single cluster spectrum which is shown in figure 5.

REFERENCES

[1] Schmitt-Rink, S., Miller, D. A. B., and Chemla, D. S. *Phys. Rev.* **B35**(1987), 8113
[2] Steigerwald, M. L. and Brus, L. E. *Annu. Rev. Mater. Sci.* **19** (1989), 471
[3] Bawendi, M. G., Kortan, A. R., Steigerwald, M. L., and Brus, L. E. *J. Chem. Phys.* **91**(11), [1989], 7282
[4] Brus, L. E. *J. Chem. Phys.* **80**(9), [1984], 4403
[5] Hu, Y. Z., Lindberg, M., and Koch, S. W.
Submitted Phys. Rev. **B**
[6] Hilinski, E., Lucas, P., and Wang, Y. *J. Chem. Phys.* **89**(1988), 3435

Picosecond Imaging of Photoexcited-Carrier Transport in GaAs/AlGaAs Multiple Quantum Wells

D.R. Wake, H.W. Yoon, J.S. Preston, H. Morkoç, and J.P. Wolfe*

Materials Research Laboratory, University of Illinois at Urbana-Champaign, Urbana, IL61801, USA

*Present address: McMaster University, Hamilton, Ontario, Canada

Abstract. Spatial and time-resolved induced bleaching at the heavy-hole exciton in a multiple quantum well structure is used to study ambipolar diffusive transport of a quasi-2D electron-hole plasma. The diffusivity is found to depend on excess carrier energy, lattice temperature, and carrier density. We attribute the density dependence to effects of screening and degeneracy at high density.

Although charge transport has been studied extensively in 3 and 2 dimensional semiconductor structures, questions remain as to the relative contribution of mechanisms to the scattering rate and energy relaxation process. Since the presence of electrons in the conduction band (or holes in the valence band) in a semiconductor quantum well reduce the absorption at the heavy-hole exciton, the reduction in absorption may be used to map out the expansion of a spatially localized electron-hole plasma following picosecond optical excitation [1]. In this work, an electron-hole plasma is excited in a 4 μm spot in a 210Å/100Å GaAs/AlGaAs multiple quantum well structure. A probe pulse is then raster-scanned across a 50 μm square field to produce a fixed-delay image of the subsequent expansion with a time resolution of 5 picoseconds. (Fig. 1.) We have measured the rate of

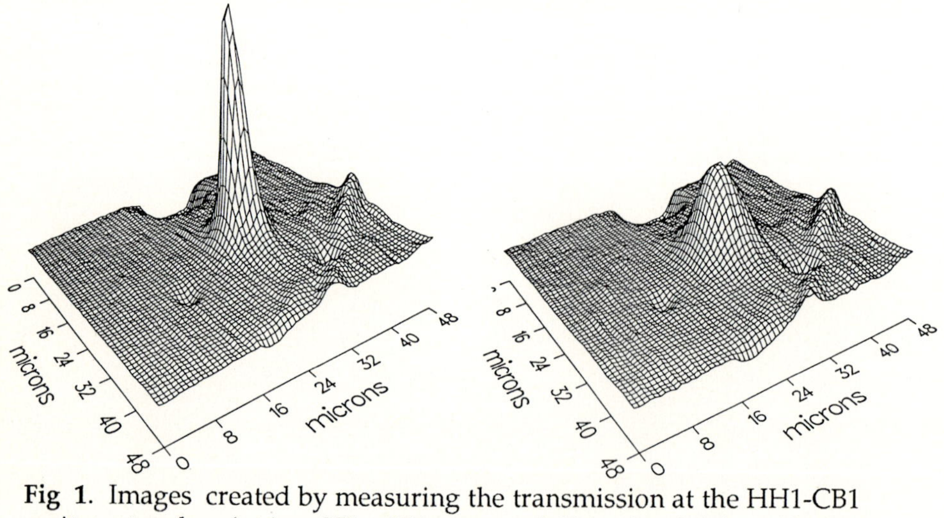

Fig 1. Images created by measuring the transmission at the HH1-CB1 exciton at each point in a 256 x 256 array. a) 20 ps fixed delay. b) 1000 ps.

Springer Series in Chemical Physics, Vol. 53 **Ultrafast Phenomena VII**
Editors: C.B. Harris · E.P. Ippen · G.A. Mourou · A.H. Zewail

diffusive expansion and its dependence on excitation density, excess energy (including resonant pumping), and a lattice-temperature from 8K to 65K.

The ambipolar diffusion of the two-component semiconductor plasma of density n is modeled by the transport equation modified for a recombination time τ:

$$\frac{\partial n}{\partial t} = \vec{\nabla}\cdot\left(D\vec{\nabla}n\right) - \frac{n}{\tau}$$

For a density-independent diffusion coefficient, D, this implies that the FWHM^2, Δ^2, of a Gaussian distribution increases linearly in time as:

$$\Delta^2 = 11.08Dt + \Delta_i^2$$

Although D decreases during the expansion, we approximately describe the motion with an effective D, D_{eff}, given by the slope of Δ^2 vs. time.

Experimental Results

Fig. 2 shows a plot of Δ^2 vs. time for a range of initial excess energies, ΔE, showing D_{eff} decreasing with time. The early time expansion is more rapid with greater ΔE, showing that, in part, D_{eff} decreases with time because the kinetic energy decreases. (The later-time expansion for large ΔE where D_{eff} is relatively unchanging, has been reported previously [1].) However the density also decreases with time, complicating the analysis.

In order to reduce the hot-carrier effect on D_{eff} and examine the density dependence, we concentrate on the resonant excitation. We again deduce D_{eff} by measuring the time dependence of the FWHM^2 of the distribution. D_{eff} decreases with time for initial n, n_o, in excess of 2×10^{10} cm^{-2}. (See Fig. 3a.) The n-dependence is complicated by the fact that the carrier distribution is initially degenerate and then becomes non-degenerate as the carriers recombine. The relative carrier density "n" as a function of time can be estimated by the peak of the spatial distribution in Fig. 1. We thus infer D_{eff} as a function of "n" as shown in Fig. 3b. D_{eff} is relatively independent of "n" at low excitation density, and **increases** with "n" above a few times 10^{10} cm^{-2}. We find D_{eff} increases with temperature, T, from 8K to 65K, in agreement with luminescence studies [2].

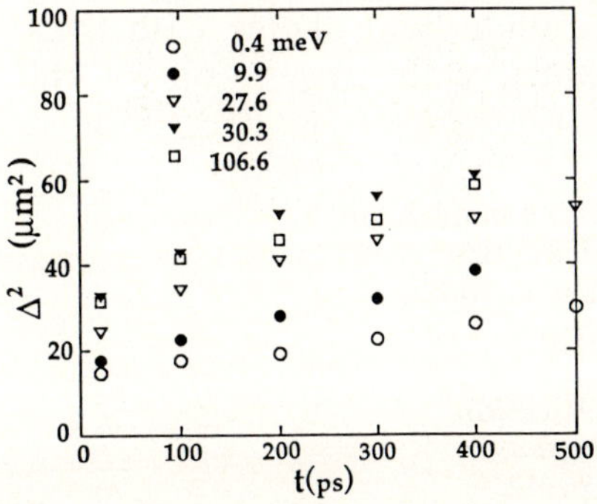

Fig 2. Time dependence of the FWHM^2 at 8.5K for $n_o \sim 2 \times 10^{10}$ cm^{-2}.

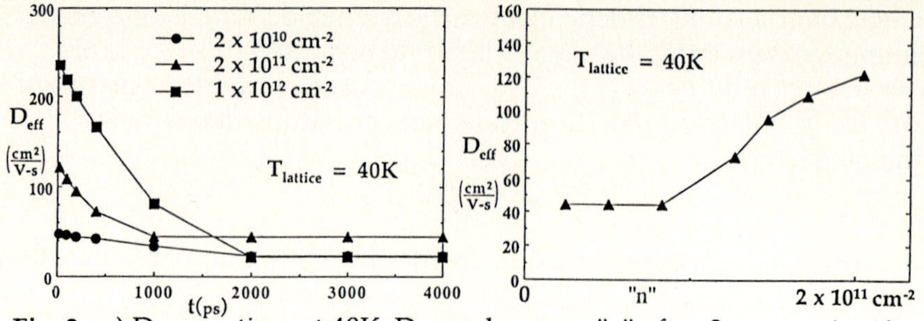

Fig 3. a) D_{eff} vs time at 40K. Dependence on "n" after 2 ns not significant.
b) D_{eff} vs estimated density "n" for middle series of Fig. 3a.

Ambipolar Diffusivity

The ambipolar diffusivity [3] is determined by the mobility, μ, and the diffusivity, D, for equal densities of electrons, n, and holes, p:

$$D_{eff} = \frac{\mu_p D_n + \mu_n D_p}{\mu_p + \mu_n}$$

Since μ and D are linearly related, the smaller hole D dominates D_{eff}. Coulomb scattering from background impurities dominates scattering in modulation-doped GaAs quantum wells under corresponding conditions and is likely to dominate the scattering of holes for the range of T studied [4].. The ΔE and T dependences indicate that increased kinetic energy contributes to increased D_{eff}. In a degenerate system, the average kinetic energy increases with n due to $T_{Fermi} > T_{lattice}$. Also, the higher velocity should reduce Coulomb-impurity scattering, which falls off as $1/Q^2$, the momentum transferred in the scattering. Calculations show that the mobility is either constant or increasing with temperature due to impurity scattering [4]. The diffusivity will then increase at least linearly in T according to the modified Einstein relation [5]. Additional enhancement of D_{eff} is expected at high density due to screening of holes by electrons.

In conclusion, the ambipolar diffusive transport in a GaAs quantum well is strongly dependent on carrier density and on carrier kinetic energy for the ranges described. The diffusivity is enhanced with increasing carrier energy either by increasing the excess excitation energy or by increasing the lattice temperature. The underlying theme with these trends is an ambipolar diffusivity correlated with carrier velocity and enhanced at high carrier densities.

References.

1. L. M. Smith, D. R. Wake, J. P. Wolfe, D. Levi, M. V. Klein, J. Klem, T. Henderson, and H. Morkoç, Phys. Rev. B 38, 5788 (1988)
2. H. Hilmer, A. Forchel, S. Hansmann, M. Morohashi, E. Lopez, H. P. Meier, and K. Ploog, Phys Rev. B 39, 10901 (1989).
3. W. van Roosbroeck, Phy. Rev. 91, 282 (1953).
4. K. Hirakawa and H. Sakaki, Phys. Rev. B 33, 8291 (1986).
5. N. W. Ashcroft and N. D. Mermin, *Solid State Physics*, (Holt, Rinehart, and Winston, New York, 1976), p. 612.

Fast Recovery of Excitonic Absorption Peaks in Tunneling Bi-Quantum-Well Structures

A. Tackeuchi, S. Muto, T. Inata, and T. Fujii

Fujitsu Laboratories, Ltd., 10-1, Morinosato-Wakamiya, Atsugi 243-01, Japan

The fast onset of optical nonlinearity of excitons in multiple quantum wells (MQWs) is of wide current interest with respect to its future applications to optical switching and computing[1]. It is especially remarkable that nonlinear absorption and refraction changes occur with a time constant shorter than 1 ps due to the photo-excitation of the exciton peak[2]. However, the slow recovery of optical constants such as absorption coefficient is a serious problem. The typical absorption recovery time in a GaAs/AlGaAs MQW is 30 ns at 300 K[3] which is determined mainly by the radiative recombination of carriers. This limits the time response of devices using the excitonic optical nonlinearity. One way of reducing recovery time involves the ac-stark effect which occurs only during pump excitation. However, the ac-stark effect is known to have low efficiency in modifying the optical properties[4].

To reduce the recovery time without sacrifice of optical efficiency, we proposed a new superlattice structure using carrier tunneling, which we call a "Tunneling Bi-Quantum-well (TBQ)"[5]. The TBQ consists of the repeated structure of narrow and wide wells as shown in Fig. 1. This structure uses the optical nonlinearity of the exciton level in the narrow well. Since the energies in the narrow and wide wells are different, photo-excited carriers in the narrow well can escape toward the wide well by tunneling and relax into the lowest energy level here. This process recovers the optical constants of the narrow well. It's also possible to control the recovery time by changing the tunneling time.

The studied TBQ structures consisted of narrow GaAs quantum wells 4.5 nm thick and wide quantum wells 9.0 nm thick. The thickness of $Al_{0.51}Ga_{0.49}As$ barriers, L_B, varied

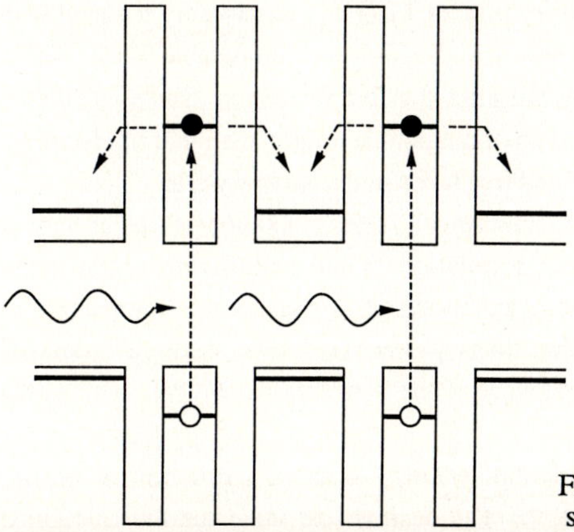

Fig. 1. Energy band diagram of TBQ structures.

Editors: C.B. Harris · E.P. Ippen · G.A. Mourou · A.H. Zewail

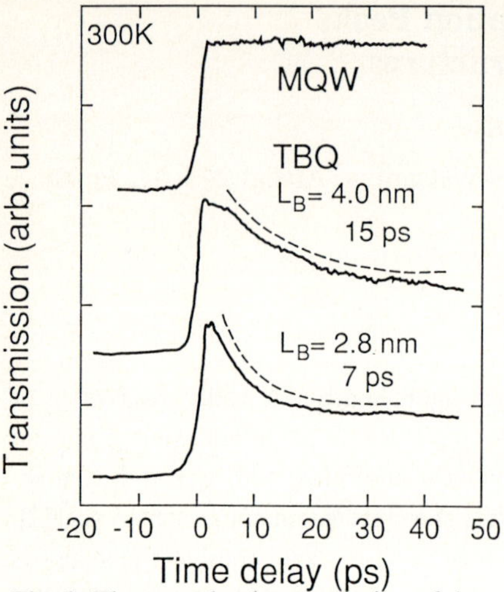

Fig. 2. Time resolved transmission of the excitonic absorption peak.

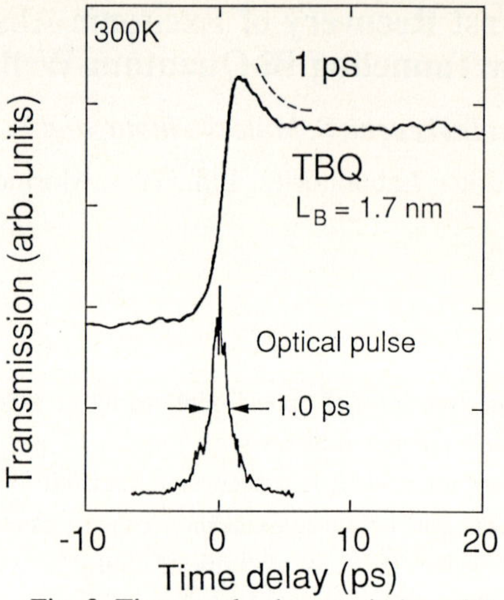

Fig. 3. Time resolved transmission of the excitonic absorption peak for the TBQ with 1.7 nm thick barrier and auto-correlation of optical pulses.

between 5.1 nm (18 MLs) and 1.7 nm (6 MLs) to investigate the dependence of the recovery time on barrier thickness. Here, we measured the time dependence of the absorption as one optical constant . In the experiments we used the ordinary time resolved pump-probe absorption measurement technique which uses a lock-in amplification system. Optical pump and probe pulses of 1 ps duration at 12 ns intervals were obtained from a tunable mode-locked styryl dye laser synchronously pumped by the compressed YAG laser pulses.

Figure 2 shows the observed transmission time dependence at the lowest heavy-hole exciton peak in the narrow wells. After photo-excitation, the conventional MQW did not show a clear absorption recovery in this time region. In contrast, the TBQs showed fast recovery times of 15 ps and 7 ps for 4.0 nm thick barrier and 2.8 nm thick barrier respectively. The absorption changes of the TBQ are about 50% of the conventional MQW for the same pumping power. Note that the recovery time of TBQ was remarkably reduced with the barrier thickness. Also note that the TBQ have a high efficiency comparable to the MQW. In Fig. 2, we can also see the long recovery tail after the fast relaxation. Since calculated tunneling time of holes is about three to six orders of magnitude longer than that of electrons, the long recovery tail is attributed to the remained free holes in the narrow wells.

For the barrier thickness of 1.7 nm, we observed much faster relaxation of 1ps as shown in Fig. 3. It was confirmed by a numerical simulation that the reduction of the peak transmission change compared with the tail occurred because this relaxation is comparable to the optical pulse duration of 1 ps. Note that the 1 ps recovery time is three orders of magnitude faster than the few nanoseconds recovery of the conventional MQW. This is the fastest tunneling process ever observed.

The semilogarithmic plot of the experimental recovery time as a function of barrier thickness is shown by a solid circles in Fig. 4. The dashed line indicates the calculated

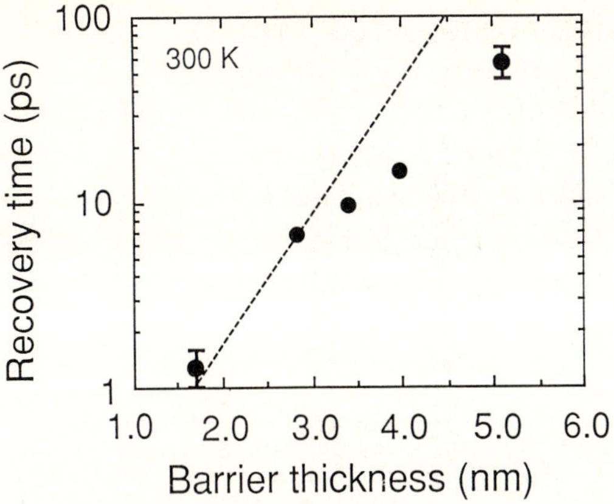

Fig. 4. Recovery time dependence on the barrier thickness. The dashed line indicates the calculated tunneling time.

tunneling time of electrons in consideration of LO phonon emission. The proportionality between the observed recovery time and the calculated tunneling time shows that the absorption recovery is governed mainly by electron tunneling. By extrapolating experimental and calculated results, we can easily estimate that the TBQ with the barrier thickness less than 1.7 nm will have a sub-picosecond recovery time

In conclusion, fast recovery of the excitonic absorption peak in Tunneling Bi-Quantum-well structures is proposed and demonstrated. With the use of electron tunneling from the narrow well toward the wide well, the recovery time was reduced to as short as 1 ps, which is three orders shorter than the few nanoseconds recovery time of conventional multiple quantum wells. This is the fastest tunneling process ever observed. We believe this result proves the high potential of the TBQ as the breakthrough for slow carrier recombination of semiconductors.

We thank M. Yamashita of Electrotechnical Laboratory for basic advice on short pulse dye laser. We also thank O. Ohtsuki, T. Yamaoka, and M. Kobayashi, and T. Misugi for their encouragement.

REFERENCES

1. A. Migus, A. Antonetti, D. Hulin, A. Mysyrowicz, H.M. Gibbs, N. Peyghambarian, and J.L. Jewell: Appl. Phys. Lett. 46 (1985) 70.
2. W.H. Knox, R.L. Fork, M.C. Downer, D.A.B. Miller, D.S. Chemla, C.V. Shank, A.C. Gossard, and W. Wiegmann: Phys. Rev. Lett. 54 (1985) 1306.
3. D.S. Chemla, D.A.B. Miller, P.W. Smith, A.C. Gossard, and W. Wiegmann: IEEE J. Quantum Electron. QE-20 (1984) 265.
4. A. Von Lehmen, D.S. Chemla, J.E. Zucker and J.P. Heritage:Opt. Lett. 11 (1986) 609.
5. A. Tackeuchi, S. Muto, T. Inata, and T. Fujii: Jpn. J. Appl. Phys. 28 (1989) L1098.

Degenerate Four Wave Mixing in Staggered Type-II $Al_xGa_{1-x}As/AlAs$ Multiple Quantum Well Structures

T. Mishina, F. Sasaki, and Y. Masumoto

Institute of Physics, University of Tsukuba, Tsukuba, Ibaraki 305, Japan

Recently much attention has been paid to carrier dynamics in staggered type-II multiple quantum well structures. In these systems, the energy level of the lowest Γ conduction electron state in a well layer is higher than that of the X-electron state in a barrier layer. Thus the photoexcited Γ-electrons in the well layer are easily scattered into the X-state in the barrier layer. The electrons scattered into the X-state in the barrier layer are expected to form X-Γ indirect excitons with holes in the well layer. We and several authors have measured these interlayer scattering times using femtosecond pump-and-probe technique with white light continuum [1,2,3]. However, the dephasing and momentum relaxation processes of the carriers in staggered type-II systems have not been understood. To clarify these relaxation processes, we have performed the time-resolved degenerate-four-wave mixing experiments in staggered type-II multiple quantum well structures.

The sample mainly studied in this work is 100 alternate layers of 9.2nm $Al_{0.34}Ga_{0.66}As$ and 2.7nm AlAs. Figure 1 shows absorption and luminescence spectra of the sample. Absorption spectrum measured at 2K shows heavy and light hole exciton peaks at 2.04 and 2.06 eV, respectively. Indirect X-Γ exciton at 1.88 eV and direct Γ-Γ exciton at 2.03 eV are observed in luminescence spectrum at 4.2K. Since the electrons and holes are separated in both real and momentum spaces, the

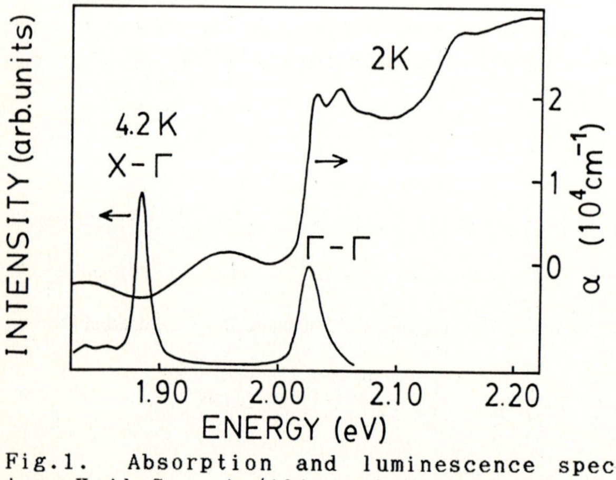

Fig.1. Absorption and luminescence spectra of the staggered type-II $Al_xGa_{1-x}As/AlAs$ (x=0.34) multiple quantum well structures measured at 2.0K and 4.2K, respectively.

Springer Series in Chemical Physics, Vol. 53 **Ultrafast Phenomena VII**
Editors: C.B. Harris · E.P. Ippen · G.A. Mourou · A.H. Zewail

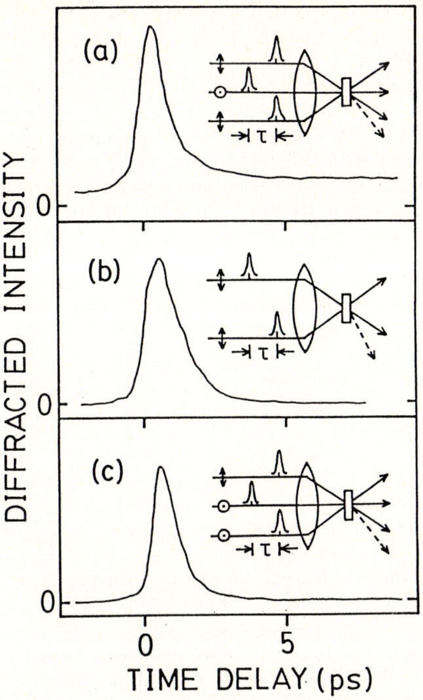

Fig.2. Degenerate four wave mixing signals of the staggered type-II $Al_xGa_{1-x}As/AlAs(x=0.34)$ multiple quantum well structures. Population grating (a), photon echo (b) and orientational grating (c).

recombination of the X-Γ indirect exciton occurs in a microsecond time scale. Actually, we have confirmed the microsecond decay of the Γ-X indirect luminescence by the time-correlated single photon counting technique. We have also confirmed the very good correspondence between the decay of X-Γ indirect luminescence and the recovery of the Γ-Γ absorption saturation in a microsecond time scale [4].

The typical forms of degenerate four wave mixing are population grating, photon echo and orientational grating. These experiments give us various information of the carrier dynamics.

The experiments were carried out using a cavity dumped Rh6G dye laser tuned to the heavy hole exciton peak. The spectral width and the pulse duration were 2nm and 450 fs ($sech^2$ shape), respectively. The repetition rate of the laser pulses was 4.1MHz.

Figure 2(a) is the temporal trace of diffracted intensity from population grating. The flat tail part corresponds to the slow decay component. Moreover, the considerable baseline appearing in Fig.2(a) indicates that the decay is slower than 240ns which is an inverse of 4.1MHz pulse repetition. The slow decay component is the direct consequence of the saturation of the hole state. This result is consistent with the slow decay of Γ-X luminescence. The initial peak is attributable to both the rapid grating decay due to the Γ-X interlayer scattering and a coherent artifact. Thus the time constant of the Γ-X interlayer scattering cannot be estimated from this result.

The photon echo experiment gives the optical phase coherence time T_2. The result of photon echo experiment is shown in Fig. 2(b). The excitation density was weak enough so that carrier-

carrier scattering does not destroy the optical coherence. The obtained decay time is 0.7ps. This indicates that T_2 is 2.8ps, because the spectral width of the heavy exciton is inhomogeneously broadened [5]. For the sample where the excited life time T_1 is very short compared to all the other relaxation processes, T_2 is expected to be twice as large as T_1. Thus the life time of the Γ-electron estimated from the photon echo experiment is 1.4ps. This is somewhat larger than the decay constant of 1.2ps derived from previous experiment[3] The difference of the decay constants can be explained by the different excitation conditions and the limited accuracy of the experiment. It should be noted that the dephasing time is purely determined by the T_1 mechanism, the interlayer Γ-X scattering.

Orientational grating is the grating of optical anisotropy formed by two orthogonally polarized pulses. The optical anisotropy is closely related to the momenta of the photo-generated carriers. Thus the decay of orientational grating measures the carrier scattering in momentum space. The temporal trace is shown in Fig. 2(c). The decay time constant of the diffracted intensity is 0.7ps and the corresponding relaxation time of the orientational grating is 1.4ps. Thus the relaxation time of the orientational grating and the life time of the Γ-electrons are the same. This simply means that Γ-X interlayer scattering randomizes the carrier momentum parallel to the layer.

Similar results were also obtained for another sample consisting of 100 alternate layers of 9.6nm $Al_{0.41}Ga_{0.59}As$ and 2.8nm AlAs.

In conclusion, we have studied the staggered type-II $Al_xGa_{1-x}As$/AlAs (x=0.34, 0.41) multiple-quantum-well structures with time-resolved degenerate-four-wave mixing. The obtained results clearly reflect the short life time of the Γ-electron state of the well layer due to the interlayer Γ-X scattering.

References

[1] P. Saeta, J.F. Federici, R.J. Fischer, B.I. Greene, L. Pfeiffer, R.C. Spitzer, and B.A. Wilson, Appl. Phys. Lett. 54, 1681 (1989).

[2] J. Feldmann, R. Sattmann, E.O. Göbel, J. Kuhl, J. Hebling, K. Ploog, R. Muralidharan, P. Dawson, and C.T. Foxon, Phys. Rev. Lett. 62, 1892 (1989).

[3] Y. Masumoto, T. Mishina, F. Sasaki, and M. Adachi, Phys Rev. B40, 8581 (1989).

[4] T. Mishina, F. Sasaki, and Y. Masumoto, Submitted to J. Phys. Soc. Jpn.

[5] T. Yajima and Y. Taira, J. Phys. Soc. Jpn. 47, 136 (1979).

B-Exciton Bleaching in Bulk CdSe

D. Braun [1], *W.W. Rühle* [1], *and J. Collet* [2]

[1] Max-Planck-Institut für Festkörperforschung, Heisenbergstr. 1
D-7000 Stuttgart 80, Fed. Rep. of Germany

[2] Laboratoire de Physique des Solides, Associé au CNRS,
Avenue de Rangueil, F-31077 Toulouse Cedex, France

Abstract The energy separation between B and A-excitons in CdSe equals the longitudinal optical phonon energy and thus allows a fast relaxation of B into A-excitons. The transition is investigated by ps-luminescence and fs-reflection measurements at ~10K and a transition time of 0.9ps is measured. A theoretical estimate yields 0.4ps. B-excitons are bleached by A-excitons as well.

1. Introduction

The interest in excitons in semiconductors has been renewed in recent years since optical nonlinearities in the excitonic regime have potential applications in photonic devices.[1,2] However, only few time-resolved investigations on excitons in CdSe have been performed,[3,4] although its special band-structure offers an efficient transformation of B into A-excitons.[5]

2. Experimental

The ps-luminescence is performed using a synchronously pumped, tunable dye laser (~ 4ps FWHM pulse width, 80MHz repetition rate) for excitation and a 2D streak camera for detection (~7ps 1/e decay). The pump and probe reflection experiments are carried out using a tunable dye laser with 600fs pulse width (FWHM) which is synchronously pumped by the second harmonic of a compressed Nd:YAG laser. The polarization of the light is always set parallel to the c-axis, since only optical transitions of the B-excitons, which we want to investigate, are then allowed (see insert in Fig.1).

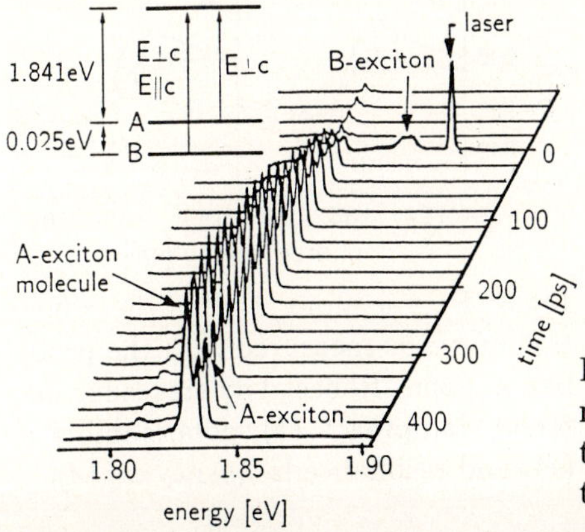

Fig.1: Photoluminescence with ps-time resolution revealing the fast decay of the B-exciton. $E_{laser}||c$, $E_{lumi.}||c$, excitation density $100W/cm^2$ at 1.873eV.

Springer Series in Chemical Physics, Vol. 53 **Ultrafast Phenomena VII**
Editors: C.B. Harris · E.P. Ippen · G.A. Mourou · A.H. Zewail

3. Results

The luminescence, depicted in Fig.1, is still dominated by the free A-exciton and the A-exciton molecule although these transitions are forbidden for E||c. The B-exciton luminescence is weak and the decay cannot be resolved with the streak camera. On the other hand, the linewidth from the lower polariton branch sets a lower decay-time limit of 0.25ps.

The reflectivity, depicted in Fig.2, reveals two time constants: a fast one of 0.9ps and a slow component of approximately 50ps that varies with photon energy. The fast component decreases the reflectivity at all energies, whereas the slow component decreases the reflectivity below resonance and increases it above. In order to explain the result, a usual oscillator model (neglecting spatial dispersion) is applied for the dielectric constant ϵ:

$$\epsilon = \epsilon_b + \frac{4\pi\,\alpha\,\omega_0^2}{\omega_0^2 - \omega^2 - i\Gamma\omega},$$

where ϵ_b denotes the background dielectric function, $4\pi\alpha$ the polarizability, ω_0 the transverse exciton frequency, and Γ the damping.

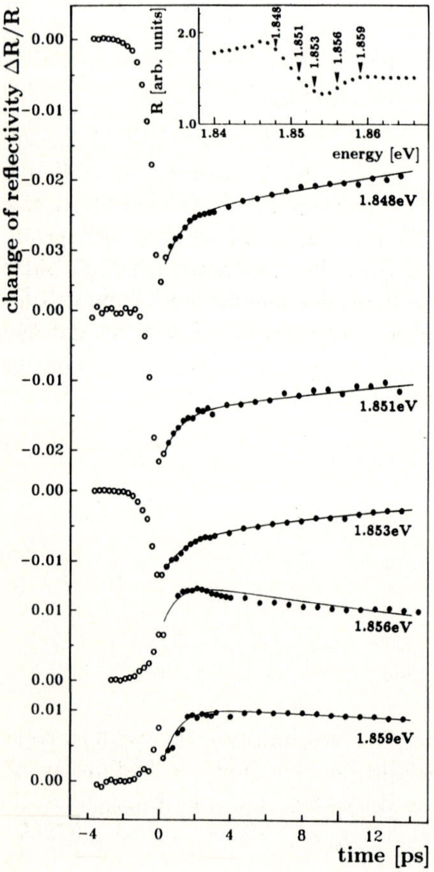

Fig.2: Time-resolved change of the reflectivity at different wavelengths. Excitation: $E_{probe}||E_{pump}||c$. The inset shows a reflection curve measured with the probe beam.

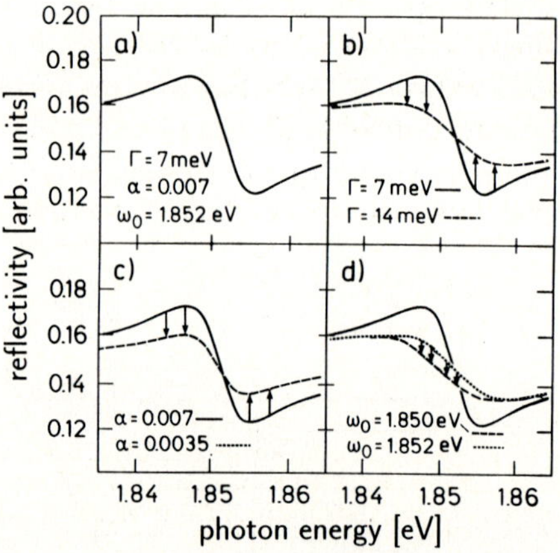

Fig.3: Part *a* shows a fit to the reflection curve as measured with the probe beam alone. Reflectivity traces for different damping Γ (*b*), polarizability α (*c*), and resonance frequency ω_0 (*d*).

Figures 3b-d illustrate the effect of Γ,α, and ω_0 on the reflectivity. An increase of Γ decreases the reflectivity below ω_0 and increases it above (Fig.3b). In addition, the reflectivity trace is broadened. A decrease of the polarizability (Fig.3c) acts similarly on the reflectivity: the reflection trace changes toward the background reflectivity. Only a shift of ω_0 in the presence of broadening (Fig.3d) reduces the reflectivity at all energies around ω_0 as observed for the fast component in Fig.2. A shift of 0.1meV is sufficient to explain the amplitude of the fast component. The effect of a change in ω_0 is shown in Fig.3d by the dashed and the dotted lines for a large Γ, i.e. an energetically broad reflection curve, as observed in our data, since the shift occurs at high densities where broadening is already significant.

The shift can be caused by two effects. First, the band gap is renormalized due to exchange and correlation effects leading to a red shift. Second, the exciton binding energy is reduced by screening, causing a blue shift. Fehrenbach *et al* [6] verified experimentally in GaAs that the two effects cancel almost exactly. We measure an initial negative change of the reflectivity which we attribute to the B→A transition which implies that the net shift between bandgap renormalization and screening is reduced in this process. As expected, we observe no fast component in the A-exciton line.

We estimate the B to A-exciton transition time by emission of an LO-phonon. With Fermi's golden rule, we obtain a polar optical scattering time of 0.4ps, if we assume a distribution of the B-exciton momentum corresponding to the luminescence linewidth. The time is consistent with the experiment.

The slow component is due to damping caused by the A-excitons and eventually due to a reduction of the polarizability induced by phase space filling (PSF). At later times, the excitons diffuse into the bulk and hence damping and PSF is reduced. PSF should occur at our exciton densities of $1.2 \cdot 10^{17}\text{cm}^{-3}$, since the excitons with K~0 cover 3.5% of real space. Consequently, α should be reduced. However, the fast component does not change amplitude with ω, i.e. PSF should not change in the B→A transformation. The reason is that PSF is merely caused by the much lighter electrons and the B→A transition only involves the holes. As a consequence, PSF of B-excitons could be caused by A-excitons, an interesting effect for switching devices.

Finally, we want to acknowledge the growth of the high quality crystal in the Institute by G.O. Müller.

1. S. Schmitt-Rink, D.S. Chemla, and D.A.B. Miller, Advances in Phys. **38**, 89 (1989)
2. C. Klingshirn and H. Haug, Phys. Reports **70**, 315 (1981)
3. Yoshida, H. Saito, and S. Shionya, J. Phys. Soc. Jpn. **50**, 881 (1981)
4. J.G. Fujimoto, S.G. Shevel, and E.P. Ippen, Solid State Commun. **49**, 605 (1984)
5. P.Y.Yu, Solid State Commun. **19**, 1087 (1976)
6. G.W. Fehrenbach, W. Schäfer, J. Treusch, and R.G. Ulbrich, Phys. Rev. Lett. **49**, 1281 (1982)

Femtosecond Room Temperature Measurements of Transport over Nanometer Dimensions in Quantum-Confined Heterostructures

D.P. Norwood [1], *H.-E. Swoboda* [1], *M.D. Dawson* [1], *T.F. Boggess* [1], *A.L. Smirl* [1], *and T.C. Hasenberg* [2]

[1] Center for Laser Science and Engineering, 124 AMRF Oakdale Campus, University of Iowa, Iowa City, IA 52242, USA
[2] Hughes Research Laboratories, 3011 Malibu Canyon Road, Malibu, CA 90265, USA

Abstract. Tunneling and over-barrier hopping are measured over nm spatial dimensions, with 100-fsec temporal resolution, in GaAs/AlGaAs multiple quantum wells at room temperature.

1. Introduction

We discuss 100-fsec, short-period (~120 nm) laser-induced transient grating measurements of perpendicular transport over nm-dimensions in $GaAs/Al_{0.3}Ga_{0.7}As$ quantum-confined heterostructures. By systematically varying the barrier width, we have examined the transition from dominance of tunneling via the confined sub-levels to that of over-barrier hopping. We show that the extracted transport parameters are consistent with a simple rate-equation model of these processes. Improved (100-fsec) resolution and room temperature measurements contrast with previous work on perpendicular transport using time-resolved photoluminescence [1].

2. Experimental

Multiple quantum well samples each with 100 Å wells but with differing barrier thicknesses (14 Å, 30 Å, 50 Å, and 100 Å) were grown by MBE. All samples showed clearly-resolved excitonic absorption features at room temperature. An ultrafast, cavity-dumped, Styryl 9/IR140, dye laser was used to optically excite each sample near the fundamental absorption edge. The laser produced 70-100 fsec pulses, of energy ~ 4 nJ and repetition rate of 1 MHz, tunable in the range 840-860 nm. The laser output was split into three beams, two equal-intensity, nearly-counter-propagating pumps, which formed the ~ 120 nm grating, and a weak (pump-probe ratio 40:1) cross-polarized probe beam, which was arranged to be temporally and spatially coincident with the pumps. Fluences of up to 15 $\mu J/cm^2$ at the sample were obtained from each

Springer Series in Chemical Physics, Vol. 53 **Ultrafast Phenomena VII**
Editors: C.B. Harris · E.P. Ippen · G.A. Mourou · A.H. Zewail

pump beam. The grating decay was measured by monitoring the diffraction efficiency of the probe beam versus probe time delay, and the diffracted signal was detected and analyzed with lock-in techniques.

Streak camera measurements of the carrier recombination times (at room temperature) determined that in each case the carrier lifetime was longer than the grating decay time. Carrier transport was, therefore, primarily responsible for the grating decay. For arbitrary angles α between the grating wavevector and the MQW growth direction, both intra- and inter-well diffusion can contribute to the decay, as was shown by Miller, et al. [2] Our data were, therefore, taken for α much below 1°, such that perpendicular transport dominated.

3. Results and Discussion

The decay of the diffraction efficiency (D.E.) was observed to increase from ~ 1.5 psec for the thinnest barrier (14 Å, Fig. 1a) sample, to ~ 80 psec for the 100 Å barrier structure (Fig. 1b). Our simple rate-

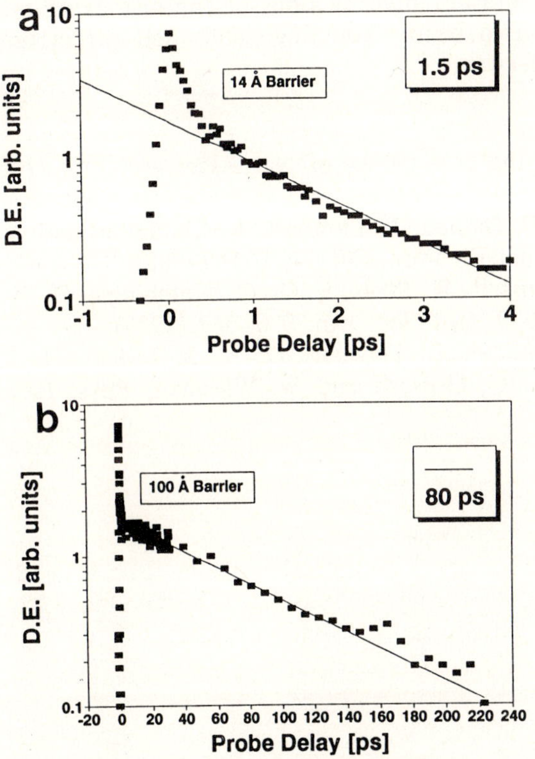

Fig. 1. Grating decay for a) 14 Å barrier and b) 100 Å barrier samples. Solid lines indicate exponential decays with indicated time constants.

equation model, which includes thermionic (over-barrier) emission from the wells and tunneling through the n=1, n=2 and n=3 confined electron levels, gives good agreement with these values. It indicates that the over-barrier thermal hopping process dominates for barrier widths approaching 100 Å, and above, while tunneling is increasingly dominant for the thinner barrier structures.

We draw attention to the initial enhanced peak in the data (Fig. 1) with a resolved decay of 200-300 fsec. This peak is much reduced, relative to the longer lived signal, when we tune the laser above the excitonic absorption features (not shown) and can, therefore, be associated with excitonic effects. One possible contribution to this could come from excitonic orientational gratings (e.g., formed between the cross-polarized forward pump and probe near zero delay and read by the backward pump beam). We have obtained evidence for such gratings at room temperature in preliminary 2 psec measurements of cross-polarized forward pump and probe self-diffraction, in the background-free direction. Another possible source for this initial fast decay is the rapid partial recovery of the excitonic absorption [3], following resonant bleaching, as the excitons are ionized by LO-phonons (time constant ~ 300 fsec). Further investigations of the origins of this peak using femtosecond two-beam coupling and self-diffraction techniques are currently underway.

This work was supported by the U.S. Office of Naval Research.

[1] B. Deveaud, J. Shah, T. C. Damen, B. Lambert, A. Chomette, and A. Regrency, IEEE J. Quantum Electron. QE-24, 1641 (1988).

[2] A. Miller, R. J. Manning, P. K. Milson, D. C. Hutchings, D. W. Crust, and K. Woodbridge, J. Opt. Soc. Am. B 6, 567 (1989).

[3] W. H. Knox, R. L. Fork, M. C. Downer, D. A. B. Miller, D. S. Chemla, C. V. Shank, A. C. Gossard and W. Wiesman, Phys. Rev. Lett 54, 1306 (1985).

Femtosecond Nonlinear Optical Properties of GaAs/AlAs Type-II Superlattices

G.R. Olbright[1], *W.S. Fu*[2], *J.F. Klem*[1], *T.E. Zipperian*[1], *R. Binder*[3], *and S.W. Koch*[3]

[1]Sandia National Laboratories*, Albuquerque, NM 87185, USA
[2]Solid State Electronics Laboratory, Stanford University, Stanford, CA 94305, USA
[3]Optical Sciences Center, University of Arizona, Tucson, AZ 85721, USA
*Supported by the Dept. of Energy, contract No. DE-ACO4-76DP00789

1. Abstract

We present femtosecond nonlinear absorption spectra of highly-excited type-II GaAs/AlAs multiple quantum wells that reveal a pronounced optically-induced blue shift and bleaching of the heavy-hole exciton absorption peak. The optical nonlinearity arises from many-body effects associated with dense electron and hole plasmas that are separated in both coordinate and momentum spaces. The blue shift disappears as we approach the indirect-to-direct crossover region by application of an axial electric field, thus verifying our assertion that the blue shift arises from type-II behaviour.

2. Experiment

In bulk semiconductors, high excitation leads to pure bleaching, with no observable shift of the exciton resonance. The absence of a shift is due to compensation between bandgap shrinkage and the concomitant reduction in the exciton-binding energy due to screening.[1] Spatial separation of the electron and hole plasmas leads to significantly different nonlinear optical properties in type-II structures[2] compared with bulk or type-I-alignment quantum wells. Recently we reported a blue shift of the heavy-hole exciton in type-II superlattices in the quasi steady-state.[3] The major part of this shift is due to the increase in the hole chemical potential in type-II systems. Here we report the study of femtosecond optical nonlinearities associated with GaAs/AlAs type-II superlattices. We discuss these phenomena in terms of modifications of the many-body plasma effects caused by electron and hole distributions functions which vary with electric-field-dependent Γ-X splitting.

Our femtosecond experiments were performed on a GaAs/AlAs photodiode. The MBE-grown p-i-n structure was processed into individually isolated and contacted mesa photodiodes. The intrinsic region consists of 90 periods of GaAs(30 Å) and AlAs(80 Å). CW PL spectra reveal Γ-X alignment and a concomitant ~ 100-fold increase in the integrated PL intensity at the crossover region as expected.[4] We acquired data in the form of time-resolved differential transmission spectra (DTS) using pump/continuum-probe spectroscopy. From electric-field-dependent DTS we obtained field- and time-dependent heavy-hole exciton shifts. We present these heavy-hole exciton shifts in Figure 1 for delays of +200 fs and +500 fs as a function of applied field ranging from the flat-band to the indirect-to-direct crossover

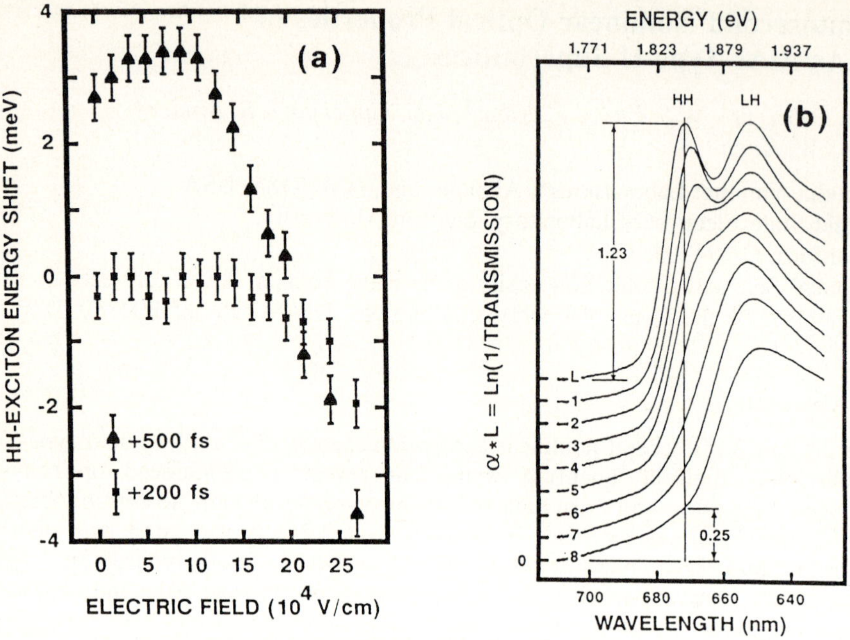

Figure 1. (a) Field-dependent Heavy-Hole exciton peak energy shift for the case of +200 fs and +500 fs pump/probe delays. The pump energy and pulsewidth are 1.96 eV and 100 fs, respectively. T = 15 K. (b) Excitation-intensity-dependent absorption spectra, α(I). L = linear spectra (I_{linear} = 1 W/cm²), I_{1-8} = 5, 8, 10, 12, 15, 25, 35, 45 kW/cm² respectively, and E = 1.96 eV. The sample consists of 75 periods of GaAs(29 Å) and AlAs(90 Å). T = 15 K.

region. In the vicinity of the indirect/direct crossover region (the high-field case) we observe the disappearance of the blue shift (see Fig. 1a). (The field-dependent Γ-X splitting is given approximately by, $\Delta_{\Gamma\text{-}X}(F) = \Delta_{\Gamma\text{-}X} - eFP/2$, where F is the field, $\Delta_{\Gamma\text{-}X}$ is the Γ-X splitting for the flat-band case, and P is the length of one superlattice period.) At very high fields the heavy-hole exciton shows a transient red shift. Conversely, at low fields, we observe a blue shift of the heavy-hole-exciton peak and partial recovery of the absorption saturation which develop after a few hundred femtoseconds. This blue shift can be regarded as evidence of the formation of electron and hole space-charge layers[2] arising from subpicosecond transfer of electrons from the GaAs to AlAs layers.[5] This blue shift exists until electrons in the AlAs layers recombine with holes in the GaAs layers via indirect recombination on microsecond timescales.

We have also obtained quasi-steady state absorption spectra on another sample consisting of 75 periods of GaAs(29 Å)/AlAs(90 Å) using modulated-pump and probe spectroscopy. Pump-intensity-dependent absorption spectra are shown in Figure 1b. The absorption spectra (see Fig. 1b) show that the heavy-hole-exciton resonance shifts significantly toward higher energies and bleaches with increasing excitation intensity. We explain the blue shift in terms of non-compensating many-body effects. The many-body effects are two-fold: (i) the spatial separation of the electrons and the holes yields a macroscopic space-charge potential (the so-called Hartree-potential) and (ii) each plasma renormalizes the bandstructure of the corresponding layer via exchange and correlation interactions and leads to a screening of the Coulomb potential. Phase-space filling effects are accounted for by considering the influence of the two-component Fermi gas on the absorption spectrum. For this case the absorption will be modified by a factor of the form $N_k=1-f_e(k)-f_h(k)$ where N_k is the phase-space filling factor and the f's are the Fermi distribution functions for the carriers. Here only the heavy holes contribute to phase-space filling, thus the f's are non-zero only for the heavy holes and the system cannot support a population inversion and can not exhibit gain. Therefore, we expect the heavy-hole energies to be renormalized by the screened Hartree-Fock self energies and the electron/heavy-hole Coulomb attraction is reduced by the screening caused by the heavy holes.

The blue shift is a consequence of the fact that the increase of the chemical potential of the hole-plasma overcompensates the reduction of the bandgap, and that this overcompensation increases with increasing density. The strong bleaching of the heavy-hole resonance is mainly caused by the phase-space filling of the heavy-hole states. We expect this nonlinearity to dominate after electrons transfer to the AlAs layers. We verify this by applying an electric field thereby reducing the Γ-X splitting and resulting in a reduction of the density of electrons at the AlAs X-point. As the field approaches a level corresponding to Γ-X alignment we expect and observe the disappearance of the blue shift (the field-dependent blue shift of the X-point in AlAs layers is $\simeq$ 100 meV/30 kV).

3. Summary

We have presented femtosecond experimental details of large optical nonlinearities in type-II multiple quantum wells. The primary effect is a blue shift of the heavy-hole exciton accompanied by significant bleaching after a few hundred femtoseconds for low fields. We observe the absence of the blue shift when we apply a large axial electric field. The absence of the blue shift for this case is due to the indirect to direct anti-crossing which results in negligible charge transfer.

1. H. Haug and S. W. Koch, "Quantum Theory of the Optical and Electronic Properties of Semiconductors," World Scientific, Singapore (1990).

2. G. R. Olbright, J. Klem, A. Owyoung, T. M. Brennan, R. Binder, and S. W. Koch, To appear in JOSA B, August (1990).

3. G. R. Olbright, J. Klem, G. R. Hadley, W. S. Fu, R. Binder, and S. W. Koch, paper PD16 Optical Society of America Annual Meetings (1989).

4. M.- H. Meynadier, R. E. Nahory, J. M. Worlock, M. C. Tamargo, J. L. de Miguel, and M. D. Sturge, Phys. Rev. Lett. 60, 1338 (1988).

5. J. Feldmann, R.Sattmann, E. O. Göbel, J.Kuhl, J.Hebling, K. Ploog, R. Muralidharan, P.Dawson, and C.T.Foxon, Phys. Rev. Lett. 62, 1892 (1989).

Part VII

Semiconductors; Solid-State Physics

Ultrafast Scattering Times in Amorphous Silicon

D. Hulin[1], *A. Mourchid*[1], *R. Vanderhaghen*[2], *and P.M. Fauchet*[3]

[1]Laboratoire d'Optique Appliquée, ENSTA, Ecole Polytechnique, Batterie de l'Yvette, F-91120 Palaiseau, France
[2]Laboratoire de Physique des Interfaces et Couches Minces, Ecole Polytechnique, F-91120 Palaiseau, France
[3]Department of Electrical Engineering, Princeton University, Princeton, NJ 08544, USA

ABSTRACT. Disorder in amorphous semiconductors leads to unusual and interesting optical properties. Using femtosecond time-resolved spectroscopy, we find that the scattering time of carriers in the extended states is subfemtosecond and that the electronic susceptibility is well described by a Drude model. These free carriers recombine non-radiatively in a few picoseconds and the increase in lattice temperature follows the carrier recombination with no measurable delay.

In optical experiments, an ultrashort coherence time of the excited states leads to an interesting and unusual optical response. In pump and probe femtosecond experiments performed in crystalline semiconductors [1] or dyes [2], a homogeneous spectral hole burning can be observed. In most semiconductors, the electron-optical phonon scattering time is of the order of 100 fsec [3] and the electron-electron scattering time has been measured to be as short as 3.5 fsec at large ($N{\sim}10^{18}cm^{-3}$) carrier densities [4]. In the limit of extremely short coherence times, momentum is no longer conserved in optical transitions and spectral information is lost. We report experiments performed in amorphous silicon (a-Si:H), a material in which subfemtosecond carrier scattering due to static disorder fluctuations relaxes completely the k-vector conservation rule.

In amorphous semiconductors, the electronic states are divided in localized and extended states, depending on the position of the energy level with respect to the mobility edge. Electrons excited above the mobility edge have a finite mobility and can thermalize in the extended states within a fraction of 1 psec or be trapped into the localized states in $\sim$ 1 psec. Thus, to study the properties of carriers in the extended states, femtosecond time resolution is required. In contrast to crystalline semiconductors, their optical properties are strongly influenced by free carrier absorption, a process which becomes first-order since momentum conservation is not relevant. In our experiments, a 100 fsec pump pulse at 2 eV injects free carriers in thin a-Si:H films having a bandgap $\sim$ 1.7 eV at room temperature. The reflectivity and transmission of a tunable probe pulse generated from a white light continuum are measured simultaneously as a function of time delay and injected carrier density (from $N = 10^{19}$ to $10^{21}cm^{-3}$). By inversion of the Fabry-Perot formulae, we obtain directly the absolute values of the real and imaginary parts of the index of refraction [5]. In this paper, we focus on the carrier momentum scattering time in the extended states and on the energy transfer time between the electronic and phonon baths.

In the first experiment, we measure the complex electronic susceptibility due to free carriers immediately after their injection, i.e before trapping or recombination. Both Δn and $\Delta\alpha$ vary linearly with N. This shows that carrier-carrier scattering is not dominant. If we select the probe photon energy to be below the bandgap, interband transitions are forbidden and the susceptibility corresponds to intraband or free carrier absorption. In Figure 1, we plot the measured ratio $\Delta k/\Delta n$ ($k=\lambda\alpha/4\pi$) vs. λ. Our results suggest that the Drude model is an adequate description of the free carrier susceptibility. It is easy to show that if electrons and holes have similar properties, $\Delta k/\Delta n = 1/\omega\tau$, where τ is the carrier momentum scattering time [6]. The value of τ that fits the data is $\sim$ 0.5 fsec, independent of the carrier density. For the average carrier injected $\sim$ 150 meV above the mobility edge, this short scattering

Springer Series in Chemical Physics, Vol. 53 **Ultrafast Phenomena VII**
Editors: C.B. Harris · E.P. Ippen · G.A. Mourou · A.H. Zewail

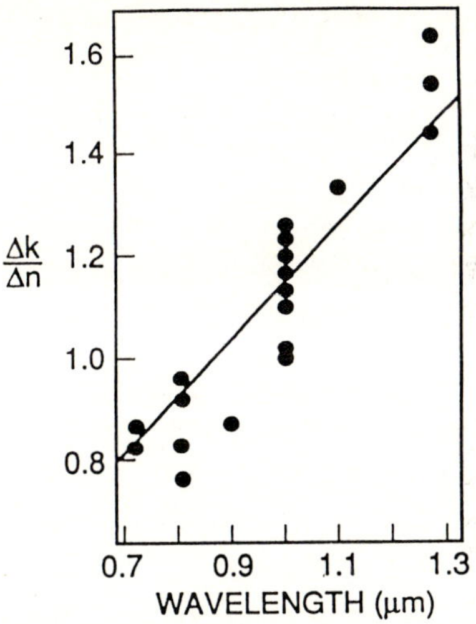

Figure 1 $\Delta k/\Delta n$ vs λ for probe photon energies below the bandgap and fit using the Drude model with $\tau = 0.46$ fsec

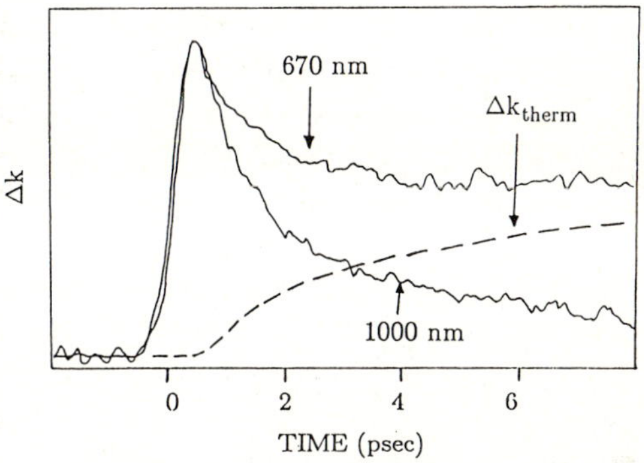

Figure 2 Induced absorption measured **(a)** below the bandgap and **(b)** above the bandgap after injection of N=2.5 $10^{20}cm^{-3}$. The difference between the two normalized curves gives the thermal induced absorption (dashed line).

time is consistent with a mean free path ~ 0.1 nm and thus a momentum randomizing collision at every atomic site. The radius of the wavefunction for an electron at the mobility edge is such that it always samples one localized state [7]. These localized states produce static potential fluctuations which act as scattering centers and thus are responsible for the subfemtosecond scattering time we measure. Note that this result also represents the first direct proof that momentum is totally randomized in a-Si:H.

When we repeat our measurements as a function of time delay [8], we observe a rapid decrease of the free carrier absorption (Figure 2.a). This could result from carrier recombination or from carrier trapping into states having a smaller optical cross-section. When the probe photon energy is above the bandgap, we observe that within a few psec Δk reaches a plateau (Figure 2.b). This plateau is the signature of bandgap shrinkage due to lattice heating that develops as the carriers disappear. The recovery of the index of refraction (below and above the bandgap) above its initial value after the carriers have disappeared is also an indication of thermal bandgap shrinkage [9]. By measuring Δn and Δk vs temperature by ellipsometry, we find that each pair that disappears gives ~ 2 eV of energy to the lattice. Thus, for large N, most of the carriers do not get trapped in localized states below the mobility edge but rather recombine across the gap by a non-radiative mechanism.

We can measure the energy transfer time from the carrier system to the phonon bath. The rate of carrier recombination in Figure 2.b is given by Figure 2.a, scaled using the Drude model. The difference between those two curves is proportional to the lattice temperature obtained from the bandgap shrinkage. We can then compare the time evolutions of the lattice temperature and of the total energy released by the carriers as they recombine. Figure 3 shows that within the uncertainty of 100 fsec those two curves can be superimposed. We conclude that the energy transfer time must be less than 100 fsec. It is energetically favorable for the carriers to produce optic phonons as they recombine. On the other hand, thermal bandgap shrinkage is usually associated with production of acoustic phonons [10]. It follows that the decay time of optic phonons into acoustic phonons must be less than 100 fsec. This time is much shorter than in crystalline semiconductors [11] and also results directly from the relaxation of the momentum conservation rule in a-Si:H.

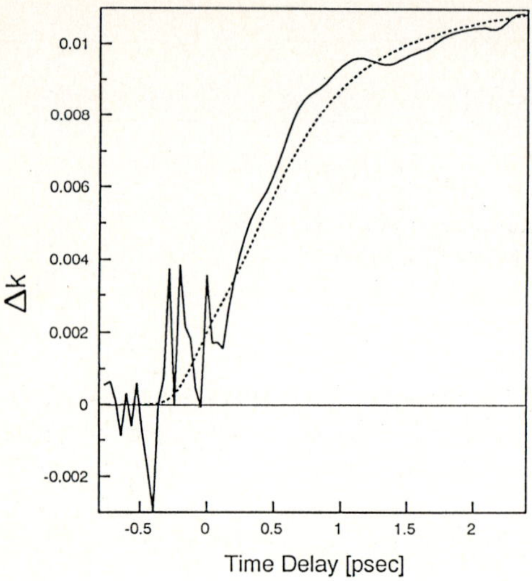

Figure 3 Induced absorption proportional to the lattice temperature increase (full line) and energy released by the carriers (dotted line). There is no delay within the 100 fsec experimental uncertainty.

In conclusion, static potential fluctuations due to disorder in amorphous silicon relax completely the momentum conservation rule. We find that the momentum scattering time for electrons in the extended states is < 1 fsec and thus the mean free path is comparable to the interatomic distance. The picosecond non-radiative recombination observed for $N > 10^{19}cm^{-3}$ produces an instantaneous lattice temperature increase, from which it appears that the decay of optic phonons into acoustic phonons takes less than 100 fsec.

P.M. Fauchet's research was supported by NSF, ONR and the Alfred P. Sloan Foundation. D. Hulin is also at Ecole Normale Supérieure and R. Vanderhaghen is also at CNRS.

References

1. J.L.Oudar et al, Phys. Rev. Lett. **55**, 2074 (1985)
2. C.H. Brito Cruz et al, Chem. Phys. Lett. **132**, 341 (1986)
3. J. Shah, IEEE Journ. Quantum Electron. **QE-24**, 276 (1988)
4. P.C. Becker et al, Phys. Rev. Lett. **61**, 1647 (1988)
5. C. Tanguy et al, Appl. Phys. Lett. **53**, 880 (9188)
6. P.M. Fauchet et al, J. Non-Cryst. Solids **114**, 564 (1989)
7. P.M. Fauchet et al, J. Non-Cryst. Solids **114**, 277 (1989)
8. A. Mourchid et al, J. Non-Cryst. Solids **114**, 582 (1989)
9. P.M. Fauchet and D. Hulin, J. Opt. Soc. Am. **B6**, 1024 (1989)
10. J.A. Van Vechten, in *Handbook on Semiconductors*, vol 3, North-Holland, 1980, Ch. 1
11. D. von der Linde, J. Kuhl and H. Klingenburg, Phys. Rev. Lett. **54**, 2151 (1985)

k-Space Carrier Dynamics in GaAs

M.T. Portella[1], *J.-Y. Bigot*[1], *R.W. Schoenlein*[1], *C.V. Shank*[1], *and J.E. Cunningham*[2]

[1]Lawrence Berkeley Laboratory, University of California, 1 Cyclotron Road, M/S:70-110A, Berkeley, CA 94720, USA
[2]AT&T Bell Laboratories, Holmdel, NJ 07733, USA

Abstract. We investigate the scattering dynamics of carriers in **k**-space for different carrier densities in GaAs, by analyzing the polarization of pump-probe signals produced by 9-femtosecond infrared pulses. The relaxation time is found to vary from 25 to 60 fs for a carrier density range of $8x10^{16}$ to $8x10^{17}$ cm^{-3}. The dependence of the relaxation time on the carrier density shows that the main scattering process is the carrier-carrier interaction.

1. Introduction

We report here the first measurements of **k**-space carrier scattering times in GaAs as a function of the density of carriers. Studies were performed by measuring the dynamics of anisotropic momentum distribution of carriers created at the band edge of the semiconductor, at room temperature. These measurements were made by analyzing the polarization of pump-probe signals produced by linearly polarized 9 femtosecond infrared pulses [1] centered at 820 nm.

For band to band photoexcited electrons, the momentum distribution function (following [2]) is given by $F(\theta) \alpha [1+aP_2(\cos\theta)]$, where $P_2(\cos\theta)=(3x^2-1)/2$ is the second order Legendre polynomial, θ is the angle between the polarization vector of the light pulse and the momentum **k** of the electron, and a is the anisotropic parameter. In GaAs, a=-1 for transitions from the heavy-hole valence band (hh), and a=+1 for transitions from the light-hole valence band (lh) to the conduction band.

The band to band transition probability depends on the density of initial states in the valence band and density of the available (empty) final states in the conduction band. Since the density of states (and effective mass [3]) is much greater in the hh-subband than in the lh-subband, for wave vectors k smaller than $5.8x10^6$ cm^{-1}, the transition probability is larger for electrons originating from the hh-subband than from the lh-subband. Thus, when the GaAs is excited by a 9 femtosecond infrared pulse, centered at 820 nm (1.51 eV), the majority of photoexcited electrons come from the hh-valence band. Electrons with an excess of energy larger than 110 meV will originate only from hh-subband.

As a result, more than 90% of the photoexcited electrons will have wave vectors **k** perpendicular to the polarization of the electric field of the laser pulse. Consequently, an anisotropic momentum distribution is created at the band edge of the GaAs changing its absorption coefficient anisotropicaly.

Springer Series in Chemical Physics, Vol. 53 **Ultrafast Phenomena VII**
Editors: C.B. Harris · E.P. Ippen · G.A. Mourou · A.H. Zewail

2. Experiment

We have used the pump-probe technique, with the pump and probe beams linearly polarized, and the probe polarization axis oriented at a 30-degree angle with respect to the pump. At zero time delay between the two pulses, the pump pulse creates an anisotropic distribution of carriers in **k**-space which anisotropically perturbs the absorption coefficient of the sample. As a consequence, the polarization of the transmited probe pulse rotates and then relaxes to its original direction as the carriers scatter to an isotropic distribution in **k**-space. The **k**-space redistribution time is determined by measuring the polarization rotating angle of the probe at different delay times. Measurements are performed in the following manner. After the sample, the probe beam is split into two parts which are analyzed by glan polarizers at 45 and 135 degrees with respect to the pump pulse. The ratio R(t) between these two signals $S_{135}(t)$ and $S_{45}(t)$ is used to determine the small angle probe polarization rotation according to [4] $\varepsilon(t)=(R(t)/R_o-1)\sin(2\varphi)/4$. $R_o=\tan(\varphi)$ is the ratio between S_{135} and S_{45} in the absence of any anisotropic distribution, where φ is the angle between the probe pulse polarization and the analyzer (15 degrees).

Changes in the probe beam polarization are on the order of one degree. The relaxation times of this polarization rotation are measured for different excitation intensities, thereby providing a direct measure of the anisotropic carrier dynamics in **k**-space as function of the carrier density.

Relaxation times as a function of carrier density are shown in Fig. 1. The measured times range from 24 fs to 58 fs for carrier densities varying between 7.7×10^{17} to 8.3×10^{16} cm^{-3}. The time dependence of the relaxation times with the density of carriers N (solid curve in Fig. 1)was found to be $T\alpha N^{-0.36\pm0.03}$. This dependence of the relaxation time on density indicates that carrier-carrier interaction is the dominant process in the anisotropic scattering. These results can be understood by considering the interaction between carriers with the nearest neighbors that are distributed in a volume proportional to N^{-1}. This yields a scattering time T, which depends on the carrier density according to $T\alpha N^{-1/3}$. These relaxation times of carrier-

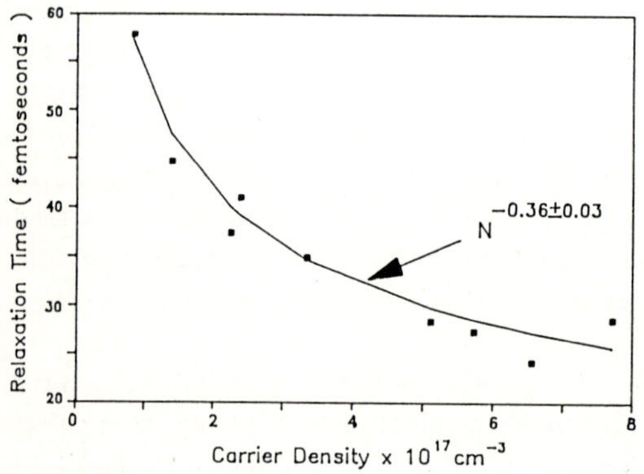

Fig. 1. **k**-space carrier scattering times as function of the density of carriers in GaAs.

carrier scattering and their dependence on the density are in agreement with the previous measurements of the dephasing times performed by photon echoes experiments [5]. One should notice that the relaxation times reported here are somehow shorter than reported in [4] using longer pulses at low temperature.

In conclusion, we have investigated the dependence on **k**-space carrier scattering time on carrier density in GaAs. Studies were performed by measuring the anisotropic distribution dynamics of carriers at the band edge of the semiconductor. The times measured vary from 25 to 60 femtoseconds for the density range $8x10^{17}$ - $8x10^{16} cm^{-3}$. This is consistent with a nearest neighbor carrier-carrier interaction.

3. References

1. P.C. Becker, H.L. Fragnito, R.L. Fork, F.A. Beisser, C.V. Shank, *Appl. Phys. Lett.* 54, 411 (1989).
2. B.P. Zakharchenya, D.N. Mirlin, V.I. Perel, I.I. Reshina, *Sov. Phys. Usp.* 25, 143 (1982).
3. J.S. Blakemore, *J. Appl. Phys.* 53, R123 (1982).
4. J.L. Oudar, A. Migus, D. Hulin, G. Grillon, J. Etchepare< A. Antonetti, *Phys. Rev. Lett.* 53, 384 (1984).
5. P.C. Becker, H.L. Fragnito, C.H. Brito Cruz, R.L> Fork, J.E. Cunningham, J.E. Henry, C.V. Shank, *Phys. Rev. Lett.* 61, 1647 (1988).

This work was supported by the U.S. Department of Energy
Contract No. DE-AC03-76SF00098

Time-Resolved Picosecond Photon Echo Studies of Phase Relaxation in CdSSe Mixed Crystals

U. Siegner, G. Noll, S.G. Shevel, and E.O. Göbel*

Philipps-Universität Marburg, Fachbereich Physik und Zentrum für Materialwissenschaften, Renthof 5, D-3550 Marburg, Fed. Rep. of Germany
*Permanent address: Institute of Physics,
Ukrainian Academy of Sciences, Kiev, USSR

Phase relaxation of optical excitations in condensed matter can be studied by coherent four–wave–mixing experiments in the time domain without requiring assumptions on the spectral properties of the material. For inhomogeneously broadened transitions, three laser pulses at times $t = 0$, $t = t_{21}$ and $t = T$ with wave vectors $\mathbf{k_1}$, $\mathbf{k_2}$ and $\mathbf{k_3}$, respectively, generate a stimulated photon echo pulse, emitted time–delayed at $t = T+t_{21}$ in the momentum–matching direction according to $\mathbf{k_3}+\mathbf{k_2}-\mathbf{k_1}$. As a function of t_{21} the echo amplitude decays due to scattering events, described by the dephasing time constant T_2 [1].

We have investigated phase relaxation of resonantly excited localized excitons in $CdS_{0.40}Se_{0.60}$ mixed crystals at low temperatures. Photon echoes due to intrinsic excitations in a semiconductor are demonstrated unambigiously, i.e. resolved in time. The results show the strong influence of disorder–induced localization on the dynamical optical properties of excitons.

Streak camera traces of the delayed echo pulse and the excitation pulses are depicted in Fig. 1 for different time spacings t_{21}

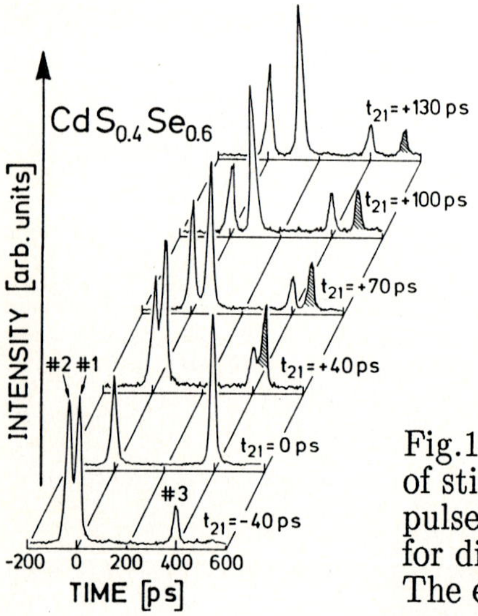

Fig.1 Streak camera traces of stimulated photon echo pulses and excitation pulses for different time delays. The echo pulse is shaded.

Springer Series in Chemical Physics, Vol. 53 **Ultrafast Phenomena VII**
Editors: C.B. Harris · E.P. Ippen · G.A. Mourou · A.H. Zewail

between pulse #1 and pulse #2 and a fixed time delay T = 400 ps of the third pulse. No echo pulse is emitted into the momentum–matching direction if pulse #2 precedes pulse #1. For $t_{21} > 0$, the time–delayed emission of the echo pulse can be clearly observed, demonstrating unambigiously the inhomogeneous broadening of the localized exciton transitions. The dephasing time evaluated from the data amounts to 400 ps.

Figure 2 shows the increase of the dephasing times (triangles, left scale) with decreasing excitation photon energy for a constant excitation intensity. The dephasing time varies between 260 ps and 500 ps. These values for resonantly excited localized excitons are longer by an order of magnitude than the dephasing times reported for free excitons in binary compound semiconductors [2,3]. Consequently, the slow dephasing has to be due to localization, which reduces the rates of exciton–phonon and exciton–exciton scattering. In particular, exciton–exciton scattering is strongly reduced as the T_2 times do not depend on the exciton density in the density range investigated (about 10^{15} cm^{-3}) [4]. The stronger localization at lower photon energies results in the increase of the T_2–times.

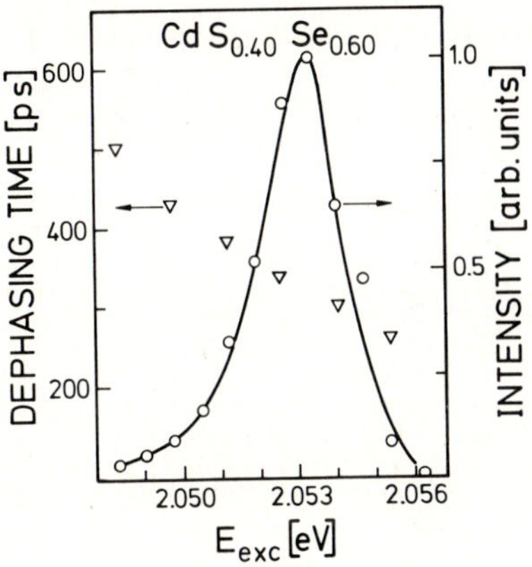

Fig.2 Dephasing times (triangles) and photon echo intensity (circles) versus excitation photon energy

Measurements of the echo intensity as a function of the excitation intensity reveal that the echo intensity saturates with increasing excitation. The saturation of the echo intensity is accompanied by a superlinear increase of the transmitted intensity of pulse #2 due to absorption bleaching. Bleaching due to filling of the optically coupled states is easily accomplished as a consequence of the reduced scattering. We determine from our results the lower limit for the intraband energy relaxation time T_{rel}, describing hopping within the distribution of the localized states, to be in the range 130 ps to 250 ps ($T_{rel} \geq T_2/2$), depending on photon energy. These values well agree with data from time–resolved luminescence experiments [5], thus giving evidence that energy relaxation is the main dephasing process.

The echo intensity measured at $t_{21} = 100$ ps, T = 400 ps and a constant excitation intensity is also plotted versus photon energy in Fig. 2 (circles, right scale). The spectral range where photon echoes can be observed coincides with the exponential tail of the absorption spectrum corresponding to localized exciton transitions. The strong decrease of the echo intensity at higher photon energies reveals the decrease of the strengths of the third order nonlinearity, which generates the echo signal. The observation of absorption bleaching indicates that state filling is the microscopic process giving rise to the nonlinear optical response and in fact, state filling is less efficient at higher photon energies where the energy relaxation rate becomes faster. The decrease of the echo intensity at longer wavelengths is attributed to the decreasing absorption in this spectral region.

In conclusion, our time–resolved photon echo experiments reveal the dynamical behaviour of localized excitons, which is characterized by reduced scattering compared to free excitons.

References

1. A.M. Weiner, S. De Silvestri, E.P. Ippen, J. Opt. Soc. Am. **B2**, 654 (1985)
2. L. Schultheis, J. Kuhl, A. Honold, C.V. Tu, Phys. Rev. Lett. **57**, 1635 (1986)
3. C. Dörnfeld, J.M. Hvam, IEEE J. Quantum Electron. **QE–25**, 904 (1989)
4. G. Noll, U. Siegner, E.O. Göbel, H. Schwab, R. Renner, C. Klingshirn, to be published in Journ. Cryst. Growth
5. S. Shevel, R. Fischer, E.O. Göbel, G. Noll, P. Thomas, C. Klingshirn, Journ. Lum. **37**, 45 (1987)

Direct Observation of the Electron and Hole Contributions in the Impulse Response of a Metal-Semiconductor-Metal Schottky Diode

M. Lambsdorff[1], *J. Kuhl*[1], *M. Klingenstein*[1], *C. Moglestue*[2], *J. Rosenzweig*[2], *A. Axmann*[2], *J. Schneider*[2], *H. Leier*[3], *and A. Forchel*[3]

[1]Max-Planck-Institut für Festkörperforschung, Heisenbergstr. 1, D-7000 Stuttgart 80, Fed. Rep. of Germany
[2]Fraunhofer Institut für Angewandte Festkörperphysik, Eckerstr. 4, D-7800 Freiburg, Fed. Rep. of Germany
[3]IV. Physikalisches Institut, Universität Stuttgart, Pfaffenwaldring 57, D-7000 Stuttgart 80, Fed. Rep. of Germany

Interdigitated metal-semiconductor-metal (MSM) Schottky diodes [1] are attractive high-performance photodetectors characterized by a fast response, high quantum efficiency, and small leakage current. The key features providing the broad bandwidth of these devices are the rapid carrier sweep-out obtained by the narrow electrode spacing and the low capacitance associated with the small active area of the structure. Recently van Zeghbroeck *et al.* [2] have demonstrated an overall bandwidth of an MSM-GaAs-Schottky diode as large as 105 GHz. The time response has directly been analyzed in the time domain using a silicon-on- sapphire (SOS) photoconductive switch controlled by the frequency doubled output of a pulse-compressed mode-locked YAG-laser (1.8ps).

We present here a comprehensive study of the impulse response of GaAs MSM-Schottky diodes. The investigation includes systematic variations of the diode parameters (active area and finger spacing) and the experimental conditions (external voltage and excitation intensity) over broad ranges. The experimental results are compared to ensemble Monte Carlo calculations. For the first time, we show experimentally and theoretically that electron and hole contributions to the response of a photodetector are separated in time.

The diodes are manufactured on bulk GaAs using $Au/Ti/Pt$ as the Schottky metal. The diode structure under investigation has an active area of $10\mu m \times 10\mu m$, $1\mu m$ wide fingers with 1.2 μm spacing.

For the measurement of the impulse response the GaAs chip with the photodiode is bonded via 300 μm long wires to a radiation-damaged SOS chip (ion dose $3 \times 10^{14} Si^{+}/cm^{2}$) containing a coplanar transmission line with 50 μm linewidth and spacing connnected to a dc bias voltage (see inset of Fig.1). The diode is excited with a 70fs optical pulse at a repetition rate of 115 MHz and an average power of 140 μW. The shape of the generated current pulse propagating on the coplanar transmission line is sampled by illuminating the fast sampling gap by a synchronized 70fs pulse of the same laser.

The cross-correlation width of the SOS switch (1.5ps) demonstrates a time resolution of 1ps for the sampling gap. Figure 1 depicts the sampling signal versus delay between the two optical pulses for two different bias voltages. Both

Springer Series in Chemical Physics, Vol. 53 **Ultrafast Phenomena VII**
Editors: C.B. Harris · E.P. Ippen · G.A. Mourou · A.H. Zewail

curves reveal two temporally separated exponential decays. For the 1V bias we observe a signal rise time of 3.7 ps, a full width half at maximum of 11.4 ps, an initial rapid decay to about 20% of the peak amplitude with a time constant of 6.8ps and a second exponential decay with 40ps. The decrease of the bias to 0.25 V results in an increase of the rise(4.5ps) and decay times (12.1 and 100ps) of the main peak whereas the relative amplitude of the slow tail increases by roughly a factor of 2. Similar variations are observed if the excitation density is increased at a fixed bias. The small oscillations on the pulse tail are most probably due to the impedance mismatch at the wire bonds.

The initial rapid decay of the signal can be attributed to the sweep-out of photoexcited electrons and the slower tail to the removal of the lower mobility holes. This explanation is confirmed by self-consistent two-dimensional Monte Carlo calculations of the carrier transport in these MSM-photodiode structures, taking into account optical- and acoustic phonon, intervalley, and impurity scattering. The model includes the Γ, X, and L valley of the conduction band and the heavy-hole, light-hole, and split-off valence band. The carrier drift under the influence of the electric field is obtained from the Boltzmann transport equation. The electric field is calculated in steps of 25 fs by solving Poisson's equation for the given spatial carrier distribution. This procedure explicitly includes dynamical screening of the externally applied electric field. Calculations performed for different values of finger spacing, external voltage and excitation intensity reveal that this field screening becomes particularly important at low voltages (< 0.5 V), large electrode spacing (1.5 μm) and high excitation density ($> 10^{17}\text{cm}^{-3}$). Strong screening of the external field results in a faster decay of the main peak and the formation of a long tail to which both electrons and holes contribute. Only a fraction of the photoexcited carriers contribute to the external photocurrent for this condition. The rest remains inside the device as a stationary electron-hole plasma which decays via recombination on a nanosecond timescale. Figure 2 depicts a temporal evolution of the electron and hole current predicted by this model for an MSM diode with a finger separation of 1.5 μm, a voltage of 2V, and an excitation density of 4.8×10^{16} electron-hole pairs per cm^3. The time dependence of the total current shows qualitatively good agreement with the measured diode response function.

The different field dependence of the fast and slow signal decay is explained by the drift velocity saturation for electrons due to intervalley scattering in the studied field regime whereas the hole drift velocity still increases with increasing field. The slight variations of the rise and decay time of the main peak with field strength may be attributed to small variations in the relative distribution of electrons in the Γ– and L-valley of the conduction band and/or the variation of dynamic screening with the applied electric field.

We do not expect a significant influence of an inherent diode capacitance on the transient photocurrent. The capacitance of the structure was estimated to be 6 fF, corresponding to an RC time of 300 fs, more than one order of magnitude below the measured rise and fall times. Moreover, the risetime of a

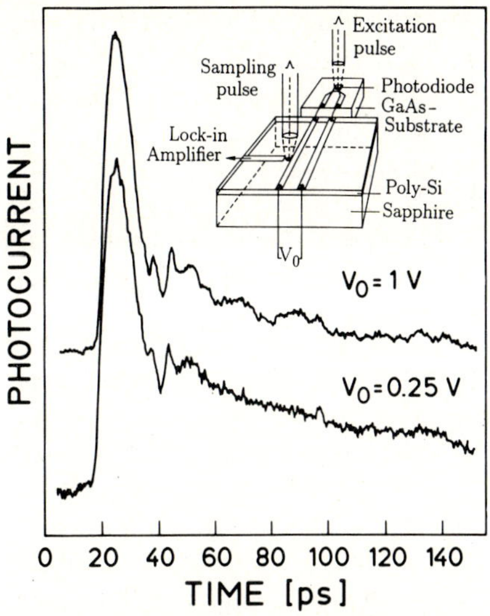

Fig.1: Experimental impulse response of the MSM-GaAs diode with 1.2 μm finger spacing for two different bias voltages. The inset shows the experimental set-up.

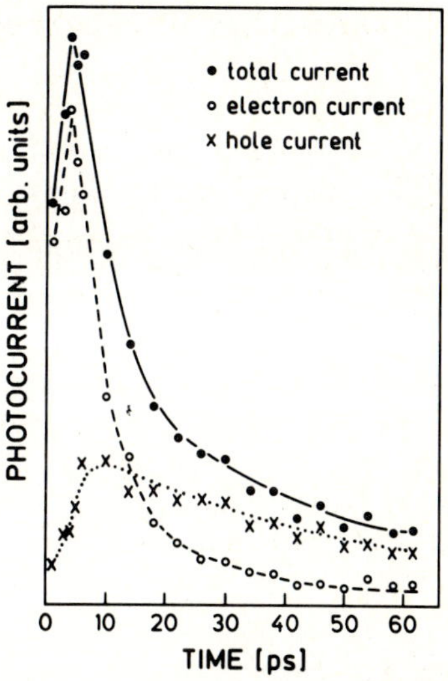

Fig.2: Theoretical time evolution of the electron, hole, and total current in a MSM photodiode as predicted by Monte Carlo calculations.

diode with 7.5× larger active area and measured 23 fF capacity only slightly increased to 4.8 ps.

Further experiments on diodes with a smaller electrode spacing (0.5μm) and on new structures where the photodiode and the photoconductive switch applied for sampling of the signal shape are integrated on the same GaAs-chip are currently in progress.

References

1. S.M. Sze, D.J. Coleman, and A. Loya, *Sol.Stat.Elect.* **14**, 1209 (1971).
2. B.J. van Zeghbroeck, W. Patrick, J.-M. Halbout, and P. Vettiger, *IEEE Electron Device Lett.* **9**, 527 (1988).

Ultrafast Recombination in Ion-Damaged InP Studied by Femtosecond Luminescence

K.F. Lamprecht [1], *R.A. Höpfel* [1], *and L. Palmetshofer* [2]

[1]Institute of Experimental Physics, University of Innsbruck, A-6020 Innsbruck, Austria
[2]Institute of Experimental Physics, University of Linz, A-4040 Linz, Austria

Time–resolved photoluminescence in H^+ bombarded InP reveals decay times down to 415 fs due to ultrafast recombination processes which are in the same order of magnitude as the intraband relaxation processes. As a consequence extreme nonequilibrium carrier distributions are observed in the luminescene spectra.

1. Introduction

Ion–damaged III–V semiconductors with ultrashort carrier lifetimes are important materials for applications in ultrafast optoelectronics [1]. Recently, in addition to photoconductivity and far infrared techniques, optical methods such as time–resolved reflectivity [2] and subpicosecond luminescence [3,4] have been applied to study the recombination dynamics.

In the work reported here, time–resolved photoluminescence in InP is studied to measure the carrier lifetimes. The results show decay times shorter than 500 fs for the highest damage doses applied. The luminescence spectra give evidence for extreme nonequilibrium carrier populations as a consequence of optical injection at high energies and rapid recombination. The possibility of *intraband* population inversion in the conduction band is discussed.

2. Experiment

The samples used in the present work are made from semiinsulating InP. They are H^+–ion bombarded with an energy of 200 keV and doses up to 5×10^{15} cm^{-2}. One sample is undamaged.

The time–resolved photoluminescence experiment is based on the nonliner correlation method as introduced by Rosen *et al.* and von der Linde *et al.* [5]: Two electron and hole populations $n_1(k,t)$, $p_1(k,t)$, $n_2(k,t+\tau)$, $p_2(k,t+\tau)$ are generated by two laser pulses with a variable delay τ, which are polarized perpendicular to each other. The cross correlation of the two populations is obtained by measuring the nonlinear luminescence signal, in our experiment, by chopping the two beams at two different frequencies (ν_1,ν_2) and measuring the signal component of the difference frequency $\nu_1-\nu_2$. For the excitation we use the focused laser pulses of a balanced colliding pulse mode–locked dye laser (gain and absorber: rhodamine 6G and DODCI) [6], giving pulses at $\lambda \approx 620$ nm, pulse durations of 100 fs, and 20 mW average power (100 MHz repetition rate). The excitation level is of the order of 10^{17} cm^{-3}. All measurements are performed at room temperature.

3. Results

In figure 1 the luminescence correlation signal for the sample with the maximum radiation damage is shown. The observed lifetime for this sample is 415

Springer Series in Chemical Physics, Vol. 53 **Ultrafast Phenomena VII**
Editors: C.B. Harris · E.P. Ippen · G.A. Mourou · A.H. Zewail

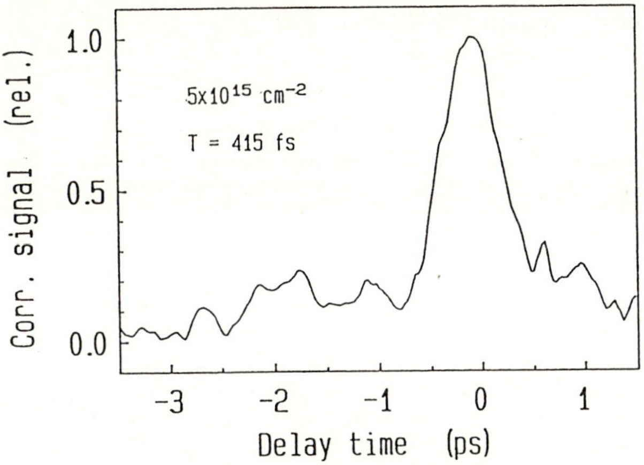

Fig. 1: Luminescence correlation signal as a function of delay time for the sample with the maximum radiation damage dose. The plot is for a sample temperature of 300 K and a wavelength of $\lambda = 860$ nm.

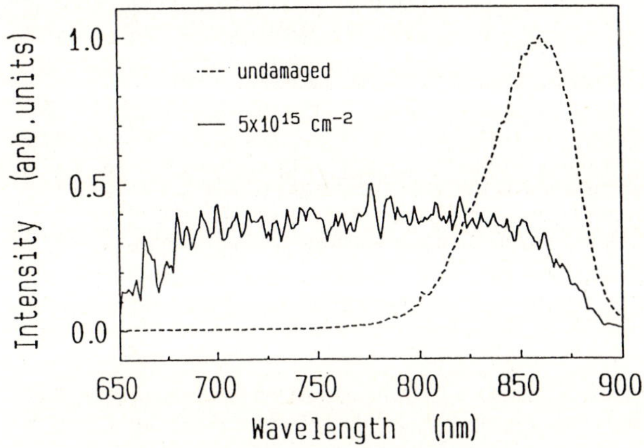

Fig. 2: Time–integrated luminescence spectra of undamaged and 5×10^{15} cm^{-2} bombarded material excited by CPM–laser at $\lambda = 620$ nm.

fs. This value, which is the overall correlation time – thus containing risetime *and* falltime – documents one of the shortest luminescence events ever experimentally observed [7].

In figure 2 the time–integrated luminescence spectra are shown, as obtained by excitation with a CPM–laser. As detector a GaAs–photomultiplier at 250 K is used. As expected, the undamaged sample shows a thermalized carrier distribution with a carrier temperature of 400 K close to the crystal temperature of 300 K. The spectra of the sample with a damage of 5×10^{15} cm^{-2} with a lifetime of 415 fs shows an intense high–energy tail.

We interpret the extremely hot luminescence spectra as a consequence of the ultrafast recombination, which is a faster process than the cooling of the carrier system. Since the luminescence spectra reflect the product of the energy distributions of electrons *and* holes, and since the hole–optical–phonon interaction is by an order of magnitude faster than the electron–phonon interaction, we conclude that the electron energy distribution is "hotter" than the spectra of the combined distribution. From the ultrashort lifetime of 415 fs we expect the electron distribution to be even "inverted" [3,8], in contrast to the hole distribution which should be not far from the room temperature. Furthermore, the spectra is very sensitive to the excitation intensity through the effects of carrier–carrier interaction on the thermalization process. With a decrease of the excitation intensity the high energy–tail becomes more distinct in the shape of the luminescence spectra.

We acknowledge collaboration with Robert Christanell. This work has been supported by the "Fonds zur Förderung der wissenschaftlichen Forschung, Österreich", project number P 6184.

References

1. D.H. Auston, in *Ultrashort Laser Pulses and Applications*, edited by W. Kaiser (Springer, Berlin, 1988), pp. 183–233.
2. F.E. Doany, D. Grischkowsky, and C.C. Chi, in *Picosecond Electronics and Optoelectronics II*, edited by F.J.Leonberger, C.H.Lee, F.Capasso, and H.Morkoc, Springer Series in Electronics and photonics (Springer, Berlin, 1987), Vol. 24, p.228.
3. R.A. Höpfel and N. Sawaki, Appl. Phys. Lett. *55*, 460 (1989).
4. M.B. Johnson and T.C. McGill, Appl. Phys. Lett. *54*, 2424 (1989).
5. D. Rosen, A.G. Doukas, Y. Budansky, A. Katz, and R.R. Alfano, Appl. Phys.Lett. *39*, 935 (1981).
   D.von der Linde, J.Kuhl, and E.Rosengart, J.Lumin. *24/25*, 675 (1981).
6. A. Valdmanis, R.L. Fork, and J.P. Gordon, Opt. Lett. *10*, 131 (1985)
7. For a review on time–resolved luminescence, see: J. Shah, IEEE J. Quantum Electronics *QE–24*, 276 (1988).
8. R.A. Höpfel, Proc. 6th Int. Conf. "Hot Electrons in Semiconductors", Scottsdale, AZ, published in Solid State Electronics (1990, in print).

Subpicosecond Photoconductivity in III-V Compound Semiconductors Using Low Temperature MBE Growth Techniques

S. Gupta, J. Pamulapati, J. Chwalek, P.K. Bhattacharya, and G. Mourou

Department of Electrical Engineering and Computer Science,
University of Michigan, EECS Bldg., 1301 Beal Avenue,
Ann Arbor, MI48109, USA

Abstract - Molecular beam epitaxial growth of III-V semiconductors at very low temperatures results in ultrashort carrier lifetimes. We have characterized this by photoconductive switching and time-resolved reflectivity measurements.

Picosecond electrical transients are commonly generated by laser activated photoconductive switches. Doping with compensating impurities, ion-damaging the material[1] and growth of polycrystalline and amorphous materials[2] have been undertaken to decrease the carrier lifetime, thus shortening the pulse. Sub-picosecond electrical transients have been generated using ion-implanted silicon-on-sapphire (SOS) thin films[1] and CdTe grown by MOCVD techniques. Smith et al.[3] have recently used the technique of low temperature molecular beam epitaxial (MBE) growth of GaAs buffer layers to improve the isolation properties of field effect transistors (FET). The good crystal structure combined with the very high density of point defects[4] , indicates that sub-picosecond photoconductive response should be possible in this material, with good responsivity[5]. We have studied the photocarrier response in low temperature MBE grown : GaAs, InAlAs (lattice matched to InP), and AlGaAs/GaAs materials by measuring their picosecond time-resolved photoconductive and reflectance properties.

The GaAs epilayers were 1-2 μm thick and grown at a rate of 1μm/hr at substrate temperatures ranging from 200-400°C. Some of the grown layers were annealed in-situ under As overpressure, by raising the substrate temperature to about 600°C for about 10 minutes after the growth was completed. The 1 μm thick $In_{0.52}Al_{0.48}As$ layers were grown at a rate of 0.6 μm/hr at substrate temperatures ranging from 150-500°C. Some of the layers grown at 150°C were annealed in-situ by raising the temperature to 500°C for 10 minutes. Even at these low growth temperatures, the material has very good crystallinity as evidenced by TEM and X-ray data. EPR and AES measurements have indicated As-rich compositions for the LT-GaAs layers, giving rise to a very high density of point defects[4]. Hence ultrafast carrier recombination times are predicted with good photoconductive switch response.

The response of a photoconductive gap switch was measured by external electro-optic sampling using a balanced CPM dye laser[6]. The measured subpicosecond signals are shown in Fig.1 for the annealed 200°C grown GaAs, unannealed & annealed 150°C InAlAs samples. To the best of our knowledge, the response of the 200°C MBE grown GaAs (FWHM=600 fs) is the shortest photoconductively generated electrical pulse to date. This measurement is clearly limited by the resolution (~300fs) of the elctro-optic sampling system. From the measured photocurrent and system parameters an approximate value of the mobility of the carriers can be deduced according to the formula as given in [2]. The annealed 200°C

Springer Series in Chemical Physics, Vol. 53 **Ultrafast Phenomena VII**
Editors: C.B. Harris · E.P. Ippen · G.A. Mourou · A.H. Zewail

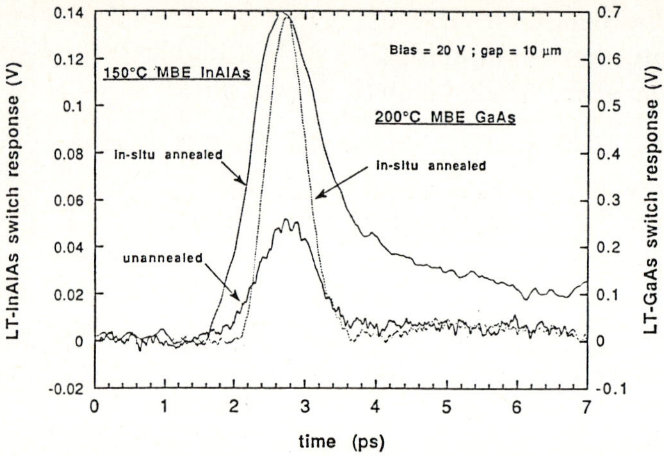

Fig. 1 Photoconductive switch response of low temperature MBE grown layers of GaAs and InAlAs, measured by external electro-optic sampling.

grown GaAs layer has a very good responsivity (μ~50-120 cm^2/V-s) for a sub-picosecond material. The 150°C unannealed $In_{0.52}Al_{0.48}As$ layer show a poorer response (μ~5 cm^2/V-s), although they have a fairly high dark resistance to make a measurement, unlike unannealed GaAs layers. A previous study indicates that clustering effects[7] are important in this material due to the different migration rates of the cations, In and Al. Hence the reduction in mobility could arise from the scattering at the boundaries between the clusters of slightly different phases. In-situ annealing improved the response by almost a factor of three, but caused the carrier lifetime to increase to about 1.2 ps.

The above CPM laser was also used to do an all optical pump-probe type of experiment to better understand the photo-generated carrier evolution. A pump beam of ~ 4-8 mW average power focussed to about 30 μm spot, was used to generate an electron-hole plasma of density ~ 3-5 x 10^{17} /cm^3. This e-h plasma modifies the refractive index[3], and thus causes a change in the reflection coefficient R(t), which is monitored by a delayed probe beam attenuated to ~10% of the pump beam power.

Fig.2a shows the time-resolved reflectivity signals measured at room temperature for unannealed $In_{0.52}Al_{0.48}As$ layers grown by MBE at temperatures ranging from 150-500°C. A reduction in the carrier trapping/recombination time is seen as the growth temperature is lowered. Note the drastic reduction in the lifetime for the material grown at 150°C as well as the reversal in the sign. A similar trend is observed for GaAs samples, as well as the reversal in the sign for the 200°C MBE grown GaAs. For the annealed GaAs material, the time-resolved reflectivity is of the usual sign, as shown in Fig.2b. The carrier trapping/recombination time, is seen to be faster than ~200 fs as was also suggested by the switching data of Fig.1.

Some preliminary time-resolved reflectivity results on MBE grown $Al_{0.3}Ga_{0.7}As$ layers show that the carrier capture times decrease to ~2.5ps from ~30ps, by growing the samples around 580°C instead of the usual temperatures of around 650°C.

In conclusion, we have demonstrated sub-picosecond photoconductivity in a number of III-V compound semiconductor systems by low temperature MBE growth techniques. The photocarrier response in these materials as a function of the growth and measurement temperatures, gives important clues about the mechanism

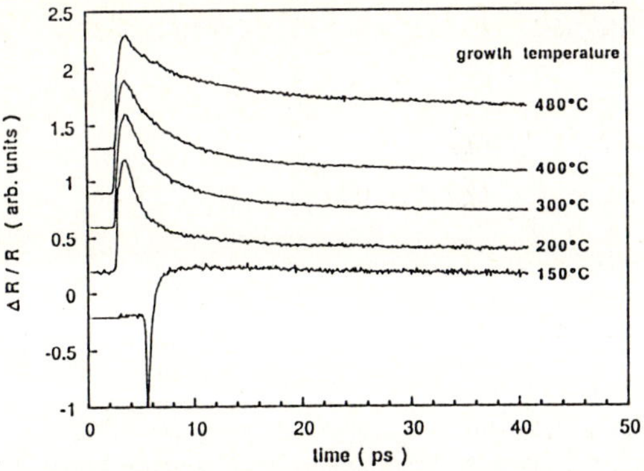

Fig. 2a Time-resolved reflectivity of InAlAs / InP layers, grown by MBE at lower growth temperatures.

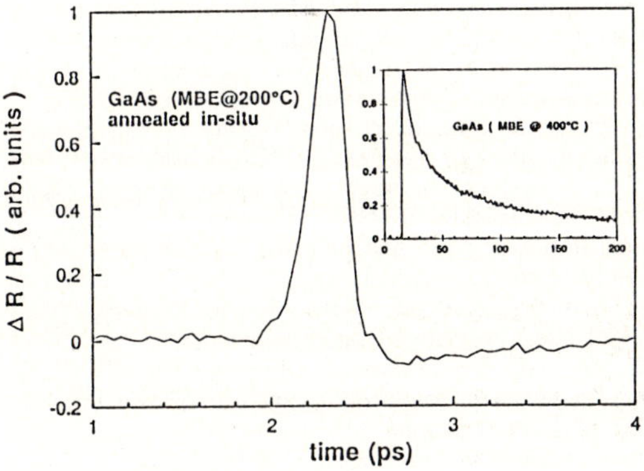

Fig. 2b Time-resolved reflectivity of GaAs layers, grown by MBE at low growth temperatures.

responsible for this ultrafast behavior. GaAs, InAlAs and AlGaAs are all used as buffer layers in various electronic/optoelectronic devices, and devices have been grown on these low temperature buffer layers. Therefore these photoconductive switches can be easily integrated for high speed device testing.

References

1. F.E. Doany, D. Grischkowsky, C.-C. Chi, Appl. Phys. Lett. **50** , 460 (1987)
2. D.H. Auston, P. Lavallard, N. Sol, D. Kaplan, Appl. Phys. Lett. **36**, 66 (1980)
3. F.W. Smith, A.R. Calawa, C.-L. Chen, M.J. Manfra, L.J. Mahoney, IEEE Electron Dev. Lett. **9** , 77 (1988)
4. M. Kaminska, E.R. Weber, Z. Liliental-Weber, R. Leon, Z.U. Rek, J. Vac. Sci. Technol. **B 7** , 710 (1989)
5. F.W. Smith, S. Gupta, H.Q. Le, M. Frankel, V. Diadiuk, M.A. Hollis, A.R. Calawa, D.R.Dykaar, G.A.Mourou, T.Y.Hsiang, Appl.Phys.Lett.**54**, 890(1989)
6. J.A. Valdmanis, Electron. Lett. **23** , 1308 (1987)
7. J.E. Oh, P.K. Bhattacharya, Y.C. Chen, O. Aina, M. Mattingly, Jour. of Electron. Mater. **19** , 435 (1990)

Nonequilibrium Carrier Dynamics in GaAs Semiconductors Induced by Energetic Ions

D. McMorrow, A.R. Knudson, and A.B. Campbell

Naval Research Laboratory, Washington, DC20375, USA

Energetic ions interact with semiconductor devices to produce ionization tracks that consist of dense plasmas of free carriers. These ion tracks generally penetrate through the active regions of the device into the substrate, as is shown schematically in Figure 1. The penetration depth depends on the material, the energy of the ion, and the characteristic rate of energy transfer from the ion to the material (dE/dx, commonly referred to as the stopping power). The charge created by such interactions can collect on various circuit nodes and produce transient errors in integrated circuit (IC) memories and latches that are commonly referred to as single event upsets (SEU's), or soft fails. Single event upsets pose a serious reliability problem for space-based digital circuitry because of the large flux of energetic ions in cosmic rays and the earth's radiation belts, and become more significant in later generation IC's which are faster and more densely packed than their predecessors. A better understanding of basic charge collection mechanisms and their relationship to SEU phenomena is thus of considerable interest. One of the important parameters not yet fully understood is the temporal dependence of the ion-induced current transients.

To improve our understanding of the physical processes associated with SEU phenomena we have initiated a study[1,2] of the ion-induced current transients produced in high-frequency GaAs devices that are representative of the current MESFET technology. We present here a picosecond time-resolved investigation of the current transients produced by several different high energy ions (2–20 MeV) in micrometer sized digital GaAs MESFET's. Because of the random arrival of the ions from the accelerator, the powerful optical sampling techniques common in laser studies of semiconductor devices and IC's are not applicable. Instead, the current transients are measured with a HYPRES high-bandwidth, superconducting sampling oscilloscope. The unique advantage of this instrument for the current experiments is its built-in, effectively dispersion-free, lossless, superconducting delay line that allows the sampling of random transients, while maintaining its full 70 GHz bandwidth. These measurements thus represent the state-of-the-art in the area of ion/semiconductor interactions.

Measurements were performed on depletion and enhancement mode GaAs test structures fabricated by Vitesse Semiconductor as standard electrical test structures for parametric analysis of their enhancement/depletion process. Figure 1 gives a schematic diagram of a 1 μm gate length depletion-mode device. The devices were fabricated by ion implantation into un-compensated, semi-insulating material with a p^- implant for threshold control, n^+ implants for source, drain, and channel, and a

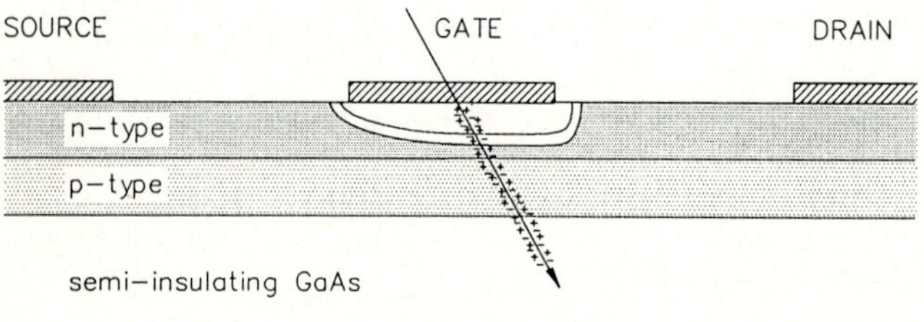

Figure 1. Schematic represention of ion track in depletion-mode GaAs MESFET

Springer Series in Chemical Physics, Vol. 53 **Ultrafast Phenomena VII**
Editors: C.B. Harris · E.P. Ippen · G.A. Mourou · A.H. Zewail

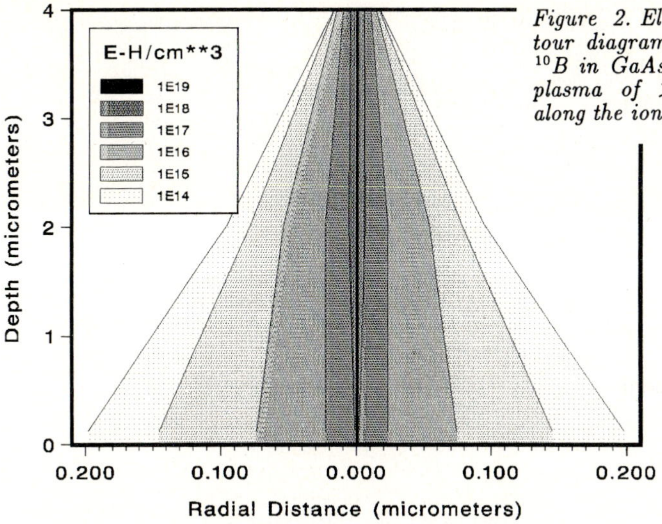

Figure 2. Electron-hole density contour diagram for incident 7.8 MeV ^{10}B in GaAs. Note the high-density plasma of $>10^{19}$ e-h pairs/ cm^3 along the ion's path.

Schottky barrier gate. The active depth of these devices is approximately 0.2 μm, with gate lengths ranging from 1 μm to 30 μm. The pinch off voltages for the depletion-mode devices are typically −1 V. The devices were wirebonded to 50 Ω microstrip lines mounted in high-frequency microwave packages, which were mounted in a vacuum chamber on the accelerator beam line. The signal was launched into semi-rigid co-axial cable, fed through the vacuum wall (~6 inches), and into the HYPRES input (via K-connectors).

The passage of an energetic ion through a semiconductor device produces a very localized and highly dense column of electron-hole pairs on a time scale of less that one picosecond. The resultant electron-hole plasma screens electric fields from its interior, resulting in dramatic distortions of the field lines so that regions of the device that were initially field-free can contain significant fields. A representative electron-hole density contour diagram[3] for a 7.8 MeV ^{10}B ion in GaAs is presented in Figure 2. This figure illustrates graphically the initial radial distribution of charge produced along the ion trajectory, with a high-density plasma of $>10^{19}$ *e-h* pairs/cm^3 being generated along the ion's path. This figure was calculated by a computer code[4] based on an analytic theory which evaluates the energy deposited in a cylindrical shell of unit length and some mean radius as a function of incident ion charge, speed and electron density of the medium.

Representative current transients measured at the drain of the 30 μm x 60 μm depletion-mode device are given in figure 3 for four different ions produced by the Pelletron accelerator at the NRL Microbeam facility.[5] This facility allows control over the location of the ion strike to within ~3 μm. The signal-to-noise ratio of this data is determined primarily by the limited acquisition time due to radiation damage effects. For each of these curves the gate was biased at −1.2 V, the drain at 3.0 V, and the source grounded, with the largest signals occuring for ion strikes in the high-field region between the gate and drain. The tracks produced by the different ions can be dramatically different, with each depositing a different amount of energy per unit length (He~540; B~2300; O~4300; Si~9500 KeV/μm), giving rise to significant differences in both penetration depth and radial extent (for most ions of interest the track diameter is on the order of 0.1 μm, *cf.*, Figure 2). It is thus somewhat surprising that the dynamics observed for the four very different ions are so similar, with each exhibiting a pulse rise-time of ~25 ps. This observation is in contrast to earlier measurements performed on large area silicon diodes which showed dramatic pulse shape variations as a function of the incident ion, with the transients produced by the heavier ions being significantly slower than those of the lighter ions.[1] Preliminary considerations suggest that this result is a consequence of the very thin (~0.2 μm)

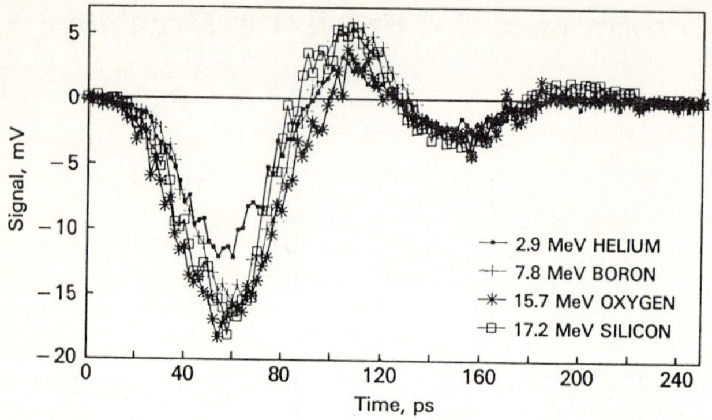

Figure 3. Ion-induced current transients measured at the drain of a 30 μm x 60 μm depletion mode GaAs MESFET as a function of the incident ion.

active region of these high-speed devices, and the buried p-layer, which isolates the active region from the semi-insulating substrate. It appears that charge deposited in the active region is the main component of charge collected at the various circuit nodes. The suprising lack of dependence of the signal on the ion type (and consequently, track structure) is perhaps due to this thin charge collection region.

A major difficulty of charge collection measurements with energetic ions is the occurrence of radiation damage. High-speed GaAs devices have been shown to be particularly sensitive to such effects. One attractive alternative to accelerator experiments involves the use of picosecond and femtosecond laser pulses. This technique possesses an inherently higher temporal resolution, is much less expensive, more convenient, and does not produce the cumulative radiation damage that is characteristic of ion bombardment. We have demonstrated[6] that the use of focussed picosecond laser pulses providing above band-gap excitation is a viable method for locating SEU sensitive areas in complex IC's. There are, however, significant differences in the mechanisms of charge deposition and collection for laser and ion beam irradiation. To better quantify both the similarities and differences between the two cases, and we have initiated a parallel program to investigate the dynamics of the charge collection process initiated by ultrashort laser pulses as a function of the laser pulse width and excitation frequency. Preliminary measurements have been performed on the same GaAs devices described above using ~10 ps pulses at various wavelengths above and approaching the GaAs band gap. The dynamical waveforms produced are qualitatively and quantatively very similar to the ion data of figure 3, illustrating the feasibility of simulating ion-induced charge collection dynamics with ultrashort optical pulses. Further experiments are in progress.

References

1. R.Wagner, N. Bordes, J. Bradley, C. Maggiore, A. Knudson, A. Campbell, IEEE Trans. Nuc. Sci. NS-35, 1578 (1988).
2. A. Campbell, A. Knudson, D. McMorrow, W. Anderson, J. Roussos, S. Espy, S. Buchner, K. Kang, D. Kerns, and S. Kerns, IEEE Trans. Nuc. Sci. NS-36, 2292 (1989).
3. W.J. Stapor and P.T. McDonald, J. Appl. Phys. **64**, 4430 (1988)
4. Figure 2 provided courtesy of W.J. Stapor of NRL.
5. A. Knudson and A.B. Campbell, IEEE Trans. Nuc. Sci. NS-28, 4017 (1981).
6. S. Buchner, D. Wilson, K Kang, D. Gill, J. Mazer, W. Raburn, A. Campbell, and A. Knudson, IEEE Trans. Nuc. Sci. NS-34, 1228 (1987).

Ultrafast Relaxation Processes of Hot Carriers in Graphite

K. Seibert[1], *G.C. Cho*[1], *W. Kütt, H. Kurz, D.H. Reitze*[2], *M.C. Downer*[2], *and A.M. Malvezzi*[3]

[1]Institute of Semiconductor Electronics, RWTH Aachen, Sommerfeldstrasse, D-5100 Aachen, Fed. Rep. of Germany
[2]Physics Department, University of Texas at Austin, Austin, TX78712, USA
[3]Department of Electronics, University of Pavia, I-27100 Pavia, Italy

Because of the simple quasi-two-dimensional structure [1] and small overlap of the valence- and conduction band states in highly oriented pyrolytic graphite (HOPG) the relaxation dynamics of nonequilibrium electronic carriers should differ distinctly from that already observed in semiconductors. Due to its short absorption depth (30 nm) hot carrier populations with a density above $10^{21}cm^{-3}$ can be easily generated with amplified fs-pulses at $\lambda = 625$ nm. We report for the first time femtosecond details of the relaxation process of extremely dense carrier gases in HOPG at room temperature over a wide range of optical excitation densities.

The source of the femtosecond pulses with a duration of 50 fs was a colliding pulse mode-locked dye laser and a multipass dye amplifier pumped by a 7 kHz copper vapor laser covering the low and high excitation regime. Experiments were done in a conventional pump/probe geometry. A part of the amplified output was used to permit a wide spectral probing range by generating white light continuum probe pulses in an ethylenglykol cell.

Fig.1 presents femtosecond transmission and reflectivity data (λ_{probe}=625nm) recorded simultaneously from a 35nm thick HOPG film mounted on a transparent substrate. High excitation (F= 7.5mJ/cm^2) and low excitation results (F= 5μJ/cm^2) are compared in Fig. 1a-b. The corresponding absorption and refractive index changes (Δk, Δn) for the two fluence levels are evaluated from thin film optics (Fig. 1c-d, bottom panels). In the case of high excitation the initial Δk-change indicates a maximum bleaching towards the end of the excitation pulse with a recovery to the unperturbated value within 2ps. A careful fit of both curves to an exponential recovery convolved with Gaussian pump and probe temporal profiles yields a decay constant $\tau = 130 \pm 15$fs for $\Delta T/T_o$ and $\Delta R/R_o$. At a three orders of magnitude lower excitation level the $\Delta T/T_o$ -transients are significantly affected by changes in the refractive index as shown in Fig.1b and d. The absorption saturation decays much faster yielding a time constant of $\tau = 30 \pm 10$fs close to the pulse

Springer Series in Chemical Physics, Vol. 53 **Ultrafast Phenomena VII**
Editors: C.B. Harris · E.P. Ippen · G.A. Mourou · A.H. Zewail

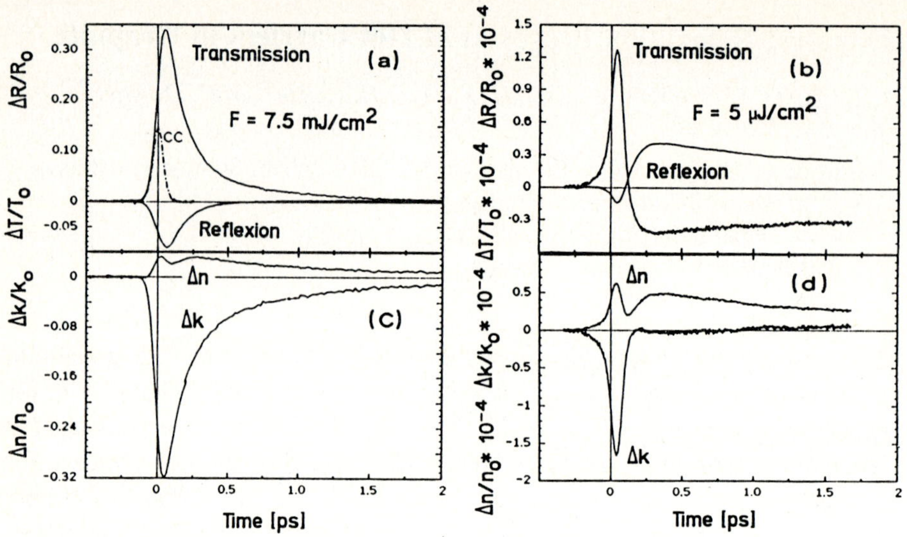

Fig. 1: Femtosecond reflectivity and transmission (λ_{probe}=625nm) changes of a 35nm thick HOPG film recorded simultaneously for the high (case a) and low excitation (case b) regime. The corresponding changes in absorption ($\Delta k/k_O$) and refractive index ($\Delta n/n_O$) for the two fluences are shown in the bottom panels (case c,d).

width parameter indicating the ultrarapid scattering of carriers out from the optically coupled states.

Since at high fluence level (F = 13 mJ/cm) the reflectivity changes ($\Delta R/R_O$) are mainly due to Δk-variations, the subpicosecond bandfilling due to rapid internal thermalization and subsequent carrier cooling can be monitored by reflectivity measurements at different wavelengths. Although the pump pulse couples valence and conduction band states only within its spectral bandwidth, carrier-carrier thermalization immediately produces a broader distribution responsible for the absorption saturation in states outside the excitation region (see Fig. 2).

Beyond the first picosecond the positive plasma induced reflectivity change relaxes with a time constant of 5ps to the unperturbated value. After 20 ps the reflectivity change turns slightly negative due to lattice heating in accordance with the experimental data of previous ps-reflectance measurements[3].

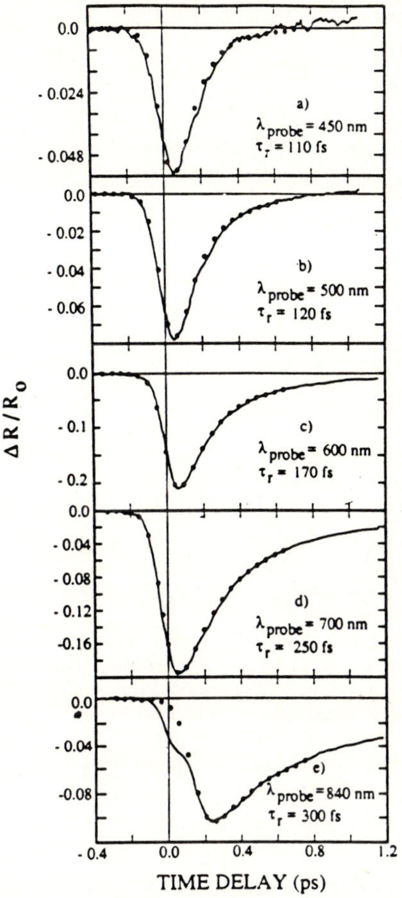

Fig.2: Wavelength dependence of reflectivity response of HOPG using amplified femtosecond pulses ($13 mJ/cm^2$) at λ_{pump} =625nm and white light continuum probing. Between 450 nm and 700 nm the maximum bleaching is reached within 100 fs indicating an ultrafast spreading of the initial carrier distribution. The recovery time τ_r derived from fitting a convolved exponential response to the data (dots) increases monotonically with λ_{probe}. For $\lambda_{probe} < \lambda_{pump}$ the distribution relaxes towards the band minima via electron-phonon interaction indicated by a rapid recovery of absorption saturation. The carriers relax more rapidly out of states higher in the band. For $\lambda_{probe} > \lambda_{pump}$ the absorption saturation shows a delayed recovery because initially high energy carriers relax into low energy states coupled by these wavelengths.

This work is supported entirely by the "Krupp Foundation".

References:

[1] B.T. Kelly, Physics of graphite (London, Applied Science Publisher, Ltd. 1981)

[2] K. Seibert, G.C. Cho, W. Kütt, H. Kurz, D.H. Reitze, M.C. Downer, and A.M. Malvezzi, to be published in Phys. Rev. B15

[3] J. Heremans, C.M. Olk, G.L. Eesley, J. Steinbeck and G. Dresselhaus, Phys. Rev. Letters 60, 452, (1988), A.M. Malvezzi, G. Reverberi, and N. Bloembergen, Mat. res. Doc. Symp. Proc. 100, 453, (1988)

Femtosecond Dynamics of Photo-Generated Carriers in Amorphous Hydrogenated Germanium

M. Wraback[1], *J. Tauc*[1], *D. Pang*[2], *W. Paul*[2], *and Z. Vardeny*[3]

[1]Department of Physics and Division of Engineering, Brown University, Providence, RI 02912, USA
[2]Division of Applied Sciences, Harvard University, Cambridge, MA 02138, USA
[3]Department of Physics, University of Utah, Salt Lake City, UT 84112, USA

Abstract. Femtosecond pump and probe technique with 2eV photons was applied to study ultrafast dynamics of photocarriers in high quality a-Ge:H varying the carrier densities over three orders (10^{18}-10^{21}cm^{3}). Photoinduced changes in the complex dielectric constant $\epsilon_1 + i\epsilon_2$ were calculated from the measured changes in reflectivity ΔR and transmission ΔT. The results are interpreted using a model based on absorption saturation of interband transitions and polarization produced by free carriers. The intensity dependence of the decays in $\Delta\epsilon_1$ and $\Delta\epsilon_2$ suggests that they are primarily due to a non-monomolecular recombination process.

1. Introduction

Amorphous hydrogenated germanium (a-Ge:H) is an attractive physical system for application of the femtosecond pump-probe technique with 2eV photons because its small optical gap (1.24eV) allows one to study the dynamics of photoexcited carriers with large excess energies in the extended states. While amorphous hydrogenated silicon (a-Si:H) has been extensively studied by this technique [1-3], it possesses a substantially larger optical gap (1.8eV), and consequently such experiments primarily probe photocarriers with little excess energy and their time evolution close to the mobility edge. Recent advances [4] in the preparation of high quality a-Ge:H have resulted in the production of a material exhibiting significantly improved optical, electrical, and structural properties. In this paper, we present a femtosecond study of this new material which, through comparison with similar studies on a-Si:H, offers new insights into the time evolution of photocarriers in extended states.

2. Experimental

The sample was a 1250Å film of a-Ge:H deposited on Corning 7059 glass and produced on the 150°C powered electrode of a r.f. glow discharge reactor [4]. Pulses of 120 fsec duration were obtained from a colliding-pulse mode-locked ring dye laser [5] for the low intensity measurements and amplified to a pulse energy of 3μJ in a Cu vapor laser-pumped amplifier [6] for the high intensity experiments. The conventional pump-probe geometry was used. For the high excitation measurements, the pump pulse was focused to a diameter of 110μm on the sample while the probe pulse was mechanically delayed and focused inside the pump area to ensure the probing of a uniform carrier density. For the low intensity measurements both pump and probe pulses were focused to 30μm. Photoinduced changes in reflectivity ΔR and transmission ΔT were recorded for initial photocarrier densities between 10^{18} and 10^{21}cm^{-3} and the time evolution of the complex dielectric constant $\epsilon_1 + i\epsilon_2$ in the first 45 psec was calculated from a first order expansion of the changes in the complex refractive index $n + i\kappa$ in terms of ΔR and ΔT [7].

3. Results and Discussion

Figs. 1 and 2 show the calculated $\Delta\epsilon_1$ and $\Delta\epsilon_2$ respectively in the first 6 psec as a function of excitation density. Both $\Delta\epsilon_1$ and $\Delta\epsilon_2$ exhibit a sharp negative onset at $t = 0^+$. At the highest carrier density (Fig. 1a) $\Delta\epsilon_1$ decays to zero and becomes positive within the first 2 psec. As the carrier density is decreased (Figs. 1b and 1c) the decays become slower and approach zero at longer times. Although the decays in $\Delta\epsilon_2$ (Fig. 2) also become slower with decreasing intensity,

Springer Series in Chemical Physics, Vol. 53 **Ultrafast Phenomena VII**
Editors: C.B. Harris · E.P. Ippen · G.A. Mourou · A.H. Zewail

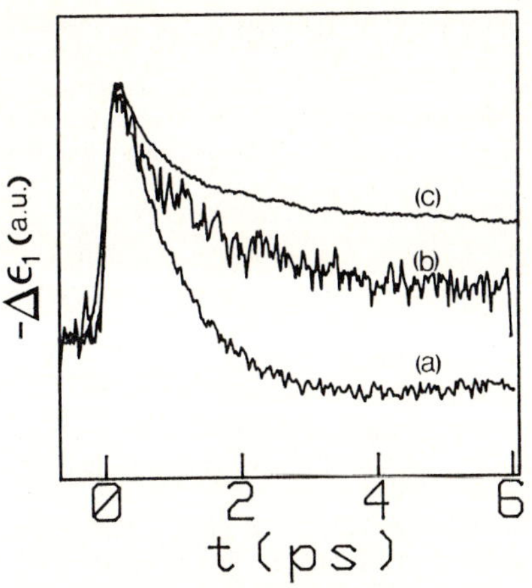

Fig. 1. Comparison of decays in $\Delta\epsilon_1$ at three carrier densities (a) 7×10^{20}; (b) 7×10^{19}; (c) $3 \times 10^{18} cm^{-3}$.

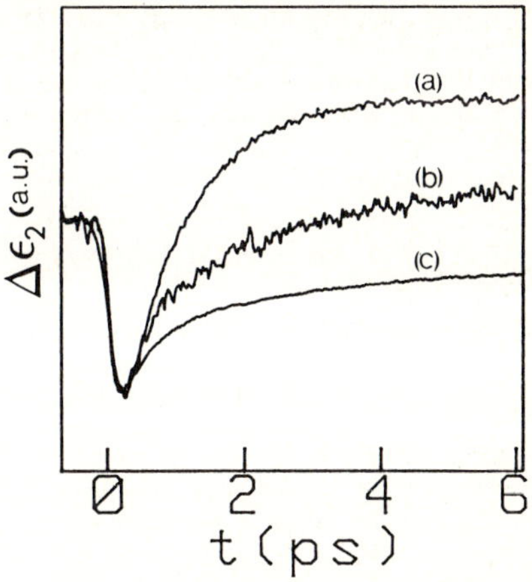

Fig. 2. Comparison of decays in $\Delta\epsilon_2$ at three carrier densities (a) 7×10^{20}; (b) 7×10^{19}; (c) $3 \times 10^{18} cm^{-3}$.

$\Delta\epsilon_2$ becomes positive for all intensities at which it was measured, but at much longer times (~20 psec) at the lowest density than indicated by Fig. 2c.

In order to attain insight into these data we compare them to typical results in a-Si-H. Both materials exhibit sharp negative onsets in $\Delta\epsilon_1$ at $t = 0^+$; in a-Si:H, this has been attributed to the polarization produced by photo-generated free carriers [2,3]. At high excitation densities the polarizability decays because of recombination and heat generation as described in a quantitative model by Esser et al. [2], while at low intensities it decays by trapping [3]. Our data indicate that a similar model for $\Delta\epsilon_1$ decay may apply to a-Ge:H, but we have not yet compared our data to a quantitative model. Since the thermalization process in the band dissipates a much larger fraction of energy in a-Ge:H than in a-Si:H, appropriate changes in the model have to be made which account for the dependence of the polarizabilities on energy, especially at lower carrier densities, where the effect of recombination on the initial decays may be weaker than that of thermalization.

The most striking difference between these two materials at short times is that a-Ge:H exhibits an initial onset of photoinduced bleaching ($\Delta\epsilon_2 < 0$), while a-Si:H is characterized by an onset of photoinduced absorption. At low excitation levels, this induced absorption in a-Si:H is attributed to photoinduced interband electroabsorption [8], which is strong near the band edge. Therefore, we do not expect the electroabsorption effect to be significant in a-Ge:H at 2eV. We propose that the observed bleaching is due to the filling of band states similarly as in crystalline semiconductors. This bleaching is strong if the k-vector is conserved; nevertheless, the bleaching due to band filling should be observable even if k-vector is not conserved, although it must be much weaker. A simple calculation of the relative change in the fundamental absorption gives an estimate of the order of the bleaching effect:

$$- \Delta \alpha/\alpha = N\left\{\left[\int_0^{h\nu-E_g} D_c(E)dE\right]^{-1} + \left[\int_0^{h\nu-E_g} D_v(E)dE\right]^{-1}\right\} \tag{1}$$

where $D_c(E)$ and $D_v(E)$ are the densities of states of the conduction and valence bands, respectively, and N is the carrier density. The integration gives $(1/D_c^{av} + 1/D_v^{av})(h\nu-E_g)^{-1} \sim 10^{-22} cm^3$, and for $N = 10^{20} cm^{-3}$ we obtain $-\Delta\alpha/\alpha \simeq 10^{-2}$, in agreement with our experimental results. In this model we assume that while the free-carrier contribution dominates the changes in the polarizability in a-Ge:H at 2eV, the band filling effect prevails over the free-carrier absorption.

The magnitude of the bleaching is significantly smaller than what would be expected for a crystal at the same excitation density. The difference is due to the broader energy distribution of optically coupled states in an amorphous semiconductor, which also tends to smear out the details of the thermalization process. Bleaching is expected to last longer in an amorphous semiconductor than in a crystal because the changes in the occupation of different states by scattering and thermalization is of less consequence. The decay of bleaching is primarily produced by recombination and heating, as for $\Delta\epsilon_1$. The decay time t_d of the bleaching into photoinduced absorption depends on excitation density as $t_d \sim N^{-0.6}$. This indicates that the recombination is not monomolecular.

4. Conclusion

In conclusion, we have determined the photoinduced changes of $\Delta\epsilon_1$ and $\Delta\epsilon_2$ in high quality a-Ge:H and interpreted them using a model based on interband transitions (band filling) and polarization of free carriers. We report the first observation of absorption saturation of interband transitions between extended states in an amorphous semiconductor. Measurement of the intensity dependence of the femtosecond decays in the range of 10^{18} - 10^{21} photo-excited carriers/cm^3 suggests that they are primarily determined by recombination and heat generation.

The authors thank T. R. Kirst and J. A. Moon for assistance and R. I. Devlen for helpful discussions. We are also grateful to J. K. Lee and E. A. Schiff, who provided a sample for some of the preliminary measurements. This work was principally supported by NSF grant DMR8706289.

References

1. Z. Vardeny and J. Tauc, in Disordered Semiconductors, eds. M. A. Kastner et al., (Plenum Press, N.Y., 1987) p. 339.
2. A. Esser, K. Seibert, H. Kurz, G. N. Parsons, C. Weng, B. N. Davidson, G. Lucovsky and R. J. Nemanich, J. Non-Crystalline Solids 114 573(1989); A. Mourchid, R. Vanderhaghen, D. Hulin, C. Tanguy and P. M. Fauchet, ibid. 114 582(1989).
3. J. Kuhl, E. O. Gobel, Th. Pfeiffer and A. Jonietz, Appl. Phys. A 34 105(1984).
4. W. A. Turner, S. J. Jones, D. Pang, B. F. Bateman, J. H. Chen, Y.-N. Li, F. C. Marques, A. E. Wetsel, P. Wickboldt, W. Paul, J. Bodart, R. E. Norberg, I. El Zawawi and M. L. Theye, to be published in J. Appl. Phys., June 1990.
5. R. L. Fork, B. I. Greene and C. V. Shank, Appl. Phys. Lett. 38 671(1981).
6. W. H. Knox, M. C. Downer, R. L. Fork and C. V. Shank, Opt. Lett. 9 552(1984).
7. H. T. Grahn, C. Thomsen and J. Tauc, Opt. Commun. 58 226(1986).
8. R. I. Devlen, Ph.D. Thesis, Brown University, 1990.

Ultrafast Pump–Probe X-Valley Absorption Spectroscopy in GaAs

W.B. Wang, N. Ockman, M.A. Cavicchia, M. Yan, and R.R. Alfano

Institute for Ultrafast Spectroscopy and Lasers, Physics Department, The City College of New York, New York, NY 10031, USA

The X valley plays an important role in hot carrier transport in GaAs [1,2]. In this paper, we present pump–probe IR absorption measurements to determine the time evolution of electrons in the X_6 valley, the upper limit of the $X_6 \rightarrow \Gamma_6$ scattering time $t_{X\Gamma}$, the energy separation of the X_6 and X_7 bands at their minima Δ_{7-6}, and the density of states effective mass of the X_z valley.

IR picosecond probe pulses at different wavelengths were used to monitor several processes, including the change of electron population in the X_6 valley subsequent to excitation by a 527 nm pump pulse [3].

The measured changes in optical density for GaAs as a function of delay time are displayed in Fig. 1. The curve for 3.9 μm (Fig. 1.b) only has a flat decay, which is attributed to free carrier absorption (FCA) [4]. The curve for 3.1 μm (Fig. 1a) shows a multicomponent decay; the flat decay portion is attributed to the sum of FCA and intervalence band absorption (IVA), while the fast decay portion is attributed to the $X_6 \rightarrow X_7$ interconduction band absorption [4]. A curve similar to Fig. 1b but with the flat decay value of $(\Delta\mathrm{OD})_{\mathrm{FC+IV}}$ for 3.1 μm is subtracted from the curve of $(\Delta\mathrm{OD})_{\mathrm{tot}}$ (Fig. 1a) to give the change in optical density attributed only to the $X_6 \rightarrow X_7$ absorption at 3.1 μm.

The change in $(\Delta\mathrm{OD})_{X_6 \rightarrow X_7}$ with delay times at $\lambda_{\mathrm{probe}} = 3.1\ \mu$m was used to determine the time evolution of the population density of electrons in the X_6 valley, N_X, using the equation

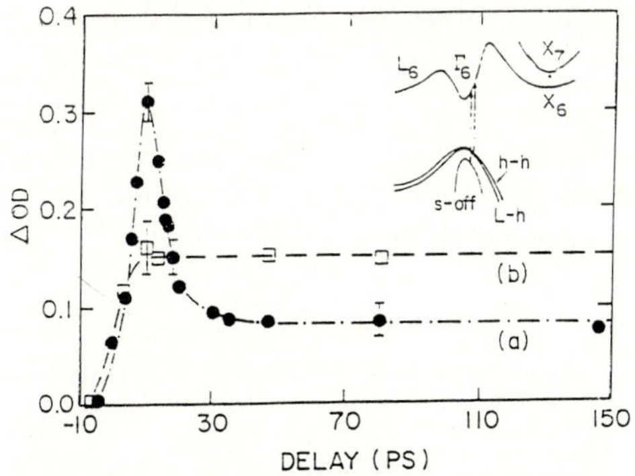

Fig. 1. Time evolution of the total change in induced optical density at **(a)** 3.1 μm (•), and **(b)** 3.9 μm (□) for GaAs at room temperature

Springer Series in Chemical Physics, Vol. 53 **Ultrafast Phenomena VII**
Editors: C.B. Harris · E.P. Ippen · G.A. Mourou · A.H. Zewail

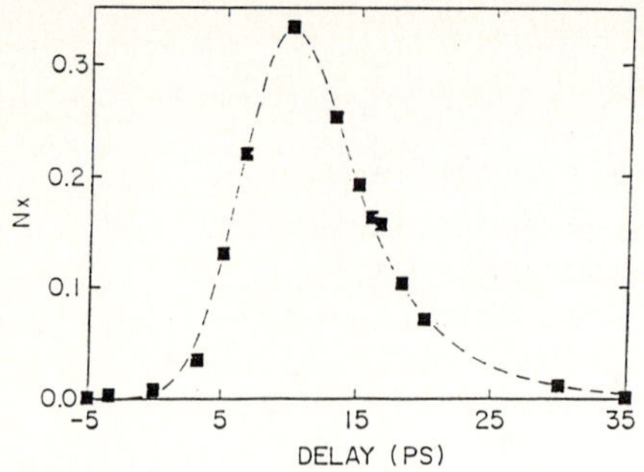

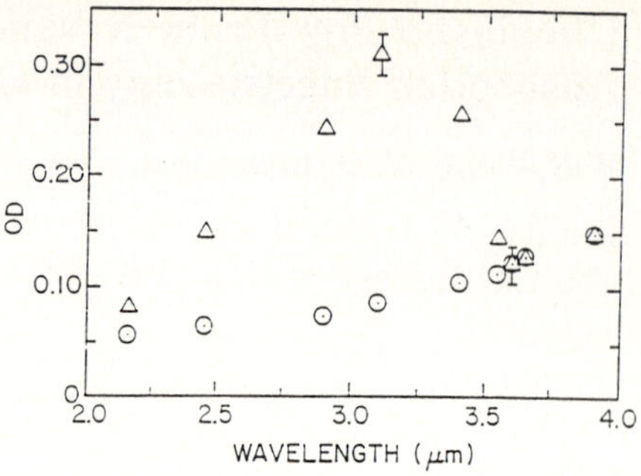

Fig. 2. The time evolution of N_X. The squares represent the experimental data. The dashed line is the profile

Fig. 3. Absorption spectra for (△) total induced absorption at the peak position ($t_d \sim 10$ ps), and (○) sum of FCA and IVA at the flat decay region

$$T(t_d) = \int_{-\infty}^{\infty} dt\, I(t - t_d)\, e^{-N_X(t)\sigma_X d} \Big/ \int_{-\infty}^{\infty} dt\, I(t)\, . \tag{1}$$

The curve of N_X as a function of t_d is shown in Fig. 2.

An upper limit of $t_{X\Gamma}$ can be extracted from the time evolution of N_X by making the following approximations: (1) neglect $\Gamma_6 \rightarrow L_6$ scatter [2]; (2) assume one half of the electrons from $X_6 \rightarrow \Gamma_6$ return to the X_6 valley since $t_{\Gamma X} \sim t_{\Gamma\Gamma}$ [2]. With these approximations, it is estimated that there will be five passes of the back-and-forth scattering between the X_6 and Γ_6 valleys near the bottom of the X_6 valley by both emission and absorption of phonons. Consequently, the $X_6 \rightarrow \Gamma_6$ backscattering time is determined to be < 2.6 ps.

To determine the band structure properties of the X_7 valley, measurements were extended to another 7 probe wavelengths from 3.65 μm to 2.16 μm. The decay behavior at different probe wavelengths is similar to that of 3.9 μm for $\lambda_{\text{probe}} > 3.6$ μm, and to that of 3.1 μm for $\lambda_{\text{probe}} < 3.6$ μm. The values of $(\Delta\text{OD})_{\text{tot}}$, for all of the probe wavelengths at $t_d = 10$ ps, and $(\Delta\text{OD})_{\text{FC+IV}}$, obtained in a similar manner to their determinations for 3.1 μm, are shown in Fig. 3. Subtracting $(\Delta\text{OD})_{\text{FC+IV}}$ from $(\Delta\text{OD})_{\text{tot}}$ gives the $X_6 \rightarrow X_7$ absorption, $(\Delta\text{OD})_{X_6 \rightarrow X_7}$, shown in Fig. 4. The long wavelength threshold of 3.6 μm in Fig. 4 indicates $\Delta_{7-6} = 0.345$ eV.

The $X_6 \rightarrow X_7$ absorption spectrum shown in Fig. 4 can be used to estimate the density of states effective mass for the X_7 band. The change of the optical density for the $X_6 \rightarrow X_7$ transition was derived to be

$$(\Delta\text{OD})_{X_6 \rightarrow X_7} = C\left(\hbar\omega - \Delta_{7-6}\right)^{1/2} \times \frac{(2m_r)^{3/2}}{\exp[(m_r/m_6)(\hbar\omega - \Delta_{7-6}) - (E_F)_6]/k_B T + 1}\, , \tag{2}$$

where $C = l|H|^2/(2.3 N_0 \pi \hbar^4)$ is a constant, $\hbar\omega$ is the photon energy, $m_r =$

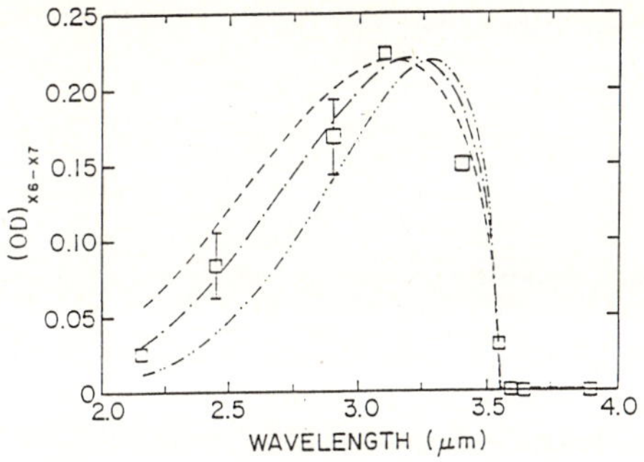

Fig. 4. The fitting result. The squares are the experimental data. The dash–dash, dot–dash, and dot–dot–dash curves are the computer fits corresponding to density of states effective masses of $0.44m_0$, $0.48m_0$, and $0.54m_0$ for the X_7 band, respectively

$m_6m_7/(m_6 - m_7)$ is the reduced effective mass, and $(E_F)_6$ is the Fermi energy for the X_6 electrons.

To fit our experimental data, we substituted into (2) values of Δ_{7-6}, $(E_F)_6$, and other parameters [5] for GaAs. The density of states effective mass for the X_7 band was treated as a variable parameter to fit the calculated changes of the optical density to the experimental data. The best fit, shown in Fig. 4, gives the value of $0.48m_0$ for the density of states effective mass in the X_7 band.

Our work provides three parameters, the upper limit of $t_{X\Gamma}$, the value of Δ_{7-6}, and the density of states effective mass for the X_7 band, which are important both in semiconductor physics and in hot carrier transport studies.

This research is supported by AFOSR and NSF.

References

1. K. Berthold, A.F.J. Levi, J. Walker, R.J. Malik: Appl. Phys. Lett. **54**, 813 (1989)
2. R.G. Ulbrich, J.A. Kash, J.C. Tsang: Phys. Rev. Lett. **62**, 949 (1989)
3. N. Ockman, R. Dorsinville, W.B. Wang, R.R. Alfano: IEEE J. QE-**23**, 2008 (1987)
4. W.B. Wang, N. Ockman, M. Yan, R.R. Alfano: Solid-State Electron. **32**, 1337 (1989)
5. J.S. Blakemore: J. Appl. Phys. **53**, R 123 (1982)

Subpicosecond Nonlinearities of Semiconductor Near Infrared Bragg Reflectors

G. Le Saux [1], *J.P. Pocholle* [2], *P. Georges* [3], *J.P. Schnell* [2], *F. Salin* [1], *M. Papuchon* [2], *and A. Brun* [1]

[1] Institut d'Optique Théorique et Appliquée, CNRS UA 14, B.P. 14, F-91403 Orsay Cedex, France
[2] Thomson-CSF, Laboratoire Central de Recherches, Domaine de Corbeville, B.P. 10, F-91401 Orsay, France
[3] Ecole Polytechnique Féminine, 3 bis rue Lakanal, F-92230 Sceaux, France

Band filling effects are used to modify the absorption of layers in a GaAlAs/GaAs Bragg reflector in the spectral region between the band gap of the two materials. Signifiant reflectivity modifications are observed in a subpicosecond regime.

Bragg reflector devices have been built in semiconductor materials. Many applications covering the realization of surface emitting laser diodes and mirrors activated in an electrical field have been demonstrated [1] . We describe in this paper experimental results showing the optical non-linearities induced by light carrier interactions in $Ga_{1-x}Al_xAs$ / $Ga_{1-y}Al_yAs$ Bragg reflectors, in the subpicosecond regime.

The samples are fabricated using the molecular beam epitaxy technique . This technique allows to control very well the layer thicknesses and the aluminium concentration.The linear reflection spectrum of the sample is calculated using the matrix transfer technique.In our calculations we take into account the refraction index dispersion and the variation of the absorption coefficient with wavelength for both materials.As shown in figure 1 , for two particular examples , the measured spectra are well fitted by the theoretical predictions.

In a first experiment we have studied the behavior of the GaAs/$Ga_{.54}Al_{.46}As$ structure.In this structure the layer thicknesses are respectively 49.8 nm and 51.5 nm. These values are measured by double X-ray diffraction techniques. The reflectivity maximum of the mirror is 78 % at 675 nm. In this spectral domain GaAs layers are absorbant, so the peak reflectivity is not optimized. By focusing an intense femtosecond pulse on the mirror, we saturate the absorption of the GaAs layers.The pump beam spectrum covers a wide spectral range (from 640nm to 800nm) to insure the

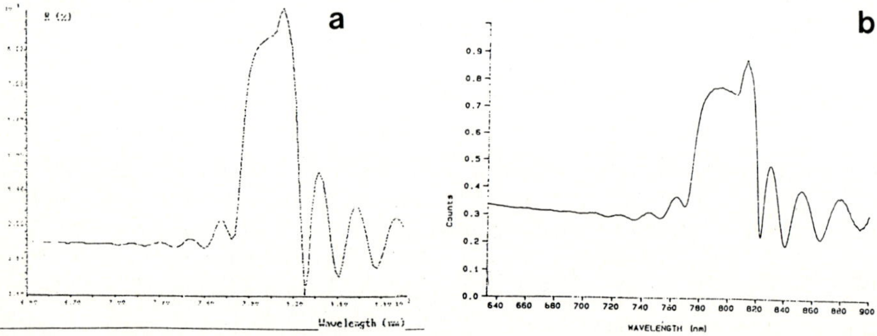

Figure 1 : Calculated (a) and measured (b) spectra for a sample made of 30 periods of alternating $Ga_{.92}Al_{.08}As$ and $Ga_{.6}Al_{.4}As$ layers.

Springer Series in Chemical Physics, Vol. 53 **Ultrafast Phenomena VII**
Editors: C.B. Harris · E.P. Ippen · G.A. Mourou · A.H. Zewail

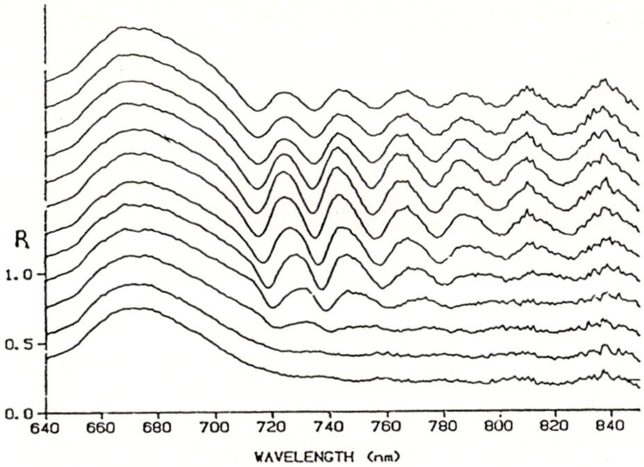

Figure 2 : Evolution of the reflectivity spectrum of the sample as a function of time .Two successive curves are separated by 166 fs.

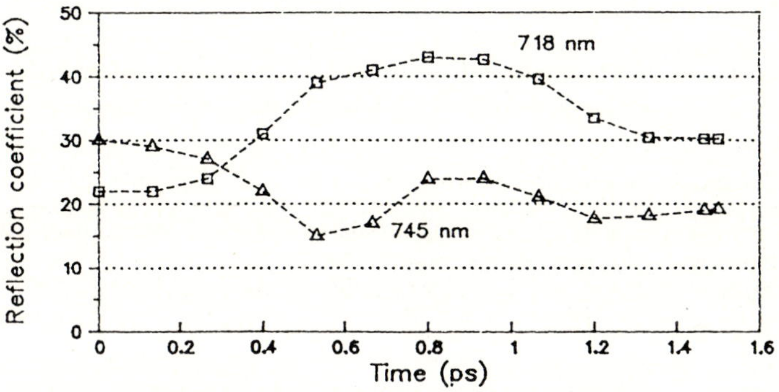

Figure 3 : Evolution of the reflectivity as a function of time for two different wavelengths:
△ - (a) 745 nm
□ - (b) 718 nm

saturation of the GaAs absorption even near the maximum reflectivity peak.The pump beam energy is about 10 microjoules. Then we obtain coherent interactions between the light reflected by the front surface of the sample and the light reflected by the deep layers. These interferences lead to a displacement of the secondary maxima towards the main peak; and to a better contrast of the interference pattern. In the spectral region between 720 nm and 800 nm the contrast of the fringes varies from about 6% in the linear case to 50% when the GaAs layers are saturated.. The kinetics of the non-linearity is measured on a femtosecond time scale and shows a great dependence of the response time with the wavelength.

In a second experiment, we have studied a structure made of alternating $Ga_{.92}$ $Al_{.08}$ As and $Ga_{.6}$ $Al_{.4}As$.The layer thicknesses are respectively 55 nm and 60 nm. In this case the band gap of $Ga_{.92}$ $Al_{.08}As$ is situated around 810 nm. We have realized a near resonant pumping of the sample by using a near infrared femtosecond laser emitting 100 fs pulses around 800 nm with a 100 MHz repetition rate. This laser is described in ref [2]. In this case the non-linearity is enhanced and signifiant modifications of the reflection coefficient are observed with low pump power. For example a 8 mW beam focused with a x20 microscope objective on the sample (giving a fluence of about 10 $\mu J/cm^2$) produces a relative increase $\Delta R/R$ of + 10 %. The temporal behavior of the modification, measured with a classical pump and probe experiment, is shown in figure 4.

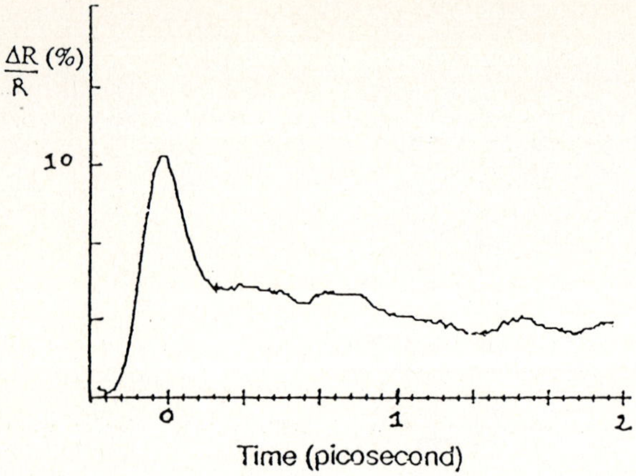

Figure 4 : Evolution of the reflectivity around 800 nm for the $Ga_{.92}Al_{.08}As$ / $Ga_{.6}Al_{.4}As$ structure as a function of time .

References :

[1] G.W. Yoffe,D.G. Schlom, J.S. Harris Jr
Appl. Phys. Lett. 51 ,1395 (1987).

[2] P.Georges,F.Salin, G. Le Saux, G. Roger and A.Brun
Optics Letters (in press).

Femtosecond Thermomodulation of Single-Crystalline and Polycrystalline Gold Films

H.E. Elsayed-Ali[1], *T. Juhasz*[2], *G.O. Smith*[2], *and W.E. Bron*[2]

[1]Laboratory for Laser Energetics, University of Rochester, Rochester, NY 14623, USA
[2]Department of Physics, University of California, Irvine, CA 92717, USA

If a metal is excited by laser pulses for which the pulse duration is less than or comparable to the hot electron energy loss lifetime (τ_e), a transient inequality between the effective electron and lattice temperatures (T_e and T_ℓ, respectively) occurs [1]. Previous femtosecond thermomodulation of metals have temporally resolved the decay of the hot electrons [2]. Although results for τ_e showed agreement with calculations [3], there remains unresolved the role of lattice imperfections on hot electron energy loss and transport.

We report femtosecond time-resolved thermoreflectivity ($\Delta R/R$) and thermotransmissivity ($\Delta T/T$) for thin (100–800 Å) polycrystalline Au films (PCF), with an average grain size of a few hundred angstroms, and single-crystalline (SCF) Au films. Pump-probe experiments are performed using a femtosecond dye-laser (~150 fs FWHM at $\lambda = 615$ nm). Time-resolved $\Delta T/T$ of a 200-Å single-crystalline Au film is shown in Fig. 1. The initial sharp rise is due to electron heating by the laser pulse. Hot electrons thermalize mostly through electron-electron (e-e) scattering and transfer their energy to the lattice through electron-phonon (e-ph) scattering. These are interrupted by scattering of the electrons at imperfections (e-i).

If we assume that e-e scattering and ph-ph scattering maintain the electron and the phonon subsystems in pseudoequilibrium at separate temperatures T_e and T_ℓ, then the usual expressions for the temporal evolution of T_e and T_ℓ are[1]

$$C_e(T_e)\, \partial T_e/\partial t = \nabla \cdot (\kappa \nabla T_e) - G(T_e - T_\ell) + P_o(r,t), \tag{1}$$

$$C_\ell\, \partial T_\ell/\partial t = G(T_e - T_\ell)\,. \tag{2}$$

The electronic heat capacity, $C_e(T_e)$, is assumed proportional to T_e, C_ℓ is the lattice heat capacity, κ is the electronic thermal conductivity, G is the e-ph coupling parameter, and $P_o(r,t)$ represents laser energy deposition in the film. In Eqs. (1) and

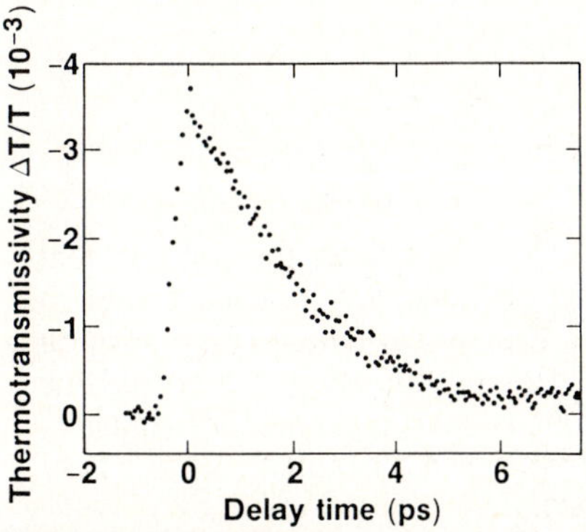

Fig. 1 Temporal evolution of the transient thermotransmissivity ($\Delta T/T$) of a 200-Å thick single-crystalline gold film. Heating laser fluence is 4.1 mJ/cm^2.

Springer Series in Chemical Physics, Vol. 53 **Ultrafast Phenomena VII**
Editors: C.B. Harris · E.P. Ippen · G.A. Mourou · A.H. Zewail

(2), the effect of lattice imperfections (in both the PCF and SCF) are not explicitly included, except as they affect G. Modifications to these simple expressions may be needed if e-i must be explicitly included. For films of thickness of the order of (or less than) the photon skin depth (~150 Å), electron diffusion can be ignored.

In Fig. 2 the experimental results for the decay time of the fast transient are displayed for various values of the pump laser fluence along with the theoretical curves for τ_e for different values of G. These results clearly show that (i) the dependence of $\Delta T/T$, and that of $\Delta R/R$, on time is the same, (ii) the decay rate in PCF is faster than that of SCF, (iii) its dependence on laser fluence essentially follows the predictions of Eqs. (1) and (2), (iv) $G = 4 \times 10^{16}$ W/m^3K overlaps with τ_e obtained experimentally in polycrystalline ~200-Å films, and (v) G for SCF is somewhat smaller than that for PCF. The observed difference in τ_e in PCF versus SCF could not be due to variations in their optical properties. Since the maximum $\Delta T/T$ for the PCF and SCF show better than 5% agreement. Therefore, the shorter transient decay time observed for the PCF, is due to faster τ_e. Enhanced electron-acoustic phonon coupling in the presence of imperfections has been postulated to be a result of the partial removal of constraints on (crystal) momentum conservation as the result of the randomization of the electron wave vector due to e-i scattering [4,5]. If we assume that the concentration of point and line defects are the same in the PCF and SCF, then the observed difference could lie in the grain boundaries for which the scattering power for electrons is thought to be very strong [6]. On the other hand, for the 200-Å films, the average grain size is comparable to film thickness and it becomes difficult to separate the role of grain boundaries from that of surfaces. Previous resistivity studies on Au have indicated that the surface roughness of PCF causes diffuse scattering of electrons. This is in contrast to the atomically smooth surface of SCF which specularly reflect the electrons [7].

We next consider transient thermomodulation experiments on films of thickness greater than the skin depth. We display in Fig. 3 the results from an 800-Å film. In this case, hot electron transport becomes important. Thus, one would expect that the

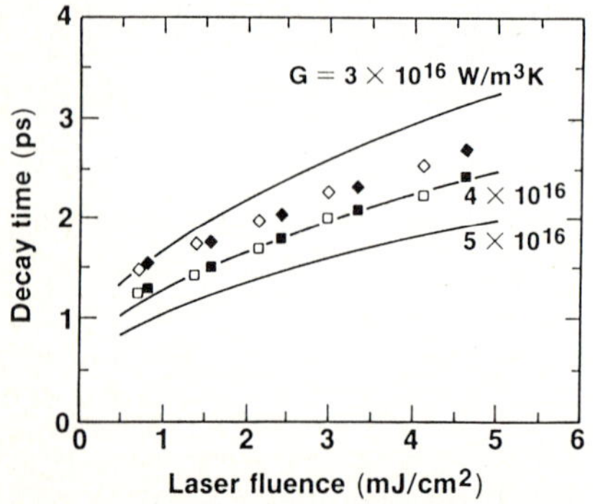

Fig. 2 Decay time of fast transient in thermoreflectivity (closed symbols) and thermotransmissivity (open symbols) of 200-Å single-crystalline (diamonds) and polycrystalline (squares) gold films as a function of heating laser fluence. Solid lines represent theoretical τ_e .

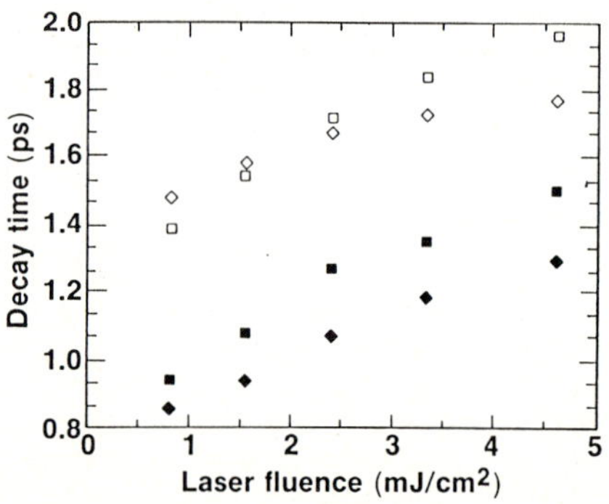

Fig. 3 Decay time of the fast transient in thermoreflectivity (closed symbols) and thermotransmissivity (open symbols) of 800-Å single-crystalline (diamonds) and polycrystalline (squares) gold films as a function of heating laser fluence.

transient decay rate of $\Delta R/R$ be faster than observed in 200-Å films. Results for $\Delta T/T$ show that it is affected less by the increased film thickness as expected since transmissivity probes across the film. In contrast to the 200-Å films, in the 800-Å films the decay time of the fast transient in $\Delta R/R$ is observed to be slower for PCF as compared to the SCF. This result is consistent with the conclusion that grain boundaries impede electron transport. An interesting sign reversal is observed in the difference between the decay time of the fast transient in $\Delta T/T$ for PCF and SCF as a function of laser fluence. Although at this stage we have no direct proof of the source of this effect, we offer the following qualitative explanation. Hot electron transport and energy relaxation are two competing processes. For low laser fluences, where diffusion is less important, the increased τ_c for the PCF results in a faster decay as observed for the 200-Å films. At the higher fluences, diffusion becomes more important. This affects the spatial distribution of hot electrons across the film causing it to be less uniform for the PCF than for the SCF due to lower κ. Since τ_c increases with the effective T_c, we would expect a slower decay time for the PCF.

In conclusion, we have demonstrated that femtosecond thermoreflectivity and thermotransmissivity of polycrystalline and single-crystalline thin Au films indicate that the loss of energy of an initially hot electron distribution is slightly faster in PCF as compared to SCF, and that hot electron transport is slower in PCF.

The work conducted at the University of Rochester was supported in part by the U. S. Department of Energy under contract No. DE-FG02-88ER45376. Additional support was provided by the Sponsors of the Laser Fusion Feasibility Project at the Laboratory for Laser Energetics. The work conducted at the University of California-Irvine was supported through NSF DMR 8603888 and DMR 8913289.

[1] S. I. Anisimov, B. L. Kapeliovich, and T. L. Perelman, Sov. Phys. JETP 39, 375 (1974).

[2] H. E. Elsayed-Ali, T. B. Norris, M. A. Pessot, and G. A. Mourou, Phys. Rev. Lett. 58, 1212 (1987); R. W. Schoenlein, W. Z. Lin, J. G. Fujimoto, and G. L. Eesley, Phys. Rev. Lett. 58, 1680 (1987); S. D. Brorson, J. G. Fujimoto, and E. P. Ippen, Phys. Rev. Lett. 59, 1962 (1987); S. D. Brorson, A. Kazeroonian, J. S. Moodera, D. W. Face, T. K. Cheng, E. P. Ippen, M. S. Dresselhaus, and G. Dresselhaus, Phys. Rev. Lett. 64, 2172 (1990).

[3] P. B. Allen, Phys. Rev. Lett. 59, 1460 (1987)

[4] For a review, see G. Bergmann, Phys. Rep. 107, 1 (1984).

[5] B. Keck and A. Schmid, J. Low Tem. Phys. 24, 611 (1976).

[6] J. M. Ziman, Electrons and Phonons (Clarendon Press, Oxford, 1962).

[7] K. L. Chopra, L. C. Bobb, and M. H. Francombe, J. Appl. Phys. 34, 1699 (1963).

Vibrational Exciton Decay in Uranyl Crystals by Picosecond Coherent Raman Scattering: Energy Bottlenecks and High Order Relaxation Mechanisms

E.L. Chronister and R.A. Crowell

Department of Chemistry, University of California, Riverside, CA92521, USA

1. Introduction

Anharmonicity in the crystal potential leads to phonon interactions which result in vibrational relaxation and dephasing. For many crystal systems a satisfactory description of vibrational relaxation is obtained by considering the first anharmonic correction to the crystal potential, namely a cubic interaction. However, recent results on vibrationally sparse systems have shown that the next higher order anharmonic relaxation term (i.e quartic) is necessary to account for the observed thermally induced dephasing [1-3]. In this work we focus on the importance of anharmonic relaxation mechanisms by creating a large vibrational energy gap. One expects the mode immediately above the gap to have an increased lifetime relative to other vibrational modes of the molecule due to the absence of any cubic decay channels.

The vibrational mode above the energy gap is found to have a lifetime orders of magnitude longer than any of the other modes of the molecule at low temperature. We show that the decay of this vibrational mode must occur by 5th and 6th order anharmonic phonon interactions. Temperature dependent changes in the decay of this vibrational mode, however, point to a cubic relaxation mechanism involving the absorption of a $90\mathrm{cm}^{-1}$ phonon mode to up-convert the population to the asymmetric ν_3 (U-O) stretching mode.

2. Experimental

The picosecond CARS experiments were performed using a home-built mode-locked, Q-switched, Nd^{3+}:YAG laser which synchronously pumps two cavity dumped dye lasers. The repetition rate of the laser system is 800Hz and it produces tunable dye laser pulses (λ=555nm-630nm) with pulse energies of 5 to 10μJ. The excitation and probe pulses are all focussed through the same 20cm lens into the sample which is held in thermal contact with the cold tip of a closed cycle He refrigerator (9-300K). The crystals were all oriented with respect to the optic axis and the laser geometry consisted of all laser polarizations parallel to the b crystallographic axis. The CARS signal was dispersed with a monochromator, detected with a photomultiplier, signal averaged and digitized using a lock-in amplifier, a fast A/D and an IBM/AT.

3. Anharmonic Relaxation Mechanisms

The lowest anharmonic perturbation (i.e. cubic terms in the potential) evaluated to second order yield the contribution to the linewidth due to three phonon scattering processes [2, 3],

$$\Gamma_l^{(3)} = 18\pi\hbar^{-2} \sum_{m,n} |B^{(3)}_{lmn}|^2 \left[(n_m+n_n+1)\, \delta(\omega_l-\omega_m-\omega_n) + 2(n_m-n_n)\, \delta(\omega_l+\omega_m-\omega_n) \right]$$

where $B^{(3)}$ is the anharmonic cubic coefficient, $n = [\exp(\hbar\omega/kT) - 1]^{-1}$ is the phonon occupation number, $\delta(x)$ yields a two phonon density of states and l, m, n, are indexes representing both the phonon branch and the phonon wavevector. The first term on the right hand side represents a down conversion process (i.e. $\omega_l \rightarrow \omega_m+\omega_n$), while the second term describes an up-conversion process (i.e. $\omega_m+\omega_l \rightarrow \omega_n$).

For the uranyl crystals in this study the large energy gap below the symmetric uranium oxygen stretch causes the two-phonon density of states to be zero at the energy of this vibrational mode (i.e. there are no two modes of the solid which add up to this energy) and therefore, higher order anharmonic terms in the crystal potential must be considered. The quartic contribution to the linewidth is given below [1-3],

$$\begin{aligned}\Gamma_l^{(4)} = 96\pi\hbar^{-2} \sum_{l,m,n} |B^{(4)}_{lmnp}|^2 \{ & [(n_m+1)(n_n+1)(n_p+1) - n_m n_n n_p]\, \delta(\omega_l-\omega_m-\omega_n-\omega_p) \\ & + 3[n_m(n_n+1)(n_p+1) - (n_m+1)n_n n_p]\, \delta(\omega_l+\omega_m-\omega_n-\omega_p) \\ & + 3[n_m n_n(n_p+1) - (n_m+1)(n_n+1)n_p]\, \delta(\omega_l+\omega_m+\omega_n-\omega_p) \}\end{aligned}$$

where $B^{(4)}$ is the quartic anharmonic coefficient. For homogeneously broadened lines the relaxation rate (T_1) is related to the linewidth by $\Gamma = (2\pi c T_1)^{-1}$.

Springer Series in Chemical Physics, Vol. 53 **Ultrafast Phenomena VII**
Editors: C.B. Harris · E.P. Ippen · G.A. Mourou · A.H. Zewail

4. Vibron Relaxation at Low Temperature: an Energy Bottleneck

A sketch of the vibrational modes and lattice phonons in these molecular crystals are shown in Figure 1. The symmetric U-O stretch ν_1, located at the top of the vibrational energy gap, has a vibrational decay time several orders of magnitude longer than all other Raman active modes of the molecule as shown in Table 1 and Figure 2. Spontaneous decay of the ν_1 mode at low temperature must occur by a high order anharmonic mechanism since a cubic mechanism (i.e. $\nu_1 \rightarrow \nu_m + \nu_n$) cannot conserve energy. In fact, since the nearest lower energy vibrational mode is ν_2 at 263cm^{-1}, it is clear that even a quartic mechanism (i.e. $\nu_1 \rightarrow 3\nu_2$) cannot conserve energy (i.e. 3 X 263cm^{-1} = 789cm^{-1} is still much less than the required 831cm^{-1}). Thus, the spontaneous decay of ν_1 in $Cs_2UO_2Cl_4$ must occur by at least a fifth order anharmonic interaction. These same features also exist for the ν_1 mode in $Cs_2UO_2Br_4$.

In general the relaxation rate decreases with increasing number of phonons according to, $1/\tau_{relax}$ = exp(-A*N) [5], where N is the number of modes involved in the fission process (i.e. the order). Thus it is

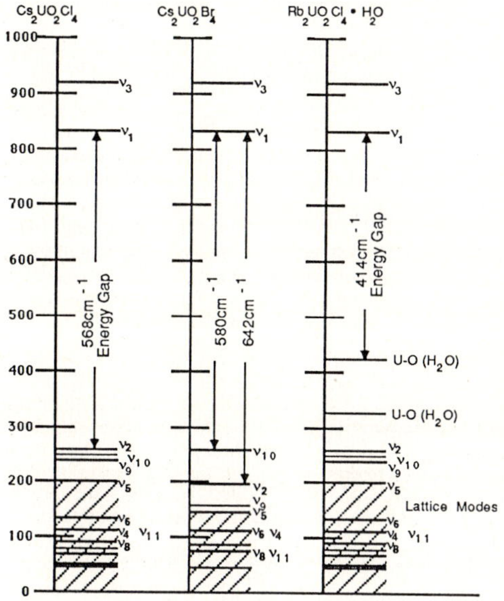

Figure 1. Shown are all of the vibrational and lattice phonon modes [4] of three different ionic uranyl crystals. There are several salient features: The large energy gap below the ν_1 mode for Cesium Uranylchloride and Cesium Uranylbromide and the introduction of water bands into this energy gap area for the case of Rubidium Uranylchloride. In addition, the phonon band region is cross hatched to indicate the dispersion of the phonon modes.

Table 1. Vibrational Frequencies and Decay Times for the Raman Active Modes of Uranyl Crystals

	$Cs_2UO_2Cl_4$		$Cs_2UO_2Br_4$		$Rb_2UO_2Cl_4 \cdot H_2O$	
Mode	ν(cm^{-1})	τ(ps)	ν(cm^{-1})	τ(ps)	ν(cm^{-1})	τ(ps)
ν_1	831	3400±400	836	510±30	838	≤17
ν_2	263	≤17	-	-	-	-
ν_5	230	≤17	-	-	-	-
ν_{11}	206	≤17	-	-	-	-

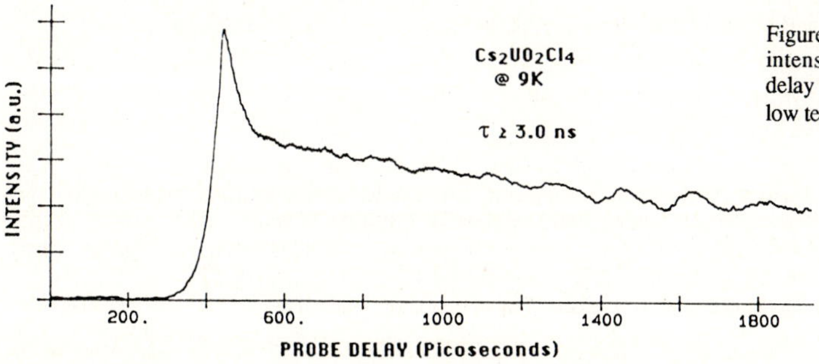

Figure 2. Picosecond CARS intensity as a function of delay for the ν_1 vibration at low temperature.

not surprising that 5th and 6th order relaxation processes have decay times which are several orders of magnitude longer than modes which have cubic relaxation pathways available for spontaneous decay.

Picosecond CARS measurements on the ν_1 mode in $Rb_2UO_2Cl_4 \cdot H_2O$ yielded a lifetime much shorter than in the other Uranyl crystals, as shown in Table 1. The increased relaxation rate for ν_1 can be attributed to cubic relaxation channels involving vibrational modes which have appeared within the energy gap region due to hydrogen bond interactions with the water molecule, as seen in Figure 1.

5. Thermally Induced Relaxation of Vibrational Excitons: High-Order Relaxation Processes

At elevated temperatures vibrational relaxation is enhanced by thermally induced mechanisms such as up-conversion to a higher energy vibration and down-conversion by stimulated phonon emission. It is also possible to lose coherence by a quartic mechanism which does not cause a loss of population [6]. However, when quartic anharmonic interactions become important it is found that quartic relaxation mechanisms dominate the pure dephasing mechanism [1-3].

The temperature dependence of dephasing in the high temperature limit ($kT>h\omega$) reduces to a linear dependence for the cubic case (i.e. $n \sim kT/\hbar\omega$) and a T^2 dependence for the quadratic case. However, if the low temperature limit is more accurate, one obtains an arrhenius type dependence.

Figure 3 yields phonon "activation" frequencies of 110±15cm^{-1} and 115±15cm^{-1} for ν_1 in both $Cs_2UO_2Cl_4$ and $Cs_2UO_2Br_4$ respectively. Assuming depopulation is responsible for the temperature effects, then the phonon "activation" energy either represents the mode energy needed to up-convert to a higher lying vibration or it represents the phonon energy responsible for stimulated phonon emission.

If we make the approximation that the down-conversion fission involves N phonons of similar energy then the product of the phonon occupation numbers in the low temperature limit yields an arrhenius equation of the form, $(k - k_o) = k_1 \exp(N*(h\nu_a/kT))$. The observed phonon activation energy is closer to the 90cm^{-1} phonon energy needed to up-convert to the ν_3 vibration than it is to the multiphonon energy $N\nu_a$. We conclude that the dominant thermal depopulation mechanism is almost purely a cubic up-conversion of ν_1 to ν_3 by the absorption of a 90cm^{-1} phonon.

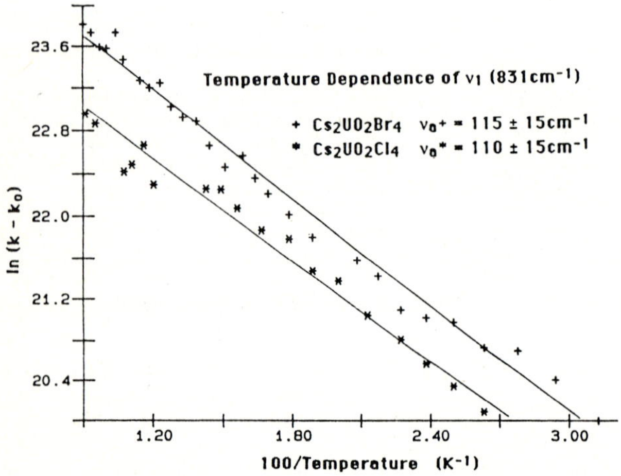

Figure 3. The temperature dependence of the decay rate for the ν_1 vibration in $Cs_2UO_2Cl_4$ and $Cs_2UO_2Br_4$ is plotted as an arrhenius rate in order to determine the approximate phonon "activation" energy for the thermal depopulation process. We conclude that thermal depopulation occurs predominantly by up-conversion to ν_3 by the absorption of a 90cm^{-1} phonon.

Acknowledgment to The Petroleum Research Fund, Klinger Scientific, and Newport Corp. for support.

References

1. S. Califano and V. Schettino, Inter. Rev. Phys. Chem. 7 19 (1988) and references within
2. L. Angeloni, R. Righini, E. Castellucci, P. Foggi, and S. Califano, J. Phys. Chem. 92 983 (1988)
3. V.K. Jindal, R. Righini, and S. Califano, Physical Review B 38 4259 (1988)
4. K.Ohwada, Spect.Acta 31A 973 (1975); J.Inorg.Nucl.Chem. 41 1145 (1979); Appl.Spec. 34 327 (1980).
5. Nitzan, S. Mukamel and J. Jortner, J. Chem. Phys. 60 3929 (1974); 63 200 (1975)
6. S. Marks, P. Cornelius, C.B. Harris, J. Chem. Phys. 73 3069 (1980)

Second-Harmonic Efficiency and Reflectivity of GaAs During Femtosecond Melting

J.-K. Wang, P. Saeta, Y. Siegal, E. Mazur, and N. Bloembergen

Division of Applied Sciences and Department of Physics,
Harvard University, Cambridge, MA 02138, USA

1. Abstract

Second-harmonic and reflectivity probes of photoexcited GaAs demonstrate melting on a 200-fs timescale and partial recrystallization of the molten material.

2. Introduction

Studies of the ultrafast melting of semiconductor surfaces under intense laser irradiation at frequencies above the bandgap reveal changes in the dielectric properties on the sub-picosecond timescale. Reflectivity measurements of silicon with femtosecond pulses demonstrate significant changes within 300 fs [1]. Second-harmonic measurements on silicon further show that the top 70–130 Å of the silicon surface loses its cubic symmetry within 150 fs [1]. In this paper we report on simultaneous reflectivity and second-harmonic generation (SHG) measurements during the melting of GaAs with 160-fs laser pulses. The results show bulk melting to a depth of at least 90 atomic layers within the pulsewidth.

3. Experimental

Amplified 200-μJ pulses at 620 nm are split into a pump and a probe beam with an energy ratio of 20:1. The *p*-polarized probe pulse is incident upon a (100) GaAs surface at 45°; the pump is orthogonally polarized and enters along the normal. The probe interrogates the central one-tenth of the pumped area. We determine the origin of time by attenuating the pump beam and maximizing the reflection sum frequency signal of the pump and probe beams. For each laser shot, the reflected intensity and the second-harmonic intensity are measured. In addition, we check for the presence of broadband ultraviolet radiation from a surface plasma. The pump beam by itself generates no spurious reflection or second-harmonic signal.

4. Results and Discussion

Figure 1 summarizes the reflectivity measurements normalized to the appropriate value for negative times. The dip in reflectivity at low fluence in the temporal range

Springer Series in Chemical Physics, Vol. 53 **Ultrafast Phenomena VII**
Editors: C.B. Harris · E.P. Ippen · G.A. Mourou · A.H. Zewail

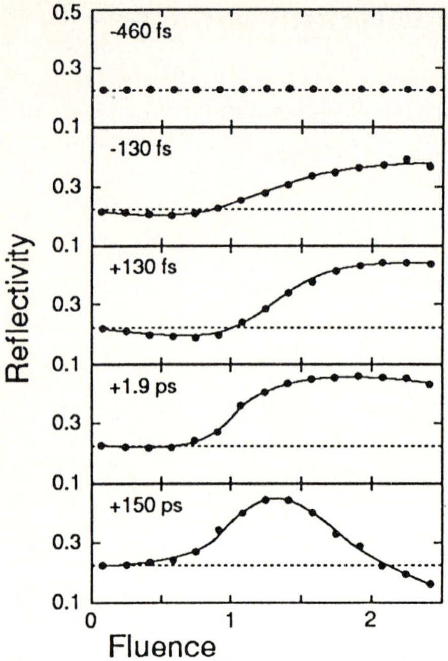

FIG 1. GaAs reflectivity vs fluence for various time delays. A fluence of 1 indicates where the SH efficiency first drops to zero.

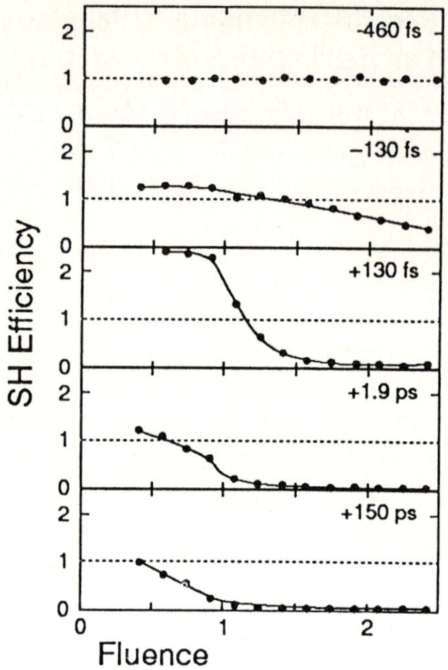

FIG 2. GaAs second-harmonic efficiency vs fluence for various time delays. A fluence of 1 indicates where the SH efficiency first drops to zero.

of −130 fs to +1.9 ps can be attributed to a dense free-carrier plasma created by photoexcitation [2]. This plasma decreases the real part of the dielectric constant, leading to a drop in reflectivity. By +130 fs the reflectivity at high fluence rises to roughly twice the crystalline value, consistent with the value predicted by a Drude model for metallic molten GaAs. After 50 ps the reflectivity at high fluence drops dramatically. This drop is probably due either to heating of the molten material far above the melting temperature [3], or to the delayed formation of a plasma cloud in front of the surface [4,5].

The second-harmonic signal is divided by the square of the probe intensity to obtain the second-harmonic (SH) efficiency, normalized to unity for unpumped material. Figure 2 shows a rapid, dramatic decrease to zero in SH efficiency as early as +130 fs, coincident with the observed plateau in reflectivity. The absence of any second-harmonic signal in this region indicates melting to a depth of at least 90 atomic layers, the absorption depth of the second harmonic.

Concurrent with the observed dip in reflectivity at short times and low fluence one finds a marked increase in SH efficiency. This increase is caused by a change in index of refraction due to the presence of the free-carrier plasma. First, lowered reflectivity results in an increased probe intensity beneath the surface. Second, with a decrease in the real part of the index, the angle between the probe *E*-field inside the material and the $\langle 0\bar{1}0 \rangle$ crystal axis increases from 10° to about 15°, resulting in approximately a factor 2 increase in SH efficiency.

We also measured the reflectivity and SH efficiency after complete equilibration of the material. The reflectivity rises again and settles at a value of 0.24, slightly higher than the value of crystalline GaAs (0.22), but lower than that of amorphous GaAs (0.28); the SH efficiency recovers to roughly half its original value. The net increase in reflectivity rules out any ablation of surface material, while the second harmonic from the resolidified material suggests a partial recrystallization.

5. Conclusion

In summary, we have performed femtosecond pump-probe measurements of the second-harmonic efficiency and reflectivity of GaAs during melting. Within 200 fs the second-harmonic signal drops to zero at high fluence, indicating melting to a depth of at least 90 atomic layers.

PS acknowledges a Ph.D. Fellowship from AT&T Bell Laboratories. YS acknowledges a Ph.D. Fellowship from the National Science Foundation. This work was supported by the Joint Services Electronics Program and the Material Research Laboratory Program under contract with Harvard University.

References

1. H. W. K. Tom, G. D. Aumiller, and C.H. Brito-Cruz, *Phys. Rev. Lett.*, **60**, 1438 (1988)
2. C. V. Shank, R. Yen, and C. Hirlimann, *Phys. Rev. Lett.*, **50**, 454 (1983)
3. P.M. Fauchet and K.D. Li, *Mat. Res. Soc. Symp. Proc.*, 100, 477 (1988)
4. M. C. Downer, R. L. Fork, and C. V. Shank, *J. Opt. Soc. Am. B*, **2**, 595 (1985)
5. J. M. Liu, A. M. Malvezzi, and N. Bloembergen, *Appl. Phys. Lett.*, **49**, 622 (1986)

Picosecond Time-Resolved Measurements of Photorefractive Grating Dynamics in Zincblende Semiconductors

A.L. Smirl[1], *M.D. Dawson*[1], *W.A. Schroeder*[1], *T.S. Stark*[1], *T.F. Boggess*[1], *and G.C. Valley*[2]

[1]Center for Laser Science and Engineering, 124 AMRF Oakdale Campus, Iowa City, IA 52242, USA
[2]Hughes Research Laboratories, 3011 Malibu Canyon Road, Malibu, CA 90265, USA

Abstract. The ultrafast evolution of photorefractive gratings in GaAs:EL2 and CdTe:V is separated from that of several co-existing gratings by a novel time-resolved polarization-rotation technique.

1. Introduction

We describe a novel nondegenerate, polarization-sensitive, transient-grating technique, which we use to isolate, to probe the symmetry of, and to time resolve (with resolution < 5-ps) the individual dynamics of several co-existing gratings written in GaAs:EL2 and CdTe:V. Specifically, we utilize optically-induced anisotropy to allow the first temporal resolution of the formation and decay of photorefractive gratings in these semiconductors on picosecond timescales. The measurements allow us to identify various contributions to the photorefractive effect, including the first direct evidence of a photorefractive grating arising from the Dember field associated with free electron hole pairs. The technique also allows the instantaneous electronic and the cumulative free-carrier gratings (which normally obscure the weaker photorefractive grating) to be simultaneously monitored. Measurements were conducted with a unique, amplified, Styryl 13 dye laser system that was developed for these studies.

2. Experimental Results and Discussion

As illustrated schematically in Fig. 1, the gratings were written by two 1-psec pulses at 960 nm with fluences of between 6 $\mu J/cm^2$ and 3 mJ/cm^2. The writing beams were s-polarized and had equal fluences to eliminate transient energy transfer between the two beams. The pulses were produced by a Styryl 13 dye laser (synchronously pumped by the frequency-doubled, fiber-grating-compressed output of a mode-locked Nd:YAG laser) followed by a three-stage amplifier (which was pumped by the frequency-doubled output from a 10 Hz regenerative

Springer Series in Chemical Physics, Vol. 53 **Ultrafast Phenomena VII**
Editors: C.B. Harris · E.P. Ippen · G.A. Mourou · A.H. Zewail

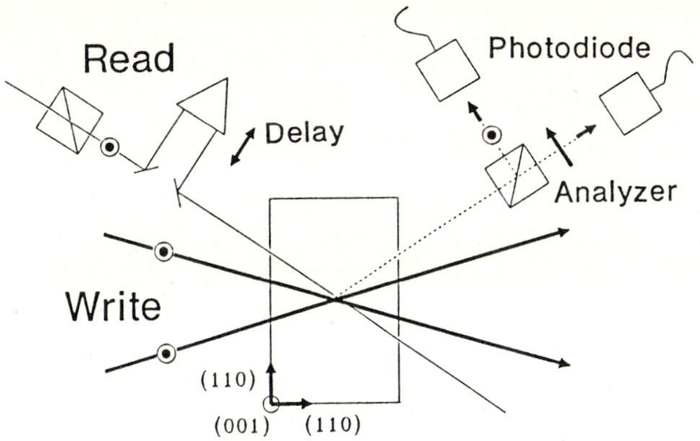

Fig.1 Experimental configuration

amplifier). The s-polarized "readout" beam was generated by Raman-shifting a fraction of the 960 nm radiation in benzene to produce a 1.06 μm pulse with a measured duration < 5-ps. A frequency-shifted probe is necessary in the Bragg regime to allow separation of the diffracted signal from the transmitted pump beams. Furthermore, since backward readout configurations are limited in temporal resolution by the sample thickness, a forward traveling probe was used. The diffracted signal was measured as a function of time delay, pump fluence, crystal orientation, and analyzer angle. The "photorefractive-cut" samples were oriented with the (100) crystallographic axis orthogonal to the grating wave vector, which for zincblende symmetry dictates that only the photorefractive grating will rotate by 90^0 the polarization of the diffracted probe. Consequently, the photorefractive contribution can be isolated from stronger free-carrier and bound-electronic gratings by an analyzer oriented to transmit p-polarization.

Representative measurements of the photorefractive and non-photorefractive gratings, recorded for three similar fluences (in mJ/cm^2) in a 3.17 mm-thick GaAs:EL2 crystal, are displayed in Fig. 2. The non-photorefractive signal, which was obtained by measuring the s-polarized component of the diffracted probe, displays three distinct features, all of which were independent of crystal orientation. The initial peak, which is symmetric about zero delay and has a χ^3 power law dependence, is associated with the instantaneous electronic index change (n_2). This peak is superposed on a broader feature arising from the accumulation and partial decay (by ambipolar diffusion) of a free-carrier grating, which for fluences above 0.25 mJ/cm^2 is primarily an electron-hole grating produced by two-photon and step-wise (through EL2 midgap levels) two-photon absorption. The feature has a fifth-order ($\chi^1\chi^3$) power law dependence so that it dominates the lower order χ^3 instantaneous electronic index change at fluences above ~ 2 mJ/cm^2. The free carrier grating is prevented from decaying completely by the space-charge field associated with the ionized EL2. This is manifested in a much weaker signal, that is observed to persist for delays greater

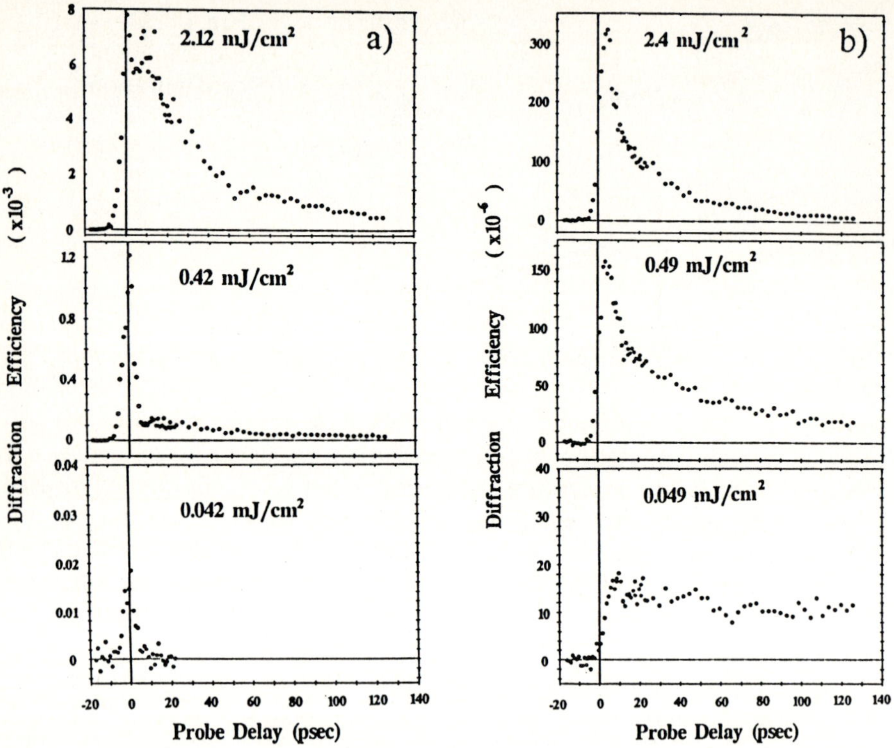

Fig. 2 a) Non-photorefractive (bound-electronic and free-carrier) and b) photorefractive grating diffraction efficiencies versus probe delay.

than 100 ps. The photorefractive signal, which was measured by detecting the p-polarized component of the diffracted probe, is also characterized by three distinct features. At fluences above ~ 0.2 mJ/cm², there is an initial transient response (the origin of which is still under investigation), which has a rise time determined by the temporal integration of the probe pulse and a decay shorter than our temporal resolution. This is followed by a signal that is consistent with diffraction from a photorefractive grating associated with the space charge field between electron-hole pairs (i.e. the Dember field), since it decays with a time constant determined by ambipolar diffusion. At longer delays and lower fluences, we observe a weaker persistent signal associated with the space charge field between free electrons and ionized EL2 donors. We emphasize that no p-polarized signal was observed when the crystal was oriented with the (100) axis parallel to the grating wave vector. Similar results have been obtained for CdTe:V.

This work was supported by DARPA and the Office of Naval Research.

Part VIII

Coherent Phonons

Hot Phonon Spectra of Polar Semiconductors Studied via Picosecond Free Carrier Absorption Between 5 and 10 μm

T. Elsaesser, R.J. Bäuerle, R. Klann, and W. Kaiser

Physik-Department E11, Technische Universität München,
D-8000 München 2, Fed. Rep. of Germany

Excess LO phonons created by carrier cooling play an important role for the relaxation of hot electrons and holes in polar semiconductors. Recently, we have introduced a novel method for the direct study of hot phonons. Transient changes of the intraband absorption of electrons were investigated [1]. Intraband absorption represents a three particle interaction of electron, photon and phonon, where the coupling to the lattice provides the conservation of the k-vector [2]. In our experiments, the electron gas in an n-doped sample is heated to a temperature of approximately 500 K by an intense infrared pump pulse. The resulting change of free carrier absorption was monitored by probe pulses at the same wavelength [1]. In InAs, a significant enhancement of intraband absorption was found, which relaxes on a time scale of several tens of picoseconds. The theoretical analysis of the data demonstrates, that this effect is mainly due to the occurrence of excess LO phonons, which couple in the absorption process and increase the absorption coefficient. In this paper, we discuss new frequency resolved measurements of the transient free carrier absorption in InAs and GaAs. Data taken at different wavelengths between 5 and 10 μm give information on the spectral distribution of the excess phonons in k-space.

Picosecond infrared pulses of a duration of 8 ps are generated by parametric frequency mixing in two nonlinear $AgGaS_2$ crystals [3]. The pump and the probe pulses are independently tunable in the wavelength range from 4 to 10 μm. Different n-doped GaAs and InAs samples with electron concentrations of approximately 10^{18} cm^{-3} are investigated at a lattice temperature of 10 K.

In Fig. 1, we present data for an InAs crystal of an electron concentration of 1.5×10^{18} cm^{-3}. After heating the carriers by a pump pulse at $\lambda_{ex}=7\mu m$, a fast rise of free electron absorption is found, which decays - depending on the specific wavelength λ_{pr} - on a time scale of several tens of picoseconds. The maximum change of absorption increases from $\Delta A=0.17$ at 5 μm to $\Delta A=0.3$ at 9 μm. The complete relaxation of the signals at later times gives evidence, that no excess carriers are created by the excitation pulse.

The absorption of hot electrons in n-type GaAs ($N=1.5\times10^{18}$ cm^{-3}) shows a similar behavior. Here the maximum absorption change is smaller ($\Delta A=0.05$ at 8 μm) and decays on a faster time scale of 20 ps.

Different parts of the distribution of excess phonons in k-space are monitored at the various probe wavelengths. At short wavelengths, phonons with large k-vectors predominantly contribute to the signal, whereas the phonon spectrum at small k

Springer Series in Chemical Physics, Vol. 53 **Ultrafast Phenomena VII**
Editors: C.B. Harris · E.P. Ippen · G.A. Mourou · A.H. Zewail

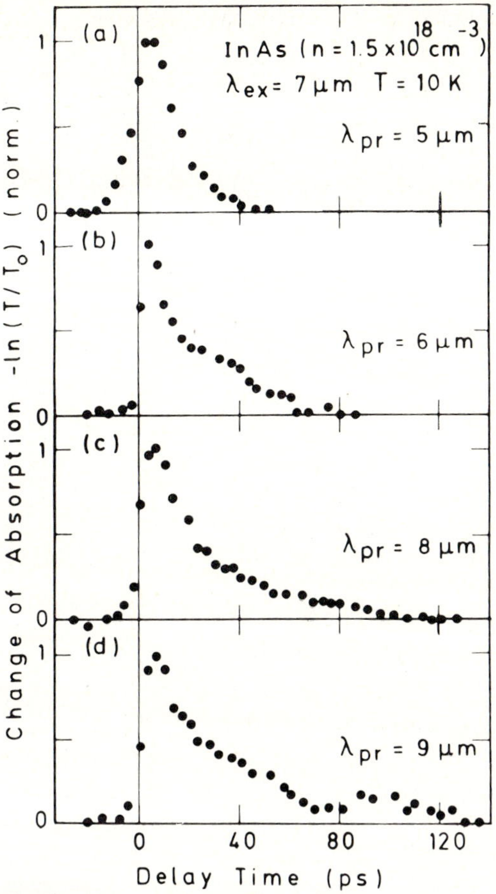

Fig. 1
Transient rise of free-carrier absorption in n-doped InAs after excitation by a picosecond pulse at λ_{ex}=7μm (T_0, T : transmission of the sample before and after excitation). The absorption change was monitored at four wavelenghts λ_{pr}. Lattice temperature of the sample : 10 K.

gives rise to the absorption change at long wavelengths. The shape of the phonon spectrum is obtained from a theoretical simulation of the experiments. We calculate the transient electron temperature and the distribution of nonequilibrium LO phonons from the time dependent excess energy of the carriers and the phonon Boltzmann equation [4]. Screening of the polar interaction in the dense Fermi sea is important for the time scale of carrier cooling and the shape of the phonon spectrum [1]. A non-thermal excess population of LO phonons occurs in the k-interval between 5×10^5 and 1.5×10^7 cm^{-1} with a broad maximum around k=2×10^6 cm^{-1}. The maximum phonon population reaches a value of N_q=0.2. Infrared photons at 5 μm mainly probe excess phonons with k-vectors between 3×10^6 and 7×10^6 cm^{-1}. At a wavelength of 9 μm, a broader k-interval from 9×10^5 to 6×10^6 cm^{-1}

contributes to the enhancement of absorption. The theoretical model is in good agreement with the experimental data.

[1] T. Elsaesser, R.J. Bäuerle, and W. Kaiser, Phys. Rev. B 40 (1989), 2976

[2] V.L. Gurevich, I.G. Lang, and Yu. A. Firsov, Sov. Phys. Sol. State 4 (1962), 918

[3] T. Elsaesser, H. Lobentanzer, and A. Seilmeier, Opt. Commun. 52 (1985), 355

[4] W. Pötz, and P. Kocevar, Phys. Rev. B 28 (1983), 7040

Resonant Excitation of Coherent Optic Phonons by Femtosecond Multiple-Pulse Impulsive Stimulated Raman Scattering

A.M. Weiner[1], *D.E. Leaird*[1], *G.P. Wiederrecht*[2], *and K.A. Nelson*[2]

[1]Bellcore, 331 Newman Springs Road, Red Bank, NJ 07701, USA
[2]Department of Chemistry, Massachusetts Institute of Technology, Cambridge, MA 02139, USA

Generation of coherent optic phonons through impulsive stimulated Raman scattering (ISRS) can occur whenever a sufficiently short optical pulse passes through a Raman-active medium [1-3]. In this paper we report the application of specially shaped femtosecond pulse trains to achieve improved control over coherent vibrational motion. In particular, we demonstrate "multiple-pulse" ISRS, in which terahertz-rate trains of fsec pulses are used for repetitive impulsive excitation of optic phonons. By matching the pulse repetition rate to a particular phonon frequency, we are able to select and amplify individual vibrational modes [4]. Each successive pulse in the sequence arrives at the sample in phase to amplify the coherent vibrational motion, analogous to repetitively pushing a child on a swing.

Our multiple-pulse ISRS experiments were performed on samples of the organic molecular crystal α-perylene at <10 K. "Single-pulse" ISRS studies of this material have revealed simultaneous excitation of several low frequency phonons [1]. In our work a CPM dye laser and copper vapor laser dye amplifier provide 75-fsec, 0.62-μm pulses (in α-perylene's transparency range), which are converted into Thz-rate trains of fsec pulses by spectral phase filtering within a pulse shaping apparatus [5]. We have used both transient grating and pump-probe measurements to demonstrate selective amplification of individual vibrational modes.

In the grating experiments, two crossed pulse trains were used to excite a phonon standing wave, and the vibrational response was monitored via diffraction of a single probe pulse. Figure 1 shows (a) the excitation pulse train which resonantly drives α-perylene's 80 cm^{-1} librational mode and (b) the diffracted signal vs. probe delay. Although the signal is initially dominated by the crystal's electronic response to each pulse in the sequence, the ISRS signal builds up throughout the train, eventually becoming as intense as the strongest electronic scattering peak. The diffracted signal subsequent to the excitation pulses oscillates at a single frequency (twice the vibrational frequency, since the scattering intensity is proportional to the square of the coherent polarization). These data clearly demonstrate selective amplification of a single vibrational mode [4].

Further experiments, with the repetition rate of the excitation pulse train detuned from the phonon frequency, were also performed. Figure 1(c) shows data for a relatively small detuning of ≃2 cm^{-1}. Although some vibrational amplification is still achieved, the efficiency is clearly reduced. The frequency of the observed oscillations remains at twice the phonon frequency, independent of the excitation rate, as

Springer Series in Chemical Physics, Vol. 53 **Ultrafast Phenomena VII**
Editors: C.B. Harris · E.P. Ippen · G.A. Mourou · A.H. Zewail

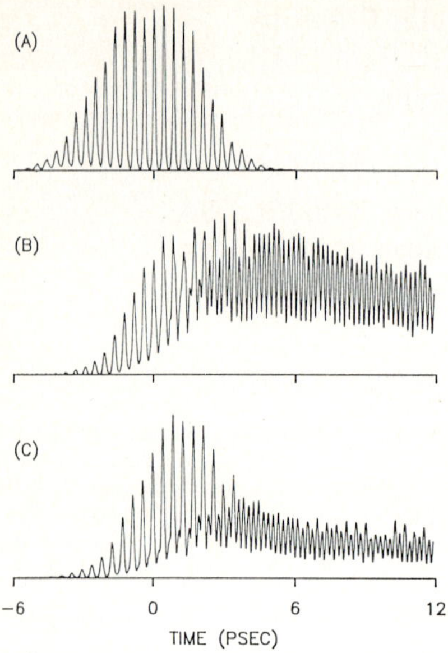

Fig. 1 (A) Intensity profile of a 2.39-THz pulse train used for driving the 80 cm^{-1} libron in α-perylene. (B) Diffracted signal as a function of probe delay. (C) Diffracted signal for slightly off-resonant excitation.

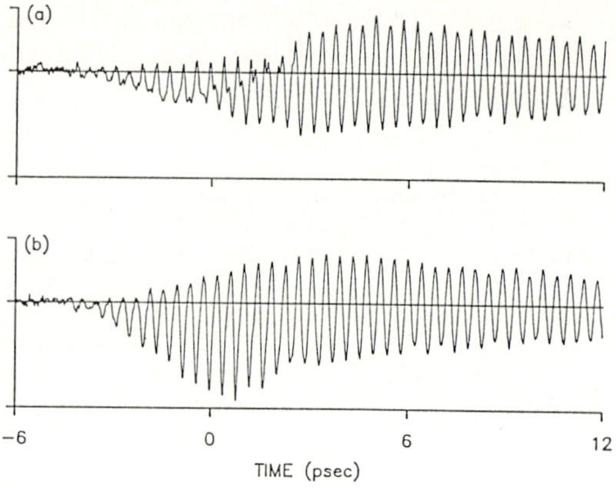

Fig. 2 (a) Pump-probe data with only the red portion of the spectrum detected. (b) Pump-probe data with only the blue portion of the spectrum detected.

expected. For significantly larger detunings, the vibrational amplification is suppressed.

Pump-probe measurements, with the same resonant pump pulse sequence as above, were also performed. Depending on its delay, the probe either amplifies or opposes the coherent vibrational motion induced by the pump, and therefore the transmitted probe pulse is "Doppler-shifted" either to the red or to the blue. Figure 2(a,b) shows pump-probe data obtained when only the red portion or only the blue portion of the probe spectrum are detected, respectively. The oscillations which take place after the pump pulse sequence are exactly out of phase for traces 2(a) and 2(b), as expected for a pure spectral shift. The differences seen in the plots for $t < 3$ psec arise due to interference with pump-induced two photon absorption. The details of this interference indicate that probe pulses are red-shifted when they arrive coincident with a pump pulse and reinforce the vibrational motion and blue-shifted when they arrive between pump pulses and oppose the vibrational motion.

By comparing transient grating data and pump-probe data, we can elucidate the timing dynamics of ISRS. In the grating measurements the peak diffraction signal occurs at times of maximum phonon-induced index change, corresponding to maximum vibrational displacement. In contrast, the extrema of the pump-probe signal correspond to a maximum time derivative of the index, and this coincides with maximum vibrational velocity. A detailed comparison of pump-probe and transient grating data, shown in Fig. 3, demonstrates the expected phase

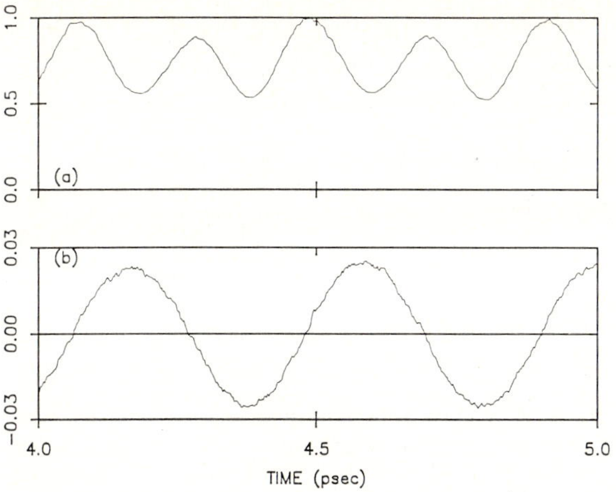

Fig. 3 (a) Transient grating data. (b) Pump-probe data.

shift. In particular, we note the extrema of the pump-probe signal coincides with minima in the diffraction data, while the peaks in the diffraction data coincide with zero crossings of the pump-probe data. Precise knowledge of the timing of phonon oscillations induced by ISRS, as illustrated here, can make possible time-resolved spectroscopy of the electronic energy states of vibrationally distorted samples.

In summary, we have described multiple pulse impulsive stimulated Raman scattering, an initial application of fsec pulse shaping techniques for optical manipulation of molecular and collective motion. Further possible applications of specially shaped fsec pulses have been discussed in the literature [6,7], and one can anticipate dramatic advances in this area in the 1990s.

G.P. Wiederrecht and K.A. Nelson acknowledge support in part by JSEP grant no. DAAL03-89-C-0001 and contributions from DuPont and Perkin-Elmer.

References

1. S. DeSilvestri, J.G. Fujimoto, E.P. Ippen, E.B. Gamble, L.R. Williams, and K.A. Nelson, Chem. Phys. Lett. 116, 146 (1985).
2. J. Chesnoy and A. Mokhtari, Phys. Rev. A38, 3566 (1988).
3. F.W. Wise, M.J. Rosker, and C.L. Tang, J. Chem. Phys. 86, 2827 (1986).
4. A.M. Weiner, D.E. Leaird, G.P. Wiederrecht, and K.A. Nelson, Science 247, 1317 (1990).
5. A.M. Weiner, J.P. Heritage, and E.M. Kirschner, J. Opt. Soc. Am. B5, 1566 (1988); A.M. Weiner and D.E. Leaird, Opt. Lett. 15, 51 (1990).
6. H. Rabitz, in Atomic and Molecular Processes with Short Intense Laser Pulses, A. Bandrauk, ed. (Plenum, New York, 1988).
7. W.S. Warren, Science 242, 878 (1988).

Impulsive Phase Coherent Femtosecond Spectroscopy of Atomic and Molecular Transients with Phase-Locked Pulse Pairs

A.J. Ruggiero, N.F. Scherer, M. Du, and G.R. Fleming

Department of Chemistry and James Franck Institute, University of Chicago, 5735 S. Ellis Ave., Chicago, IL 60637, USA

Abstract. Advances in the development of experimental techniques for coherent femtosecond time-resolved spectroscopy of atomic and molecular transients using phase-locked pulse pairs are reported.

1. Introduction

Control over the relative phase relationship between optical pulses in a multiple pulse sequence permits the exploitation of quantum interference effects in the experimental study of atoms, molecules, and chemical reaction dynamics [1]. The interference cross-terms found in the transition probabilities for excitation with multiple phase coherent pulses are analogous to the optical field interference effects found in a Young's double slit experiment. In principle, the time dependent suppression of quantum interference due to the coupling of the interfering degrees of freedom with the environment should provide a sensitive probe of these interactions.

Generation and use of phase coherent pulse pairs in high resolution laser spectroscopy [2] and time-resolved optical coherent transient spectroscopy [3] is well established. The primary constraint on the successful implementation of these approaches results from the technical difficulties associated with maintaining the phase difference between the pulses during the course of the experiment. We report on a new experimental method for setting and maintaining the relative phase of femtosecond laser pulses in multiple pulse experiments. The ultrashort duration and broad frequency bandwidth of these pulses allows the phase and amplitude evolution of coherent superposition states in multilevel systems to be probed. Molecular iodine is used as an example.

2. Experimental Methods and Results

Ultrashort transform limited pulses used in these experiments were generated in a novel cavity dumped antiresonant ring femtosecond dye laser [4]. The excitation pulse pair is created via a phase-locked scanning Mach-Zender interferometer. One arm of the interferometer serves as a course scanning stepper motor delay line, while the other arm contains a position regulated piezo electric transducer (PZT). The PZT is used to

Springer Series in Chemical Physics, Vol. 53 **Ultrafast Phenomena VII**
Editors: C.B. Harris · E.P. Ippen · G.A. Mourou · A.H. Zewail

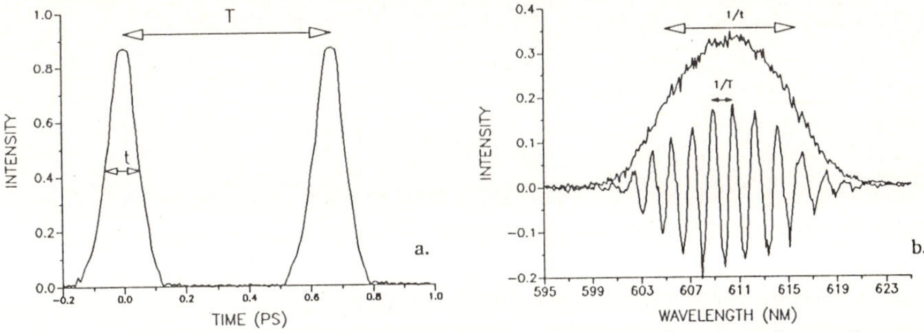

Fig. 1. Single pulse and double pulse temporal and spectral envelopes. **a.** Phase-locked femtosecond pulse autocorrelations **b.** Top curve: single pulse frequency spectrum; bottom trace: double pulse frequency spectrum.

make small corrections to the relative phase of the pulses following each stepped advance of the scanning delay line. The novel aspect of this technique for setting and maintaining the relative phases of the pulses is in the generation of the error signal for the PZT phase locked loop. The basic idea of this new technique is to use the interference structure in the frequency spectrum of the pulse pair to fix the relative phase of the pulses at a given delay. Single pulse and double pulse temporal and spectral envelopes are shown with their corresponding relationships to the delay time, T, and the pulsewidth, t, in Figure 1. The interference fringes can be maintained for delay periods more than 1000 times greater than the pulse durations by storing the fields in either a monochrometer via geometrical dispersion [5] or in the excitation probability amplitude of a two level coherence in an atomic vapor discharge (hollow cathode lamp, HCL). The maximum phase-locking time interval is dictated by the spectrally selected bandpass or in the case of the HCL by the dephasing time of the atomic coherence. Relative phase differences and time-delays between pulses can be chosen with regard to the coherent transient measurement of interest.

In a generalized pump-probe experiment with phase coherent pulses, the first pulse creates a dipole moment in the system with a phase related to the phase of the exciting field. During the delay period the dipole precesses freely at the Bohr frequency, w_0 of the transition and accumulates a phase w_0T, while the field at the carrier frequency w_c acquires a phase w_cT. The interaction with the second pulse will either further excite the system or deexcite it by stimulated emission, depending on the relative phase between the electric field and the dipole oscillations. The probability for excitation by the second pulse is related to the probability of excitation by the first pulse through the quantum interference cross-term. The direct measurement of the interference cross-term is made by monitoring the total emission from the sample in a right angle geometry using a double

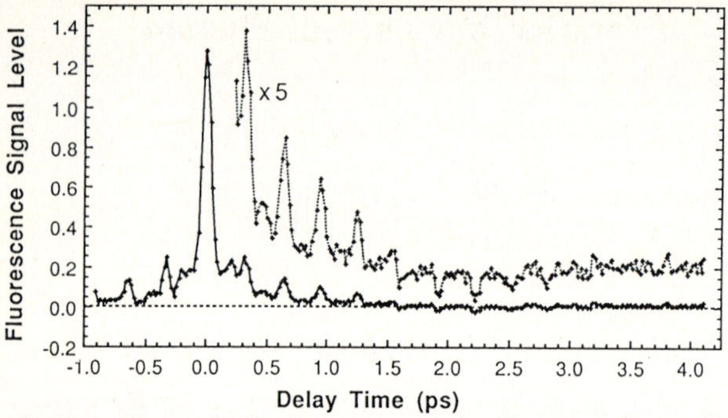

Fig. 2. Phase-locked pump-probe fluorescence signal from iodine vapor at 300K. The excitation pulses are locked with the same phase.

modulation technique with phase sensitive detection. An example of the application of this general technique to measurements of free-induction-decay dynamics of the ro-vibrational coherent superposition state in molecular iodine are shown in Figure 2. The peaks in the iodine transient occur at the excited state vibrational period (300fs). A notable feature of the transient signal is the sign reversal after 5 vibrational periods. This phase reversal is believed to result from detuning effects and anharmonicities in the potential [6]. Full experimental details and interpretation of the measured transients are given elsewhere [7].

3. References

1. R. Kosloff, S.A. Rice, P. Gaspard, S. Tersigni and D.J. Tannor: Chem. Phys. **139**, 201 (1989); T. Seideman, M. Shapiro and P. Brumer: J. Chem. Phys. **90**, 7132 (1989).
2. M.M. Salour: Rev. Mod. Phys. **50**, 667 (1978), and references therein.
3. W.S. Warren and M.S. Silver: Adv. Mag. Reson. **12**, 247 (1988); W.S. Warren and A.H. Zewail: J. Chem. Phys. **78**, 2279 (1983); **78**, 2298 (1983).
4. N.F. Scherer, A.J. Ruggiero and G.R. Fleming: Optics Letters, in prep.
5. N.H. Schiller and R.R. Alfano: Opt. Commun. **35**, 451 (1989).
6. J. Cina: Private communication.
7. N.F. Scherer, A.J. Ruggiero, M. Du and G.R. Fleming: J. Chem. Phys. in press.

Direct Time-Resolved Observation of Coherent-Phonon Oscillations in III-V Compounds

G.C. Cho, W. Kütt, M. Strahnen, A. Esser, U. Lemmer, and H. Kurz

Institute of Semiconductor Electronics, RWTH Aachen,
Sommerfeldstrasse, D-5100 Aachen, Fed. Rep. of Germany

Improvements in time resolution and signal-to-noise-ratio of reflective electro-optic sampling[1] allows for the first time the direct visualization of lattice vibrations in III-V-compounds in time domain. Individual cycles of phonon oscillations as well as vibrational dephasing are demonstrated via modulations of time resolved reflectivity changes through the electro-optic effect. By the same effect LO-phonons are launched synchronously by an impulsive electric field transient associated with the ultrafast separation of the optically injected carriers in the surface space-charge field of III-V-compounds.

The experiments have been performed with the 100 MHz pulse train of a CPM-ring dye laser and a novel "fast-scan pump-probe"-technique[2] capable of resolving transient reflectivity changes $\Delta R/R_0 \sim 10^{-7}$, corresponding to surface electric field changes of 50V/cm. The probe beam is linear polarized and oriented parallel to the [010]-crystal axis of the (100)-oriented samples. By sampling the difference between the orthogonally polarized [0$\underline{1}$1] and [011] components reflected from the sample via a polarizing cube beam splitter the time evolution of the electric field change E(t) is monitored directly.

Fig. 1a shows the temporal evolution of the longitudinal field changes on a (100)-GaAs:Cr surface. At an optically excited carrier density of 10^{18} cm^{-3} the ultrafast initial field rise to values of appr. 40 KV/cm is closely correlated to the pump-probe correlation signal (AC). The rise time (10%-90%) of 80 fs corresponds to an actual 50 fs after probe pulse deconvolution. For positive time delays the reflectivity signature is modulated by a periodic signal with a frequency of 8.9 THz, matching exactly the frequency of the Γ_{15}-LO-phonons in GaAs[3].

Regarding the driving force which launches the coherent phonons, impulsive optical rectification ($\chi^{(2)}$) and stimulated Raman scattering (ISRS) [4] have been discussed. For both processes an instantaneous response to the fs-excitation pulse is expected and they are governed by polarization symmetry considerations. There is, however, no change in the magnitude of the oscillations, when the polarization of the pump

Springer Series in Chemical Physics, Vol. 53 **Ultrafast Phenomena VII**
Editors: C.B. Harris · E.P. Ippen · G.A. Mourou · A.H. Zewail

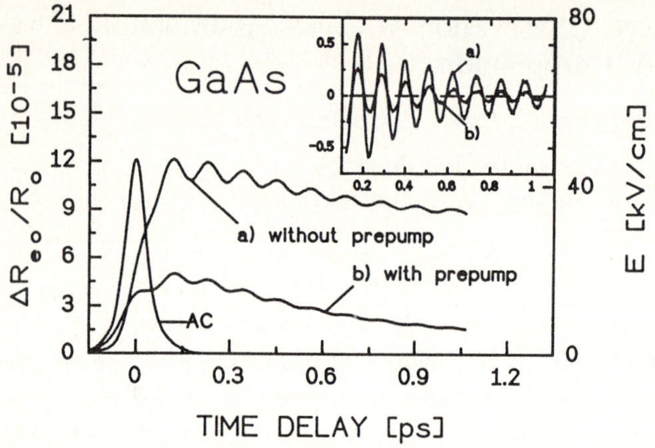

Fig.1: Reflectivity changes and corresponding electric field changes at the surface of an intrinsic GaAs-(100)-crystal a) at an excitation density of 10^{18} carriers/cm^3 and b) with prepump excitation ($\sim 10^{18} cm^{-3}$). The insert shows the reflectivity oscillations due to vibrations of coherent LO-phonons

beam is rotated for fixed orientations of sample and probe beam polarization. Thus by symmetry arguments we can rule out $\chi^{(2)}$ and $\chi^{(3)}$ processes acting as driving force.

In our opinion the coherent phonons are rather generated by impulsive electric field transients manifested in Fig.1. The rapid rise of the electric field transients is short enough to excite LO-phonons synchronously through electro-optic coupling. The ultrafast electric field change is caused by the drift of the optically injected carriers within the surface space charge field of III-V-compounds. This is confirmed by lowering the initial driving field through optical injection in a three-pulse experiment. Ten picoseconds before the actual pump-probe experiment the sample was excited by a prepulse extracted from the other arm of the CPM laser. As expected and shown in Fig. 1b the initial field change as well as the magnitude of the phonon induced oscillations is strongly reduced. This is consistent with the screening of the surface space charge field by photoexcited carriers. The strict proportionality between initial electric field amplitude and reflectivity oscillation holds for a wide range of excitation levels. The basic features of the electro-optic field transients agree well with Monte-Carlo results of carrier transport in surface space charge fields[5] at high excitation intensities.

Similar oscillations and field transients have been also observed in n- and p-doped GaAs, AlGaAs and intrinsic and p-doped InP. The corresponding energies of 34.2 meV ($Al_{0.21}Ga_{0.79}As$) and 42.2 meV (InP) are in good agreement with published literature values of the Γ-LO-phonon energy.

Details of the dephasing of the coherent-phonon oscillations can be obtained from the time resolved measurements as shown in the insert of Fig. 1. The decay is attributed mainly to vibrational dephasing due to coupling with incoherent phonons generated by the cooling of photoexcited hot carriers. The phonon induced oscillations decay on a time-scale of 1.7 ps for low excitation values of a few $10^{17}cm^{-3}$. For higher excitation the dephasing time decreases to appr. 0.7 ps. However, for strong excitation above a few $10^{18}cm^{-3}$, preliminary experiments show strong deviations from a monoexponential decay possibly due to anharmonic coupling to TO-phonons. In n- and p-doped GaAs the coherent phonons dephase much more rapidly. The coupling due to the free carriers is obvious. In addition a striking correlation between dephasing rate and carrier cooling rate has been found in intrinsic and p-type InP.

In conclusion we have observed subpicosecond coherent phonon oscillation via electro-optic sampling. The optimized signal-to-noise-ratio of 10^{-7} and high time resolution of 50 fs opens the way to detailed studies of the interaction of coherent phonons with other subsystems such as incoherent phonons and plasmons.

This work was supported entirely by the "Deutsche Forschungsgemeinschaft".

References:

/1/ Lixing Min and R.J. Dwayne Miller, Appl. Phys. Lett 56, 524 (1990)
/2/ M. Strahnen, Ph.D. thesis, RWTH 1990
/3/ Sadao Adachi, J. Appl. Phys. 58, R1 (1985)
/4/ W.E. Bron, J. Kuhl, and B.K. Rhee, Phys. Rev B34, 6961 (1986)
/5/ Xiang Zhou, Thomas Y. Hsiang, and R.J. Dwayne Miller, J. Appl. Phys. 66, 3066 (1990)

Excitation of Low Frequency Polaritons in $PbTiO_3$ with Femtosecond Pulses

J. Etchepare[1], *G. Grillon*[1], *A. Antonetti*[1], *J.C. Loulergue*[2], *M.D. Fontana*[3], *and G.E. Kugel*[3]

[1]L.O.A. CNRS URA 1406, ENSTA, Ecole Polytechnique, F-91120 Palaiseau, France
[2]Institut d'Optique Théorique et Appliquée, CNRS UA 14, F-91406 Orsay, France
[3]C.L.O.E.S. Université de Metz et Sup. Elec., F-57078 Metz, France

Abstract. Four wave mixing in a forward arrangement has been used to directly drive and probe definite phonon modes in a single crystal of tetragonal $PbTiO_3$. Polariton frequency variation as a function of the wave vector can be easily measured by this technique.

1. Introduction

Time resolved transient grating experiments have been developed for several decades; advances in ultrashort pulses allow us nowadays to excite and directly probe nuclear vibrational motions [1]. This technique [2] is well suited to the study of lattice dynamics of $PbTiO_3$, due to the existence of a few well-defined and underdamped Raman bands in the low frequency region [4,4]. In the ferroelectric phase of this perovskite, which is the stable form at room temperature, the phonon modes, of symmetry A_1 and E, are simultaneously IR and Raman active: we are thus moreover concerned with polar modes, which may experience a polariton behavior.

Using a forward nondegenerate four wave mixing, we demonstrate the ability to achieve the polariton regime, to describe its dispersion in the $\omega - \boldsymbol{K}$ space and in conclusion to deduce the static dielectric constant value.

2. Theoretical Background

Two ultrashort laser pulses, $(\omega_{p1}, \boldsymbol{K}_1)$ and $(\omega_{p2}, \boldsymbol{K}_1)$, issued from a femtosecond-pulse laser system and centered at 620 nm are crossed inside the crystal to form a standing wave of difference wave vector $\boldsymbol{K} = (\boldsymbol{K}_1 - \boldsymbol{K}_2)$. The temporal behavior of the transient processes involved [5] is monitored by coherent scattering of variable delayed probe pulses of frequency ω_t.

The temporal width of the exciting pulses being much shorter than the oscillating period of the phonon Ω derived from mixing of frequency components contained within the broad excitation pulse width, the measured nuclear response function will appear as a damped sinusoid.

Other third-order processes, with different temporal characteristics, may be generated during the nonlinear interaction: electronic third-order polarizability

Springer Series in Chemical Physics, Vol. 53 **Ultrafast Phenomena VII**
Editors: C.B. Harris · E.P. Ippen · G.A. Mourou · A.H. Zewail

gives rise to an instantaneous response, with regard to the pulse temporal width; relaxational motions will be described as usual by an exponential function.

3. Experimental Results

Figure 1 shows the diffraction efficiency as a function of the temporal delay between pump and probe pulses. The diffracted signal polarization has been analyzed along y (or z), the polarization of the probe pulse directed along y (or z) and that of the pump along z (c axis of the crystal); the relevant nonlinear susceptibility tensor element generated corresponds therefore to $X^{(3)}_{yyzz}$ (or $X^{(3)}_{zzzz}$). The potentially driven phonons will be of A_1 (TO) type, extraordinary in character, and no geometrical dispersion is expected [6]. In addition to an instantaneous response, the signal reveals in each case an interference between an oscillatory mode and a long relaxation component. As a result of the interaction with the long lived process, we can see that, first, the signal is linearly proportional to the oscillatory contribution, and second, we measure directly the phonon frequency value Ω and not 2Ω.

We have studied the efficiency of the long lived process as a function of the pump pulse intensity. Figure 2 shows that a fourth power law dependence is measured with a good confidence. We can therefore conclude that creation of a free carrier index grating has occurred by a two-photon absorption process.

TO components may interact with transient induced dipoles, which give rise to a dispersive character, the polariton regime, for values of the wave vector in the 10^2–$10^3\,\mathrm{cm}^{-1}$ range. Such a magnitude for the wave vector is far smaller than that which can be expected when using conventional right angle Raman scattering or phase conjugate four wave mixing; it is on the other hand easily reached in our forward geometry where the angle φ between the pump pulses inside the crystal is around 1°.

Figure 3 displays the evolution of the polariton frequency as a function of the $\boldsymbol{K}_{\mathrm{ph}}(\varphi)$ wave vector values. The experimental points have been fitted with an undamped single oscillator model [6].

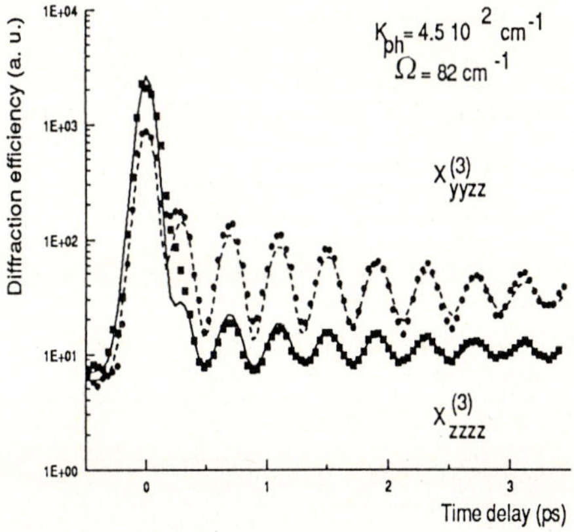

Fig. 1. Diffraction efficiency, on a logarithmic scale, as a function of the time delay between pump and probe pulses; the two curves correspond to different polarization directions of the probe and diffracted beam with respect to that of the pump beams; the z direction corresponds to the c axis of the crystal

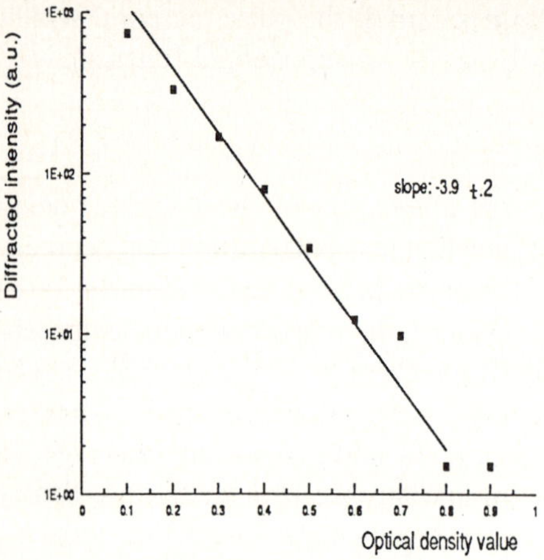

Fig. 2. Variation of the diffracted intensity, at 3 ps time delay, versus optical densities put in the common path of the pump pulses

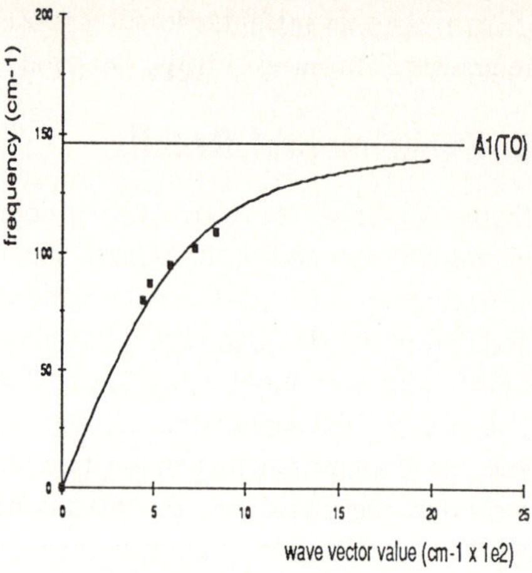

Fig. 3. Evolution of the polariton frequency versus phonon wave vector value and fit to the experiment using a single oscillator model

4. Conclusion

Using femtosecond pulses in a forward arrangement, coherent optical polariton oscillations and decay have been observed for the first time in the ferroelectric phase of $PbTiO_3$. From its dispersion behavior, an accurate value of $\varepsilon_{||}(0)$ has been deduced.

Complementary work in the vicinity of the phase transition temperature is in progress: it will take advantage of a time-domain approach over a frequency-domain one to distinguish between an overdamped oscillation and a relaxational motion [7].

Acknowledgement. The authors are grateful to K. Wojcik of the Institute of Physics in Katowice (Poland) for supplying the $PbTiO_3$ single crystal.

References

1. Y.X. Yan, E.B. Gamble, Jr., K.A. Nelson: J. Chem. Phys. **83**, 5391 (1985)
2. J. Etchepare, G. Grillon, A. Antonetti, J.C. Loulergue, M.D. Fontana, G.E. Kugel: Phys. Rev. B, to be published
3. G. Burns, B.A. Scott: Phys. Rev. B **7**, 3088 (1973)
4. R.A. Frey, E. Silberman: Helv. Phys. Acta **49**, 1 (1976)
5. J. Etchepare, G. Grillon, J. Arabat: Appl. Phys. B **49**, 425 (1989)
6. R.Claus, L. Merten, J. Brandmuller: *Light Scattering by Phonon-Polaritons* (Springer, Berlin, Heidelberg 1975)
7. M.D. Fontana, H. Idrissi, K. Wojcik: Europhys. Lett. **11**, 419 (1990)

Time-Resolved Electron–Hole Plasma and Optical Phonon Interaction in GaP

G.O. Smith, T. Juhasz, and W.E. Bron

Department of Physics, University of California, Irvine, CA92717, USA

Abstract. The interaction of an optically generated non-stationary electron-hole plasma with coherently excited optical phonons has been investigated through TR-CARS. A non-exponential phonon decay is observed in the presence of a non-stationary two-component plasma.

1. Introduction

Gallium phosphide is of increasing importance in the generation of ultrashort laser pulses due to its high index of refraction [1]. Thus the response of this crystal to pulses of high irradiance becomes increasingly important. Toward this end, we have investigated the interaction of a non-stationary electron-hole plasma (NEHP), injected by two-photon absorption (TPA) of intense picosecond laser pulses, with coherently excited longitudinal optical (LO) phonons.

2. Experimental

Time-resolved coherent anti-Stokes Raman scattering (TR-CARS) is used to carry out the experiment [2]. Pulses of 6ps duration from a dual synchronously pumped dye laser system are amplified to the micro-Joule energy level using a Nd:YAG regenerative amplifier as a pumping source [3]. The difference in output frequencies of the system $(\omega_l - \omega_s)$ is tuned to the LO phonon frequency and part of the ω_l beam is used to probe the coherence of the LO phonons, thus $\omega_p = \omega_l$. Approximately 90% of the ω_s beam is used to inject the NEHP via TPA into the interaction volume. The intensities of the ω_l, ω_s , and ω_p beams are kept sufficiently weak to avoid additional TPA and intense LO phonon generation that could cause non-exponential dephasing of the LO phonons [4]. The four beams are focused into a pure GaP crystal kept at 5K. Data was collected alternately with and without NEHP injection at a set of fixed probe delay times in order to avoid the effects of long term laser fluctuations.

3. Results and Discussion

Figure 1 contains a typical experimental result for the intensity of the TR-CARS signal as a function of the probe delay. The upper curve, which rep-

Springer Series in Chemical Physics, Vol. 53 **Ultrafast Phenomena VII**
Editors: C.B. Harris · E.P. Ippen · G.A. Mourou · A.H. Zewail

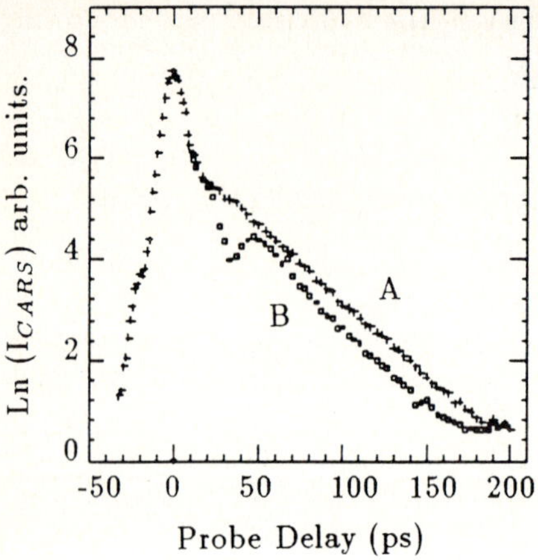

Fig. 1. Ln of the TR-CARS signal as a function of the delay time. A: without NEHP, B: with NEHP.

resents the LO phonon dephasing without NEHP injection, clearly shows an exponential decay. The lower curve displays the LO phonon dephasing with NEHP injection. The dip at $\sim$ 30ps delay is consistent with the timing of the NEHP injection pulse, but its origin is not as yet understood. An increase in the dephasing rate due to the presence of the NEHP occurs after the dip in a $\sim$ 100ps long time interval and appears to be non-exponential. This is indicated in figure 2 which plots the difference (with and without NEHP) between the two dephasing curves. The solid line is a second order polynomial fit to the data.

We offer the following explanation of the results. The excited LO phonons and the plasma oscillations can be treated as two damped vibrational systems coupled through the electric field. Due to the low density of the electron-hole gas ($n \simeq 5 \times 10^{16}\text{cm}^{-3}$), the plasma frequency is much smaller than that of the LO phonons. Under these conditions, the solution to the coupled system gives an expression for the temporal evolution of the coherent

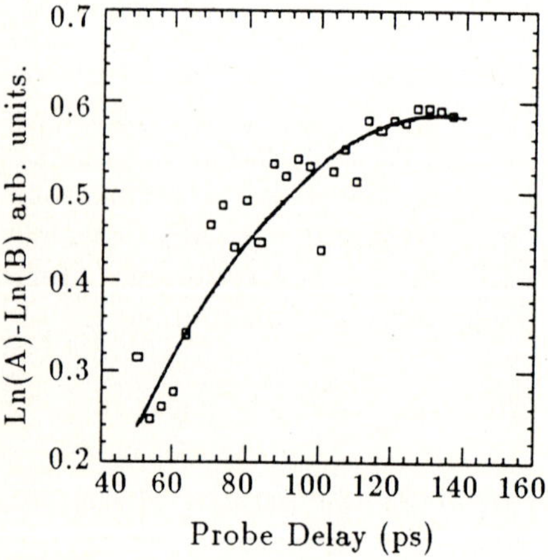

Fig. 2. The difference in the Ln of the two decay curves (A and B) as a function of the delay time.

intensity of the dephasing optical phonons which can be written in the following form [5]:

$$|A(t)|^2 = |A(0)|^2 \exp(-\Gamma t) \times \exp(-\beta \int_0^t \gamma(t')dt')$$

where $A(0)$ is the phonon amplitude at $t = 0$, Γ is the phonon dephasing constant, β is the coupling constant between the LO phonons and the plasmons, $\gamma(t)$ is the damping of the plasma oscillations.

In order to understand the behavior of the optical phonon dephasing, the evolution of the NEHP, leading to an expression for $\gamma(t)$, must be considered. After their injection, the carriers lose energy though a cascade of steps associated with single LO phonon emission [6]. This results in the formation of a NEHP near the conduction band minimum after $\sim$ 7ps following the excitation pulse [5]. We note that, in the forward scattering geometry used here, single carriers do not interact with the coherently excited LO phonons due to wave vector mismatch [7]. After formation, the estimated plasma temperature is $\sim$ 180K. The plasma continues to cool over the next $\sim$ 60ps time interval [5]. At the begining of this interval, the plasma damping is relatively strong due to the short scattering time of the relatively hot carriers. This results in an increase of LO phonon dephasing through the coupling of the two vibrations. As the plasma cools down, however, the plasma damping decreases, resulting in a decreasing effect on the LO phonon dephasing.

4. Summary

In conclusion, we have shown experimentally that a low density ($n \simeq 5 \times 10^{16} cm^{-3}$) NEHP, injected by TPA of a strong laser pulse, influences the dephasing of LO phonons through the coupling of the two vibrational systems. The observed non-exponential dephasing can be explained by the decreasing plasma damping as the NEHP cools down.

The authors acknowledge support through NSF DMR-86-03888, DMR-89-13289 and ARO DAAL 03-89-k-0060.

References

[1] T. Sizer II, Appl. Phys. Lett. **55**, p. 2694 (1989).
[2] W. E. Bron, J. Kuhl and B. K. Rhee, Phys. Rev. B. **34**, p. 6961 (1986).
[3] T. Juhasz, J. Kuhl, and W. E. Bron, Opt. Lett. **13**, p. 577 (1988).
[4] W. E. Bron, T. Juhasz and S. Mehta, Phys. Rev. Lett. **62**, p.1655 (1989).
[5] I. B. Levinson, Private communication.
[6] J. A. Kash, J. C. Tsang, and J. M. Hvam, Phys. Rev. Lett. **54**,p. 2151 (1985).
[7] W. E. Bron, in "Phonon Scattering in Condensed Matter V", ed. by A. C. Anderson and J. P. Wolfe, (Springer, Berlin), 1986.

Part IX

Superconductors

Far-Infrared Spectroscopy of High Temperature Superconductors with Terahertz Electrical Pulses

M.C. Nuss, K.W. Goossen, P.M. Mankiewich, M.L. O'Malley, J.L. Marshall, and R.E. Howard

AT&T Bell Laboratories, Crawfords Corner Road,
Room 4C-330, Holmdel, NJ 07733, USA

In recent years, a variety of techniques have been developed to generate and detect ultrafast electromagnetic transients with bandwidths up to 1 THz [1]. The spectrum of these transients spans part of the microwave, submillimeter wave and far-infrared range of the electro-magnetic spectrum. The spectroscopic benefits include the enormous spectral bandwidth covered, the coherent detection that enables to extract both real and imaginary part of the dielectric response function, and a sensitivity that far exceeds the sensitivity of conventional far-infrared spectrometers in the same spectral range.

Here, we use subpicosecond electrical pulses to measure the dielectric response of high-quality $YBa_2Cu_3O_7$ high-temperature superconducting films for frequencies below the superconducting bandgap. A transmission line geometry is employed to obtain a large interaction length between the electrical pulses and the superconductor (Fig. 1). Electrical pulses of ~750 fs duration are generated with a photoconductive switch integrated on a coplanar stripline. A flip-chip contact launches this transient on the superconducting transmission line patterned from a 2000 Å thick $YBa_2Cu_3O_7$ film epitaxially grown on a $LaAlO_3$ substrate. Input and propagated pulses are then probed by electro-optic sampling in reflection mode with a cryogenic electro-optic prober.

This experiment is capable to measure the complex dielectric response of the superconductor. The electrical transient suffers both absorption and dispersion as it

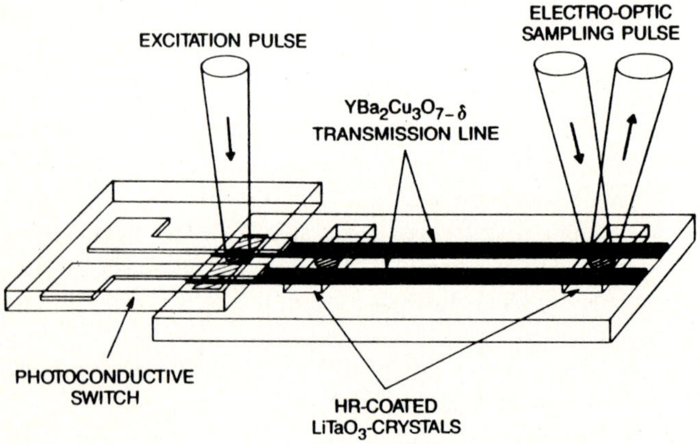

Fig. 1 Transmission line geometry to measure absorption and phase shifts.

Springer Series in Chemical Physics, Vol. 53 **Ultrafast Phenomena VII**
Editors: C.B. Harris · E.P. Ippen · G.A. Mourou · A.H. Zewail

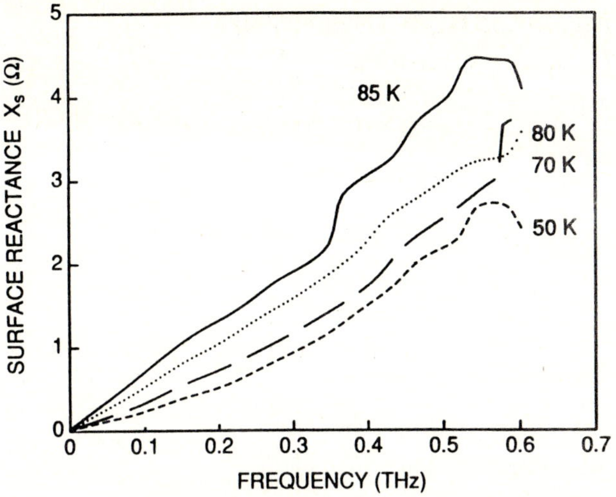

Fig. 2 Measured imaginary part X_s of the surface impedance.

propagates down the line; $E(\omega) = E_0(\omega) \cdot e^{-(\alpha + i\beta)x}$. Both α and β can be reduced to either the complex surface impedance Z_s or alternatively to the complex conductivity $\sigma(\omega)$ of the intrinsic superconductor if the geometrical parameters of the transmission line are known [2,3].

Fig. 2 shows the imaginary part X_s of the surface impedance (obtained from the phase shift $\beta_s x$) over the broad frequency range from 10 GHz to 0.5 THz for several temperatures below the transition temperature $T_c = 91\,K$. The important feature of X_s is that it is only determined by the magnetic penetration depth and $X_s = \omega\mu_0\lambda$ independent of the theoretical model used for the superconductor. Evidently the linear relation of X_s with ω is satisfied for T=85 K and 80 K. For the lower temperatures, the curves are slightly sublinear, pointing to a frequency dependent penetration depth $\lambda(\omega)$. From a linear fit at low frequencies, we obtain $\lambda(0) = 1800\,\mathring{A}$ and $\lambda(77\,K) = 3000\,\mathring{A}$, both roughly 20% higher than the penetration depth measured in single crystals from muon spin resonance [4].

The real part R_s of the surface impedance (which is proportional to the absorption coefficient) over the same frequency range is plotted in Fig. 3 as bullets (•••). All curves have a slope $\omega^{1.8}$ below 200 GHz. This is different from the 2-fluid model, which predicts a strictly quadratic dependence on frequency, but close to the Mattis-Bardeen (MB) model [5] (solid line in Fig. 2). The offset in magnitude between experiment and MB-theory is only due to the slightly larger penetration depth measured in our films compared to $\lambda(0) = 1400\,\mathring{A}$ assumed for the MB-calculations. The MB-theory also predicts the reduction in losses observed at frequencies above 200 GHz.

The good agreement of our data with the Mattis-Bardeen theory is surprising, as it is only applicable to either the extreme anomalous limit or the dirty limit, which are both not satisfied in $YBa_2Cu_3O_7$ superconductors because of the extremely short coherence length ξ_0. On the other hand, full BCS calculations [6] (□□□ in Fig. 3)

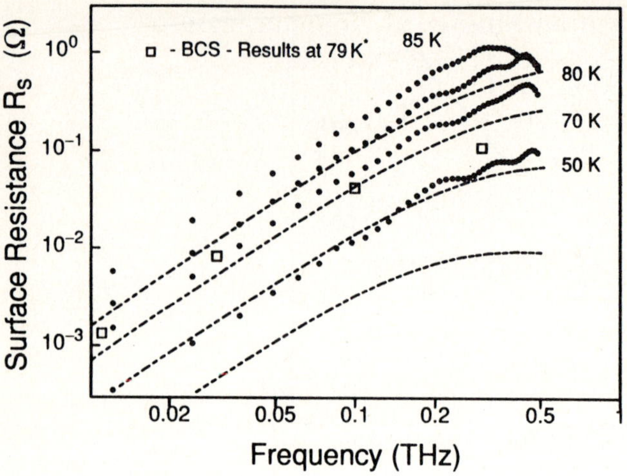

Fig. 3 Measured surface resistance R_s (•••) compared to Mattis-Bardeen model (solid line) and BCS-model (□□□).

agree well with our data at low frequencies but deviate at the high frequency end. We attribute these findings to the two-dimensionality of the $YBa_2Cu_3O_7$ superconducting films which has been neglected in the calculations.

1. D. H. Auston, in *Ultrashort Laser Pulses and Applications*, edited by W. Kaiser, Springer-Verlag (1988)
2. M. C. Nuss, and K. W. Goossen, IEEE Journ. Quant. Electr. **25**, 2596 (1989)
3. M. C. Nuss, K. W. Goossen, P. M. Mankiewich, R. E. Howard, B. L. Straughn, G. W. Berkstresser, and C. D. Brandle, IEEE Electr. Dev. Lett., **11**, 200 (1990)
4. D. R. Harshman et al., Phys. Rev. **B 39**, 851 (1989)
5. D. C. Mattis, J. Bardeen, Phys. Rev. Lett. **111**, 412 (1958)
6. J. Halbritter, priv. comm. and Z. Physik, **266**, 209 (1974)

Femtosecond Absorption Studies of Nonequilibrium Electronic Processes in High-T_c Superconductors

J. Chwalek[1], *C. Uher*[2], *S. Gupta*[1], *J. Whitaker*[1], *G. Mourou*[1], *J. Agostinelli*[3], *and M. Lelental*[3]

[1]Ultrafast Science Laboratory, University of Michigan, Ann Arbor, MI48109, USA

[2]Department of Physics, University of Michigan, Ann Arbor, MI48109, USA

[3]Corporate Research Laboratories, Eastman Kodak Co., Rochester, NY 14650, USA

Abstract. Femtosecond laser pulses are used to probe nonequilibrium carrier distributions in high-T_c superconducting films. An increase in the relaxation time and peak fractional transmissivity change was observed when the films became superconducting. No such changes were observed for a nonsuperconducting film.

The ability to generate and time resolve nonequilibrium electron distributions using short laser pulses (laser pulse durations less than or comparable to electron-phonon energy relaxation time) has produced important information regarding the thermal properties of the carriers and phonons in metals [1,2]. Short duration laser pulses have also been used to study the carrier dynamics in nonsuperconducting [3] and superconducting [4] cuprate compounds at room temperature and below the critical temperature [5]. In this paper, we present evidence of nonequilibrium carrier heating on a subpicosecond time scale in thin-film samples of the superconducting compounds $YBa_2Cu_3O_{7-x}$ ($x \sim 0$) and $Bi_2Sr_2Ca_2Cu_3O_{10+\delta}$ ($\delta \sim 0$) and nonsuperconducting $YBa_2Cu_3O_{6+y}$ ($y < 0.4$).

From recent optical studies the broad absorption feature centered near 2.5 eV common to the copper-oxide perovskite materials is believed to be due to an interband transition [6]. If the optically probed states are near the Fermi level a thermomodulation process can occur.

Our experimental set-up is similar to transient thermomodulation pump-probe experiments reported in Ref. 1. In our case the short pulse light source was a colliding-pulse mode-locked dye laser that produced pulses of ~ 100-fs duration at a 100-MHz repetition rate and a wavelength of 630 nm (2 eV). The pump beam provided ultrafast heating which in turn produced a change in the refractive index. This was directly measured with a second probing beam (polarized perpendicularly to the pump beam) by reflecting it off the surface of the film, or by passing it directly through the semi-transparent film. The resultant change in reflectivity or transmissivity was sensed by a photodiode and recorded versus the pump-probe delay time. The measurements could be performed from ~ 7 K to room temperature.

The c-axis oriented $Bi_2Sr_2Ca_2Cu_3O_{10}$ film used for our study was produced using a metallo-organic decomposition (MOD) technique on a MgO substrate [7]. Ac susceptibility measurements of the sample indicated an onset to the superconducting state at 108 K. Shown in Fig. 1 is the normalized transmissivity change versus pump-probe delay time obtained for the BSCCO film at several sample temperatures. The essential features of the data which provide evidence for a nonequilibrium carrier effect include a rapid transient near 0 ps which follows the

Editors: C.B. Harris · E.P. Ippen · G.A. Mourou · A.H. Zewail

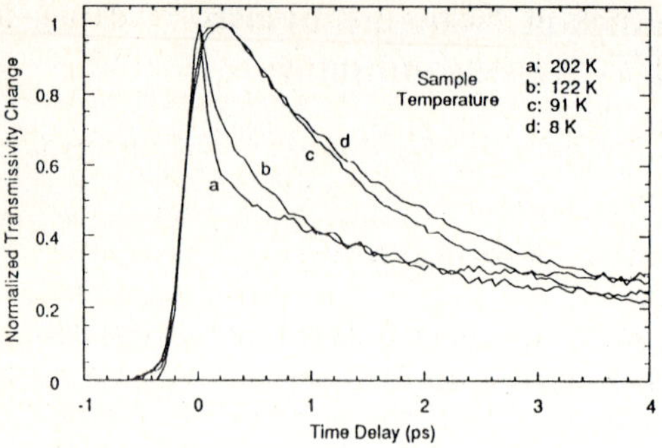

FIG. 1. The transmissivity change normalized to unity is plotted relative to the pump-probe delay time for the BSCCO film at a few sample temperatures. The 0 ps delay time was chosen to coincide with the maximum transmissivity change of the 202 K trace.

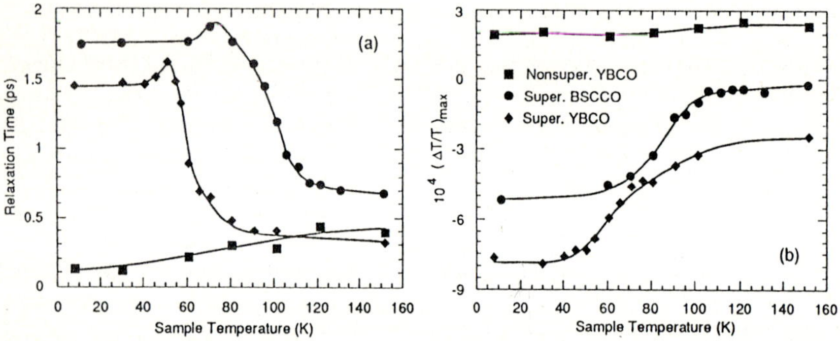

FIG. 2. (a) The relaxation time as a function of sample temperature is shown for the superconducting BSCCO, superconducting YBCO, and nonsuperconducting YBCO samples. (b) The peak fractional transmissivity change vs. sample temperature for the films in (a). The lines through the data points serve *only* as a guide for the eye.

cross-correlation of the pump and probe pulses and a consequent relaxation which occurs on a sub-picosecond time scale. In measured d-band metals this type of signature can be interpreted as an initial heating of the conduction electrons which thermalize rapidly via electron-electron scattering. This hot electron distribution consequently thermalizes with the lattice by energy transfer through electron-phonon scattering. For later delay times heat diffusion out of the illuminated volume of the film will occur. The most interesting feature of the data is the distinct, dramatic increase in the relaxation time that occurs for sample temperatures below the critical temperature of the film. Another interesting feature for sample temperatures below T_c is a delay in the rising edge near 0 ps.

This dependence of the relaxation time with sample temperature for the BSCCO film is given in Fig. 2a. As shown in the figure, the relaxation time actually increases slightly with increasing temperature as T_c is approached before dropping. Accompanying the relaxation time change is a substantial increase in the peak fractional transmissivity change, $(\Delta T/T)_{max}$, as the sample is cooled below T_c (see Fig. 2b). Neither of these abrupt changes are expected in conventional metals. Also shown in Figs. 2a and 2b is data obtained from superconducting and nonsuperconducting YBCO samples. Both samples came from the same c-axis oriented thin-film deposited on a MgO substrate using a standard laser ablation technique. The superconducting YBCO sample revealed an onset to diamagnetism at

87 K. The oxygen content of the second sample was reduced rendering it nonsuperconducting by heating it in flowing argon. Similar behavior is observed for the superconducting YBCO sample with the exception that the temperature dependence was shifted accordingly by the critical temperature. These changes were not observed for the nonsuperconducting YBCO sample. The only change observed is a gradual increase of the relaxation time with increasing sample temperature eventually surpassing the relaxation time of either superconducting sample.

From Fig. 2b it can be seen that the peak fractional transmissivity change is positive for the nonsuperconducting sample; this is opposite to the change measured for the superconducting samples. From simultaneous measurements of the fractional transmissivity and reflectivity changes, the change in the imaginary part of the complex dielectric function, $\Delta\varepsilon_2$, can be deduced. This measurement was performed at room temperature. A positive $\Delta\varepsilon_2$ consistent with probing unoccupied states below the Fermi level was found for the superconducting YBCO and BSCCO samples while a negative $\Delta\varepsilon_2$ was found for the nonsuperconducting YBCO sample Removing oxygen in the superconducting YBCO sample will reduce the number of carriers which can depress the Fermi level. If the Fermi level is reduced sufficiently so that occupied states lying above the Fermi level are being probed, a polarity reversal will ensue. Although the Fermi surface for these materials is not well known, a sign reversal is consistent with probing states lying near the Fermi level, lending further support to the interpretation of the measured transients being a direct result of a nonequilibrium carrier distribution.

Because no structural phase transition is believed to accompany the entry into the superconducting domain in these high-T_c compounds and due to the extremely fast time scales involved, we believe the observed behavior is electronic in nature and intimately related to the superconductivity. It is well known for BCS superconductors that as the material is cooled below its critical temperature the density of states exhibits an increase just below the Fermi level and a decrease at the Fermi level. This behavior corresponds to the opening of the energy gap in the superconductor. This change in the density of states close to the Fermi level should then alter the Fermi smearing, although further studies are needed.

This work was supported by the AFOSR, University Research Initiative under contract number F49620-87-C-0016 and the Eastman Kodak Company.

1. H. E. Elsayed-Ali, T. B. Norris, M. A. Pessot, and G. A. Mourou, Phys. Rev. Lett. **58**, 1212 (1987).
2. R. W. Schoenlein, W. Z. Lin, J. G. Fujimoto, and G. L. Eesley, Phys. Rev. Lett. **58**, 1680 (1987).
3. G. L. Doll, G. L. Eesley, S. D. Brorson, M. S. Dresselhaus, G. Dresselhaus, A. Cassanho, H. P. Jenssen, and D. R. Gabbe, Appl. Phys. Lett. **55**, 402 (1989).
4. S. D. Brorson, A. Kazeroonian, D. W. Face, T. K. Cheng, G. L. Doll, M. S. Dresselhaus, G. Dresselhaus, E. P. Ippen, T. Venkatesan, X. D. Wu, and A. Inam, 1990 American Physical Society Meeting, Anaheim,CA, 15 March 1990.
5. J. M. Chwalek, C. Uher, J. F. Whitaker, G. A. Mourou, J. Agostinelli, and M. Lelental, submitted to Appl. Phys. Lett.
6. M. K. Kelly, P. Barboux, J.-M. Tarascon, and D. E. Aspnes, Phys. Rev. B **40**, 6797 (1989).
7. J. A. Agostinelli, G. R. Paz-Pujalt, and A. K. Mehrotra , Physica C **156**, 208 (1988).

Femtosecond Thermomodulation Study of Conventional and High-T_c Superconductors

S.D. Brorson[1], *A. Kazeroonian*[2], *J.S. Moodera*[3], *D.W. Face*[1], *T.K. Cheng*[1], *E.P. Ippen*[1], *M.S. Dresselhaus*[2], *G. Dresselhaus*[3], *G.L. Doll*[2], *T. Venkatesan*[4], *X.D. Wu*[5], *and A. Inam*[5]

[1]Department of Electrical Engineering and Research Laboratory of Electronics, Massachusetts Institute of Technology, Cambridge, MA 02139, USA
[2]Department of Physics and Center for Material Science and Engineering, Massachusetts Institute of Technology, Cambridge, MA 02139, USA
[3]Francis Bitter National Magnet Laboratory, Cambridge, MA 02139, USA
[4]Bellcore, Red Bank, NJ 07701, USA
[5]Department of Physics, Rutgers University, Piscataway, NJ 08854, USA

Abstract. We report femtosecond thermomodulation measurements of the electron-phonon coupling constant λ in metallic and high-T_c superconductors. In metallic superconductors, excellent agreement with other measurements of λ is obtained. For high-T_c materials, we find qualititive agreement with theoretical expectations.

The Eliashberg generalization of BCS theory includes a parameter λ describing the strength of the electron – phonon coupling constant.[1] In 1987, P. B. Allen made the suggestion that λ might be measured in the metallic superconductors using femtosecond thermomodulation (pump – probe) spectroscopy.[2] Such experiments [3,4] have been performed on Cu and Au, although not in the context of measuring λ. Here we report the results of systematic femtosecond thermomodulation measurements of λ in metallic thin films of Cu, Au, Cr, Ti, W, Nb, V, Pb, NbN, and V_3Ga. We find excellent aggreement between our measured λ values and those obtained by other means, thus confirming Allen's theory. This method to measure λ has several advantages over other techniques: it is the most direct measure of $\lambda\langle\omega^2\rangle$, it works at room temperature, it can be applied to both superconducting and non-superconducting samples, and it is not affected by interactions competing with superconductivity such as the antiferromagnetism present in Cr.

In our experiment, the metal is pumped with a 60 fs pulse of $\sim$ 2 eV light from a CPM laser. Immediately after the arrival of the pump pulse, the electronic temperature is higher than that of the lattice, since the electron energy loss rate is not fast enough to maintain equilibrium with the lattice. Then, the electrons lose energy to the lattice via phonon emission; the electron temperature equilibrates to the lattice temperature on a time scale determined by the electron-phonon coupling strength. In many metals, the optical reflectivity depends on the electron temperature.[5] This allows the electron's temperature recovery process to be mapped out in time by monitoring the sample's change in reflectivity (ΔR) using a variably delayed probe pulse. Qualitatively, the central feature of Allen's theory is that metals with large T_c have strong electron-phonon coupling, and hence will show faster relaxation times than those with weaker coupling and lower T_c. Quantitatively, his theory gives an expression for the relaxation

Springer Series in Chemical Physics, Vol. 53 **Ultrafast Phenomena VII**
Editors: C.B. Harris · E.P. Ippen · G.A. Mourou · A.H. Zewail

rate having only $\lambda\langle\omega^2\rangle$ as a free parameter; the measured relaxation rate can be fit to directly determine $\lambda\langle\omega^2\rangle$.

In Fig. 1 is shown ΔR *vs.* delay t data for several representative metals. A computer generated least squares fit to the relaxation is shown as the solid line. From the fit we extract $\lambda_{exp}\langle\omega^2\rangle$; values of this quantity are presented in Table I. Using literature values of $\langle\omega^2\rangle$ we derived the value of λ_{exp}. For comparison are shown literature values of λ_{lit} determined by other methods (where they exist). As can be seen, the agreement is excellent.

We have also extended this technique to the new high-T_c copper oxide based superconductors. Three oriented superconducting thin-films were studied: $YBa_2Cu_3O_{7-\delta}$, $Bi_2Sr_2CaCu_2O_{8+x}$, and $Bi_2Sr_2Ca_2Cu_3O_{10+y}$. This time, by monitoring changes in both reflection and transmission, we measure the change of the imaginary part of the dielectric function, $\Delta\epsilon_2$. Measuring $\Delta\epsilon_2$ is important since very little is known about the electronic structure of these materials. The measured $\Delta\epsilon_2$ *vs.* delay dynamics are shown in Fig. 2. In the inset are shown the raw ΔR and ΔT data. In all cases, the observed recovery time is fast, occurring on a $\approx$ 100 fs time scale.

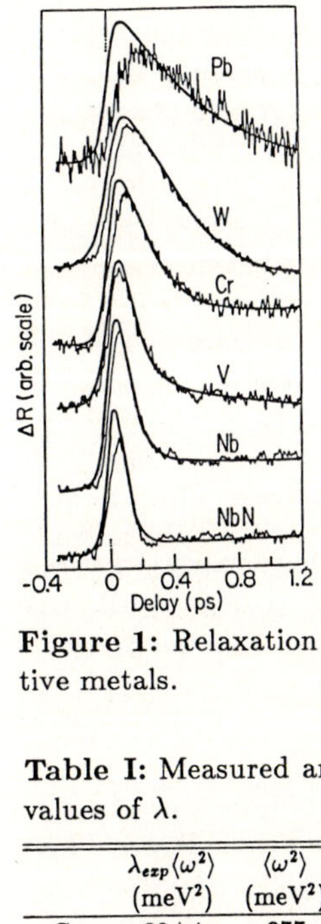

Figure 1: Relaxation data in representative metals.

Table I: Measured and literature values of λ.

	$\lambda_{exp}\langle\omega^2\rangle$ (meV^2)	$\langle\omega^2\rangle$ (meV^2)	λ_{exp}	λ_{lit}
Cu	29±4	377	0.08±0.01	0.10
Au	23±4	178	0.13±0.02	0.15
Cr	128±15	987	0.13±0.02	-
W	112±15	425	0.26±0.04	0.26
V	280±20	352	0.80±0.06	0.82
Nb	320±30	275	1.16±0.11	1.04
Ti	350±30	601	0.58±0.05	0.54
Pb	45±5	31	1.45±0.16	1.55
NbN	640±40	673	0.95±0.06	1.46
V_3Ga	370±60	448	0.83±0.13	1.12

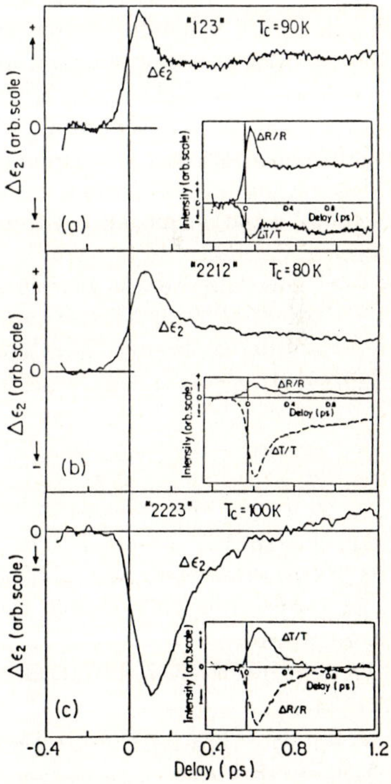

Figure 2: Relaxation data in high-T_c materials.

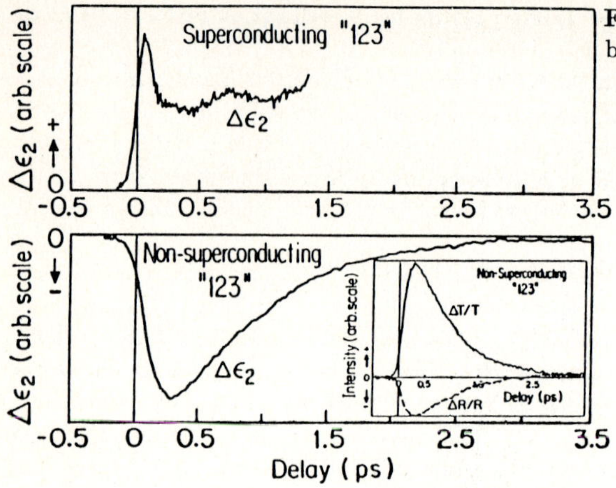

Figure 3: Relaxation dynamics before and after deoxygenation.

The time development of $\Delta\epsilon_2$ can be related to the dynamics of the Cu d to O p band charge transfer excitation occurring in the Cu – O planes[6]. Since the optically probed states lie near the Fermi level, the dynamic change of ϵ_2 is possibly the result of a thermomodulation process similar to that occurring in Cu and Au. In this picture, the observed fast recovery of $\Delta\epsilon_2$ can again be attributed to non-equilibrium carrier temperature relaxation.

To confirm that a thermomodulation process is indeed occurring, we deoxygenated a 123 sample. Deoxygenation removes carriers from the sample (thus decreasing the Fermi level) and simutaneously depresses T_c. $\Delta\epsilon_2$ *vs.* delay t data are shown in Fig. 3 for a superconducting and a non-superconducting $YBa_2Cu_3O_{7-\delta}$ sample. The sign of $\Delta\epsilon_2$ reverses in the non-superconducting sample, consistent with our experiment probing states near the Fermi level, and the observed relaxation time increases by over an order of magnitude, qualitatively consistent with the expectation that smaller T_c implies weaker coupling and slower relaxation.

This work was supported in part by JSEP contract DAAL03-89-C-0001, AFOSR contract F49620-88-C-0089, AFOSR contract F19628-85-C-0002, Air Force agreement BX2961 (administered through MIT Lincoln Laboratory and NSF contract DMR87-FG05-85ER45151.

[1] W. L. McMillan, Phys. Rev. **167**, 331 (1968); P. B. Allen and R. C. Dynes, Phys. Rev. B **12**, 905 (1975).

[2] P. B. Allen, Phys. Rev. Lett. **59**, 1460 (1987).

[3] H. E. Elsayed-Ali, T. B. Norris, N. A. Pessot, and G. A. Moreau, Phys. Rev. Lett. **58**, 1837 (1987).

[4] R. W. Schoenlein, W. Z. Lin, J. G. Fujimoto, and G. L. Eesley, Phys. Rev. Lett. **58**, 1680 (1987).

[5] W. J. Scouler, Phys. Rev. Lett. **18**, 445 (1967); R. Rosei and D. W. Lynch, Phys. Rev. B **5**, 3883 (1972).

[6] M. K. Kelly, P. Barboux, J.-M. Tarascon, and D. E. Aspnes, Phys. Rev. B. **40**, 6797 (1989).

Rate of Electron–Phonon Relaxation in Niobium

K.M. Yoo[1], *X.M. Zhao*[1], *M. Siddique*[1], *R.R. Alfano*[1], *D.P. Osterman*[2], *M. Radparvar*[2], *and J. Cunniff*[2]

[1]Institute for Ultrafast Spectroscopy and Lasers, Department of Electrical Engineering, City College of the City University of New York, New York, NY 10031, USA
[2]HYPRES Inc., 175 Clearbrook Road, Elmsford, NY 10523, USA

Thermal modulation measurements using 120 fs / 620 nm laser pulses were performed on niobium films. The electron-phonon relaxation time was found to be ultra fast and measured to be 360 fs, and the electron-phonon coupling constant was estimated to be $17\times 10^{17}\ Wm^{-3}K^{-1}$.

Niobium is an important superconducting metal for the fabrication of Josephson junctions and other superconducting devices. One of the parameters which governs the ultimate switching speed of a Josephson junction is the electron-phonon relaxation time[1] τ_{ep}. The dynamics of the nonequilibrium superconducting state is governed in part by τ_{ep}. Nonequilibrium superconductivity[2] has been studied in the past by driving the junction into the nonequilibrium state. Using ultrafast laser excitation on the Josephson junction or weak link[3-4], the transition from superconducting to normal and back to superconducting state is monitored by the transient change in voltage across the junction[3-6]. The rise time in the transient change of voltage is mainly governed by the electron-phonon relaxation time which has been estimated to be on the order of a picosecond. This ultrafast time scale cannot be resolved with the fastest voltage measurement instrumentation presently available. In this paper, the electron-phonon relaxation time and electron-phonon coupling constant of niobium was measured by a time resolved thermal modulation[7-11] optical transmission technique.

The experiments were performed using ultrafast laser pulses which were generated from a colliding pulse mode-locked (CPM) dye laser and amplifier system. A copper vapor laser at a repetition rate of 6.6 kHz amplified the output of the CPM laser. The full width at half maximum of the amplified laser pulse is 120 fs and the spectrum of the laser is centered at 620 nm. The laser beam was split into a strong pump and a weak probe beam. The energy of the pump beam is 50 nJ per pulse and the energy of the probe beam was set to about 4% of the pump beam energy. The pump-probe delay is controlled by a motorized translational stage with 1 μm accuracy. The probe beam is polarized either parallel or perpendicular with respect to the polarization of the pump beam. Both the pump and probe beam are focused to a spot of about 34 μm diameter on the niobium film at room temperature of 300 K. Niobium films of 20, 50, and 100 nm thickness were deposited on fused quartz substrates by DC magnetron sputtering. The pump beam was chopped so that the difference between pumped and unpumped in probe beam transmission was measured by the lock-in amplifier.

The change in transmission at various pump-probe delay times ($\Delta T/T$ *vs* t) is plotted in Figs. 1. Fig. 1(a) shows the salient feature when the polarization of the probe beam is perpendicular to the

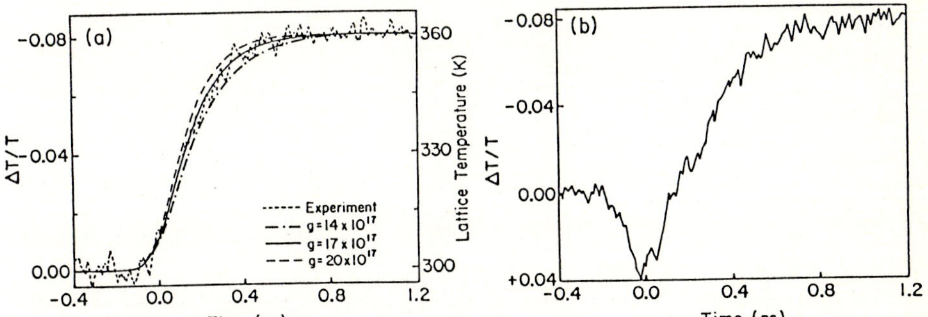

Fig. 1 Time-resolved thermal modulation change of transmittivity of 20 nm Nb film. (a) Experiment - probe beam is polarized perpendicular to the pump beam. The three numerical curves are computed from equations (1) and (2) for the lattice temperature. (b) Experiment - probe beam is polarized parallel to the pump beam.

Springer Series in Chemical Physics, Vol. 53 **Ultrafast Phenomena VII**
Editors: C.B. Harris · E.P. Ippen · G.A. Mourou · A.H. Zewail

pump beams. An initial fast change in transmission is observed, characterized by a 380 fs rise-time measured from 10% to 90% of the total rise. The actual rise time is estimated to be 360 fs which is obtained by deconvolution of the measured rise-time with the laser pulse width, i.e. $(380^2-120^2)^{1/2}$. The maximum change in transmission is about 8%. Fig. 1(b) shows the presence of a coherent artifact when the probe beam is polarized parallel to the pump beam. The profile of the coherent artifact is observed as a dip, with temporal profile similar to the laser profile. The change in transmission curve has a long decay time of about a nanosecond. This long decay time may be related to the time taken for the lattice to cool back to room temperature; that is, the time taken for the excess phonons created in the film to escape to the substrate. Thus, the observed change in transmission may be attributed to the change in lattice temperature. The fast relaxation in transmission due to a falling hot electron temperature[8,9] (Fermi level smearing) is not observed because the sharp d-band to Fermi level transition for niobium does not fall within the probing laser energy.

In a simple physical model, the electrons absorb the laser energy and get excited to a higher energy state, i.e. to a higher temperature[8-10]. These hot electrons cool down to the lattice temperature through electron-phonon scattering. The time evolution of the hot electron and lattice temperatures may be approximated by the following energy rate equations:

$$C_e \frac{\delta T_e}{\delta t} = \kappa \nabla^2 T_e - g(T_e - T_l) + S(r,t) \quad , \tag{1}$$

$$C_l \frac{\delta T_l}{\delta t} = g(T_e - T_l) \quad , \tag{2}$$

where $C_e = 720 T_e \; Jm^{-3}K^{-2}$ is the heat capacity of the electrons, $C_l = 2.3\times 10^6 \; Jm^{-3}K^{-1}$ is the heat capacity of the phonons (lattice), g is the electron-phonon coupling constant, κ is the diffusion coefficient for the electrons, and $S(r,t)$ is the source term. The diffusion term in equation (1) for the case of transmission geometry may be neglected on a picosecond time scale. The source term $S(r,t)$ is equivalent to the laser pulse profile which is approximated to be Gaussian. The peak power of the laser is $1.3\times 10^{15} W/m^2$ for 50 nJ pulse at 120 fs. The niobium film is estimated to absorb 1.5% of the laser energy. These two coupled equations are solved numerically for the electron temperature T_e and lattice temperatures T_l. The curve for the lattice temperature is fitted to the experimental curve in Fig. 1(a). The fitting parameter used is the electron-phonon coupling constant g. This fitting is valid assuming that the transmission of the niobium film is linearly related to the lattice temperature of the film. The rise time of the lattice temperature becomes shorter with stronger electron-phonon coupling or lower pumping energy $S(r,t)$, or mathematically with larger value of g or smaller value of $S(r,t)$. The best fit of the computed curve for lattice temperature to the experimental data in Fig. 1(a) is $g = 17\times 10^{17} \; Wm^{-3}K^{-1}$. This numerical data is traced by the solid line in Fig. 1(a). The dash-dash and dot-dash curves are the computed lattice temperature for $g = 20\times 10^{17} \; Wm^{-3}K^{-1}$ and $g = 14\times 10^{17} \; Wm^{-3}K^{-1}$, respectively.

The computed hot electron and lattice temperatures for Nb and Cu are plotted in Fig. 2(a) and (b) for the same pump intensity, respectively. The solid curve is for niobium and the dot-dash curve is for copper. The hot electron temperature of the copper rises to a peak of 1660K which is significantly higher than the peak electron temperature of niobium which is $580K$. The peak lattice temperature for niobium is $360K$ which is higher than copper (344K). The electron-phonon cou-

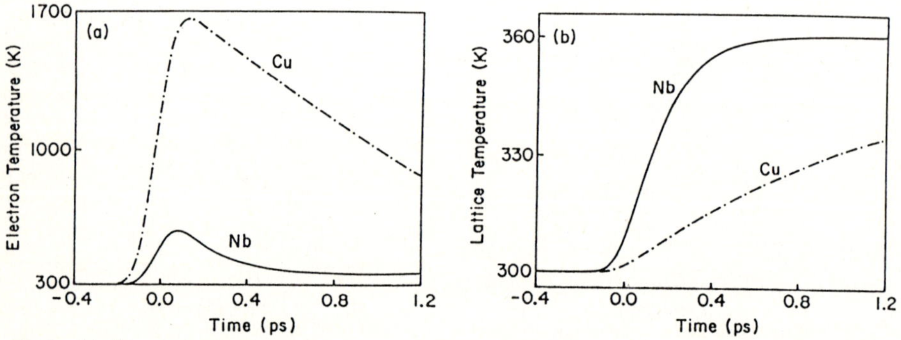

Fig. 2 (a) Computed hot electron temperature in time for copper and niobium. (b) Computed phonon temperature in time for copper and niobium.

pling constant of copper[9] has been estimated to be $1\times10^{17}\ Wm^{-3}K^{-1}$. Thus, the electron-phonon coupling constant of niobium is 17 times higher than copper. This large electron-phonon coupling constant for niobium increases the rate of electron transfer of its excess energy to the lattice. Thus, reducing the time for the hot electron to transfer its excess energy to lattice through electron-phonon scattering in niobium metal.

The time taken for the hot electrons and lattice to achieve thermal equilibrium may be estimated in the following crude picture. The electron excited to $2\ eV$ releases its excess energy through a series of scattering with the lattice. Each scattering generates a phonon of energy approximated by $k_B T$, where k_B is the Boltzmann constant and T is the lower of either the Debye or average sample temperature. The total number of scattering n_s for the hot electron to release all its excess energy of E_{exc} (excitation photon energy = 2 eV) is estimated by $E_{exc}/k_B T$, where $k_B T$ is the energy of hot electron releases to the lattice with the creation of a phonon during each scattering. The time taken for the electron to release all its energy is given by $n_s\tau$, where τ is the time interval between each scattering which can be approximated by Drude relaxation time. For niobium, τ = 4.2 fs. The electron-phonon relaxation time computed by $n_s\tau = (2eV/275k_B)(4.2\,fs)$ is 354 fs. This computed time is surprising close to our measured value of 360 fs. For copper, τ = 27 fs, $n_s\tau = (2eV/315k_B)(27\,fs) = 2.1\,ps$; and gold, τ = 30 fs, $n_s\tau = (2eV/170k_B)(30\,fs) = 4.1\,ps$; These computed relaxation times of 2.1 and 4.1 ps are also in good agreement with the measured electron-phonon relaxation time of 1-4 ps for copper[9] and 2-3 ps for gold[10], respectively from the time-resolved thermal modulation measurement. The above indicates that the electron-phonon relaxation time for metal may be estimated directly from the Drude relaxation time. We must emphasize that the picture presented above is crude. The relaxation of hot electrons may be dominated by electron-electron scattering at high energy, and there is a wide distribution of phonon energy instead of just the Debye energy.

In conclusion, the electron-phonon relaxation time and coupling constant for niobium have been measured to be 360 fs and $17\times10^{17}\ Wm^{-3}K^{-1}$, respectively. The electron-phonon relaxation time may be estimated from $(E_{exc}/k_B T)\tau$.

Helpful discussions with A. Katz and G. L. Eesley are gratefully acknowledged. We thank C. Cohen for fabricating the niobium films. This research is supported by HYPRES Inc..

References

[1] S. B. Kaplan, C. C. Chi, D N. Langenberg, J. J. Chang, S. Jafarey, and D. J. Scalapino, Phys. Rev. B 14, 4854 (1976); 15, 3567 (E) (1977)

[2] V. F. Elesin and Yu. V. Kopaev, Sov. Phys. Usp. 24, 116 (1981)

[3] L. R. Testardi, Phys Rev. B 4, 2189 (1971)

[4] C. C. Chi, M. M. T. Loy, and D. C. Cronemeyer, Phys Rev. B 23, 124 (1981)

[5] C. Vanneste, A. Gilabert, P. Sibillot, and D. B. Ostrowsky, Appl. Phys. Lett. 38, 941 (1981)

[6] R. Sobolewski, D. P. Butler, T. Y. Hsiang, C. V. Stancampiano, and G. A. Mourou, Phys. Rev. B 33, 4604 (1986)

[7] S. I. Anisimov, B. L. Kapeliovich, and T. L. Perel'man, Zh. Eksp. Teor. Fiz. 66, 776 (1074) [Sov. Phys. JEPT 39, 375 (1975)]

[8] G. L. Eesley, Phys. Rev. Lett. 61, 2140 (1983); G. L. Eesley, Phys. Rev. B 33, 2144 (1986)

[9] H. E. Elsayed-Ali, T. B. Noris, M. A. Pesot, and G. A. Mourou, Phys. Rev. Lett. 58, 1212 (1987)

[10] R. W. Schoenlein, W. Z. Lin, J. G. Fujimoto, and G. L. Eesley, Phys. Rev. Lett. 58, 1680 (1987)

[11] J. G. Fujimoto, J. M. Liu, E. P. Ippen, and N. Bloembergen, Phys. Rev. Lett. 53, 1837 (1984)

Part X

Interfaces and Surfaces

Surface Vibrational Energy Relaxation: $c(2 \times 2)$-CO/Cu(100)

S.F. Shane, L. Rothberg, L.H. Dubois, N.J. Levinos, M. Morin, and A.L. Harris

AT&T Bell Laboratories, 600 Mountain Ave., Murray Hill, NJ 07974, USA

The first excited vibrational level lifetime of CO in a c(2x2) overlayer on the Cu(100) surface is measured to be 3 ps.

We report the lifetime for the first excited vibrational state of CO in the ordered c(2x2) overlayer on Cu(100), using transient sum frequency generation (SFG) vibrational spectroscopy to probe the population dynamics. This system has become a prototype for study of surface vibrational relaxation as a result of extensive theoretical work on its vibrational damping by coupling to electron-hole pair excitations in the metal.[1]

To measure the excited state vibrational lifetime, an intense, resonant infrared pulse pumps the vibrational mode. An infrared-visible probe pulse-pair monitors the vibrational level populations by transient, vibrationally resonant SFG.[2,3] The normalized difference of square roots of the SFG signal with and without pump gives the change in the fractional populations in the ground, n_a, and excited, n_b, vibrational levels:

$$\Delta[n_a(\tau_d)-n_b(\tau_d)] = \{[S_0]^{1/2} - [S(\tau_d)]^{1/2}\}/[S_0]^{1/2}$$

where S_0 is the signal with no pump and $S(\tau_d)$ is the signal at delay τ_d after the pump.

The adsorbate/metal system is prepared in an ultrahigh vacuum chamber by standard methods.[4] An amplified picosecond dye laser and cesium heat pipe Raman shifter provide ~3.5 ps pulses of visible (~600 nm) and tunable infrared (2000-2200 cm^{-1}) radiation.[2] The pump infrared pulse (3 μJ) is focused to a 200 μm diameter at the sample surface and is spatially overlapped with the probe infrared (1 μJ) and visible (3 μJ) pulse-pair. The SFG signal is monitored by a PMT at a reflection angle determined by k-vector matching of the probe pulse-pair at the surface.

Fig. 1a shows the probe SFG signal as a function of infrared frequency for the c(2x2)-CO overlayer on the Cu(100) surface, with the pump blocked, taken with the sample at 110K.

Springer Series in Chemical Physics, Vol. 53 **Ultrafast Phenomena VII**
Editors: C.B. Harris · E.P. Ippen · G.A. Mourou · A.H. Zewail

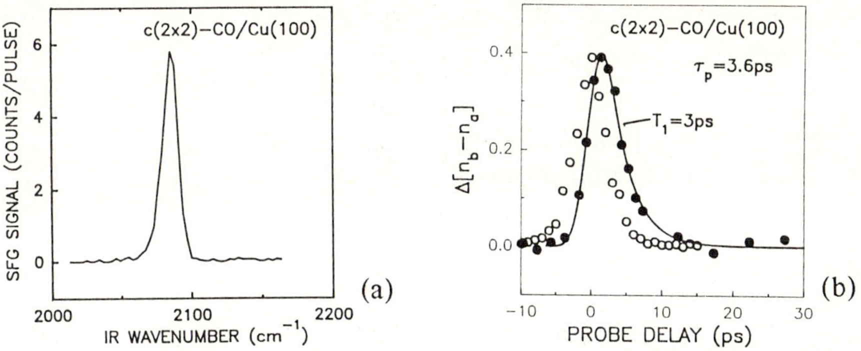

Fig. 1. a) SFG vibrational spectrum of c(2x2)-CO/Cu(100); b) Solid circles: transient vibrational population change; open circles: infrared-visible cross correlation; solid line: calculated transient response assuming parameters in text.

The vibrational resonance SFG line position at 2085 cm^{-1} is consistent with IR absorption measurements.[4] The observed linewidth, after deconvoluting the 6 cm^{-1} spectral width of the probing infrared laser pulse, is 12 cm^{-1}, substantially broader than the the narrowest observed IR absorption linewidth [4] of 4.6 cm^{-1}. The SFG resonance is probably inhomogeneously broadened. Careful surface preparation should eventually provide an SFG linewidth as narrow as that observed in IR absorption.

The normalized transient SFG response as a function of pump-probe delay, obtained with the pump and probe infrared pulses tuned to the vibrational resonance, is shown as the solid points in Fig. 1b. The cross correlation of the infrared pulse with the visible probe pulse is shown by the open circles (5.1 ps autocorrelation width FWHM, 3.6 ps pulse width assuming equal visible and infrared Gaussian pulses). The excited state lifetime is near the experimental time resolution. It can be estimated by calculating the transient vibrational response using a five-wave mixing description, which includes coherent pulse mixing effects as well as the population dynamics.[3] Gaussian pulses of 3.6 ps width, and an effective vibrational dephasing time T_2=0.9 ps given by the vibrational linewidth, are assumed. The best fit to the transient data is shown by the solid line in Fig. 1b, with T_1=3ps. While this lifetime is somewhat dependent upon assumptions about the pulse shape, it cannot be much longer than 4 ps. The lifetime is similar to the 4.9 ps CO lifetime recently measured on Pt(111) by picosecond infrared absorption saturation spectroscopy.[5]

The measured 3 ps lifetime is somewhat longer than the linewidth-based estimate of T_1=1.2ps (from the temperature independent 4.6 cm^{-1} linewidth)[4], suggesting that inhomogeneous broadening may contribute to the measured linewidth. Theoretical estimates [1] of the CO/Cu(100) lifetime based upon vibrational coupling to electron-hole pair excitation in the metal surface fall in the range 2-14 ps. These models thus yield a reasonably good prediction of the experimentally measured lifetime, supporting a model in which the damping is determined by coupling to the electronic degrees of freedom of the metal. Further studies will be needed to confirm and elucidate the nature of the coupling.[1]

In summary, the vibrational relaxation rate of c(2x2)-CO/Cu(100) has been measured at 120K by transient SFG vibrational spectroscopy. The excited state lifetime is short, ~3 ps, but is somewhat longer than the linewidth-determined limit of 1.3 ps. The excited state lifetime is consistent with theoretical models based upon vibrational coupling to the electronic response of the metal surface.

References

1. A. Liebsch, Phys. Rev. Lett. **54**, 67 (1985); A. G. Eguiluz, Phys. Rev. **B30**, 4366 (1984); P. Avouris and B. N. J. Persson, J. Phys. Chem. **88**, 837 (1984); and references therein.
2. A. L. Harris, L. Rothberg, L. H. Dubois, N. J. Levinos, and L. Dhar, Phys. Rev. Lett. **64**, 2086 (1990).
3. A. L. Harris, L. Rothberg, L. H. Dubois, N. J. Levinos, and L. Dhar, to be published.
4. R. Ryberg, Phys. Rev. **B32**, 2671 (1985); and references therein.
5. J. D. Beckerle, M. P. Casassa, R. R. Cavanagh, E. J. Heilweil, and J. C. Stephenson, Phys. Rev. Lett. **64**, 2090 (1990).

Femtosecond Laser-Induced Structural Changes in Gallium Arsenide

H. Schulz, J. Bialkowski, K. Sokolowski-Tinten, and D. von der Linde

Institut für Laser- und Plasmaphysik, Universität-GHS-Essen, D-4300 Essen 1, Fed. Rep. of Germany

Abstract. A decrease of the reflected second harmonic and an increase of the optical reflectivity is observed in GaAs following intense femtosecond laser excitation. Our data indicate that a structural phase transformation from the crystalline state to a disordered metallic liquid takes place in less than one picosecond.

It is generally accepted that solid-liquid phase transformations induced by picosecond laser pulses can be described as a rapid thermal melting process [1]. However, a thermal picture is no longer adequate to explain the interaction of laser pulses with the material when the pulse width is shorter than the thermalization time of the optical energy, which is approximately 10^{-12}s. Although previous femtosecond experiments on silicon [2-4] revealed changes of the optical reflectivity and of the reflected second harmonic which appeared to be consistent with a solid-liquid phase transition occurring in less than one picosecond, the physical mechanism of the subpicosecond phase transformation is not well understood as yet.

We present here some new data on femtosecond structural changes in GaAs. This material is interesting for the study of phase transformations [5, 6] because crystalline GaAs is non-centrosymmetric and possesses a large $\chi^{(2)}$. Upon melting $\chi^{(2)}$ is expected to drop to zero because molten GaAs is an isotropic metallic liquid.

In our experiments we used 100 fs s-polarized pump pulses at 620 nm to induce the phase transformation. The surface was probed with p-polarized pulses at the same wavelength. Pump and probe pulses were incident at 30 degrees and 49 degrees, respectively. The sample was a GaAs crystal with (100) surface orientation. The (010) was axis normal to the plane of incidence. In this configuration the reflected second harmonic of the probe pulse is s-polarized, whereas second harmonic generation of the pump pulse is symmetry-forbidden. Sum frequency mixing of pump and probe leads to p-polarized second harmonic in a direction approximately given by the bisector between the reflected pump and probe beam. Measurement of the p-polarized second harmonic provides the pump-probe cross correlation function.

In our experiment both the reflected fundamental and the reflected second harmonic of the probe pulse are measured simultaneously as a function of the delay time between pump and probe. Figure 1 shows results for a relatively low pump fluence (0.18 J/cm^2). We observe an initial very fast decrease of the reflectivity from 20% to 18% followed by an increase to 22% in about 5 ps. The second harmonic appears to follow a mirror-image pattern of the reflectivity: there is an initial fast rise and a subsequent decrease of the second harmonic on a time scale of several picoseconds.

The changes of the reflectivity can be attributed to intraband transitions of the photoexcited carriers and population changes in the valence band and the conduction band. The increase of the reflectivity at long times is probably due to lattice heating.

The changes of the reflected second harmonic at low pump fluences can also be attributed to the induced changes of the linear optical constants. No changes of $\chi^{(2)}$ have to be assumed. It can be shown that the variations

Springer Series in Chemical Physics, Vol. 53 **Ultrafast Phenomena VII**
Editors: C.B. Harris · E.P. Ippen · G.A. Mourou · A.H. Zewail

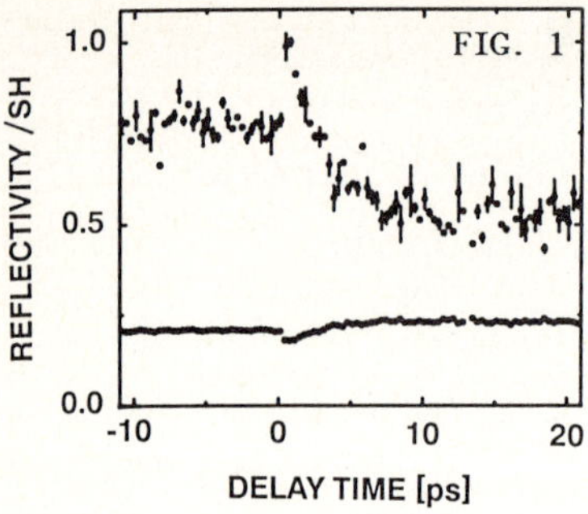

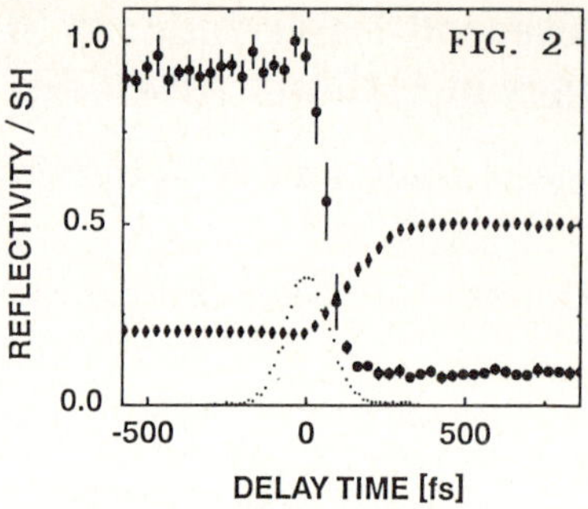

Reflectivity (lower curve in Fig. 1, diamonds in Fig. 2) and reflected second harmonic (upper curve in Fig. 1, full circles in Fig. 2) as a function of delay time. Fig. 1: 0.18 J/cm^2 (below threshold); Fig. 2: 0.75 J/cm^2 (above threshold). The dotted curve in Fig. 2 is the cross correlation of the pump pulses and the probe pulses.

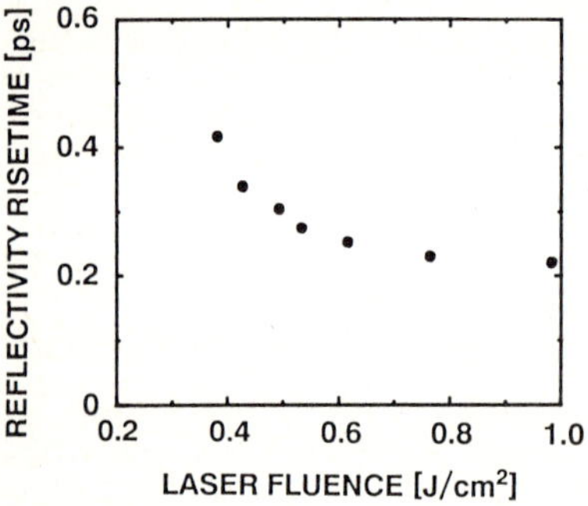

FIG. 3

Measured rise time of the optical reflectivity as a function of laser fluence.

of the optical constants provide a consistent explanation of both the second harmonic and the optical reflectivity.

The situation is quite different for higher pump fluences. Figure 2 shows the measured second harmonic and the optical reflectivity of the probe pulse for a pump fluence of 0.75 J/cm^2. The principal feature of the reflectivity is a large, step-like increase from 20% to 50% in 230 fs. The magnitude of the observed reflectivity change is in good agreement with the expected reflectivity of liquid GaAs, which we estimated on the basis of a Drude model. The rise of the optical reflectivity is accompanied by a drastic decrease of the second harmonic. The measured fall time is 110 fs, which is close to the time resolution limit of our experiment.

Measurements of the optical reflectivity and of the second harmonic were performed for a variety of pump fluences. From these data we established a threshold fluence of 0.25 J/cm^2 for the onset of the step-like increase of the reflectivity. The rise time of the optical reflectivity was observed to decrease with pump fluence. As shown in Fig. 3 the rise time is 400 fs at 0.35 J/cm^2, falling to 230 fs for fluences greater than 0.75 J/cm^2.

The observed features of the reflectivity and of the second harmonic for high fluences are consistent with the picture of a phase transformation from the solid to the liquid state:

1) Second harmonic generation is expected to disappear when the non-centrosymmetric crystal is melted;

2) Formation of a metallic liquid is expected to lead to an increase of the reflectivity approximately by a factor of two.

In fact, the measured decay time of the second harmonic is consistent with the measured rise time of the reflectivity when the formation of a liquid overlayer is assumed, which absorbs a portion of the incident fundamental and attenuates the reflected second harmonic generated at the liquid-solid interface. However, to attribute the observed rise and fall times to a growing liquid overlayer a propagation velocity of the phase boundary in excess of 10^7 cm/s had to be assumed. This value is very much larger than the speed of sound, which is believed to be an upper limit of the melting velocity [7].

The optical reflectivity was also measured at long times. For fluences well above the threshold we find a decrease of the reflectivity after the initial subpicosecond rise. For example, at 0.75 J/cm^2 we observed a drop from 50 % to 35 % in 10 ps. The time constant of the decay decreases with fluence. The apparent relaxation of the reflectivity is probably caused by the formation of a plasma from ablated surface material.

In conclusion, we have observed changes of the optical reflectivity and of the reflected second harmonic which indicate a structural phase transition of GaAs occurring in a few hundred femtoseconds. This time is shorter than the thermalization time of the optical energy, and much shorter than the expected time for a melting process involving the propagation of a phase boundary from the surface into the bulk. We believe that our data as well as recent results of others [8] contribute further evidence that other models are needed to explain the observed ultrafast phase transformations.

Support of this work by the Deutsche Forschungsgemeinschaft is gratefully acknowledged.

References

1. N. Bloembergen, Mat. Res. Soc. Symp. Proc. 51, 3 (1986).
2. C. V. Shank, R. Yen, C. Hirlimann, Phys. Rev. Lett. **50**, 454 (1983).
3. C. V. Shank, R. Yen, C. Hirlimann, Phys. Rev. Lett. **51**, 900 (1983).
4. H. W. K. Tom, G. D. Aumiller, C. H. Brito-Cruz, Phys. Rev. Lett. **60**, 1438 (1988).
5. A. M. Malvezzi, J. M. Liu, and N. Bloembergen, Appl. Phys. Lett. **45**, 1019 (1984).
6. S. A. Akhmanov, Sov. Phys. Uspekhi **28**, 1084 (1985).
7. F. Spaepen, D. Turnbull, in *Laser Annealing of Semiconductors*, ed. by J. M. Poate, J. W. Mayer (Academic Press, New York, 1982) p. 15.
8. J. K. Wang, P. Saeta, Y. Siegal, E. Mazur, and N. Bloembergen, elsewhere in this volume.

Ultrafast Relaxation of Electrons Probed by Surface Plasmons at a Thin Silver Film

R.H.M. Groeneveld, R. Sprik, M. Wittebrood, and A. Lagendijk

Natuurkundig Laboratorium der Universiteit van Amsterdam, Valckenierstraat 65, 1018 XE Amsterdam, The Netherlands

We employed the surface plasmon resonance of a thin silver film to induce and detect nonequilibrium heating of the electron gas. We measured an increase of the electron temperature of 30 - 90 K and an electron-phonon energy relaxation time of 670(70) fs.

In recent years, femtosecond thermomodulation experiments[1,2,3] have been performed to study the relaxation dynamics of electrons in metals. The electrons were heated to a high (> 1000 K) nonequilibrium temperature. At these extreme conditions, the dynamics depends strongly on the electron temperature, resulting in a nonexponential decay.

We present[4] a novel method that enables us to follow *perturbative* transients in the electron temperature (30 - 90 K) with femtosecond time resolution. It employs the surface plasmon (SP) resonance of a thin (45 nm) silver film evaporated on a prism[5]. The SP resonance appears as a sharp dip in the reflectivity $R(\Phi)$ of p-polarized light that has been reflected by the fused-silica-silver interface at a critical angle of incidence Φ_{sp}, the plasmon angle. The position Φ_{sp} and width $\Delta\Phi_{sp}$ depend on the complex dielectric constant $\epsilon_{Ag} = \epsilon_r + i\epsilon_i$ of the film. The SPs decay into hot electron-hole pairs. The heated electron gas cools by transferring its energy to phonons in a characteristic electron-phonon energy relaxation time τ_E ($\approx$ a picosecond). Our optical pump-probe setup consists of a colliding pulse modelocked laser which provides 150 fs pulses for a weak probe beam and a doubly (500 Hz $\otimes$ 8.8 MHz) amplitude modulated pump beam that is focused to an intensity as low as 180 Wcm^{-2} onto the film. Part of the reflected probe beam falls on a phase-sensitive photodetector via a movable slit which determines the angle of detection Φ_d.

Measuring the time-resolved change of reflectivity ΔR, we observe (Fig. 1.a) a rapid increase of -ΔR to a maximum ("peak") value, followed by an *exponential* decay to a constant level with a time constant τ_E = 670(70) fs (the electron-phonon energy relaxation time.) The final ("plateau") level lies significantly higher than the initial level. Theoretically, we expect that thermomodulation of the absorption edge at $\lambda = 310$ nm by hot electrons appears as a dispersive modulation $\Delta\epsilon_r$ around the laser wavelength (620 nm). We observed that the measured "peak" values at different angles of detection (Fig. 2.a) matches the calculated first derivative $dR/d\epsilon_r$ (Fig. 2.b). It follows that the instantaneous electron temperature rise amounted to ΔT_e = 30 - 90 K.

Springer Series in Chemical Physics, Vol. 53 **Ultrafast Phenomena VII**
Editors: C.B. Harris · E.P. Ippen · G.A. Mourou · A.H. Zewail

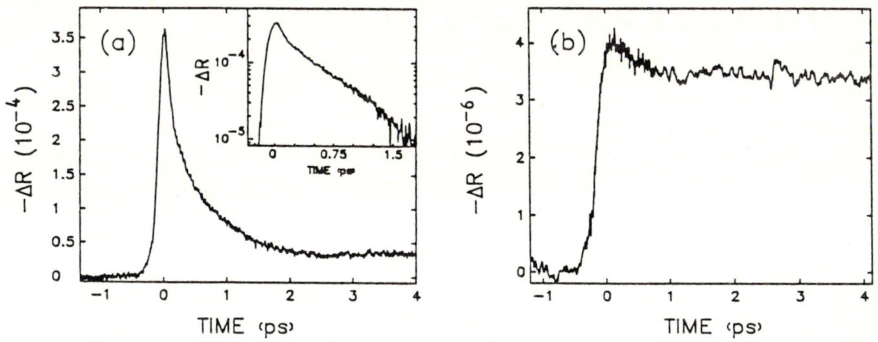

Figure 1: Measured change of the reflectivity ΔR as a function of delay time: **a)** when surface plasmons (SP) are used to excite and probe the silver film. The inset shows the logarithmical plot. **b)** When only the pump pulse excites SP.

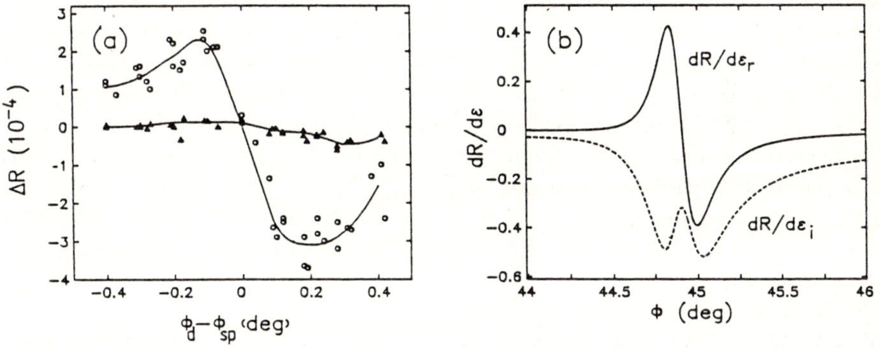

Figure 2: **a)** Measured change of reflectivity ΔR versus the angle of detection Φ_d. The circles mark ΔR around zero delay time; the triangles mark ΔR at ≈ 3 ps. **b)** Calculated derivative of the reflectivity R with respect to the real and imaginary part of the dielectric constant of the silver, versus the angle of incidence.

By comparing the change of reflectivity under SP-resonant (Fig. 1.a) and non-SP-resonant (Fig. 1.b) probing conditions, we can discriminate between the electron and lattice heating (corresponding to the "peak" and "plateau" levels), and estimate the lattice temperature rise to be $\Delta T_l = 0.10(5)$ K.

In order to understand the nature of the extra sharp structure on the time resolved curve (Fig. 1.a) around zero delay time, we measured the exact moment of zero delay time (Fig. 3). After a time resolved scan was taken, the role of the two beams as pump and probe beams was swapped and a second, 'reversed' scan was taken. The point of intersection of the two scans is the zero delay time. We see that the maximum of ΔR occurs at a delay time of 30 fs. This shows that the sharp structure is not an instantaneous artefact, and that

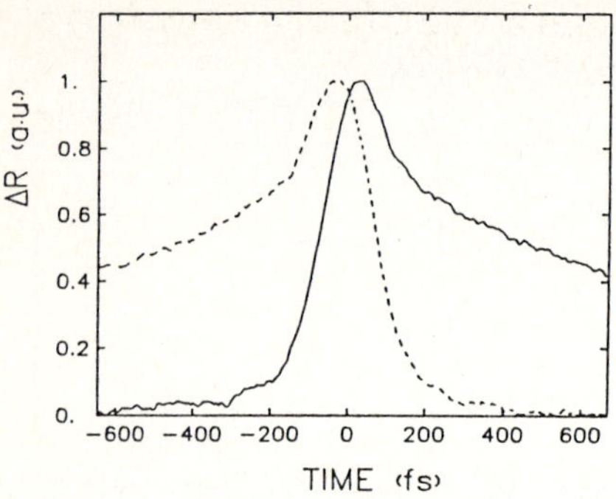

Figure 3: Determination of the zero delay time. The full line shows a 'normal' time resolved scan. The dashed line shows the'reversed' scan after the role of the beams as pump and probe beams has been swapped.

an initially very rapid relaxation process, such as electronic heat transport in the film, may be involved.

In conclusion, we observed a nonequilibrium electron temperature rise of 30 - 90 K and an electron-phonon energy relaxation time of 670(70) fs by the application of the surface plasmon resonance of a thin silver film. We expect the energy relaxation to slow down below the Debye temperature (215 K) and will extend the experiment to the low-temperature regime.

References

[1] H.E. Elsayed-Ali, T.B. Norris, M.A. Pessot, and G.A. Mourou, Phys. Rev. Lett. **58**, 1212 (1987).
[2] R.W. Schoenlein, W.Z. Lin, J.G. Fujimoto, and G.L. Eesley, Phys. Rev. Lett. **58**, 1680 (1987).
[3] S.D. Brorson, W.Z. Lin, J.G. Fujimoto, and E.P. Ippen, Phys. Rev. Lett. **59**, 1962 (1987).
[4] R.H.M. Groeneveld, R. Sprik, and A. Lagendijk, Phys. Rev. Lett. **64**, 784 (1990).
[5] E. Kretschmann, Z. Physik **241**, 313 (1971).

Picosecond Transient Surface Temperature Measurement by Reflection High-Energy Electron Diffraction

H.E. Elsayed-Ali and J.W. Herman

Laboratory for Laser Energetics, University of Rochester, Rochester, NY 14623, USA

In the study of the interactions of ultrafast lasers with surfaces, a key parameter is the time evolution of the surface temperature. Typically, this is determined for simple systems using a heat-diffusion model. It has been previously demonstrated that second-harmonic generation at surfaces could be used, when resonantly enhanced, as a surface temperature probe [1]. However, this approach does not offer a general technique suitable for different materials. Here we demonstrate the utilization of picosecond time-resolved reflection high-energy electron diffraction (RHEED) as a lattice-surface temperature probe.

Electron diffraction is a well-developed surface-structural probe and has been utilized for many decades in surface studies [2]. The development of a picosecond transmission electron diffraction system was reported several years ago [3]. More recently, we have demonstrated picosecond RHEED [4]. Details of our ultrahigh vacuum picosecond RHEED system are given in Ref. [5].

The basic idea of this technique (Fig. 1) is the utilization of a picosecond laser pulse to create a focused and collimated electron pulse with a comparable time duration. We use a Nd:YAG laser ($\lambda = 1.06$ μm) to irradiate the sample while the electron probing pulse is generated by irradiation of a photocathode with the frequency quadrupled Nd:YAG laser ($\lambda = 0.266$ μm). By varying the spatial delay between the ultraviolet laser pulse and the fundamental, it is possible to obtain time-resolved RHEED patterns. The sensitivity of the diffraction pattern to temperature comes from the fact that as the temperature is raised, there is an increased atomic vibrational amplitude which reduces the number of elastically scattered electrons (Debye-Waller effect) [6].

We utilize the surface Debye-Waller effect to temporally resolve the surface temperature of a Pb(110) crystal. The system is operated at 8 Hz repetition rate. The

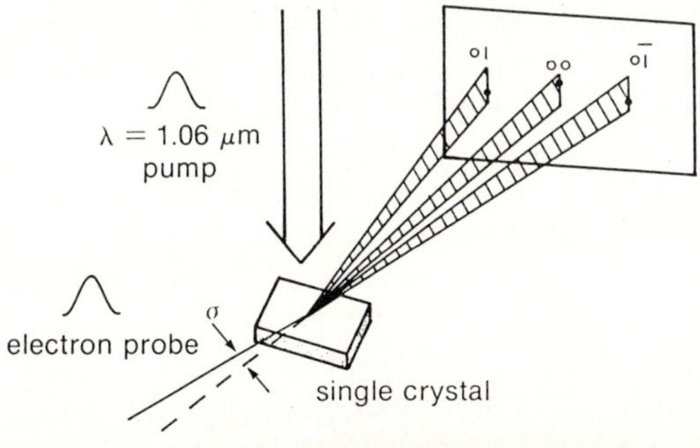

Fig. 1 Picosecond time-resolved reflection high-energy electron diffraction (RHEED). The laser pump and electron probe are well synchronized.

Springer Series in Chemical Physics, Vol. 53 **Ultrafast Phenomena VII**
Editors: C.B. Harris · E.P. Ippen · G.A. Mourou · A.H. Zewail

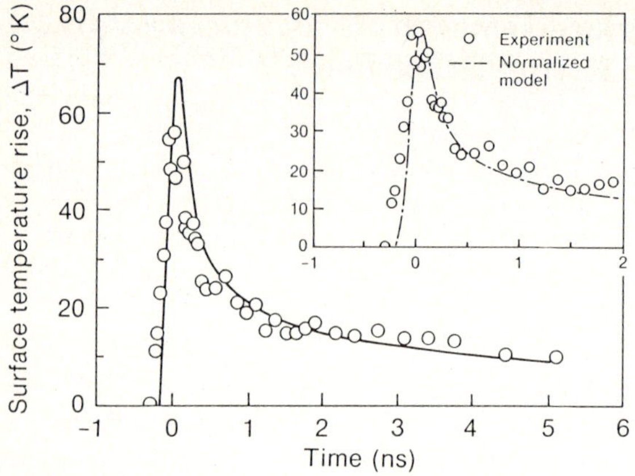

Fig. 2 Transient surface temperature rise of Pb(110) irradiated with 3.5×10^7 W/cm^2 peak intensity, 167 ps FWHM, λ = 1.06 µm laser pulses. Experimental results are obtained from the transient surface Debye-Waller effect on RHEED intensity using a static heating calibration. The solid line is obtained from a numerical heat diffusion model. In the inset the peak temperature from the model is normalized to that from the experiment.

laser heating pulse is incident normal to the sample and has a maximum energy of ~20 mJ per pulse.

The time-resolved results are related to the lattice temperature by observing the diffraction intensity changes under static heating conditions and comparing with the intensity changes due to laser heating. The validity of this temperature calibration is not directly tested. However, this calibration should hold if (i) the considered time scale is longer than that needed for equilibrium to be established between the electrons and the lattice. (In metals, at room temperature or higher, this typically occurs within a few picoseconds) [7], (ii) no surface phase transformation occurs, and (iii) no accumulated surface modifications occur by repeated laser heating. These conditions appear to be fulfilled in the experiments reported here.

An example of an experimental run is shown in Fig. 2 along with the decay of the surface temperature rise obtained from a heat diffusion model [8]

$$C[dT(z,t)/dt] = K[d^2T(z,t)/dz^2] + I(1 - R)\, \alpha\, e^{-\alpha z} f(t),$$

where C is the heat capacity per unit volume, T(z,t) is the temperature profile at distance z normal to the sample surface (z = 0), t is time, K is the thermal conductivity, I is the laser peak intensity, α is the absorption per unit length, and f(t) is the time dependence of the laser pulse which is assumed to be Gaussian.

The surface temperature rise obtained using the transient surface Debye-Waller effect on the picosecond RHEED pattern shows general agreement with that predicted by a heat diffusion model (Fig. 2). An uncertainty in the model, however, is the value of the reflectivity and absorptivity for which we relied on published results for the complex index of refraction. Experiments performed at different heating laser peak intensities show trends consistent with those predicted by the heat diffusion model. That is, for a constant heating laser pulse width, the maximum surface temperature is directly proportional to the laser peak intensity, while the decay rate of the surface temperature is independent of the laser peak intensity. Nonlocal effects on the thermal conductivity which arise if the temperature gradient is significant within a mean free path of the heat carriers [9], i.e., hot electrons, are not included.

We have demonstrated the use of picosecond RHEED as a time-resolved surface temperature probe. The temperature measurement is based on the surface Debye-Waller effect which results in a reduction of the number of elastically scattered electrons with lattice heating. Time-resolved measurements of the picosecond laser heated Pb(110) surface show general agreement with a heat diffusion model.

The technique of picosecond RHEED offers both surface selectivity and temperature sensitivity. In addition to its utilization as a time-resolved structural and temperature probe of fast surface processes, it can also be used to study the thermal properties of extremely thin films (up to a few monolayers).

ACKNOWLEDGMENT

The authors would like to thank G. A. Mourou for his input to the various stages of this work. This work was supported by the U. S. Department of Energy under contract No. DE-FG02-88ER45376. Additional support was provided by the Laser Fusion Feasibility Project at the Laboratory for Laser Energetics which has the following sponsors: Empire State Electric Energy Research Corporation, New York State Energy Research and Development Authority, Ontario Hydro, and the University of Rochester.

1. J. M. Hicks, L. E. Urbach, E. W. Plummer, and H.-L. Dai, Phys. Rev. Lett. 61, 2588 (1988).
2. For example, D. Lichtman, Methods of Surface Analysis, Vol. I, edited by A. W. Czanderna (Elsevier Sci. Pub., New York, 1975), p. 39.
3. G. Mourou and S. Williamson, Appl. Phys. Lett. 41, 44 (1982).
4. H. E. Elsayed-Ali and G. A. Mourou, Appl. Phys. Lett. 52, 103 (1988).
5. H. E. Elsayed-Ali and J. W. Herman, Rev. Sci. Instrum. 61, 1636 (1990).
6. C. Kittel, Introduction to Solid State Physics, 5th edition (Wiley, New York, 1976), p. 63.
7. H. E. Elsayed-Ali, T. B. Norris, M. A. Pessot, and G. A. Mourou, Phys. Rev. Lett. 58, 1212 (1987).
8. J. H. Bechtel, J. Appl. Phys. 46, 1585 (1975).
9. F. Claro and G. D. Mahan, J. Appl. Phys. 66, 4213 (1989).

Lifetime of an Adsorbate Substrate Vibration: H on Si(111)

P. Guyot-Sionnest[1], *P. Dumas*[2], *Y.J. Chabal*[3], *and G.S. Higashi*[3]

[1]LURE, Centre Universitaire Paris-Sud, Bât. 209D, F-91405 Orsay, France
[2]LASIR, 2 rue Henri Dunant, F-94320 Thiais, France
[3]AT&T Bell Laboratories, 600 Mountain Ave., Murray Hill, NJ07974, USA

Abstract. We present the measurement of the lifetime of the adsorbate-substrate vibration for hydrogen on silicon (111) by Sum-Frequency Generation. The lifetime is 0.8 ns.

The study of dynamical processes on the ultrafast time scale is among the recent developments in surface science. In the past, there has been a great effort in the study of the various relaxation processses for molecular vibrations in gas, liquid and solid phases that stems from an interest in the fundamental nature of the various coupling mechanisms as well as in the potentially selective reactivity of molecules in an excited vibrational mode. For molecules adsorbed on surfaces, essentially the same questions can be raised but sensitivity problems require improved or new techniques.

Until recently, the vibrational linewidth was the available experimental data for learning about the adsorbate-substrate coupling. In the case of a homogeneous broadening, as can be the case for well ordered adsorbate monolayers on single crystal substrate, one can obtain significant information from the temperature dependence of the linewidth, essentially on the nature of the dephasing process[1]. However, the lifetime of the vibration cannot be extracted from the linewidth and therefore direct time resolved measurements are needed.

Recently, there have been several experiments using linear reflectivity or Sum-Frequency Generation (SFG) techniques to probe the lifetime of adsorbate vibrations. The various mechanisms put forward for explaining the measured lifetimes have been via intramolecular[2] electronic,[3] or phonon[4] excitations, thus already covering a broad spectrum of the potential mechanisms.

In our work, we have measured the lifetime of the stretching vibration of hydrogen adsorbed on silicon. We have chosen to study this simple and nearly ideal system where ab initio calculations are in principle feasible. The sample studied is a single crystalline surface of silicon (111) that has been covered by a very ordered monolayer of atomic hydrogen[5,6]. Prepared in an appropriate HF:NH_4F etching solution the hydrogen saturates the otherwise reactive dangling bonds of the silicon atoms, leading to a very stable and unreconstructed surface. The hydrogen presents a single stretching vibration at 2084 cm^{-1} characteristic of a monohydride termination[5].

In addition to band structure calculations for this highly ideal system[7], the dephasing has been well characterized from temperature dependence infrared spectroscopy and linewidth studies showing that the quartic coupling to Si phonons is the dominant mechanism[6]. Finally, Molecular Dynamics simulations performed for the similar system of H:Si(100) predict a long lifetime of 20 ns due to the high frequency gap between the Si-H stretch and the other surface vibrations[8]. The lifetime measurement is therefore an important test for future theoretical studies.

The surface time resolved measurement is carried out by combining infrared transient bleaching and the surface nonlinear optical technique of SFG. In SFG, a visible and an infrared pulse illuminate the surface[9]. Coherent sum-frequency photons are emitted from the surface where the inversion symmetry is broken. When the infrared is tuned to a resonance of the adsorbate (the mode has to be infrared and raman active) the second order nonlinearity is enhanced, yielding an increased output at the sum-frequency.

Springer Series in Chemical Physics, Vol. 53 **Ultrafast Phenomena VII**
Editors: C.B. Harris · E.P. Ippen · G.A. Mourou · A.H. Zewail

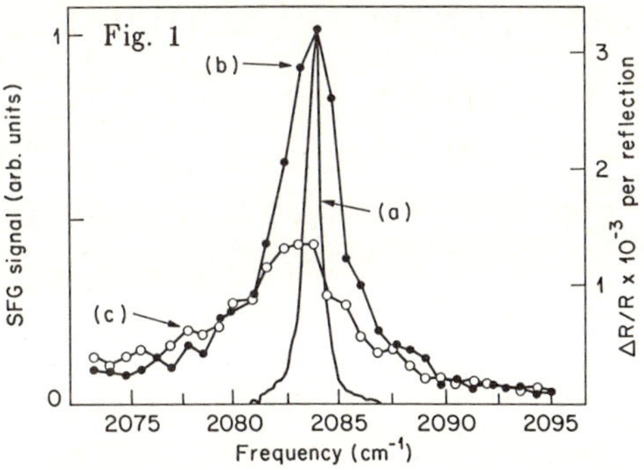

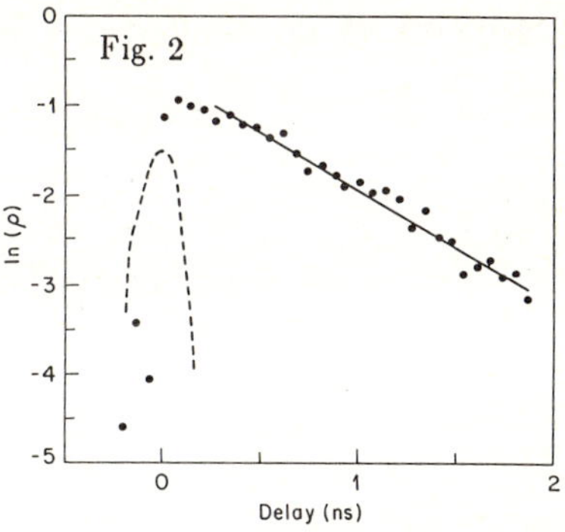

Figure 1: Spectra of the Si-H stretch. a) Linear absorption obtained by multiple internal reflection with a 0.3 cm^{-1} resolution. b) SFG spectrum taken with the probe infrared beam (with a 3 cm^{-1} laser width). c) Same spectrum taken 200 ps after the arrival of the pump infrared beam.

Figure 2: Excited state population as a function of time. The dotted line is the cross-correlation of the infrared pump and the visible pulse. The straight line is a least squares fit giving T_1=795ps.

In our laser system, the infrared radiation is generated by difference frequency mixing of the fundamental Nd:YAGradiation and an infrared dye laser[10]. Outputs of about 80 μJ around 2000 cm^{-1} are obtained. The pulse width is around 200 ps. Using 8% of the infrared over 150×400 μm and 300 μJ of the doubled Nd:YAGradiation, the Si-H stretch vibration gives a large SFG as shown in Fig. 1. The remaining IR can then be used for pumping the vibration.

Neglecting the small nonresonant contribution, the SFG signal is just proportional to the square of the population difference between the ground state and the excited state[11]. Following the pump pulse, the SFG is then reduced by 60% corresponding to an 18% population in the first vibrational level. Varying the delay allows therefore the direct measurement of the ground state recovery time. As shown in Fig. 2, we observe a straight line on a logarithmic scale that indicates a single exponential decay giving a 0.8 ns lifetime. This corresponds to a 0.007 cm^{-1} contribution to the linewidth.

For the Si-H stretching mode, electronic mechanisms are not possible since the vibrational quantum is much smaller than the band gap. The mode can only decay to the bending mode (637 cm^{-1}) or substrate phonons ($<500 cm^{-1}$). The large energy difference implies a mutiphonon process and therefore a slow decay, as observed. These can happen via anharmonic coupling terms of a high order. Among the various anharmonic processes allowed by symmetry, one could think, for example, of the Si-H stretch vibration decaying into two bending modes and two substrate phonons. At present, we do not know the specific phonon-decay mechanisms, but some insight will be obtained by measuring the lifetime as a function of temperature.

Due to our 200 ps time resolution, we could not detect faster decay mechanisms. We can expect that, on the same time scale as the dephasing time (10 ps at room temperature), the highly excited q=0 Si-H stretch phonon will scatter into the weakly dispersive phonon band. Indeed, this could happen via the same anharmonic coupling to the Si phonons that has been found to be dominant for the dephasing[6]. This should be a general feature for ordered monolayers and we are presently working on improving our time resolution. The laser system is also tunable around 1500 cm^{-1} so that we plan to study the Si-D stretch.

The Sum-Frequency Generation technique therefore opens a new line of experimental research in the field of vibrational dynamics of surfaces. The multiplicity of relaxation mechanisms that are possible justifies the study of simple systems at first such as the Si-H ordered monolayer system.

References.

[1]. E.g. B. N. J. Persson, F. M. Hoffman, and R. Ryberg, Phys. Rev. **B34**, 2266 (1986), and references therein.

[2]. A. L. Harris. L. Rothberg, L. H. Dubois, N. J. Levinos, and L. Dhar, Phys. Rev. Lett. **64**, 2086 (1990).

[3]. J. D. Beckerle, M. P. Casassa, R. R. Cavanagh, E. J. Heilweil, and J. C. Stephenson, Phys. Rev. Lett. **64**, 2090 (1990).

[4]. P. Guyot-Sionnest, P. Dumas, Y. J. Chabal, and G. S. Higashi, Phys. Rev. Lett. **64**, 2156 (1990).

[5]. G. S. Higashi, Y. J. Chabal, G. W. Trucks and Krishnan Raghavachari, Appl. Phys. Lett. **56**, 656 (1990).

[6]. P. Dumas, Y. J. Chabal, and G. S. Higashi (to be published).

[7]. L. Miglio et al., Physica Scripta **37**, 768 (1988); and U. Harten et al., Phys. Rev. **B38**, 3305 (1988).

[8]. J. C. Tully, Y. J. Chabal, K. Raghavachari, J. M. Bowan, and R. R. Lucchese, Phys. Rev. **B31**, 1184 (1985).

[9]. Y. R. Shen, Nature (London) **337**, 519 (1989), and references therein.

[10]. T. Elsaesser, H. Lobentanzer, and A. Seilmeir, Opt. Commun. **52**, 355 (1985).

[11]. J. H. Hunt, P. Guyot-Sionnest, and Y. R. Shen, Chem. Phys Lett. **133**, 189 (1987).

Pump–Probe Investigation of Femtosecond Desorption

F. Budde, T.F. Heinz, M.M.T. Loy, J.A. Misewich, and B.D. Smith

IBM Research Center, T.J. Watson Research Center, Box 218,
Yorktown Heights, NY 10598, USA

Abstract: Desorption of NO molecules from a Pd(111) surface has been induced by femtosecond visible laser pulses. Time-resolved measurements of desorption were obtained by means of a correlation technique. The desorption process was found to be characterized by a subpicosecond response, consistent with an electronic mechanism.

It has recently been demonstrated that strongly nonequilibrium excitation can be induced in a metal irradiated by a femtosecond laser pulse [1]. The laser energy is initially absorbed as electronic excitation, which rapidly thermalizes. Only on a subpicosecond or picosecond timescale is energy transferred to lattice vibrations of the metal. Because of the low heat capacity of the electrons, the peak electronic temperature may far exceed that of the lattice. This distinctive nonequilibrium regime may give rise to new physical processes. In our laboratory, we are examining the nature of desorption from a metal surface under such excitation. In this paper, we report direct time-resolved measurements of the desorption of nitric-oxide molecules (NO) from a Pd(111) surface induced by femtosecond laser pulses.

In recent studies of the NO/Pd(111) system, the yield and final-state energy distributions were determined for NO molecules desorbed by 200-fs pulses of 620 nm radiation [2]. Among the striking results of these measurements were the high yield and vibrational temperature of the desorbed NO molecules, as well as the markedly nonlinear dependence of the desorption yield on the laser fluence. These findings could not be understood in terms of a conventional thermal desorption mechanism, nor could they be explained by a simple photochemical process. It was suggested that a novel desorption mechanism was operative as a consequence of the high density of nonequilbrium electronic excitation expected on the femtosecond time scale.

In the present work, the desorption process is probed on an ultrafast time scale by means of a pump-probe technique. The approach is based on a two-pulse correlation measurement of the desorption yield, similar in concept to the method applied by Fujimoto *et al.* [3] to the electron emission. In the experiment, orthogonally polarized pairs of pulses are formed by sending the output

Springer Series in Chemical Physics, Vol. 53 **Ultrafast Phenomena VII**
Editors: C.B. Harris · E.P. Ippen · G.A. Mourou · A.H. Zewail

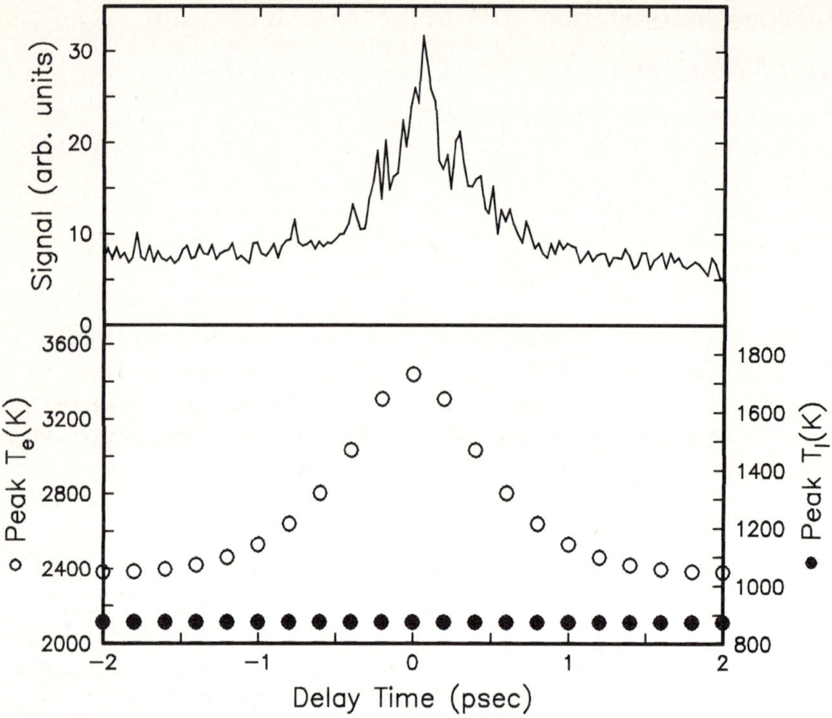

Fig. 1. (a) Yield of desorbed NO as a function of the delay between the equivalent pump and probe pulses. (b) The peak electronic temperature (open symbols and left scale) and lattice temperature (filled symbols and right scale) corresponding to the indicated pump-probe separation.

from an amplified CPM laser system through a Michaelson interferometer. The pulses are of 400-fs duration when they reach the Pd surface held in an ultrahigh vacuum chamber. The Pd surface is maintained at saturation coverage of NO at a base temperature of 400 K. The desorbed NO molecules are detected through multi-photon ionization [2] and the yield is established as a function of the time separation between the pairs of femtosecond pulses. Since the desorption yield is nonlinear in the laser intensity, we expect that the desorption signal for pulses at small separation will exceed that for well separated pulse pairs. The resulting correlation signal should reveal the characteristic time response of the desorption process.

Fig. 1(a) displays such correlation data for the NO desorption yield. A pronounced fast response can be seen. Under the present experimental conditions, the width of the correlation peak in the desorption signal is similar to that of the laser pulse autocorrelation and no attempt has been made to extract a finite relaxation rate. Any coherent contribution to the correlation signal is expected to be weak, since orthogonally polarized pulses were used to excite the system. This conclusion is further substantiated by the fact that the observed desorption correlation trace is appreciably wider than the electric-field autocorrelation of the laser pulses. The symmetric form of the desorption correlation signal is obtained only for conditions under which equal energy is absorbed from each laser pulse. For inequivalent excitation pulses, a systematic asymmetry develops [4].

The results of Fig. 1(a) indicate that the excitation responsible for molecular desorption has a lifetime $\lesssim$ 1 ps. This suggests immediately that an electronic, rather than a thermal desorption mechanism must be operative. To examine this issue more quantitatively, we have evaluated the electronic (T_e) and lattice (T_l) temperatures of the surface by means of a solution of the set of coupled diffusion equations for these two quantities [1-2,4]. Fig. 1(b) displays the peak electronic and lattice temperatures obtained as a function of the time separation between the two excitation pulses for our experimental conditions. The observed desorption signal is found to follow the behavior of the peak electronic temperature (solid symbols), but to differ dramatically from the peak lattice temperature (open symbols), which only varies for pulse separations of many picoseconds. Thus, in contrast to the findings for laser-induced desorption with pulses of nanosecond duration [5], a thermal desorption mechanism does not apply for femtosecond laser pulses. The desorption mechanism on the ultrafast time scale apparently requires the high density of electronic excitation present for a fraction of a picosecond following irradiation of the surface. This electronic excitation may lead to desorption either through excitation of the molecule-surface vibration or of higher-lying electronic states of the adsorbate complex [2].

References

1. H. E. Elsayed-Ali, T. B. Norris, M. A. Pessot, and G. A. Mourou, Phys. Rev. Lett. **58,** 1212 (1987); R. W. Schoenlein, W. Z. Lin, J. G. Fujimoto, and G. L. Eesley, *ibid.* **58**, 1680 (1987); P. B. Corkum, F. Brunel, N. K. Sherman, and T. Srinivasan-Rao, *ibid.* **61**, 2886 (1988).
2. J. A. Prybyla, T. F. Heinz, J. A. Misewich, M. M. T. Loy, and J. H. Glownia, Phys. Rev. Lett. **64**, 1537 (1990).
3. J. G. Fujimoto, J. M. Liu, E. P. Ippen, and N. Bloembergen, Phys. Rev. Lett. **53**, 1837 (1984).
4. F. Budde, T. F. Heinz, M. M. T. Loy, J. A. Misewich, and B. D. Smith, to be published.
5. J. A. Prybyla, T. F. Heinz, J. A. Misewich, and M. M. T. Loy, Surface Sci., in press.

Part XI

Electron or Proton Transfer and Solvation

Femtosecond Dynamics of Single Electron Transfer in Aqueous Media

Y. Gauduel, S. Pommeret, N. Yamada, A. Migus, and A. Antonetti

Laboratoire d'Optique Appliquée – INSERM U275, Ecole Polyte
ENS Techniques Avancées, F-91120 Palaiseau, France

Abstract. In this paper we develop the contribution of femtosecond spectroscopy to our understanding of the primary events involved in single-electron reactions in homogeneous aqueous solutions. Electron localization and hydration, electron–cation recombination and radical pair formation have been investigated at this time scale.

1. Introduction

The reactions of electrons and protons in polar liquids and in particular in liquid water are of fundamental importance in chemical physics and biology. It is known that the polar nature of water molecules, the energetics and the time dependence of the charge–water molecule interactions can impede or assist oxidation–reduction reactions. A general discussion is continuing on the role of the solvent friction (collisional and dielectric components) in the frequency factor for the reaction dynamics [1]. The primary reactions involving relaxed electrons with prototropic radicals represent a fundamental aspect of single-electron transfer linked to radiation chemistry and a clear understanding of them would be an important step in improving our knowledge of free radical reactions occurring at short times after the energy deposition. In the case of excess subexcitation electrons in an aqueous environment, the elementary charge (electron) implicates fascinating issues on its coupling with the solvent or with other chemical entities (ionic and biological acceptors).

2. Primary Species in Neat Liquid Water

2.1 Initial Energy Deposition and Ionization Process

The time scale of the physical and chemical events related to the absorption of energy by water molecules extends from 10^{-16} s (formation of the ionized state of water) to 10^{-12} s (thermal orientation of water molecules) and to 10^{-7}s (formation of molecular products). The energy exchanges result in the formation of nonequilibrium configurations including quasi-free or dry electrons (e^-_{qf}), thermalized (e^-_{th}), localized (e^-_{loc} or e^-_{trap}) and hydrated electrons (e^-_{hyd}) and prototropic radicals (Fig. 1). Femtosecond spectroscopy of neat liquid can be conducted by using high intensity pump light pulses at 310 nm at intensities in the 10^{11} W cm^{-2} range. Time resolved induced absorption kinetics taken over a large set of probe

Springer Series in Chemical Physics, Vol. 53 **Ultrafast Phenomena VII**
Editors: C.B. Harris · E.P. Ippen · G.A. Mourou · A.H. Zewail

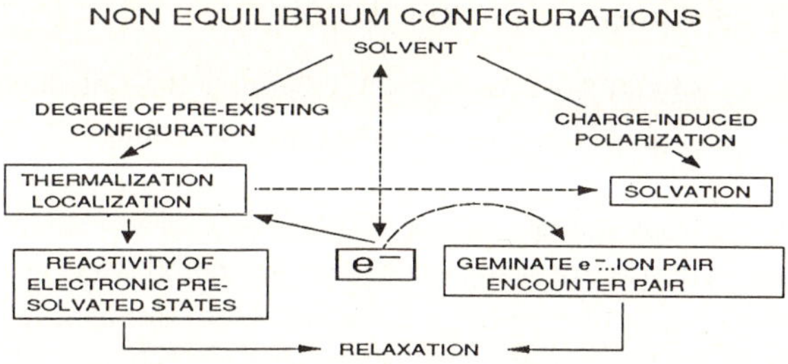

Fig. 1. Representation of different channels involved in the behaviour of non-equilibrium electron configurations in aqueous solutions

wavelengths allow investigation of the formation of primary species following the photoionization process [2,3].

2.2 Towards an Identification of H_2O^+

Immediately after the ionization of H_2O molecules the cationic radical (H_2O^+) undergoes, within a few vibrational periods, an ultrafast proton transfer to an adjoining water molecule to give a cationic ion H_3O^+ and a radical OH. This charge transfer is believed to take place in less than $5 \times 10^{-15} - 10^{-14}$ s [4], i.e. the time required for a water molecule vibration.

$$H_2O + 2\,h\nu(310\,\text{nm}) \rightarrow [H_2O^*] \rightarrow H_2O^+ + e^-_{qf}\,, \tag{1}$$

$$H_2O^+ + n(H_2O) \rightarrow (H_3O^+ + OH)nH_2O\,. \tag{2}$$

Femtosecond investigations in the near ultraviolet clearly identify an ultrafast induced absorption which disappears following a monoexponential law.

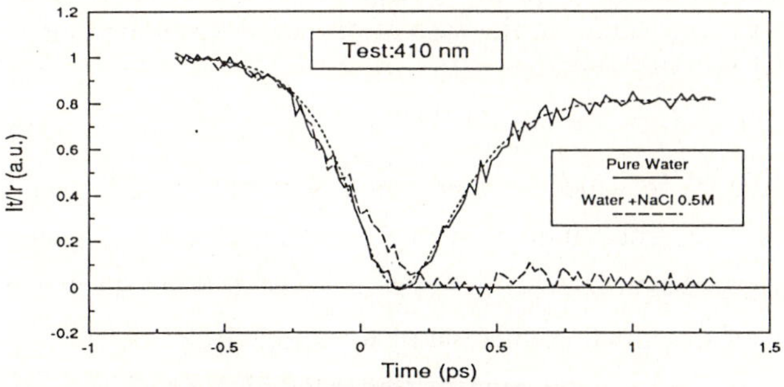

Fig. 2. Time dependence of the induced absorbance at 410 nm following the femtosecond ionization of pure liquid water or ionic aqueous solutions

Data analysis gives an instantaneous risetime, i.e. within the pulse duration, and a time relaxation of 130 ± 10 fs. This transient species is not observed in ionic aqueous media. From the spectral data literature [5,6], we attribute this absorption to the formation of the water cation H_2O^+ and its subsequent reaction through the formation of the hydronium ion H_3O^+ (equation 2).

2.3 Electron Relaxation in Neat Water

Information on the dynamics of electron localization and solvation is of considerable interest for understanding electron transfer processes in liquids. The e^-_{hyd} is often represented as a bound state and characterized by a binding energy between 1 and 2 eV and a charge distribution radius of 2.5–3 Å. Our investigations have shown that electron hydration in water proceeds through at least one intermediate state (e^-_{prehyd}) absorbing in the infrared (IR) and whose lifetime is in the femtosecond regime. The measured initial electron trapping time (110 fs) is longer than the estimates for the thermalization time of electrons in liquid water: $T_{th} = (2.4-4) \times 10^{-14}$ s [7]. This would suggest that if the early energy loss rate of electrons in water is about 10^{13} eVs^{-1}, electrons would be thermalized before being localized. The trapping time ratio ($T_{1(D_2O)}/T_{1(H_2O)} = 1.09$) is lower than (i) the ratio of the energetic vibrational modes (OH vs OD), which is $\sqrt{2}$ times greater in H_2O than in D_2O, and (ii) the estimate of the ratio of the energy loss rate [8]. It can be concluded that the thermalization step in D_2O is not the limiting factor of the prehydration process. The absence of a significant time-dependent spectral shift from the IR to the visible clearly indicates that the relaxation of water molecules around an electron involves an extremely small motion without a dominant dielectric response of the polar medium [2]. Recent simulations suggest that the final hydration step in neat water involves a nonadiabatic transition from the excited state to the ground state of e^-_{hyd} [9,10].

3. Ultrafast Electron-Cation Reaction

The initial spatial distribution of electron–ion pairs ($e^-_{hyd} \cdots X_3O^+$ or $e^-_{hyd} \cdots OX$, X = H or D) can influence the early behavior of e^-_{hyd}:

$$H_3O^+ + e^-_{hyd} \rightarrow H_2O + H \quad (k = 2.3 \times 10^{10}\,M^{-1}\,s^{-1})\,, \tag{3}$$

$$OH + e^-_{hyd} \rightarrow OH^- \quad (k = 3 \times 10^{10}\,M^{-1}s^{-1})\,. \tag{4}$$

We have considered a model in which an electron–radical pair ($XO_3^+ \cdots e^-_{hyd}$ or $OX \cdots e^-_{hyd}$) executes a recombination in a one-dimensional process (jump process). The analytical expression is $N_{e^-_{hyd(t)}} = \gamma N \cdot_{e^-_{hyd}} \mathrm{erf}(T_d/t)^{1/2}$, in which γ represents the fraction of the initial population of hydrated electrons ($N \cdot_{e^-_{hyd}}$) which will recombine and T_d the jump time of the recombination. The jump rate is found to be significantly influenced by a H/D isotope substitution ($0.45 \times 10^{12}\,s^{-1}$ in D_2O and $0.83 \times 10^{12}\,s^{-1}$ in H_2O). Considering the high mobility of the hydronium ion compared to e^-_{hyd} ($D_{H_3O^+} = 9 \times 10^{-5} cm^2 s^{-1}$, $D_{e^-_{hyd}} = 4.75 \times 10^{-5} cm^2 s^{-1}$), the model favours the role of the dynamical structure of the hydrated X_3O^+ including the dynamics of hydrogen-bond polarization.

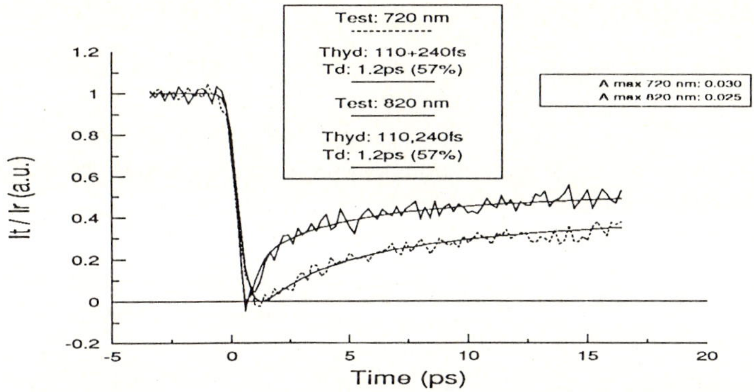

Fig. 3. Time dependence of the induced absorbance at 720 and 820 nm following the femtosecond photoionization of light water. At 720 nm, the early decay is attributed to a recombination process ($e^- \cdots H_3O^+$; $e^- \cdots OH$). At 820 nm, the early decay is due to a fast relaxation of the non-fully hydrated electrons [2]

4. Effect of Ionic Strength on Electron Reactivity

4.1 Generation and Relaxation of Electrons

Femtosecond photochemistry in ionic aqueous solutions allows investigation of the early electron reactivity in the presence of a solute and electron acceptor. Electron photodetachment from the chloride anion, through CTTS, was achieved by pumping ionic aqueous solutions at 310 nm: $Cl^- + h\nu(310\,nm) \rightarrow [Cl^-]^* \rightarrow Cl + e^-$. In dilute ionic aqueous solutions (NaCl = 0.4 M), the photodetachment of an electron from Cl^- and its subsequent hydration proceed through at least one intermediate step (localization step) absorbing in the infrared and having a lifetime of 250 fs. Figure 4 summarizes the influence of ionic strength on electron trapping and hydration dynamics. The most significant ionic strength effects are observed on the electron hydration dynamics but the dynamics of electron hydration is faster than the formation time (T_r) of the Na^+ or Li^+ atmosphere around e^-_{hyd}.

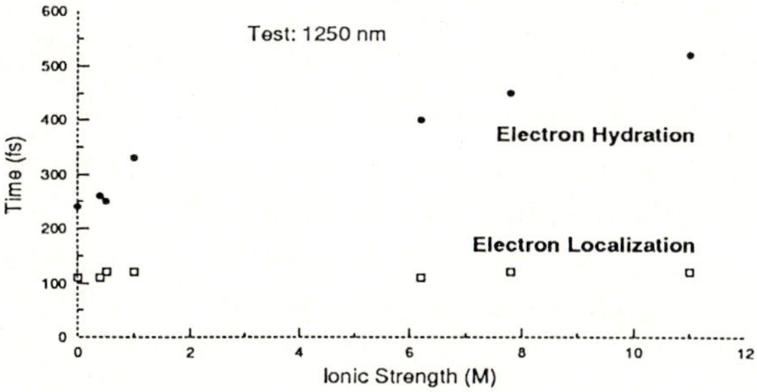

Fig. 4. Effect of the ionic strength on the dynamics of electron localization and hydration in aqueous media

4.2 Radical Pair Formation

It has been suggested that in concentrated ionic solutions reaction of excess electrons could occur through the formation of an electron–ion pair [11]. In the presence of a high concentration of hydrated protons ($[H_2O]/[H_3O^+] = 5$), femtosecond data exhibit an important structureless spectral distribution peaking around 920 nm ($E = 1.35$ eV), which we assign to an electron–cation pair $(H_3O^+ : e^-)_{hyd}$. The formation of this pair is found to follow a rate constant of $4 \times 10^{12}\,s^{-1}$ and has a lifetime of 850 fs [12]. This pair is not observed with Li^+ as a counterion. This is the first demonstration that the single electron transfer $(H_3O^+)nH_2O + e^- \rightarrow (H_3O^+ : e^-)nH_2O$ occurs *prior to* the electron hydration step. A significant fraction of electrons are therefore directly localized in the reaction radius of the hydronium ion, i.e. are either present in the hydration shell or tunnel there through a long range. In the evolution of the coupling between the electron–ion pair and water molecules we suggest that the limiting factor of the deactivation dynamics of the encounter pair corresponds to the activation energy of the radical–ion bond cleavage reaction which includes either a proton migration from H_3O^+ to neighbouring water molecules or a local polarization effect on H bonds.

References

1. E.W. Castner, G.R. Fleming, B. Bagchi, M. Maroncelli: J. Chem. Phys. **89**, 3519 (1988)
2. Y. Gauduel, A. Migus, A. Antonetti: In *Chemical Reactivity in Liquids: Fundamental Aspects*, ed. by M. Moreau, P. Turcq (Plenum, New York 1988) pp. 15–32
3. H. Lu, F.H. Long, R.M. Bowman, K.B. Eisenthal: J. Phys. Chem. **93**, 27 (1989)
4. N.V. Klassen: In *Radiation Chemistry*, ed. by Farhataziz, Rodgers (VCH, New York 1987) pp. 29–64
5. H. Lew, I. Heiber: J. Chem. Phys. **58**, 1246 (1973)
6. O. Dutuit, A. Tabche-Fouhaile, I. Nenner, H. Frolich, P.M. Guyon: J. Chem. Phys. **83**, 584 (1985)
7. A. Mozumder: Radiat. Phys. Chem. **32**, 287 (1988)
8. V.V. Konokalov, A.M. Raitshmring, Y.D. Tsvetkov: Radiat. Phys. Chem. **32**, 623 (1988)
9. J.E. Aldrich, M.J. Bronskill, R.K. Wolff, J.W. Hunt: J. Chem. Phys. **55**, 530 (1971)
10. R.B. Barnett, U. Landman, A. Nitzan: J. Chem. Phys. **90** (1989)
11. M.C. Messmer, J.D. Simon: J. Phys. Chem. **94**, 1220 (1990)
12. Y. Gauduel, S. Pommeret, A. Migus, N. Yamada, A. Antonetti: J. Am. Chem. Soc. **112**, 2925 (1990)

Femtosecond Studies of Electrons in Water

F.H. Long, Hong Lu, and K.B. Eisenthal

Department of Chemistry, Columbia University, New York, NY 10027, USA

We have measured the dynamics of an electron photoionized from a neutral water molecule. Our results suggest that the electron solvation dynamics can be viewed as an internal conversion process.

Although the equilibrium properties of the aqueous electron have been well-studied, the solvation dynamics and other non-equilibrium properties are an area of active research. A fundamental question concerning the electron solvation dynamics is the nature of the transition from the wet electron to the solvated or equilibrium electron. Theoretical studies have predicted dynamics involving a continous change in the electron absorption spectra.[1] Early experimental work has claimed that only two-states were involved, i.e. the solvation dynamics were not continous.[2] In order to resolve these issues we have done experiments with improved time resolution and have looked for an isosbestic wavelength, which if found would be a stringent test of a two state solvation dynamics.[3]

The two state model means that a wet electron, identifiable by its characteristic absorption spectra, undergoes a transition to the solvated electron. In other words, there are only two electron species present in the kinetics rather than a series of wet electron species, having their own spectra, that evolve into the equilibrium solvated electron. In the two state model, if the absorption of the two species (states) overlap then there must be a wavelength where the two spectra intersect (the isosbestic wavelength) and therefore have the same absorption coefficient ε. A consequence of the presence of an isosbestic wavelength can be seen by considering the absorption A(t) at a time t and a wavelength λ.

$$A(t)=\varepsilon_{wet}N_{wet}(t) + \varepsilon_{sol}N_{sol}(t)$$

where $N_{wet}(t)$ and $N_{sol}(t)$ are the concentrations of wet and solvated electrons. At the isosbestic wavelength, $\varepsilon_{wet}=\varepsilon_{sol}$ and the kinetics will therefore appear to be characteristic of a single species, with one risetime and a long decay, since the total electron concentration, $N_{wet}(t) + N_{sol}(t)$, is conserved.

To find the isosbestic wavelength we scanned from the infrared to the visible. At 820 nm the isosbestic point was observed in the wet to solvated electron transition, Fig. 1. A small decay present due to the geminate recombination of the electron has been corrected for using our previous results. We conclude from the observation of the isosbestic point that the wet to solvated electron transition involves only two states.

Springer Series in Chemical Physics, Vol. 53 **Ultrafast Phenomena VII**
Editors: C.B. Harris · E.P. Ippen · G.A. Mourou · A.H. Zewail

Accordingly, we used a simple kinetic model to fit the data.

$$e_{quasi\text{-}free} \xrightarrow{k_1} e_{wet} \xrightarrow{k_2} e_{sol}$$

This model was used because it is physically reasonable and allows for an unambiguous fit to the data. By the discovery of the isosbestic point, the two state model is verified. Our experimental results show that the wet electron formation time, $1/k_1$, is 180 ± 40 fs. This time is interpreted as the total time needed for the electron to localize in a pre-existing trap followed by a further deepening of the well to a configuration appropriate to the wet electron. The lifetime of the wet electron, $1/k_2$, was found to be 540 ± 50 fs. In the visible region, an ultrafast rise and decay was also found. We believe this might be the water cation, H_2O^+ or the excited state of neutral water.

We propose that the existance of only two states can be understood if that the wet electron is really an *excited state* of the solvated electron. This would quite naturally lead to the two states observed in the dynamics. The excited state is likely to be lowest excited state of the solvated electron corresponding to a p-like electron wavefunction. We therefore see that the rate-limiting step in the solvation dynamics of the electron is really an energy relaxation process referred to as a radiationless transition in molecules. This physical picture is supported by recent non-adiabatic simulations.[4] One anticipates that the internal conversion process can be described by the terms neglected in the Born-Oppenheimer approximation, as is the case with molecules. For the solvated electron, it is reasonable to say that the relevant motions might be water librations and translations.

Using the known values for the solvated electron absorption, the wet electron absorption spectrum is obtained, Fig. 2. This was done by comparing the signal at each wavelength to the absorption at long times which is due totally to the solvated electron. Both absorption spectra shown are normalized to the

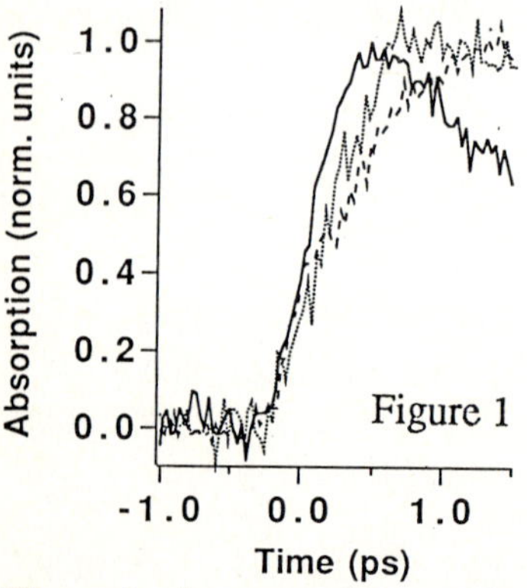

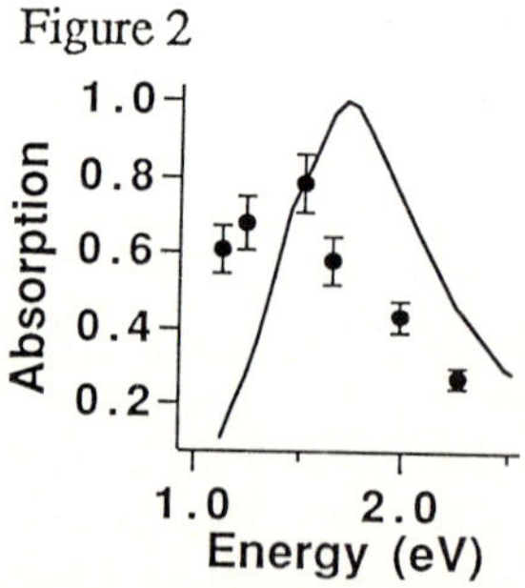

Figure 1. Aqueous electron dynamics at 625 (dashed line), 1000 nm (solid line) and 820 nm (dotted line).The appearance of an isosbestic point at 820 nm is evident.

Figure 2. Experimental wet (pre-solvated) electron (dots) and solvated electron (line) absorption spectra.

maximum of the solvated electron absorption. It can be seen that the wet electron absorption has a blue tail extending out past 2.5 eV indicating that the potential well the electron sits in is already quite deep after 180 fs. The absorption has a broad maximum at around 900 nm or 1.4 eV and that the overall absorption strength is similar to that of the solvated electron.

In summary the experimental verification of an isosbestic wavelength, suggests that the dielectric relaxation picture of electron solvation is fundamentally wrong. We propose that the existence of two-states can be interpreted by a physical picture in which the wet electron is an excited state of the solvated electron. Thus the rate-limiting step in the solvation dynamics is an internal conversion process from the excited state (wet electron) to the ground state (the solvated electron).

Acknowledgements

We would to thank NSF, AFOSR and the Petroleum Research Fund for their generous support. FHL would like to thank IBM for a graduate student fellowship.

References

1. P.J. Rossky and J. Schnitker J. Phys. Chem. **92**, 4277, (1988).
2. A. Migus, Y. Gaudel, J. L. Martin and A. Antonetti Phys. Rev. Lett **58**, (1987), 1559.
3. F.H. Long, H. Lu, K. B Eisenthal Phys. Rev. Lett. **64**, 1469, (1990).
4. F. Webster, J. Schnitker, R. Friesner, P. J. Rossky private communication.

Femtosecond Optical Spectroscopies of Solvated Polyatomic Molecules

S. Mukamel[1] *and Y.J. Yan*[2]

[1]Chemistry Department, University of Rochester, Rochester, NY 14627, USA
[2]Chemistry Department, University of California at San Diego, La Jolla, CA 92093, USA

Abstract. A multimode Brownian oscillator model is developed for third order nonlinear optical spectroscopies in condensed phases, taking into account molecular dynamics of coherent nuclear motions of the absorber and the dephasing and reorganization processes of solvent molecules. Calculations of femtosecond photon echo are discussed.

In this paper we present a generalized Brownian oscillator model which illustrates various aspects of molecular dynamical processes during time resolved nonlinear optical measurements.[1-5] Our description of molecular dynamics is based on the time evolution of density matrix in Liouville (or phase) space.

Let us consider an optical measurement of a solvated chromophore. We assume that only two electronic states, |g> and |e>, of the chromophore are relevant for the optical excitation. All four wave mixing spectroscopies of this molecular system can be expressed in terms of the molecular response function which is the sum of four-time correlation functions of the chromophore dipole moment μ. The typical form of these four-time correlation functions is $\mathrm{Tr}\,[G_{eg}(t_3)G_{mm}(t_2)\,G_{ge}(t_1)\rho_0]$ with m=g or e. Here ρ_0 describes the initial thermal distribution of the material system in the ground electronic state |g>. The molecular dynamics during the optical measurement are completely described by the Liouville space Green function $G_{mn}(t)$, which is defined by its action on an arbitrary dynamical variable A:

$$G_{mn}(t)\,A = \exp(-iH_m t/\hbar)\,A\,\exp(iH_n t/\hbar); \qquad m,n=e \text{ or } g. \tag{1}$$

The diagonal Green function $G_{mm}(t)$ describes the molecular dynamics in the potential surface V_m (corresponding to H_m) and is governed by the single molecular nuclear Hamiltonian H_m. Its phase space representation $G_{mm}(p,q,t;\,p_0,q_0)$ has a well defined classical analogue, i.e. the conditional probability of finding the system in phase space point (p,q) at time t, assuming it evolved in the potential surface V_m from (p_0,q_0) at time t=0. Since G_{mm} describes the coupled dynamics of the chromophore and the solvent, the complete phase space has a macroscopic number of degrees of freedom. However, we consider explicitly only a few relevant degrees of freedom, denoted as the system modes, by using the generalized Langevin equation:

$$\dot{q}(t) = p(t) \tag{2a}$$

Springer Series in Chemical Physics, Vol. 53 **Ultrafast Phenomena VII**
Editors: C.B. Harris · E.P. Ippen · G.A. Mourou · A.H. Zewail

$$\dot{p}(t) = \partial V/\partial q - \int_0^t \hat{\gamma}(t-\tau)\, p(\tau)\, dt + f(t). \tag{2b}$$

Here V is the effective potential surface for the system modes in the m = e or g electronic state. The friction kernel and the random force satisfy the fluctuation-dissipation relation: $\langle f(t)f(0)\rangle = k_B T\, \gamma(t)$. The quantum dynamics in state $|m\rangle$ can be obtained by considering eqs.(2) as operator equations. A Fokker-Planck equation for $G_{mm}(p,q,t;\, p_0,q_0)$ can then be obtained from eqs.(2). The off-diagonal Green function, e.g. $G_{eg}(p,q,t;\, p_0,q_0)$, describes the molecular dynamics during an optical transition and depends on both of the electronic state Hamiltonians, H_g and H_e. This Green function has no classical analogue due to its quantum nature which involves two electronic states.[3]

It is obvious that all the optically active modes of the chromophore should be considered explicitly in eqs.(2). In addition, the *solvation coordinate* should also be considered explicitly. Physically, the solvation coordinate originates from the collective nuclear response of the solvent upon optical excitation of the chromophore. This solvation coordinate is described by a strongly overdamped oscillator governed by the Smoluchowski equation of motion. In a frequency-resolved experiment, the optically active modes contribute to the Franck-Condon progressions and the solvation coordinate contributes to the line broadening. In a time-resolved experiment, the optically active modes show quantum beats with the vibronic dephasing and relaxation induced by the solvent friction and fluctuations, whereas the strongly overdamped solvation coordinate is responsible for spectral diffusion, e.g. the time-dependent Stokes shift and broadening, as well as homogeneous and inhomogeneous electronic dephasing.

Let us consider an ideal stimulated photon echo experiment in which the system is subject to three impulsive pulses with wavevectors $\underset{\sim}{k}_1$, $\underset{\sim}{k}_2$ and $\underset{\sim}{k}_3$, separated by τ_{21} and τ_{32}, subsequently. The echo signal is then generated after the third pulse at the phase matching direction of $\underset{\sim}{k}_3+\underset{\sim}{k}_2-\underset{\sim}{k}_1$ as a function of τ_{21} and τ_{32}:

$$S_{SPE}(\tau_{21},\tau_{32}) = |\mathrm{Tr}\{G_{eg}(\tau_{21})\, [G_{ee}(\tau_{32})+G_{gg}(\tau_{32})]\, G_{ge}(\tau_{21})\, \rho_0\}\,|^2. \tag{3}$$

The physical picture of the stimulated photon echo described by eq.(3) as follows: At time t=0, the initially equilibrated system ρ_0 is excited by the first pulse to an optical coherence ρ_{ge}, which then evolves freely, as described by the Green function $G_{ge}(\tau_{21})$, until $t=\tau_{21}$. The system is then subjected to the second impulsive pulse, which transfers the system to either the electronic ground ρ_{gg} or the excited ρ_{ee} state. The system then evolves freely as described either by $G_{gg}(\tau_{32})$ or by $G_{ee}(\tau_{32})$ until it interacts with the third impulsive pulse. After the third pulse the system is again in an optical coherence and the echo arises due to the free rephasing processes described by the Green function $G_{eg}(\tau_{21})$, which is the Hermitian conjugate of $G_{ge}(\tau_{21})$. The ordinary (two-pulse) photon echo is a limiting case of the stimulated echo when $\tau_{32}=0$ and $\underset{\sim}{k}_3=\underset{\sim}{k}_2$. In this case all the relevant molecular dynamics are related to the optical coherence and described by[4]

$$S_{SPE}(\tau_{21}) = |\mathrm{Tr}\{G_{eg}(\tau_{21})G_{ge}(\tau_{21})\, \rho_0\}\,|^2. \tag{4}$$

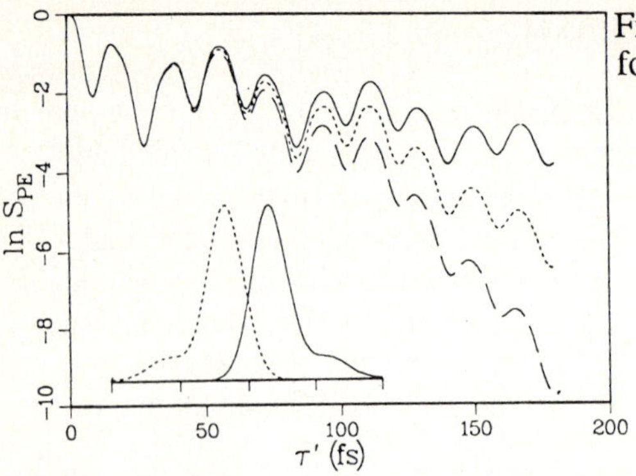

Fig.1: The impulsive two pulse echo for a three mode system [1b].

Figure 1 presents the ordinary photon echo signal for a three mode system (two coherent vibrations and a single solvation coordinate). The various curves correspond to a different choice of solvent timescale. Details of the calculations are give in Reference 1b. The calculations demonstrate the capacity of the present Brownian oscillator model to describe the coherent nuclear motion as well as the solvation dynamics which couple to optical transitions, as measured recently.[6]

The present Brownian model provides a unified description of the molecular dynamics, which include the coherent nuclear motion, electronic and nuclear dephasing and solvent relaxation, as probed in nonlinear optical measurements of polyatomic molecules in solution. The electronic solvation dynamics (spectral diffusion, homogeneous and inhomogeneous electronic dephasing) is described by a macroscopic solvation coordinate, which undergoes an overdamped motion. The stochastic lineshape model[7] is a special limit case of the present model obtained by considering an overdamped oscillator and neglecting the time dependent Stokes shift. The present Brownian solvation model can describe the complete picture of solvent reorganization processes, which include spectral (Stokes) shift as well as line broadening.

ACKNOWLEDGEMENTS

The support of the Air Force Office of Scientific Research, the National Science Foundation and the Petroleum Research Fund, administered by the American Chemical Society, is gratefully acknowledged.

REFERENCES

1. a) S. Mukamel, Adv. Chem. Phys. **70**, Part I, 165 (1988);
   b) S. Mukamel, Ann. Rev. Phys. Chem. **41** (in press) (1990); W. Bosma, Y.J. Yan and S. Mukamel (to be published).
2. R.F. Loring, Y.J. Yan and S. Mukamel, J. Chem. Phys. **87**, 5840 (1987).
3. Y.J. Yan and S. Mukamel, J. Chem. Phys. **89**, 5160 (1988).
4. R.F. Loring and S. Mukamel, Chem. Phys. Lett. **114**, 426 (1985).
5. Y.J. Yan and S. Mukamel, Phys. Rev. A (in press) (1990); W.B. Bosma, Y.J. Yan and S. Mukamel, J. Chem. Phys. (submitted) (1990).
6. P.C. Becker, H.L. Fragnito, J.Y. Bigot, C.H. Brito Cruz, R.L. Fork, and C.V. Shank, Phys. Rev. Lett. **63**, 505 (1989).
7. R. Kubo, Adv. Chem. Phys. **15**, 101 (1969).

Ultrafast Studies on Intramolecular Charge Transfer and Solvation

P.F. Barbara

Department of Chemistry, University of Minnesota, Minneapolis, MN 55455, USA

Abstract. Ultrafast fluorescence spectroscopy has been used to study the solvent mediated excited state intramolecular electron (charge) transfer of 9,9'–bianthryl. The excited state electron transfer (et) kinetics are well modeled by an electronically adiabatic approach, employing an "outer sphere" generalized Langevin equation description of motion along the et reaction coordinate. The relationship between solvation and electron transfer dynamics is discussed.

1. Introduction

There is a growing appreciation [1] that fast electron transfer reactions are not well characterized by traditional models (such as Marcus–Hush theory or related non–adiabatic theories) which are based on a quasi–equilibrium approximation for the population of thermally excited states of the solute/solvent systems. Contemporary theories treat solvation dynamics in detail. A key result is that the rate of a charge transfer reaction should be a function of the microscopic dynamics of the specific solvent. In fact, in the case of a very small intrinsic charge transfer activation barrier, the rate is predicted to be roughly equal to the rate of solvation. The predictions are considerably more complex for solvents with a distribution of relaxation times and electron transfer barrier energies that are comparable or larger than the available thermal energy.

2. A Prototype Charge Transfer Molecule: Bianthryl

Excited state charge transfer has been extensively investigated for the molecule BA by a number of research groups [2,3]. The fluorescence spectrum of BA is strongly solvent polarity dependent. In polar solvents the emission can be roughly interpreted as being due to the sum of bands from two isomers of S_1; a non–polar LE (locally excited) form and CT (charge transfer) form, as proposed over a decade ago by Nakashima et al. and Grabowski et al. [2,3].

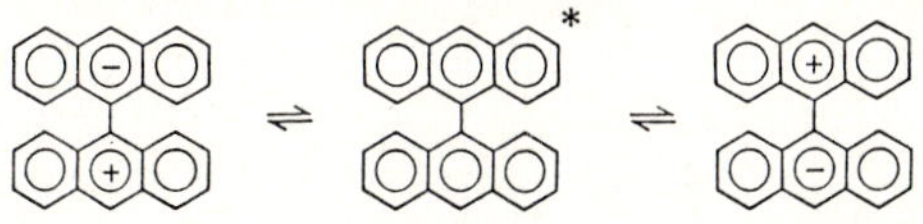

CT' LE CT

The effect of solvent on the total free energy of a BA/solvent system is represented in figure 1 for a polar solvent such as propylene carbonate. These zero order (uncoupled) curves for the various configurations correspond to a Marcus theory like outer sphere model for the electron transfer reaction coordinate. The parameters employed in constructing figure 1a were determined empirically as described elsewhere. [4] The various energy curves in figure 1a ignore mixing (configuration interaction) between the various zero order states. In fact, the LE/CT mixing is quite substantial and the adiabatic energy curves of figure 1b are a more accurate representation of S_1 BA than are the diabatic curves of figure 1a. These curves come from mixing the zero order curves of figure 1a.

Springer Series in Chemical Physics, Vol. 53 **Ultrafast Phenomena VII**
Editors: C.B. Harris · E.P. Ippen · G.A. Mourou · A.H. Zewail

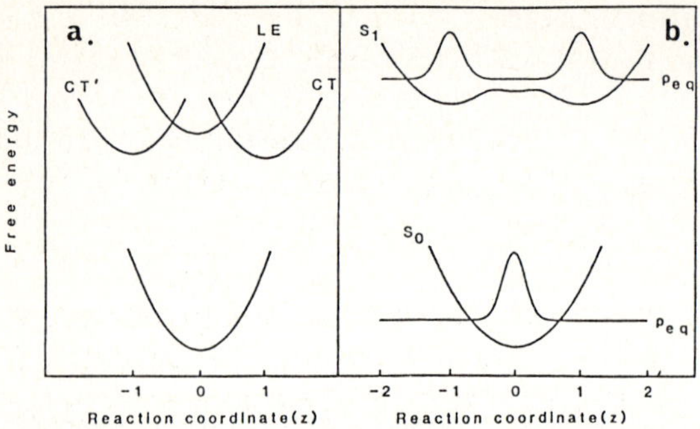

Figure 1. (a) The zero order energies for the S_0, LE, CT and CT' states as a function of the solvent coordinate z. The parameters have been adjusted in order to bring agreement between predicted and observed spectra for BA in propylene carbonate. (b) The adiabatic energy for the first excited singlet state (S_1) of BA in propylene carbonate. The equilibrium distribution function is denoted by ρ_{eq}.

The physical significance of figure 1b is firmly established by considering how emission spectra are calculated from this model. First it can be assumed that the solvent is stationary on the time scale of emission so that for each specific value of z we can make the Franck–Condon approximation for the emission intensity.

$$I(\omega) \propto \beta_{el}^2 [\langle X|X'\rangle]^2 \tag{1}$$

Here β_{el} is the electronic part of the transition moment. The total spectrum is the integral.

$$I(\omega,t) \propto \int_{-\infty}^{\infty} g(\omega_0, \omega - \omega_0)\, \rho(z, t)\, \beta_{el}^2(z)\, dz \tag{2}$$

We have been able to reproduce, see Figure 2, the equilibrated emission spectra of BA remarkably well by using eq. 2, the potential and $\rho(z,t)$ in figure 1b. The emission shape $g(\omega_0, \omega - \omega_0)$ function was determined empirically by assuming that it is simply equal to the emission spectra of BA in hexane where the S_1 of BA is dominated by LE in the relevant region of the potential and ω_0 is not a function of z because LE and S_0 are both nonpolar.

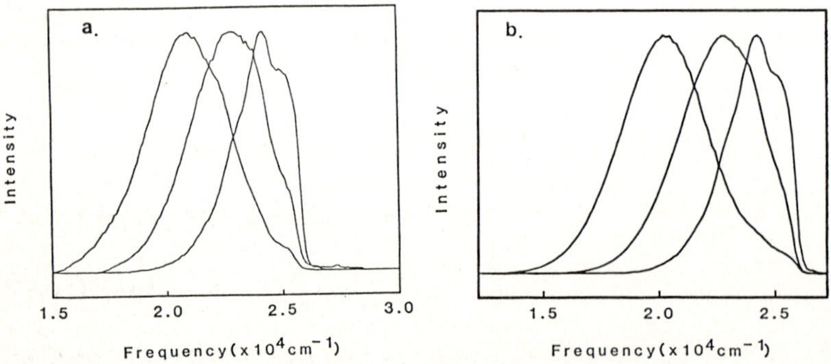

Figure 2 (a) Steady state fluorescence spectra of BA in hexane (right), glycerol triacetate (middle) and propylene carbonate (left). (b) Simulated emission spectra for BA in a nonpolar, moderately polar and very polar solvent. The potential energy parameters are optimized to represent the solvents indicated in figure 1a. In both figures, the peak intensities are normalized to the spectrum in hexane.

In summary, an adiabatic model for the first excited state of BA is able to account well for the equilibrium absorption and emission properties of BA.

3. Simulation of the Charge Transfer Dynamics of Bianthryl

Recently, we developed a theoretical model for charge transfer dynamics of BA. The starting point is the generalized Langevin Equation, GLE, as adopted for stochastic motions involving coupling to a solvent coordinate, as follows.

$$m\ddot{z} = -\frac{\partial V(z)}{\partial z} - m\int \eta(t-\tau)\,\dot{z}(\tau)\,d\tau + F(t) \tag{3}$$

The GLE is a stochastic equation of motion for the coordinate z, see figure 1b. The l.h.s. of eq. 3 is the inertial force along z in terms of the effective polarization mass m of the solvent and the acceleration $\ddot{z}$. The term $-\frac{\partial V[z(t)]}{\partial z}$ is associated with the force due to the potential $V(z) = E_{S_1}(z)$. The term F(t) is a random force due to fluctuations of the solvent. The remaining term on the r.h.s. of eq. 3 accounts for the friction (or retarding motion) along the reaction coordinate due to the lag of the solvent motion. The quantity $\eta(t)$ is the time–dependent friction kernel. It characterizes the dissipation effects of the solvent motion along the reaction coordinate. We assume that the underlying dynamics of the dielectric function for BA and other molecules are similar to the dielectric transient Stokes shift dynamics for the coumarins, i.e. as quantified by the average solvation time τ_s.

The key results of the simulations are (i) the predicted time dependent spectra are in good agreement with experiment and (ii) the predicted electron transfer times are predicted to be similar to the average solvation time τ_s, see Table 1.

Table 1. A comparison of the average electron transfer time $<\tau_{et}>$ of 9,9'–bianthryl(BA) in various polar solvents to the average solvation times $<\tau_s>$ of coumarin probes, as determined by the transient Stokes shift method [4].

solvent[a]	compound	τ_1, ps	τ_2, ps	$<\tau_{et}>$[b], ps	$<\tau_s>$[c], ps
propanol (273 K)	BA	91(62 %)	276(38 %)	161	138±21
butanol (273 K)	BA	115(64 %)	486(36 %)	249	
pentanol (273 K)	BA	183(41 %)	635(59 %)	450	
pentanol	BA	138(75 %)	366(25 %)	195	142±2
acetone	BA	0.86			0.83
benzonitrile	BA	1.48(46 %)	8.75(54 %)	5.4	4.5–5.9
DMS	BA	0.67(73 %)	2.70(27 %)	1.2	1.2–1.4
PC	BA	0.72(66 %)	3.59(34 %)	1.7	2.4–3.4

a. Measurements were made at ambient temperature if not specified.

b. Average times obtained by weighting each time constant. For alcohol solvents, measurements were made by time–correlated single photon counting. The rest of the measurements were made by fluorescence up–conversion system. The transient data of acetone were fitted by a single exponential function and the rest of the transient data were modeled with a bi–exponential function.

c. Average solvation times were obtained with several coumarin probes (see ref. 4).

4. Experiments on 9,9'–Bianthryl

The time and wavelength resolved fluorescence dynamics of bianthryl has been investigated by several groups [4]. The results are generally in accord with the theoretical model just discussed. One of the challenges of studying the photodynamics of BA is that the LE and CT interconversion is so rapid (i.e. on the time scale of solvation) that it is necessary to employ UV subpicosecond and even femtosecond fluorescence spectroscopy which has only recently been available.

We have studied the time evolution of the fluorescence spectrum of BA in acetone[4]. The results are in qualitative agreement with the simulated time dependent spectra using the simulated $\rho(z,t)$ and eq. 2. This strongly supports the validity of the adiabatic, GLE model for the charge transfer of S_1 BA. Many of our measurements on BA have been confined to single emission wavelength near the short wavelength edge of the BA fluorescence. The emission at these wavelengths monitors population in the LE region of figure 1b. The results show that population in this region evolves with non–mono–exponential charge transfer kinetics due to solvent kinetic control, and in turn, the underlying non–mono–exponential solvation dynamics. Fitting parameters for the emission dynamics are compared in Table 1 with the fitting parameters of C(t) measured with coumarin probes in the same solvent. It is remarkable how closely the population dynamics of LE parallel dynamics of C(t). Thus experimental and theoretical data strongly indicate that the excited state LE/CT interconversion of BA is controlled by solvation dynamics and that the barrier for the reaction is small compared to k_BT.

The relationship between the intensity dynamics near the peak wavelength of LE (≃420 nm) and the conventional description of charge transfer rate constants is clearly established with GLE simulations. We observe that the simulated *intensity* dynamics for a given solvent are nearly identical to the simulated *survival* dynamics, which supports the notion that the intensity dynamics on the short wavelength edge of the spectrum does indeed give an accurate measure of the underlying charge transfer dynamics.

5. Acknowledgements

Acknowledgement is made to the National Science Foundation (Grant CHE–8251158) and the Office of Naval Research for support of this research. I would like to thank my coworkers who actually accomplished most of the work described in this paper: Tai Jong Kang, W. Jarzeba, and Teresa Fonseca.

6. References

1. J. T. Hynes, *J. Phys. Chem.* **90**, 3701 (1986); D. F. Calef and P. G. Wolynes, *J. Phys. Chem.* **87**, 3387 (1983); H. Sumi and R. A. Marcus, *J. Chem. Phys.* **84**, 4894 (1986); W. Nadler and R. A. Marcus, *J. Chem. Phys.* **86**, 3906 (1987); I. Rips and J. Jortner, *J. Chem. Phys.* **87**, 2090 (1987); ibid **87**, 6513 (1987); **88**, 818 (1988); *Chem. Phys. Lett.* **133**, 411 (1987); M. Sparpaglione and S. Mukamel, *J. Chem. Phys.* **88**, 3266 (1988); ibid **88**, 1465 (1988); M. D. Newton and H. L. Friedman, *J. Chem. Phys.* **88**, 4460 (1988); I. Rips and J. Jortner, *Chem. Phys. Lett.* **133**, 411 (1987); G. E. McManis, A. K. Mishra, and M. J. Weaver, *J. Chem. Phys.* **86**, 5550 (1987).
2. T. J. Kang, M. A. Kahlow, D. Giser, S. Swallen, V. Nagarajan, W. Jarzęba and P. F. Barbara, *J. Phys. Chem.* 92, 6800 (1989) and references therein.
3. N. Nakashima, M. Murakawa, and N. Mataga, *Bull. Chem. Soc. Japan,* **49**, 854 (1976);W. Rettig and M. Zander, *Ber. Bunsenges. Phys. Chem.* **87**, 1143 (1983); M. Zander and W. Rettig, *Chem. Phys. Lett.* **110**, 610 (1984); N. Mataga, H. Yao, T. Okada, and W. Rettig, *J. Phys. Chem.* **93**, 3383 (1989).
4. T. J. Kang, W. Jarzeba, T. Fonseca, P. F. Barbara, to be submitted; P. F. Barbara, Advances in Photochemistry, 15, 1(1990).

Dynamics of Intramolecular Electron Transfer in Polar Solvents

T. Okada, S. Nishikawa, K. Kanaji, and N. Mataga

Department of Chemistry, Faculty of Engineering Science, Osaka University, Toyonaka, Osaka 560, Japan

1. Introduction

When the electronic interaction between an electron donor (D) and acceptor (A) is very weak, the electron transfer (ET) process is considered to be non-adiabatic. When the interaction becomes fairly strong, the reaction will become adiabatic. If the electronic interaction becomes sufficiently strong and the energy gap relations are also favorable, the ET process will become barrierless. In such a case it is believed that the ET process is governed mainly by the orientational motions of polar solvent molecules or polar groups in the environment surrounding D and A, and the longitudinal dielectric relaxation time τ_L will be important as a factor controlling the ET rate. In the limit of strong interaction between D and A groups combined by a rigid spacer or single bond, its excited singlet state can be regarded as a very polar single molecule and we can observe a large fluorescence Stokes shift due to the solvation in polar solvents. For the elucidation of the above mechanisms, especially the interactions of D and A with solvent including its dynamical effects on the ET process, systematic femtosecond-picosecond laser photolysis studies on various combined D and A systems with different degrees of electronic interaction between them are of crucial importance.

2. Results and Discussion

(A) Photoinduced electron transfer in 1-butanol

The compounds examined here are listed below. In these compounds, there is no appreciable electronic interaction between two groups both in the ground state and in the initial state immediately after laser excitation and the electron transfer can take place very easily when the surrounding polar solvent orientations are favorable. Actually, the transient absorption spectra of these compounds show the change in the course of electron transfer from the $S_n \leftarrow S_1$ band localized in the pyrene or anthracene part to the superposition of donor cation and acceptor anion absorption bands [1,2].

The fluorescence decay curves monitored at the fluorescence band of the locally excited state of pyrene or anthracene moiety were nonexponential as indicated in Fig. 1. We have analyzed the decay curves by means of an extended exponential function,

$$F(t) = A\exp[-(t/\tau)^{\beta}] + B\exp(-kt),$$

where the second term is due to the contribution from the photoproducts, the amount of which was about 0.1% in a fresh sample. The obtained values of β and τ depend strongly on the

Springer Series in Chemical Physics, Vol. 53 **Ultrafast Phenomena VII**
Editors: C.B. Harris · E.P. Ippen · G.A. Mourou · A.H. Zewail

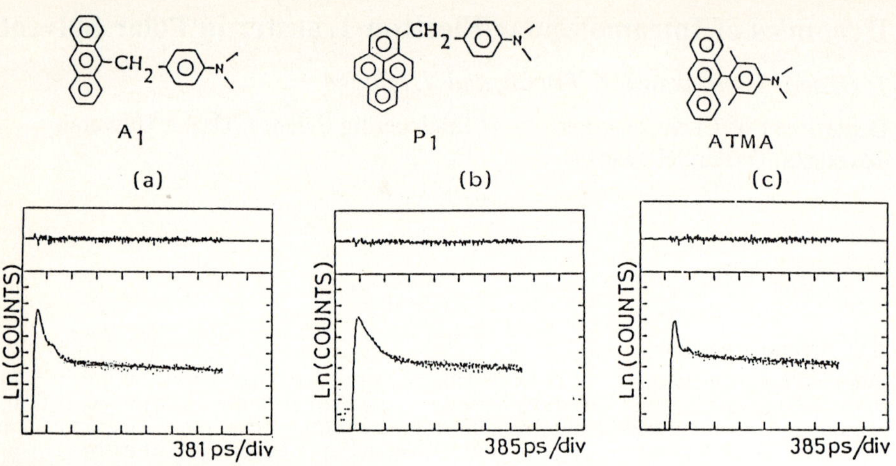

Fig. 1. The fluorescence decay curves monitored at the band of the locally excited state of A_1(a), P_1(b), and ATMA(c) in 1-butanol at room temperature.

Table 1. β and τ values of A_1, P_1, and ATMA in 1-butanol at 298 k.

compound	β	τ /ps
A_1	0.62	8.2
P_1	0.95	87
ATMA	0.52	3.6

compounds as indicated in Table 1. The rise curve of the ion-pair state obtained by transient absorption measurement was reproduced very well even in the picosecond time regime by using the same values of τ and β obtained by the fluorescence measurement as shown in Fig. 2, although the time resolution of our single photon counting system was about 10 ps.

It is well known that the dielectric behaviors of alcoholic solvents do not show a simple Debye dispersion. 1-Butanol has three different dielectric relaxation times. The initial slope of the observed fluorescence decay curve in the case of A_1 and ATMA is rather close to the fastest longitudinal dielectric relaxation time of 1-butanol, $\tau_3 \sim 2$ ps, but close to $\tau_2 \sim 25$ ps in the case of P_1. One of the possible mechanisms responsible for the extended exponential behavior of the fluorescence decay curve may be saturation effects of solvation dynamics in nearby solvents strongly coupled to the charge distributions in D-A systems in the excited state of the solute molecule. Such a solute-solvent interaction will give a radial distribution of the orientational relaxation time of the solvent, leading to the observed fluorescence behavior.

From the above viewpoints, it seems very important to study precisely the dynamics of electron transfer processes by using simpler polar solvents such as alkanenitriles. We have measured time resolved absorption spectra of A_1 and P_1 in acetonitrile and butyronitrile solutions [2]. The rise curves of the ion-pair state absorbance can be expressed approximately by an exponential function, although the dynamic range of our

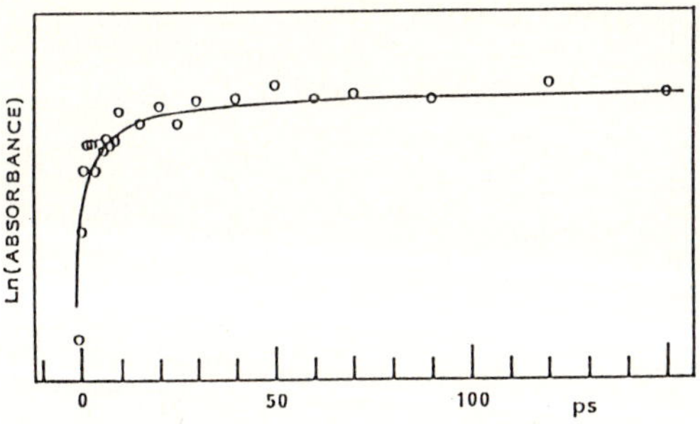

Fig. 2. Rise curve of ion-pair state of A_1 in 1-butanol.
—— : calculated with β=0.62, τ=8.2 ps.

detection system is about one order of magnitude. The obtained rise times in acetonitrile and butyronitrile are 0.65 ps and 1.0 ps for A_1 and 1.7 ps and 2.5 ps for P_1, respectively [2]. The rise times are much longer than the longitudinal dielectric relaxation time of the solvent; τ_L(acetonitrile)=0.2 ps and τ_L (butyronitrile)=0.5 ps. The electron transfer process will not be barrierless but will be controlled mainly by the magnitude of the D, A electronic interaction and the Franck-Condon factor of the reaction [2]. These results obtained in alkanenitrile suggest that the nonexponential fluorescence decay in 1-butanol should be explained as due to the characteristic nature of the alcoholic solvent.

(B) Dynamical behavior of 9,9'-bianthryl in the excited state

Since two anthryl planes of bianthryl are mutually perpendicular in the ground state, the fluorescence spectrum in low temperature matrices is structured and quite similar to the emission of monomeric anthracene. In nonpolar solvent at room temperature, the fluorescence band is broadened and slightly red shifted compared with that detected at 77 K as indicated in Fig. 3a, which indicates that the excited equilibrium state at room temperature is delocalized between two equal anthracene subunits and probably has a form a little inclined from a perpendicular configuration. The $S_n \leftarrow S_1$ absorption spectra are also different from each other between two temperatures corresponding to the difference of the fluorescence bands as shown in Fig. 3a. Time-resolved absorption spectra of bianthryl in isooctane at room temperature are indicated in Fig. 3b. The absorption spectra detected at 0 ps and 1 ps may be assigned mainly to the spectrum due to the ground state geometry. With increase of the delay time, the spectrum changes toward that of equilibrium state. We can estimate the approximate decay time of the initial configuration by plotting the absorbance at wavelengths where the time-dependent changes are large. The obtained decay times, τ , in some alkane solvents depend only slightly upon the solvent viscosity, η , namely, τ (hexane) $\sim$ 5 ps, η=0.31 cp; τ (isooctane)$\sim$9 ps, η =0.50 cp; τ (cyclohexane)$\sim$10 ps, η=0.98 cp; τ (trans decalin) $\sim$ 6 ps, η=2.1 cp; and τ (liquid paraffin) $\sim$16ps, η =85 cp. Rather small dependences on the solvent viscosity and excitation wavelength in the isomerization process of cyanine dye were reported by Akesson et al. [3]. They analyzed their obtained results by a model assuming a barrierless motion on a potential surface with position-dependent sink [4]. The conformational relaxation in the excited state of bianthryl in nonpolar solvents may be a similar process. In the present experimental conditions, the excited

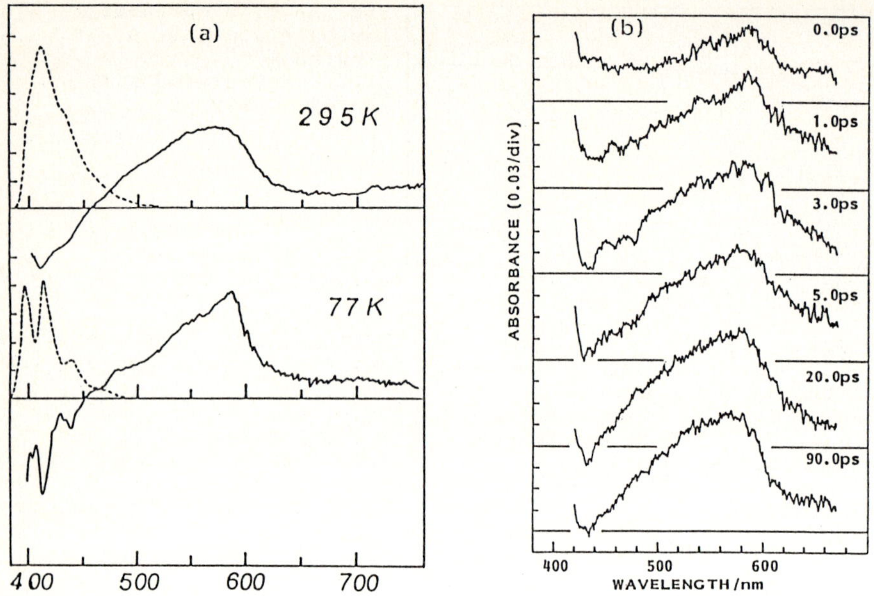

Fig. 3. (a) $S_n \leftarrow S_1$ absorption (solid line) and fluorescence (dotted line) spectra of bianthryl in methylcyclohexane-isopentane mixed solvent. (b) Time resolved absorption spectra in isooctane at room temperature.

state bianthryl just after laser irradiation has an excess vibrational energy of about 2400 cm^{-1}. Therefore, it might be possible that the torsional relaxation of the excited bianthryl by the above described mechanisms[4] is affected also by the relaxation process of this excess vibrational energy, leading to the observed viscosity dependence on τ values by some mechanisms which are not very clear at the present stage of investigation.

The formation of the charge transfer state in a viscous polar solvent, 1-pentanol, has been reported elsewhere [5]. The charge transfer took place from the twisted form (relaxed S_1 state observed in nonpolar solvent) and the observed approximate charge transfer time was similar to the longitudinal relaxation time for 1-pentanol at 25°C (174 ps). By examining the time-resolved spectra of bianthryl in hexanenitrile solution we have concluded that the charge transfer process seems to be competing with the relaxation process to the inclined form. The observed spectra of the charge transfer state are not the same as the superposition of absorption bands of anion and cation radicals of anthracene and/or 9-methylanthracene, indicating some delocalization interaction between two identical components.

From an analysis of the transient absorption spectra, approximate rise times of the charge transfer state have been estimated to be 1.8 ps, 3.2 ps, and 7.5 ps in acetonitrile, butyronitrile, and hexanenitrile, respectively. The obtained rise times are considerably longer than the fluorescence decay times detected at ca. 400 nm reported by Barbara et al. [6]. One of the possible explanations for this disagreement may be that the charge transfer proceeds via multiple intermediate states with different configurations and with slightly different electronic structures to the equilibrium state. In fact, 4-(9-anthryl)-N,N-dimethylaniline (A_0) shows multiple intermediate states in the course of the solvent induced change of the electronic and geometrical structures in the excited state [1,2].

3. References

[1] T. Okada, N. Mataga, W. Baumann, and A. Siemiarczuk: J. Phys. Chem. 91, 4490 (1987).
[2] N. Mataga, S. Nishikawa, T. Asahi, and T. Okada: J. Phys. Chem. 94, 1443 (1990).
[3] E. Akesson, H. Bergstrom, V. Sundstrom, and T. Gillbro: Chem. Phys. Lett. 126, 385 (1986).
[4] B. Bagchi, G. R. Fleming, and D. W. Oxtoby: J. Chem. Phys. 78, 7375 (1983).
[5] N. Mataga, H. Yao, T. Okada, and W. Rettig: J. Phys. Chem. 93, 3383 (1989).
[6] M. A. Kahlow, T. J. Knag, and . F. Barbara: J. Phys. Chem. 91, 6452 (1987).

Femtosecond Proton Transfer Reactions in Electronically Excited Aromatic Molecules

T. Elsaesser[1], *F. Laermer*[1], *W. Frey*[1], *W. Kaiser*[1], *M. Wiechmann*[2], *and H. Port*[2]

[1]Physik-Department E 11, Technische Universität München, D-8000 München 2, Fed. Rep. of Germany
[2]3. Physikalisches Institut, Universität Stuttgart, D-7000 Stuttgart 80, Fed. Rep. of Germany

Intramolecular proton transfer represents an elementary chemical reaction occurring in numerous aromatic molecules after excitation to a higher singlet state. The motion of the proton towards its new location in the molecular geometry was not temporally resolved in previous picosecond measurements suggesting subpicosecond time constants of the structural change. In this paper, we present a study of proton transfer with a temporal resolution of 70 fs. The transfer rates are directly measured for different benzothiazole and -triazole compounds and the mechanisms of the reaction are discussed.

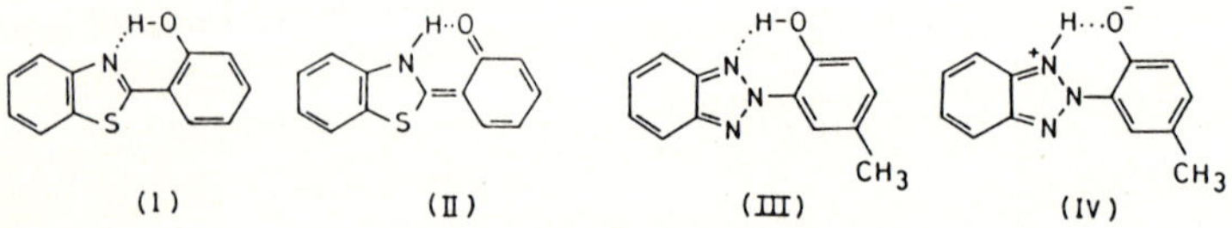

(I) (II) (III) (IV)

The enol tautomer (I) represents the predominant ground state structure of 2-(2'-hydroxyphenyl)benzothiazole (HBT) dissolved in nonpolar solvents. Excitation of this species via the S_0–S_1 absorption band in the ultraviolet around 350 nm results in a weak fluorescence emission with maximum at 530 nm. The emission band of a Stokes shift of 11000 cm^{-1} is due to the excited keto tautomer (II) with the proton now linked to the nitrogen atom. The formation of the latter structure is evident from picosecond infrared measurements, where the stretching vibrations of the new N-H and C=O groups were observed in the excited electronic state [1].

The speed of the enol $\rightarrow$ keto transformation is measured by monitoring the femtosecond rise of the keto emission [2]. After excitation of the enol tautomer to the S_1 state by a pump pulse at 310 nm, the probe pulse induces stimulated emission from the keto S_1 state to the keto ground state. The rise time τ_1 of the gain directly reveals the delayed formation of the keto species. Independent of the specific wavelength of the probe pulses, we find a very small value of τ_1=170 $\pm$ 20 fs, representing the transfer time of the proton. After the femtosecond transfer reaction, the keto S_1 state decays to the keto ground state by internal conversion on a picosecond time scale (S_1 lifetime 300 ps for HBT in C_2Cl_4 [3]). Finally, the keto $\rightarrow$ enol back reaction proceeds in the ground states on a slow microsecond time scale. Here a barrier exists between the keto and the enol minima of the potential energy surface.

Springer Series in Chemical Physics, Vol. 53 **Ultrafast Phenomena VII**
Editors: C.B. Harris · E.P. Ippen · G.A. Mourou · A.H. Zewail

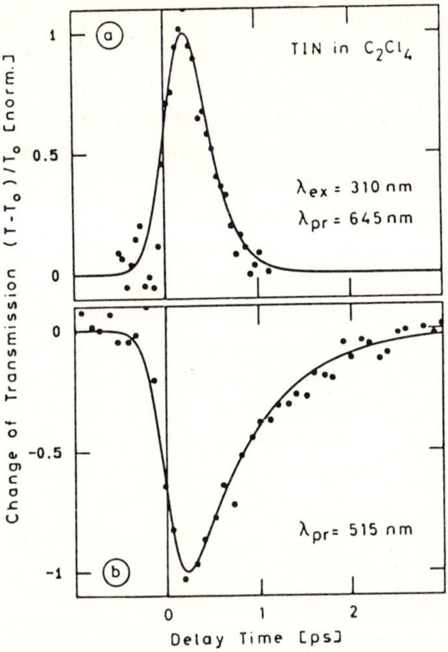

Fig. 1
a) Femtosecond gain of TIN at λ_{pr}=645nm (points, T_0, T : transmission of the sample before and after excitation at λ_{ex}=310nm).
b) Transient absorption of TIN at λ_{pr}=515nm.

The data suggest the following mechanism of femtosecond proton transfer in the excited state : Excitation of the enol tautomer to the S_1 state results in a redistribution of electronic charge, establishing a barrierless potential energy surface with a minimum for the keto configuration. As a consequence, the motion of the proton along the energy surface occurs with the time constant τ_1=170 fs, which corresponds to the period of a low frequency vibration of large amplitude.

Next we discuss the photophysics of 2-(2'-hydroxy-5'-methylphenyl)benzotriazole (trade name : TINUVIN P or TIN), a widely used photostabilizer of polymers. The planar enol tautomer (III) with an intramolecular hydrogen bond represents the main ground state species in nonpolar solution [4]. The strongly Stokes shifted emission between 600 and 850 nm, which is indicative of proton transfer in the excited state, shows an extremely low quantum yield of approximately 10^{-5}. The enol species undergoes a reaction scheme comparable to that of HBT. In TIN, however, the reaction cycle comprising the formation of the keto-type structure (IV), internal conversion in this molecule and the keto → enol back reaction is completed within less than 1 ps. In Fig. 1 a, the time dependent gain at 645 nm, i.e. in the range of the keto emission band, is plotted vs delay time. The signal rises with a delay time of approximately 100 fs related to the speed of the initial proton motion. This step is followed by the very rapid depopulation of the keto S_1 state with a time constant of 150 fs (decay in Fig. 1 a). Low frequency librations of the phenyl ring relative to the triazole moiety of TIN seem to be involved in the highly efficient S_1-S_0 internal conversion [4].

The data of Fig. 1 b represent the transient absorption of TIN at a probe wavelength of 515 nm, where the absorption of the keto ground state dominates for delay times longer than 500 fs. The absorption change decays with a time constant of 600 fs giving evidence of the fast back reaction from the keto to the enol ground state. The subpicosecond back reaction suggests an essentially barrierless potential surface in the electronic ground state favored by the zwitterionic character of the keto-type tautomer (IV). The speed of proton transfer may be additionally enhanced by the high vibrational temperature in the keto ground state, which is a result of the femtosecond S_1-S_0 internal conversion in the keto-type species and of the concomitant rapid redistribution of energy over the vibrational degrees of freedom.

Very recently, we have studied the proton transfer of TIN in a polystyrene film. Similar to the liquid phase, the complete reaction cycle takes approximately 1 ps making TIN a very effective stabilizer of polymers.

[1] T. Elsaesser, and W. Kaiser, Chem. Phys. Lett. 128 (1986), 231

[2] F. Laermer, T. Elsaesser, and W. Kaiser, Chem. Phys. Lett. 148 (1988), 119

[3] T. Elsaesser, B. Schmetzer, M. Lipp, and R.J. Bäuerle, Chem. Phys. Lett. 148 (1988), 112

[4] M. Wiechmann, H. Port, F. Laermer, W. Frey, and T. Elsaesser, Chem. Phys. Lett. 165 (1990), 28

Solvent/Solute Interactions Probed by Transient Absorption and Transient Raman Spectroscopy

T.L. Gustafson, K. Iwata, and W.L. Weaver

Department of Chemistry, The Ohio State University,
120 West 18th Avenue, Columbus, OH 43210, USA

Abstract. We use transient absorption spectroscopy and picosecond transient Raman spectroscopy to probe the effect of solvent on the excited state structure of *trans*-stilbene, its isotopomers, and sulfur and selenium analogues of DODCI.

The importance of *cis-trans* photoisomerization in the polyenes as a model system for isomerization is well established and has been reviewed.[1,2] The structure and ordering of the excited electronic states in these systems remain important areas of study. The polyenes also provide an important test case for the study of the fundamental interaction between solvent and solute in a chemical reaction in solution. Of particular interest is the *dimensionality* of the isomerization pathway.[3-6] We are using transient absorption spectroscopy to probe the dynamics of isomerization and transient Raman spectroscopy to probe the influence of friction on the *structure* of reacting species in solution.

The apparatus we are using is based on a MHz amplified synchronously pumped dye laser.[7] We have recently improved our amplifier scheme to include a three passes through the amplifier. This has increased our output energy by ~30%. In Figure 1 we show the output energy as a function of input energy for the triple pass amplifier; the inset shows the optical design of the triple pass amplifier. The output powers are sufficient for obtaining a broadband probe pulse for transient absorption spectroscopy using stimulated Raman scattering in an optical fiber.[8] In Figure 2 we show the transient absorption spectrum of DODCI at delay times of -25, 25, 100 and 500 ps. Each spectrum in Figure 2 was obtained in just 4 s, demonstrating the high signal-to-noise available with the apparatus. The additional output power also permits the acquisition of high quality transient Raman spectra. In Figure 3 we show the picosecond transient Raman spectra of *trans*-stilbene and α,α'-^{13}C-*trans*-stilbene in hexane over the range from 900 to 1700 cm^{-1}. An important feature of these data is the

Springer Series in Chemical Physics, Vol. 53 **Ultrafast Phenomena VII**
Editors: C.B. Harris · E.P. Ippen · G.A. Mourou · A.H. Zewail

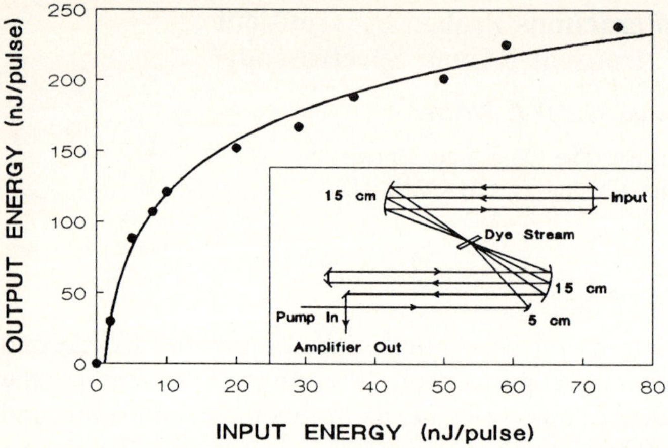

Figure 1: Output energy as a function of input energy for the triple pass amplifier. Pump energy: 2.5 μJ/pulse; Repetition rate: 1 MHz. Inset: Optical design of triple pass amplifier.

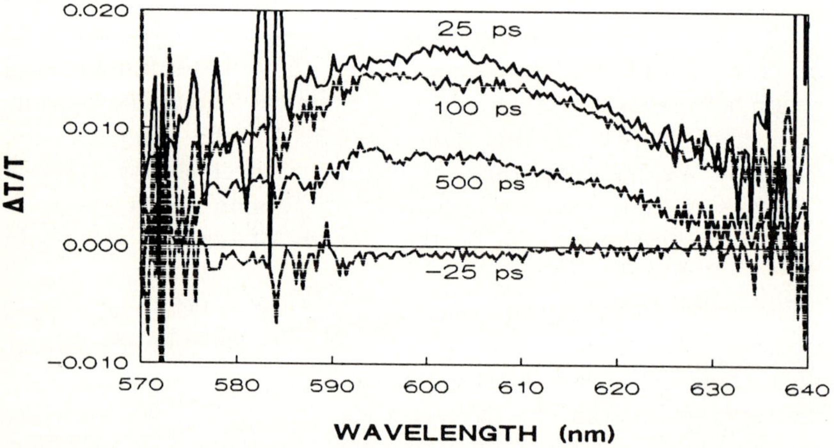

Figure 2: Picosecond transient absorption spectra of DODCI at -25, 25, 100, and 500 ps: 583 nm pump; 4 s accumulation.

observation of a broad (~45 cm^{-1}) band at ~1546 cm^{-1}. We attribute the breadth of this band to conformations in the excited state that involve rotations about the olefinic carbon-phenyl carbon single bond. This torsional motion is the same molecular coordinate that has been proposed to give the multi-dimensional character to the isomerization process.[3,4]

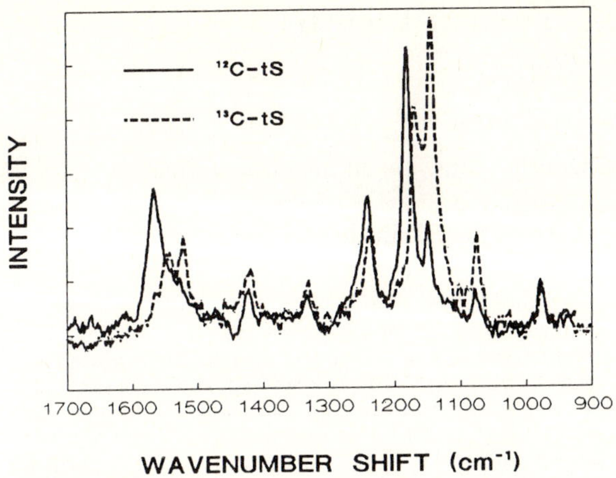

Figure 3: Picosecond transient resonance Raman spectra of *trans*-stilbene (solid line) and α,α'-^{13}C-*trans*-stilbene (dashed line) in hexane; Delay: 25 ps; Pump: 296.3 nm; Probe: 592.7 nm.

References

1. (a) J. Saltiel, J. D'Agostino, E. D. Megarity, L. Metts, K. R. Neuberger, M. Wrighton, and O. C. Zafiriou, *Org. Photochem.* **3**, 1 (1973); (b) J. Saltiel and J. L. Charlton, "Rearrangements in Ground and Excited States, Vol. 3," P. de Mayo, Ed. (Academic Press, New York, 1980) pp 25-89.
2. R. M. Hochstrasser, *Pure Appl. Chem.* **52**, 2683 (1980).
3. N. S. Park and D. H. Waldeck, *Chem. Phys. Lett.*, submitted for publication.
4. N. Agmon and R. Kosloff, *J. Phys. Chem.* **91**, 1988 (1987).
5. B. J. Matkowsky, A. Nitzan and Z. Schuss, *J. Chem. Phys.* **88**, 4765 (1988).
6. M. Lee, J. N. Haseltine, A. B. Smith III, and R. M. Hochstrasser, *J. Am. Chem. Soc.* **111**, 5044 (1989).
7. (a) T. L. Gustafson, D. M. Roberts, and D. A. Chernoff, *J. Chem. Phys.* **79**, 1559 (1983); *J. Chem. Phys.* **81**, 3438 (1984).
8. K. Iwata, W. L. Weaver, and T. L. Gustafson, unpublished results.

Solvation and Barrierless Electron Transfer: How Different Are the Dynamics?

B. Bagchi [1], *A. Chandra* [1], *and G.R. Fleming* [2]

[1] Solid State and Structural Chemistry Unit, Indian Institute of Science, Bangalore-560012, India

[2] Department of Chemistry and James Franck Institute, University of Chicago, Chicago, IL60637, USA

Recent experiments on photoinduced electron transfer reactions have revealed that in many conjugated aromatic molecules the electron transfer in the excited state can proceed without the intervention of a significant activation barrier to the reactive motion [1]. Experiments also indicate that these reactions are rather strongly coupled to solvent polarization relaxation, indicated by a strong dependence of the rate on the longitudinal relaxation time, τ_L of the dipolar solvent. Here we present a microscopic theory of dynamic solvent effects on a barrierless electron transfer reaction and compare the dynamics with that of solvation of the newly formed charge transfer state. We also investigate the effects of solvent translational modes on barrierless electron transfer reactions.

For an explicit microscopic calculation, we consider the following simple model of an intramolecular electron transfer reaction

$$A\text{-}B \xrightarrow{h\nu} (A^{\delta+}-B^{\delta-})^* \longrightarrow A^+ - B^- , \qquad (1)$$

where A and B are the two segments of the molecule and the intermediate state is the locally excited state. The reason for the absence of a sizeable activation barrier in the present problem is that there is no intermediate state of incomplete solvation of the electron in the reaction process, in contrast to a self-exchange reaction of the type Fe^{2+}/Fe^{3+} which faces a sizeable barrier.

We assume a one dimensional reaction free energy surface with the following definition of the reaction coordinate, X(t)

$$X(t)= -\int dr\, \Delta D(r)\cdot P(r, t) , \qquad (2)$$

where $\Delta D(r)$ is the change in the displacement vector as a result of the charge redistribution arising from the electron transfer and P(r,t) is the instantaneous solvent polarization at position r at time t. If the locally excited state is non-polar then X(t) is the time dependent solvation energy of the charge transfer state. We assume that the motion of the reaction coordinate can be described by a modified Smoluchowski equation for the time evolution of the probability density, P(X,t) of the reaction coordinate, with a position dependent sink term to account for the electron transfer reaction at certain specific values of the reaction coordinate. A similar equation of motion has been used earlier in the description of barrierless isomerization reactions [2]. The important parameters of this equation are the effective friction (ζ) on the reaction coordinate, the frequency (w) of the (assumed) harmonic surface, a molecular rate (k_o) of electron transfer

Springer Series in Chemical Physics, Vol. 53 **Ultrafast Phenomena VII**
Editors: C.B. Harris · E.P. Ippen · G.A. Mourou · A.H. Zewail

from the origin (where the rate is maximum) and a sink function (S(X)) which gives the position dependence of the rate. Sumi and Marcus [3] pointed out that $k_oS(X)$ is determined by the nature of the coupling between the two diabatic electronic surfaces. If the coupling is strong (the adiabatic limit), then $k_oS(X)$ may be approximated by an absorbing barrier at the position of crossing (X_c). For weak coupling (the non-adiabatic limit), $k_oS(X)$ is a finite sink at X_c. For the former case, the time dependent survival probability, $P(t) = \int dX\, P(X,t)$, can be obtained analytically if we assume that initial vibrational relaxation on the locally excited state is so rapid that P(X,t=0) can be approximated by the equilibrium distribution. The resulting simple expression is given by

$$P(t)= (2/\pi)\, \sin^{-1}(\exp(-w^2t/\zeta)) \quad . \qquad (3)$$

The electron transfer dynamics is non-exponential with a long time rate given by $K_L=w^2/\zeta$. Eq. 3 is an important result of this work. The parameters ζ and w are determined by the reaction time correlation function, Q(t), defined by [4]

$$Q(t) = \langle X(0)X(t)\rangle / \langle X^2(0)\rangle \quad . \qquad (4)$$

Note that Q(t) is identical with the solvation energy time correlation function (usually denoted by $C_S(t)$) of the charge transfer state. If we neglect the distortion of the solvent by the reactant, (that is assume a linear response of the solvent), then Q(t), and in turn ζ and w, are determined by the dynamics of the dipolar liquid. From an analysis of the relations between these quantities, the following important expression can be derived

$$w^2/\zeta = [\int_0^\infty dt\, Q(t)]^{-1} \quad . \qquad (5)$$

Since Q(t) is also the solvation time correlation function, Eqs. 3 and 5 show that <u>the long time rate of a barrierless adiabatic electron transfer reaction is exactly equal to the inverse of the average solvation time</u>. Eqs. 3 and 5 also provide an interesting relationship between the dynamics of electron transfer and solvation which can be tested against experiments.

The correlation function Q(t) can be evaluated by following the method developed in Ref. 5. For explicit (microscopic) numerical calculation, we approximate the field **D**(r) by that of a point dipole at the center of a sphere that encompasses the reaction system and the solvent by a system of dipolar hard spheres for which an analytical solution of the equilibrium two-particle direct correlation function is available. The importance of the translational contribution to the solvent polarization relaxation is gauged by the parameter p', given by $D_T/2D_R\sigma^2$, where D_T and D_R are the solvent translational and rotational diffusion coefficients and σ is the solvent molecular diameter.

The results of numerical calculations are shown in figures 1 and 2 for a liquid of dielectric constant 18 and reduced density 0.8. The reactant and the solvent molecules are of the same size. In Fig. 1, the dynamics of electron transfer is compared with the dynamics of solvation for two values of p'. Although the two processes are rather similar in the absence of translational contribution (p' = 0), they deviate from each

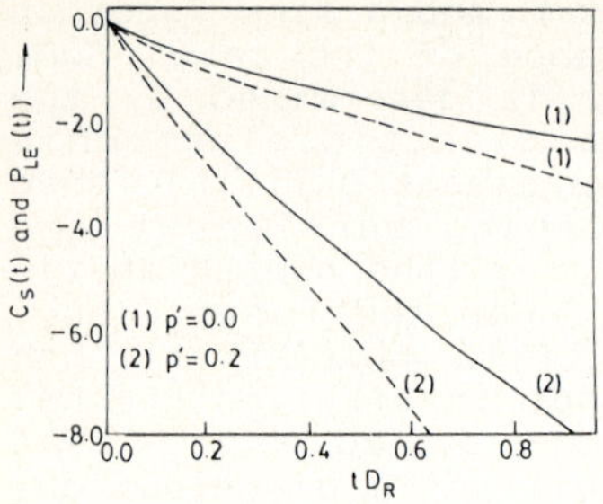

Fig.1 Comparison between the dynamics of electron transfer (P(t): dashed line) and solvation ($C_S(t)$: solid line).

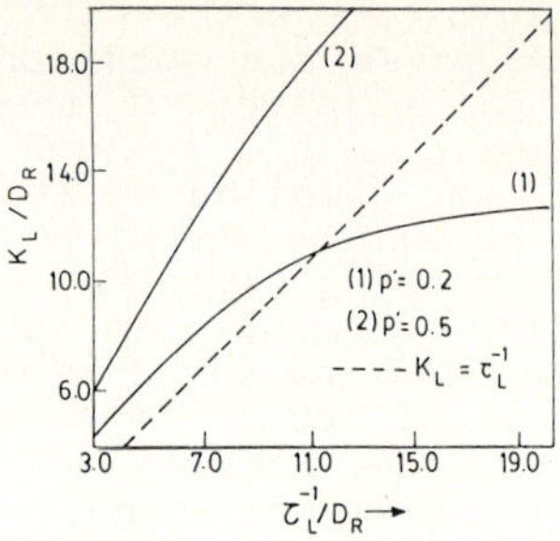

Fig.2 Dependence of electron transfer rate (k_L) on the longitudinal relaxation time, τ_L.

other as p' is increased. The electron transfer dynamics is faster than the solvation dynamics at all times. In Fig. 2, the long time rate k_L is plotted against the longitudinal time τ_L for two different values of p'. The prediction of Sumi-Marcus theory ($k_L = \tau_L^{-1}$) is also shown for comparison. It can be seen from this figure that the dependence of k_L on τ_L is non-trivial.

For a non-adiabatic reaction, the Smoluchowski equation has been solved numerically with a Gaussian sink function. The results have been discussed elsewhere [6].

In conclusion, we have presented a theory which treats both electron transfer and solvation dynamics within a consistent theory of polar solvent dynamics. The theory makes several predictions which can be tested against experiments.

REFERENCES

1. E.W. Kosower and D. Hupert, Annu. Rev. Phys. Chem. 37, 127 (1986); P.F. Barbara and W. Jarzeba, Acct. Chem. Res. 21, 195 (1988); Adv. Photochem (in press).
2. B. Bagchi, G.R. Fleming and D.W. Oxtoby, J. Chem. Phys. 78, 7375 (1983); B. Bagchi and G.R. Fleming, J. Phys. Chem. 94, 9 (1990).
3. H. Sumi and R. Marcus, J. Chem. Phys. 84, 4894 (1986).
4. J.T. Hynes, J. Phys. Chem. 90, 3701 (1986).
5. A. Chandra and B. Bagchi, J. Phys. Chem. 93, 6996 (1989); ibid 94, 1874 (1990).
6. B. Bagchi, A. Chandra and G.R. Fleming, J. Chem. Phys. (Communicated)

Low Temperature Ultrafast Charge Separation: Rate Versus Free Energy

M.P. O'Neil[1], *G.L. Gaines, III*[2], *W.A. Svec*[1], *M.P. Niemczyk*[1], *and M.R. Wasielewski*[1]

[1]Chemistry Division, Argonne National Laboratory, Argonne, IL60439, USA
[2]Biological and Medical Research Division, Argonne National Laboratory, Argonne, IL60439, USA

Abstract. The rate of electron transfer from the lowest excited singlet state of a porphyrin to an acceptor was measured in 2-MTHF glass at 77 K for a series of molecules. Analysis of the rate data as a function of free energy of reaction reveals that the ion-pair state of the oxidized donor and reduced acceptor is destabilized in solid solution by about 0.7 eV relative to its energy in polar liquids.

1. INTRODUCTION

A long-standing problem in the study of porphyrin and chlorophyll based donor-acceptor compounds designed to mimic photosynthesis is the failure of most of these molecules to undergo high quantum efficiency photoinitiated charge separation in the solid state at very low temperatures. The rate constants for oxidation of the lowest excited singlet states of porphyrin donors by acceptors decrease sharply at the freezing points of the media in which the compounds are dissolved [1]. This results in a small or negligible quantum yield of charge separation because the rate of electron transfer cannot compete with the decay of the porphyrin excited singlet state to ground state. In low temperature solid solutions dipole rotations of the solvent are eliminated as a source of stabilization of the incipient ion pair. Thus, the ion pair energy is increased, the free energy of the charge separation reaction is less negative, and the reaction rate is slower.

2. RESULTS

We have prepared the series of 32 donor-acceptor molecules depicted in Figure 1. Picosecond transient absorption and fluorescence measurements along with fluorescence quantum yield measurements were used to determine the rates of electron transfer for each pair using published techniques [1]. The one-electron oxidation and reduction potentials for the donor and acceptor, E_{ox} and E_{red}, respectively, and the lowest excited

Springer Series in Chemical Physics, Vol. 53 **Ultrafast Phenomena VII**
Editors: C.B. Harris · E.P. Ippen · G.A. Mourou · A.H. Zewail

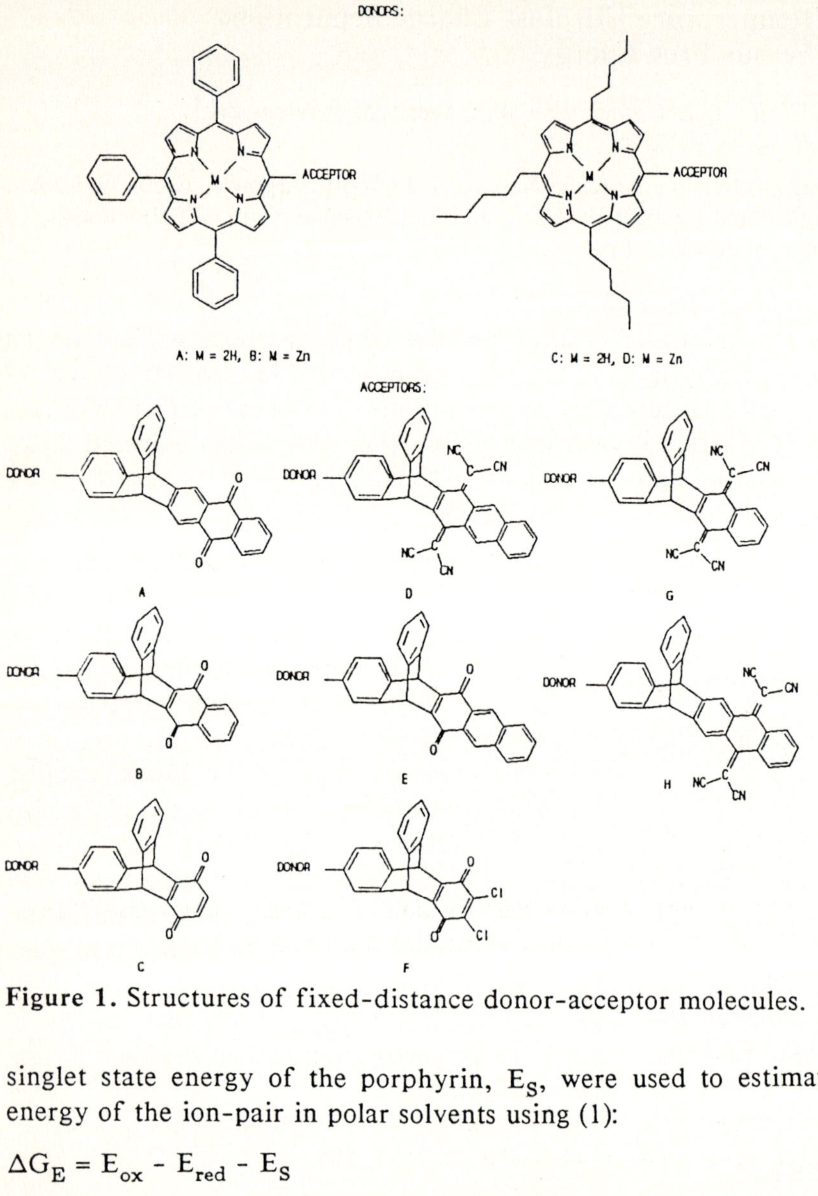

Figure 1. Structures of fixed-distance donor-acceptor molecules.

singlet state energy of the porphyrin, E_S, were used to estimate the energy of the ion-pair in polar solvents using (1):

$$\Delta G_E = E_{ox} - E_{red} - E_S \qquad (1)$$

The log of the electron transfer rate constant is plotted *vs* $-\Delta G_E$ in Figure 2. The plot shows that the rate of photoinduced charge separation does not compete effectively with excited singlet state decay until the free energy as measured in polar media is about -0.7 eV. The natural excited state decay times for the free base and Zn containing porphyrins are indicated by the abscissa and the horizontal bar in Figure 2 at 10^8 s^{-1} and 4 x 10^8 s^{-1}, respectively.

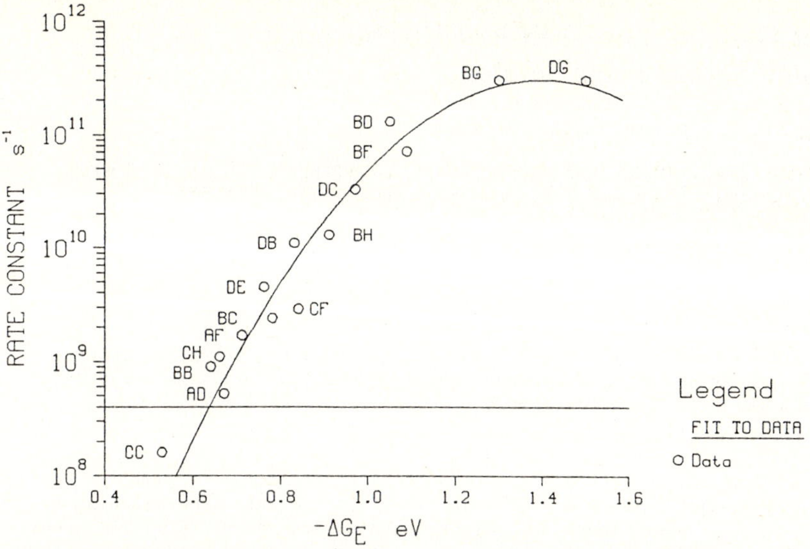

Figure 2. Rate constant *vs* free energy data. The first letter in the labeled points = donor, and the second letter = acceptor.

3. CONCLUSION

The data show that elimination of solvent dipole rotations in the solid state result in about a 0.7 eV destabilization of the $D^{+}A^{-}$ ion-pair. This energy must be taken into account when porphyrin-based donor-acceptor molecules are designed for applications requiring efficient charge separation in the solid state.

(This work was supported by the Division of Chemical Sciences, Office of Basic Energy Sciences, U. S. Department of Energy.)

4. REFERENCE

1. M. R. Wasielewski, D. G. Johnson, W. A. Svec, K. M. Kersey, and D. W. Minsek, *J. Am. Chem. Soc., 110*, 7219, (1988).

Solvent Dynamics by Transient Holeburning: Nonpolar Versus Polar Solutes

Tai Jong Kang, Jongwan Yu, and M. Berg

Department of Chemistry and Biochemistry, University of Texas at Austin, Austin, TX 78712, USA

Solvent interactions with the electronic states of a non-polar solute are measured with transient holeburning. Large interactions much more rapid than dielectric relaxation are observed.

1. Introduction

Shank and coworkers have used ultrafast techniques to extend holeburning to the liquid phase, where they have observed the effects of rapid intramolecular relaxation [1]. Recently, there have been predictions that holeburning can also be used to study intermolecular interactions of solvents with solute electronic states [2-4]. We present the first transient holeburning results attributable to intermolecular interactions. They show that large non-polar interactions can develop much more rapidly than the relatively well-studied dipolar interactions.

2. Experimental

We have studied dimethyl-s-tetrazine (DMST) dissolved in glycerol triacetate (GTA). DMST is a small, rigid molecule with a simple, well studied vibronic structure [5]. As a result, the 0-0 transition can be cleanly excited, precluding any intramolecular relaxation effects. The form of the time evolving holeburning spectrum can also be easily predicted as shown in Fig. 1. This model assumes that excited state and ground state broadenings and frequency shifts all proceed at the same rate. Both bleaching and stimulated emission contribute to the spectrum, overlapping at the 0-0 transition. Before solvent motion occurs (Fig. 1a), the spectrum is only broadened by the laser spectrum, giving narrow, well resolved lines. After complete solvent relaxation (Fig. 1d), the holeburning spectrum is given by the overlap of the steady-state absorption and stimulated emission spectra. Note that even the fully relaxed spectrum retains some structure due to this overlap.

GTA is a viscous (18 cP), polar, non-hydrogen bonding solvent, with a long polarization relaxation time (700 ps) previously measured by time-resolved Stokes shifts [6]. For a highly polar solute, electronic state perturbations are expected to be dominated by this interaction. The holeburning spectrum would then relax on the same polarization relaxation timescale measured by time-resolved Stokes shifts [3]. However, DMST has no dipole moment by symmetry so these experiments provide a previously unavailable measure of non-polar solvation.

The experiments have been performed with a synchronously-pumped dye-laser pumped by the compressed output of a CW mode-locked Nd:YAG laser and amplified in a Q-switched Nd:YAG pumped amplifier. The experiment is performed as a standard pump-probe experiment using a continuum probe generated in a water cell and polarized at the magic angle relative to the pump beam. Data is presented only after at least a 1 ps delay to avoid any stimulated Raman or coherence effects [1]. A small solvent contribution to the earliest data (1 ps) has been subtracted. The excited state absorption spectrum has been independently determined from measurements in hexane solution and its contribution removed from the data.

Springer Series in Chemical Physics, Vol. 53 **Ultrafast Phenomena VII**
Editors: C.B. Harris · E.P. Ippen · G.A. Mourou · A.H. Zewail

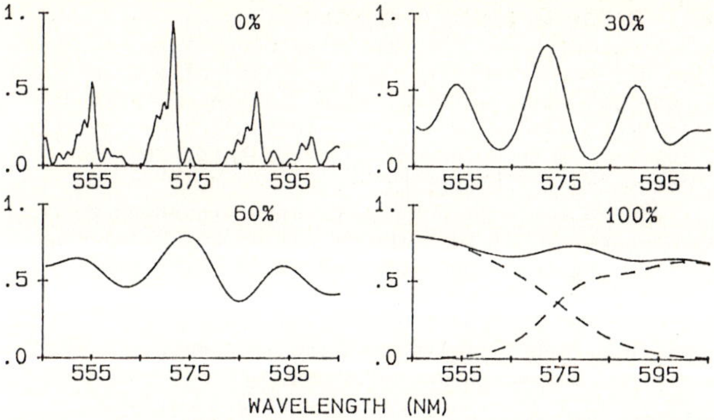

Fig. 1 Calculated holeburning spectrum of DMST vs. percentage of solvent relaxation. Bleaching and stimulated emission contributions to the 100% relaxed spectrum are shown (dashed).

3. Results

Figure 2 shows holeburning results on DMST in GTA over several time ranges. The 1.2 ns data shows the weak structure expected from a fully relaxed spectrum. However, comparison to the 100 ps data shows that there has been no significant relaxation of the electronic states on the time scale of the solvent polarization relaxation. This is a direct contrast to the expectations for a polar solute.

An alternate mechanism for solute electronic state relaxation is the reorientation of the relatively small solute molecule within a static solvent structure. The solute reorientation rate has been measured by the decay of the excited state absorption anisotropy. The solute reorientation is substantially faster than solvent polarization relaxation and is non-exponential. Fit to a biexponential, the decay gives time constants of 20 ps (60%) and 140 ps (40%). However, comparing the 100 ps spectrum to the 10 ps spectrum (Fig. 2) shows that there is no relaxation on the time scale of the solute reorientation.

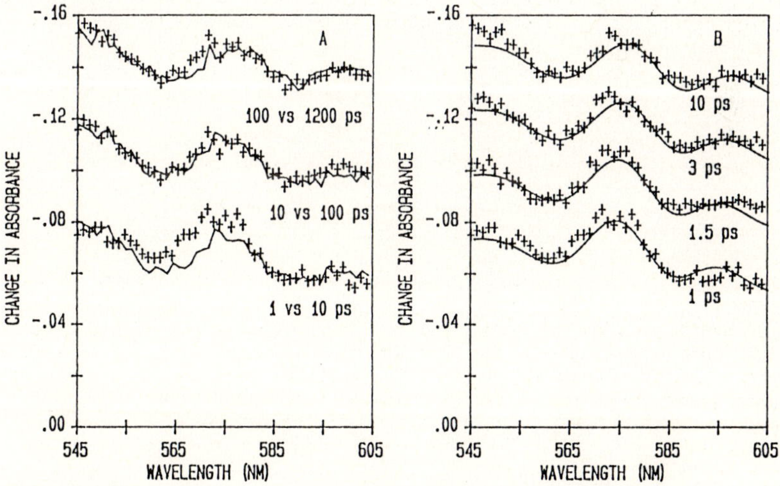

Fig. 2 Holeburning spectra of DMST in GTA. A) 1.2 ns, 10 ps - solid; 100 ps, 1ps - (+). B) Short time data (+) vs. model (line) of a 100 cm^{-1} frequency shift with a 2 ps time constant.

A comparison of the 1 ps and 10 ps spectra (Fig. 2a), shows a small evolution of the spectrum. Figure 2b examines this time range in more detail. The nature of the relaxation has been elucidated with the aid of a model based on broadening and shifting the gas-phase spectra. The results of the model are shown superimposed on the data in Fig. 2b. The model indicates that the excited state undergoes a 100 cm^{-1} shift (40% of the total shift) but no broadening. The shift occurs with a 2 ps time constant.

Most important, however, is the large amount of solvent relaxation which occurs within the first picosecond. This consists of all of the broadening (700 cm^{-1} FWHM) and much of the excited state frequency shift (150 cm^{-1}). This result is surprising considering the high viscosity of the solvent and the two orders of magnitude slower relaxation found for polar solutes.

In summary, holeburning effects due to solvent-solute interactions have been observed. The results show that for a non-polar solute, solvent interactions relax much more rapidly than the dipolar interactions which dominate the solvation of polar solutes. They also show two timescales contributing to the relaxation of the solute electronic state, suggesting that more than one type of interaction is important.

Acknowledgements

This work was supported by the National Science Foundation (Grant No. CHE-8806381) with additional support from the Robert A. Welch Foundation.

References

1) C.H. Brito Cruz, R.L. Fork, W.H. Knox and C.V. Shank, Chem. Phys. Lett. 132, 341 (1986); C.H. Brito Cruz, J.P. Gordon, P.C. Becker, R.L. Fork and C.V. Shank, IEEE J. Quant. Electron. QE-24, 261 (1988); W.T. Pollard, C.H. Brito Cruz, C.V. Shank and R.A. Mathies, J. Chem. Phys. 90, 199 (1989).
2) W. Vogel, D.-G. Welch and B. Wilhelmi, Chem. Phys. Lett. 153, 376 (1988).
3) R.F. Loring, Y.J. Yan and S. Mukamel, J. Chem. Phys. 87, 5840 (1987).
4) S. Kinoshita, J. Chem. Phys. 91, 5175 (1989).
5) G.G. Asmuth, K.K. Innes and V.A. Job, J. Mol. Spectrosc. 83, 267 (1980).
6) V. Nagarajan, A.M. Brearley, T.J. Kang and P.F. Barbara, J. Chem. Phys. 86, 3183 (1987).

Part XII

Chemical Dynamics

Femtosecond Dynamics from c.w. Spectroscopy

E.J. Heller

Departments of Chemistry and Physics, University of Washington,
Seattle, WA 98195, USA

Abstract. The conventional c.w. spectroscopies contain much explicit dynamical information. We discuss electronic absorption and Raman scattering. C.w. spectroscopy is considered as a limit of multiple femtosecond pulses. Once it is understood that a c.w. laser is performing a Fourier transform on femtosecond pulsed dynamics, much insight can be gained into intramolecular dynamics and the complementary pulsed experiments.

Many people have been drawn into molecular spectroscopy by the new femtosecond technology. My purpose here is not to recount the new successes [1-5], which will be done by others in this book. Neither is there space to present a good account of the theory of the femtosecond pump-probe pulsed experiments (See Refs.[6-13]).

I wish to present a view of the relationship of the pulsed experiments with the ancient technique of c.w. spectroscopy, and to call attention to the *dynamical* information contained in c.w. Raman and absorption spectra. It is important to know what can be achieved via c.w. techniques, even if one is a committed femtosecond practitioner.

The starting point is the simple observation that a c.w. laser beam can be thought of as an overlapping set of femtosecond pulses, as in

$$\begin{aligned} e^{i\omega t} &= 1 \cdot e^{i\omega t} \\ &= \eta \Big\{ \sum_n e^{-\beta(t-t_n)^2} \Big\} e^{i\omega t}. \end{aligned} \tag{1}$$

If the Gaussian widths and spacings are arranged appropriately then the Gaussians in the curly brackets are only modestly overlapping yet the sum is nearly a constant, and η, a normalization, keeps the whole sum very nearly equal to 1 at all times. The constant β is chosen to give the femtosecond pulses.

We consider electronic absorption (or emission) spectroscopy. The time dependent Schrödinger equation for the molecule in the presence of the field reads

$$i\hbar \frac{\partial \Psi(t)}{\partial t} = [H + V(t)]\Psi(t) \tag{2}$$

which can be written

$$i\hbar \frac{\partial \Psi(t)}{\partial t} = [H + \sum_n V_n(t)]\Psi(t) \tag{3}$$

where $\sum_n V_n(t) = V(t)$ is the sum of femtosecond pulses.

Springer Series in Chemical Physics, Vol. 53 **Ultrafast Phenomena VII**
Editors: C.B. Harris · E.P. Ippen · G.A. Mourou · A.H. Zewail

Within the usual perturbation treatment, the first order wavefunction becomes a sum of terms, one for each femtosecond pulse:

$$\Psi^{(1)}(t) \approx \sum_n \psi_n(t) \tag{4}$$

where each term $\psi_n$is the result of the femtosecond pulse arriving at time t_n. This in turn is simple to recast as

$$\Psi^{(1)}(t) \approx \sum_n e^{i\omega t_n - iE_0 t_n}\phi(t - t_n) \tag{5}$$

where $\phi(0)$ is the promoted part of the wavefunction that was electronically excited by the short pulse at $t = 0$.

Using the Born-Oppenheimer approximation and taking the case of a diatomic we have the situation shown schematically in Fig. 1. We have indicated the c.w. laser by a grey sinusoid; one of the pulses is shown darker. When the pulse interacts with the molecule, it promotes a part of the ground state wavefunction to the excited electronic state. This is shown in the upper left panel. That piece simply continues to evolve on the upper Born-Oppenheimer surface, returning to the Franck-Condon region after one vibrational period. It acquires a phase which here changes sign upon returning.

Figure 2 shows the two pulse case. If the second pulse is timed to arrive when the the wavepacket $\phi(t)$ returns, then either constructive or destructive interference results, depending on the phase of the freshly promoted wavepacket. If the fresh packet is in phase, then constructive interference re-

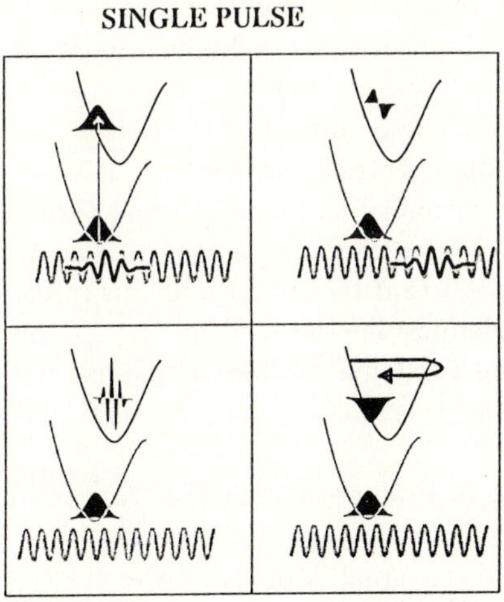

Figure 1. A single femtosecond pulse excites an electronic transition in a diatomic molecule.

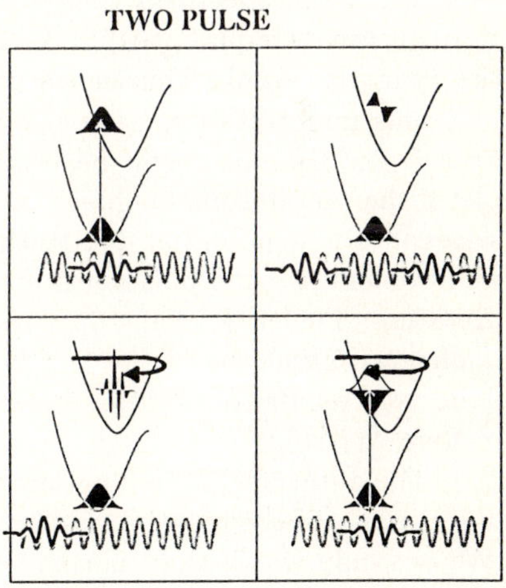

Figure 2. Two femtosecond pulses excite an electronic transition in a diatomic molecule; they can interfere destructively or constructively.

sults and the probability of being in the excited state is enhanced. If it is out of phase, then destructive inteference results and the probability of being in the excited state is diminished. The simplest way to control the phase is with the frequency of the light ω. One can also phase shift the pulses with respect to each other; this has been done recently by Fleming [14].

In the c.w. case, pieces of the ground state are being continuously promoted with a phase given via the laser frequency. The interference of the fresh and aged wavepackets is the issue of peaks or troughs in the spectrum. One can treat an arbitrary laser field very similarly. The most general case of an arbitrary pulse shape for one and two photon electron spectroscopy was first given in Ref. [6].

If we ask: in the c.w. case, how much probability is in the excited Born-Oppenheimer surface after a long time, we find that it is proportional to

$$S(E) \equiv \frac{1}{2\pi\hbar} \int_{-\infty}^{\infty} e^{iEt/\hbar} \langle \phi(0)|\phi(t)\rangle \, dt$$

$$= \sum_{n=0}^{\infty} p_n^{\phi} \delta(E_n - E) , \qquad (6)$$

where the p_n^{ϕ} are Franck-Condon factors. This is a somewhat different slant on things, casting the absorption spectrum as the Fourier transform of the autocorrelation function of the initially produced wavepacket and the time evolved wavepacket [6]. Note that we are now talking of a packet $\phi(t)$ produced by a imaginary femtosecond pulse contained within the c.w. laser field. The features of the spectrum thus accrue from the dynamics of $\phi(t)$ as viewed through the "window" $\phi(0) = \phi$. The spectral features are a direct handle on the dynamics via the Fourier transform.

This interpretation has been available for some time [15]. It is the way Franck and Condon might have cast electronic spectroscopy, because it completes the picturesque language of the sluggish nuclei suddenly finding themselves in a new potential due to the agile electrons changing state. Now that we have femtosecond pulses it is even more important to understand this interpretation. Frederick et al.[16], were able to deduce the detailed history of the nonlinear interaction of the two benzene ring torsional motions of benzophenone over ten or so vibrational periods. Other relevant papers may be found in Refs. [17-23].

The autocorrelation function $\langle \phi(0)|\phi(t)\rangle$ in Eq. (6) is analogous to watching a movie of the molecule following electronic promotion–with one big hitch. We can only watch that motion through a single small window...the state ϕ (Fig. 3,top).

Limitations of space prevent a fuller discussion of two photon Raman spectroscopy, which is vastly richer than absorption spectroscopy. Suffice it to say that it is like watching the movie through many small windows, one for each final state accessed in the Raman process (Fig. 3,bottom).

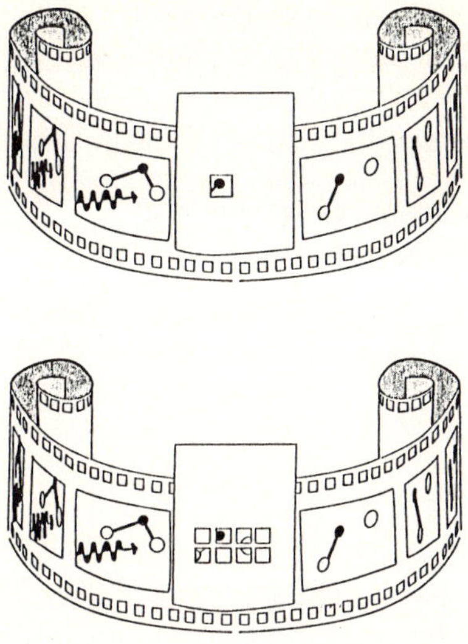

Figure 3. The analogy between absorption (top) and Raman (bottom) spectroscopy and a view of the dynamics from small windows.

References

1. L. R. Williams and K. A. Nelson, J. Chem. Phys., **87**, 7346(1987).
2. M. Dantus, R. M. Bowman, and A. H. Zewail, Nature, **343**, 737(1990).
3. W. Lin, R. W. Schoenlein, , J. G. Fujimoto, and E. P. Ippen, IEEE J. Quantum Electron., QE-24, 267(1988).
4. M. Maroncelli, and G. R. Fleming, J. Chem. Phys., **92**, 3251(1990).
5. F. H. Long, H. Lu, and K. B. Eisenthal, J. Chem. Phys., **91**, 4413 (1989).
6. (a) Soo-Y. Lee and Eric J. Heller, J. Chem. Phys., **71**, 4777 (1979).
7. D.J. Tannor and S. A. Rice, J. Chem. Phys. **82**, 1073 (1985); Adv. Chem. Phys. **70**, 441 (1988).
8. S. O. Williams and D. G. Imre, J. Phys. Chem. **92**, 6648(1988).
9. V. Engel and H. Metiu, J. Chem. Phys., **91**, 1596 (1989); **90**, 6116(1989).
10. S. O. Williams and D. G. Imre, J. Phys. Chem. **92**, 6648(1988).
11. H. Kono, and Y. Fujimura, J. Chem. Phys., **91**, 5960 (1989).
12. B. Fain, S. H. Lin, and N. Hamer, J. Chem. Phys., **91**, 4485 (1989).
13. W. T. Pollard, Soo Y. Lee, and R. A. Mathies, J. Chem. Phys., **92**, 4012 (1990).
14. G. Fleming, to be published.
15. K. Kulander and E.J. Heller,J. Chem. Phys. **69**, 2439 (1978); E.J. Heller, J. Chem. Phys., **68**, 3891(1978).
16.J.H. Frederick, E.J. Heller, J.L. Ozment and D.W. Pratt, J. Chem. Phys., **88**, 2169 (1988).
17. M. O. Trulson, G. D. Dollinger, and R. A. Mathies, J. Chem. Phys., **90**, 4274 (1989).
18. E.J. Heller, Accts. Chem. Res., **14**, 368 (1981).

19.S.Y. Lee and E.J. Heller, J. Chem. Phys., **76**, 3035 (1982).
20. E.J. Heller, R.L. Sundberg and D.J. Tannor, J. Phys. Chem., **86**, 1822 (1982).
21. A.B. Meyers, R.A. Mathies, D.J. Tannor and E.J. Heller, J. Chem. Phys., **77**, 3857 (1982).
22. A.B. Meyers, B. Li and X. Ci, J. Chem. Phys., **89**, 1876 (1988).
23. R.L. Sundberg, D. Imre, M. O. Hale, J.L. Kinsey, and R. D, Coalson, J. Phys. Chem. , **90**, 5001 (1986).

Ultrafast Molecular Dynamics of Liquid Carbon Disulfide Observed at High Pressure in a Diamond Anvil Cell

B. Kohler and K.A. Nelson

Department of Chemistry, Massachusetts Institute of Technology, Cambridge, MA 02139, USA

Femtosecond time-domain light scattering was performed on liquid carbon disulfide as a function of pressure. At high densities intermolecular vibration is observed.

1. Introduction

During the past few years we have used an entirely time domain form of light scattering known as impulsive stimulated scattering (ISS) which has distinct advantages for the study of fast intermolecular motions in simple liquids [1]. By measuring the evolution of the macroscopic polarizability with femtosecond time resolution we are able to investigate elementary dynamical processes such as molecular reorientation and translation. For time intervals shorter than a few picoseconds such motion is expected to be nondiffusional, reflecting correlated interactions among near neighbors. As the density of the liquid is increased the collective relaxation of local structure may become sufficiently hindered that molecular motions become oscillatory for short time periods. Such rapid intermolecular motions are believed to play an important role in condensed phase chemical reaction dynamics.

2. Experimental Procedures and Results

ISS scans were obtained from liquid carbon disulfide contained in a diamond anvil cell at 293K. In figure 1 the ISS signal is presented as a function of pressure. The ISS experiments were conducted in a transient four-wave mixing or transient grating geometry, as has been described extensively elsewhere [1]. Excitation and probe pulses were 70 fs in duration at a central wavelength of 620 nm. As our interest here is in the dynamical motions of the nuclei, the polarizations of all beams were chosen to minimize the contribution to the signal from the liquids instantaneous electronic optical nonlinearity [2]. The scans have been scaled so that the maximum of the nuclear response of each scan has equal magnitude.

All scans show a rise in the signal after zero delay time corresponding to inertial reorientational motion of the heavy CS_2 molecules. As the pressure is increased a weak oscillation appears at delay times near 500 fs. The data for CS_2 at high pressure are qualitatively similar to data recorded at low temperature in our previous study of the liquid at constant pressure [3]. We suggested then that the heavily damped, oscillatory character of the ISS signal is a manifestation of intermolecular motion which is vibrational in character at times < 1 ps. This was analyzed in

Springer Series in Chemical Physics, Vol. 53 **Ultrafast Phenomena VII**
Editors: C.B. Harris · E.P. Ippen · G.A. Mourou · A.H. Zewail

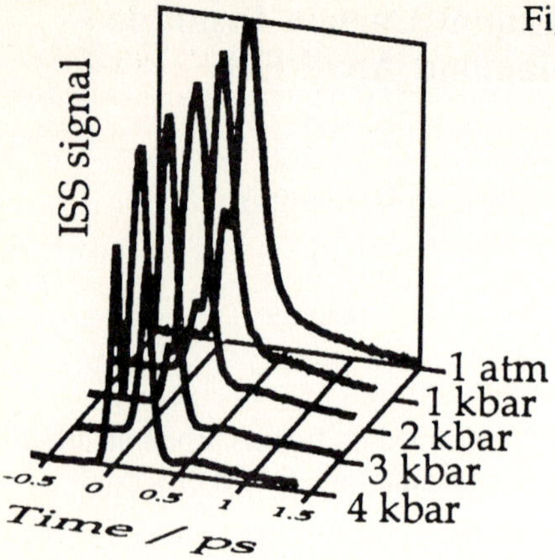

Fig.1. ISS signal from liquid CS_2 as a function of pressure.

terms of librational motion of CS_2 molecules within transient potential wells created by cages of near neighbors.

3. Molecular Dynamics Simulations

Increasing liquid density has two principal consequences for a light scattering experiment. First of all, the reorientation of single molecules becomes highly hindered, especially in liquids of anisotropic molecules which may exhibit strong local order. Secondly, molecules in a dense liquid interact strongly such that the collective polarizability, the observable in a light scattering experiment, can no longer be expressed as a sum of polarizabilities of individual molecules. Thus, the light scattering experiment is sensitive not only to dynamical motions of single molecules but also to motions of pairs, triplets and higher configurations of molecules [4]. The contribution of this interaction-induced scattering complicates the assignment of any light scattering signal to simple dynamical motions.

While we expect there to be an interaction-induced light scattering component to the ISS signal from carbon disulfide, it is nonetheless useful to investigate whether single molecule dynamics can reproduce the main features of the data. With this goal we have performed molecular dynamics simulations of a computer model of liquid carbon disulfide at a wide range of densities and temperatures. We used a Lennard-Jones interaction site potential proposed by Madden and Tildesley to represent the effective pair interactions [5].

From the MD trajectory data we have computed single molecule orientational correlation functions and used these to simulate our time-domain data. Figure 2 displays the simulated ISS signal due to single molecule reorientation at two densities. The simulated responses correctly reproduce the inertial peak in the signal and the subsequent rapid and then slower signal decays. However, they fail to show any oscillatory character. Even at the highest densities simulated single molecule reorientation in the computer liquid never becomes sufficiently hindered to show librational oscillations.

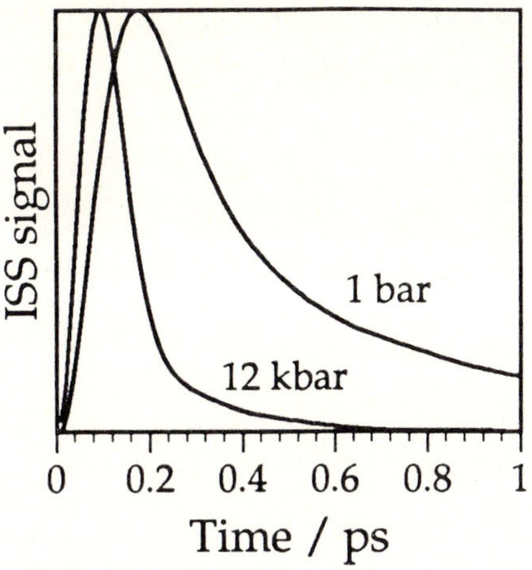

Fig.2. MD simulation of ISS signal evaluated from single molecule reorientation.

4. Conclusions

We have measured the ultrafast nonlinear optical response of liquid carbon disulfide as a function of pressure at constant temperature in a diamond anvil cell. We observe a subpicosecond signal which shows heavily damped oscillations as the density of the liquid is increased, corresponding to a low frequency mode of about 40 cm^{-1}. Molecular dynamics simulations performed over a wide range of densities suggest that single molecule reorientation is not sufficiently librational to account for our observed signal. Since subpicosecond librational motion in the liquid should be quite sensitive to quickly relaxing, local structure in the liquid it is possible that a better intermolecular pair potential for CS_2 would produce libration. Despite the challenges in separating the single molecule polarizability anisotropy contribution from interaction-induced contributions, we believe that the ISS signal provides evidence of subpicosecond intermolecular dynamics in dense liquids involving molecular translations or rotational motions, or possibly both, which are vibrational in character.

This work was supported in part by NSF grants DMR-8704352 and DMR-9002279 and by contributions from DuPont, Perkin-Elmer and the Alfred P. Sloan Foundation.

References

1. S. Ruhman, A.G. Joly, B. Kohler, L.R. Williams and K.A. Nelson, *Revue de Phys. Appl.* **22**, 1717 (1987).
2. J. Etchepare, G. Grillon, J.P. Chambaret, G. Hamoniaux and A. Orszag, *Optics Comm.* **63**, 329 (1987).
3. S. Ruhman, B. Kohler, A.G. Joly and K.A. Nelson, *Chem. Phys. Lett.* **141**, 16 (1987).
4. See e.g. P.A. Madden, Ultrafast Phenomena IV, ed. by D.H. Auston and K.B. Eisenthal (Springer, Berlin, 1984).
5. D.J. Tildesley and P.A. Madden, *Mol. Phys.* **42**, 1137 (1981).

Femtosecond Transition-State Absorption Spectroscopy of Bi Atoms Produced by Bi_2 Photodissociation

J.A. Misewich, J.H. Glownia, R.E. Walkup, and P.P. Sorokin

IBM T.J. Watson Research Center, P.O. Box 218,
Yorktown Heights, NY 10598, USA

Abstract: Two classes of Bi transient absorption spectra observed in the 308-nm photolysis of gaseous Bi_2 molecules are identified and discussed.

A recent femtosecond transient absorption study [1] focussed on the gas-phase photodissociation of Bi_2. With excitation at 308nm, there are at least two active dissociative channels, each producing both a ground state atom ($^4S°_{3/2}$) and an excited state atom ($^2D°_{3/2}$ or $^2D°_{5/2}$). Formation of $^2D°_{5/2}$ (slow) atoms was observed to be delayed in time by ~0.5 ps with respect to formation of $^2D°_{3/2}$ (fast) atoms, in rough accord with the ~0.5 eV energy separation between the channels.

For one of the lines monitoring the fast channel (Fig. 1a), the spectra strongly resemble transient spectra we have observed in other systems (e.g., TlI, Tl-cyclopentadienyl). We term these Class I spectra. The features common to Class I spectra are (1) the transient spectra are all centered about the free atom line positions, and (2) the spectral profiles at early times appear to be more dispersive than absorptive. In general, Class I transient spectra are satisfactorily explained (e.g., Fig. 1b) in terms of a model [1] based upon the transient behavior of the polarization induced by the probe continuum pulse as it interacts with a growing population of two-level atoms whose resonance frequencies shift in time. The salient features of Class I spectra can also be readily understood in terms of a simplified model known as the *impact approximation* [2]. In the impact approximation, it is assumed that during the half collision the induced polarization is simply shifted by a phase equal to the integral of the frequency shift from the point of excitation to infinity, and that the polarization only radiates after the half collision.

In marked contrast to the spectra shown in Fig. 1a are the transient spectra observed for transitions monitoring the slow channel, e.g. Fig. 2a. The "Class II" spectra shown in Fig. 2a are characterized by extremely broad absorption features at early times, with wings to one side of the atomic lines which evolve into the unperturbed asymptotic atomic absorption profiles without showing the dispersion-like spectral signatures of Class I spectra. Class II spectra can also be explained with the use of a model involving the transient behavior of the induced polarization. However, as a result of the smaller repulsive energy associated with the slow channel, a much broader wavepacket is produced on

Springer Series in Chemical Physics, Vol. 53 **Ultrafast Phenomena VII**
Editors: C.B. Harris · E.P. Ippen · G.A. Mourou · A.H. Zewail

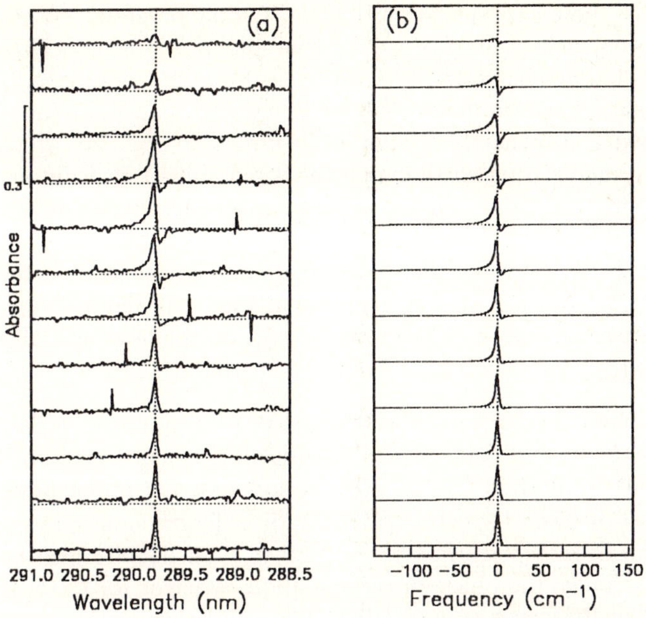

Fig. 1. (a) Transient Bi spectra monitoring the fast dissociation channel, 100-fs steps. (b) Modelling of experimental data.

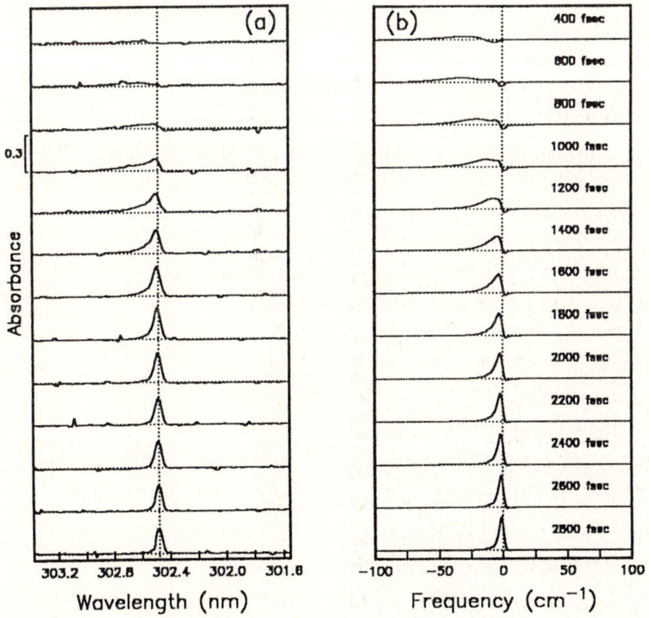

Fig. 2. (a) Transient Bi spectra monitoring the slow dissociation channel, 200-fs steps. (b) Modelling of experimental data.

its dissociative potential than is produced on the fast channel potential. When one averages the transient polarization over the broad distributions associated with the slow channel, the characteristic Class I features are dramatically changed. The negative absorption feature is greatly reduced, and the relative contribution of the wing is increased. This is illustrated in Fig. 2b which contains a series of simulated spectra obtained for a red shifted R^{-3} dipolar difference potential appropriate for a homonuclear system. The spectra of Fig. 2b represent averages over the expected thermal distribution of velocities. For numerical reasons, averages over the distribution of initial positions or the pump pulse duration were not made. We expect that the small residual negative absorption feature associated with the early spectra would disappear with more complete averaging.

In summary, the observed spectra can be categorized as "atomic-like" (Class I) or "molecular-like" (Class II). The above classification is associated with the origin of the radiation emitted by the polarization that the probe pulse induces in the system. For Class I spectra, the radiation is largely emitted over a long period when the Bi is unperturbed by its partner, with the effect of the half collision being to change the phase of the radiation, whereas for Class II spectra, the spectrum in the wings is associated with a polarization radiating while the Bi is still very much in the force field of its partner.

Support from the Army Research Office is acknowledged.

REFERENCES

1. J.H. Glownia, J.A. Misewich, and P.P. Sorokin, J. Chem. Phys. 92, 3335 (1990), and references therein.

2. R.E. Walkup, J.A. Misewich, J.H. Glownia, and P.P. Sorokin, to be published.

Femtosecond Infrared Spectroscopy of Complex Molecules Using CW IR Lasers

R.M. Hochstrasser, P.A. Anfinrud, R. Diller, C. Han, M. Iannone, T. Lian, and B. Locke

Department of Chemistry, University of Pennsylvania, 231 South 34th Street, Philadelphia, PA 19104, USA

Abstract: Recent applications to vibrational relaxation, protein (hemoglobin) dynamics and molecular motions of the CW-IR laser gated detection method are described.

1. Introduction

The transient vibrational spectra of polyatomic molecules can provide accurate structural information that refines our understanding of their dynamics. Recently, we have introduced a new method of obtaining vibrational spectra with a time resolution of 250 fs [1,2]. This infrared gating system was used in optical pump/IR probe experiments and recently was extended to an IR pump/probe configuration [3]. This latter experiment is very useful since it permits the study of vibrational dynamics at the low conversions and high repetition rates suitable for dynamical and orientational studies of complex, dilute systems. All the experiments reported in this paper rely on gating a CW IR laser with ps or fs-pulses.

2. Experimental

The infrared gated detector used in the present experiments uses a visible 200 fs pulse to upconvert slices of the CW laser (diode or CO laser) to be detected by a photomultiplier. The gating pulse is passed through a variable delay thereby enabling spectra to be obtained by tuning the CW laser at fixed delay. This configuration also removes the problem of variations in the up-converted signal at different delays since the upconverter is run at twice the frequency that the sample is pumped and the difference transmission is obtained with two lock-in amplifiers. The detectivity of this gated system can be better than 10^{-5} OD units (RMS) which enables the study of vibrational transitions having average strength. Signals from metal carbonyls are in the range of 0.03 OD for pumping energies of ca. 2 μJ of visible energy (0.2 μJ in the IR). The free carbon monoxide signal is ca. 50 times weaker, giving peak signals of 6×10^{-4} OD for the same pump energies.

A system has been developed to generate infrared pulses at high repetition rates (Figure 1a) based on a mode-locked, Q-switched Nd:YAG oscillator which synchronously pumps two dye lasers and seeds a regenerative amplifier. The single-pulse 600 μJ, frequency-doubled output of the YAG regenerative amplifier is used to amplify one of the dye lasers and to pump a $LiIO_3$ parametric amplifier, which results in the generation of tunable infrared pulses (25 ps and 300 nJ at 500 Hz) which are used to excite a vibrational mode of the sample. The sample is probed by the output of a CW laser (CO for the present experiment), which is upconverted by a pulse from the second dye laser in a second $LiIO_3$ crystal as discussed above. The probe may be set to the wavelength of the ground state absorption, the excited vibrational state absorption, a different vibrational mode of the molecule or scanned to obtain a spectrum.

Springer Series in Chemical Physics, Vol. 53 **Ultrafast Phenomena VII**
Editors: C.B. Harris · E.P. Ippen · G.A. Mourou · A.H. Zewail

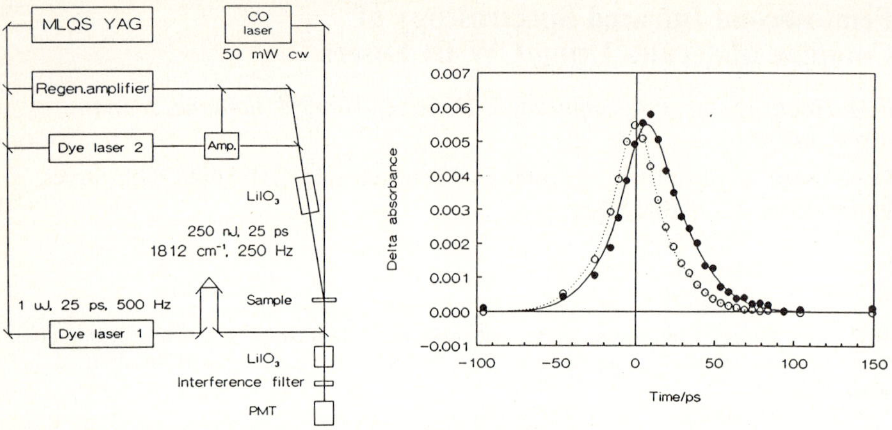

Figure 1: (a) Diagram of high rep rate IR pump / IR probe apparatus (b) Transient IR absorption of acetylbromide in $CC\ell_4$. Open circles: normalized instrument response; solid line: convolution of response with 10 ps decay of V = 1 in C = 0 stretch.

Tunable IR pump/Tunable IR probe experiments:

In preliminary work with the system shown in Figure 1(a), we have determined population lifetimes of organic carbonyl vibrations (ca. 1800 cm^{-1}) and metal carbonyls. For example, Figure 1(b) shows the result of an experiment in which the CO vibration of acetyl bromide (CH_3COBr) in CCl_4 was excited by a 250 nJ pulse at 1812 cm^{-1} and was probed by the CO laser at 1787 cm^{-1}. A transient absorption was observed with a lifetime of approximately 10 ps. This absorption is assigned to the v=2 ← v=1 transition consistent with the location of a transition in the overtone region of the infrared spectrum. The data shown were obtained in 100 s/point; the detectivity of the system is thus better than $5x10^{-4}$ optical density change for 1 s of data acquisition. Studies aimed at determining relaxation pathways in this and similar molecules are underway. Improvements to the apparatus are planned which will substantially improve the time resolution. Projected applications of this versatile system include measurement of T_1 times, studies of relaxation pathways in organic molecules, and determination of rotational reorientation times by measurement of the anisotropy decay of transient infrared signals.

Photodissociation of Hemoglobin-CO and Protein Dynamics

The photodissociation of hemoglobin-CO (HbCO) provides a way of studying protein dynamics since the equilibrium structures of Hb and HbCO are different. With transient IR spectroscopy the bound, ..Fe-C≡O localized, and "free" carbon monoxide spectra can be studied to provide direct information on the excited states of Hb and HbCO, geminate rebinding and the rate at which CO leaves the protein [2].

Ultrafast IR Studies of the photogenerated CO from hemoglobin-CO

The ^{13}CO infrared absorption at ca. 2090 cm^{-1} is ca. 50 times weaker than the bound CO absorption and was only seen by using very long integration times (ca. 100 hours). The spectra are narrower than found for CO in nonpolar solvents (4) and there are no significant spectral changes seen after 10 ps. The spectra at very early times are superimposed on a broad signal extending over at least 100 cm^{-1}. Figure 2(a) shows the time dependence of this signal at a frequency where no free CO absorption is occurring; 2(b) shows the

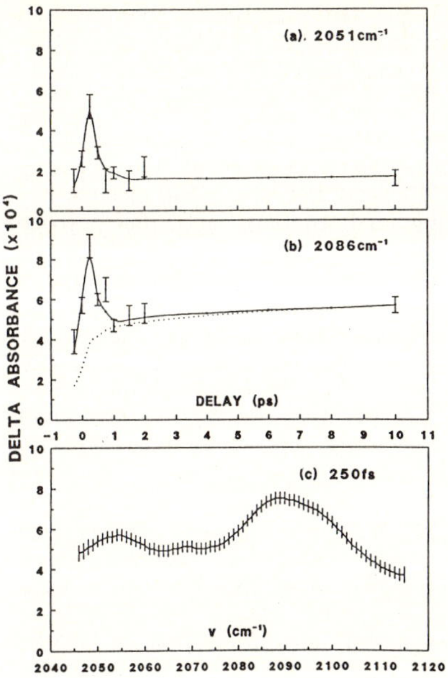

Figure 2: Photodissociating $Hb^{13}CO$: (a) Transient absorption at 2051 cm^{-1}. (b) Transient absorption at 2086 cm^{-1} (peak of free CO absorption). (c) IR spectrum at 250 fs delay (peak of (b)).

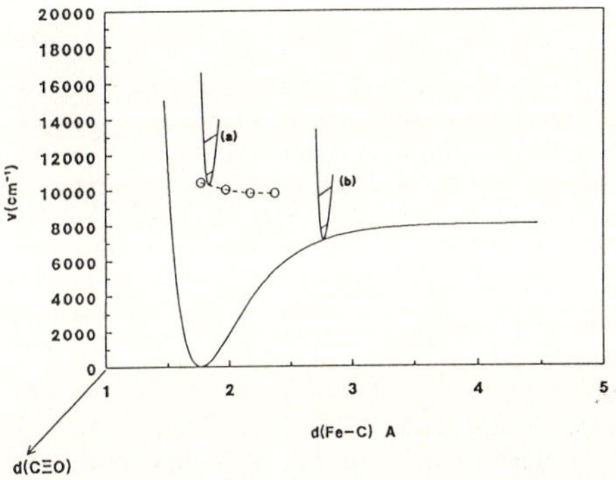

Figure 3: Morse potential for FeCO dissociation (ω_e=507cm^{-1}, D_e=8000cm^{-1}). Circles: Dissociative potential energy [scaled from Waleh and Loew, J.A.C.S. 104, 2346 (1982)]. (a) Bound potential at transition state. (b) Nearly "free" CO.

kinetics at 2090 cm^{-1} where the free CO absorbs; and 2(c) shows the spectrum at 250 fs, exactly at the peak of the kinetics in Figure 2(b). The data in Figure 2(c) permit us to measure the amount of CO present at 250 fs as ca. 60% that found at later times, known from the asymptote in Figure 1(b). The rise time of the CO signal, shown as a dashed line through the datum at 250 fs in Figure 2(b), could therefore be determined. The best fit rise times are less than 100 fs. We are therefore certain that the CO is released faster than can be measured with our pulses: The best fits are for release times considerably less than 100 fs.

Such rapid expulsion of CO is consistent with an Fe-CO vibration frequency of 507 cm^{-1}. According to the Morse curve for ground state HbCO along this coordinate in Figure 3, the CO leaves the field of the Fe when the bond length reaches ca. 3.29A° or after the CO has moved only 1A°. Our picture of the dynamics is that the dissociation originates from hot ligand field states reached by rapid internal conversion and IVR. The available CO kinetic energy is less than 0.5eV but more likely 0.16eV. The exit velocity of 1-2A° per 100 fs is sufficient to release essentially free CO on the observed time scale. The lower value is preferred because IVR energy sharing is likely to be very fast. The relocation of the amino acids having van der Waals contacts with the bound CO will make the rebinding of dissociated CO more difficult, perhaps accounting for the absence of a fast geminate recombination.

Nature of the Ultrafast IR transient:

The ultrafast transient shown in Figure 2(a) is found throughout the spectral regions explored - 1920 to 2130 cm^{-1} for Hb and $Hb^{12}CO$. Two possibilities are considered to explain this result: heating and cooling; and the presence of an excited state transient.

A sudden change in the IR spectrum in the overtone region is expected to occur when the molecules absorb the energy of the photons but it is surprising that the cooling should occur so rapidly (sub-picosecond) based on the 10's of ps cooling times observed for other aromatic molecules [5-7]. A simulation [8] predicted an initial rapid 50% drop in temperature (1-4 ps) followed by a slower (20-40ps) relaxation and Raman spectra [9] showed evidence of cooling over the 1 to 10 ps regime. Possibly our observation refers to the fast first stage of cooling only but the whole range expected is not evident in the signals.

A short-lived electronic state of Hb [9,10] having a changed IR spectrum could generate a signal as in Figure 2(a). For HbCO, the sub-100fs bound state would have a strong IR absorption distributed over >50 cm^{-1} that would be absent for Hb excitation. This 'transition state' signal from structure 3(a)would rise and fall within our time resolution just as found in Figures 2(a) and 2(b).

Infrared Anisotropy:

The anisotropy r(t) measures the correlation function $\langle P_2[\hat{\mu}_{PUMP}(o)\cdot\hat{\mu}_{PROBE}(t)]\rangle$ where $\hat{\mu}_{PUMP}$ and $\hat{\mu}_{PROBE}(t)$ are body fixed unit vectors corresponding to the directions of the transition moments in the pump and probe process. Optical pump / IR probe or IR pump / IR probe are particularly useful because $\hat{\mu}_{IR}$ can be chosen along well defined molecular axes (local modes) even in the case of unsymmetric molecules. Such measurements can yield the assignment of electronic wavefunctions, the overall rotational dynamics of the molecule or the alignment of reactants and products.

As an example of optical/IR anisotropy the photolysis of a di-iron tetracarbonyl (see Figure 4) is considered [11]. The IR transition of the bridge CO is polarized along the 2-fold axis. Observation of an anisotropy greater than zero in this case indicates the IR and optical transition dipoles are parallel. As Figure 4 shows, r(t) at 1789 cm^{-1} has already dropped well below the theoretical maximum value of 0.4 by zero delay. This implies an ultrafast loss of orientational order -- a time constant of ca. 400 fs is required. The signal at 1789 cm^{-1} is a bleach of the bridging CO absorption which is recovering by geminate rebinding. To account for the anisotropy decay a reorientation of the bridge CO direction, probably as a result of bridge-terminal carbonyl interchange and isomerism in the excited state,

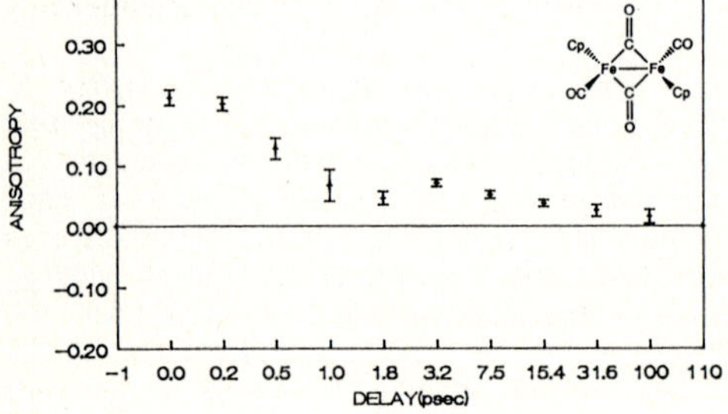

Figure 4: Anisotropy of the IR absorption of Fe_2CO_4 Cp_2 after optical excitation.

should occur prior to geminate recombination. This motion is so rapid as to correspond to a forced motion -- that is, after bond breakage the potential along the reactive coordinate drops steeply. The recurrence in r(t) at ca. 1.5 ps is perhaps a result of the caging of the fragments generated by the optical pulse.

Acknowledgements: This research was supported by grants from NSF and NIH.

References:

(1) P.A. Anfinrud, C. Han, J.N. Moore, P.A. Hansen and R. M. Hochstrasser in Ultrafast Phenomena VI [Editors: T. Yajima, K. Yoshihara, C.B. Harris and S. Shionoya] Springer, 1988; p 442.

(2) P.A. Anfinrud, C. Han and R.M. Hochstrasser, Proc. Natl. Acad. Sci. USA 86, 8387 (1989).

(3) R. Diller, M. Iannone and R.M. Hochstrasser, to be published.

(4) D. Richon and D. Patterson, Chem. Phys. 24, 235 (1977).

(5) F.E. Doany, B.I. Greene and R.M. Hochstrasser, Chem. Phys. Lett. 75, 206 (1980).

(6) A. Sellmeier, P.O. Scherer and W. Kaiser, Chem. Phys. Lett. 105, 140 (1984).

(7) N.H. Gottfried, A. Sellmeier and W. Kaiser, Chem. Phys. Lett. 111, 326 (1984).

(8) E.R. Henry, W.A. Eaton and R.M. Hochstrasser, Proc. Natl. Acad. Sci. USA 83, 8982 (1986).

(9) J.W. Petrich, J.L. Martin, D. Houde, C. Poyart and A. Orszag, Biochem. 26, 7914 (1987).

(10) P.A. Cornelius, A.W. Steele, D.A. Chernoff and R.M. Hochstrasser, Proc. Natl. Acad. Sci. USA 78, 7526 (1981).

(11) P.A. Anfinrud, C. Han, T. Lian and R.M. Hochstrasser, "Subpicosecond Infrared Spectra: Condensed Phase Photochemistry of Iron Carbonyls", see paper in this Volume.

The Observation of Vibrational Relaxation and Solvent Reorientation Following Ultrafast Photodissociation of $Cr(CO)_6$

Soo-Chang Yu, Xiaobing Xu, R. Lingle, Jr., and J.B. Hopkins

Department of Chemistry, Louisiana State University,
Baton Rouge, LA70803, USA

Abstract. The photodissociation of $Cr(CO)_6$ at 266nm has been studied in cyclohexane and n-propanol by using picosecond Raman spectroscopy. Vibrational relaxation on a time scale of 100ps has been observed in cyclohexane in both the Stokes and anti-Stokes spectra. Raman band shifts have been observed in n-propanol within 100ps which are assigned to solvent coordination and reorientation from the alkane coordinated intermediate to the more stable oxygen-metal coordinated complex.

1. Introduction

There has been great interest in the ultrafast phenomena following photodissociation of $Cr(CO)_6$ as a result of the recent controversy concerning the role of solvent coordination to the vacant bonding site. Simon et al.[1,2] first reported that photodissociation of $Cr(CO)_6$ results in the formation of solvated $Cr(CO)_5$ which appears in a time of < 0.8ps in hydrocarbon solvents. Their picosecond transient absorption results have been supported by similar femtosecond experiments by Nelson et al.[3] However,the appearance time of solvated $Cr(CO)_5$ has been questioned by subsequent investigations using both IR[4-5] and uv-visible[6-7] transient absorption spectroscopy. Spears et al.[4-5] have claimed a much longer solvation time of 100ps. Lee and Harris[6-7] suggested that vibrational relaxation controls solvation dynamics on a 17ps time scale. The subsequent migration dynamics of the coordinated solvent species from H-Cr bonding to the thermodynamically more stable O-Cr bonding in THF and alcohols has also been a matter of controversy.

2. Experimental

The experimental apparatus has been previously described in detail.[8] Two 5ps pulses at 266nm were used as pump and probe pulses in this experiment. Both pulses were produced in the same laser by injecting two seed pulses into the regenerative amplifier. The delay between the pulses depends on the cavity length mismatch between the injection (seed) laser and the regenerative amplifier. This delay was measured using standard autocorrelation techniques. Two spectra were obtained by alternately exciting the sample with a single pulse or a double pulse sequence. A pure transient difference spectrum was computed by subtracting these two spectra. Laser intensity was 20μJ/pulse at 2kHz in a 0.2mm beam waist. The concentration was 10mM in both cyclohexane and n-propanol solvents.

Springer Series in Chemical Physics, Vol. 53 **Ultrafast Phenomena VII**
Editors: C.B. Harris · E.P. Ippen · G.A. Mourou · A.H. Zewail

3. Vibrational Energy Relaxation of Solvated $Cr(CO)_6$ in Cyclohexane

We have observed the dynamics of vibrational cooling which occur over 100ps in both the Stokes and anti-Stokes regions of the spectrum.[8] When vibrational relaxation occurs, the intensity of the Stokes signal is expected to grow while that of the corresponding anti-Stokes signal diminishes in time. Thus the complementary observation of a growing Stokes band and a decaying anti-Stokes band gives unambiguous evidence of cooling in that particular vibrational mode.

Comparing frames A-D in Figure 1, it is readily apparent that a single transient appears in roughly 100ps. The metal- CO stretch at 381 cm^{-1} and the CO vibration at 1935 cm^{-1} indicate that the transient is a metal carbonyl complex which is assigned to $Cr(CO)_5$. The negative going peaks are ground state $Cr(CO)_6$ bands which appear in the transient spectrum as a result of population bleaching.

Figure 2 illustrates the transient anti-Stokes spectrum in the region of the 381 cm^{-1} band assigned to $Cr(CO)_5$. The observation an initially strong signal which decays in time indicates that the anti-Stokes spectrum originates from the hot vibrational states which decay in 100ps. The fact that the anti-Stokes and Stokes spectrum have complementary dynamics is consistent with vibrational relaxation which occurs on the time scale of 100ps.

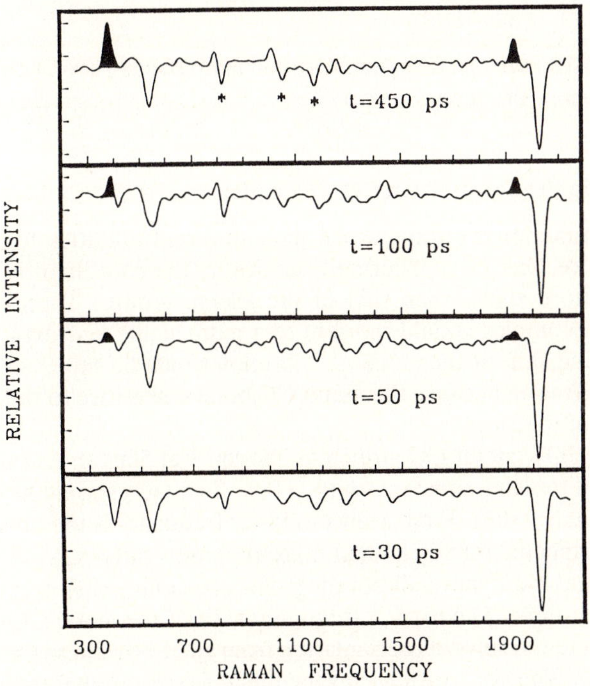

Figure 1:Transient Stokes Raman spectrum in cyclohexane. Colored-in bands are assigned to $Cr(CO)_5$. Asterisks denote noise from solvent Raman bands which have been subtracted out of each spectrum. Frequency in cm^{-1}.

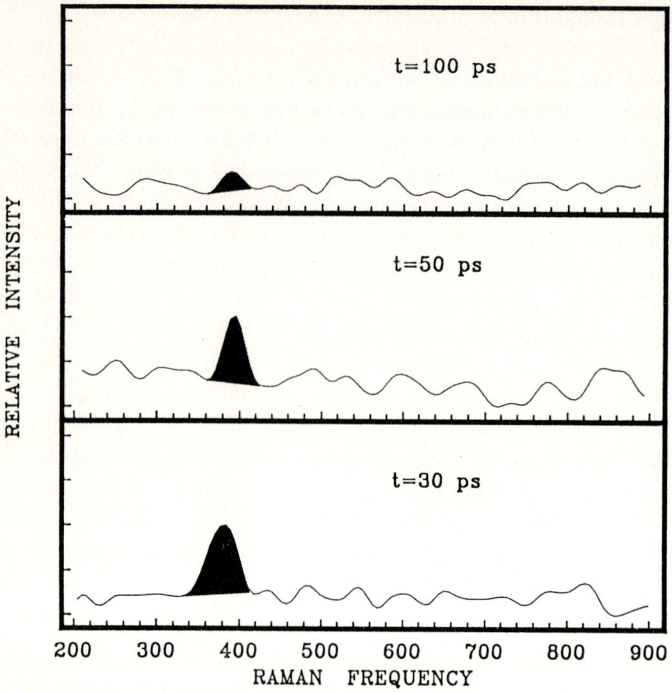

Figure 2:Transient anti-Stokes Raman spectrum. Ground state bands have been subtracted out of the spectrum. Frequency in cm^{-1}.

4. Solvent Reorientation in the Solvated $Cr(CO)_5$ Complex

N-propanol has two functional groups which have different affinities for bonding to the vacant site of $Cr(CO)_5$. Thermodynamically, the coordination of the hydroxyl group is more stable than that of the alkane group. Therefore an alkane-coordinated complex should reorient to a more stable hydroxyl-coordinated complex. A band shift of the metal-Co vibration would be expected as a result of this solvent migration because the metal-CO bond is sensitive to the coordination environment of the metal.

In n-propanol the $Cr(CO)_5$ metal-CO stretch at 384 cm^{-1} at 50ps shifts to 378 cm^{-1} in 100ps and then remains constant up to 500ps. Two facts should be noted in assigning this frequency shift. First, a shift to lower frequency is the opposite behavior of what is expected for vibrational relaxation. Secondly, no shift is observed at all for cyclohexane (where all bonding sites are equivalent) over the entire 500ps observation time. The 6 cm^{-1} frequency shift of the metal-CO vibration is therefore attributed to solvent reorientation from an H-bonded to O-bonded coordinated solvent. The observation that the metal-Co vibrational frequency is sensitive to solvent coordination together with the fact that no frequency shift is observed in cyclohexane indicates that the coordination time in cyclohexane is exceedingly fast ($\leq$10ps).

References

[1.] J.D.Simon and X.Xie,J.Phys.Chem.,90,6751(1986).
[2.] J.D.Simon and X.Xie,J.Phys.Chem.,93,291(1989).
[3.] A.G.Joly and K.A.Nelson,J.Phys.Chem.,93,2876(1989).
[4.] L.Wang,X.Zhu and K.G.Spears,J.Am.Chem.Soc.,110,8695(1988).
[5.] L.Wang,X.Zhu and K.G.Spears,J.Phys.Chem.,93,2(1989).
[6.] M.Lee and C.B.Harris,J.Am.Chem.Soc.,111,8963(1989).
[7.] M.Lee and C.B.Harris,J.Am.Chem.Soc.,Submitted for Publication.
[8.] S.C.Yu,X.Xu,R.L.Lingle,and J.B.Hopkins.J.Am.Chem.Soc.,112,3668(1990).

Ultrafast Rotational Dynamics of Electronically Excited Aniline Molecules in Solution

M.A. Pereira, P.E. Share, M.J. Sarisky, and R.M. Hochstrasser*

Department of Chemistry, University of Pennsylvania, 231 South 34th Street, Philadelphia, PA 19104, USA

*Present address: Department of Chemistry, University of Rochester, Rochester, NY 14627, USA

1. Introduction

The study of rotational dynamics of molecules in solution is a source of valuable information on the forces that mediate intermolecular interactions. Although hydrodynamic approachs are moderately successful in describing the motion of large molecules in many liquids, the regime where the solvent induced reorientation and the free, inertial motion of the solute occur on the same time scale is not yet well understood. Examples of this category is the motion of aniline in isopentane [1] and liquid CS_2 [2]. In the present work the fluorescence anisotropy of aniline is studied with ca. 250 fs time resolution, greatly extending previous investigation [1.a] of this system.

2. Experimental Methods

In order to obtain the necessary time resolution and to generate the needed excitation wavelength [302 nm], a dye laser was built following the design of Dawson et al [3] which comprises a double-jet dye laser with intracavity dispersion correction realized by two prisms in double pass configuration. These prisms disperse the light transversely allowing tunability just by the insertion of a slit in the optical path, controlling both the frequency and the bandwidth of operation. The pump laser is a continuously pumped, acousto-optically mode-locked Nd:YAG laser (Antares), frequency doubled using KTP. The dye laser routinely produced 40 mW of average power at 604 nm with a pulse autocorrelation width of 150 fs. The pulses were then pre-chirped before amplification in such a way that the process of amplification also reshaped them.

The dye laser pulses were amplified at 20 Hz by a Q-switched Nd:YAG-pumped three-stage amplifier (Spectra-Physics PDA-1) with sulforhodamine 640 in methanol/water as the amplifier dye up to levels of ~ 400 μJ/pulse and an approximately Gaussian autocorrelation width of 200 fs. Pulse broadening in the amplifier was reduced by carefully adjusting the concentration in the three stages of the dye amplifier. The amplified beam was sent through a beam splitter and half of it was focused with a long focal length lens and passed through a 0.5 mm KDP crystal to generate the second harmonic needed for excitation. The cross-correlation of the amplified dye fundamental and the second harmonic, determined by sum frequency generation in BBO, was ~250 fs in width and Gaussian in shape.

The 302 nm light (10-20 μJ/pulse) was focused into the sample with a 10 cm f/l lens after passing through a polarizer and a quarter-wave plate to allow excitation with either vertically or horizontally polarized light.

Springer Series in Chemical Physics, Vol. 53 **Ultrafast Phenomena VII**
Editors: C.B. Harris · E.P. Ippen · G.A. Mourou · A.H. Zewail

Fluorescence was collected collinearly with the pump by a reflective microscope objective (Beck reflecting objective x 74 magnification, focal length 2.6 mm, aperture 0.65), collimated for filtering purposes, then focused again by a Cassegrian telescope (reflective telescope) into a BBO crystal 200 μm thick. In the region where the fluorescence was collimated, a solution filter of naphthalene/hexadecane was placed to reject the residual pump beam which could leak through the sample cell. The purpose of collinear fluorescence collection was to minimize the transit time spread in the fluorescence originating from different parts of the illuminated volume.

The remaining fundamental light at 604 nm traversed a variable delay line and was weakly focused and overlapped with the fluorescence in the mixing BBO crystal after reflection from a mirror positioned very close to the fluorescence path.

The signal generated at the sum frequency (~209 nm) was passed through a small monochromator (PTR tunable filter TGF-200-02) and detected with a solar blind photomultiplier (Hamamatsu R-1356) and a gated integrator. For each pulse the pump beam was sampled and the ratio between up-converted fluorescence and pump beam was recorded. The signal was rejected if the pump beam was not within some predefined fraction of the standard deviation from the average pump beam signal.

3. Results and Discussion

An exponential fitting yielded a rotational decay time of ca. 0. 95ps, only 55% longer than the free rotor rotational time (0. 61 ps). The free rotor anisotropy decay curve is also shown in Figure 1.

A careful measurement of the anisotropy at time zero yielded r(0) ~ 0.26, which corresponds to 13% long in-plane axis component in both absorption and emission. Earlier measurements [1.a] resulted in similar anisotropies and which are in reasonable agreement with the fluorescence polarization work of Albrecht [4].

Extended diffusion theory [5] attempts to bridge the gap between free rotation and diffusion . Collisions are assumed to randomize the angular momentum and the fit of theory to experiments requires specifying a collision frequency. Collision frequencies can be obtained for simple liquids in a standard manner [6] and compared with the predictions from the theory. The fitting in this case for aniline in isopentane yielded a collisional period of 205 fs implying that the aniline molecule performs, on average unhindered rotations of 20^0. The calculated Enskog collisional frequencies [1.b] increases for solvents through the series isopentane, isooctane, hexadecane. The extended diffusion collision rate decreased through the same series indicating that this theory does not adequately predict trends in the orientation dynamics of aniline. Fig. 1 shows the free rotor, extended diffusion and experimental results.

Hydrodynamics modeling failed to predict the observed anisotropy. Fig. 2 presents the fitting using slip and stick boundary conditions. Fig. 2 also shows the fitting for an exponential decay having a time constant equal the sum of the free rotation correlation time and the calculated rotational diffusion time using slip boundary conditions. This is an often used artifice to match the zero viscosity rotational time, although it lacks a legitimate theoretical basis.

Molecular dynamics simulations (1.55 ns total trajectory, 125 molecules per cell, Lennard-Jones potentials) were performed to evaluate the suitability of this approach to predict the orientational dynamics of aniline in isopentane. The resulting

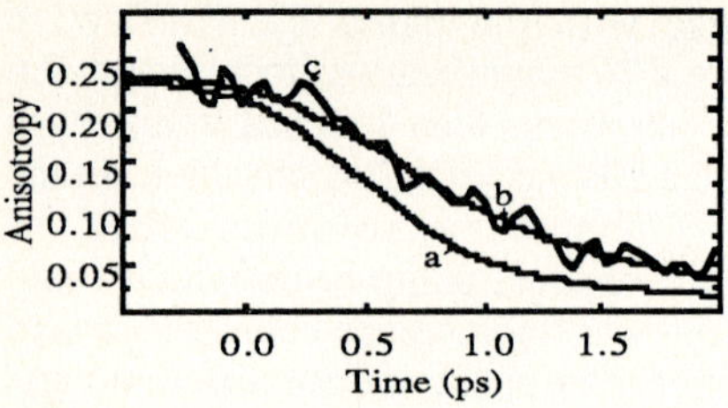

Fig. 1: Fluorescence Anisotropy of Aniline in Isopentane: (a) the anisotropy considering aniline as a free rotor; (b) extended diffusion fitting and (c) the experimental anisotropy.

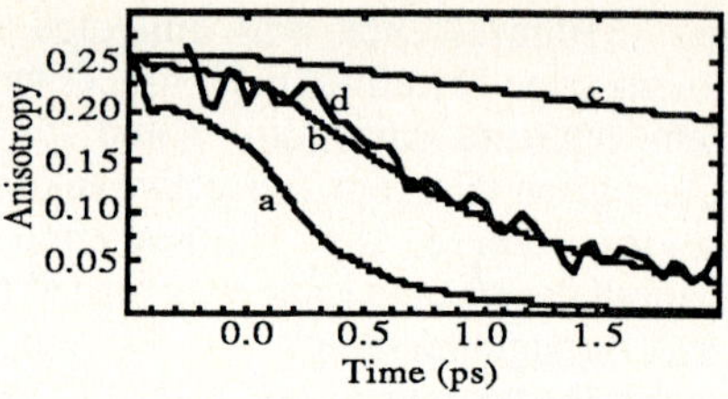

Fig. 2: Fluorescence Anisotropy of Aniline in Isopentane: (a) slip hydrodynamics, (b) slip + free rotation, (c) stick hydrodynamics, (d) the experimental data.

calculated anisotropy decayed too fast, indicating that the Lennard-Jones potentials used to mimic the intermolecular interaction did not reproduce the experimental friction. The torque autocorrelation obtained from this simulation was then scaled to yield the correct anisotropy decay in a generalized Langevin calculation. Further details on the simulations will be presented elsewhere [1.b].

4. Conclusions

UV fluorescence anisotropy decay measurements of aniline in various solvents were carried out. In isopentane there is a significant amount of inertial character in the aniline motion. Extended diffusion calculations indicate that the molecule is rotating through ca. 20 degrees between collisions, which occur at intervals of ca. 205 fs. This result is consistent with the inadequacy of the hydrodynamics predictions (slip and stick).

References

1. (a) A. B. Myers, M. A. Pereira, P. L. Holt and R. M. Hochstrasser, J. Chem. Phys., 86, 5146 (1987); (b) M. A. Pereira, P. E. Share, M. J. Sarisky, J. C. Shelley, M. L. Klein and R. M. Hochstrasser, in preparation, (1990).
2. (a) S. Ruhman, B. Kohler, A. G. Joly and K. A. Nelson, Chem. Phys. Let., 141, 16 (1987); (b) C. Kalpouzos, D. McMorrow, W. T. Lotshaw and G. A. Kenney-Wallace, J. Phys. Chem., 91, 2028 (1987); Chem. Phys. Lett., 150, 138 (1988).
3. (a) M. D. Dawson, T. F. Borges, D. W. Garvey and A. L. Smirl, Opt. Lett., 11, 721 (1986).
4. A. H. Kalantar and A. C. Albrecht, Ber. Bunsenges. Phys. Chem., 68, 361 (1964).
5. (a) R. G. Gordon, J. Chem. Phys., 44, 1830 (1966); (b) D. Chandler, J. Chem. Phys.; (c) 60, 3508 (1974); (d) 60, 3500 (1974); (e) 62, 1358 (1975).
6. K. S. Schweizer and D. Chandler, J. Chem. Phys., 76, 2296 (1982).

Dynamics of cis-Stilbene Isomerization in Solution

J.M. Jean, D.C. Todd, S.J. Rosenthal, A.J. Ruggiero, and G.R. Fleming

Departments of Chemistry and Physics and James Franck Institute,
University of Chicago, 5735 S. Ellis Ave., Chicago, IL 60637, USA

Fluorescence lifetime measurements of cis-stilbene show weak dependence on solvent viscosity and suggest that the reaction coordinate involves substantial ethylenic hydrogen motion.

1. Introduction

The influence of the solvent on the dynamics of reactive processes continues to be of great interest [1,2]. The past decade has seen a substantial increase in our understanding of the effect of solvent friction on activated barrier crossing processes. The trans-cis isomerization of stilbene has served as a useful prototype for understanding these types of reactions [3,4]. Less well understood is the role of the solvent in reactions for which there is little or no potential energy barrier separating reactant from products [2]. An example of the latter type of process is isomerization of stilbene from the cis side. The lack of a substantial barrier along the reaction coordinate leads to a rapid internal twisting motion that results in internal conversion ultimately leading to formation of the trans isomer.

A fundamental difference between barrierless processes and activated barrier crossing arises from the lack of a clear separation of time scales between motion along the reaction coordinate and vibrational relaxation of the remaining orthogonal degrees of freedom in the former case. The resulting dynamics can be quite complicated and, in the case of stilbene, can be viewed as a result of the competition between the intrinsic nonradiative rate from the twisted configuration and the viscous drag exerted by the solvent on the reaction coordinate.

2. Results and Discussion

We have carried out measurements of the excited state lifetime in both alcohols and alkanes using the fluorescence upconversion method [5]. In addition we have measured the fluorescence anisotropy in 2-propanol. A summary of the lifetimes is shown in Table1. A typical decay curve is displayed in Fig. 1. We find weak viscosity dependence in both types of solvents. Additionally, we find that the lifetime is shorter by almost a factor of two in 2-propanol than in tetradecane though the viscosity of

Springer Series in Chemical Physics, Vol. 53 **Ultrafast Phenomena VII**
Editors: C.B. Harris · E.P. Ippen · G.A. Mourou · A.H. Zewail

Table 1. Best single exponential fits to cis-stilbene fluorescence decay curves. T = 293 K.

Solvent	Viscosity (cp)	Emission Wavelength 400 nm	430 nm
h12 cis-stilbene			
2-propanol	2.35	0.76 ± .10 ps	0.87 ± .05 ps
decanol	15.51	1.27 ± .10 ps	1.47 ± .10 ps
hexane	0.32	1.03 ± .15 ps	----
tetradecane	2.43	1.38 ± .10 ps	1.52 ± .05 ps
d2 cis-stilbene			
tetradecane	2.43	1.71 ± .15 ps	1.75 ± .10 ps

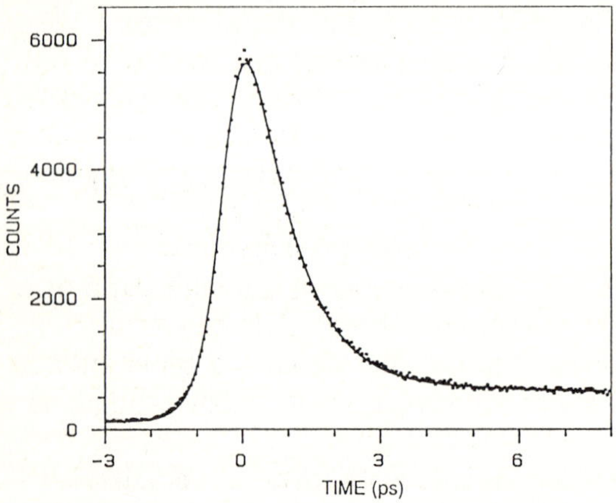

Fig.1 Fluorescence decay curve of cis-stilbene in 2-propanol at 293 K. Emission wavelength is 430 nm.

these two solvents differ by only 3% at room temperature. It is suspected that these are manifestations of the breakdown of the Stokes-Einstein relation connecting the microscopic friction with the shear viscosity of the solvent. Extensive hydrogen bond formation in alcohols leads to a larger free volume providing less resistance to large amplitude twisting of the molecule. An alternative explanation can be based on the effect of the polar environment of the alcohol solvents on the reaction coordinate. If the potential surface along which the nuclei relax to achieve the twisted configuration arises from an avoided crossing of the S_1 and S_2 surfaces, then differential solvation effects on these two states would lead to a reaction coordinate with a solvent dependent frequency.

An important question to consider is the nature of the reaction coordinate. We have found that deuteration of the ethylenic protons leads to a lifetime that is approximately 20% slower than that for the nondeuterated species indicating the reaction coordinate for this process is indeed more complicated than simple ethylenic torsion. Similar conclusions have been reached by Myers and Mathies [6] and Abrash et al.[7]. A recent spectroscopic study by Petek et al.[8] on cis-stilbene analogs that are prevented from isomerizing has concluded that ethylenic-phenyl bending and phenyl torsional modes are strongly displaced in the S_1 state. The viscosity dependence of these motions should be less pronounced than that for ethylenic torsion and should lead to an isomerization rate that depends weakly on viscosity. This is in accord with our observations.

Regardless of the nature of the reaction coordinate, motion along this the avoided crossing due to solvent polarity. It is known that in trans-stilbene the barrier to isomerization is substantially lower in alcohols than in alkanes due to preferential solvation of the S_2 state [10].

3. Conclusion

The weak viscosity dependence of the fluorescence lifetime of cis-stilbene along with the observation of ~ 20% increase in the lifetime in the deuterated species suggest that the reaction coordinate is more complicated than ethylenic torsion and involves a substantial amount of ethylenic hydrogen motion. The rate of isomerization in alcohols is faster than that in alkanes at constant viscosity, though it is not known whether this is a result of a dynamic or static solvent effect. The anisotropy result, though tentative, suggests that the isomerization process in 2-propanol is barrierless and that we are directly measuring motion along the reaction coordinate.

References

1. J.T. Hynes: in Theory of Chemical Reaction Dynamics, Vol IV. ed. M. Baer, CRC Press, Boca Raton, FL (1985).
2. B. Bagchi and G.R. Fleming: J. Phys. Chem., **94**, 9 (1990).
3. S.H. Courtney and G.R. Fleming: J. Chem. Phys. **83**, 215 (1985).
4. G. Rothenberger, D.K. Negus, and R.M. Hochstrasser: J. Chem. Phys. **79**, 5360 (1983).
5. A. J. Ruggiero,D.C. Todd, and G.R. Fleming: J. Am. Chem. Soc. **112**, 1003 (1989).
6. A.B. Myers and R.A. Mathies: J. Chem. Phys. **81**, 1552 (1984).
7. S. Abrash, S. Repinec, and R.M. Hochstrasser: J. Chem. Phys., accepted.
8. H Petek,K. Yoshihara,J. Frey, J. Penn, and J. Frederick: private comm.
9. H. Petek, Y. Fujiwara, D. Kim, and K. Yoshihara: J. Am. Chem. Soc. **110**, 6269 (1988).
10. S. Kim, S.H. Courtney, and G.R. Fleming: Chem. Phys. Lett, **159**, 543 (1989).

Isomerization of cis-Stilbene in Ar and Kr Clusters: Direct Measurements of trans-Stilbene Formation Rates on a Picosecond Time Scale

H. Petek[1], *K. Yoshihara*[1], *Y. Fujiwara*[2], *and J.G. Frey*[3]

[1]Institute for Molecular Science, Myodaiji, Okazaki 444, Japan
[2]Faculty of Pharmaceutical Sciences, Kanazawa University, Takara-machi, Kanazawa 920, Japan
[3]Department of Chemistry, Southampton University, Southampton SO9 5NH, UK

1. Introduction.

cis-Stilbene photochemical isomerization rate and quantum yields in solution phase have been the focus of many studies on the effect of solvent on barrierless isomerization [1-6]. Femtosecond fluorescence upconversion [2], multiphoton ionization, and transient absorption measurements [3] of the cis-stilbene S_1 state lifetime show a weak dependence on the solvent: the S_1 state decays in 0.7 -1.5 ps [2,3]. The solvent dependence of the S_1 state lifetime has been attributed to the effect of solvent friction on the rapid torsion around the central C = C bond to the 90° twisted geometry [2,3], which is thought to be the common intermediate for both cis- and trans-stilbene S_1 state decay, and has historically been designated as the "phantom" state [5]. However, another isomerization channel, the dihydrophenathrene (DHP) formation requires 1.2 kcal/mol activation energy, and occurs with as high as 30 % quantum yield at room temperature [6,7]. The analysis of LIF spectra of cis-stilbene homologs shows that the initial motion on the cis-stilbene S_1 state surface is along the DHP formation reaction coordinates [7]. The effect of solvent viscosity on the trans-stilbene formation is measured in supersonic expansions, where it is possible to form clusters of cis-stilbene with solvents such as rare gases, alcohols, or hydrocarbons, at temperatures near the freezing point of the solvent. In the low temperature cluster environment the DHP formation is minimized and the effect of solvent friction is clearly observed.

2. Experimental.

The experimental apparatus has been previously described [4,7,8]. Briefly, the experiment consists of: i) a molecular beam apparatus; ii) a picosecond laser system for pump-probe measurements; and iii) fluorescence detection and signal processing instrumentation. The clusters are formed in supersonic expansion by mixing either Ar or Kr with cis-stilbene within a pulsed nozzle and expanding the sample through a 0.5 mm orifice. For cis-stilbene:organic solvent clusters, expansions are generated by mixing the solvent vapor at room temperature with 1 600 Torr of He.

The picosecond laser system is an amplified synchronously pumped dye laser which is doubled to produce ~100 μJ at ~300 nm with <1 ps pulse width and 10 Hz repetition rate. The experiment uses a standard pump-probe configuration in which the pump pulse induces cis-trans isomerization and an identical time-delayed probe pulse excites the trans-stilbene product fluorescence. trans-Stilbene emission intensity is selectively detected through a 329 nm interference filter, and it is measured as a function of delay between pump and probe pulses.

3. Results and analysis.

The trans-stilbene rise time following photolysis of cis-stilbene solvated in Ar and Kr clusters is shown in Figure 1. The signal is symmetric with respect to the time origin because the pump and probe pulses are identical. The rise times are fit using the following expression:

$$A_f(t) = A_1 + A_2 - A_1\exp(-t/\tau_1) - A_2\exp(-t/\tau_2). \quad (1)$$

The fitting parameters for measurements at 297.0 nm are presented in Table I. The calculated fast rise time (τ_1) for trans-stilbene formation in Ar clusters is probably longer than the true rise time because the finite excitation pulse width is not considered in the fitting procedure.

The dependence of trans-stilbene rise time on rare gas stagnation pressure, cis-stilbene concentration, laser power, and excitation wavelength is investigated; however, there is no significant deviation from the

Springer Series in Chemical Physics, Vol. 53 **Ultrafast Phenomena VII**
Editors: C.B. Harris · E.P. Ippen · G.A. Mourou · A.H. Zewail

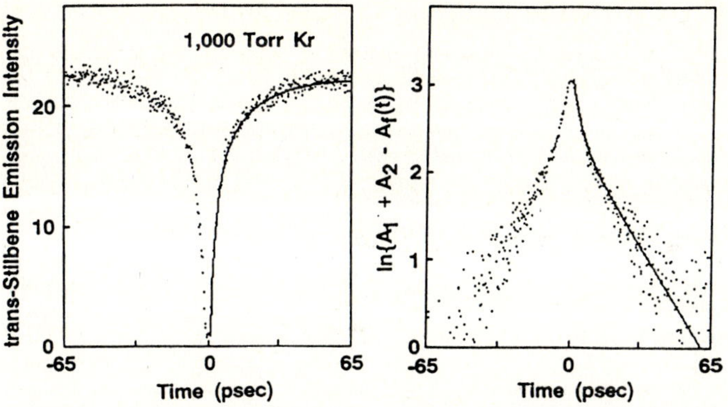

Figure 1. The trans-stilbene emission intensity as a function of delay between the pump and probe pulses in Ar clusters. The linear plot is shown on the left, and the corresponding logarithmic plot on the right side. The best fit to the biexponential rise is shown with the solid line.

Table I The parameters from fit of trans-stilbene rise times in solvent clusters to the Eq. 1.

Solvent	A_1/%	τ_1/ps	A_2	τ_2
Ar	58	1.7	42	15.9
Kr	60	3.0	40	15.9
n-hexane	100	8.7		
methanol	100	5.7		
ethanol	100	12.2		

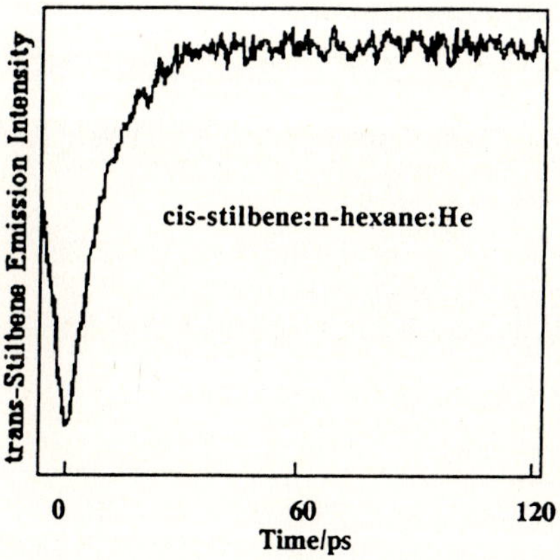

Figure 2. The emission intensity as a function of delay between pump and probe pulses of nascent trans-stilbene isomerization product in n-hexane clusters. The clusters are generated by coexpanding He, n-hexane, and cis-stilbene from total stagnation pressure of 1 600 Torr.

fitting parameters in Table I. Increasing the stagnation pressure, nozzle temperature, and laser power increases the total signal level.

The production rates of trans-stilbene in organic solvents is single exponential, and slower than the fast production rate in Ar and Kr (Figure 2). The data are fit by setting the A_2 amplitude in Eq. 1 to zero. The measured production rates decrease with increasing He stagnation pressure. This is consistent with decreasing cluster temperature, and therefore increasing cluster viscosity as the stagnation pressure increases. This temperature effect is confirmed by observing changes in the emission spectra with stagnation pressure.

4. Discussion.

The trans-stilbene rise time shows previously unobserved biexponential production rate with a fast component that depends on the cluster environment (solvent), and a slow component that is the same for Ar and Kr. The slower trans-stilbene formation in Kr than in Ar clusters is consistent with the smaller trans-stilbene formation quantum yield, and higher cis-stilbene emission quantum yield in Kr than in Ar clusters [8]. The slow production rate is the same for both Ar and Kr clusters, suggesting that it is due to a solvent independent dynamical process.

There are several dynamical processes which could be contributing to the biexponential production of trans-stilbene, for instance: i) twisting to 90° configuration on cis-stilbene S_1 surface; ii) intramolecular or intracluster vibrational relaxation on ground and excited state surfaces; iii) internal conversion from the phantom state; and iv) isomerization to dihydrophenanthrene. Furthermore, the biexponential production rate could reflect different environments in clusters, such as surface vs. bulk, or ordered vs. disordered phases. It is tentatively proposed that the fast component is the viscosity dependent torsion to the vibrationally hot phantom state, and the slow component is decay of the relaxed phantom state. It is assumed that the decay of the phantom state strongly depends on the temperature. Therefore the rate limiting step in the fast component is the production rather than the decay of the vibrationally hot phantom state. Organic solvent clusters are expected to be considerably warmer than the rare gas clusters, resulting in unobservably rapid phantom state decay.

Although the relative trans-stilbene production rates for various solvents in Table I are consistent with a viscosity dependence at a given temperature, the agreement is not quantitative. This is expected since the temperature of clusters is different, because it strongly depends on the solvent and the expansion conditions. Elucidation of the mechanism for trans-stilbene formation, and quantitative measurements of of the relationship between solvent friction and the trans-stilbene formation rates will be subjects of future work.

References

1.) G.R. Fleming, Chemical Applications of Ultrafast Spectroscopy, Oxford University Press, Oxford (1986); S. K. Kim, Ultrafast Time-resolved Studies of Stilbene Dynamics in Solution, Ph.D. Thesis, University of Chicago (1988).
2.) D.C. Todd, J.M. Jean, S.J. Rosenthal, A.J. Ruggiero, D. Yang, and G.R. Fleming, J. Chem. Phys. (submitted).
3.) F.E. Doany, R.M. Hochstrasser, B.I. Greene, and R.R. Millard, Chem. Phys. Lett. **118**, 1 (1985); B.I. Greene and T.W. Scott, Chem. Phys. Lett. **106**, 399 (1984).
4.) H. Petek, K. Yoshihara, Y. Fujiwara, and J.G. Frey, J. Opt. Soc. Am. B, (in press).
5.) G.S. Hammond, J. Saltiel, A.L. Lamola, N.J. Turro, J. S. Bradshaw, D.O. Cowan, R.C. Counsell, V. Vogt, and C. Dalton, J. Am. Chem. Soc. **86**, 3197 (1964).
6.) T. Wismonski-Knittel, G. Fischer, and E. Fischer, J. Chem. Soc. Perkin I, 1930 (1974).
7.) H. Petek, K. Yoshihara, Y. Fujiwara, Z. Lin, J.H. Penn, and J.H. Frederick, J. Phys. Chem. (submitted).
8.) H. Petek, Y. Fujiwara, D. Kim, and K. Yoshihara, J. Am. Chem. Soc. **110**, 6269 (1988).

Rotational Dynamics in Simple and Nematogenic Liquids: Transient Grating Optical Kerr Effect Investigations

F.W. Deeg[1] *and M.D. Fayer*[2]

[1]Institut für Physikalische Chemie, Universität München, Sophienstrasse 11, D-8000 München 2, Fed. Rep. of Germany
[2]Department of Chemistry, Stanford University, Stanford, CA 94305, USA

Abstract. We have studied the reorientational dynamics of neat biphenyl (BP) and pentylcyanobiphenyl (5CB) as well as solutions of these compounds in n-heptane over an extended temperature range with a polarization selective transient grating (TG) optical Kerr effect (OKE) approach. Whereas neat BP and the solution samples show a behavior typical for a simple liquid and hydrodynamic models, the dynamics of neat 5CB (in the isotropic phase) is dominated by the specific liquid structure.

Results and Discussion

The TG nuclear OKE signals of biphenyl in the neat liquid and in solution on the ultrafast timescale $\tau \leq 1$ ps are due to librational molecular dynamics (and collision-induced processes) as has been recently discussed by other groups [1]. For $\tau > 1$ ps one finds diffusional reorientational relaxation characterized by a biexponential decay. The time constants show a linear dependence on η/T, the ratio of viscosity and temperature, as expected from hydrodynamics and the Debye-Stokes-Einstein (DSE) model. The comparison with a depolarized light scattering (DLS) investigation of the reorientational dynamics of biphenyl [2], which revealed only the slower component, illustrates the complementary features of frequency and time domain experiments, as pointed out by Yan and Nelson [3].

The results of TG OKE experiments on 5CB in its isotropic phase above the nematic-isotropic phase transition are in strong contrast to the BP data and indicate large deviations from hydrodynamic theory. The TG signals in fig. 1 illustrate the complexity of the rotational dynamics of 5CB spanning five orders of magnitude in time and incorporating processes with time constants from hundreds of fs to tens of ns. The plots suggest three different dynamic ranges: There is the ultrafast timescale $\tau <$ 1-2 ps (fig. 1a), which again can be associated with intermolecular librational and translational motions, an intermediate timescale 1 ps $< \tau <$ 1 ns (fig. 1b) and a distinctly slower process on the ns-timescale (fig. 1c).

The dynamics on the slower timescales (fig. 1b and c) are determined by the specific structure of the 5CB liquid. Even in the isotropic phase, above the phase transition temperature, there exists a locally orientationally ordered liq-

Springer Series in Chemical Physics, Vol. 53 **Ultrafast Phenomena VII**
Editors: C.B. Harris · E.P. Ippen · G.A. Mourou · A.H. Zewail

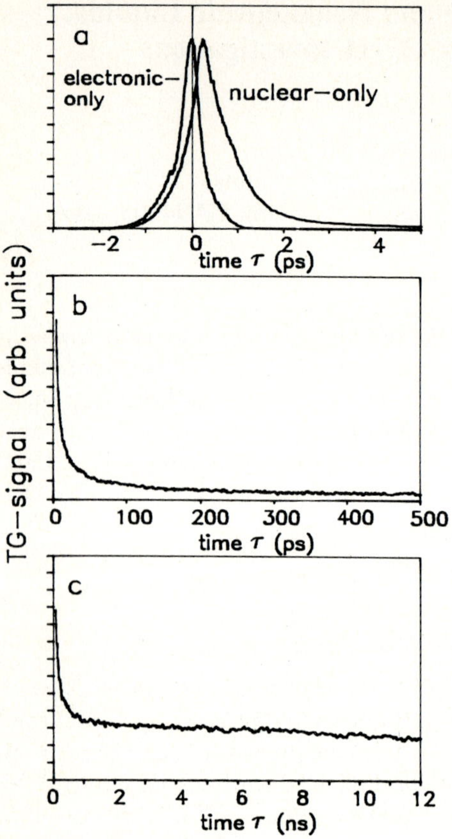

Figure 1:
TG OKE signals for a neat pentylcyanobiphenyl sample at 41.7 °C and three different timescales

uid. That is, one can qualitatively picture the liquid as consisting of pseudonematic domains, characterized by a coherence length $\xi(T)$ much larger than the molecular size [4]. The ns-timescale process in fig. 1c has monoexponential character and can be attributed to the well-known collective reorientational dynamics of the pseudonematic domains. The Landau-deGennes theory and the pseudonematic domain picture provide a satisfactory description of the temperature dependence of this collectice rotational dynamics up to 35 degrees above the phase transition.

The intermediate timescale decay (fig. 1b) is distinctly nonexponential with time constants from 1 to 300 ps. We associate these relaxation channels with local molecular rotational dynamics *within* the pseudonematic domains. Our results present the first detailed time domain investigation of these intradomain dynamics which had been postulated before [5]. In strong contrast to the ns-timescale data these intermediate timescale dynamics are temperature *independent* up to 35 degrees above the nematic-isotropic phase transition. This is illustrated in fig. 2 which shows in a double-logarithmic plot the intermediate timescale decay at 41.7°C and 69.4°C. Between these two temperatures the

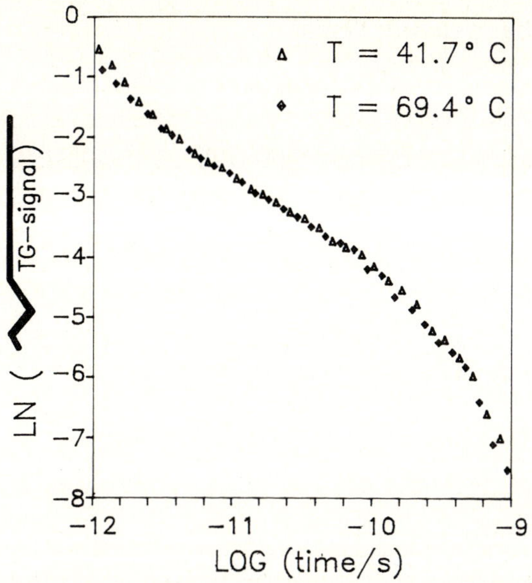

Figure 2: Intermediate timescale dynamics of neat 5CB at 41.7°C and 69.4°C.

bulk viscosity changes by nearly a factor of three. The negligible change of the observed dynamics reflects a decoupling of the local molecular relaxation within the structured domains from the bulk viscosity. Only at higher temperatures when the pseudonematic domains fall apart, the local dynamics start to exhibit a temperature dependence. These results demonstrate an intimate relationship between liquid structure and reorientational dynamics in neat 5CB.

References

1. see e.g. S. Ruhman, B. Kohler, A.G. Joly and K.A. Nelson: IEEE J. Quantum Elec. **24**,470 (1988)
2. Th. Dorfmüller in: *Rotational Dynamics of Small and Macromolecules*, ed. by Th. Dorfmüller and R. Pecora (Springer, Berlin 1987).
3. Y.X. Yan and K.A. Nelson: J. Chem. Phys. **87**, 6240 and 6257 (1987)
4. P.G. deGennes: *The Physics of Liquid Crystals* (Clarendon, Oxford 1974)
5. C. Flytzanis and Y.R. Shen: Phys. Rev. Lett. **33**, 14 (1974)

Rotational Diffusion of Phenoxazine Dyes: Characterization of Molecular Friction

D.S. Alavi, R.S. Hartman, and D.H. Waldeck

Department of Chemistry, University of Pittsburgh, Pittsburgh, PA 15260, USA

Abstract: Relative contributions of mechanical, dielectric, and specific solute/solvent interactions to molecular friction in liquids are evaluated by measuring solute reorientation times for phenoxazine dyes.

Measurement of rotational diffusion of molecules in liquid solutions is a sensitive probe of molecular friction. This friction can arise from a variety of physical processes, including solvent displacement, dielectric interactions, and specific solute/solvent interactions. It is often difficult to disentangle the relative contributions of these interactions since in general they are present together. Our goal is to isolate and evaluate these contributions by measuring reorientation times for a set of probe molecules whose physical properties can be varied in a systematic way.

The second-order rotational correlation function r(t) is measured in a state selective manner using time resolved optically heterodyned polarization spectroscopy (OHPS) [1]. Briefly, a linearly polarized probe pulse creates anisotropic distributions of both ground and excited state molecules. A copropagating, noncollinear probe pulse, linearly polarized 45° from the pump polarization, is monitored through an analyzer polarizer. If δ_D is the angle (radians) between the analyzer transmission axis and the crossed position, then the fraction of probe light transmitted is

$$T \propto \exp\left(\frac{-\sigma dK}{3}\right) \left| \frac{dK(t)r(t)\sigma}{2} + i\phi(t) + \delta_D\left(2 + \frac{d\sigma K(t)}{3} + \frac{d\sigma K(t)r(t)}{6}\right) \right|^2$$

where σ is the absorption cross section, K is the solute concentration, d is the sample thickness, K(t) is the excited state population, and $\phi(t)$ is the birefringent part of the signal. The term which varies linearly with δ_D is the signal of interest. A refinement of this technique is used to isolate this term in an unambiguous way. First a decay with the analyzer at $+\delta_D$ is collected and then a second decay at $-\delta_D$ is recorded. Both decays are normalized with respect to laser intensities, and are then subtracted. Only the linear term survives, and fitting this decay to a single exponential yields the decay time τ_m. Excited state lifetimes (τ_f) are measured using time correlated single photon counting, and the reorientation time is computed as $\tau_{or} = (1/\tau_m - 1/\tau_f)^{-1}$. The pump probe apparatus has been described [1].

The molecules studied are three phenoxazine dyes, resorufin (anionic), resorufamine (neutral but polar), and oxazine 118 (cationic). They have similar structures so the effects of solute size and shape are constant (see Figure 1). This allows independent evaluation of dielectric effects in DMSO, which is polar but not hydrogen bonding. In order to quantify the effects of overall charge and dipole moment on the friction. Reorientation times were measured over a range of temperatures for all three dye molecules in both ground (S_0) and excited (S_1) electronic states. The results are shown in Figure 2. A plot of τ_{or} vs. η/T should, according to

Springer Series in Chemical Physics, Vol. 53 **Ultrafast Phenomena VII**
Editors: C.B. Harris · E.P. Ippen · G.A. Mourou · A.H. Zewail

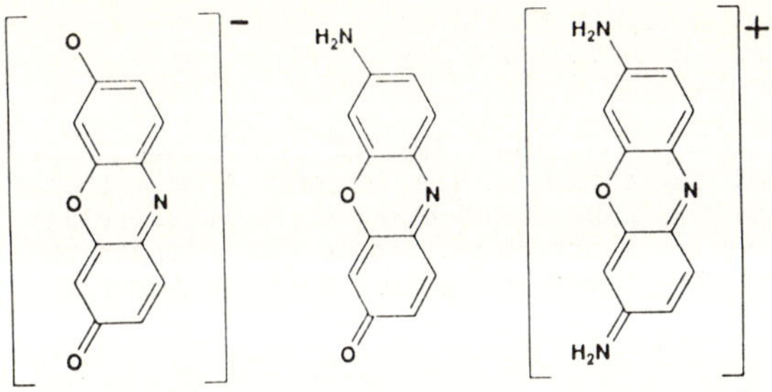

Figure 1. Structures of the test solute molecules are shown.

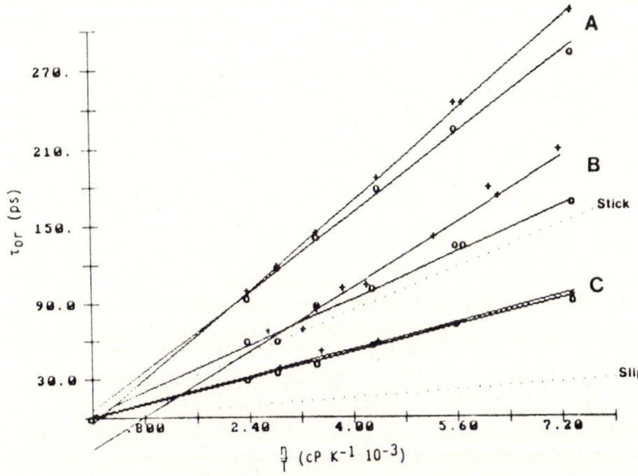

Figure 2. A DSE plot is shown for oxazine (A), resorufamine (B) and resorufin (C) in DMSO. Rotation times for both electronic states are shown, (+) is excited state and (o) is ground state.

Debye-Stokes-Einstein (DSE) theory, yield a straight line through the origin. Also shown in the figure are the lines calculated for these molecules with slip and stick hydrodynamic boundary conditions. The best fit slopes and intercepts are listed in Table 1.

Both steady state spectroscopy and electronic structure calculations for resorufamine indicate a ground state dipole moment of 4-5 D and an increase to 8-10 D in the excited state. Longer reorientation times are observed in the excited state than in the ground state. Qualitatively, this trend is expected because of the larger dipole moment in the excited state, which would couple the solute more strongly to the solvent. Attempts to quantify the observed reorientation times with several models of dielectric friction,

TABLE I. Best fit parameters of τ_{or} vs. η/T to a line.

Solute	Slope(ns-K/cp)	Intercept(ps)	τ_{or} at 294 K ps
resorufin (S_0)	13.0±0.4	-0.5±1.5	95
resorufin (S_1)	13.1±0.7	1.2±2.5	98
resorufamine (S_0)	22.7±2.0	2.4±7.2	170
resorufamine (S_1)	31.6±2.3	-24.5±9.2	209
oxazine (S_0)	38.5±1.0	5.9±3.8	290
oxazine (S_1)	43.4±0.4	-4.3±1.4	316

have been unsuccessful. Continuum models of dielectric friction [2] overestimate the dipole moment (i.e., underestimate the friction) by a factor of five to ten.

These continuum models fail qualitatively when applied to the ionic dyes. A continuum theory predicts that any effect on the friction from a charge should be the same regardless of the sign of the charge. In fact, relaxation times for the cation oxazine are approximately three times longer than for the anion, resorufin. A continuum model alone cannot explain this difference. Furthermore, the friction for either ionic dye should be larger than that of resorufamine, but as Figure 2 shows, resorufin has the fastest relaxation and oxazine the slowest, with resorufamine intermediate between the two.

A major assumption in these analyses is that DMSO does not interact in any specific way with the solute molecules. This assumption may be incorrect. The lone electron pairs on the oxygen of DMSO could form hydrogen bonds with amino groups on oxazine (two amino groups) and resorufamine (one group), while no hydrogen bonding interaction can occur with resorufin. While solvent attachment alone does not quantitatively account for the differences in reorientation times, the qualitative trend between the three dyes is in the right direction.

Resorufamine has also been studied in isopropanol, where changes in local hydrogen bonding interactions between ground and excited state may be superimposed on the dielectric friction effects. The observed decay profile is nonexponential. Preliminary results indicate a stronger state dependent reorientation, with a 50 to 75% increase in τ_{or} from S_0 to S_1. Further studies will compare resorufamine to the charged dyes, so that effects of overall charge and dipole moment can be compared.

In conclusion, a quantitative study of dielectric friction effects on solute molecule rotation was performed by studying three mechanically similar dye molecules with differing electrical properties in the polar, aprotic solvent DMSO. Continuum theories of dielectric friction fail to account for the experimental data, both qualitatively and quantitatively. It appears that molecularity cannot be neglected in the description of microscopic friction, and that explicit treatment of at least the first solvation shell is required.

The support of the National Science Foundation (CHE-8613488) is appreciated.

[1] D.S. Alavi, R.S. Hartman, D.H. Waldeck, J. Chem. Phys. *92*, 4055 (1990).

[2] a) T. Nee, R. Zwanzig, J. Chem. Phys. *52*, 6353 (1970).
b) J.B. Hubbard, P.G. Wolynes, J. Chem. Phys. *69*, 998 (1978).
c) G. van der Zwan, J.T. Hynes, J. Phys. Chem. *89*, 4181 (1985).

Linear Electron–Phonon Interaction in Dye-Doped Polymers Studied by Femtosecond Accumulated Photon Echo

S. Saikan

Department of Physics, Faculty of Science, Osaka University, Toyonaka, Osaka 560, Japan

The linear electron-phonon interaction in dye-doped polymers has been investigated with Fourier-transform spectroscopy based on femtosecond accumulated photon echo [1]. This technique enables us to extract the homogeneous spectrum from the inhomogeneously broadened absorption spectrum. The obtained homogeneous spectrum, in particular, the phonon sideband spectrum and its temperature dependent variation has been analyzed according to the theory which takes into account the linear electron-phonon interaction [2,3]. The relation between the phonon sideband spectrum and the low-frequency Raman spectrum of the host polymer has also been discussed from fracton picture [4].

The sample employed in this experiment is tetraphenylporphine (TPP) doped in PMMA. The $Q_x(0,0)$ band was excited by a synchronously pumped DCM dye laser. Observed temperature dependence of the echo signal and its Fourier-transformed spectrum are shown in Fig.1(a) and (b), respectively.

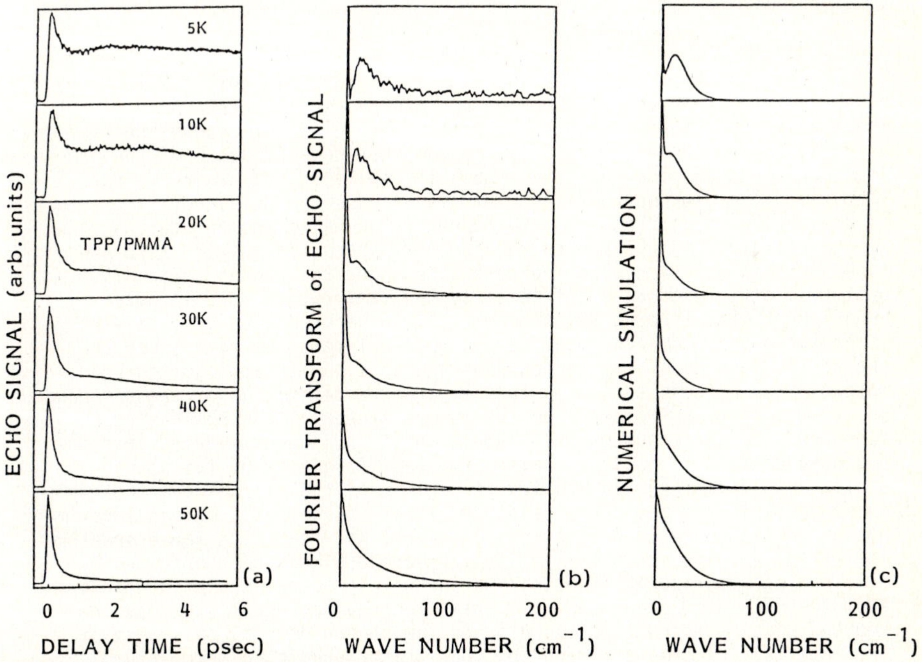

Fig.1 (a) Temperature dependence of the heterodyne-detected accumulated photon echo in tetraphenylporphine(TPP) doped in PMMA. (b) Fourier cosine transformed spectrum of (a). (c) Numerical simulations.

Springer Series in Chemical Physics, Vol. 53 **Ultrafast Phenomena VII**
Editors: C.B. Harris · E.P. Ippen · G.A. Mourou · A.H. Zewail

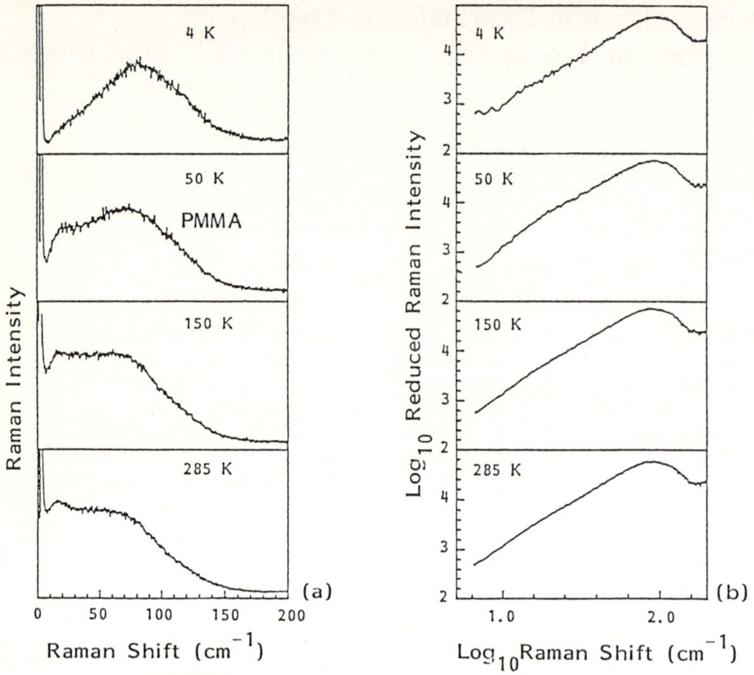

Fig.2 (a) Temperature dependence of low frequency Raman spectrum in PMMA polymer. (b) Reduced Raman spectrum in log-log coordinates.

The Fourier spectrum indicates that the echo signal in this sample is mainly generated from the zero-phonon line and the phonon sideband. According to the theory which takes into account the linear electron-phonon interaction, we tried to reproduce numerically the temperature dependence of the phonon sideband spectrum. The result is shown in Fig.1(c), from which we noticed that the ultrafast decay of the echo signal observed at higher temperatures is due to the excitation from the thermally occupied phonon state in the ground state and is not due to the spectral broadening of the zero-phonon line. This calculation also indicates that once the phonon state distribution is determined at a sufficiently low temperature, the temporal behaviour of the echo at any temperature can be predicted precisely [2].

From the measurements in different host polymers, we have found that the peak frequency of the phonon sideband spectrum is an important parameter, which affects seriously the temperature dependence of the homogeneous spectrum. The peak frequency of the phonon sideband spectrum has been measured in a large number of polymers and it has been revealed that the hydrogen bonding in polymers raises the peak frequency of the phonon sideband spectrum [4]. To clarify the origin of the vibrational mode which is responsible for the phonon sideband spectrum, we performed an experiment of low frequency Raman scattering in several polymers. The Raman spectrum in PMMA polymer is shown in Fig.2(a). Figure 2(b) has been obtained by eliminating numerically the contribution of the boson occupation factor from the observed Raman spectrum. This spectrum has usually called the reduced Raman spectrum and it reflects the phonon density of states of the material. Figures 2(a) and (b) show that there is no temperature dependent variation of the phonon density of states although the Raman spectrum varies dependent on the temperature. The characteristic feature of the Raman spectrum of polymers is that there appears two spectral edges in the spectrum. The low frequency peak has usually been called the boson peak since this peak becomes less prominent with decreasing temperature, as seen in Fig.2(a). The reduced Raman spectrum (Fig.2(b)) reveals that the phonon density of states in polymers obeys ω^{μ} dependence between the two spectral edges. The above mentioned two characteristic features can be explained well by considering the fractal structure in polymers. In view of the fracton picture, the boson peak is attributable to the low frequency fracton edge. From the comparison between the Raman spectrum and the phonon sideband spectrum observed in the Fourier-transformed spectrum of the echo signal, we find that the peak frequency of the phonon sideband spectrum corresponds to the low frequency fracton edge [4]. Therefore, it can be concluded that the homogeneous spectrum, in particular, the phonon sideband spectrum is affected seriously by the fractal structure in polymers.

References

[1] S.Saikan, T.Nakabayashi, Y.Kanematsu and N.Tato Phys.Rev. B38 7777 (1988)
[2] S.Saikan, A.Imaoka, Y.Kanematsu and T.Kishida Chem.Phys.Lett. 162 217 (1989)
[3] S.Saikan, A.Imaoka, Y.Kanematsu, K.Sakoda, K.Kominami and M.Iwamoto Phys.Rev. B41 3185 (1990)
[4] S.Saikan, T.Kishida, Y.Kanematsu, H.Aota, A.Harada and M.Kamachi Chem.Phys.Lett. 166 358 (1990)

Time-Resolved Decay of Absorption Anisotropy due to Rotation of Free Molecules in the Gas Phase

N.A. Borisevich[1], *E.V. Khoroshilov*[1], *I.V. Kryukov*[1], *P.G. Kryukov*[1], *G.B. Tolstorozhev*[2], *M.Yu. Shakhbazyan*[1], *and A.V. Sharkov*[2]

[1]P.N. Lebedev Physics Institute, USSR Academy of Sciences, 117924 Moscow, USSR

[2]B.I. Stepanov Institute of Physics, Belorussian Academy of Sciences, 220603 Minsk, USSR

Abstract. A photoinduced orientational anisotropy of an ensemble of free molecules has a complex time dependence, and differs significantly from the relaxation in condensed media. We have studied the time-resolved $S_1^* \rightarrow S_n^*$ absorption anisotropy decay resulting from rotation of free POPOP molecules in vapour at 600 K and PPO molecueles at 483 K after $S_0 \rightarrow S_1^*$ excitation by femtosecond linearly polarized pulses at 308 nm. The maximum absorption anisotropy r=0.4 recorded at the moment of excitation speaks in favour of parallel dipoles of $S_1^* \rightarrow S_n^*$ and $S_0 \rightarrow S_1^*$ transitions of POPOP and PPO molecules. Later,value r decays approximately to zero, then increases and gets a stationary nonzero level due to nonchaotic rotation of free molecules.

1. Introduction

A molecular absorption of linearly polarized light gives origin to an ensemble of anisotropically excited molecules. As a result of reorientation (rotation) of molecules the anisotropy (absorption or luminescence) changes with time. The theoretical calculations performed classically by Blokhin et al. [1,2] and later quantum mechanically by Myers and Hochstrasser [3] and Felker and Zewail [4] showed that the relaxation of optically induced orientational anisotropy of the ensemble of free molecules in gas phase has a complex time dependence, and differs significantly from the relaxation in condensed media. An experiment with stilbene was taken by Scherer et al. [5] at temperature 2K in a molecular beam. The anisotropy decay time was about 50 ps. But the application of a picosecond technique did not make it possible to measure in detail the kinetics of decay of luminescence anisotropy in the experiment with the stilbene vapour at 463 K (Myers et al. [6]).

2. Experimental

We have studied decay of a photoinduced absorption anisotropy in vapour (of several Torr pressure) of 1,4-Di-[2- (5-phenyloxazolyl)] -benzene (POPOP) molecules at T= 600 K and 2,5-Diphenyloxazol (PPO) molecules at 483 K. The pulses from colliding pulse mode-locked dye laser with duration of 70 femtoseconds were amplified in a multipass

Springer Series in Chemical Physics, Vol. 53 **Ultrafast Phenomena VII**
Editors: C.B. Harris · E.P. Ippen · G.A. Mourou · A.H. Zewail

jet amplifier designed by Khoroshilov et al. [7]. The amplified radiation was transformed to the second harmonic (308 nm). This linearly polarized radiation served for the excitation of $S_0 \rightarrow S_1^*$ transition of POPOP and PPO molecules. The probing was performed by the pulses at 616 nm. This wavelength corresponds to $S_1^* \rightarrow S_n^*$ transition of the molecule. The values of photoinduced absorbance changes, $\Delta A_{\parallel}(t)$ and $\Delta A_{\perp}(t)$, were measured under the conditions of mutually parallel and perpendicular polarization of the exciting and probing radiation. From those data we calculated the value of anisotropy, $r(t)=(\Delta A_{\parallel}(t)-\Delta A_{\perp}(t)) \;/\; (\Delta A_{\parallel}(t)+2\Delta A_{\perp}(t))$.

3. Results and Discussion

Figure 1 illustrates the experimental dependences $\Delta A_{\parallel}(t)$ and $\Delta A_{\perp}(t)$ and anisotropy r(t) calculated on the basis of those results for POPOP (left side) and PPO (right side). Figure 1 exhibits also the dependence of anisotropy on time after excitation calculated by Blokhin [8] for the POPOP molecule. Immediately after excitation we recorded maximum possible value of anisotropy r = 0.4 for both molecules. This is indicative, first, of parallel dipoles of $S_0 \rightarrow S_1^*$ and $S_1^* \rightarrow S_n^*$ transitions. Second, the maximum anisotropy can be observed only in the case when the pulse duration is much less than the measured decay time.

As a result of molecular rotation, at collision-free conditions, the anisotropy decreases abruptly and reaches minimum within the time

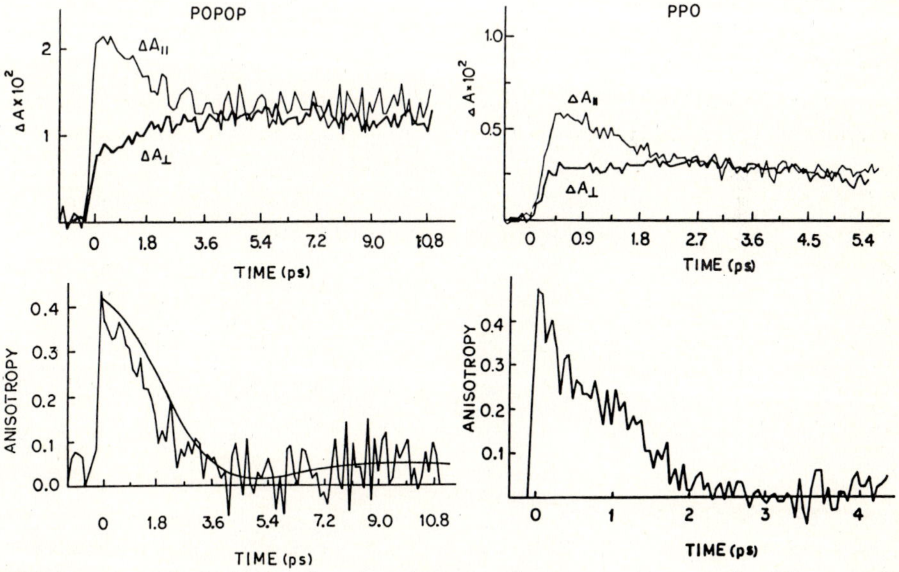

Figure 1. Photoinduced absorbance changes, $\Delta A_{\parallel}$ and $\Delta A_{\perp}$, and anisotropy, r(t), for POPOP molecules at 600 K (left) and PPO molecules at 483 K (right).

close to $\tau = \sqrt{I/kT}$, where I is the momentum of inertia of a molecule. This process takes about 4 ps for POPOP molecules and about 2 ps for more light PPO molecules. According to the calculations [8] one should expect that the anisotropy achieves a stationary nonzero level (r=0.05 for POPOP) after a fast decay to a minimum value. Such a deviation from zero is due to a nonchaotic rotation of free molecules. As seen from Fig.1 this stage has been observed in experiment too.

Thus, from the experimental results obtained it follows that we have observed, in a real time scale, the rotation of free molecules in the gas phase manifested as the decay of induced absorption anisotropy due to the dephasing of the molecules which rotate with different rates.

References

1. A.P.Blokhin, N.A.Borisevich, I.I.Kalosha, V.A.Tolkachev: Dokl.AN SSSR (Russian) **238**, 123 (1978)
2. A.P.Blokhin, V.N.Knyukshto, V.A.Tolkachev: Izv. AN SSSR (Russian) **42,** 359 (1978)
3. A.B.Myers, R.M.Hochstrasser: J.Chem.Phys. **85,** 6301 (1986)
4. P.M.Felker, A.H.Zewail: J.Chem.Phys. **86,** 2460 (1987)
5. N.F.Scherer, L.R.Khundkar, T.S.Rose, A.H.Zewail: J.Phys.Chem. **91,** 6478 (1987)
6. A.B.Myers,P.L.Holt, M.A.Pereira, R.M.Hochstrasser: Chem.Phys.Lett. **132**,585 (1986)
7. E.V.Khoroshilov, I.V.Kryukov, P.G.Kryukov, A.V.Sharkov: *Ultrafast Phenomena VI* (Springer, Berlin, Heidelberg, 1988) pp.22-23
8. A.P.Blokhin: Vest. AN BSSR (Russian) **4**, 77 (1986)

Ultrafast Predissociation Dynamics of Isolated Small Molecules from UV Resonance Raman Depolarization Ratios

A.B. Myers and Bulang Li

Department of Chemistry, University of Rochester, Rochester, NY 14627, USA

Abstract. Resonance Raman depolarization ratios and bandshapes are described as a probe of ultrafast dissociation dynamics of small gas phase molecules. The role of a finite laser bandwidth in inducing incoherent "fluorescence" emission is discussed. Applications to the predissociative S_3 state of CS_2 vapor are shown.

1. Introduction

In resonance Raman scattering in solution, the electronic dephasing is usually fast compared with molecular rotation, and the depolarization ratios may be calculated as originating from a fixed isotropic distribution. For small molecules in the vapor phase, however, the electronic dephasing is determined by some combination of collisions, photochemistry, and "light-induced" dephasing (vide infra), none of which are necessarily fast compared with rotation [1]. Thus the depolarization ratios as well as the bandshapes depend on the electronic dephasing time, which determines the "time" that may elapse between destruction of the incident photon and creation of the scattered photon. If collisions are infrequent, the dephasing may be dominated by pure lifetime decay, and analysis of the resonance Raman depolarization ratios and bandshapes may be used to determine pico- and subpicosecond excited state lifetimes. Additionally, if rapid phase fluctuations in the applied field cause its bandwidth to be significant relative to the molecular resonance widths, the resulting emission will consist of both coherent (Raman-like) and incoherent (fluorescence-like) components [2], with pronounced effects on the near-resonant emission polarization.

The S_3 state of CS_2 near 200 nm is an excellent candidate for such studies. Previous work employing photofragment spectroscopy and fluorescence quantum yields suggests that the predissociation-limited excited state lifetime is comparable with rotational periods [3] and depends strongly on excess vibrational energy [4].

2. Theoretical and experimental methods

We start from eqs. (18)-(19) of ref. [2], a density matrix treatment of resonantly excited spontaneous emission for an arbitrary applied field. We assume that the incident field is given by $E(t)=\phi(t)\exp(i\omega_L t)+c.c.$, where $\phi(t)$ is a randomly fluctuating phase with a correlation function given by $\langle\phi^*(\tau)\phi(t)\rangle=\exp(-\Gamma'|t-\tau|/\hbar)$. Generalizing from a 3-level to a multilevel system [5], performing the time integrals, and separating the matrix elements into rotational and vibrational parts leaves the following expression for the "Raman" plus "fluorescence"emission intensity in the absence of material pure dephasing:

$$I_{i\,f}^{z\alpha}(\omega_L,\omega_S) \propto \sum_{v,v'} \langle f|v\rangle\langle v|i\rangle\langle i|v'\rangle\langle v'|f\rangle \sum_{a,b,b',c} P_a \langle a|C_z|b\rangle\langle b|C_\alpha|c\rangle$$

$$\langle c|C_\alpha|b'\rangle\langle b'|C_z|a\rangle \frac{1}{i(\Delta_S+\omega_{v'v}+\omega_{b'b})+\gamma/2}\left\{\frac{1}{-i\Delta_L+\Gamma'+\gamma 2}\;\frac{1}{i(\Delta_S-\Delta_L)+\Gamma'}\right.$$

$$\left.+\frac{1}{i(\omega_{v'v}+\omega_{b'b})+\gamma}\left[\frac{1}{-i\Delta_L+\Gamma'+\gamma/2}+\frac{1}{i\Delta_L'+\Gamma'+\gamma/2}\right]\right\}+\text{c.c}$$

Here z and α are the polarizations of the incident (ω_L) and scattered (ω_S) light, {|i>, |v>, |f>} and { |a>, |b>, |c>} are the initial, intermediate, and final vibrational and rotational states, the C's are direction cosines, Δ_S and Δ_L are the detunings from resonance, $\gamma/2$ is the linewidth (HWHM) due to excited state population decay, and Γ' is the laser linewidth.

For CS_2, we evaluate the direction cosine matrix elements for a parallel transition of a linear rotor in the high-J limit. Integration over ω_S for α=z and α=x give expressions for the depolarization ratio, which become particularly simple when a single vibronic state is on resonance or the emission is summed over all final vibrational states [6,7]. The explicit expressions for the rotational band contours are more complicated [7].

Resonance Raman spectra and depolarization ratios of CS_2 vapor were obtained as described previously [6]. The nominal temporal and spectral pulse widths were 8 ns and 0.7 cm^{-1}, respectively. The spectral resolution of the detection system was ~10 cm^{-1} for the Raman bandshape measurements.

3. Results and discussion

Figure 1 shows experimental depolarization dispersion curves in two excitation regions as well as calculated curves as a function of excited state lifetime and bandwidth of the driving field. The depolarization ratios are quite sensitive to the lifetime, while "light-induced dephasing" has a pronounced effect only in the preresonance region. This is because the Raman component of the emission approaches ρ= 1/3 as the laser is tuned away from resonance and the "time" spent in the excited electronic state becomes negligible, while the fluorescence polarization is independent of detuning. The excited state lifetime is strongly dependent on excess vibrational energy, falling from ~1.5 ps near 208 nm to 0.44 ps near 200 nm.

Figure 2 shows emission bandshapes excited near 208 nm. With excitation on the low-frequency side of a vibronic band, P and R branches of different J states are on resonance, and three lobes (O, Q, and S-branches) are observed; on the high-frequency side, only R branch transitions are on resonance, and only Q and S branches appear. The reasonably good calculated fits using a constant γ suggest a relatively

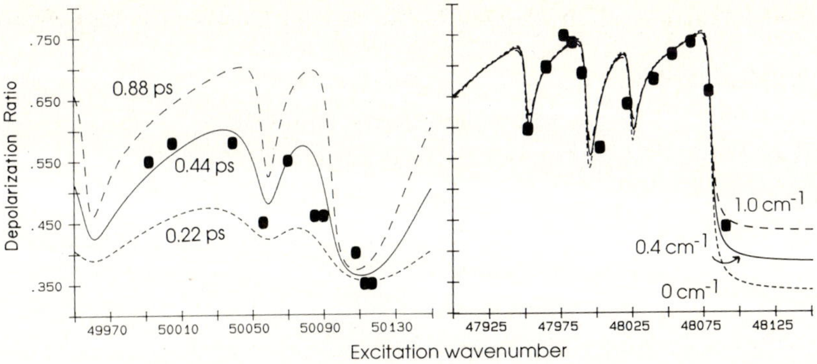

Figure 1. Experimental (points) and calculated emission depolarization ratios for CS_2. Left: Excitation near 200 nm, Γ'=0, excited state lifetime as indicated. Right: Excitation near 208 nm, excited state lifetime = 1.5 ps, laser linewidth as indicated.

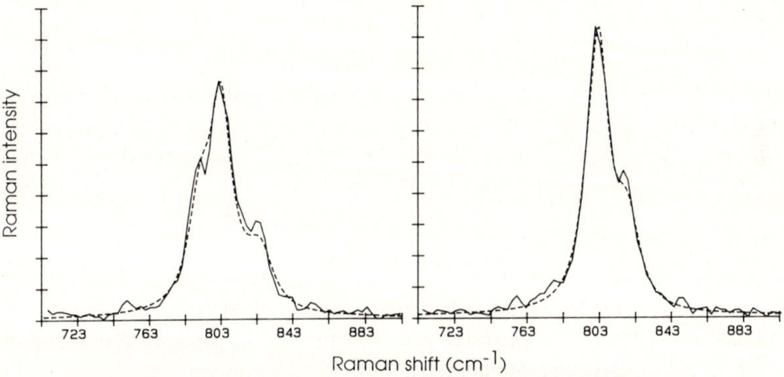

Figure 2. Experimental (solid) and calculated (dashed) CS_2 emission bandshapes with excitation at 47976 cm^{-1} (left) and at 47983 cm^{-1} (right).

weak J dependence of the lifetime, but definite conclusions await a more detailed analysis of the data.

References

1. A. B. Myers and R. M. Hochstrasser, J. Chem. Phys. **87**, 2116 (1987).
2. J. S. Melinger and A. C. Albrecht, J. Chem. Phys. **84**, 1247 (1986).
3. I. M. Waller and J. W. Hepburn, J. Chem. Phys. **87**, 3261 (1987).
4. D. Hara and D. Phillips, J. Chem. Soc. Faraday Trans. 2 **74**, 1441 (1978).
5. S. Mukamel, J. Chem. Phys. **82**, 5398 (1985).
6. B. Li and A. B. Myers, J. Chem. Phys. **89**, 6658 (1988).
7. B. Li and A. B. Myers, manuscript in preparation.

Vibrational Dynamics of Chloroform, Bromoform and Iodoform

H.J. Bakker, P.C.M. Planken, L. Kuipers, and A. Lagendijk

FOM Institute for Atomic and Molecular Physics, Kruislaan 407, 1098 SJ Amsterdam, The Netherlands

Abstract We present a study of the energy relaxation processes of chloroform, bromoform and iodoform in solution after excitation of the C-H stretch vibration. The experiment is performed in a polar and a nonpolar solvent in order to study the effect of the polarity of the solvent on the relaxation. We observe that in both solvents the relaxation leads to ultrafast changes in the absorption band of the C-H stretch vibration.

Introduction

In this study we investigate the vibrational lifetimes and the energy-transfer processes of the haloforms in solution with infrared saturation spectroscopy. In this technique a significant fraction of the molecules is excited with an intense picosecond infrared pulse (pump). This induces a bleaching of the sample. The time dependence of the bleaching can be probed by measuring the transmission of a weak delayed second pulse (probe).[1]

Experimental

The infrared pulses are generated by means of parametric generation and amplification in $LiNbO_3$ crystals pumped by a Nd:YAG laser.[2] The generated pulses are tunable between 2200 and 7200 cm^{-1} and have a pulse duration of approximately 19 picoseconds. A pulse at 3000 cm^{-1} has a typical energy of 200 μJ and a bandwidth of 15 cm^{-1}.

Results and discussion

In figure 1, the relative transmission of the probe pulse as a function of the delay between pump and probe is presented for $CHCl_3$ dissolved in CCl_4 (carbontetrachloride). In figure 1a, the central frequency of the pulses is higher than the maximum of the absorption band. We observe that when the probe overlaps with the pump, the transmission of the probe increases, because of the bleaching of the sample due to the pump. For large delay times, all molecules have relaxed and the transmission returns to its initial value.

In figure 1b, the central frequency of the pulses is lower than the maximum of the absorption band. We observe that the transmission decreases to a value that is lower than the one before excitation by the pump. For $CHBr_3$ we observe a similar behaviour. For CHI_3 we do not observe such an overshoot, but we do observe a steepening of the transmission decrease.

We deduce from these results that for all three haloforms the relaxation takes place via two consecutive relaxation processes. In the first process, the vibrational energy is transferred from the C-H stretch vibration to other vibrations within the molecule. After this process, the transition frequency of the C-H stretch vibration is lowered due to anharmonic coupling with the other molecular vibrations. If we choose the central frequency of the laser pulse lower than the maximum of the absorption band (fig. 1b),

Springer Series in Chemical Physics, Vol. 53 **Ultrafast Phenomena VII**
Editors: C.B. Harris · E.P. Ippen · G.A. Mourou · A.H. Zewail

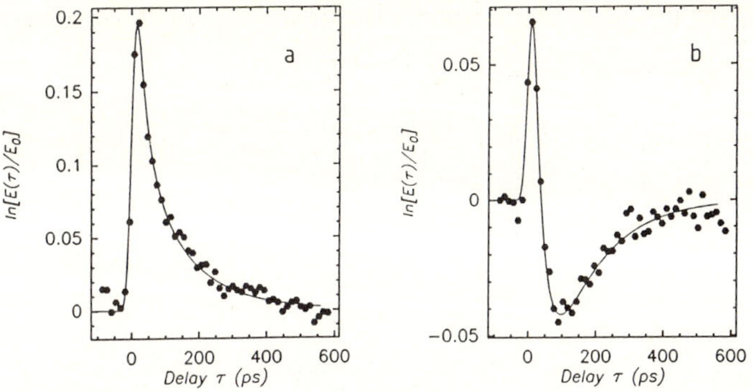

Fig. 1: Relative transmission of the probe for $CHCl_3$ dissolved in CCl_4 (ν_{max} = 3021 cm^{-1}) for two different central frequencies. a) 3024 cm^{-1} b) 3014 cm^{-1}

Table 1 : Exponential relaxation time constants of the IVR and the IET process.

solvent	CCl_4		CD_3COCD_3	
process	IVR	IET	IVR	IET
$CHCl_3$	30±2 ps	140±5 ps	6±1 ps	15±2 ps
$CHBr_3$	56±2 ps	260±10 ps	5±1 ps	19±2 ps
CHI_3	95±4 ps	310±10 ps	54±2 ps	

the transition shifts better into resonance as a result of this effect. In the second process, the other vibrations transfer the energy to the solvent molecules.

The time constants of both the internal vibrational relaxation (IVR) process and the intermolecular energy transfer (IET) process can be deduced by comparing the experimental results with numerical calculations in which the exponential relaxation time constants are used as fit parameters. The resulting time constants are presented in table 1. Both relaxation processes become slower as the energy gap between the C-H stretch vibration and the other vibrations in the molecule increases. For $CHBr_3$, the time constants of the relaxation processes have been measured previously with other techniques.[3]

With CD_3COCD_3 as solvent (fig. 2), we observe a significant acceleration of the two relaxation processes due to the strong interaction of the haloforms with the large dipole moment of the CD_3COCD_3 molecules. This interaction leads to large fluctuations of the vibrational energies so that the rate of relaxation is no longer determined by the energy gaps between the C-H stretch vibration and the other vibrations. For CHI_3, we observe that the IVR process is relatively slow because the small dipole moment of CHI_3 leads to a weak interaction with CD_3COCD_3 at the C-H side of the CHI_3 molecule. The IET process is relatively fast because the large polarizability of the iodine atoms leads to a strong interaction with CD_3COCD_3 at the CI_3 side of the molecule. In CHI_3, the shift of the absorption band cannot be observed and the time constant of the IET process cannot be determined because the molecules that have relaxed via the IVR process relax too fast via the IET process to have a significant influence on the transmission of the probe.

The transfer of the vibrational energy to the solvent molecules in the IET process leads to a rise in temperature. In CD_3COCD_3, this rise in temperature results in a decrease of the absorption coefficient at the low-frequency side of the absorption band of the C-H stretch vibration. This causes the increased transmission level in figure 2b. From the study of the temperature dependence of the absorption band it follows that the increased transmission level corresponds with a rise in temperature of 30±5 K.

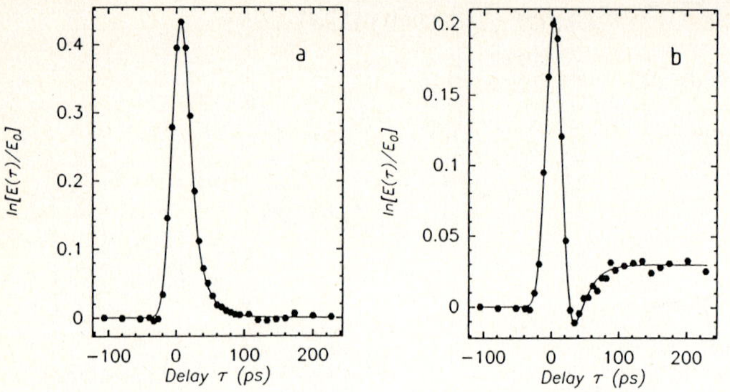

Fig. 2: Relative transmission of the probe for $CHBr_3$ dissolved in CD_3COCD_3 (ν_{max} = 3028 cm^{-1}) for two different central frequencies. a) 3033 cm^{-1} b) 3016 cm^{-1}

[1] J. Chesnoy and D. Ricard, Chem. Phys. Lett. **73**, 433 (1980)

[2] A. Seilmeier, K. Spanner, A. Laubereau and W. Kaiser, Optics Comm. **24**, 237 (1978)

[3] H. Graener and A. Laubereau, Chem. Phys. Lett. **102**, 100 (1983)
H. Graener, R. Dohlus and A. Laubereau, Chem. Phys. Lett. **140**, 306 (1987)
H. Graener, R. Dohlus and A. Laubereau in *Ultrafast Phenomena V*, 458 (1987)
H. Graener, Chem. Phys. Lett. **165**, 110 (1990)

Photoisomerization Dynamics of Methylstilbenes

N.S. Park and D.H. Waldeck

Department of Chemistry, University of Pittsburgh, Pittsburgh, PA 15260, USA

Abstract: Different conformers for meta-methylstilbenes cause nonexponential decay of the trans excited state. The decay kinetics of these molecules is used to explore the nature of the reaction coordinate.

1. Introduction

The dynamics of stilbene isomerization has been studied by a number of workers [1-3]. Comparisons of the observed friction dependence with theoretical models presumes a reaction coordinate which is well defined, independent of solvent conditions (temperature and viscosity). The appropriateness of this assumption is tested by studying meta substituted methylstilbenes in a single solvent (decane) as a function of temperature. These data show that more than one internal coordinate is important to the reaction dynamics.

Spectroscopic studies of methylstilbenes in supersonic jets are important to this work. The presence of a methyl group in the meta and para positions has little effect on the electronic character of the trans excited state. In particular Majors et al. [4] has found little or no change in the barrier to isomerization for 4-methylstilbene. Spangler et al. [5] find a small shift in the excitation origin (< 200 cm^{-1}) for the different meta conformers. The latter workers also find that the methyl rotor couples with low frequency modes present in the unsubstituted stilbene (i.e. ν_{37}).

In solution the nonexponentiality of the decay of 3,3'-dimethylstilbene (3,3'-DMS) has been reported and attributed to different conformers [2,6]. Herein kinetic data for 3-methylstilbene (3-MS) and 3,3',5-trimethylstilbene (3,3',5-TMS) is reported and a quantitative analysis of the competition between conformer interconversion and trans to cis isomerization is undertaken. Figure 1 shows the simple kinetic scheme used in this analysis. The interconversion rates are assumed the same here for simplicity, in actuality they may be different. This scheme can be solved and is found to give a double exponential decay.

2. Qualitative Results

The isomerization kinetics of different methylstilbenes have been studied and the characteristics of the decay curves at room temperature (298 +/- 2 K) are given in Table 1. This data shows that when all of the para ring positions and all of the meta ring positions are substituted the decay remains single exponential, as is the unsubstituted species. When the meta ring postions are not all substituted the decay becomes nonexponential. In the case of 3,3'-DMS the decay requires more than two exponentials for a good fit at temperatures below 273K [2], but is adequately fit by two exponentials at elevated temperatures. The decays of 3-MS and 3,3',5-TMS are adequately fit by two exponentials at all temperatures.

Springer Series in Chemical Physics, Vol. 53 **Ultrafast Phenomena VII**
Editors: C.B. Harris · E.P. Ippen · G.A. Mourou · A.H. Zewail

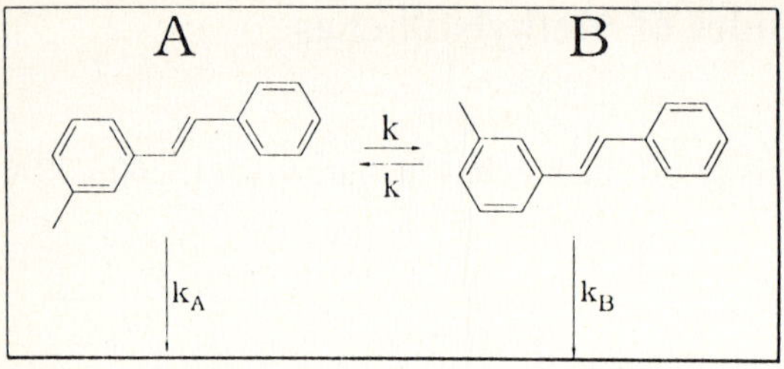

Figure 1. The kinetic scheme for the excited state decay of 3-MS is shown. The same scheme applies to 3,3',5-TMS and a slightly more complicated one to 3,3'-DMS [2].

TABLE 1: Decay characteristics of methylstilbenes in decane.

solute	number of exponentials	τ_1(ps)	A_1(%)	τ_2(ps)	A_2(%)
stilbene	1	100	100	-	-
4,4'DMS	1	309	100	-	-
3,3'-DMS	≥2	105	46	230	54
3-MS	2	125	93	220	7
3,3',5-TMS	2	200	50	300	50
3,3',5,5'-TMS	1	278	100	-	-

4,4'-dimethylstilbene (4,4'-DMS); 3,3'-dimethylstilbene (3,3'-DMS); 3-methylstilbene (3-MS); 3,3',5-trimethylstilbene (3,3',5-TMS); 3,3',5,5'-tetramethylstilbene (3,3',5,5'-TMS).

The single exponential decays reveal the importance of displaced solvent volume on the isomerization rate. The different displaced volumes for the different conformers of meta stilbenes causes the differing decay rates and nonexponential kinetics. The rates of stilbene and 3,3',5,5'-TMS can be used to set a bound on the individual decay rates of the different conformer decay rates shown in figure 1.

3. Quantitative Results

The isomerization kinetics of 3-MS has been studied in decane as a function of temperature. A plot of the percent of fast component versus temperature is shown in figure 2. If the two species have equal populations in the ground electronic state and the conformer interconversion is slow compared to trans to cis isomerization in the excited state, then the weight of the fast component should not exceed 50%. If the excited state interconversions were rapid compared to trans to cis isomerization then the decay should be single exponential and the percentage would reach 100%. A preliminary analysis of the fitting parameters in terms of the kinetic scheme shown in figure 1 yields the decay rate constant for each conformer

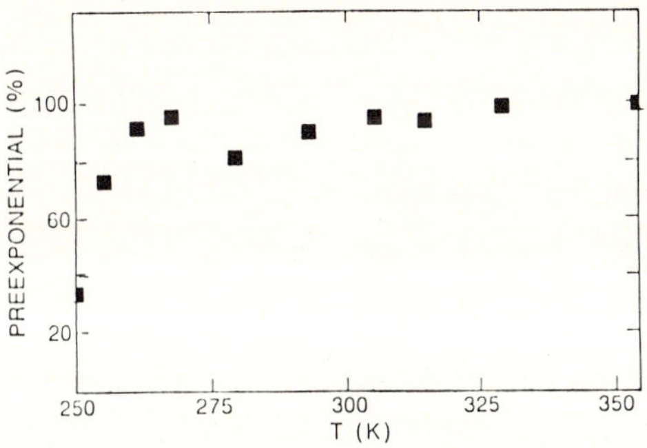

Figure 2. A plot of the percent of fast component versus temperature is shown for 3-MS.

and the interconversion. For 3-MS at room temperature an interconversion rate of 0.4 ps^{-1} is found compared to trans to cis isomerization rates of 5.0 ns^{-1} and 10. ns^{-1}. The importance of phenyl torsional motion in trans to cis isomerization is demonstrated by monitoring the kinetics of meta-methylstilbenes.

4. Acknowledgements

The support of the National Science Foundation (CHE-8613468) is appreciated.

5. References

1. N. S. Park and D. H. Waldeck, J. Chem. Phys. *91*, 943 (1989) and references therein.
2. N. S. Park and D. H. Waldeck, Chem. Phys. Lett., accepted.
3. N. Agmon and R. Kosloff, J. Phys. Chem. *86*, 5146 (1987).
4. T. J. Majors, U. Even and J. Jortner, J. Chem. Phys. *81*, 2330 (1984).
5. a) L. H. Spangler, R. van Zee and T. S. Zwier, J. Phys. Chem. *91*, 2782 (1987); b) L. H. Spangler, W. B. Bosman, R. D. van Zee and T. S. Zwier, J. Chem. Phys. *88*, 6768 (1988).
6. a) E. Fischer, J. Mol. Struc. *84*, 219 (1982); b) A. C. Matthews, R. Sakuvovs and K. P. Ghiggino, J. Photochem. *19*, 235 (1982).

Ultrafast Energy Trapping in Mercuric Iodide

*L.R. Williams, R.J. Anderson, and M.J. Banet**

Sandia National Laboratories, Livermore, CA 94551, USA
*Present address: Massachusetts Institute of Technology,
Rm 2-069, Cambridge, MA 02139, USA

Abstract. Subpicosecond time-resolved transient grating experiments performed in mercuric iodide reveal the formation of trapped carrier states with a 2 ps risetime and a 1.8 ns decay time. Time-resolved fluorescence measurements show similar rise and decay times. In addition, we demonstrate time-resolved observation of a coherently excited optic phonon mode in a semiconductor.

Mercuric iodide is a wide bandgap semiconductor which is of technological interest because it can be used as a room temperature x-ray, gamma-ray and optical detector. In order to understand the physical processes involved in the optical response of mercuric iodide detectors, we have conducted transient grating and time-resolved photoluminescence experiments. Time-resolved experiments allow us to explore the dynamics of optically excited species and possibly to deduce the role played by defect states in excited state relaxation.

The laser system consisted of a colliding-pulse-modelocked (CPM) ring dye laser amplified by a copper vapor laser resulting in 2 μjoule of energy in a pulse centered at 628 nm, with a pulse duration of 80 femtoseconds (FWHM), and at a repetition rate of 5.3 kHz. Experiments were performed in thin (500 μm to 1mm) single crystal samples of tetragonal mercuric iodide. The bandgap in mercuric iodide is 2.13 eV at room temperature, and shifts to 2.33 eV at 78 K. Mercuric iodide can be excited by two-photon absorption at the laser fundamental, or by one-photon absorption of the frequency-doubled light (314 nm).

The amplified laser output was split into three beams for the transient grating experiments. Two beams are crossed in time and space in the sample to create an optical interference pattern. The resultant index of refraction grating is probed by diffraction of the third, variably delayed beam. Figure 1 shows typical transient grating scans at 78 K in mercuric iodide. The diffraction efficiency is recorded as a function of the time delay between the probe and the two excitation pulses. After the zero delay coherent coupling spike, the signal consists of oscillations in intensity at a regular frequency superimposed on a rising background. A simulation of the data in figure 1 is shown with a dashed line and was generated by adding a sine function to a rising background. The frequency of the oscillations determined from the simulation is 3.4 THz which matches the 114 cm^{-1} line in the Raman spectrum of mercuric iodide.[1] The oscillations in signal intensity are due to the creation of a coherent optic phonon population through stimulated Raman

Springer Series in Chemical Physics, Vol. 53 **Ultrafast Phenomena VII**
Editors: C.B. Harris · E.P. Ippen · G.A. Mourou · A.H. Zewail

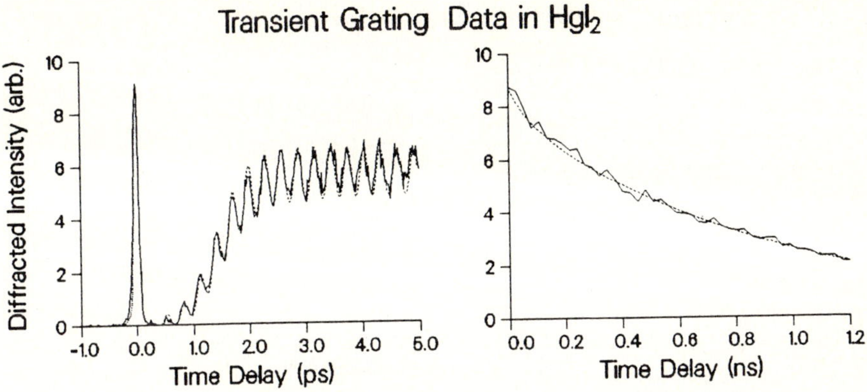

Figure 1. Transient grating data in a single crystal of mercuric iodide at 78 K. Left side: Oscillations in signal intensity due to coherently excited optic phonons are superimposed on a rising signal level due to trapping of carriers at defects. An autocorrelation of the pulse is shown with the dotted line and a simulation of the data is shown with the dashed line. Right side: Long-time decay of the transient grating signal (solid line) and biexponential simulation (dashed line).

scattering where the mixing frequencies are contained within the bandwidth of the ultrashort pulse.[2]

Underlying the oscillations in the transient grating signal shown in figure 1 is a background signal which rises over a few picoseconds to a constant level. Pump-probe transient absorption experiments show no corresponding change in the absorption coefficient at the CPM laser wavelength. The right hand side of figure 1 shows a transient grating scan taken out to much longer delay times. The dashed line shows a simulation of the long-time scan generated with a biexponential functional form. The two decay times are 130 ps and 1.8 ns.

At 78 K, photoluminescence of mercuric iodide can be excited by two photon absorption at the CPM fundamental and by one photon absorption of the doubled CPM. We have followed the time evolution of this photoluminescence using a Hamamatsu M1952 streak camera with a nominal time resolution of 2 ps. Figure 2 shows typical rise and fall times of 315 nm excited photoluminescence at 78K. Simulations of the data are shown with dashed lines, using a risetime of 3 ps and a biexponential decay with decay times of 125 ps and 0.9 ns. Decay curves for the photoluminescence generated by two-photon absorption at 628 nm are similar.

The similarity of the general form of the transient grating background and the time-resolved photoluminescence is striking. In both cases a few ps rise from essentially zero signal level is followed by a biexponential decay with decay times about 100 ps and 1 ns. The background in the transient grating data is most likely due to two-photon absorption at the maxima of the optical interference pattern. 4 eV excitation leads to the creation of translationally hot carriers in the conduction band which cool rapidly and are trapped at defects or impurity sites. Low temperature, cw excited photoluminescence spectra have shown a number of bands red-shifted from the

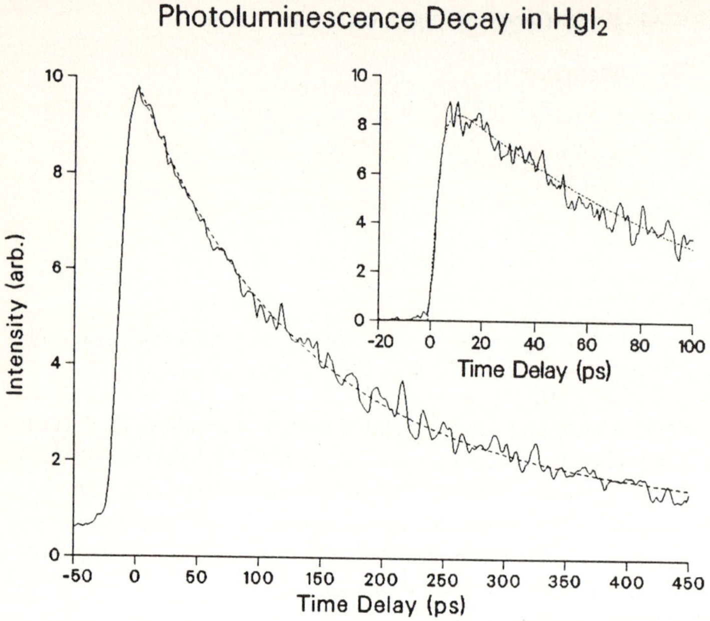

Figure 2. Fluorescence decay curves in mercuric iodide at 78 K. Data curves are shown with a solid line and simulations are shown with dashed lines.

band-gap which have been attributed to defects and impurities[3], and trapping times on the order of 100 fs to a few ps have been inferred from thermally stimulated current measurements[4]. Thus, the few picosecond risetime we observe is consistent with the timescale for electron cooling and trap formation. Correlating the decay times of the transient grating signal and the photoluminescence signal must be done with care. The transient grating signal reflects a change in the index of refraction which could be due to the luminescent species (trapped carriers) or could be due to species which decay through nonradiative processes.

Acknowledgments: This work supported by U.S.D.O.E. Office of Basic Energy Sciences, Division of Materials Sciences, under contract DEAC04-76DP00789.

References

1. A. J. Melveger, R. K. Khanna, B. R. Guscott and E. R. Lippincott, Inorg. Chem. 7, 1630-4 (1968).

2. S. De Silvestri, J. G. Fujimoto, E. P. Ippen, E. B. Gamble, Jr., L. R. Williams, and K. A. Nelson, Chem. Phys. Lett. 116, 146 (1985). Y.-X. Yan, E. B. Gamble, Jr. and K. A. Nelson, J. Chem. Phys. 83, 5391-9 (1985).

3. R. B. James, X. J. Bao, T. E. Schlesinger, J. M Markakis, A. Y. Cheng and C. Ortale, J. Appl. Phys. 66, 2578 (1989).

4. R. H. Bube, Phys. Rev. 106, 703-17 (1957).

Non-Markovian Dynamics of Azulene in Solution

E.T.J. Nibbering, K. Duppen, and D.A. Wiersma

Ultrafast Laser and Spectroscopy Laboratory, University of Groningen, Nijenborgh 16, 9747 AG Groningen, The Netherlands

The aromatic hydrocarbon molecule azulene has been studied intensively in various time–domain experiments since the early days of ultrafast spectroscopy [1]. So far, the interest has focussed mainly upon the unusual population decay characteristics of this molecule. While fluorescence is easily observed from the second excited singlet state, the emission from the first excited state is quenched by very rapid internal conversion to the electronic ground state. The lifetime of this S_1 state has been determined in a solid matrix by temperature dependent lineshape analysis [2], in solution at an excess vibronic energy of 1900 cm^{-1} by two–step photo–induced fluorescence [3], and in a planar supersonic jet by absorption lineshape analysis [4].

We have found that the molecule is also perfectly suited to study the dynamics of electronic dephasing in solution [5]. The vibrational structure of both singlet transitions (S_1 and S_2) is clearly resolved in hydrocarbon solvents such as isopentane and cyclohexane. This has allowed us to study the dephasing by lineshape analysis and resonance light scattering. To support the interpretation of these results in terms of a non–Markovian model, the lifetime of the pure electronic S_1 state was measured for the first time with subpicosecond resolution.

The absorption line shape of a molecule is usually described in terms of the interaction between a "system", i.e. the ground and electronically excited state(s), and a bath that comprises the degrees of freedom of the solvent and possibly vibrational modes of the solute. In liquids, the dynamics of the bath may occur on the same time scale as the dynamics of the system. In that case the assumption of exponential phase relaxation is unwarranted, and a nonMarkovian approach should be taken. Based on a stochastic approach, Mukamel derived master equations for the density matrix of the system without invoking the Markovian assumption of an infinitely fast correlation time [6]. When the system frequency is assumed to undergo a Gaussian modulation with an exponential bath correlation function, the final result for the absorption lineshape is

$$I(\omega) \propto 2\mathrm{Re} \int_0^\infty d\tau \, \exp[i(\omega-\omega_{21})\tau]\exp[-(\Delta^2/\Lambda^2)(e^{-\Lambda\tau}-1+\Lambda\tau)] \; . \qquad (1)$$

Springer Series in Chemical Physics, Vol. 53 **Ultrafast Phenomena VII**
Editors: C.B. Harris · E.P. Ippen · G.A. Mourou · A.H. Zewail

When the line profiles of the electronic transitions of azulene were fit to this expression, correlation times Λ^{-1} of 26 fs and 11 fs were obtained for the $S_1 \leftarrow S_0$ and the $S_2 \leftarrow S_0$, respectively, with amplitudes Δ of 36 THz and 83 THz. This means that the solution dynamics falls in the intermediate modulation regime (Λ and Δ comparable), where the Markovian approximation is indeed not applicable. In the fast modulation limit ($\Delta << \Lambda$), the lineshape would reduce to a Lorentzian with a pure dephasing time constant $T_2^* = \Lambda\Delta^{-2}$.

Information on stochastic fluctuations can also be obtained by resonance light scattering experiments. With the same master equation as above, the ratio of the integrated fluorescence over Raman intensities can be given as a continued fraction [7]:

$$R = -2\,\mathrm{Im}\,\cfrac{\Delta^2 T_1}{2\pi\delta + i(\Lambda + 1/(2T_1)) - \cfrac{2\Delta^2}{2\pi\delta + i(2\Lambda + 1/(2T_1)) - \cfrac{3\Delta^2}{2\pi\delta + i(3\Lambda + 1/(2T_1))\ldots}}} \qquad (2)$$

For fast modulation and small detuning ($\Delta, \delta << \Lambda$) the Markovian limit would be obtained, where the ratio between the fluorescence and Raman yields is given by $R = 2T_1/T_2^*$.

The resonance light scattering spectrum of the $S_1 \leftarrow S_0$ transition of azulene in cyclohexane is shown in fig.(1), which is corrected for instrumental response and Raman signals of the solvent. The sharp features constitute the Raman spectrum and the broad features the dephasing-induced fluorescence spectrum of azulene. With Δ and Λ in principle known from the absorption line profiles, the lifetime of the S_1 state remains as the only unknown parameter in eq. (2). When the ratio of the integrated fluorescence over Raman yield as a function of detuning is fitted to eq.(2), a value of $T_1 = 1.4 \pm 0.2$ ps is obtained.

The lifetime T_1 of eq.(2) was measured with short amplified continuum pulses at the $S_1 \leftarrow S_0$ transition (700 nm), generated with a CPM-laser. A first pulse excites azulene molecules in the S_1 origin. A second, delayed, probe pulse then further excites these molecules into the second singlet S_2. The fluorescence from this state is detected. As shown in fig.(2) a symmetric trace results that directly gives the lifetime as $T_1 = 1.6 \pm 0.2$ ps. The same experiment with pulses at 625 nm gave a value slightly different from that of ref. [3]: 1.2 ± 0.2 ps.

The results of the resonance light scattering data and the lineshape analysis can now be combined. In fig.(3) the experimental ratio of the integrated fluorescence over the Raman intensity is given as a function of detuning. The solid line is calculated from eq. (2) with the stochastic parameters Δ and Λ from the absorption line profile eq.(1), and T_1 determined as discussed above. *There is no fitting parameter*

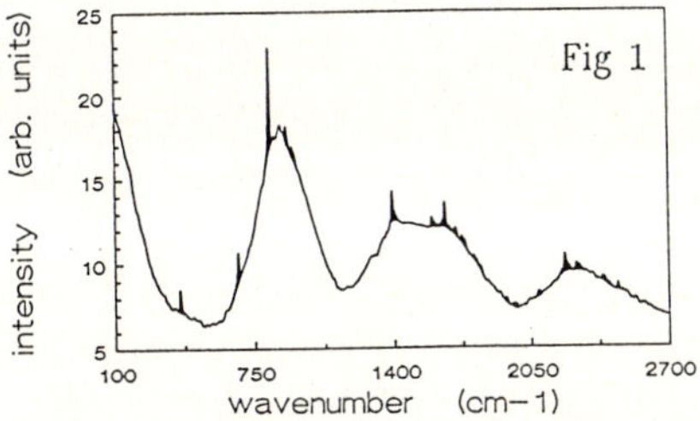

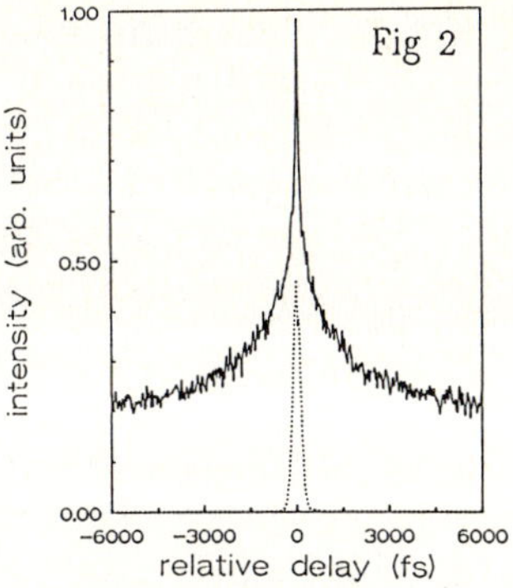

Fig 1: Resonance light scattering spectrum for the $S_1 \leftarrow S_0$ electronic transition of azulene in cyclohexane. Raman and fluorescence are clearly distinct.

Fig 2: Two–step photo–induced fluorescence from the S_2 state of azulene in cyclohexane. The decay on both sides gives $T_1(S_1)$= 1.6 ps. The dotted trace indicates the time resolution.

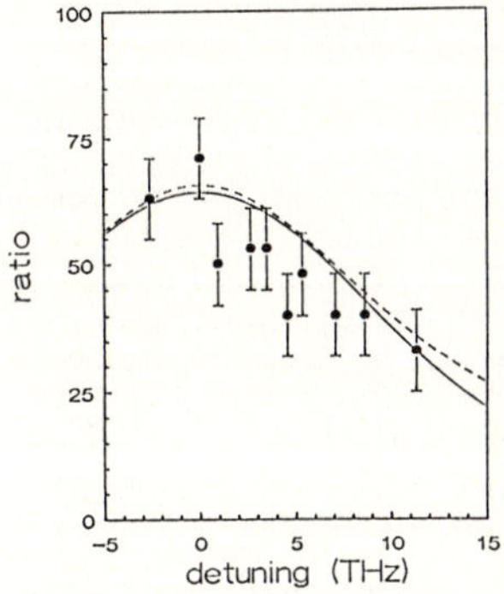

Fig 3:

Ratio R of the integrated fluorescence to Raman yield as a function of detuning of the excitation laser from $S_1 \leftarrow S_0$. The predicted functional dependence is given by the solid line.

involved in this picture. The broken line is based on a different model for the stochastic fluctuations.

We conclude that a non–Markovian master equation is needed to describe the optical dynamics of azulene in solution. Lineshape analysis and resonance light scattering give consistent results: the correlation time of the S_1 state is about 26 fs, and the stochastic fluctuations occur in the intermediate modulation regime where Δ and Λ are comparable.

References

1. R.M. Rentzepis, Chem. Phys. Lett. **2**, 117 (1968)
2. R.M. Hochstrasser and T.Y. Li, J. Mol. Spectrosc. **41**, 297 (1972)
3. E.P.Ippen, C.V.Shank and R.L.Woerner, Chem. Phys. Lett. **46**, 20 (1977)
4. A. Amirav and J. Jortner, J. Chem. Phys. **81**, 4200 (1984)
5. E.T.J. Nibbering, K. Duppen and D.A. Wiersma, J. Chem. Phys.
6. S. Mukamel, Chem. Phys. **37**, 33 (1979)
7. Y.J. Yan and S. Mukamel, J. Chem. Phys. **86**, 6085 (1987)

Resonance Raman Scattering as a Probe of Ultrafast Electronic Surface Hopping and Nuclear Vibrational Motion

J.Z. Zhang [1], *E.J. Heller* [2], *and D.G. Imre* [2]

[1] Department of Chemistry, University of California at Berkeley, Berkeley, CA 94720, USA

[2] Department of Chemistry, University of Washington, Seattle, WA 98195, USA

We present a study of the dynamics and spectroscopy of a model system with two identical, dissociative chromophores using a quantum-mechanical time-dependent wave packet theory. We demonstrate that resonance Raman spectra carry distinct signatures of both the dissociation dynamics in the femtosecond regimes and the non-adiabatic electronic coupling between the two chromophores.

1. Introduction

Study of ultrafast molecular reaction dynamics have been of great importance in chemistry. With the development of ultrafast laser techniques, molecular dynamics in the picosecond and femtosecond regimes have been studied in the time domain directly in recent years. Typically a pump-probe scheme is employed [1]. An alternative approach to probe the short time dynamics is to use spectroscopic methods with CW or long pulse lasers; dynamical information is then extracted from the observed spectrum by resorting to some theoretical methods. Resonance Raman spectroscopy is a good example. Especially when interpreted in a time dependent picture using a wave packet theory, it has been demonstrated to be a powerful tool in probing photodissociation dynamics of molecules [2,3].

2. Model and Method

An interesting system in molecular dynamics and spectroscopy is provided by molecules with more than one chromophores [4-6]. They can be treated as having multiple interacting diabatic electronic states, one for each chromophore. In a two chromophore system, there are two degenerate, coupled, diabatic electronic states. We focus on the fast events (several femtosecond) and observe the consequence of the exchange of electronic excitation between the two chromophores on the early time dynamics. We have chosen the model, represented as A-B-A, such that each chromophore A-B is dissociative if left electronically uncoupled to the other. The calculation is based on propagating the wave packet (dynamics) by solving the time-dependent Schrödinger equation numerically and then the spectrum is calculated directly from the dynamics according to a time-dependent formalism [2]. The detailed procedure is given in Ref.6.

3. Results and Discussion

We carried out the study from very weak to strong couplings and found that resonance Raman spectrum is sensitive to both the coupling strengths and the dissociation dynamics. The dynamics and spectra, on the other hand, depend on the relative orientations of the two chromophores. For parallel dipoles, for example, we have found that one of the adiabatic surfaces (lower,dissociative)

Springer Series in Chemical Physics, Vol. 53 **Ultrafast Phenomena VII**
Editors: C.B. Harris · E.P. Ippen · G.A. Mourou · A.H. Zewail

Fig.1

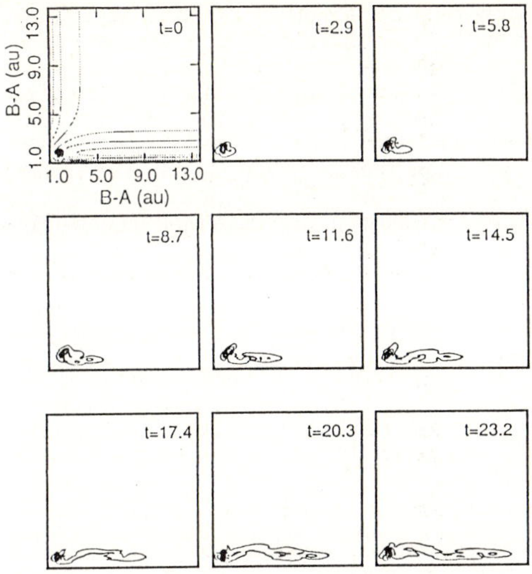

Fig.2

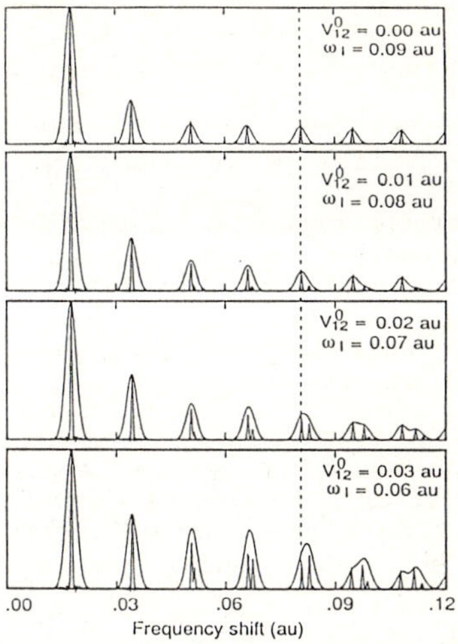

Fig.1. Evolving wave packet on one of the diabatic surfaces for coupling equal to 0.020au and parallel dipoles. The gray contour lines in the first frame represent one of the diabatic surfaces. The dynamics on the other surface is exactly symmetric to what happens on this surface. Time is in fs.

Fig.2. Resonance Raman spectra for different couplings V^0_{12} with different excitation frequencies ω_I as indicated in the figure. The frequencies are shifted in such a way that we always probe the lower adiabatic state.

becomes electronically forbidden but vibronically allowed [6]. Figure 1 shows the dynamics on one of the diabatic surfaces for parallel dipoles with coupling equal to 0.02au. The effect of the non-adiabatic coupling on the dynamics has been such that the stronger the coupling is, the more slowly the dissociation happens, the faster the electronic excitation hops, and the further the wave packet moves along the symmetric stretch. For the example shown in Fig.1, it appears that the wave packet is moving more likely on a bound surface rather than on a dissociative surface. In this case, the bound state is the upper adiabatic surface. The part of the wave function which appears to be predissociated is due to the lower dissociative adiabatic surface. Therefore, strong coupling corresponds to the adiabatic limit while weak coupling corresponds to the diabatic regime. The electronic hopping happens on the time scale of about 7-8 fs in this case.

The Raman spectra for different coupling strengths are shown in Fig.2, both a high and a low resolution spectra are shown in each case. The effect of the coupling on the spectra is obvious. The zero-coupling spectra show a single progression in both high and low resolutions. In this case, we know that the motion of the wave packet on the excited state is along the local mode A-B, and we expect the emission spectrum to show a pure local mode progression since the ground state surface is known to be of local mode character. Thus the motion on the excited state lights up the pure A-B progression in the ground state. When we use this simple spectrum to obtain a dissociation limit using anharmonicities we find dissociation limit for an A-B bond. When the coupling V_{12}^0

increases and motion along the symmetric stretch begins, the high resolution spectrum becomes more complex. This is because it takes a superposition of a few local mode eigenstates to create some symmetric stretch motion. Aside from an increase in the intensities and number of combination modes we observe changes in the low resolution spectra. The spacing of the low resolution peaks increases with coupling. When we use these peaks to obtain a dissociation limit for the ground state we find almost twice the energy required to break an A-B bond. Thus, the motion along the symmetric stretch on the excited state is expressed in the low resolution Raman spectra through smaller anharmonicity and thus larger dissociation limit.

As a test of the model developed here, we have calculated the Raman spectra of CH_2I_2 and H_2O and found the coupling strengths to be 0.024au and 0.020au respectively by making direct comparisons with available experimental results [7,8]. These strong couplings suggest that both molecules are in the adiabatic limit. We found that CH_2I_2 dissociates in about 75 fs while it takes about 25 fs for H_2O to dissociate completely.

In conclusion, we have demonstrated that resonance Raman scattering is sensitive to the non-adiabatic electronic coupling for degenerate two chromophore systems; and therefore it can be used as a powerful tool to probe the ultrafast electronic surface hopping and the dissociation dynamics.

Acknowledgement

This work was supported by the National Science Foundation and PRF.

References

1. T. Rose, M.J. Rosker, and A.H. Zewail, J. Chem. Phys., 88, 6672(1988).
2. E.J. Heller, Acc. Chem. Res., 14, 368(1981).
3. J.Z. Zhang and D.G. Imre, J. Chem. Phys. 89, 309(1988).
4. R. Heather and H. Metiu, J. Chem. Phys., 90, 6903(1989).
5. S. Das and D. Tannor, J. Chem. Phys., 91, 2324(1989).
6. J.Z. Zhang, D. Huber, E.J. Heller and D.G. Imre, J. Chem. Phys., submitted.
7. R.J. Sension, R.J. Brudzynski, and B.S. Hudson, 61, 694(1988).
8. J.Z. Zhang, D. Huber, E.J. Heller and D.G. Imre, J. Chem. Phys., 89, 3602(1988).

Nickel(II) Octaethylporphyrin: Conformational Dynamics, Ligand Binding and Release

S.H. Courtney[1,*], *T.M. Jedju*[1], *J.M. Friedman*[1,*], *L. Rothberg*[1], *R.G. Alden*[2], *M.S. Park*[2], *and M.R. Ondrias*[2]

[1]AT&T Bell Laboratories, 600 Mountain Ave., Murray Hill, NJ 07974, USA
[2]Department of Chemistry, University of New Mexico, Albuquerque, NM 87131, USA
* Present address: Department of Chemistry, New York University, New York, NY 10003, USA

We report time resolved Raman and absorption experiments to characterize excited state vibrational frequencies and ligand binding in Nickel(II) porphyrins.

Metalloporphyrins are important in a variety of biological functions, such as oxygen transport, electron transfer, and energy conversion. To understand their ability to bind ligands and participate in electron transfer reactions, the excited state dynamics of these complex molecules must be elucidated. In non-coordinating solvents, photoexcitation of NiOEP leads to the formation of a relatively long lived excited state ($^*B_{1g}$) that decays to the $^1A_{1g}$ ground state in ≈300ps[1,2,3]. To further investigate the excited state dynamics, we undertook a subpicosecond transient absorption study in chloroform and methylene chloride. We observed the appearance of transient absorption bands in <2 ps and spectral evolution of the transient bands (see figure 1).

NiOEP is known to crystallize in three distinct conformations; two triclinic planar crystal forms and a tetragonal form. The tetragonal form exhibits a contracted metal core and a highly distorted non-planar macrocycle relative to the planar forms. In solution, the band width of ν_{10}, an in plane skeletal mode, is unusually large and studies using resonance Raman spectroscopy have shown that multiple conformers in solution are responsible for this effect[4,5]. Higher temperatures appear to favor a ruffled conformation and the skeletal modes above 1400 cm^{-1} are at higher frequency in the planar form. The core of the macrocycle expands upon photo-excitation due to population of the $d_{x^2-y^2}$ orbital in the $^*B_{1g}$ state and calculations suggest this should favor a planar conformation. The spectral evolution could then be accounted for by the conversion of ruffled conformers to the favored planer form. To date we have obtained Raman spectra of the 'bottleneck' excited state on a 30 ps time scale (see figure 2). The ground state (center spectrum) shows the band profile of ν_{10} peaking at 1653cm^{-1} and the upper spectrum in figure 2 shows the formation of the transient d-d excited state when the sample is irradiated with a 532 nm pump pulse spatially and temporally overlapped with the 'probe' pulse. Figure 2 also shows the excited state Raman spectrum as constructed by subtraction of the ground state spectrum (ie probe only). Unfortunately, it will be quite complex to analyze band

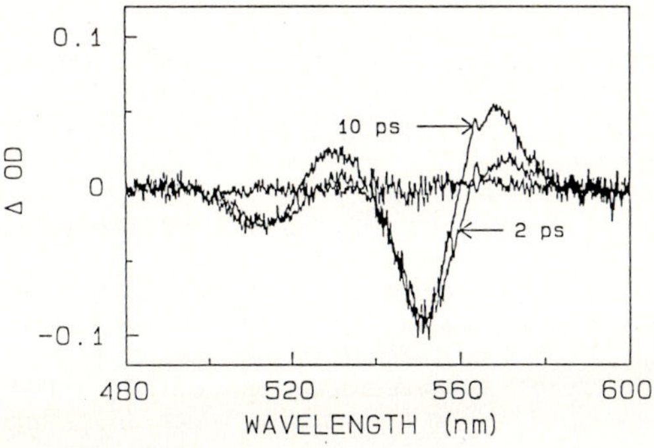

Figure 1) Transient absorption spectra NiOEP in methylene chloride as a function of time (-19, +2, and +10 ps). There is a slight distortion at the laser excitation wavelength (563 nm) and the spectra was assembled from two overlapping intervals. Notice the lack of an isosbestic point at approximately 560 nm.

Editors: C.B. Harris · E.P. Ippen · G.A. Mourou · A.H. Zewail

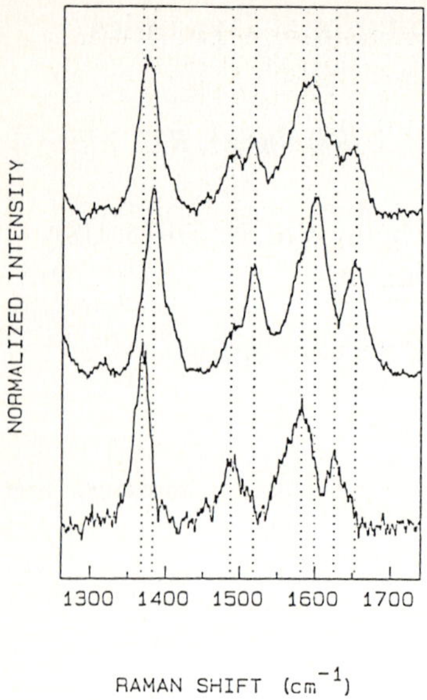

Figure 2) Resonance Raman spectra of NiOEP in chloroform. The center spectrum is that of the ground state obtained using only the 436 nm pulse and the upper spectrum was obtained with a time delay between the pump (532 nm) and the probe (436 nm) pulse of 30 ps. The lower trace is the excited state spectrum obtained by subtracting the residual ground state fraction (0.65 in this case) from the upper spectrum. The fluorescent background was removed before subtracting and there is a broad solvent band at approximately 1490 cm^{-1} (the shoulder in the center spectrum). Note how ν_{10} down shifts to 1625cm^{-1}.

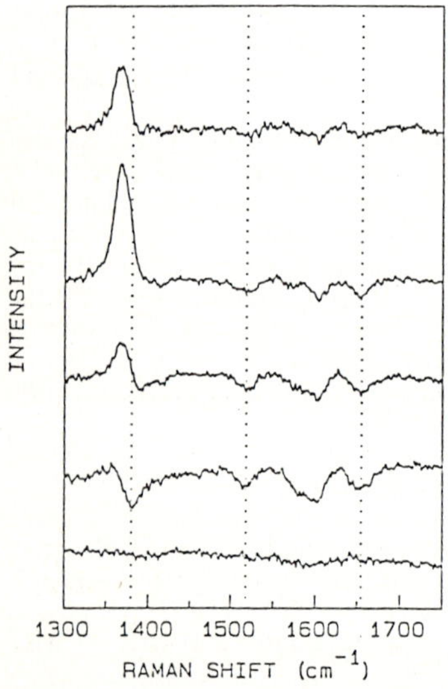

Figure 3) Time dependent resonance Raman difference spectra of NiOEP in pyridine. The appearance of the large Raman band near 1370 cm^{-1} is due to solvent binding in an excited electronic state. The species is probably six coordinate and the spectra were taken at at -200 ps, +33 ps, +200 ps, +1 ns, and +5 ns time delays (from bottom to top). Negative going peaks are due to depletions of the ground state. The 'probe' wavelength was 436 nm to resonantly enhance the red shifted bound species. Depletion of the ground state species and new Raman bands are present and at 'negative time' we obtain featureless difference spectra.

profiles/positions on faster time scales where multiple excited states, vibrational heating, and conformational changes may all play a role.

In coordinating solvents, we confirm that solvent ligand binding and release occurs in the excited states of these molecules. For example, NiOEP in pyridine exists primarily as a 4-coordinate ground state species (A_{1g}) and binds a solvent molecule(s) when photoexcited. The appearance time of the bound species in figure 3 is approximately 400 ps.

Re-establishing the original ground state equilibrium occurs on a 10 ns time scale. This process is expected to be temperature dependent and it should be possible to determine the activation energy. Similar studies of excited state ligand release are also possible by using Raman 'probe' wavelengths to resonantly enhance the 4-coordinate species. The system of choice for this experiment is probably NiPPDME in piperidine in which the ground

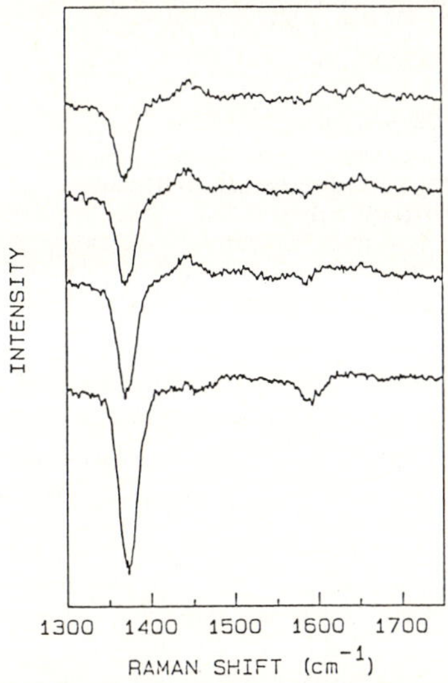

Figure 4) Time resolved resonance Raman difference spectra of NiOEP in piperidine. These spectra were produced by direct subtraction of a 'negative time' spectrum from spectra obtained at other delays (33 ps, 300 ps, 600 ps, and 1 ns from bottom to top). In this case, a mixture of ground state 4-coordinate and bound species has been excited by the 532 nm pump pulse. Careful studies of the relative Raman cross sections are required to further study such cases.

state equilibrium favors the six coordinate species. This case may exhibit other interesting photochemistry such as ligand release from upper electronic states competing with fast electronic relaxation to lower bound states (followed by ligand release).

References

1) D.H. Kim, C. Kirmaier and D. Holten, Chem. Phys. **75,** 305 (1983).

2) S.H. Courtney, T.M. Jedju, J.M. Friedman, R.G. Alden and M.R. Ondrias, Chem. Phys. Lett., **164,** 39 (1989).

3) A.Yu. Chikishev, V.F. Kamalov, N.I. Koroteev, V.V. Kvach, A.P. Shkurinov, and B.N. Toleutaev, Chem. Phys. Lett., **144,** 90 (1988).

4) R.G. Alden, B.A. Crawford, R. Doolen, M.R. Ondrias, and J.A. Shelnutt, J. Am. Chem. Soc., **111,** 2070 (1989).

5) R.S. Czernuszewicz, Xiao-Yuan Li, T.G. Spiro, J. Am. Chem. Soc., **111,** 7024 (1989).

Application of Ultrafast Broadband Infrared Spectroscopy to Measurement of Metal-Carbonyl Dynamics

S.A. Angel, P.A. Hansen, E.J. Heilweil, and J.C. Stephenson

Molecular Physics Division, Room B268, Building 221, National Institute of Standards and Technology, Gaithersburg, MD 20899, USA

Abstract. Population lifetimes and energy transfer dynamics obtained from transient multichannel infrared spectra of CO and NO(v=1) vibrations for gas and liquid phase metal-carbonyl systems are presented.

1. Introduction

Recent developments in ultrafast time-resolved infrared (IR) spectroscopy enable one to directly observe transient structures, mode-specific energy content and chemical reaction products generated in condensed-phase molecular systems on the picosecond or shorter timescale [1]. This paper reports our first broadband IR transient spectral measurements for vibrational energy decay (T_1) and population transfer (VET) of NO and CO-stretching vibrations in metal-carbonyl molecules and cluster compounds.

2. Experimental Technique

The method employed in this work uses picosecond broadband IR probe pulses to extract the maximum spectral information from a single laser shot [2]. By using multichannel detection rather than tuning a probe laser across infrared-active transient spectral features, time-resolved spectral information is rapidly gathered with a minimum of external control. This technique is similar to nanosecond TRISP developed by P. Sorokin and coworkers [3] where metal heat pipe frequency converters were used.

For these vibrational population and energy transfer studies, a tunable narrow-band visible dye laser is downconverted to produce tunable IR pumping pulses (ca. 5 μJ, 18 ps, 5 cm^{-1} FWHM at 5 μm). Efficient downconversion of a second synch-pumped picosecond broadband dye laser pulse (no tuning elements, 18 ps FWHM, 150-350 cm^{-1} FWHM at 10 Hz) to the IR (5 μm or 2000 cm^{-1} region, IR probe pulse) and upconversion back to the visible (ca. 480 nm) is performed by frequency mixing with YAG second harmonic (532 nm) in $LiIO_3$. Detection is accomplished with a spectrograph and OMA detector covering ca. 120 cm^{-1} spectral regions with 3 cm^{-1} FWHM resolution. Averaging 1800 to 2400 laser shots (6-8 minutes) without normalization is sufficient to obtain a single time-delay spectrum with 2% sensitivity.

3. Results and Discussion

Transient spectra obtained for $Rh(CO)_2$(acac) in n-hexane solution measured the decay of the CO-stretching symmetric $\nu_1(0\rightarrow1)$ at 2086 cm^{-1} and asymmetric $\nu_2(0\rightarrow1)$ at 2014 cm^{-1} bleaching features and their v=1→2 overtone and combination absorptions [4]. Following ν_1 excitation (Fig 1), a bleach for ν_2 at 2014 cm^{-1} and absorptions for ν_2 (1→2) at 2002 cm^{-1} and $\nu_1=1\rightarrow\nu_1=1+\nu_2=1$ at 1987 cm^{-1} are observed. The 2014 cm^{-1} bleach is due to ground state depletion. The 2002 cm^{-1} absorption is caused by population transfer from $\nu_1=1$ to $\nu_2=1$ which occurs in ≤10 ps. The coupled ν_1 and ν_2 modes relax together with a decay time of $T_1=64 \pm 10$ ps.

Intramolecular VET among the ground electronic (S_o) state NO and CO-stretch modes of gas phase $Co(CO)_3NO$ was also studied. Under collisionless

Springer Series in Chemical Physics, Vol. 53 **Ultrafast Phenomena VII**
Editors: C.B. Harris · E.P. Ippen · G.A. Mourou · A.H. Zewail

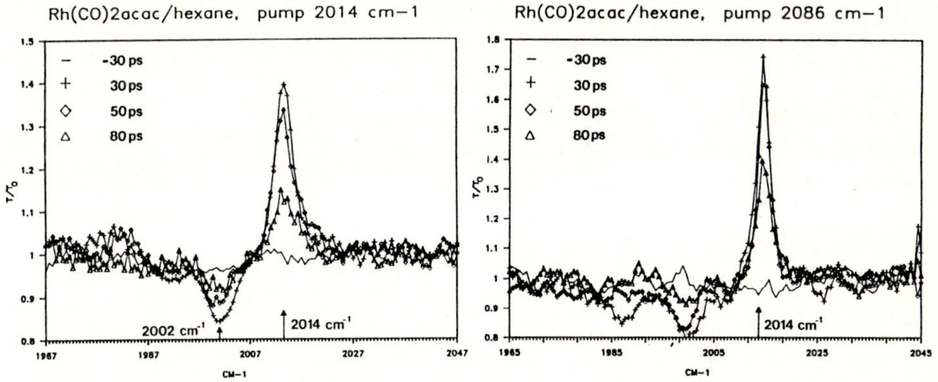

Fig. 1 - Broadband IR spectra for $Rh(CO)_2acac$ in hexane at 298 K showing the transient $\nu_2(0\rightarrow1)$ bleach $(T/T_o>1)$ at 2014 cm^{-1} and absorptions $(T/T_o<1)$ for $\nu_2(1\rightarrow2)$ at 2002 cm^{-1} and $(\nu_1=1)\rightarrow(\nu_1=1+\nu_2=1)$ at 1987 cm^{-1}.

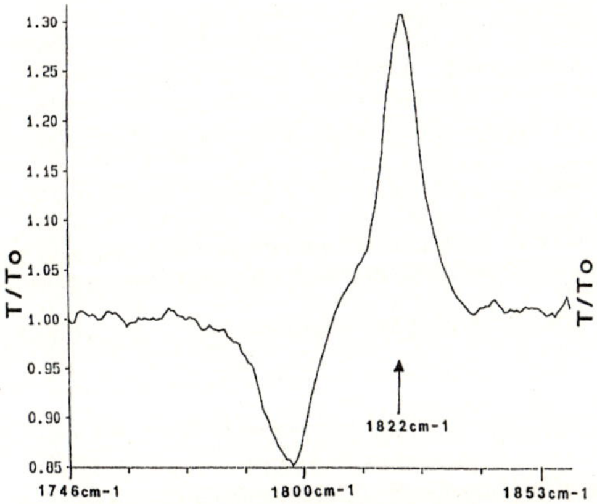

Fig. 2 Transient ν_2 bleach at 1822 cm^{-1} and $\nu_2(1\rightarrow2)$ absorption at 1796 cm^{-1} after 1822 cm^{-1} excitation of $Co(CO)_3NO$ gas.

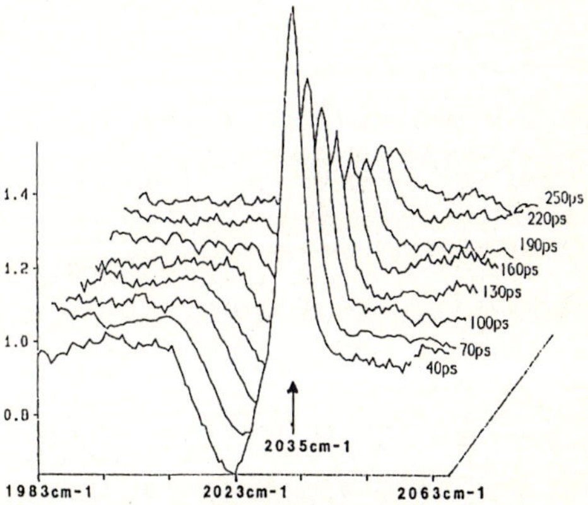

Fig. 3 Time-resolved IR spectra for ν_8 $Co(CO)_3NO$/hexane at 298 K

room temperature (20 Torr) conditions, time resolved measurements of the bleaches and v=1→2 absorptions for the directly excited asymmetric ν_8 CO-stretch (2048 cm^{-1}) and the ν_2 NO-stretch (1822 cm^{-1}) (see Fig. 2) reveal $T_1(\nu_8)=870\pm105$ ps and $T_1(\nu_2)=415\pm51$ ps, respectively. When pumping ν_8, an absorption at 2101 cm^{-1} corresponding to the $\nu_1(1\rightarrow2)$ symmetric CO-stretch appears within the pulse duration and decays with a rate comparable to that of ν_8, indicating rapid equilibration between these two modes. Conversely, transients were not observed at early delay times in the ν_2 region when exciting the ν_8 CO-stretch. However, **delayed** appearance (> 100 ps) of a small ν_2 bleaching feature suggests weak CO-NO coupling and population transfer in several hundred ps from the CO to NO-stretch without collisions.

Time dependent spectra of $Co(CO)_3NO$ in room temperature hexane solution gave $T_1(\nu_8)$ = 148±26 ps (Fig. 3) and $T_1(\nu_2)$ = 37±9 ps. The T_1 relaxation of the NO(v=1) stretch is faster than that observed for any related CO-stretch for metal-carbonyls in hexane [4].

Studies of the high frequency terminal (2055, 2064 cm^{-1}) and bridging (1866 cm^{-1}) CO-stretching modes of the $Co_4(CO)_{12}$ cluster compound in CCl_4 produced similar results. Both terminal modes were bleached during the pump pulse (at 2066 cm^{-1}) and a strong red-shifted absorption in the region

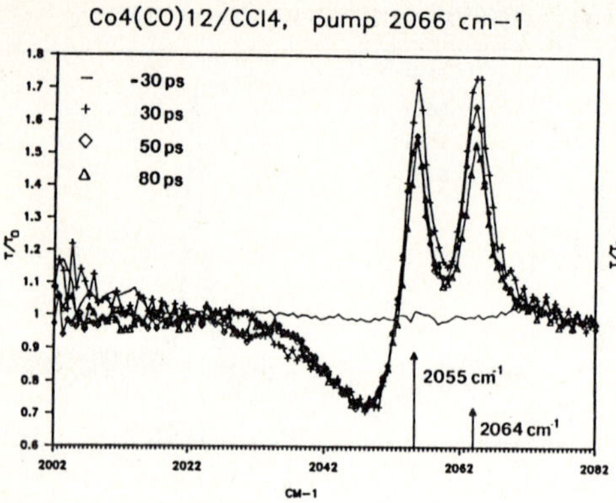

Fig. 4 Terminal CO-stretch region of $Co_4(CO)_{12}/CCl_4$, $7x10^{-4}$ M at 298K

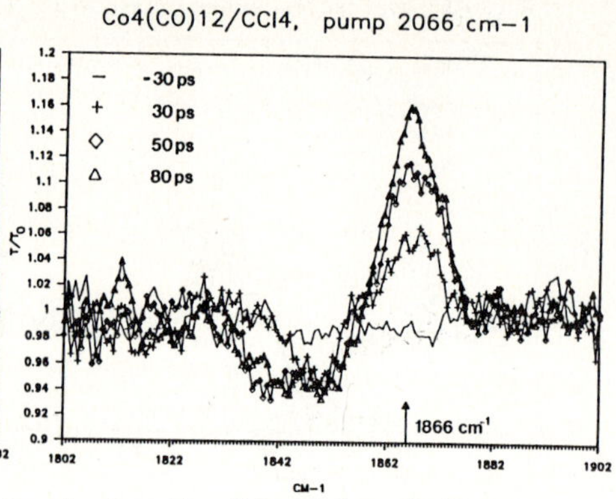

Fig. 5 Growth of bridging CO features after excitation of terminal CO-stretch

expected for the v=1→2 transition of the 2055 cm^{-1} mode persists, with its initial magnitude, for at least 90 ps (Fig. 4) then decays together with the bleaches. Kinetic runs show an initially rapid (47±5 ps) decay component of both bleaching features [4] followed by a longer (312±30 ps) component. Probing the CO bridge-bond region near 1866 cm^{-1} (Fig. 5) with 2066 cm^{-1} excitation produces no observable features near t=0, but the appearance of a bleach and absorption with a 50 ps risetime is found. Our interpretation of these observations is that the terminal CO-stretches rapidly equilibrate (few ps) followed by population transfer to the bridge-bonded CO-stretch (ca. 50 ps). The coupled terminal and bridging CO modes decay slowly with a time constant of several hundred ps.

4. **Summary** We have demonstrated a novel broadband IR probing method which allows measurement of vibrational population lifetimes and dynamics. Extensions of this approach to the femtosecond regime and using dual track normalization for studies of metal-carbonyl photochemistry are planned.

5. References

[1] P. Anfinrud, C. Han, P. A. Hansen, J. N. Moore, R. M. Hochstrasser, in "Ultrafast Phenomena VI," Springer-Verlag, Berlin, Vol. 48, p. 442, 1988.
[2] E. J. Heilweil, Opt. Lett., 14, 551 (1989).
[3] J. H. Glownia, J. Misewich, P. P. Sorokin, Opt. Lett., 12, 19 (1987)
[4] E. J. Heilweil, R. R. Cavanagh and J. C. Stephenson, Chem. Phys. Lett., 134, 181 (1987); J. Chem. Phys., 89, 230 (1988).
[5] H. Graener, R. Dohlus and A. Laubereau, Chem. Phys. Lett., 140, 306 (1987).

Broadband Ultrafast Absorption Spectroscopy in the Hard Ultraviolet: Evolution of the CF_2 Radical upon Photodissociation of CF_2Br_2

T.R. Gosnell, A.J. Taylor, and J.L. Lyman

Chemical and Laser Sciences Division, MS E543,
Los Alamos National Laboratory, Los Alamos, NM 87545, USA

Abstract. Pump-probe experiments on the photodissociation of gas-phase CF_2Br_2 have been conducted using 700-fs source pulses at 248 nm for both initiating dissociation and generating a probing ultraviolet supecontinuum. The broadband uv absorption spectra thus obtained show evidence of a long-lived (≈3 ps) highly anharmonic local mode of the CF_2Br reaction intermediate.

Time-resolved absorption spectroscopy is often rejected in favor of laser-induced fluorescence spectroscopy as a viable technique for the investigation of ultrafast photochemical phenomena primarily because the method does not allow zero-background detection. Nevertheless, the advantages of a linear spectroscopy and the ability to detect intermediate species whose lifetimes are short compared with their fluorescence lifetimes have motivated us to reexamine the absorption technique. The above advantages, however, can only be obtained at the cost of a laser system able to deliver high-energy ultrashort pulses (> 1 mJ), and by requiring the photochemical reaction under investigation to produce photoproducts with large absorption cross sections ($> 10^{-18}$ cm^2) at high quantum yield (~1).

We present here initial results of time-resolved absorption experiments conducted using 700-fs, 10-mJ pump pulses at 248 nm and a synchronized ultraviolet probe continuum [1] derived from the same KrF laser source. The reaction studied is the photolysis of CF_2Br_2 [2], which produces the spectroscopically well characterized CF_2 radical by the following two-step dissociation pathway:

$$CF_2Br_2 + h\nu \Rightarrow CF_2Br^* + Br(^2P_{3/2}), \tag{1}$$

$$CF_2Br^* \Rightarrow CF_2(\tilde{X}) + Br(^2P_{3/2}). \tag{2}$$

Step (1) of this reaction occurs by direct dissociation of the parent molecule and thus occurs on a time scale (~ 100 fs) unresolved in our

Springer Series in Chemical Physics, Vol. 53 **Ultrafast Phenomena VII**
Editors: C.B. Harris · E.P. Ippen · G.A. Mourou · A.H. Zewail

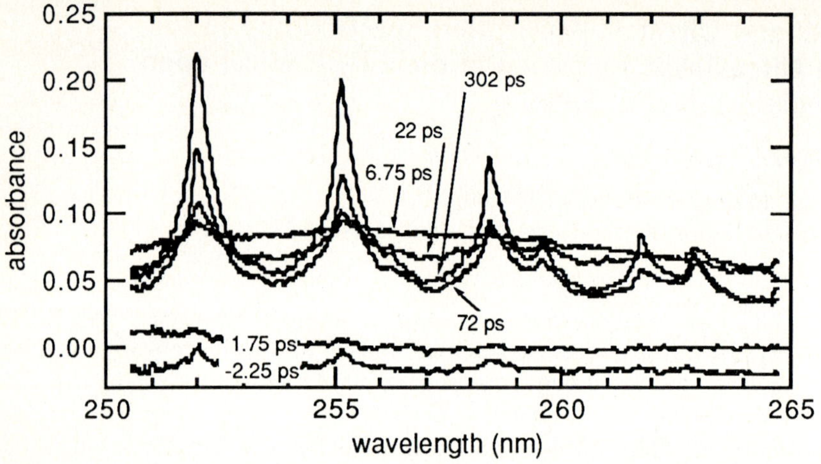

Figure 1. Time-resolved uv absorption spectra showing the evolution of ground-electronic-state CF_2 radicals in both the vibrational ground state and the first-excited vibrational state of the ν_2 bending mode.

experiment. In step (2), on the other hand, the vibrationally hot CF_2Br intermediate species undergoes unimolecular reaction on a much longer time scale to yield the final products.

The course of this second step is displayed in Figure 1, which shows the discrete CF_2 absorption spectrum as a function of time delay between the pump pulse and the probing continuum pulse. These CF_2 spectra indicate that not only the ground vibrational state of the molecule is produced, but also the first-excited state of its low-lying ν_2 bending mode. This observation is consistent with the spread of center-of-mass translational energies measured for step (1) above in Reference [2].

The most striking aspect of the spectra shown in Figure 1, however, is the rapid appearance of a broadband absorption feature upon which the discrete CF_2 spectrum is superimposed. We attribute this broadband feature to the CF_2Br reaction intermediate. Indeed, the strength of the broadband absorption decreases in parallel with the increasing strength of the CF_2 spectrum.

In greater detail, Figure 2 shows the onset of the CF_2Br absorption at small time delays. The spectra remarkably show that not only does the strength of the absorption evolve on a several-picosecond time scale, but also the spectral shape of the absorption. We believe that these results should be interpreted as the influence of intramolecular vibrational redistribution within the CF_2Br radical on the molecule's ultraviolet absorption spectrum. Moreover, the ≈3 ps time scale for redistribution is surprising long, suggesting the existence of a long-lived anharmonic local mode in the molecule.

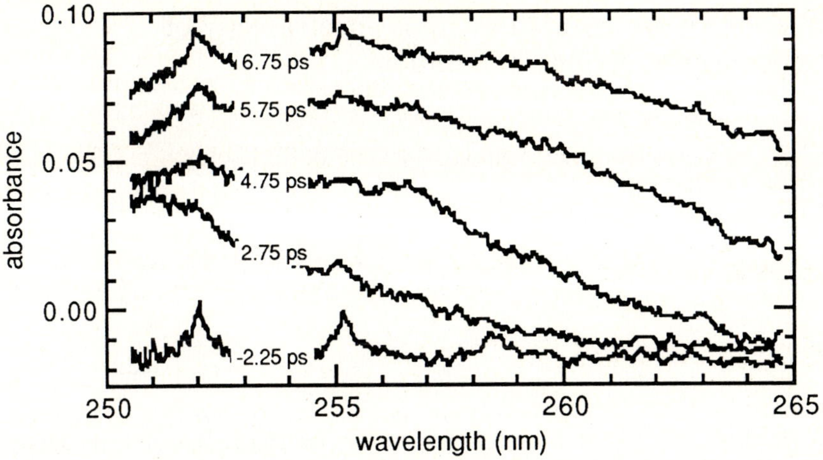

Figure 2. Evolution of broadband absorption attributed to the CF_2Br reaction intermediate.

References

[1]T. R. Gosnell, A. J. Taylor, and D. P. Greene, Opt. Lett. **15**, 130 (1990).

[2]D. Krajnovich, Z. Zhang, L. Butler, and Y. T. Lee, J. Phys. Chem. **88**, 4561 (1984).

Femtosecond Transient Absorption Spectroscopy of Photochemical Reactions in Solution

A.G. Joly and K.A. Nelson

Department of Chemistry, Massachusetts Institute of Technology, Cambridge, MA 02139, USA

1. Abstract

We present new femtosecond transient absorption data on the metal carbonyls, $M(CO)_6$ with M = Cr and W as well as preliminary resonance Raman experiments from $Cr(CO)_6$. The results indicate the dynamics of the initial ligand ejection event as well as a possible reaction coordinate.

2. Introduction

For small molecules in the gas phase, unimolecular dissociation dynamics may be determined uniquely by the structure of the potential energy surface along which reaction occurs. Femtosecond time-resolved measurements of gas-phase reaction dynamics have been analyzed quantitatively in terms of known potential energy surfaces, and can be used to determine or refine unknown ones[1]. In condensed phases, interactions between the reacting species and its environment cause dissipation of momentum along the reaction coordinate, barrier recrossing, and other events which destroy the one-to-one correspondence between reactive potential energy surface and reaction dynamics. A microscopic understanding of condensed phase chemical reactions therefore requires independent determination of both chemical reaction dynamics and reactive potential energy surface(PES).

A set of prototypical liquid-state photochemical reactions is offered by the metal carbonyls $M(CO)_6$ where M = Cr, Mo, or W. In solution, a single CO dissociates within 350 fs[2]. Within a few picoseconds of dissociation, a nearby solvent molecule coordinates to the naked site, forming the species $Cr(CO)_5S$ [S = solvent] which is stable onto at least the nanosecond time scale[2,3].

3. Experimental

Experiments are performed as described elsewhere [2] using 90 fs, 308 nm pump pulses and visible probe pulses derived from a white light continuum source. Cross correlations between ultraviolet and probe pulses indicate an instrument response of approximately 100 fs which is consistent with the fastest rise times seen in the data.

4. Results and Discussion

We have obtained the time-dependent excited-state absorption spectra of all three $M(CO)_6$ species. Figure 1 depicts the spectral changes of $W(CO)_6$ for the first 5.7 ps after excitation. Data from $Cr(CO)_6$ and $Mo(CO)_6$ are almost identical in form.

Springer Series in Chemical Physics, Vol. 53 **Ultrafast Phenomena VII**
Editors: C.B. Harris · E.P. Ippen · G.A. Mourou · A.H. Zewail

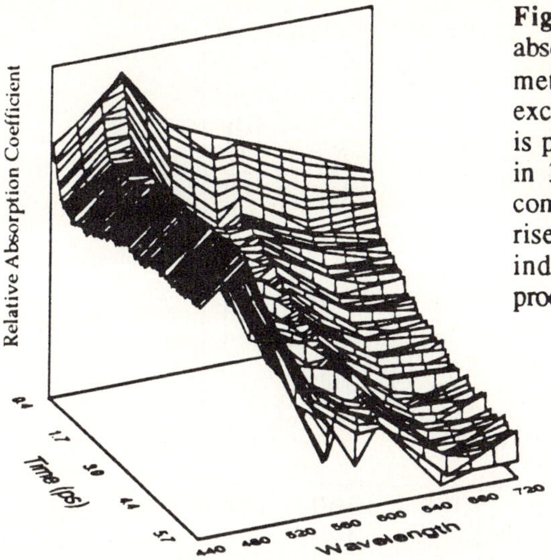

Figure 1. Broadband visible absorption spectra for $W(CO)_6$ in methanol. The excited state-excited state absorption spectrum is peaked at 500 nm and decays in 350 fs. Subsequent solvent complexation leads to a gradual rise between 440 and 520 nm indicating formation of the product $W(CO)_5(MeOH)$.

Figure 1 indicates that the excited-state absorption spectrum of $W(CO)_6$ is peaked at 500 nm and tails to the blue. During the first 500 fs, as CO loss occurs, the absorption strength decays sharply and blue shifts slightly. During the next several picoseconds the spectrum shows a more gradual increase in absorption between 440 and 500 nm. Due to its similarity with $Cr(CO)_6$, this rise is interpreted as a signature of solvent complexation[2,3], i.e. formation of the $W(CO)_5(MeOH)$ species. Further spectral evolution over a total of 70 ps after photoexcitation (not shown) is indicative of vibrational relaxation of the nascent $M(CO)_5(SOLVENT)$ species[2,4].

Transient absorption experiments performed in a variety of solvents reveal that the initial ligand ejection is not noticably altered even by viscous solvents such as ethylene glycol. It appears that the short time dynamics are determined almost solely by the intramolecular PES. Initial solvent complexation shows no preference for hydroxyl over alkyl site coordination, although the former is more stable thermodynamically and is known to occur eventually[5].

Although we know the time scale for initial CO photoejection rather well, we still know little about the intramolecular geometry changes accompanying ligand loss. To address the question of the reaction coordinate, we have performed resonance Raman (RR) experiments on $Cr(CO)_6$ in methanol. In RR, as the frequency of the excitation light approaches an electronic resonance, there is enhancement in those vibrational modes which undergo distortion in the excited state. Those vibrational modes which determine the dissociative coordinate will have large distortions in the excited state and therefore undergo large resonance enhancement.

At least one asymmetric stretch must be active in the reaction coordinate for only a single CO to dissociate. Due to symmetry considerations, enhancement of asymmetric stretches in $Cr(CO)_6$ is manifested by the appearance of even overtones in the Raman spectrum. Figure 2 shows RR data of $Cr(CO)_6$ in methanol with both off-resonance (514 nm) and pre-resonance (351 nm) excitation wavelengths. In

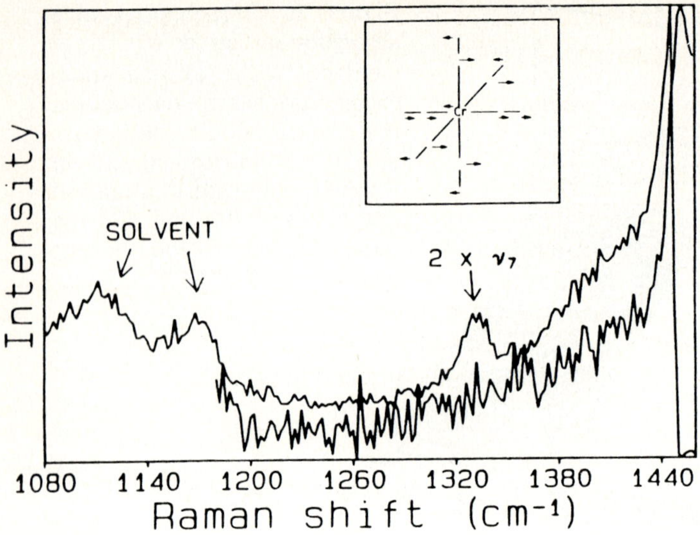

Figure 2. Resonance Raman spectra of $Cr(CO)_6$ in methanol for both off-resonant (lower trace) and pre-resonant (upper trace) excitation. Near resonance the second overtone of the ν_7 asymmetric stretch (shown in inset) appears at 1332 cm^{-1}.

the pre-resonant scan an additional line appears at 1332 cm^{-1}. This frequency corresponds to exactly twice the frequency of the ν_7 asymmetric stretch whose motion is shown in the Figure 2 inset.

The ν_7 mode also has the correct symmetry to be the vibronic promoting mode. A more complete resonance excitation profile including higher overtones should be measured to yield quantitative conclusions about the reaction coordinate.

This work was supported in part by NSF Grant No. CHE-8901722 and by contributions from DuPont, Perkin-Elmer, and the Alfred P. Sloan Foundation. Assistance by Dr. Mark Trulson in the acquisition and interpretation of Raman spectra is gratefully acknowledged.

5. References

[1] M. Gruebele, G. Roberts, M. Dantus, R. Bowman, and A. H. Zewail, *Chem. Phys. Lett.* **166**, 459 (1990).
[2] A. G. Joly and K. A. Nelson, *J. Phys. Chem.* **93**, 2876 (1989).
[3] J. D. Simon and X. Xie, *J. Phys. Chem.* **90**, 6751 (1986).
[4] M. Lee and C. B. Harris, *J. Am. Chem. Soc.* **111**, 8963 (1989).
[5] X. Xie and J. D. Simon, *J. Am. Chem. Soc.* **112**, 1130 (1990).

Subpicosecond Infrared Spectroscopy: The Condensed Phase Photochemistry of Iron Carbonyls

P.A. Anfinrud, C. Han, T. Lian, and R.M. Hochstrasser*

Department of Chemistry, University of Pennsylvania,
231 South 34th Street, Philadelphia, PA 19104, USA
* Present address: Department of Chemistry, Harvard University,
12 Oxford Street, Cambridge, MA 02138, USA

1. Introduction

The photochemistry of dinuclear transition metal carbonyls has been extensively studied in isolated matrices [1,2] and low temperature liquids [3,4] using conventional IR spectroscopic methods, and in ambient solution [5-8] using microsecond time-resolved IR techniques. While these studies have revealed long-lived photoproducts formed by visible and UV photolysis, they lack the time resolution necessary to address issues related to the primary events in the photodissociation of organometallic compounds.

A novel method for obtaining ps IR spectra [9] was recently extended to provide femtosecond time resolution [10]. This development has enabled an ultrafast investigation into the photochemistry of $[C_pFe(CO)_2]_2$ ($C_p = \eta^5$-C_5H_5) in cyclohexane where the primary photoproducts and their dynamics have been observed for the first time.

2. Experimental

A novel visible-IR pump-probe spectrometer has been developed which is based upon gated upconversion of a CW IR probe beam [9]. The IR probe originates in either a diode laser or a CO laser. The time-resolution achieved with this approach is limited only by the optical pulse duration while the spectral resolution is limited only by the probing bandwidth.

Subpicosecond optical pulses are generated by shortening pulses from a cavity dumped dye laser with an optical fiber/grating compressor. The ultrashort pulses are amplified at a high repetition rate in a multipass amplifier and split to produce the photolyzing and gating pulses. The photolyzing pulse, which is delivered to a common focus with the CW IR beam in a flowing sample, modulates the transmitted probe beam. The transmitted IR field is frequency summed (upconverted) with an ultrashort gating pulse in a nonlinear crystal of $LiIO_3$ to generate a signal which is related to the transmitted IR probe intensity at an instant in time defined by the gating pulse. The upconverted signal is detected with a photomultiplier tube, demodulated in a lock-in amplifier, and processed in a computer to determine the differential absorbance (ΔA).

The cryogenically cooled Pb-salt diode lasers (Laser Photonics, Analytics Division) generate a few milliwatts of IR intensity and can be tuned continuously over a range as broad as $\approx$ 150 cm^{-1} by adjusting the operating current and/or temperature. The CO laser (Laser Analytics) can generate a few hundred milliwatts of IR power on some transitions and is tunable over the range 1780 $\rightarrow$ 1920 cm^{-1} with lines appearing approximately every 4 cm^{-1}. Time-resolved spectra are recorded by varying the probe wavelength at a fixed delay between the pumping and gating pulses, and time-resolved absorption changes are measured by varying the delay between the photolyzing and gating pulses at a fixed probe frequency.

3. Results and Discussion

The FTIR spectrum of $[C_pFe(CO)_2]_2$ reveals three absorptions (see Fig. 1). The features at 1792, 1960, and 2004 cm^{-1} correspond to the *bridging*, *trans* and *cis* CO absorption,

Springer Series in Chemical Physics, Vol. 53 **Ultrafast Phenomena VII**
Editors: C.B. Harris · E.P. Ippen · G.A. Mourou · A.H. Zewail

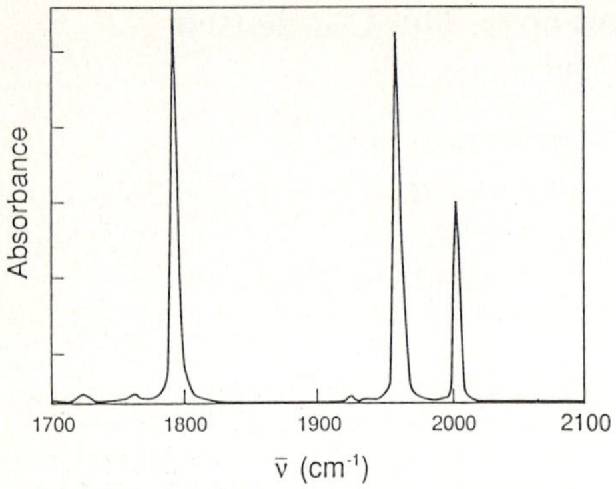

Fig. 1. The FTIR absorption spectrum of $[C_pFe(CO)_2]_2$ in cyclohexane

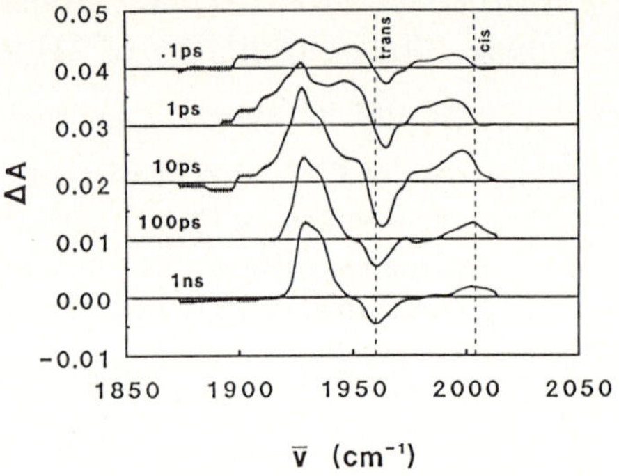

Fig. 2. The time-resolved differential IR absorption spectra of $[C_pFe(CO)_2]_2$

respectively. The *trans* and *cis* isomers, which interconvert in cyclohexane at room temperature, are present in the ratio 1.8 : 1.

The time-resolved differential IR absorption spectra in Fig. 2 were obtained following 580 nm photolysis with diode laser probes. The bridging transient IR absorptions presented in Fig. 3 were measured with a CO laser probe. Any negative-going (bleaching) features arise from ground state depletion while positive-going features reveal new transient intermediates.

The dynamics for the dissociation and reappearance of doubly bridged species is shown in Fig. 3. A transient IR absorption appears at 1797 cm^{-1} (higher energy wing of the bridging absorption) but decays within a few ps to reveal the bridging bleach. The bleach partially recovers within 20 ps but then *rebleaches* by 60 ps. The transient IR signal at 1788 cm^{-1} (lower energy wing) has no dominating absorption feature but does reveal $\approx$ 40% bleaching recovery. These results suggest the following. The absorption is likely due to an electronically excited bound dimer which persists for $\approx$ 1 ps before dissociating. The recovery of the bleach within $\approx$ 20 ps is due to geminate recombination. The rebleaching likely arises from cooling of the hot regenerated dimer and the concomitant shift of its bridging absorption toward its ground state transition at ambient temperatures (1792 cm^{-1}).

The transient IR spectra in Fig. 2 reveals features associated with terminally bound CO. The two long-lived absorptions have been observed with microsecond time resolution and were assigned to the radical resulting from homolytic cleavage of the dimer [5]. The integrated absorption and the magnitude of the trans bleach (after accounting for the background) do not evolve from 1 to 10 ps but diminish by a factor of $\approx$ 2 by 100 ps, which is consistent with the recovery of the bridging CO bleach. The integrated absorption exceeds the integrated bleach at all times, as expected, since one additional terminal CO absorption is created for each bridging bond broken. At least three transitions are evident in the early time spectra which suggests that there is more than one intermediate present. (Homolytic cleavage of the *trans* or *cis* isomers would generate identical fragments which could yield only two new absorptions in the terminal CO spectral region).

A kinetic study of the transient absorption at 2004 cm^{-1} revealed an isosbestic point from 0 to 5 ps, i.e., the change in absorption is nearly zero over this time period. The isosbestic point implies that the absorption at the lower energy side of the *cis* feature arises from an intermediate whose spectrum remains unchanged for the first 5 ps. This suggests that the intermediates cannot interconvert and their spectra cannot shift over this time period. It is proposed that the intermediates include both singly bridged and homolytically cleaved species. The singly bridged intermediate might be expected to reform its broken bridging bond and thus may account for the geminate recombination. The homolytically cleaved radicals, on the

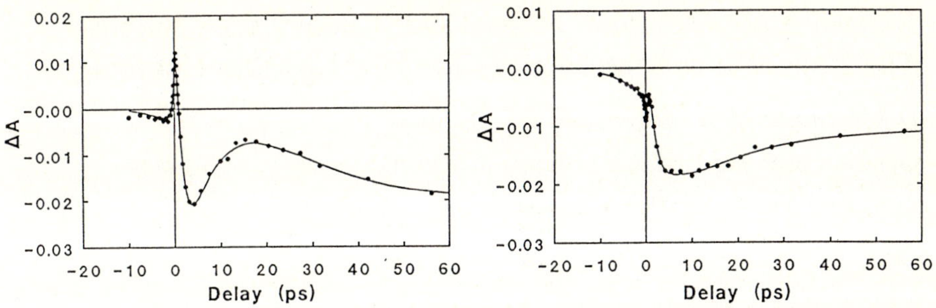

Fig. 3. The transient IR absorption of $[C_pFe(CO)_2]_2$ at 1797 cm^{-1} (a) and 1788 cm^{-1} (b). The solid line is the convolution of a multi-exponential with the instrument response function

other hand, may be destined to reform the starting material through a much slower bimolecular process. The spectral changes that occur from 5 to 10 ps likely arise from solvation dynamics; the solvation delay arises from excess thermal energy in the nascent fragments which must be partially dissipated before solvent complexation can occur.

4. Conclusions

The photodissociation proceeds through an excited bound electronic state which persists for ≈ 1 ps. The nascent photofragments may include a singly bridged intermediate and homolytically cleaved radicals which apparently become solvated between 5 and 10 ps. Roughly half of these intermediates geminately recombine to generate *hot* dimers which cool with a time constant of ≈ 20 ps.

The CO laser enhances the experimental sensitivity relative to diode laser probes. Diode lasers, however, provide broader spectral coverage and continuous tunability.

This research was supported by grants to RMH from NSF and NIH.

References

1. R.H. Hooker, K.A. Mahmoud, A.J. Rest: J. Chem. Soc., Chem. Commun. 1022 (1983)
2. A.F. Hepp, J.P. Blaha, C. Lewis, M.S. Wrighton: Organometallics 3, 174 (1984)
3. D.R. Tyler, M.A. Schmidt, H.B. Gray: J. Am. Chem. Soc. 101, 2753 (1979)
4. D.R. Tyler, M.A. Schmidt, H.B. Gray: J. Am. Chem. Soc. 105, 6018 (1983)
5. B.D. Moore, M.B. Simpson, M. Poliakoff, J.J. Turner: J. Chem. Soc., Chem. Commun. 972 (1984)
6. A.J. Dixon, M.A. Healy, P.M. Hodges, B.D. Moore, M. Poliakoff, M.B. Simpson, J.J. Turner, M.A. West: J. Chem. Soc., Faraday Trans. 2 82, 2083 (1986)
7. B.D. Moore, M. Poliakoff, J.J. Turner: J. Am. Chem. Soc. 108, 1819 (1986)
8. A.J. Dixon, M.A. Healy, M. Poliakoff, J.J. Turner: J. Chem. Soc., Chem. Commun. 994 (1986)
9. J.N. Moore, P.A. Hansen, R.M. Hochstrasser: Chem. Phys. Lett. 138, 110 (1987)
10. P.A. Anfinrud, C. Han, P.A. Hansen, J.N. Moore, R.M. Hochstrasser: In Ultrafast Phenomena VI, (Springer, Berlin 1988) p.442

Excited State Dynamics of Jet-Cooled Substituted Indoles: The Role of Polar Interactions in ^1L_a–^1L_b State Coupling

D.R. Demmer, G.W. Leach, and S.C. Wallace

Ontario Laser and Lightwave Research Center, University of Toronto, Toronto, Ontario, Canada

Abstract. New evidence has been found for the participation of the ^1L_a state in the excited state dynamics of jet-cooled alkylindoles. Picosecond time scale excited state lifetime measurements of a number of these species and their deuterated analogues afford a clear view of the ^1L_a-^1L_b state coupling.

1. Introduction

Efforts to understand the condensed phase photophysics of indole and simple indole derivatives has led to the identification of two electronic states (^1L_a and ^1L_b) that underly their lowest energy absorption bands. The characterization of the jet-cooled gas phase photophysics of these same molecules however, has involved only one electronic state, and only recently has evidence been obtained for the participation of the two electronic states which have been identified from condensed phase work. [1,2]

Estimates of the positions of the ^1L_a and ^1L_b electronic states in a variety of indoles lead to the conclusion that the primary factors affecting the position of the ^1L_a state are the location of substituents around the indole nucleus and the polarity of its immediate environment. In particular, 2- and 3-substituted indoles are expected to have relatively low energy ^1L_a states, and 2,3-disubstituted indoles the lowest of all. We present evidence that the gas phase photophysics of 2,3-dimethylindole (2,3-DMI), its hydrogen-bonded complexes and those of 3-methylindole (3-MI) is significantly perturbed by the presence of the ^1L_a electronic state which interacts vibronically with the ^1L_b through motions associated with the indole N-H bond.

2. Results and Discussion

The spectrum of 2,3-DMI is sharp and structured and has been assigned to the $^1L_b \leftarrow {}^1A_1$ transition. However the corresponding fluorescence lifetimes of individual bands within this spectrum (figure 1) decrease rapidly as one proceeds to higher energy. For 2,3-DMI(d_1) the decrease in lifetime is much more modest and characteristic of other indoles in the gas phase. The lifetime of 2,3-DMI decreases from about 6 ns at the origin to approximately 450 ps at some 900 cm^{-1} in excess vibrational energy in stark contrast to that of the deuterated analogue, which drops from 12 ns at the origin to roughly 8 ns at comparable energy. This marked decrease in quantum yield can be attributed to dissociation following internal conversion to the ^1L_a state. Our results suggest that isotopic substitution closes this dissociative pathway which we therefore ascribe to tunnelling along the N-H coordinate.

The spectra of 2,3-DMI hydrogen-bonded complexes are very broad and differ dramatically from similar clusters involving indole and some of its other derivatives. The dispersed emission spectra of these complexes are characterized by broad structureless emission that is substantially red-shifted relative to the bare molecule emission. Analysis of the time- and energy-resolved fluorescence for the hydrogen-bonded complexes of H_2O and NH_3 show multiexponential behavior as shown in figure 2 for the case of the ammonia complexes. These decays and those for the water complexes have been fitted to dual exponential lifetimes with a short component of 250 ps and a longer lifetime component of about 6-10 ns. Analysis of the time-resolved

Springer Series in Chemical Physics, Vol. 53 **Ultrafast Phenomena VII**
Editors: C.B. Harris · E.P. Ippen · G.A. Mourou · A.H. Zewail

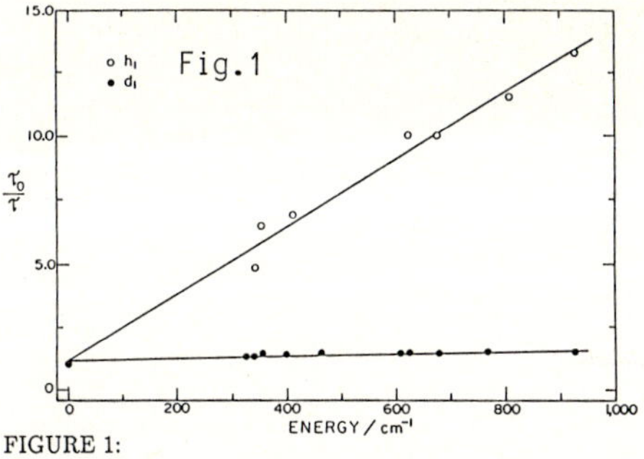

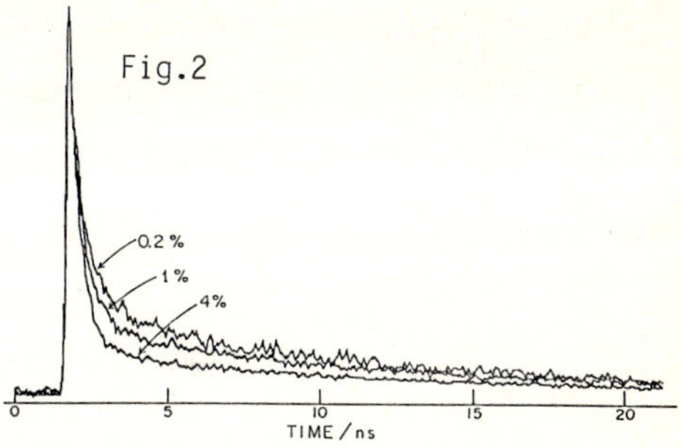

FIGURE 1:
Normalized decay rates of normal (h_1) and deuterated (d_1) 2,3-dimethylindole plotted as functions of excess vibrational energy above their respective ^1L_b origins. τ_0 is the lifetime measured at the origin: 6 ns for 2,3-DMI and 12 ns for 2,3-DMI(d_1).

FIGURE 2:
Fluorescence decays measured for expansions of 2,3-DMI containing various proportions of ammonia. The curves are labelled according to the percentage of NH_3 in the expansion mixture. The 1:1 cluster is the predominant species present only at the lowest percentage shown.

fluorescence from 2,3-DMI complexed with larger hydrogen-bonded partners show single exponential decays corresponding to fluorescence lifetimes in the 10-30 ns regime.

The complexation of indole derivatives with polar molecules can be thought of as the result of an interaction between the acidic hydrogen of the ring nitrogen atom and the complexing partner acting as a Lewis base. The strength of this interaction has been shown to correlate well with the proton affinity of the Lewis base,[3] and the resulting hydrogen bond would be expected to weaken the N-H bond of the indole nucleus in proportion to the proton affinity. Since the energy of a hydrogen bond is considerably lower than that of the covalent N-H bond the effect of this interaction on the indole moiety would normally be minimal, but in the case of the 2,3-disubstituted indoles it is this bond which is intimately involved with the coupling of the two zero order electronic potential surfaces. The unusual excited state properties of the 2,3-DMI hydrogen bonded complexes may then be viewed as the result of the interaction between the indolic hydrogen and the lone pair of the Lewis base, which enhances the coupling of the zero order ^1L_b and ^1L_a states by promoting the extension of the N-H bond.

Our results can be explained in terms of a relatively simple model by postulating that the ^1L_a states of jet-cooled 2,3-DMI complexes are bound states. The role of the N-H bond extension in enhancing the coupling between the ^1L_a and ^1L_b states would imply that the proposed minimum in the ^1L_a state potential along the N-H coordinate lies at a larger value of r_{NH} than the minimum of either the ^1L_b or ground states. If we assume this to be the case, then vertical transitions from the ground to the excited state terminating on the ^1L_a state potential surface will result in the population of excited state clusters with a relatively high degree of vibrational excitation.

In the case of 2,3-DMI-$(NH_3)_1$ and -$(H_2O)_1$ the high degree of vibrational excitation could result in a very rapid decay of population as vibrational energy is redistributed into the hydrogen bond and the clustering partner is carried away. Alternatively, if the complexing partner possesses a sufficient density of low-frequency vibrational states the cluster dissociation may be arrested by the dissipation of vibrational energy into the bath of states provided by the complexing partner. In the case of TMA and MeOH a single solvent molecule apparently provides a sufficiently dense set of bath states to inhibit this behavior, while ammonia and water do not.

The ^1L_a state of 3-MI lies to higher energy and its effects are somewhat more subtle. A measurement of the fluorescence lifetime *vs.* excitation energy, analogous to that performed for 2,3-MI in figure 1, shows that the lifetime does not vary in a regular way: 12.3 ns at the origin, it declines to roughly 6 ns at about 650 cm^{-1} before rising to 8 ns at 800 cm^{-1}. Clearly, the ^1L_b state which lends the structure to the excitation spectrum is being perturbed. We attribute this to the presence of a the nearby ^1L_a state. However, since it lies at a roughly similar energy to the ^1L_b state in the bare molecule, its vibrational manifold is relatively sparse, and coupling occurs primarily at energies

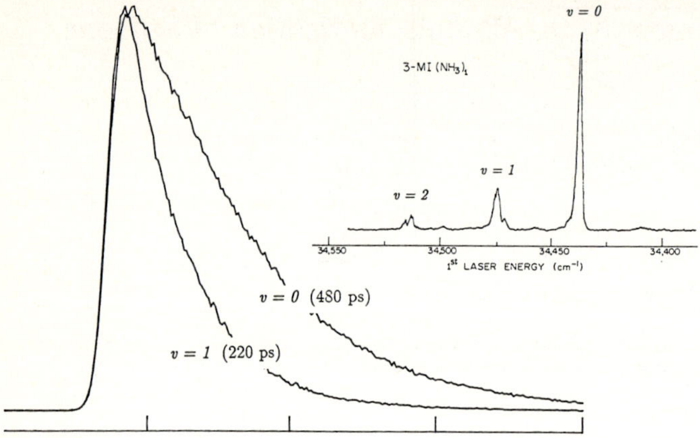

FIGURE 3:

Fluorescence decays of 3-MI:$(NH_3)_1$ for excitation at the vibrationless level of the cluster (*v=0*) and with one quantum in the van der Waals mode (*v=1*). The inset shows the excitation spectrum of this complex. The decay of the third member of the progression is too fast to measure by our current methods.

where there is a coincidental near-degeneracy between members of the two vibrational manifolds.

Figure 3 shows the excitation spectrum of 3-MI complexed with ammonia. This spectrum and those of other H-bonded complexes are dominated by sharp features (and are therefore assigned to the $^1L_b \leftarrow {}^1A_1$ transition) that can be assigned to progressions in an intermolecular vibrational coordinate. Undispersed emission resulting from excitation of the 3-MI-$(NH_3)_1$ complex shows a dual exponential lifetime, with two widely differing time constants. The short component can be isolated by monitoring the structured portion of the emission spectrum. It is seen to undergo a rapid decrease for consecutive members of the excitation progression: from 480 ps to 220 ps on going from the complex origin band to the second member of the progression, 38 cm^{-1} to higher energy (figure 3). The longer lifetime component remains constant at 14 ns.

Time-resolved measurements on the complexes of 3-MI with methanol, trimethylamine, and water show single exponential decays, with a lifetime of approximately 10 ns. By analogy with the results presented for 2,3-DMI and its hydrogen-bonded complexes, we can explain these observations in a straightforward way. The photophysics of the 3-MI complexes of methanol, trimethylamine, and water is dominated by only one electronic state, namely the ^1L_b. Complexation with ammonia however, results in an interaction of the ^1L_a and ^1L_b manifolds such that excitation to the ^1L_b state is followed by tunnelling to the ^1L_a surface where the complex undergoes dissociation. A portion of the total energy goes to relative translation of the 3-MI and NH_3 fragments and a portion to the internal excitation of each fragment, leaving the 3-MI fragment in a bound ^1L_a state where it ultimately decays by fluorescence with a 14 ns lifetime. We are currently performing experiments to elucidate the nature of the intermolecular motion that leads to such fast complex dissociation.

References:

[1.] J.W. Hager, D.R. Demmer, and S.C. Wallace, *J. Phys. Chem.* 91, 1375 (1987).

[2.] D.R. Demmer, G.W. Leach, E.A. Outhouse, J.W. Hager, and S.C. Wallace, *J. Phys. Chem.* 94, 582 (1990).

[3.] J.W. Hager and S.C. Wallace, *J. Phys. Chem.* 88, 5513 (1984).

Molecular Alignment in the Photoisomerization of Stilbene

S.T. Repinec, R.J. Sension, S.A. Abrash, and R.M. Hochstrasser

Department of Chemistry, University of Pennsylvania,
231 South 34th Street, Philadelphia, PA 19104, USA

1. Introduction

The isomerization of stilbene has been studied extensively as a model for intramolecular rearrangement. With the advent of subpicosecond lasers, measurements on *cis*-stilbene have become possible [1-3].

We have carried out a series of experiments on the isomerization of *cis*-stilbene. Transient absorption measurements were used to obtain a spectrum of the cis excited state and to measure the dynamics of *cis*-stilbene in solutions of alkane solvents [3]. Absorption anisotropy measurements, presented here, provide information on rotational dynamics and the alignment between the reactant and product molecules during isomerization.

2. Experimental

The laser system used for these studies is based on a standard design [4,5]. A CPM laser which is amplified in a four stage Nd:YAG pumped amplifier and then frequency doubled, is used to excite *cis*-stilbene at 312 nm. Probe wavelengths in the visible are made by continuum generation in a cell of D_2O. Portions of the continuum to be used as probes are selected with a set of interference filters. The probe can be further frequency doubled to obtain probe wavelengths in the ultraviolet.

A half-wave plate in the pump arm of the apparatus allows the rotation of the pump beam polarization relative to the probe to give the parallel and perpendicular geometries for the anisotropy studies.

3. Results and Discussion

Experiments were done in the *n*-alkane solvents hexane and hexadecane. The lifetime of the *cis*-stilbene excited state in these two solvents is 1.0ps and 1.5ps respectively [3].

Excited *cis*-stilbene exhibits transient absorption in the ultraviolet and visible/IR regions of the spectrum [3]. The visible/IR transient absorption is broad covering the range 400-1100 nm with a maximum at ca. 650 nm. The ultraviolet transient starts around 390 nm and increases toward higher energy. In the region around 325nm-335nm the UV transient absorption of the cis isomer overlaps the *trans*-stilbene ground state absorption sufficiently to study the alignment between the reactant and product molecules with a single probe wavelength.

Editors: C.B. Harris · E.P. Ippen · G.A. Mourou · A.H. Zewail

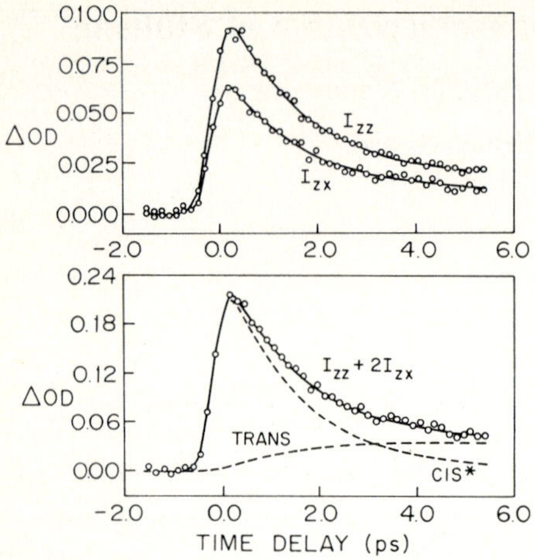

Fig. 1. Decay curves of excited *cis*-stilbene in hexadecane probed at 330nm. Upper: Parallel (I_{ZZ}) and perpendicular (I_{ZX}) pump-probe polarization geometries. Lower: Magic angle decay i.e. 57°.

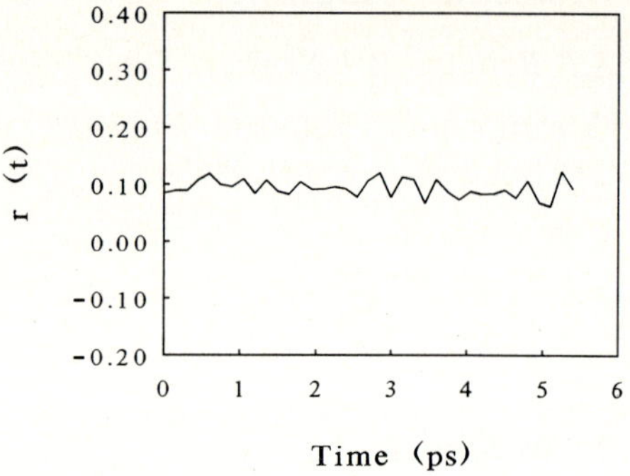

Fig. 2. Anisotropy generated from data in Fig. 1 where $r(t)=I_{ZZ}-I_{ZX}/I_{ZZ}+2I_{ZX}$. The low initial anisotropy of $r(o)=0.10\pm0.03$ is seen throughout the UV transient and is probably due to an angle of about 44° between the ground and excited state transition dipoles.

The transient absorption decay obtained when probing *cis*-stilbene in hexadecane at 330nm is shown in Figure 1. This viscous solvent was used for the alignment experiment in order to limit the overall rotation of the molecule. The spectrum does not decay down to the baseline due to a product adsorption. The spectral position and intensity of this component suggest that it might be from *trans*-stilbene ground state molecules. The photoproduct component partially decays with a lifetime of about 24ps (in hexadecane) possibly due to vibrational cooling. The different decays in the top panel of Fig.1 are the different pump-probe polarization geometries and are used to generate the anisotropy shown in Figure 2. The lower panel in Fig.1 is the magic angle decay which is shown here fit to a simple model of the product growing in with a time constant equal to the decay of excited *cis*-stilbene (τ =1.5ps). At early times the measured anisotropy in Fig.2 is dominated by excited cis isomer. When a large amount of the cis population has decayed (after approx. 1.5 ps) the anisotropy is due to product molecules. On the measured time scale of the experiment the anisotropy is essentially constant.

If the isomerization were to proceed as a rigid rotation about the ethylene bond holding one end of the molecule fixed, the observed anisotropy of generated trans molecules would be near zero. The second possibility is a symmetric rigid motion of both phenyl rings. If no other orientational relaxation effects are considered this type of motion could produce an anisotropy slightly smaller than that observed in Figure 2. If the isomerization to trans is a simple rigid rotation it is most likely a combination of these two motions. The isomerization may proceed

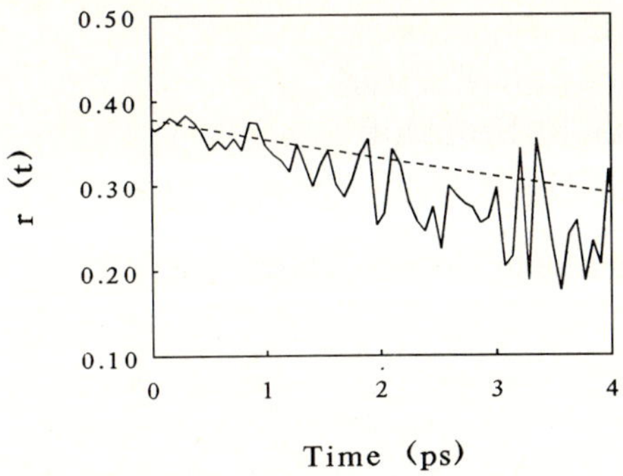

Fig. 3. Cis-stilbene anisotropy in hexane with a 650nm probe pulse.

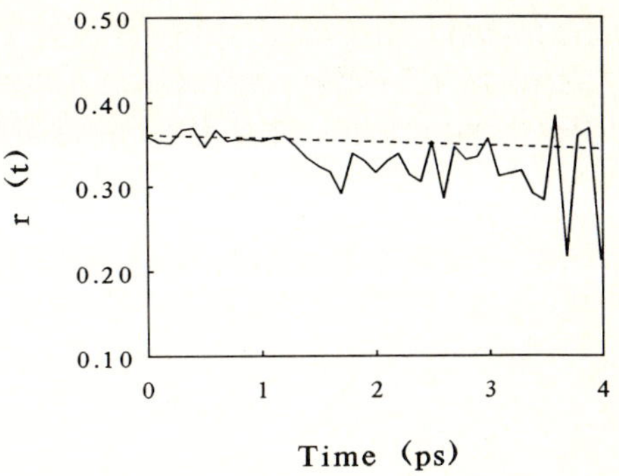

Fig. 4. 650nm anisotropy in hexadecane.

by a combination of smaller amplitude motions of many atoms. This complex pseudo-rotation involving a combination of phenyl-ethylene bond stretches and phenyl ring rotations could limit the overall rotation of the transition dipoles, and give rise to a significant trans product anisotropy as in Figure 2. The photoproduct anisotropy might also be influenced by the presence of dihydrophenanthrene and *cis*-stilbene bleach.

Anisotropy measurements on the visible transient were done at a probe wavelength of 650nm. The results are shown in Figures 3 and 4, in the solvents hexane and hexadecane respectively. The dashed lines in these figures represent the anisotropy decay times of *trans*-stilbene in these solvents (15 and 80ps). As seen in the figures, anisotropy decays for *cis*-stilbene seem to be faster than for trans. The anisotropy in hexadecane also appears to be non-exponential. The magic angle decay in this instance only is fit as two exponentials of 1.3 and 6ps. This suggests that in very viscous solvents there is a distribution of cis environments, and hence decay times, that persist for a few picoseconds.

4. Conclusions

With a 330nm probe the formation of a product molecule, assumed to be *trans*-stilbene ground state, can be observed from the isomerization of cis. The product component grows in with a time constant equal to or faster than the decay of *cis*-stilbene, and seems to partially decay due to vibrational cooling. The product anisotropy defines a reaction coordinate for the isomerization.

REFERENCES

1. B.I. Greene and R.C. Farrow, J. Chem. Phys. **78**, 336 (1983).
2. F.E. Doany, R.M. Hochstrasser, B.I. Greene and R.R. Millard, Chem. Phys. Lett. **118**, 1 (1985).
3. S.A. Abrash, S.T. Repinec and R.M. Hochstrasser, J. Chem. Phys., in press.
4. R.L. Fork, B.I. Greene and C.V. Shank, Appl. Phys. Lett. **38**, 671 (1981).
5. R.L. Fork, C.V. Shank and R.T. Yen, Appl. Phys. Lett. **41**, 224 (1982).

Femtosecond and Picosecond Laser Photolysis Studies of Photoreduction Processes of Excited Benzophenone and Its Derivatives in Solution

H. Miyasaka, K. Morita, M. Kiri, and N. Mataga

Department of Chemistry, Faculty of Engineering Science, Osaka University, Toyonaka, Osaka 560, Japan

Photoinduced electron transfer (ET) and charge transfer (CT) interaction play a crucial role in many kinds of photochemical reactions [1-3]. Hydrogen abstraction of excited benzophenone (BP^*) from the amine (AH) is one of such reactions. Although a number of investigations have been performed in order to clarify the role of CT interaction in the reaction mechanism, clear conclusion on this problem has not been established. In order to elucidate the contribution of the CT interaction and the ion pair state(s) in the reaction, the effects of the solvent, the oxidation potential of the amine, and the structure of the ion pair (IP) on the reaction mechanism were investigated by means of femtosecond-picosecond laser photolysis and transient absorption spectroscopy.

EXPERIMENTAL

Fs laser photolysis system [3] was used for the measurement of the time-resolved transient absorption spectrum in fs-several ps time region. The fundamental pulse at 710nm (0.3-0.5mJ/pulse with 200-500fs fwhm) was frequency doubled and used for the excitation. White continuum light generated by focusing the fundamental pulse into D_2O was used as the monitoring ray. In the measurement in longer time region, mode-locked YAG laser system was used.

RESULTS AND DISCUSSION

Reaction Mechanism of Hydrogen Abstraction of $^3BP^*$.

Since the S_1->T_1 intersystem crossing (ISC) time of BP is very short (ca.10ps), much attention has been concentrated on the photoreduction of BP in T_1 state ($^3BP^*$). In this study, we have examined the dynamics of the photoreduction of $^3BP^*$ with a number of amines in various solvents[3,4]. From these investigations, it has been revealed for such amines as diphenylamine (DPA) that the hydrogen abstraction and CT or IP state formation by ET are competing at encounter collision between $^3BP^*$ and DPA, and the CT or IP state relaxed with respect to D-A configurations and solvation does not contribute to ketyl radical (BPH) formation (scheme1).

$$^3BP^* + AH \rightarrow BPH + A^{\cdot} \qquad \rightarrow (BP^-AH^+) \qquad \underline{\textbf{Scheme 1}}$$

On the other hand, for some tertiary amines with low oxidation potential such as N,N-dimethylaniline (DMA) in acetonitrile, it has been revealed that the hydrogen abstraction of $^3BP^*$ is described by such a mechanism as

$$^3BP^* + AH \rightarrow (BP^{-}\cdot AH^+) \rightarrow BPH + A^{\cdot} \qquad \rightarrow BP^- + AH^+ \qquad \underline{\textbf{Scheme 2}}$$

The first step of the photoreduction is the ET at encounter collision leading to the stable IP formation $^3(BP^-DMA^+)_{enc}$, followed by the proton transfer. This difference of the mechanism depending on the nature of amines and solvent might be related to the difference of the energy

Springer Series in Chemical Physics, Vol. 53 **Ultrafast Phenomena VII**
Editors: C.B. Harris · E.P. Ippen · G.A. Mourou · A.H. Zewail

level of the IP state and the mutual configuration of BP and AH at encounter collision.

Dependence of the Proton Transfer Reactivity of IP on its Production Process

In addition to the ET in T_1 state, the excitation of the CT complex in the ground state as well as the ET reaction in S_1 state also produce IP states. From the recent investigation on the singlet IP between various donors (D) and acceptors (A) [5], it was indicated that the structure of IP such as the mutual configuration of ions and the inter-ionic distance in the pair including the surrounding solvents was different depending on the mode of its production. Hence, the elucidation of the reactivity of each IP, which consists of the same D-A pair and has a different pathway of production, may provide important information on the relation between the geometry and the reactivity of the hydrogen abstraction.

Fig.1 shows the time-resolved transient absorption spectra of BP-DMA (1.0M) in acetonitrile solution. The absorption spectra of $^1BP^*$ and $^3BP^*$ in this solvent are shown in Fig.2 and S_1->T_1 ISC time constant of BP in acetonitrile has been obtained to be 9.0ps. A transient absorption spectrum at 1ps after the excitation with fs 355nm laser pulse in Fig.1 shows absorption maxima at 475nm (DMA^+) and 575nm ($^1BP^*$). In addition to these two peaks, broad absorption whose intensity increases in wavelength region longer than 600nm (BP^-) is observed. Since the rapid increase of absorption signal due to (BP^-DMA^+) was almost identical with the response of the apparatus and BP and DMA in acetonitrile solution forms a stable weak CT complex in the ground state, the rapid appearance of IP was concluded to be due to the excitation of the ground state CT complex. From the equilibrium constant of the complex formation and the extinction coefficient of the complex, ca.30% of the excitation light at 355nm was absorbed by the CT complex. By the laser photolysis at 397nm where the CT complex can be selectively excited, it was revealed that the production of BPH was not observed in the IP produced by the excitation of the CT complex, $^1(BP^-DMA^+)_{com}$, and only the rapid recombination with the time constant of 85ps was observed.

In Fig.1, we can also see that the absorption due to (BP^-DMA^+) increases together with the decrease of $^1BP^*$ absorbance and increase in the absorbance at 525nm ($^3BP^*$) with the elapse of the delay time after the

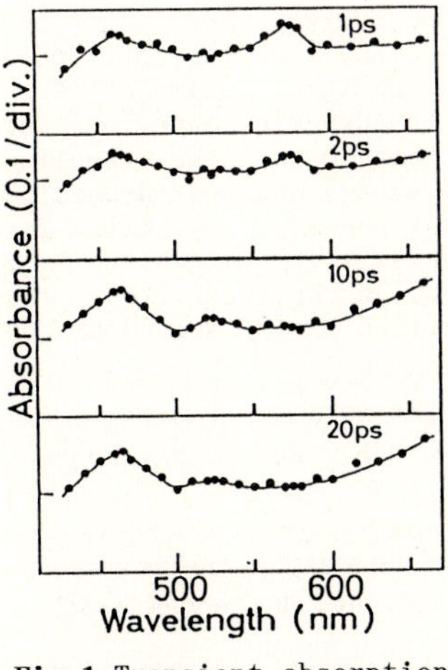

Fig.1 Transient absorption spectra of BP-DMA(1.0M) in acetonitrile excited with a 355nm fs laser pulse.

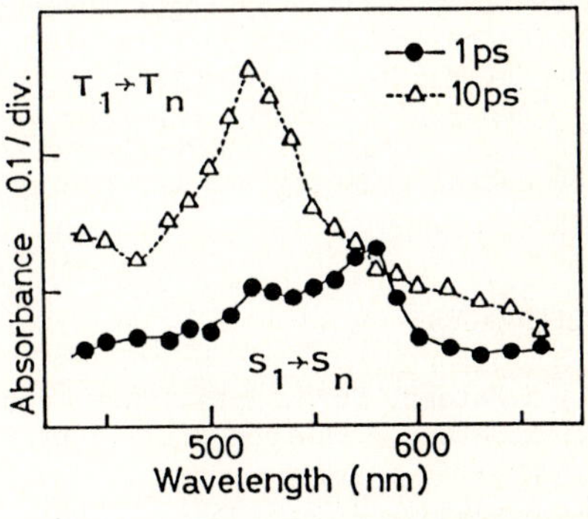

Fig.2 Transient absorption spectra of BP in acetonitrile excited with a 355nm fs laser pulse.

Table 1. Dependence of the reaction rate constants of ion pairs upon the mode of the production. Abbreviations are as follows: PT, proton transfer, ID, ionic dissociation, and CR, charge recombination.

	Lifetime	k_{PT}/s^{-1}	k_{ID}/s^{-1}	k_{CR}/s^{-1}
$^3(BP^-DMA^+)_{enc}$	140ps	$5.4x10^9$	$1.4x10^9$	-----
$^1(BP^-DMA^+)_{enc}$	460ps	$6.6x10^8$	$9.5x10^8$	$5.8x10^9$
$^1(BP^-DMA^+)_{com}$	85ps	$<<10^8$	$< 4x10^8$	$1.1x10^{10}$

excitation. The rise times of $^3BP^*$ and IP were 4.5ps and 4.8ps, respectively, and the decay time constant at 575nm ($^1BP^*$) was 5.0ps. From the coincidence of the time constants and the time evolution of the transient absorption spectrum, it may be concluded that ET reaction between $^1BP^*$ and DMA competes with the ISC process. Since the lifetime of $^1BP^*$ in amine free solution was 9.0ps, almost half of the $^1BP^*$ was quenched by DMA, which resulted in the production of $^1(BP^-DMA^+)_{enc}$, and the rest was terminated in the triplet state (which also resulted in the IP as stated above). Summarizing the above results,it is concluded that three kinds of IP are produced in the acetonitrile solution of BP-DMA (1.0M) by 355nm excitation.

By changing the concentration of DMA and the excitation wavelength, we have determined the lifetime and subsequent reaction rate constants of each IP (Table 1). As is clear from this table, the rate constant of each process is quite different depending on the mode of the production. It should be noted that the structure of each IP may be considered to be maintained at least during several hundreds of picoseconds, since the rate constants are different depending on the nature of each IP. The difference in k_{CR} between singlet and triplet IP's may be explained by the difference of the spin multiplicity. However that in k_{PT} cannot be accounted for only by this difference. According to the recent results [5], the compact IP probably with no intervening solvent molecule between A^- and D^+ would be produced in the case of the excitation of the CT complex, while more loose IP with intervening solvent molecules between A^- and D^+ would be formed in the IP produced by the ET at encounter. Along with this interpretation, it may be concluded from the present results that the factor controlling the hydrogen abstraction process is not only the mutual distance between the pair but also the initial geometry for the interaction, since the structure of each IP may be considered to be maintained at least during the lifetime of IP.

In relation to this, we have studied also the dynamic behaviors of chained molecules such as BP-O-$(CH_2)_n$-O-AH. In this case, the reaction mechanism for the hydrogen abstraction (scheme 1 or 2) changed depending on the chain length. This result also suggests that the mutual geometry for the interaction is important in the reaction mechanism.

Summarizing these experimental results, it may be concluded that the reaction scheme of the hydrogen abstraction of excited BP is dependent on the mutual geometry and the energy level of IP.

REFERENCES

[1] N.Mataga, Pure. Appl. Chem., **56**, 1255 (1984).

[2] N.Mataga, "Photochemical Energy Conversion", Elsevier, Amsterdam (1988), p.32.

[3] N.Mataga, H.Miyasaka, T.Asahi, S.Ojima, and T.Okada, "Ultrafast Phenomena VI", Springer-Verlag, Berlin, (1988), p.511.

[4] H. Miyasaka and N. Mataga, Bull. Chem. Soc., Jpn., **63**, 131, (1990).

[5] T.Asahi and N.Mataga, J.Phys.Chem., **93**, 6575 (1989).

Femtosecond Laser Photolysis Studies on the Conformation Change of Benzil in Solutions

H. Miyasaka and N. Mataga

Department of Chemistry, Faculty of Engineering Science, Osaka University, Toyonaka, Osaka 560, Japan

It is well-known that benzil (diphenylethanedione) has near-skewed form in the ground state, while the structure in the excited singlet state has been attributed to trans-planar geometry. (The terms "skewed" and "planar" used here refer to the dihedral angle between the planes containing the benzoil groups.) This structural change of benzil in solution has been considered to be very fast at room temperature. In this study, we have directly measured the structural change process by means of transient absorption spectroscopy with a femtosecond laser. On the basis of the experimental result, we will discuss the factors controlling the change of molecular structure.

Experimental

A femtosecond laser photolysis system [1] was used for the measurement of the transient absorption spectra. Second harmonics (355nm) with 500fs fwhm or 200fs fwhm of pyridine 1 dye laser (710nm) were used for the excitation. The monitoring white light was generated by focusing the fundamental pulse into D_2O. All the measurements were performed at 22°±2°C.

Results and Discussion

Figure 1 shows the time-resolved transient absorption spectra of benzil in n-hexane (a) and in cis-decalin (b). As shown in this figure, with increase of the delay time after the excitation, a broad absorption spectrum whose maximum at 545nm gradually evolves into a rather sharp spectrum with its maximum at 530nm in each solution. The former absorption spectrum with its maximum at 545nm, which was observed immediately after the excitation in each solvent, may be attributed to the absorption spectrum of excited benzil in skewed form, since the spectral shape and the absorption maximum is very close to the S_n<-S_1 absorption spectrum of benzil crystal where the structure is skewed form [2]. On the other hand, the latter spectrum is assigned to the S_n<-S_1 absorption of trans-planar form, since its absorption decay time is identical with the fluorescence lifetime. Accordingly, the spectral change observed in Figure 1 indicates that the structural change of benzil in the excited state in solution proceeds in the subpicosecond to picosecond time regions.

In order to obtain the time constant of the structural change, we plotted the time profile of the absorbance in Figure 2, where the signal around at the absorption maximum of the skewed form was exhibited. As is clear from Figure 2, the transient absorbance appears almost within the time resolution of the apparatus, followed by the rapid decrease and reaches

Springer Series in Chemical Physics, Vol. 53 **Ultrafast Phenomena VII**
Editors: C.B. Harris · E.P. Ippen · G.A. Mourou · A.H. Zewail

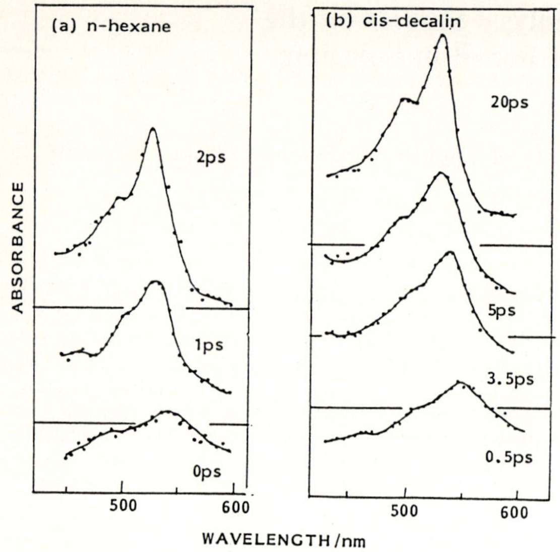

Figure 1. Transient absorption spectra of benzil in solutions, excited with 355nm femtosecond pulse. Chirping of the monitoring light was corrected.

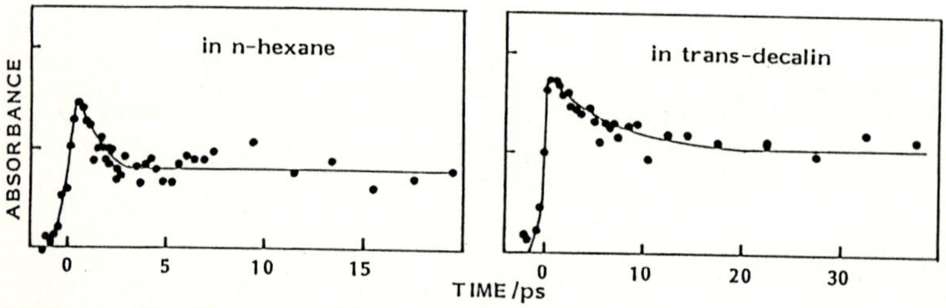

Figure 2. Time profile of the transient absorbance around at 550nm of benzil in solutions, excited with 355nm femtosecond laser pulse.

the plateau value in the time region listed in this figure. In addition, the increase of the absorbance whose time constant was similar to this decrease was observed in the absorption maximum of the trans-planar form of the excited benzil in each solution. Accordingly, this decrease of the absorbance in Figure 2 may be attributed to the time constant of the structural change from the skewed to the trans-planar form. The time for the conformation change in n-hexane was obtained to be 580fs and that in cis-decalin was 6.1ps.

In Figure 3, we show the relation between the time constant for the skewed->trans-planar change of the molecular structure and the viscosity of the solvent. This figure shows that the time for the conformation change is in proportion to the viscosity of the solvent, which indicates that the solvent friction is of crucial importance in regulating the structural change and that the dependence on the macroscopic viscosity holds even in such a short time region.

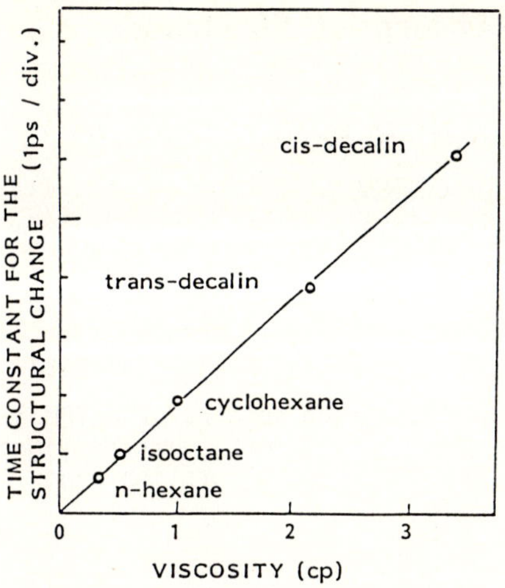

Figure 3. Solvent viscosity dependence of the time constant for the structural change of excited benzil.

References

[1] N.Mataga, H.Miyasaka, T.Asahi, S.Ojima, and T.Okada, Ultrafast Phenomena VI, edited by T.Yajima, K.Yoshihara, C.B.Harris, and S.Shionoya, (Springer-Verlag, Berlin Heidelberg, 1988),p511.

[2] N.Ikeda, M.Koshioka, H.Masuhara, and K.Yoshihara, Chem.Phys.Lett., 150(1988)452.

Fast Change in the Excited State Absorption Spectrum of TPM Dye Solutions

M.M. Martin, P. Plaza, Nguyen Dai Hung, and Y.H. Meyer

Laboratoire de Photophysique Moléculaire du CNRS,
Bât. 213, Université Paris-Sud, F-91405 Orsay, France

Abstract. Time evolution of the transient absorption spectra of a TPM dye is measured in pump-probe experiments using an unconventional subpicosecond tunable dye laser system. New transient absorption bands in the blue are reported.

1. Introduction

Relaxation dynamics of Triphenylmethane dyes in the excited state has been extensively investigated using fluorescence and pump-probe techniques on nanosecond, picosecond and femtosecond time scales (see ref. in [1-3]). It is well established that the relaxation rate is determined by the solvent viscosity but the interpretation of the transient decays measured at a few wavelengths is not completely elucidated. Time evolution of transient absorption spectra of a TPM dye is reported here on subpicosecond and picosecond time scales. New transient bands in the blue are evidenced.

2. Experimental

We measured the time resolved absorption spectra of Ethyl Violet solutions in the 320-900 nm spectral range during and after laser excitation using an original subpicosecond dye laser system at 606 nm [4] and a supercontinuum of white light as the probe in a two beam scheme. The supercontinuum was also used to obtain a tunable picosecond pump source after spectral selection and amplification. Transient spectra of Ethyl Violet in acetonitrile, ethanol and decanol, were taken after excitation in both the excited singlet states S_1 and S_2, at 606 and 560 nm. Probe beams transmitted by the excited sample and by a reference non-excited sample were simultaneously detected through a polychromator by a computer-controlled double diode array detector.

3. Results and Discussion

Fig.1 shows the change in optical density of an Ethyl Violet solution in acetonitrile, probed in the 320-520 nm wavelength range, within the - 0.83 ps to + 8.8 ps time range, when excited at 606 nm. Zero delay time was defined at the half height of the S_1-S_n absorption rise curve of the dye HIDCI in ethanol at 448 nm under the same conditions. A fast spectral shift of the excited state absorption spectrum is observed when probing during laser excitation (- 0.5 ps to + 0.5 ps). Similar results were obtained in ethanol and decanol. For delay times above + 1.5 ps, new absorption bands can be seen at 415 and around 490 nm superimposed on the ground state bleaching. In all solvents, these spectral changes occur together with the previously reported delayed bleaching [3] in the 530-580 nm wavelength range and with the decrease in the gain signal above 600 nm. At longer delays, the blue bands decay without further spectral change at

Springer Series in Chemical Physics, Vol. 53 **Ultrafast Phenomena VII**
Editors: C.B. Harris · E.P. Ippen · G.A. Mourou · A.H. Zewail

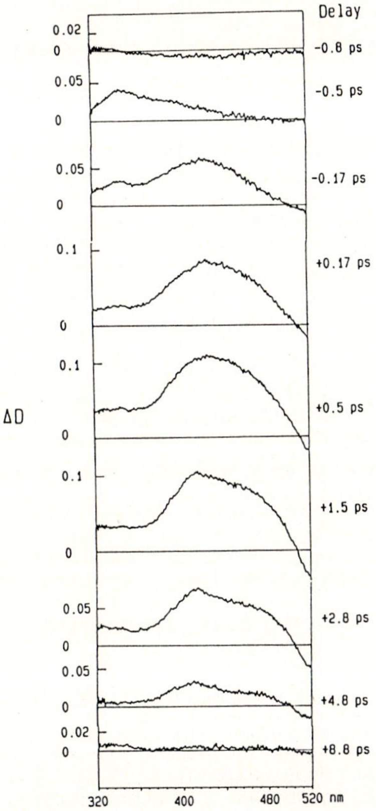

Fig.1 Time resolved transient absorption spectra of Ethyl Violet in acetonitrile under 0.6 ps laser pulse excitation at 606 nm.

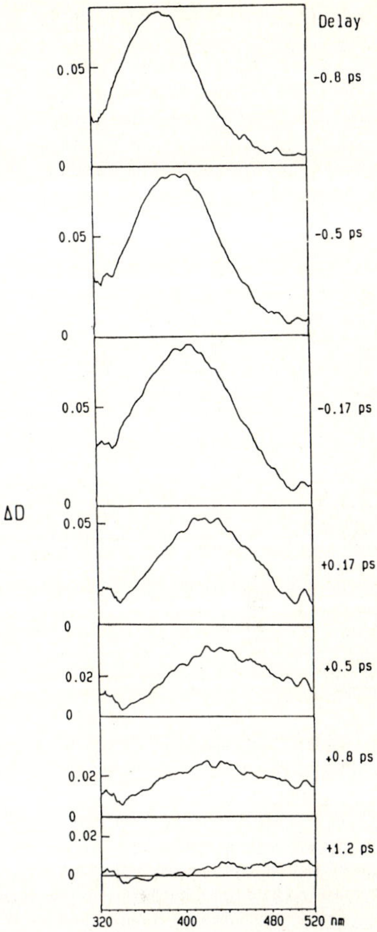

Fig.2 Two-photon absorption of 1-Chloronaphthalene measured under the same experimental conditions.

viscosity dependent rates. Similar results were obtained when tuning the excitation laser at 560nm.

We attribute the spectral shift observed when probing during the laser excitation (- 0.5 to + 0.5 ps) to the group velocity dispersion of the continuum probe. The chirp in the probe pulse is evidenced in fig.2 : two-photon absorption measurements in 1-Chloronaphthalene clearly show the probe chirp. The zero delay time corresponds to the arrival time of the probe pulse at 448 nm.

We explain the fast change in shape of the transient absorption spectrum of Ethyl Violet after excitation (fig.1, 0.5 - 2.8 ps) by electronic relaxation via an intermediate state. We previously assigned the various shapes of the rise and decay curves of the absorbance measured at selected wavelengths by a relaxation model involving the rapid formation of a transient state from the first excited state [3]. The present new results demonstrate the formation of a transient state absorbing in the same wavelength range as the S_1 state, with a peak at 415 nm, confirming our previous statement that the transient absorption decay measured in the blue wavelength range cannot be directly compared to the fluorescence lifetime or gain decay [3]. TPM dye relaxation mechanism has been described as a barrierless torsional deactivation process in a number of studies [ref in 1-3]. The BFO theoretical analysis of such a process [5] shows that the excited state decay rate should depend on the excitation wavelength. Excitation wavelength effect has not been found in this study. It was not observed either in previous experimental analysis [6].

At long delay times the blue transient peak at 415 nm, remains together with that observed in the red edge of the ground state absorption band which was attributed to the absorption of the non thermally equilibrated electronic ground state, in the case of Malachite Green [7]. In the present study similar spectral change with viscosity dependent rate were observed in acetonitrile, ethanol and decanol. Blue and red transient absorptions still persist more than 60 ps after excitation in decanol. Slow solute-solvent energy transfer was reported for Azulene in CCl_4 after $S_1 - S_0$ internal conversion [8] and explained by the lack of solvent high lying vibrational states such as those due to the C-H modes. The long decay time of the intermediate state absorption measured for Ethyl Violet in decanol, may indicate that a deactivation process such as internal conversion followed by energy transfer to the solvent may not completely explain the viscosity controlled S_1 relaxation pathway of this dye.

References

[1] G.R. Fleming: "Chemical applications of ultrafast spectroscopy", Oxford University Press, Oxford (1986), p.190.190.

[2] D. Ben-Amotz, C.B. Harris: J. Chem. Phys., 86, 4856,5433 (1987)

[3] M.M. Martin, E. Breheret, F. Nesa, Y.H. Meyer: J. Chem.Phys.,130, 279 (1989)

[4] Y.H. Meyer, M.M. Martin, F. Nesa, E. Breheret: J. Phys., 48, C7-12, 397 (1987). Nguyen Dai Hung, Y.H. Meyer, M.M. Martin, F. Nesa: VI. International Symposium on Ultrafast Phenomena in Spectroscopy, Neubrandenburg ,GDR, (1989)

[5] B. Bagchi, G.R. Fleming, D.W Oxtoby:78, 7375 (1983)

[6] U. Aberg, E. Akesson, H. Bergström, T. Gillbro, V. Sundström: Springer Series in Chemical Physics,Vol 48: Ultrafast Phenomena VI, p.520, Springer-Verlag (1988)

[7] T. Robl, A. Seilmeier, Chem. Phys. Lett.: 147, 544 (1988)

[8] A. Seilmeier, W. Kaiser: Topics in Applied Physics, Vol 60 : Ultrashort Laser Pulses and Applications, p.279, Ed W Kaiser, Springer Verlag (1988)

Temperature Dependent Superradiant Decay in Molecular Aggregates

F.C. Spano, J.R. Kuklinski, and S. Mukamel

Chemistry Department, University of Rochester, Rochester, NY 14627, USA

ABSTRACT. The effect of acoustic and optical phonons on the superradiative decay in molecular aggregates is theoretically investigated, via numerical solution of a generalized master equation for the N exciton populations. The emergence of a coherence domain, over which the aggregate cooperatively radiates, is consistent with previous experimental observations.

When a concentrated solution of certain dye molecules, such as pseudo-isocyanine bromide (PICBr), is cooled to sufficiently low temperatures, aggregates, consisting of a regular arrangement of dye molecules, are formed [1-3]. The absorption spectrum of J- aggregates is much more intense, narrower and red-shifted relative to the monomer spectrum, indicating that the excited state dynamics is governed by Frenkel excitons. This model also predicts an ultrafast radiative decay rate, which scales linearly with aggregate size N [4]. deBoer and Wiersma [1] have recently observed this superradiant decay from low temperature PIC Br J-aggregates. Between 1.5K and ~ 50K the radiative lifetime is roughly 100 times shorter than the monomer lifetime. Above 50K the lifetime increases roughly linearly with temperature.

The existence of a temperature dependent coherence domain, which contains N_{eff} ($< N$) coherently coupled molecules and hence a radiation rate proportional to N_{eff}, would explain the observed behavior in ref [1]. The domain size is determined by the magnitude of the exciton-phonon coupling and the temperature; higher temperatures, or more efficient exciton-phonon coupling, result in a smaller N_{eff}. This model has also been used to describe excited state dynamics in macroscopic semiconductor quantum wells [5] and J-aggregate monolayers [6], where N_{eff} must necessarily converge (to a value N^*) as N approaches infinity in order to remain independent of N. We have developed a microscopic theory which predicts a convergent value of N_{eff} in the presence of both acoustic (intermolecular vibrations) and optical (librations) phonon coupling to excitons.

Our starting point is an aggregate Hamiltonian which includes excitons (bandwidth = 4V), phonons (acoustical phonon bandwidth = Ω_{ac}, dispersionless optical phonon frequency = Ω_{op}) and linear exciton-phonon coupling (F_{ac},F_{op}) in the so called weak coupling form [7]. The details of our procedure can be found elsewhere [8]. Basically, a k space generalized master equation (GME) is derived for the N exciton populations by truncating an infinite hierarchy of equations of motion (generated by the Heisenberg equation) for all exciton-phonon operators .

Springer Series in Chemical Physics, Vol. 53 **Ultrafast Phenomena VII**
Editors: C.B. Harris · E.P. Ippen · G.A. Mourou · A.H. Zewail

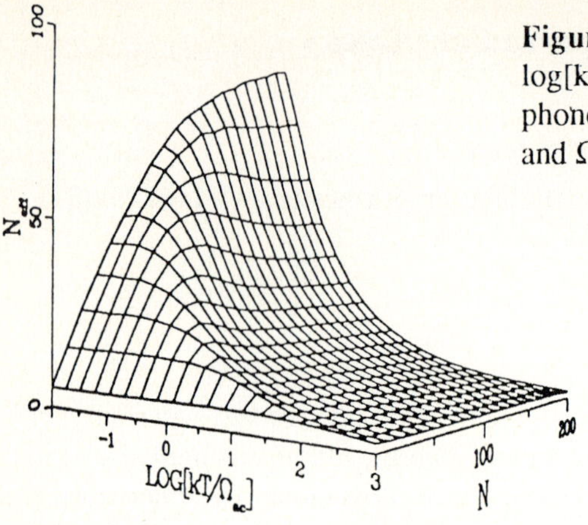

Figure 1 N_{eff} vs. N and $\log[kT/\Omega_{ac}]$, for acoustic phonons with $F_{ac} = 0.05V$ and $\Omega_{ac}=0.01V$.

The truncation is accomplished by factorizing the expectation value of bi-quadradic exciton-phonon operators into the product of the expectation values of the exciton and phonon parts, eventually leading to a set of N coupled integro-differential equations for the N exciton populations which are then solved numerically. The fluorescence intensity observable is directly proportional to the expectation value of the k=0 exciton population [4]. The resulting time dependent fluorescence decay is generally nonexponential; we define a generalized lifetime, τ_{fl} to be the time at which a fraction (1-1/e) of the total number of photons are emitted.The effective radiation rate (or coherence domain size), N_{eff}, is then simply equal to $1/\gamma\tau_{fl}$ where γ is the monomer radiation rate. Figure 1 shows N_{eff} as a function of temperature and size N for an aggregate with acoustic phonons. At fixed temperature, the aggregate radiates superradiantly as N increases ($N_{eff} = N$) but eventually levels off to a value, N^*, which remains constant as N increases. We have numerically verified that N^* is unchanged up to N=500. This is direct evidence for the convergent coherence domain hypothesis discussed above. Next, consider the effect of temperature for a given size, N. At low temperatures N_{eff} is independent of temperature until a threshold temperature is reached, at which point N_{eff} begins to decrease. For very high temperatures the aggregate lifetime approaches the monomer value (where $N_{eff} = 1$), independent of N. The temperature dependence is thermally activated because of the discrete nature of the phonon band structure, and the threshold temperature decreases as N increases because the phonon band structure approaches a continuum. Our numerical studies also demonstrate that when dispersionless optical phonons are coupled to excitons, the threshold temperature persists, and is independent of size as N increases. This leads us to conjecture that optical phonon-exciton coupling may be responsible for the observations of deBoer and Wiersma [1]. In Figure 2 we show N_{eff} vs inverse temperature for the case when $\Omega_{op} = 0.4V$ and $F_{op} = 0.35V$. (For PIC Br $4V \approx 2400\ cm^{-1}$ and $\gamma = 3.7\ nsec^{-1}$). The solid curves are numerical results for N = 150, 125 and 75 (from top to bottom). The open (blue site) and closed circles (red site) correspond to what is believed to be two separate PIC Br J-aggregate structures

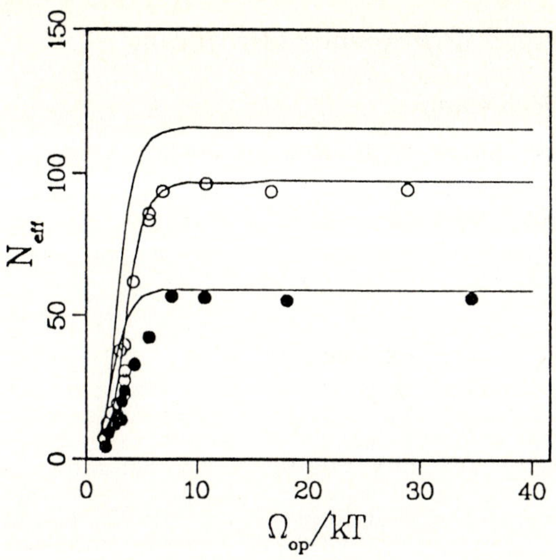

Figure 2 N_{eff} vs Ω_{op}/kT for aggregates with N=75, 125 and 150 (bottom to top) $F_{op} = 0.35V$ and Ω_{op}= 0.4V. The circles represent the experimental results of ref[1a]; the open circles correspond to the blue site and the solid circles correspond to the red site.

which absorb at different wavelengths [Fig 5 in Ref 1]. Our results agree well with experiment for N = 125 (blue site) and N = 75 (red site). The fit is however not unique; larger values for N and F_{op} can also give a satisfactory fit. The threshold temperature is relatively stable over a range of N,F_{op} values so that the frequency of the optical phonon responsible for the observed behavior is $\Omega_{op} \approx 240\ cm^{-1}$.

ACKNOWLEDGEMENTS

The support of the Air Force Office of Scientific Research, the National Science Foundation and the Petroleum Research Fund, administered by the American Chemical Society, is gratefully acknowledged.

REFERENCES

1. S. DeBoer and D.A Wiersma, Chem. Phys. Lett. **165**, 45 (1990).
2. V. Sundstrom, T. Gillbro, R.A. Gadonas and A. Piskarskas, J.Chem. Phys. **89**, 2754 (1988).
3. D.V. Brumbaugh, A.A. Muenter, W. Knox, G. Mourau and B. Wittmershaus, J. Lum. **32**, 783 (1984).
4. F. C. Spano and S. Mukamel, J. Chem. Phys. **91**, 683 (1989).
5. J. Feldmann, G. Peter, E.O. Gobel, P. Dawson, K. Moore, C. Foxon and R.J. Elliot, Phys. Rev. Lett. **59**, 2337 (1987).
6. D. Mobius and H. Kuhn , J. Appl. Phys. **64**, 5138 (1988).
7. A.S. Davydov in "Theory of Molecular Excitons" (Plenum Press, New York, 1971).
8. F.C. Spano, J.R. Kukliniski and S.M. Mukamel (in preparation).

Studies of Diffusion-Influenced Fluorescence Quenching

D.D. Eads [1], *B.G. Dismer* [2], *and G.R. Fleming* [2]

[1]Department of Chemistry, Princeton University, Princeton, NJ 08544, USA
[2]Department of Chemistry and James Franck Institute, University of Chicago, 5735 S. Ellis Ave., Chicago, IL 60637, USA

The study of diffusion limited reactions has occupied a central place in chemistry for many years. When the intrinsic reaction is sufficiently rapid the rate of mass transport becomes the rate limiting step and the reaction is considered to be diffusion controlled or diffusion influenced. Previous studies of diffusion controlled reactions [1,2] have found excellent agreement between experimental results and the classic descriptions by Smoluchowski [3] and Collins and Kimball [4]. We present data here that demonstrates a direct observation of non-Smoluchowski behavior of a diffusion influenced bimolecular fluorescence quenching reaction.

The Smoluchowski model describes a system consisting of two species, A and B, in a continuum solvent where the concentration of A particles is very low relative to the concentration of B particles. Under these conditions A can be modeled as a stationary sink and the mobile quencher B is assigned the sum of the diffusion constants of the particles A and B ($D = D_a + D_b$).

The Smoluchowski boundary condition is that of a reaction which is assumed to be infinitely fast. As relatively few bimolecular reactions in solution are truly instantaneous the utility of the Smoluchowski equation is rather limited. A significant improvement in the description can be realized by utilizing the Collins and Kimball [4] boundary conditions which assumes that the reaction rate is finite:

$$k_{\alpha}(t) = \frac{4\pi DRK_o}{4\pi DR + K_o}\left[\, 1 + \frac{K_o}{4\,\pi DR}\exp(\alpha^2 D\, t)\,\mathrm{erfc}(\alpha\sqrt{tD})\,\right]; \qquad \alpha = \frac{4\pi DR + K_o}{4\pi DR^2} \qquad \text{eqn. 1}$$

where R is the distance at which the boundary condition is evaluated, and K_o is a rate parameter that is related to the intrinsic reaction rate [4].

Although the Collins and Kimball solution ignores much of the molecular phenomenon that are expected to play a role in determining the fluorescence quenching dynamics, this description has been amazingly successful in describing the experimental data collected to date. A notable exception was noted by Nemzek and Ware [1b] who found that in order to self consistently fit their steady-state and time resolved data it was necessary to consider the static quenching. The phenomenon of static quenching occurs when there exist A-B pairs which, upon excitation of A, decay with a rate which is dominated by the intrinsic reaction rate. This can be attributed to A-B pairs that are initially separated by distances on the order of the contact distance so that diffusion is not the rate limiting step. Under these conditions the quenching dynamics will consist of a rapid transient that is characterized by the

Springer Series in Chemical Physics, Vol. 53 **Ultrafast Phenomena VII**
Editors: C.B. Harris · E.P. Ippen · G.A. Mourou · A.H. Zewail

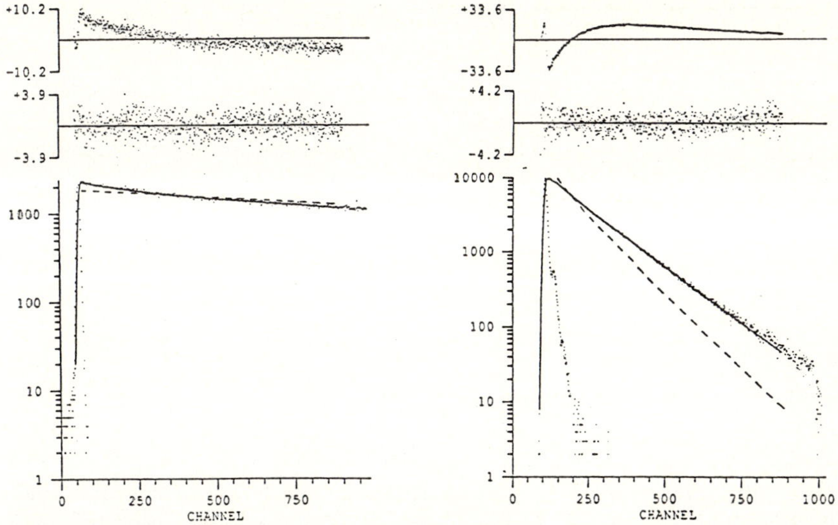

Fig 1. Time resolved fluorescence quenching of rhodamine B with 0.1 M $Fe(CN)_6^{-4}$ fit with eqns. 1 and 2.
1a) Upconversion data (0.2 ps/channel). The solid line through the data and the lower residuals are from a fit with R=1.23 nm, τ_R=36.5 ps, and D=1.25x10^{-5} cm^{-2}/s; the upper residuals and the dashed line are with the parameters of Fig. 1b.
1b) TCSPC data (4.6 ps/channel). The solid line through the data is a fit where R=1.15 nm, τ_R=87.7 ps and D=1.25x10^{-5} cm^2/s and the dashed line is for a fit with the parameters of Fig. 1a.

bimolecular rate and a term that is characterized by the diffusion controlled dynamics. To obtain an appropriate description of these dynamics we have made an ad hoc combination of the Collins and Kimball solution (eqn. 1) and a short time limit sink function that has been found by Szabo [7] to obtain the following equation:

$$I(t) = I_o \exp\left(-\left(\frac{t}{\tau_o} + \frac{4\pi c(R^3 - a^3)(1 - \exp(-t/\tau_R))}{3} + c\int_{t=0}^{t'} k_D(t)dt\right)\right) \quad \text{eqn. 2}$$

where τ_R is the bimolecular reaction rate, v is the volume of a shell surrounding the fluorophore between the van der Waals' contact distance and the Collins and Kimball radiation boundary distance, τ_o is the fluorescence lifetime of A and k_D is given by eqn. 1.

The system which we have examined is that of aqueous rhodamine B (pH 1.5) quenched by $Fe(CN)_6^{-4}$ using the techniques of time correlated single photon counting (TCSPC) fluorescence upconversion and steady state fluorescence.

We have examined several Smoluchowski type models of diffusive quenching which include simple ionic interactions (Debye-Smoluchowski equation) and distance dependent sink functions and have found that we are unable to self consistently fit the data [5,6]. Fig. 1 illustrates this by showing fits of data collected with the upconversion and TCSPC apparatus. If the TCSPC and upconversion data are fit with eqn. 2 [6,7] the optimized parameters are inconsistent between the two fits.

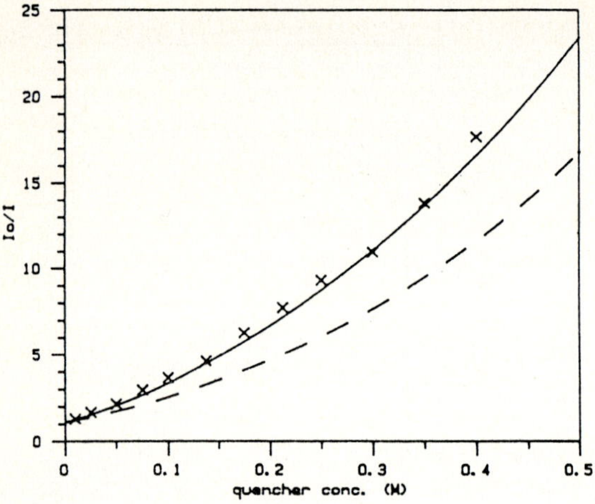

Fig 2) Steady state fluorescence intensity as a function of quencher concentration. Experimental data (x). Theoretical curves are expressed in references 6 and 7. The dashed and solid lines are based on fitted parameters obtained from eqns. 1 and 2, respectively.

Typically, the reaction rate is found to be a factor of three greater for the upconversion fit than for the TCSPC data (with each of the other variables the same for the two data sets, Fig. 1a,b).

In addition to predicting the time dependence of the data, a comprehensive model of diffusive quenching must also self consistently predict the Stern-Volmer behavior of the steady state data. In Fig. 2 we show a Stern-Volmer plot of our data and the predicted behavior according to the TCSPC and upconversion data using Szabo's formalism [7]. From these data we observe that while the data and theory are not self consistent, the upconversion and steady-state data can be described with a rate constant of $4x10^{-2}$ ps^{-1}, R=1.0 nm and $D=1.25x10^{-5}$ cm^2/s. While these comparisons clearly show a discrepancy between the data and the fits, it should be noted that the use of more primitive models which do not include a sink function results in data fits that are of significantly lower quality [6].

1. (a) R.M. Noyes, Prog. React. Kinet. 1, 129 (1961).
   (b) T.L. Nemzek and W.R. Ware, J. Chem. Phys. 62, 477 (1975).
2. (a) N. Joshi, M.L. Johnson, I. Gryczynski and J.R. Lakowicz, Chem. Phys. Lett. 135, 200 (1987).
   (b) E. Pines, D. Huppert, and N. Agmon, J.Chem. Phys. 88, 5620, 5639 (1988).
   (c) N. Periasamy, S. Doraiswamy, B. Venkataraman, and G.R. Fleming, J. Chem. Phys. 89, 4799 (1988).
3. M. Smoluchowski, Phys. Z. 17, 55 (1916).
4. F.C. Collins and G.E.Kimball, J. Colloid. Sci. 4, 425 (1949).
5. D.D. Eads, N. Periasamy and G.R. Fleming, J. Chem. Phys. 90, 3876 (1989).
6. D.D. Eads, B.G. Dismer and G.R. Fleming, J. Chem. Phys. in press.
7. A. Szabo, J. Chem. Phys. 93, 6929 (1989).

Zero-Phonon Line and Ultrafast Dephasing in Nile Blue Doped Polymer Film

S. Nakanishi, H. Ohta, and H. Itoh

Department of Physics, Kagawa University, Saiwai-Cho 1-1,
Takamatsu City, Kagawa 760, Japan

Abstract: Picosecond optical dephasing of zero-phonon line and femtosecond dephasing in Nile blue doped polymer film have been investigated over inhomogeneously broadened absorption by using incoherent photon echo technique.

Nile blue has been extensively studied in picosecond and femtosecond laser spectroscopy and several ultrafast relaxation features have been reported in solution and polymer host by using femtosecond laser pulses [1-5]. In those investigations, the relaxation of the excited state, vibration, optical coherence and sublevel coherence has been reported. However, within our knowledge, optical dephasing of zero-phonon line (ZPL) in Nile blue has not been studied. In this paper we report picosecond optical dephasing and the inhomogeneous distribution of ZPL in Nile blue doped polyvinyl alcohol (PVOH) polymer film measured with incoherent photon echo technique at 10 K. We also show the presence of femtosecond dephasing ($\sim$ 100 fs) at the top of absorption.

Experimental apparatus consists of a broadband DCM dye laser pumped by second harmonics of YAG laser, producing pulses of 8 ns duration at repetition rate of 10 Hz. To detect photon echoes from ZPL having relatively long dephasing time, we employ laser pulses with the bandwidth $\Delta\lambda$ of about 1.6 nm . These pulses give the time resolution, determined by the relation $\tau_c = \lambda^2/\pi c\Delta\lambda$, of 0.4 ps. In order to measure femtosecond dephasing, no tuning element is used in the laser cavity. In this case the bandwidth is 15.6 nm corresponding to the time resolution of about 40 fs. Output of the dye laser is split into first (wave vector $\mathbf{k}_1$) and second ($\mathbf{k}_2$) pulses. The second pulse is delayed in time. Both pulses are focussed onto the sample kept at 10 K in a cryostat. Photon echoes emitted in the direction $2\mathbf{k}_2$-$\mathbf{k}_1$ are detected and accumulated by a Boxcar integrator.

Figure 1 is a photon echo signal measured at long-wavelength side (672.0 nm) of absorption. The trace includes photon echoes from ZPL and faster dephasing components, and interference between them. Optical dephasing time T_2 of ZPL is determined to be 29.3ps at 10K, which corresponds to the homogeneous linewidth Γ_h = 10.9 GHz. When laser wavelength is scanned over the inhomogeneous broadened absorption, the dephasing time of ZPL is roughly constant , but the ratio of photon echo intensity of ZPL extrapolated to delay time $\tau = 0$ to the peak signal intensity at $\tau = 0$ changes. It is well known that the peak signal intensity involves the contributions from ZPL, phonon sideband and vibronic transitions. Therefore, the ratio is determined by many factors , such as Debye-Waller factor [6], the excitation bandwidth and the inhomogeneous broadening of ZPL and faster dephasing components. Because, during the course of experiment described above, the temperature and excitation bandwidth are kept constant, the inhomogeneous broadening of ZPL and other transitions is a major factor to dominate the variation of the ratio. We can estimate the inhomogeneous broadening of ZPL in terms of the ratio, provided that the inhomogeneous distribution of other transitions is assumed

Springer Series in Chemical Physics, Vol. 53 **Ultrafast Phenomena VII**
Editors: C.B. Harris · E.P. Ippen · G.A. Mourou · A.H. Zewail

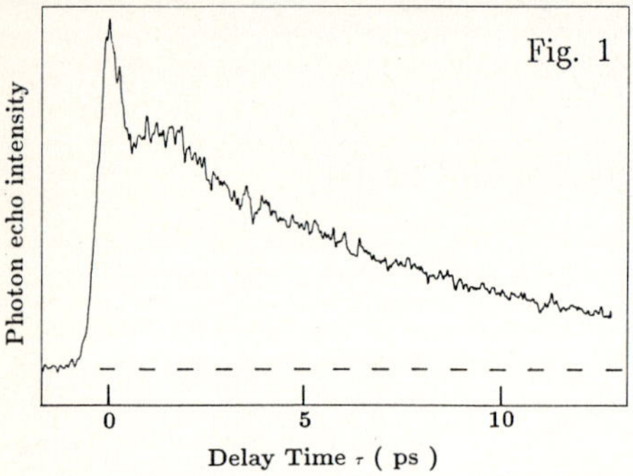

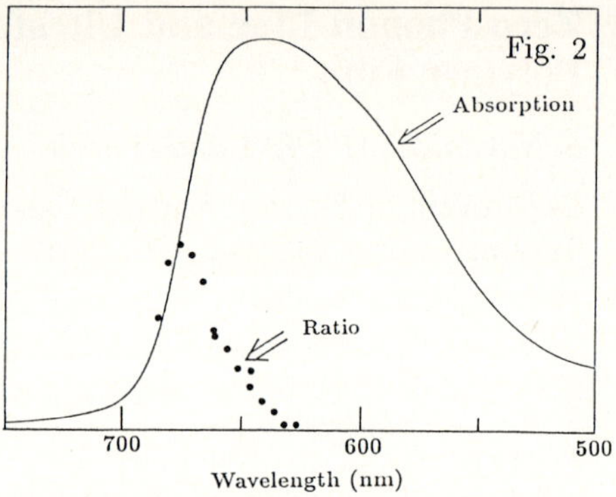

Fig. 1. Picosecond photon echo decay of ZPL measured at 672 nm and 10 K. The sharp signal spike at τ=0 consists of the contributions from ZPL, the phonon sideband and vibronic transitions.

Fig. 2. The variation of the ratio (solid circles) with the excitation wavelength, indicating the inhomogeneous broadening of ZPL. The solid line represents the absorption curve taken at room temperature.

to be much larger. We show, in Fig. 2, the ratio as a function of the excitation wavelength, indicating that the inhomogeneous distribution of ZPL has the peak wavelength of 677 nm and width (HWHM) of 15 nm. This result implies that the dephasing measurements previously performed at 625 nm [3] give the dephasing feature of the vibronic transitions rather than ZPL.

Femtosecond measurement has also been performed at the top of absorption curve. We use laser pulses with the bandwidth of 15.6 nm and center wavelength of 637 nm. We compare the photon echo decay at 10K to that at 300 K in Fig. 3, where small peak shift observed is neglected in plotting. In this case, the contribution from ZPL is too small to be detected. Femtosecond dephasing component is evidently involved in the signal at 10 K. The dephasing time is obtained in terms of the usual perturbation theory in two-level system incorporating with the stochastic nature of broadband laser light. Photon echo intensity I(τ) is simply expressed as

$$I(\tau) \propto A \int_0^\infty dT \int_0^\infty dS G(S-T) G(T-\tau) G(S-\tau) e^{-2(T+S)/T_2},$$

where $G(t)$ is a field correlation function and the extremely large inhomogeneous broadening is assumed. Result of numerical calculations (Fig. 4) shows that the dephasing time is T_2=90 fs at the top of absorption and 10 K. In the calculation, we use a function $G(t) = Cexp(-ln2t^2/\tau_c^2)$ and τ_c=40 fs. We attribute the observed femtosecond echo to the vibronic lines in Nile blue. This result contrasts with that of ref. 3, in which PMMA polymer was chosen to be a host material and no femtosecond dephasing component was detected. The reason we can detect femtosecond dephasing is probably due to the fact that our host material, PVOH, gives much less effect on the dephasing of Nile blue than PMMA, and the dephasing time is much longer in PVOH [6]. We can say that, though the excited state relaxation of the vibronic transition is supposed to be extremely fast, its optical dephasing is not lifetime-limited, but still host-dependent.

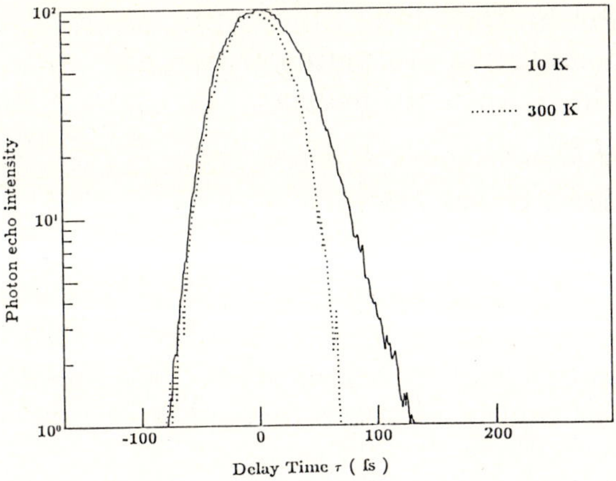

Fig. 3. Femtosecond photon echo decay measured at the top (637 nm) of absorption curve and 10 K. The dotted line is an echo decay obtained at room temperature.

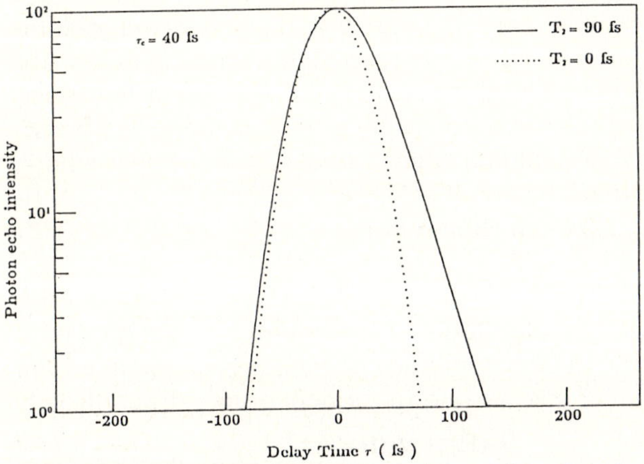

Fig. 4. Result of the numerical calculations by using the perturbation theory in two-level system. The solid and dotted lines are the calculated result with T_2=90 and 0 fs, respectively. τ_c used is 40 fs.

References

1. A.J. Taylor, D.J. Erskine and C.L. Tang, Chem. Phys. Lett. 103, 430(1984).
2. A.M. Weiner and E.P. Ippen, Opt. Lett. 9, 53(1984), Chem. Phys. Lett. 114, 456(1985).
3. A.M. Weiner, S. De Silvestri and E.P. Ippen, J. Opt. Soc. Am. B 2, 654 (1985).
4. M.J. Rosker, F.W. Wise and C.L. Tang, Phys. Rev. Lett. 57, 321 (1986).
5. P.C. Becker, H.L. Fragnito, J.Y. Bigot, C.H. Brito Cruz, R.L. Fork and C.V. Shank, Phys. Rev. Lett. 63, 505(1989).
6. S. Saikan, A. Imaoka, Y. Kanematsu, K. Sakoda, K. Kominami and M. Iwamoto, Phys. Rev. B 41, 3185 (1990).

The Impact Approximation in Transition-State Absorption Spectroscopy: On the Possibility of Obtaining Difference Potentials Directly from the Transient Spectra

R.E. Walkup, J.A. Misewich, J.H. Glownia, and P.P. Sorokin

IBM, T.J. Watson Research Center, P.O. Box 218,
Yorktown Heights, NY 10598, USA

Abstract: The impact approximation in transition-state absorption spectroscopy is introduced. Its usefulness in recovering difference potentials from transient spectra is evaluated.

For many photodissociative gas-phase systems which result in the production of atoms, transient absorption spectra taken in the vicinity of the various atomic resonance lines are observed to bear striking resemblances to the spectra shown in Fig. 1. The notable shared features are (1) the transient spectra are all centered about the free atom line positions, and (2) the spectral profiles at early times appear more dispersive than absorptive. Examples of systems which behave in this way, besides the 377-nm Tl line in TlI (Fig. 1), are the 377 and 352-nm Tl lines in Tl-cyclopentadienyl, and the 289.8-nm Bi line in Bi_2.

There is a simple physical model by which the two common features described above can be readily understood. This model, by analogy with a similar approximation made in collisional broadening theory, is called the *impact approximation*. Its use allows one (in principle) to extract directly the difference potential $\delta\omega(R) = \frac{V_1(R) - V_0(R)}{\hbar} - \omega_0$ from the transient spectra, as will be presently shown.

By way of introduction, we first consider an isolated atom that is suddenly excited at t=0. This atom will radiate with a dipole field $\propto e^{-i\omega_0 t} e^{-\gamma t}$, where ω_0 is the resonance frequency, and γ is the decay rate. The radiated power spectrum has the familiar Lorentzian shape. In transient absorption spectroscopy, the absorption spectrum arises from interference between the incident continuum pulse and the radiated wave. For an effectively white incident pulse, the signal is thus proportional to the real part of the Fourier transform of the field radiated by the excited atom. For an isolated atom, a Lorentzian shape is obtained again. However, if a collision is in progress at the time of excitation, then the radiated wave will be affected. In the simplest picture, the collision adiabatically shifts the transition frequency. In the impact approximation, one assumes that the effective duration of the collision is shorter than the characteristic

Springer Series in Chemical Physics, Vol. 53 **Ultrafast Phenomena VII**
Editors: C.B. Harris · E.P. Ippen · G.A. Mourou · A.H. Zewail

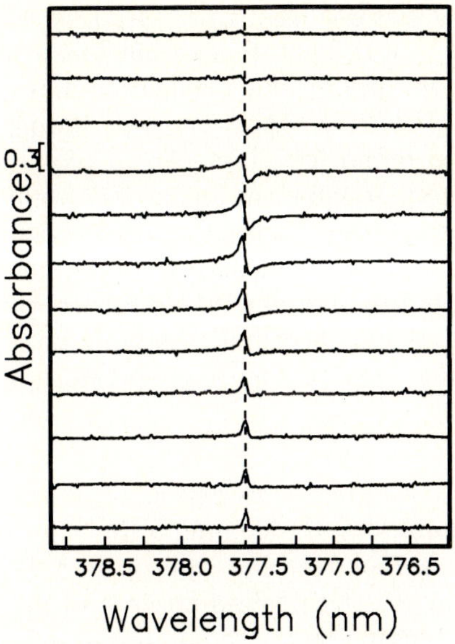

Fig. 1. Transient Tl absorption spectra resulting from 308-nm photolysis of TlI. Pump-probe separation increased by 100 fs between successive traces.

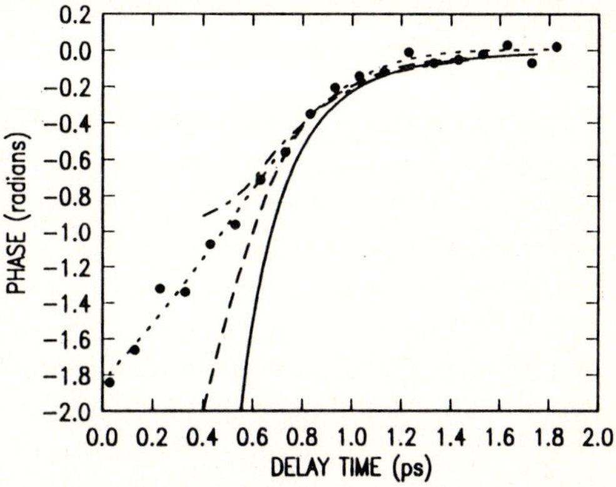

Fig. 2. Accumulated phase η vs. pump-probe separation. See text for details.

decay time. Thus most of the field is radiated after the colliding pair has separated. In this case the radiated field is approximately $e^{-i\eta} e^{-i\omega_0 t} e^{-\gamma t}$, i.e., the radiated wave is simply shifted by a phase equal to the integral of the frequency-shift from the point of excitation to $t=\infty$. The transient absorption signal, given by the real part of the Fourier transform of the radiated field, is thus $\alpha(\Delta) \propto \cos(\eta)L(\Delta) + \sin(\eta)D(\Delta)$, where L and D are Lorentzian and dispersion functions. Consider an ideal situation in which there is a unique relationship between delay time and internuclear separation. One can directly determine the phase, η, from experimental data as a function of pump-probe delay. The difference potential can then be obtained by differentiation of the phase: $\delta\omega(t_d) = -\partial\eta/\partial t$. If the relationship between delay-time and R is known, then one can determine the difference potential vs. R.

The situation for TlI is shown in Fig. 2. Transient spectra were fitted to a superposition of absorption and dispersion profiles in order to determine the phase, η. Due to noise considerations, it was necessary to fit the phase with a functional form, which could then be differentiated analytically. A global fit to the TlI data is shown as the dotted curve in Fig. 2. A good over-all fit is obtained. Differentiation of the fitted function results in an effective difference potential that is steeply attractive at large separations, but that "saturates" at $\simeq$ - 10cm^{-1}. If correct, this would imply a remarkable degree of similarity between the upper and lower potentials. There are reasons for being skeptical. The long-range van der Waals potentials for TlI should be very close to those for TlXe. From experimental and theoretical studies of TlXe, the C_6 coefficient for the 377.6-nm line is $\simeq 10^7 cm^{-1} Å^6$. One expects overlap and exchange forces (i.e. chemical effects) to be negligible for internuclear separations $> 6Å$; thus the difference potential should be dominated by van der Waals forces for $R > 6Å$. Note that at 6 Å the van der Waals potential amounts to a 200 cm^{-1} red shift; and the effective molecular sweep-rate is 2000 cm^{-1}ps^{-1}.

Are such large red shifts consistent with the data? One can answer this question by starting from an assumed potential and calculating the effective phase, including effects due to averaging over the appropriate R-distributions. The results are shown in Fig. 2. The dashed curve and the dash-dot curve show the effective phase obtained by averaging the results for a van der Waals potential with $C_6 = 10^7 cm^{-1} Å^6$. The dashed curve corresponds to a 160-fs pump pulse. The dash-dot curve was obtained with a 240-fs width. The solid curve shows the results for the same potential but with very narrow spatial distributions. The effect of averaging is quite noticeable even for rather large delay times. It is also important to note that the effective phase at small delay times is rather sensitive to the detailed shape of the pump pulse. One may conclude that for the TlI data, the effects of averaging over the distributions of R are just too severe and too uncertain to allow reliable recovery of a difference potential. Nevertheless, in principle, one could recover the difference potential if the R-distributions were sufficiently narrow. Clearly it would help to have a shorter pump pulse, and it would also help to have a flatter Franck-Condon region of the dissociative potential.

Support from the Army Research Office is acknowledged.

Part XIII

Biology

Primary Charge Separation in the Photosynthesis of Bacterial Reaction Centers

W. Zinth, W. Holzapfel, U. Finkele, C. Lauterwasser, K. Dressler, and P. Hamm

Physik-Department E 11, Technische Universität München,
Arcisstrasse 21, D-8000 München 2, Fed. Rep. of Germany

Abstract. The very rapid initial electron transfer is studied for two purple bacteria. The results favor a model of electron transfer involving a bacteriochlorophyll anion radical.

Photosynthesis is the most important photochemical reaction in nature, where light energy is converted to chemical energy. Most photosynthetic systems act according to a common principle: the absorption of a photon is followed by a first charge separation and the build-up of a potential gradient. Finally, in complex biochemical processes the synthesis of energy storing molecules takes place. The crucial part of the photosynthetic energy conversion is the primary charge separation, which occurs in so-called reaction centers (RC).

It is the purpose of this paper to show that the same reaction model well describes the primary charge separation for different purple bacteria. This fact is not obvious, since different tetrapyrols, bacteriochlorophylls (BChl) and bacteriopheophytins (BPh) are present in various reaction centers, e.g. BChl a and BPh a are essential pigments in the reaction centers of Rhodobacter (Rb.) sphaeroides while BChl b and BPh b operate in the reaction centers of Rhodopseudomonas (Rps.) viridis. For both reaction centers x-ray structure analysis has revealed the molecular arrangement /1/. It was shown that the prosthetic groups and related aminoacids are in a very similar arrangement in both reaction centers: two BChl molecules are in close contact acting as a primary electron donor P. The other pigments are arranged in two branches, A and B. Starting from the primary donor, the special pair P, one finds an accessory bacteriochlorophyll (B), a bacteriopheophytin (H), and a quinon (Q) on each branch. It was shown that the electron transfer occurs via the A-branch and after about 3 - 4 ps a radical pair $P^+H_A^-$ is formed. Another 200 ps later the electron reaches the quinon Q_A building the intermediate $P^+Q_A^-$. The role of the accessory bacteriochlorophyll B_A is still in debate /2,3/. Recent experiments on Rb. sphaeroides have proven the existence of a previously unresolved 0.9 ps kinetic which was interpreted to be related with a radical pair $P^+B_A^-$ as the real intermediate formed prior to $P^+H_A^-$ /4,5/.

The present experiments were performed using the excite and probe technique with weak subpicosecond pulses from two different laser-amplifier systems with repetition rates of 10 Hz. Excitation was made of the lowest energy band of P (at 860 nm for Rb.sphaeroides and at 950 nm for Rps. viridis). Probing was performed by a 5 nm to 20 nm wide fraction of a femtosecond light continuum. Parallel polarisation of exciting and probing pulses was used in the experiments presented here. The reaction centers were kept at room temperature under stirring.

Springer Series in Chemical Physics, Vol. 53 **Ultrafast Phenomena VII**
Editors: C.B. Harris · E.P. Ippen · G.A. Mourou · A.H. Zewail

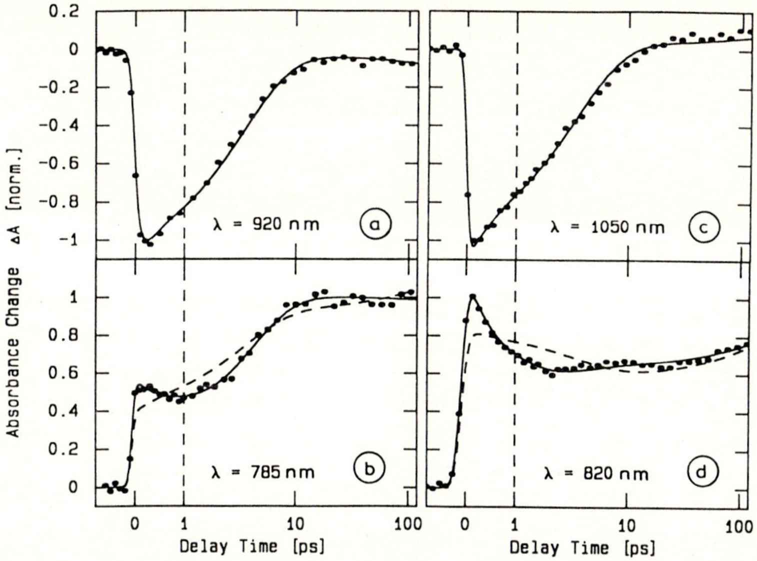

Fig.1 Transient absorption data for reaction centers from Rb. sphaeroides (a,b) and Rps. viridis (c,d). The filled circles represent the experimental data, the solid lines follow model functions using the time constants given in the text.

Time-resolved absorption data are shown in Fig.1 for different probing wavelengths. The decay of the excited electronic level of the special pair is seen at λ_{pr} = 920 nm for Rb. sphaeroides (Fig.1a) and λ_{pr} = 1050 nm for Rps. viridis (Fig.1c). Both probing wavelengths are located in the long-wave wing of the P absorption band (see Fig.2a and 2d), where the excited electronic level is visible via its stimulated emission. In Fig.1a and 1c the population of P* decays exponentially with a time constant around 3.5 ps. The situation is drastically different at some wavelengths close to the absorption band of the accessory bacteriochlorophyll. An additional fast kinetic component becomes evident. In Rb. sphaeroides at 785 nm (Fig.1b) a first absorption increase at time zero is followed by a fast relative absorption decrease before the 3.5 ps process takes over, again increasing the absorption. For Rps. viridis one finds the additional fast kinetic component quite clearly at 820 nm (Fig.1d). Extensive studies at different wavelengths gave numbers for this fast process of 0.9±0.4 ps in Rb.sphaeroides and 0.65±0.3 ps in Rps. viridis. In addition they supplied amplitudes of the various kinetic components which can be used to calculate difference spectra of the intermediate states for specific sequential reaction models.

When the fast process is assumed to precede the 3.5 ps decay, the spectral data require the existence of two excited electronic states and the electron should be transferred directly from the special pair P to the bacteriopheophytin H_A. However, this direct electron transfer gives serious difficulties for the microscopic theoretical description of the electron transfer process. In addition, recent results on mutants of reaction centers of Rb. spheroides are in contradiction to this reaction

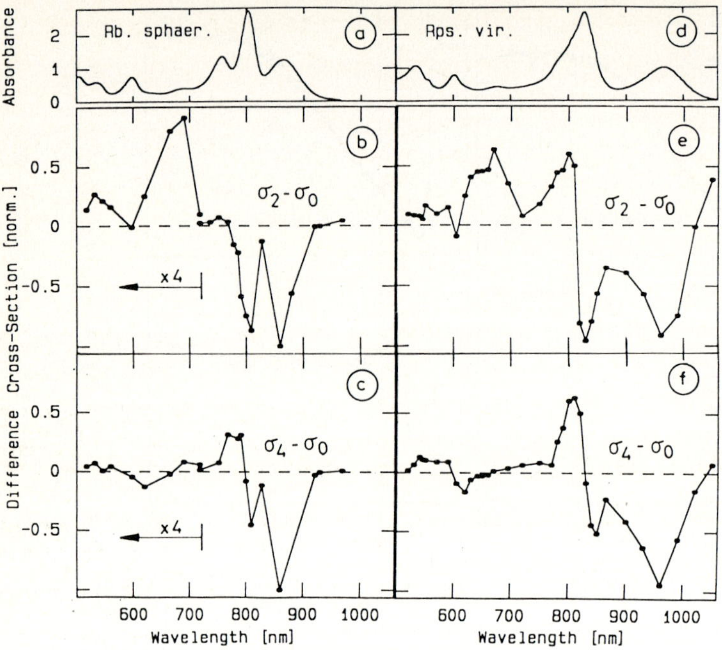

Fig.2 Spectral data for Rb. sphaeroides (a,b,c) and Rps.viridis (d,e,f). (a,d) give the absorption spectra, (b,c,e,f) show difference spectra calculated according to the second model for the intermediates I_2 and I_4.

tion model /6/. The second possibility is a model, where the 3.5 ps process precedes the 0.9ps/0.65ps decay.

$$P \xrightarrow{h\nu} \underset{P^*}{I_1} \xrightarrow{3.5ps} \underset{P^+B_A^-}{I_2} \xrightarrow[0.65ps]{0.9ps} \underset{P^+H_A^-}{I_3} \xrightarrow{220ps} \underset{P^+Q_A^-}{I_4}$$

Of interest are the difference spectra of intermediate I_2 and I_4 which are shown in Fig.2. The spectrum $\sigma_4-\sigma_0$ of the final picosecond product I_4 ($P^+Q_A^-$) does not depend on the reaction model used. The absorption changes observed are well known. They reflect the disappearance of the absorption of the special pair P and small contributions from P^+ and from electrochromic shifts. The spectrum $\sigma_2-\sigma_0$ of intermediate I_2 again exhibits some of the features known for P^+ formation. In addition, there are strong absorption changes of the Q_y band of the accessory bacteriochlorophyll (at 800 nm for Rb. sphaeroides, at 820 nm for Rps. viridis) and around 660 nm (in the bacteriochlorophyll anion band) indicative of the formation of a bacteriochlorophyll anion radical. Consequently, the calculated difference spectra give strong evidence of the existence of $P^+B_A^-$ as a short-lived intermediate in the primary electron transfer in the reaction centers of the two purple bacteria. The notion of $P^+B_A^-$ being the second intermediate is also strongly supported by additional experiments on the transient dichroism /5/.

In conclusion: We have shown that the primary electron transfer in different bacterial reaction centers proceeds according to a common reaction scheme, where a subpicosecond - previously undetected - reaction is involved. The present data support the model of sequential electron transfer. From the excited electronic state of the special pair P* an electron proceeds to the accessory bacteriochlorophyll within 3.5 ps forming the radical pair state $P^+B_A^-$, which decays more rapidly with 0.9ps/0.65ps to the radical pair state $P^+H_A^-$ where the electron has been transferred over a distance of approximately 17A to the bacteriopheophytin.

Acknowledgement

Experiments were performed in collaboration with W. Kaiser, H. Scheer, D. Oesterhelt, U. Stilz, S. Buchanan, and H. Michel.

References

/1/ J. Deisenhofer, H. Michel, EMBO J. 8 (1989) 2149

/2/ J.L. Martin, J. Breton, A.J. Hoff, A. Migus, A. Antonetti, Proc. Natl. Acad. Sci. USA 83 (1986) 957

/3/ J. Breton, J.L. Martin, A. Migus, A. Antonetti, A. Orszag, Proc. Natl. Acad. Sci. USA 83 (1986) 5121

/4/ W. Holzapfel, U. Finkele, W. Kaiser, D. Oesterhelt, H. Scheer, H.U. Stilz, W. Zinth, Chem. Phys. Lett. 160 (1989) 1

/5/ W. Holzapfel, U. Finkele, W. Kaiser, D. Oesterhelt, H. Scheer, H.U. Stilz, W. Zinth, Proc. Natl. Acad. Sci. USA, accepted for publication (1990).

/6/ U. Finkele, C. Lauterwasser, W. Zinth, K. Gray, D. Oesterhelt, to be published (1990)

Primary Charge Separation Process in Reaction Centers from *Chloroflexus aurantiacus* Bacterium

J.L. Martin[1], *J.C. Lambry*[1], *M. Ashokkumar*[1], *M.E. Michel-Beyerle*[2], *R. Feick*[3], and *J. Breton*[4]

[1]Laboratoire d'Optique Appliquée, INSERM U275,
Ecole Polytechnique–ENSTA, F-91128 Palaiseau Cedex, France
[2]Technische Universität München, D-8046 Garching, Fed. Rep. of Germany
[3]Max-Planck-Institut für Biochemie, D-8033 Martinsried, Fed. Rep. of Germany
[4]Service de Biophysique, CEN Saclay, F-91191 Gif-sur-Yvette, France

In photosynthetic bacteria the light is first absorbed by antenna proteins, which contain chlorophyll-like chromophores. The initially trapped excitation very rapidly migrates to a membrane protein called the reaction center where the primary charge separation subsequently occurs. This light-induced electron transfer reaction is extremely fast even at cryogenic temperatures (700 fs at 10 K in *Rps. viridis* [1]) and occurs with a near-unity quantum efficiency. The reaction center of different bacteria has been crystallized and high-resolution X-ray structures have been obtained for two different photosynthetic bacteria, *Rps. viridis* [2,3] and *Rb. sphaeroides* [4,5]. The primary electron donor in both species is the excited singlet state (P^*) of a dimer of bacteriochlorophyll (Bchl), the special pair P. These two Bchl molecules are strongly coupled electronically, and this coupling is responsible, at least in part, for striking optical properties, like the large red shifts of the infrared Qy states as compared to monomer spectra. The relative contributions of the protein moiety and of the pigment-pigment interaction to these shifts have still to be determined. Besides the red shift, two other unusual features are observed in the special pair absorption bands: a strong exciton splitting (850 cm^{-1} in *Rps. viridis*) and a broadening of the P band.

The X-ray structures of the reaction center have revealed an approximate C_2 symmetry in the spatial organization of the six porphyrin chromophores [the dimer (P) of Bchl molecules, the two monomeric Bchl molecules ($B_{L,M}$) and two bacteriophephytins ($H_{L,M}$) – where the subscripts L and M refer to the corresponding protein subunits]. Surprisingly, despite this symmetrical arrangement of the chromophores along two parallel branches, only one of these branches is active (the L branch) in reaction centers. The other striking observations of the last few years have come from femtosecond experiments in which, after direct excitation of the P^* state, the $P^+ B_L^-$ state appears not to be a kinetically resolvable intermediate in the electron transfer process [6–8] from P^* to H_L. As B_L is located in between P and H_L in the crystal structure, one might have expected the B_L^- state to be the first intermediate in the electron transfer reaction. Very recently [9] it has been proposed from the analysis of kinetics data obtained on the *Rb. sphaeroides* reaction center at room temperature that the state $P^+B_L^-$ could be involved if its rate of depletion is 5 times faster than its rate of appearance, as

Springer Series in Chemical Physics, Vol. 53 **Ultrafast Phenomena VII**
Editors: C.B. Harris · E.P. Ippen · G.A. Mourou · A.H. Zewail

originally proposed a few years ago by *Marcus* [10] from the analysis of earlier measurements [7].

However, at low temperatures, where the numerous overlapping bands present at room temperature are better resolved, no B_L^- intermediate has been detected [1,8,11]. In principle, cooling the reaction center down to cryogenic temperatures makes the analysis of kinetics data a more tractable problem, especially in the B region (800 – 830 nm) where different chromophores, including the higher energy exciton component Py_+ of P and the two monomeric Bchls (B_L and B_M) contribute to the absorption. The lack of evidence for a B_L^- intermediate under these conditions put new limits on the lifetime of this hypothetical intermediate [11]. One suggested possibility [12] is that the second step, in the two-step model hypothesis, is adiabatic. An alternative model has been proposed, where the B_L^- state is a virtual state which mediates the electron transfer from P to H via a superexchange mechanism [13].

In all cases, a prerequisite to an unambiguous interpretation of the kinetics data in the femtosecond experiments is an understanding of the reaction center electronic structure in the ground, excited and charge separated states, especially in the near infrared region, which is the most relevant to the study of the charge separation mechanism. At room temperature such an investigation would be tentative at best, as many underlying states are buried within the "B" band.

In an attempt to identify the different spectroscopic components involved in the transient optical spectra in the near-infrared region, we have performed experiments at 10 K with reaction centers from the green bacterium *Chloroflexus aurantiacus*. *Chloroflexus* reaction centers contain 3 bacteriopheophytins and only 3 Bchl molecules instead of 4 in *Rb. sphaeroides* or *Rps. viridis*. Polarized spectroscopy [14,15], together with the crystal structure of *Rb. sphaeroides* and *Rps. viridis* reaction centers, suggests that one of the monomeric B molecules in purple bacteria is replaced by a pheophytin in *Chloroflexus*. This implies that the remaining B molecule is responsible for most of the absorption at 810 nm; in femtosecond experiments, a transient depletion of the B state, if it exists, should be seen more clearly.

1. Materials and Methods

Reaction centers from *Chloroflexus aurantiacus* were prepared as previously described [16] and were suspended in Tris-HCl 25 mM at pH 7.5. The reaction centers were diluted to their final concentration with 60 % (v/v) glycerol. The samples were cooled in the dark using a convection cryostat with a temperature regulated flow of helium gas.

Tunable femtosecond pulses in the near infrared were generated from an amplified continuum as described elsewhere [7,8]. The excitation energy was adjusted so that no more than 20 % of the reaction centers were excited by each pump pulse. Transient spectra were recorded on a 2D multichannel detector and transient kinetics at a given wavelength were measured using a two-diode system at the output slit of a spectrometer.

2. Results and Discussion

Figure 1 shows transient difference spectra in the 740–890 nm region, taken 600 fs, 1.3 ps, 4 ps, 10 ps and 100 ps after excitation in the P absorption band. The spectrum at 600 fs clearly reveals the presence of an instantaneous bleaching of part of the band at 811 nm simultaneously with the appearance of an induced absorption at 806 nm. Such a transient spectrum can be induced by an instantaneous shift of a narrow absorption band (9 nm) from 811 nm to 806 nm, prior to electron transfer from P^*.

At longer time the ΔA spectra are shifted to the blue in a time scale corresponding to the kinetics of electron transfer from P^* to H. Such a shift corresponds to a time-dependent electrochromic shift of the B absorption band as observed in reaction centers from purple bacteria. Although the transient ΔA spectrum is progressively mixed with the overlapping ΔA signal induced by the Stark shift of the blue part of the band at 806 nm, the initial instantaneously populated state does not appear to relax significantly during the electron transfer process.

Several processes could be the cause of the apparent instantaneous shift of part of the absorption band around 810 nm. The first possibility is the instantaneous disruption of the exciton coupling within the dimer when the Py_- state is bleached, which causes a bleaching of the higher-energy excitonic transition Py_+ and the instantaneous appearance of a Bchl monomer-like band at 806 nm. This interpretation is supported by photoselection experiments [15], where the absorption at 813 nm has been assigned to the Py_+ transition. The presence of an isosbestic point in the femtosecond ΔA spectra (Fig. 2) is consistent with this view.

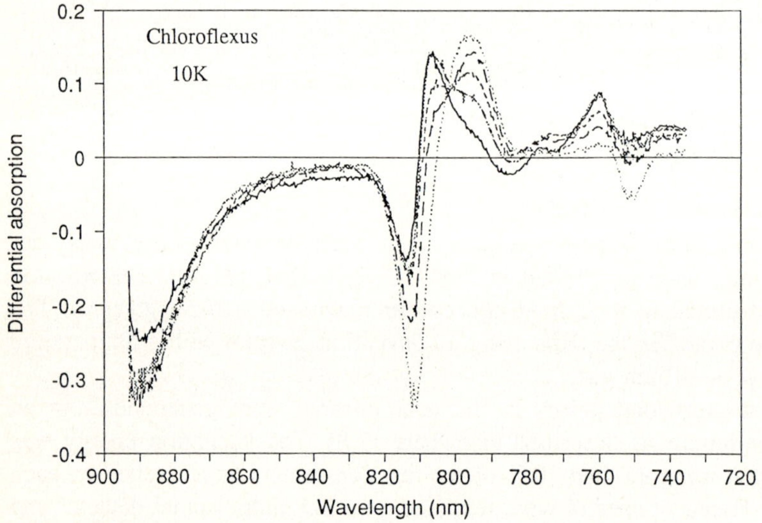

Fig. 1. Transient ΔA spectra in *Chloroflexus* reaction centers taken at 600 fs (–), 1.3 ps (– · –), 4 ps (- - -), 10 ps (— —) and 100 ps (· · ·)

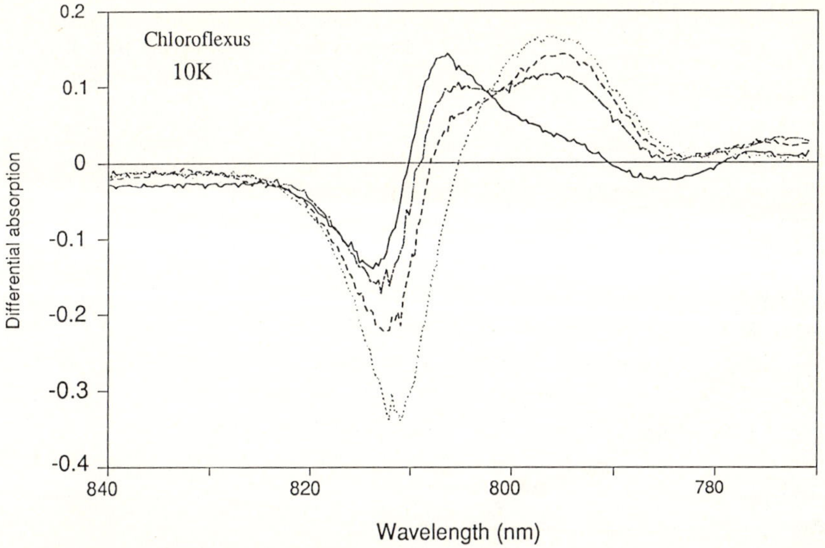

Fig. 2. Transient ΔA spectra in the 800 nm region at 600 fs (–), 4 ps (– · · –), 10 ps (- - -) and 100 ps (· · ·) after excitation at 870 nm

As the instantaneously populated Bchl-like state is still present in the P^+H^- state, it is obvious from these transient spectra that kinetic data at wavelengths in this region which reveal small deviations from a single exponential behaviour should not be a priori interpreted as indicating the presence of an intermediate in the electron transfer pathway from P^* to the bacteriopheophytin.

An alternative interpretation of an instantaneous shift in this region is that it could be an induced Stark shift of transitions which belong to B in response to the presence of P^* and P^+. It is especially noteworthy that an instantaneous shift in the B region is not what one expects from a simple two-step model of charge separation; in this case, a time-dependent bleaching of the B band is expected.

3. Rate of the Primary Electron Transfer

By monitoring the stimulated emission from P^* at 930 nm the rate of electron transfer can be accurately determined. At 10 K the kinetics are clearly not-single-exponential, as shown in Fig. 3. The best fit at 10 K using two exponentials gives 2 ps and 24 ps. At higher temperature the rate of electron transfer increases by roughly a factor of 2 but remains nonexponential. The presence of slowly interconverting conformers of reaction centers could be at the origin of a nonexponential behaviour.

An alternative model could involve an induced nuclear relaxation on the time scale of the electron transfer reaction in response to the presence of the P excited and possibly charge transfer state. In this situation the initial conformational state is more favourable to the electron transfer reaction.

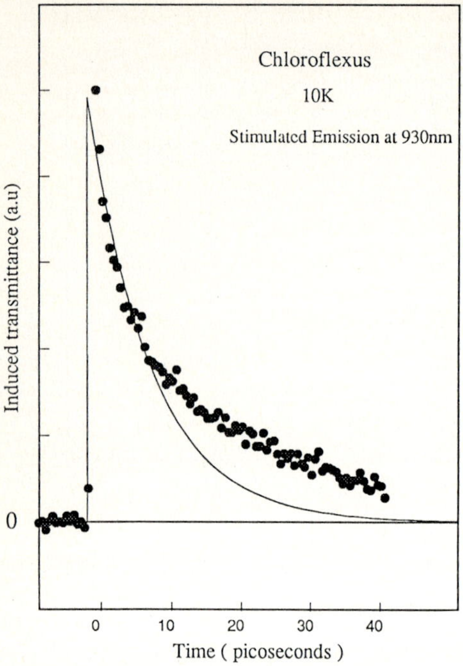

Fig. 3. Kinetic data at 930 nm corresponding to the decay of the stimulated emission at 10 K (• •) The continuous line represents a single exponential decay

References

1. G.R. Fleming, J.L. Martin, J. Breton: Nature **33**, 190–192 (1988)
2. J. Deisenhofer, O. Epp, K. Miki, R. Huber, H.J. Michel: Mol. Biol. **180**, 385–398 (1984)
3. H. Michel, O. Epp, J. Deisenhofer: EMBO J. **5**, 2445–2451 (1986)
4. J.P. Allen, G. Feher, T.O. Yeates, H. Komiya, D.C. Rees: Proc. Natl. Acad. Sci. USA **84**, 5730–5734 (1987)
5. J.P. Allen, G. Feher, T.O. Yeates, H. Komiya, D.C. Rees: Proc. Natl. Acad. Sci. USA **85**, 8487–8491 (1988)
6. N.W. Woodbury, M. Becker, D. Middendorf, W.W. Parson: Biochemistry **24**, 7516–7521 (1985)
7. J.L. Martin, J. Breton, A. Hoff, A. Migus, A. Antonetti: Proc. Natl. Acad. Sci. USA **83**, 957–961 (1986)
8. J. Breton, J.L. Martin, A. Migus, A. Antonetti, A. Orszag: Proc. Natl. Acad. Sci. USA **83**, 5121–5125 (1986)
9. W. Holzapfel, U. Finkele, W. Kaiser, D. Oesterhelt, H. Scheer, H.U. Stilz, W. Zinth: Chem. Phys. Lett. **160**, 1–7 (1989)
10. R.A. Marcus: Chem. Phys. Lett. **133**, 471–477 (1987)
11. J.L. Martin, J. Breton, J.C. Lambry, G. Fleming: In *Photosynthetic Bacterial Reaction Center: Structure and Dynamics*, ed. by J. Breton, A. Verméglio, Vol. 149 (Plenum, New York 1988) pp. 195–203
12. R.A. Marcus: Chem. Phys. Lett. **146**, 13–22 (1988)
13. A. Ogrodnik, N. Remy-Richter, M.E. Michel-Beyerle, R. Feick: Chem. Phys. Lett. **135**, 576–581 (1987)
14. C. Kirmaier, D. Holten, L.J. Mancino, R.E. Blankenship: Biochim. Biophys. Acta **765**, 138–146 (1984)
15. P. Parot, N. Delmas, D. Garcia, A. Vermeglio: Biochim. Biophys. Acta **809**,137–140 (1985)
16. J.A. Shiozawa, F. Lottspeich, D. Oesterheft, R. Feick: Eur. J. Biochem. **180**,75–84 (1989)

Picosecond Transient Phase Grating Studies of the Energetics and Structure Dynamics of Heme Proteins

L. Genberg, L. Richard, G. McLendon, and R.J.D. Miller

Department of Chemistry and Institute of Optics, University of Rochester, Rochester, NY 14627, USA

Large amplitude motions in proteins and biopolymers is a fundamental feature of the biological function of these systems. The mechanism of energy transduction from a stimulus, such as ligand binding or dissociation, to a specific protein motion associated with biological function remains a controversial issue. Large amplitude motions often involve the correlated action of a vast number of vibrational modes and energy must be directed along specific channels in order to displace a sizable number of atoms. This motion must involve energy delocalization, either over a specific vibrational coordinate as postulated by the soliton model [1], or as delocalized potential energy gradients, such as in the strain model [2]. These two possible mechanisms can be distinquished by the spatial dispersion of vibrational energy and the lifetimes of the vibrational modes that couple to the correlated motions. This problem can be ideally studied using phase grating spectroscopy by taking advantage of the high sensitivity to density changes and time resolution of this technique.

One of the most important systems exhibiting photoinduced, large amplitude, correlated structural changes is the class of heme proteins. Studies using picosecond transient phase grating spectroscopy have been performed on the ligated/deoxy structure transition for both hemoglobin (Hb) and myoglobin (Mb) using 532 nm excitation. Excitation into the Q–band dissociates the CO ligand on a 100 fsec time scale [3,4] and CO geminate recombination occurs on the 100 ns time scale [4]. Intermittent to these time scales, the protein undergoes large amplitude motions that drive the system towards the deoxy heme structure. Nonresonant probes at 1.064 μm were used to isolate the phase contribution to the grating formation in order to determine these protein dynamics through density modulations of the index of refraction.

At high excitation conditions (i.e. approximately two photons absorbed per heme), extra thermal energy is pumped into the lattice through population cycling of the short lived deoxy state, and results similar to the previous met heme studies were observed [5]. These excitation conditions allow assignment of the density change that accompanies the protein motion relative to thermal dilation. At low excitation conditions, a change in the index of refraction corresponding to an increase in the density of the surrounding water is observed, i.e. the opposite of thermal expansion. The magnitude of this response differs between the two proteins, as shown in Figure 1. In MbCO, the phase grating component is completely dominated by this protein driven wave. The HbCO has a much smaller response at the same excitation conditions per heme than the MbCO. This difference is attributed to the interaction between the subunits in hemoglobin. The subunit interaction inhibits the hemoglobin from undergoing as large an amplitude change as the myoglobin, and the amplitude of the signal due to the protein displacement is reduced relative to normal thermal expansion of the water layer. In addition, the protein wave is not observed in oxyhemoglobin, as shown in Figure 2.

Springer Series in Chemical Physics, Vol. 53 **Ultrafast Phenomena VII**
Editors: C.B. Harris · E.P. Ippen · G.A. Mourou · A.H. Zewail

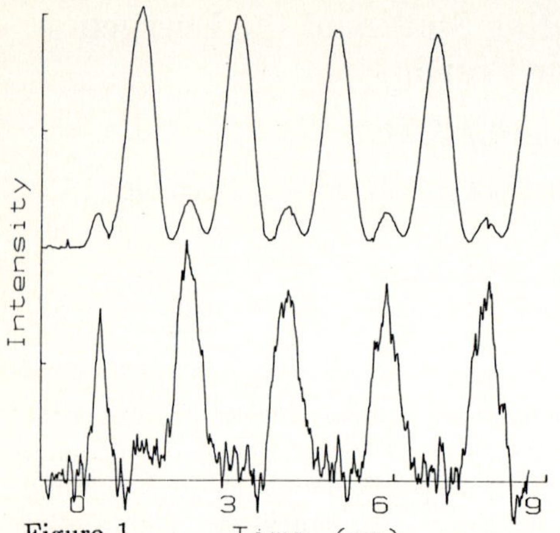

Figure 1

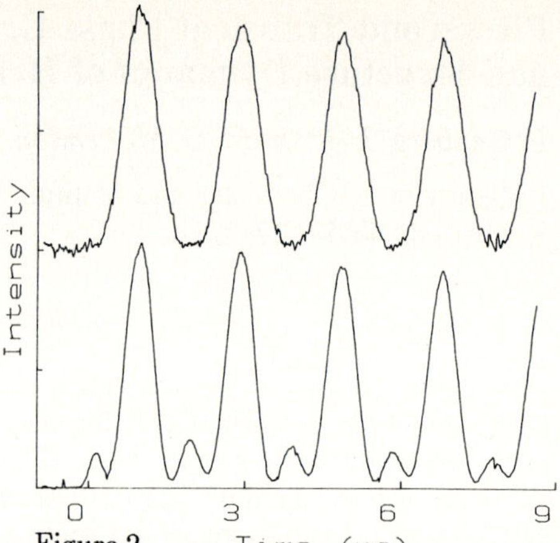

Figure 2

Figure 1. Carboxyhemoglobin (top curve) versus carboxymyoglobin (bottom curve) with 10.2^0 between 532 nm excitation pulses (low excitation). The protein driven wave dominates for myoglobin. Both a thermal and protein driven wave are clearly observable for hemoglobin, with the thermal wave having the larger amplitude.

Figure 2. Oxyhemoglobin (top) versus carboxyhemoglobin (bottom). Note the absence of the protein driven wave in the oxyhemoglobin curve.

The structure of the HbO_2 heme pocket is significantly different from the HbCO which leads to rapid O_2 recombination. The rapid O_2 recombination returns the protein to the ligated structure and turns off the protein displacement contribution to the signal. Controls were performed on the deoxy proteins. As expected, the deoxy heme proteins showed only the thermal response and no protein driven wave was observed.

It is important to note that the signal demonstrates what approximates a t=0 strain boundary condition as opposed to the normal t=0 sinusoidal stress boundary condition for a strictly thermal phase grating. This is shown in Figure 3. The longer decay of the signal at narrower angles (longer acoustic period) shows that the signal is fringe spacing dependent. This study illustrates that the signal is acoustic in origin. The acoustic periodicity relative to t=0 could also be explained by an acoustic beat against an excited state or population grating component. The population phase grating component is at least two orders of magnitude larger with a probe wavelength of 532 nm. Comparisons of absolute signal intensities at probe wavelengths of 1.064 μm and 532 nm have found that the signal at 1.064 μm is approximately one fourth that at 532 nm, as expected for a pure acoustic response. In addition, control studies at 4^0C (the zero thermal expansion point of water) find the same acoustic signal at 532 nm which rules out a thermal acoustic response. Further temperature control studies at 1.064 μm are in progress.

The temperature studies and absolute signal intensities are consistent with the assignment of the acoustic response to a nonthermal material displacement associated with the protein structure transition. These studies in combination with the dependence of the signal on the ligand, CO versus O_2 and deoxy, demonstrates that the dynamics are directly measuring large amplitude motions of the protein. The inset to figure 3 shows the expanded rise time at the narrower excitation angle. This rise time is on the order of 30 to 100 ps. Again, from speed of sound considerations, this

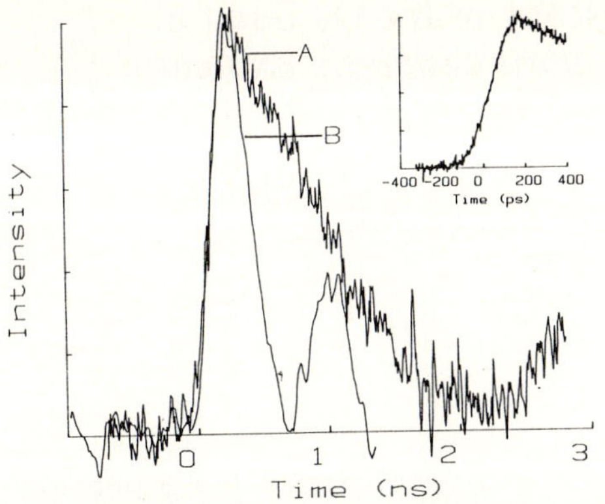

Figure 3. Fringe spacing dependence for carboxymyoglobin. Curves A and B are for excitation angles of 2^0 and 10^0, respectively. The inset shows the expanded rise time for conditions similar to curve A.

rapid rise time is only possible if the density change is a direct consequence of material displacement driven by the protein structure change. These results are consistent with a large amplitude correlated structural change occurring within 30 to 100 ps following CO dissociation.

The rapid rise time of a protein driven density change lends support to the model of stored strain energy as the driving force for protein motion. The only caveat is that the potential energy gradient would have to be extensively delocalized over the protein structure in order to account for the observed dynamics. Experiments are in progress to fully resolve this issue.

Acknowledgements: This work was supported by the NIH (1 R01 GM41909-01A1) and an NSF Presidential Young Investigator Award (RJDM). RJDM is the recipient of an A. P. Sloan Fellowship and a Camille and Henry Dreyfus Teacher-Scholar Award.

References

1. A. S. Davydov, J. Theor. Biol. **66,** 379 (1977).
2. M. F. Perutz, G. Fermi, B. Lushi, B. Shaanan and R. C. Liddington, Acc. Chem. Res. **20,** 309 (1987).
3. P. A. Anfinrud, C. Han and R. M. Hochstrasser, Proc. Natl. Acad. Sci. **86,** 8387 (1989).
4. J. W. Petrich and J. L. Martin, Chem. Phys. **131,** 31 (1989).
5. L. Genberg, Q. Bao, S. Gracewski and R. J. D. Miller, Chem. Phys. **131,** 81 (1989).

Polarized Pump–Probe Spectroscopy of the Q_Y Band of BChl a-Protein from *Prosthecochloris aestuarii*: Evidence for Ultrafast Exciton Localization

P.A. Lyle and W.S. Struve

Department of Chemistry and Ames Laboratory–USDOE, Iowa State University, Ames, IA 50011, USA

1. Introduction

The 3-dimensional structure of the bacteriochlorophyll a - protein antenna complex from the green sulfur photosynthetic bacterium Prosthecochloris aestuarii has now been known for 13 years [1]. However, little is understood about its electronic excitation transport (EET) dynamics, and the interpretation of its 809 nm Q_y absorption spectrum remains unsettled. While the organization of the BChla molecules (which are complexed by trimeric protein subunits into clusters of seven chromophores with nearest-neighbor Mg-Mg distances as short as 11 Å) mandates dipole-dipole exciton couplings on the order of 200 cm^{-1} simple exciton calculations limited to chromophores within a single subunit cannot reproduce the experimental absorption and CD spectra [2]. Spectral hole-burning studies have recently established that exciton splittings are largely responsible for the nonvibronic band structure in the 4.2 K Q_y absorption spectrum, because satellite holes appear at wavelengths corresponding to static absorption bands other than the one excited at the burn wavelength [3].

In order to address the question of the extent to which such exciton coherence is maintained during EET in this antenna complex, we have examined the dynamic linear dichroism and obtained the short-time photobleaching action spectrum for spectral evidence of energy redistribution.

2. Experimental

Solutions of BChl a - protein were dissolved in a mixture of Tris buffer and water, and adjusted for an optical density of 0.2 at 809 nm in the rotatable 800 μm path length cell used in pump-probe experiments. A hybrid mode-locked styryl 8 / IR-140 dye laser was synchronously pumped by 532 nm, 70 ps pulses from a cw mode-locked Nd:YAG laser. The repetition rate was reduced to 19 MHz with a cavity-dumper, sharpening the dye laser autocorrelation width to 2-3 ps fwhm. The pump-probe optics and the single-sideband multiple modulation scheme have been described previously [4]; the pump and probe beams were modulated at 0.5 and 6.5 MHz respectively with Isomet 1206C acoustooptic modulators, and the signal-bearing photobleaching transient was detected at the 7.0 MHz sum frequency using a Drake R-7A radio receiver.

3. Results

A Q_y photobleaching action spectrum, obtained by normalizing the isotropic photobleaching signal at 10 ps after excitation to the product of pump and probe beam intensities, is shown by the data points in Fig. 1. This action spectrum is compared with the static Q_y absorption spectrum of the antenna protein (solid curve),

Springer Series in Chemical Physics, Vol. 53 **Ultrafast Phenomena VII**
Editors: C.B. Harris · E.P. Ippen · G.A. Mourou · A.H. Zewail

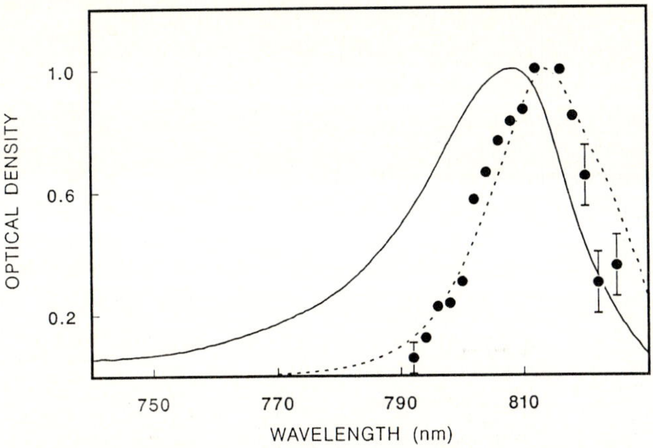

Figure 1. Room-temperature Q_y spectra for BChl a - protein from P. aestuarii: static absorption spectrum, solid curve; Boltzmann-weighted static absorption spectrum, dashed curve; and isotropic photobleaching action spectrum at 10 ps, data points.

proving that the latter is no longer uniformly bleached by this time. A similar action spectrum (not shown) was observed at 2 ps. Hence, the laser-excited exciton states have lost their coherence by this time, and the excitation has gravitated toward the lower-energy side of the Q_y spectrum. The dashed line in Fig. 1 shows the static spectrum multiplied by the Boltzmann factor $\exp(-\epsilon/kT)$, where the excitation energy ϵ is referenced to that of the lowest-energy exciton component at 827.1 nm [3]. The similarity between the thermally averaged static and photobleaching action spectra suggests that the excitation has become spectrally equilibrated by 2 ps, although some spectral redistribution must also accompany exciton decay into localized chromophore states.

We show in Fig. 2 typical polarized photobleaching profiles for the BChl a - antenna protein at 814 nm. The occurrence of an initial ratio of parallel to perpendicular signals less than 3 is significant, because this ratio physically must initialize to 3. To consider the origin of this shortfall, we consider a depolarization model in which the laser-excited state A decays unresolvably fast (e.g. via exciton localization) into a second excited state B, which then evolves with decay law s(t) into an excited state C with contrasting polarization (e.g. via EET). In this case, the form of the polarized transients is given by

$$I_{\parallel}(t) - P(t)[1 + 0.8[(a_1-a_2)s(t)+a_2]] \quad (1a)$$

$$I_{\perp}(t) - P(t)[1 - 0.4[(a_1-a_2)s(t)+a_2]] \quad (1b)$$

where P(t) is the isotropic decay law and a_1, a_2 are residual anisotropy parameters that depend on the B and C transition moment orientations relative to that of A [5]. Convolute-and-compare analyses were carried out for these and other profiles at 814, 809, and 805 nm, representing s(t) with the single-exponential function $\exp(-t/\tau)$ and floating the parameters a_1, a_2, and τ. These invariably converged to $a_1 \approx 1$, $a_2 \approx 0.15$ to 0.35, and $\tau \approx 2.3$ to 3.9 ps, indicating that essentially all of the anisotropic decay occurs in the second depolarization step with lifetime ~ 2 to 4 ps. Since the photobleaching action spectrum is thermally equilibrated (cf. Fig. 1) within 2 ps of excitation, much if not all of the depolarization

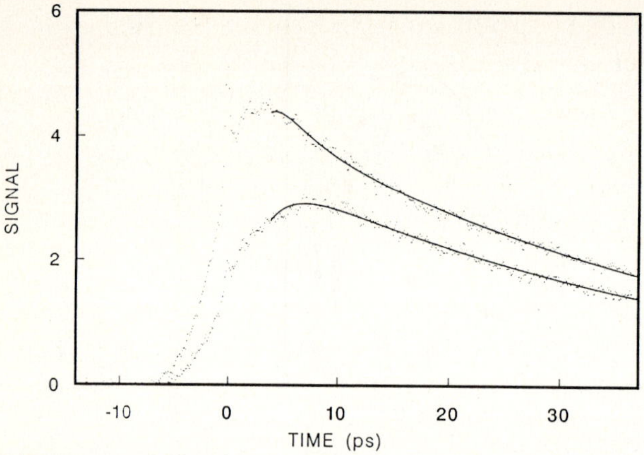

Figure 2. Polarized photobleaching profiles for BChl a - protein from P. aestuarii at 814 nm. Continuous curves are convolutions of the laser autocorrelation function with Eqs. (1).

observed in Fig. 2 corresponds to incoherent excitation hopping between localized chromophore states.

1. R.E. Fenna, L.F. Ten Eyck, and B.W. Matthews: Biochem. Biophys. Res. Commun. 75, 751 (1977).

2. R.M. Pearlstein and R.D. Hemenger: Proc. Natl. Acad. Sci. USA 75, 4920 (1978).

3. S.G. Johnson and G.J. Small: J. Phys. Chem., submitted for publication.

4. T.P. Causgrove, S. Yang, and W.S. Struve: J. Phys. Chem. 92, 6790 (1988).

5. P.A. Lyle and W.S. Struve, Biochem. Biophys., submitted for publication.

Picosecond Transient Absorption and Dichroism Studies on Excitation Transfer in Allophycocyanin

W.F. Beck, M. Debreczeny, Xinwei Yan, and K. Sauer

Department of Chemistry and Chemical Biodynamics Division,
Lawrence Berkeley Laboratory, University of California at Berkeley,
Berkeley, 1 Cyclotron Road, Berkeley, CA94720, USA

Abstract. Excitation transfer processes in allophycocyanin have been studied as a function of aggregation state using a picosecond absorption/linear dichroism spectrometer that employs a polarization and amplitude modulation technique.

1. Introduction

Allophycocyanin is a light-harvesting protein that is found in the core of the phycobilisome in cyanobacteria [1]. It directs excitation energy received from the similar phycobiliprotein C-phycocyanin towards the photosystem II reaction center. As found in the phycobilisome, allophycocyanin is organized as a ring-like homotrimer; each monomeric subunit is composed of an α and a β subunit, each of which binds a single phycocyanobilin chromophore via a thioether linkage to a cysteine residue at amino acid position 84 [1]. Interactions between the chromophores in adjacent monomers are thought to lead to a large bathochromic shift of the electronic absorption spectrum upon aggregation of monomers into trimers [2].

We have examined excitation transfer processes in allophycocyanin using a picosecond transient absorption/linear dichroism technique. We show in this paper that trimers of allophycocyanin exhibit fast linear dichroism transients that are not observed in monomers.

2. Experimental

Allophycocyanin preparations free of linker polypeptides were isolated from cultures of the AN112 mutant of *Synechococcus* 6301 by using methods similar to those described previously [3,4] and were suspended in a buffer solution at pH 7.0 containing 10 mM PO_4^{3-} and 1 mM NaN_3. Allophycocyanin monomers were prepared by dilution of the isolated trimeric allophycocyanin preparation into a buffer solution at pH 7.0 containing 1 M KSCN and 50 mM PO_4^{3-}. For picosecond spectroscopy, 15 mL samples ($A_{600\ nm} = 2\ cm^{-1}$) were pumped through a 1 mm cell; the sample reservoir was kept on ice and was purged with N_2 throughout the experiment.

Excitation transfer processes were studied with a one-color pump-probe spectrometer that employs 1.5-2.5 ps (FWHM of autocorrelation) pulses from a synchronously pumped, cavity-dumped dye laser system operated at a 3.8 MHz pulse repetition rate. The amplitude and polarization of the pump laser beam were modu-

Springer Series in Chemical Physics, Vol. 53 **Ultrafast Phenomena VII**
Editors: C.B. Harris · E.P. Ippen · G.A. Mourou · A.H. Zewail

lated by an acoustooptic and a photoelastic modulator (set as a $\lambda/2$ retarder), respectively. The probe laser transmittance signal was analyzed by lock-in amplifiers referenced to the amplitude and polarization modulation frequencies. By this technique, we recorded transient linear dichroism and isotropic ground-state absorption-recovery traces simultaneously. The absorption anisotropy was obtained by computing the ratio of the dichroism and isotropic absorption traces after proper scaling of the dichroism signal.

3. Results and Discussion

Figure 1 (left side) shows the transient absorption, dichroism, and anisotropy traces obtained at 590 nm with a monomeric allophycocyanin preparation. The anisotropy exhibits a single 85 ps exponential component that decays to a limiting anisotropy of 0.34. The absorption transient reveals a component with a similar decay time constant (100 ps) that accounts for 15 percent of the amplitude, the remaining amplitude being described by a 1400 ps time constant. At longer laser wavelengths, the intensity of the ~100 ps component decreases; at 610 nm the anisotropy is flat at the maximum value near 0.40, and the absorption decay is well described by a single decay component with a 1850 ps time constant. Thus, the ~100 ps component arises from an excitation transfer step between the α84 and β84 chromophores. The decreasing amplitude of the ~100 ps component with longer

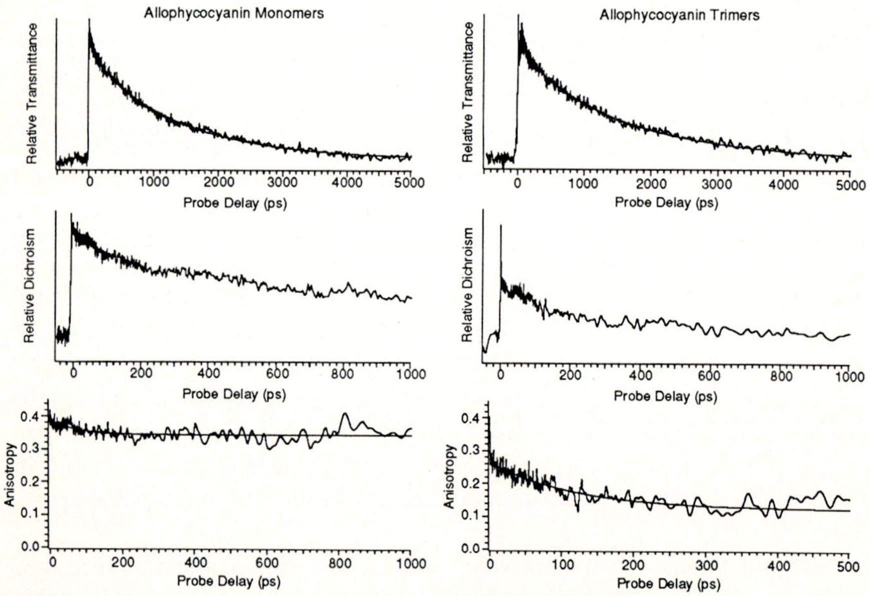

Figure 1. Isotropic transient absorption (top), linear dichroism (middle), and computed anisotropy decay traces for allophycocyanin monomers at 590 nm (left column) and allophycocyanin trimers at 595 nm (right column).

wavelengths is consistent with the slowly increasing fluorescence polarization observed in monomeric allophycocyanin by Maxson and Sauer [4].

Figure 1 (right side) shows the results obtained at 595 nm from a trimeric allophycocyanin preparation under the same conditions used for the monomers. The transient dichroism trace reveals a sharp, instrument-limited decay (< 2 ps time constant) that is not observed in monomeric preparations. The anisotropy decay exhibits an instrument-limited decay from the maximum value near 0.40 to an anisotropy of 0.28, followed by a slower, 120 ps decay to a limiting anisotropy of 0.12. Note that the < 2 ps component is not easily observed in the transient absorption trace, which is well described by a ~150 ps decay component accounting for 15 percent of the amplitude and a 1700 ps component accounting for the rest of the amplitude. While the < 2 ps anisotropy component clearly arises from interactions between chromophores in adjacent subunits, the 120 ps anisotropy component seems to be analogous to the ~100 ps anisotropy component assigned to a $\alpha 84 \leftrightarrow \beta 84$ transfer path in monomers. In the trimeric preparations, this pathway accounts for nearly three times the amount of anisotropy change as observed in the monomers.

The results described in this paper suggest that extremely fast intersubunit excitation transfer paths are established upon formation of trimeric allophycocyanin from the monomers. These paths are made easily visible with the picosecond transient linear dichroism technique that we described. Whether the observed fast dichroism and anisotropy components correspond to Förster induced-dipole resonant transfers or to exciton-state relaxation (internal conversion) events is the subject of experiments currently underway in our laboratory.

4. Acknowledgements

This work was supported by the Director, Office of Energy Research, Division of Energy Biosciences, of the U.S. Department of Energy under contract No. DE-AC03-76SF-00098. W.F.B. was the recipient of a postdoctoral fellowship from the Miller Institute for Basic Research in Science (University of California, Berkeley).

5. References

1. For reviews, see: (a) Glazer, A. N. *Annu. Rev. Biophys. Biophys. Chem.* **1985**, *14*, 47-77; (b) Glazer, A. N. *Annu. Rev. Biochem.* **1983**, *52*, 125-157.
2. MacColl, R.; Csatorday, K.; Berns, D. S.; Traeger, E. *Biochemistry* **1980**, *19*, 2817-2820.
3. Glazer, A. N.; Fang, S. *J. Biol. Chem.* **1973**, *248*, 659-662.
4. Maxson, P.; Sauer, K. in "Photosynthetic Light-Harvesting Systems," Scheer, H., and Schneider, S., Eds.; de Gruyter, Berlin, 1988; pp 440-449.

Vibrational Temperatures of Photoexcited Deoxy-Hemoglobin

S.H. Courtney[1,*], *J.M. Friedman*[1,*], *R.G. Alden*[2], *M.D. Chavez*[2], *and M.R. Ondrias*[2]

[1]AT&T Bell Laboratories, 600 Mountain Ave., Murray Hill, NJ 07974, USA
[2]Department of Chemistry, University of New Mexico, Albuquerque, NM 87131, USA
* Present address: Department of Chemistry, New York University, New York, NY 10003, USA

The anti-Stokes/Stokes ratio of several heme modes were monitored as a function of laser fluence in order to address the vibrational population changes of deoxy-hemoglobin following picosecond photoexcitation.

In recent years, time-resolved optical spectroscopies have proven extremely valuable for characterizing the sub-nanosecond dynamics of large molecular systems. This is particularly true in the field of molecular biophysics where the active sites of heme proteins and components of photosynthetic systems have been probed on increasingly short timescales using resonance Raman and absorption spectroscopies. The high peak powers required for ultrafast spectroscopies can generate non-linear effects through the interaction of the laser field and chromophore. These effects can at times cloud the interpretation of spectral data. Ultrafast laser spectroscopy invariably results in the generation of very high laser fluences at the sample (> 10^8W/cm^2). In many instances, these high laser fluences are necessary to generate substantial populations of electronic excited species. The effects of high laser fluences are magnified when the field is in near-resonance with an intense optical transition of the chromophore. The electric field can couple the resonance Raman and resonance fluorescence radiative processes of the chromophore, which will result in broadening of the spontaneous Raman and fluorescence emission. In addition, the strong driving fields used in many ultrafast optical experiments are sufficient to alter the Boltzmann populations of specific normal modes if the T_1 population relaxation time of the vibrational level is long compared to the feeding rate of the level. This study evaluates the behavior of a number of Raman active modes of deoxy-Hb under conditions of a strong resonant field.

To address the vibrational population changes of deoxy-Hemoglobin following photoexcitation(1,2), we have measured the anti-Stokes/Stokes ratio of ν_7, the strongest mode in the low frequency region of the spectrum and find an area ratio of 6.7% to 10.2%. Several observations suggest that this mode is not vibrationally hot on a 30ps timescale. 1) A two color pump/probe protocol using a strong 532nm pump pulse temporally and spatially overlapped with 436nm probe pulse resulted in no change in the relative intensities. 2) Experiments using 10ns pulses at the same wavelength yield an anti-Stokes/Stokes ratio of 8.6% for the 674 cm^{-1} mode. 3) The anti-Stokes/Stokes ratio is consistent with the value expected at 300 K. (The Boltzmann ratio of ν_7 corrected with a Kramers-Kronig transform technique is 8.2%.) Figure 1 shows the anti-Stokes and Stokes Raman spectra of deoxy-Hb taken with 30ps pulses of three heme modes, ν_7, ν_{16}, and ν_{32}, at 671, 756, and 790 cm^{-1}, respectively.

Another mode, ν_4, exhibits considerably different behavior. In this case, we observe a laser fluence dependent anti-Stokes to Stokes

Springer Series in Chemical Physics, Vol. 53 **Ultrafast Phenomena VII**
Editors: C.B. Harris · E.P. Ippen · G.A. Mourou · A.H. Zewail

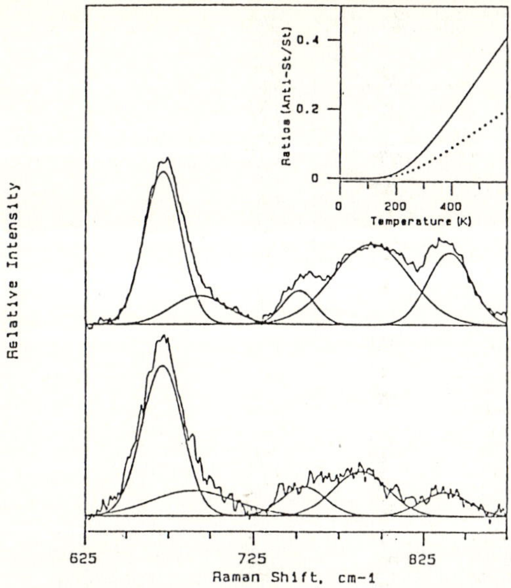

Figure 1) Resonance Raman spectra of deoxy-Hb obtained using 436nm laser pulses. The Stokes region of the spectrum (upper panel) is shown from 625-875 cm^{-1}, with fits used to extract the intensities. The best fit in an anti-Stokes/Stokes ratio of 8.6%. This value corresponds to a temperature of 300 $\pm$30K. The inset is a plot of the expected Boltzmann distribution for ν_7 in the absence of resonance and reabsorption effects (dotted line). The solid line displays the expected behavior in the presence of reabsorption effects and resonance conditions at 436nm using the Kramers-Kronig transform technique (3).

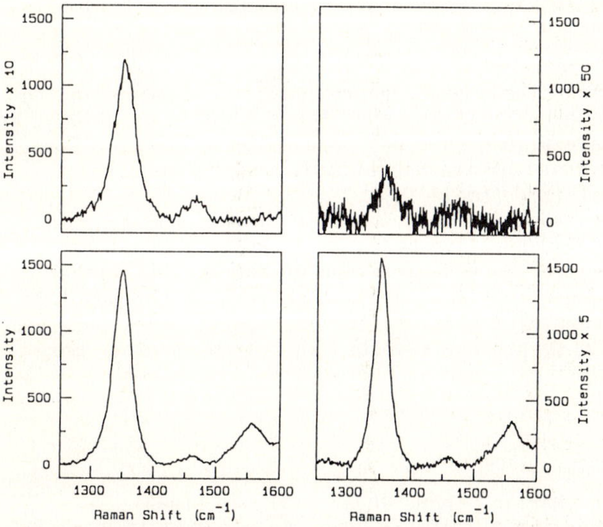

Figure 2) Resonance Raman spectra of deoxy-Hb in the high frequency region. The lower left panels are Stokes spectra, and the upper panels are anti-Stokes results. The left panels are at high laser fluence; whereas, the right panels are at approximately five times lower energy per pulse, clearly showing a decrease in the anti-Stokes intensity relative to the Stokes.

ratio. At our highest fluence, we obtain a ratio greater than 50 times the value expected from the uncorrected Boltzmann distribution.

Figure 3 shows simulations of the anti-Stokes/Stokes intensity ratios as a function of the Rabi frequency, W, with an oscillator strength of 1.1 at 300K. Comparison of these plots with the experimental values in Figure 2 shows that the population lifetime for state b> of the 1357

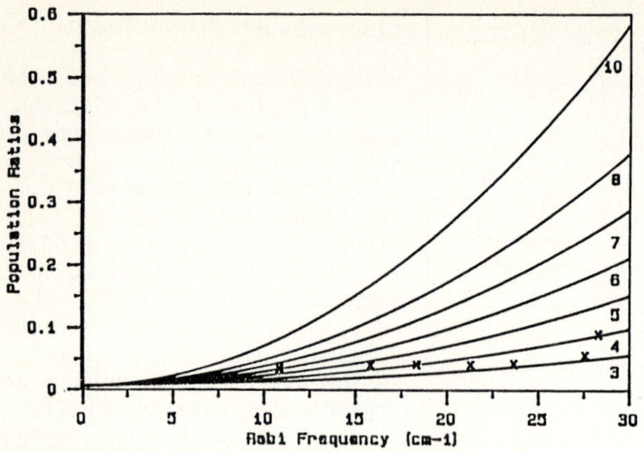

Figure 3) The solid lines are simulations of the effects of varying the population lifetime of state b> as a function of Rabi frequency. The molecular parameters used Γ_{cc}, Γ_{cb}, Γ_{ab}, and Γ_{bc} were 53.15, 106.3 10, and 106.3 cm^{-1}, respectively. The values for ω_L, ω_{ac}, ω_{ab} were 22951, 23259, and 1357 cm^{-1}, respectively. The Rabi frequency W was varied from 0-30 cm^{-1}, and Γ_{bb} was adjusted to values corresponding to 3, 4, 5, 6, 7, 8, and 10 ps. The x's correspond to the experimental values based on an oscillator strength of 1.1.

cm^{-1} mode is 4-5 ps in the steady state approximation. Changes in the relative populations of the anti-Stokes and Stokes signals as a function of Rabi frequency indicate a long (4-5 ps) population relaxation time (small Γ_{bb}), while broadening of the vibrational mode is consistent with a long dephasing lifetime (small Γ_{ab}).

This study demonstrates that the mode specific vibrational dynamics exhibited by a number of Raman active modes in deoxy-Hb can not be treated within the context of an equipartition of energy, but rather must be evaluated in terms of the relazation phenomena of the specific vibrational mode. The complex vibrational relaxation dynamics observed in the protoypic deoxy-Hb system may also be manifested in other more complicated biological systems, such as photosynthetic systems.

References

1) J.W. Petrich and J.L. Martin, Chem. Phys., 131, 31 (1989).
2) R.G. Alden, M.D. Chavez, M.R. Ondrias, S.H. Courtney, E.W. Findsen, J.M. Friedman, J. Phys. Chem. 94, 85 (1990).
3) K.T. Schomacker and P.M. Champion, J. Chem. Phys. 90, 5982 (1989).

Sub-picosecond Resonance Raman Spectroscopy of Some Biological Systems

R. van den Berg and M.A. El-Sayed

Department of Chemistry and Biochemistry, University of California at Los Angeles, Los Angeles, CA 90024, USA

We have studied the photochemical and photophysical events that occur upon excitation of two important biological molecules, bacteriorhodopsin (bR) and two differently ligated hemoglobin (Hb) molecules (HbCO and HbO_2) using a sub-picosecond resonance Raman microbeam and flow technique [1]. The laser is a Spectra Physics YAG laser which is compressed and frequency-doubled, and which synchronously pumps a cavity-dumped dye laser. The excitation wavelengths used were 587 nm for bR and 578 nm for Hb. In this one-color technique, the transients observed are those appearing within the laser pulse of 700-800 fs.

Bacteriorhodopsin is a protein found in the purple membrane of *Halobacterium halobium*. It has a retinal chromophore, which is covalently linked to the protein backbone. Upon light absorption, it goes through a cycle of different intermediates on timescales varying from fs up to ms. In the first place we have studied the C=C stretch region around 1530 cm^{-1}. The spectrum is presented in fig. 1. Apart from a strong vibration at 1518 cm^{-1}, there is a shoulder around 1510 cm^{-1}. Since the ethylenic stretch frequency has been shown to correlate linearly [2] with the absorption maximum of each intermediate, the 1518 cm^{-1} vibration is found to correspond to a species having an absorption maximum at 625 nm, i.e. the J_{625} intermediate, and the 1510 cm^{-1} mode to an even earlier

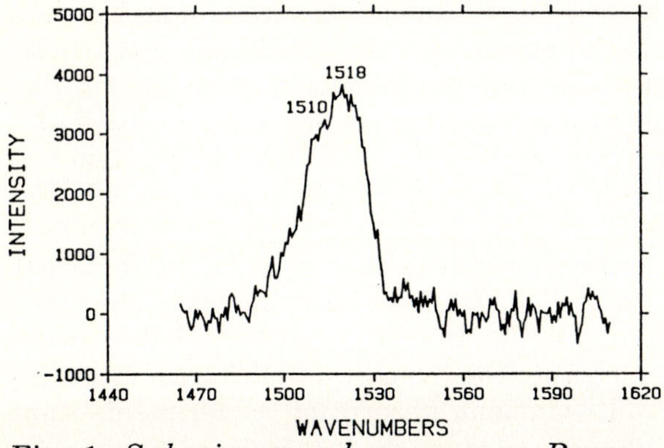

Fig. 1. *Sub-picosecond resonance Raman spectra of the earliest intermediates in the photocycle of bR in the C=C stretch region.*

Springer Series in Chemical Physics, Vol. 53 **Ultrafast Phenomena VII**
Editors: C.B. Harris · E.P. Ippen · G.A. Mourou · A.H. Zewail

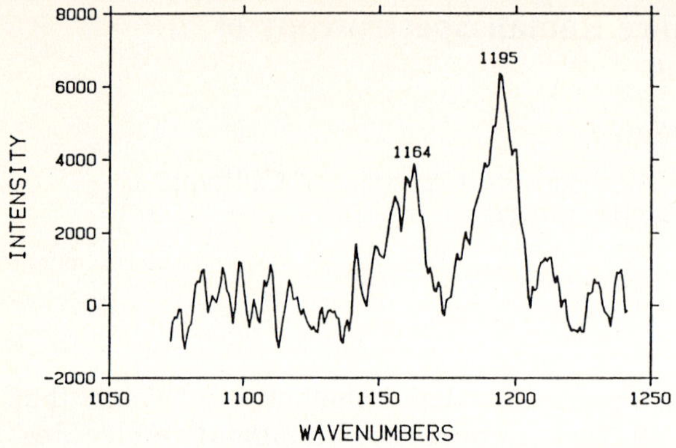

Fig. 2. *Sub-picosecond resonance Raman spectra of the earliest intermediates in the photocycle of bR in the fingerprint region.*

intermediate absorbing at 660 nm, which has been detected in the optical spectra of Mathies et al. [3].

Spectra of the fingerprint region on a ps timescale have shown [4,5] that the retinal isomerizes from an *all-trans* to *13-cis* conformation. In the present study we have observed that the J_{625} intermediate already exists in the twisted isomerized form. This can be seen in fig. 2, where the appearance of the 1195 cm^{-1} band signals the formation of the deformed *13-cis* retinal. The 1164 cm^{-1} band however, is uniquely found for the J_{625} intermediate, and shows that the isomerization at these very early times is of a different nature than at much longer times.

In a second study, we have obtained resonance Raman spectra of the photodissociation products of Hb ligated with O_2 and CO. We concentrate here on vibrations in the region between 1500 and 1650 cm^{-1}, where the so-called core-size markers are found. These vibrations are very sensitive to both the position of the Fe atom with respect to the heme plane and its spin state.

In fig. 3a we show the resonance Raman spectra obtained for the photoproduct of HbCO. From the positions of the vibrational frequencies it can be concluded that upon photodissociation a deoxy-Hb-like photoproduct is formed. The vibrational frequencies are indicative of a photointermediate in a high-spin (S=2) state, in which the Fe atom has not yet completely moved out of the plane. Similar spectra were obtained on a timescale of 40 ps [6].

For HbO_2 (fig. 3b), the vibrational modes show much larger downshifts compared to those of HbCO. This shows that the photointermediate not only has a planar heme structure, but is in an excited state. This would corroborate the hypotheses of Petrich et al. [7], on the basis of femtosecond transient absorption experiments that whereas for HbCO a deoxy-like ground state product is found, in the case of HbO_2 the principal primary photoproduct is an excited-state species.

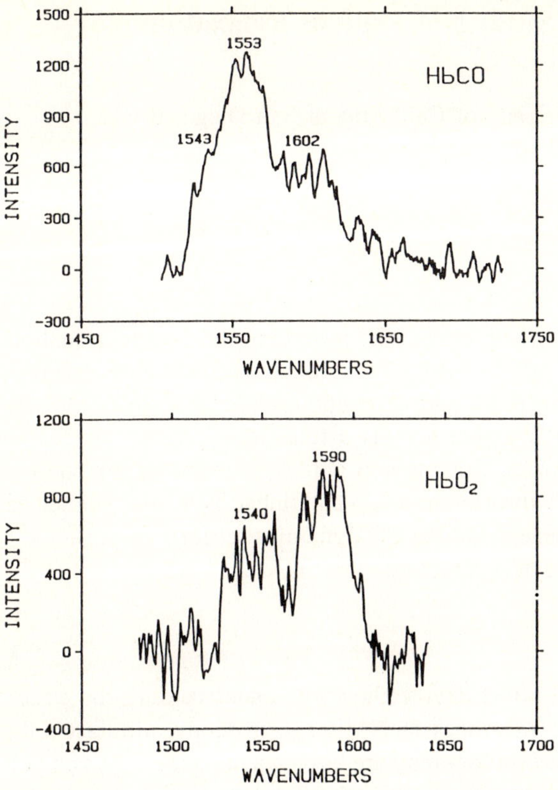

Fig. 3. *Resonance Raman spectra of the photoproducts of HbCO (fig. 3a) and* HbO_2 *(fig.3b) in the core-size region.*

References

[1] J. Terner, T.G. Spiro, M. Nagumo, M.F. Nicol and M.A. El-Sayed, *J. Am. Chem. Soc.* 102 (1980) 3238.

[2] B. Aton, A.G. Doukas, R.H. Callender, B. Becher and T.G. Ebrey, *Biochemistry* 16 (1977) 2995.

[3] R.A. Mathies, C.H. Brito Cruz, W.T. Pollard and C.V. Shank, *Science* 240 (1988) 777.

[4] J. Terner, C.-L. Hsieh, A.R. Burns and M.A. El-Sayed, *Proc. Natl. Acad. Sci. USA* 76 (1981) 3046.

[5] M. Braiman and R.A. Mathies, *Proc. Natl. Acad. Sci. USA* 79 (1982) 403.

[6] J. Terner, J.D. Stong, T.G. Spiro, M. Nagumo and M.A. El-Sayed, *Proc. Natl. Acad. Sci. USA* 78 (1981) 1313.

[7] J.W. Petrich, C. Poyart and J.L. Martin, *Biochemistry* 27 (1988) 4049.

Picosecond Polarization Studies of Protein Relaxation

Xiaoliang Xie and J.D. Simon

Department of Chemistry, University of California at San Diego, B-041,
La Jolla, CA 92093, USA

Abstract. Protein relaxation in myoglobin (Mb) following the photoelimination of CO from carbonmonoxymyoglobin (MbCO) is examined using picosecond time resolved circular dichroism (CD) and magnetic circular dichroism (MCD) spectroscopies. The transient CD signal reveals a relaxation process which requires several hundred picoseconds. Comparison with optical absorption studies on the Q-band and near infrared absorption of Mb and linear dichroism studies of the Soret optical absorptions suggest that the CD dynamics reflect a conformation relaxation of the surrounding protein structure.

1. Introduction

Currently there is a substantial effort devoted towards understanding the structure and dynamics of proteins. Advances in NMR spectroscopy, ultrafast time resolved techniques and X-ray crystallography are having a dramatic effect on our understanding of biological systems. Steady state circular dichroism and magnetic circular dichroism spectroscopy have also been extensively used to characterize structural properties of proteins. Thus, in principle, information concerning real time structural relaxations in proteins could be detected through a time resolved measurement of the CD and MCD signals.

In the last decade there have been several reports of instruments which can measure time resolved CD kinetics and spectra [1–3]. Applications of these devices to biological systems have also been reported [4,5]. Recently, we have developed a new technique for collecting time resolved CD data with picosecond resolution [6,7]. As demonstrated in the following sections, this form of spectroscopy can provide information on protein structural relaxations which are not easily detected by more conventional time resolved spectroscopies. In the present discussion, we focus on the relaxation dynamics of Mb following the photodissociation of CO from MbCO. We have also used this technique to examine the evolution of the CD spectrum of bacterial photosynthetic reaction centers and results on that system will be published elsewhere [8].

2. Experimental Details

A detailed description of the experimental apparatus has recently been published. Briefly, a high repetition rate picosecond laser system is used to generate tun-

Springer Series in Chemical Physics, Vol. 53 **Ultrafast Phenomena VII**
Editors: C.B. Harris · E.P. Ippen · G.A. Mourou · A.H. Zewail

able pump and probe pulses. The picosecond probe pulse train is polarization modulated by a piezo-electric modulator switching sequential pulses between left and right circular polarization. The difference in transmission of these probe pulses is detected using phase sensitive detection. The time delay between each pump–probe pulse pair is controlled by a computer driven optical delay line. By scanning the delay line, the transient CD signal can be recorded as a function of delay time following photolysis. On the other hand, for a fixed delay, the dye laser can be scanned in order to record a transient CD spectrum.

Transient MCD kinetics (or spectra) are recorded by placing the sample in an external magnetic field. The details of this technique will appear shortly [9].

A pH 7.0 Tris-bis buffer solution of MbCO was prepared under anaerobic conditions. The sample concentration was 0.9 mM, with an O.D. of 1.7 at 532 nm and > 4 at 355 nm. Under these conditions, the photolysis pulse (50 μJ) excites $> 90\,\%$ of the sample in the focal volume. Probe pulses on the order of 0.1 μJ were used.

3. Relaxation of Myoglobin Following Photoelimination of CO

The photodissociation of MbCO to form Mb and free CO has been the subject of many picosecond time resolved studies [10]. Transient infrared, Raman and absorption studies have revealed that photoinduced bond cleavage occurs within 350 fs. However, in considering the crystal structures of MbCO and Mb, relaxation of the tertiary structure of the protein is also expected. In particular, in the local vicinity of the reaction site, following dissociation, the heme domes, and the iron moves out of the porphyrin ring, causing the distal histidine to tilt and the F-helix to move.

To address the time scale of protein relaxation in Mb following photodissociation of CO, transient CD measurements were carried out on the N-band absorption at 355 nm. The data are shown in Fig. 1. The dashed line is a nor-

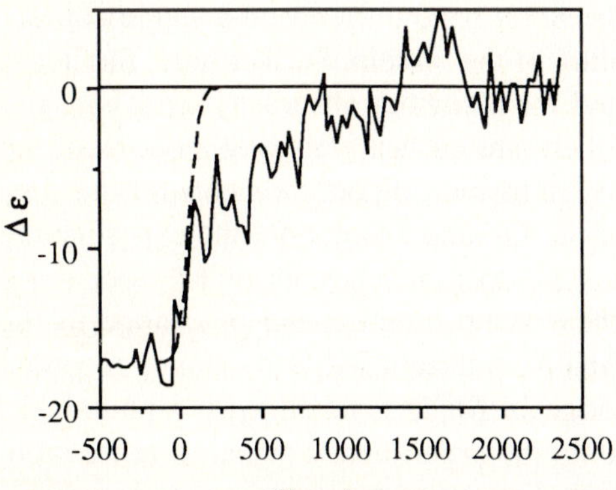

Fig. 1. Transient CD kinetics at 355 nm following photoelimination of CO from MbCO. The dashed line is the corresponding transient absorption measurement and provides a measure of the instrument response

malized transient absorption study on the same sample at the same wavelength, a measure of the instrument response.

The observed level change of the CD signal is consistent with the steady state CD values of MbCO and Mb at 355 nm. The important observation in Fig. 1 is that the evolution of the CD signal is distinctly different from the transient absorption kinetics. The CD kinetics consist of a fast initial rise within the instrument response, followed by a slower rise which reaches zero (the steady state CD value for Mb at 355 nm) at about 500 ps. It is also important to note that both MbCO and Mb have a significant absorption at 355 nm (the N-band for both is centered at 350 nm with the oscillator strength being slightly greater for Mb). The transient CD dynamics reveal a relaxation process which is two orders of magnitude slower than that associated with photoinduced bond cleavage.

In the remainder of this section, several possible interpretations of the CD dynamics are briefly discussed. A detailed study will be reported shortly [11]. Three possible mechanisms are consistent with the observed data: (1) vibrational relaxation in the vicinity of the heme, (2) a time dependent splitting of the degeneracy of the heme transitions, and (3) relaxation of the surrounding protein matrix.

Vibrational relaxation in the vicinity of the heme has been addressed both experimentally and using molecular dynamics simulations [10]. These studies indicate that the heat is dissipated into the protein within a few tens of picoseconds, significantly faster than the relaxation process revealed by the CD data. In order to address the possibility of a time dependent splitting in the degeneracy of the heme transitions, three separate experiments were performed: (1) linear dichroism studies on the Soret absorption bands, (2) transient MCD on the Q-band, and (3) detailed time resolved absorption studies of the Q-band absorption. In these latter two studies, the Q-band dynamics were probed as the Q_0 band is the absorption feature most sensitive to the lifting of the degeneracy of the e_{2g} orbitals [12].

Linear dichroism experiments on the Soret band only reveal a single component corresponding to the overall tumbling of the protein. Furthermore, the initial anisotropy observed is 0.1, that expected for a circular absorber [13]. These observations support the conclusion that there are no fast wobbling motions of the heme nor is there a rotation of the heme transition dipole moments on the time scale of the CD kinetics. In the case of the Q-band absorption, the experimental results show that the shape of the transient spectrum is unchanged from $t = 0$ to a delay of 10 ns, suggesting that there is no time dependent change in the degeneracy of the heme transitions. Finally, transient MCD dynamics recorded at 580 nm reveal an instantaneous change in MCD level following photolysis, with no ensuing dynamics observed within the experimental signal-to-noise ratio. These data support the conclusion that the iron changes its spin state extremely fast (< 1 ps). All three experiments support the conclusion that there is no time dependent evolution in the electronic structure of the heme.

The above discussion signifies that the transient CD kinetics is providing information on the relaxation dynamics of the protein structure following photodissociation. Calculations reported by *Hsu* and *Woody* [14] suggest that the CD of the N-band of Mb is primarily due to coupled oscillator interactions between the Q_x and Q_y transitions on the heme and the $\pi \rightarrow \pi^*$ transitions on the surrounding aromatic amino acids. In addition, aromatic residues as far as 12 Å away from the heme can significantly contribute to the CD spectrum.

One possible motion that could account for the evolution of the CD signal would involve the tilting of the proximal histidine. This can be independently addressed by examining the near infrared absorption band of Mb at 760 nm; this transition arises from a prophyrin-to-iron charge transfer and has been shown to be sensitive to the conformation in the immediate vicinity of the heme ring [15]. This absorption feature is characteristic of five-coordinate ferrous hemes and does not appear in the six-coordinate MbCO spectrum. Detailed dynamical studies of this absorption band reveal no change in band shape or band position from 30 ps (the instrument response) to 10 ns following photodissociation. These results suggest that the tilting of the proximal histidine as well as other structural contributions to this absorption band are not responsible for the observed CD kinetics. Studies on mutant proteins are planned in order to determine the origin of this relaxation process.

Acknowledgements. This chapter is supported by the National Institute of Health, GM-41942. JDS thanks the Alfred P. Sloan Foundation and the Camille and Henry Dreyfus Foundation for fellowship support.

References

1. P.M. Bayley, M. Anson: Biopolymers **13**, 401 (1974)
2. F.A. Ferrone, J.J. Hopfield, S.E. Schnatterly: Rev. Sci. Instrum. **45**, 1392 (1974)
3. D.S. Kliger: Rev. Chem. Intermed. **8**, 367 (1987)
4. I. Tabushi, K. Kamamura, T. Nishiya: J. Am. Chem. Soc. **101**, 2785 (1979)
5. S.J. Milder, S.C. Bjorling, I.D. Kuntz, D.S. Kliger: Biophys. J. **53**, 659 (1980)
6. X. Xie, J.D. Simon: Rev. Sci. Instrum. **60**, 2614 (1989)
7. X. Xie, J.D. Simon: J. Opt. Soc. Am. B, in press
8. X. Xie, J.D. Simon: Biophys. J., in preparation
9. X. Xie, J.D. Simon: J. Phys. Chem., in preparation
10. For a recent review see: R.M. Hochstrasser, C.K. Johnson: In *Ultrashort Laser Pulses and Applications*, ed. by W. Kaiser, Topics Appl. Phys., Vol. 60 (Springer, Berlin, Heidelberg 1988)
11. X. Xie, J.D. Simon: Biochemistry, submitted
12. W.A. Eaton, J. Hofrichter: Methods Enzymol. **76**, 175 (1981)
13. D. Magde: J. Chem. Phys. **93**, 3717 (1971)
14. M.-C. Hsu, R.W. Woody: J. Am. Chem. Soc. **93**, 3515 (1971)
15 M. Sassaroli, D.L. Rousseau: Biochem. **26**, 3092 (1987)

Subpicosecond Fluorescence Anisotropy Measurements of Tryptophanyl Residues in Proteins

J.E. Hansen, S.J. Rosenthal, and G.R. Fleming

Department of Chemistry and James Franck Institute,
University of Chicago, 5735 S. Ellis Ave., Chicago, IL60637, USA

The fluorescence anisotropy of tryptophanyl residues in proteins is used as a probe of protein motion. We have recently shown [1] that the short time (200 fs - 10 ps) region of the fluorescence anisotropy of tryptophan in solution is dominated by an electronic relaxation process, which has the potential to obscure molecular motion. The question arises as to the relevance of this result for the behavior of tryptophanyl residues in proteins. To address this question we have examined the subpicosecond fluorescence anisotropy of tryptophanyl residues in two very different enviroments - exposed to the solvent in melittin, a 26 amino acid peptide, and buried in a hydrophobic region in apoazurin purified from the bacteria Pseudomonas aeruginosa. Measurements were made by ultraviolet fluorescence upconversion described elsewhere [1]. Samples were degassed and flowed in absence of air. Concentrations were adjusted so the optical density at the excitation wavelength was between 0.6 and 1.0 for a 1 mm pathlength.

Tryptophan has two low-lying singlet-states, ^1L_a and ^1L_b. In melittin the tryptophanyl residue is exposed to water and ^1L_a is of lower energy than ^1L_b, whereas the tryptophanyl residue in apoazurin is buried in the protein, inaccessible to water, and ^1L_b is lower [2, 3]. Fluorescence emitted by the tryptophanyl residue in melittin originates primarily from the ^1L_a state. During the first few picoseconds a fast component dominates the fluorescence and anisotropy decays as observed for tryptophan in water [1]. There are remarkably similar trends in the excitation and emission wavelength dependence for melittin and tryptophan. Figure 1 compares the parallel anisotropy components of the tryptophanyl residue in melittin to those of tryptophan in water. For tryptophan in water this fast component has been attributed to rapid level kinetics between ^1L_a and ^1L_b [1]. The characteristic shape of the parallel and perpendicular (not shown) curves results from rapid decay components in both the population and anisotropy curves. We are able to simulate our results using the model described in references 1 and

Springer Series in Chemical Physics, Vol. 53 **Ultrafast Phenomena VII**
Editors: C.B. Harris · E.P. Ippen · G.A. Mourou · A.H. Zewail

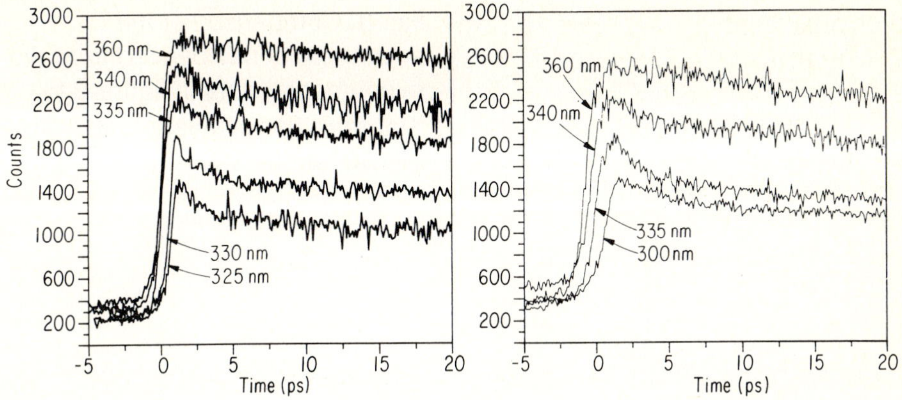

Fig. 1. Comparison of the parallel anisotropy components collected at various emission wavelengths from 300 nm excitation of the tryptophanyl residue in melittin (on the left) to those of tryptophan in water.

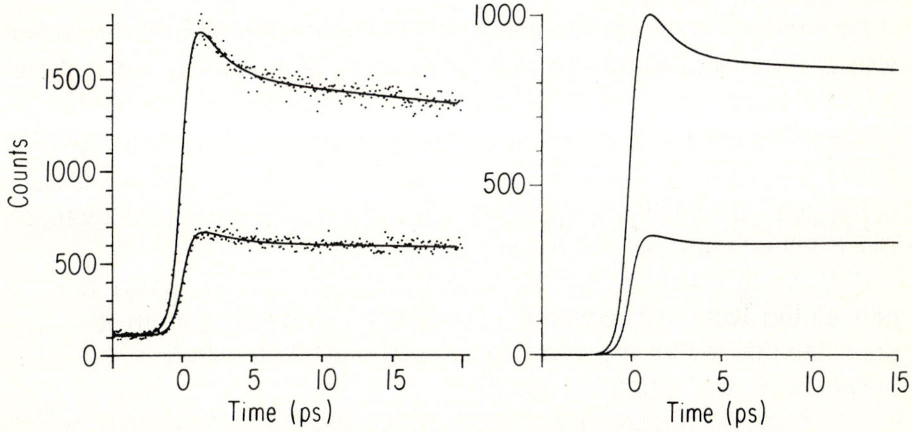

Fig. 2. Comparison of the parallel and perpendicular anisotropy components collected at 330 nm from 300 nm excitation of the tryptophanyl residue in melittin (on the left) to those from a simulation of the data generated by the model discussed in the text.

4 (see figures 2 and 3) with similar parameters as those employed for the tryptophan simulations. The only parameters that are significantly altered from those for the simulation of tryptophan in water are the excitation and detection ratios, which are defined as the fraction of absorption and emission, respectively, that is contributed by the ^1L_b state to the sum of contributions from both the ^1L_a and ^1L_b states. This is consistent with the differences observed between both

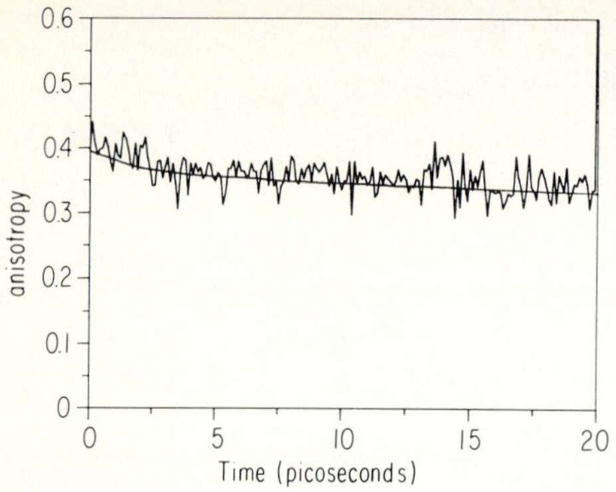

Fig. 3. Comparison of the anisotropy decay derived from the experimental data in figure 2 to that from the simulated data (shown as the solid curve).

absorption and fluorescence spectra of melittin in water and tryptophan in water. Further we report the first measurement of an initial anisotropy of 0.40 for a tryptophanyl residue in either a peptide or a protein. This result is shown in figure 3 for melittin with 300 nm excitation.

By contrast, fluorescence and anisotropy decays of apoazurin excited with 292 nm light exhibit no fast component, as observed for those of melittin, however a low initial anisotropy is observed. This is consistent with the ^1L_a and ^1L_b states being nearly degenerate (within 200 cm^{-1}, compared to the value of 500 cm^{-1} assigned to the energy separation of states for tryptophan in water), and the interconversion being faster than our resolution.

Our results demonstrate that fluorescence anisotropy decays are convoluted with nonrotational contributions originating from a population decay from one excited singlet state to another.

This work was supported by NSF.

References

1. A.J. Ruggiero, D.C. Todd, and G.R. Fleming, J. Am. Chem. Soc. **122**, 1003 (1990).
2. H. Lami and N. Glasser, J. Chem. Phys. **84**, 597 (1986).
3. J.W. Petrich, J.W. Longworth, and G.R. Fleming, Biochemistry **26**, 2711 (1987)
4. A.J. Cross, D.J. Waldeck, and G.R. Fleming, J. Chem. Phys. **78**, 6455 (1983).

Index of Contributors